4th International Symposium on Sustainable Urban Development (ISoSUD 2017)

IOP Conference Series: Earth and Environmental Science
Volume 106

Jakarta, Indonesia
9 – 10 August 2017

ISBN: 978-1-5108-5772-8
ISSN: 1755-1307

Printed from e-media with permission by:

Curran Associates, Inc.
57 Morehouse Lane
Red Hook, NY 12571

Some format issues inherent in the e-media version may also appear in this print version.

Printed by Curran Associates, Inc. (2018)

For permission requests, please contact the Institute of Physics at the address below.

Institute of Physics
Dirac House, Temple Back
Bristol BS1 6BE UK

Phone: 44 1 17 929 7481
Fax: 44 1 17 920 0979

techtracking@iop.org

Additional copies of this publication are available from:

Curran Associates, Inc.
57 Morehouse Lane
Red Hook, NY 12571 USA
Phone: 845-758-0400
Fax: 845-758-2633
Email: curran@proceedings.com
Web: www.proceedings.com

Table of contents

Volume 106

The 4th International Seminar on Sustainable Urban Development

9–10 August 2017, Jakarta, Indonesia

Accepted papers received: 18 December 2017
Published online: 31 January 2018

Preface

Papers

IOP Conf. Series: Earth and Environmental Science **106** (2017) 011001 doi:10.1088/1755-1315/106/1/011001

Preface

First of all, praise be to Allah, the Cherisher and Sustainer of the world, for His blessing for all of us. He who has provided us a chance so that we could be here. On behalf of the Organizing Committee, it is our great pleasure to welcome you to **the 4th International Seminar on Sustainable Urban Development (ISoSUD) 2017**, held by Faculty of Landscape Architecture and Environmental Technology, Universitas Trisakti, Jakarta, Indonesia at Hotel Ciputra. To the academicians, our colleagues from overseas universities, guests, participants, students, please accept our gratitude, warm welcome and appreciation.

Let me deeply express a special appreciation to the speakers:
1) Assoc. Prof. Dr. Nada Khaleefah M. A. Alrikabi from Environmental & Urban Planner, Center of Urban and Regional Planning for Post Graduate Studies, Baghdad University, Iraq
2) Dr. Yusnani Mohd Yusof from Urban Planning, Affordable City, Universiti Brunei Darussalam
3) Prof. Dr. Pasi Lehmusluoto, from Department of Limnology and Environment Protection, University of Helsinki, Finlandia
4) Prof. Dr. Mohd. Razman Salim, from Department of Environmental Engineering, Faculty of Civil Engineering, Universiti Teknologi Malaysia, Malaysia

Achieving the SDGs requires of governments, private sector, civil society and citizens alie to make sure we leave a better planet for future generations. All 17 Goals interconnect, meaning success in one goal affects success for others. In short word, this is the greatest chance we have to improve life for future generations. Some of the goals that in major issue of **4th International Seminar on Sustainable Urban Development** are:

- Goals No. 6: Clean Water and Sanitation
- Goals No. 7: Affordable and Clean Energy
- Goals No. 11: Sustainable Cities and Communities
- Goals No. 13: Climate Action
- Goals No. 14: Life Below Water
- Goals No. 15: Life on Land

The seminar gathers eminent speakers and scholars that are engaged in Sustainable Urban Development field. The theme for this event is:

THE SUSTAINABLE DEVELOPMENT GOALs: IMPROVE LIFE FOR FUTURE GENERATION

The sub-themes of the seminar include, but not limited to :
1) Ecological Disaster Mitigation and Adaptation
2) Urban Environemntal Management
3) Environmental Technology

This year, the committee received over than 150 papers both in oral or poster presentation from 15 universities in Indonesia and from UK, Japan, Malaysia, Italy, and Iraq as well. Selected papers will be proposed to be published in IOP Publisher Conferences Series: Earth and Environmental Science, indexing by Scopus and Thomson Reuters. Other selected papers will

IOP Conf. Series: Earth and Environmental Science **106** (2017) 011001 doi:10.1088/1755-1315/106/1/011001

be published in National Journal accredited by Ministry of Research and Higher Education of Indonesia, i.e: Aceh International Journal of Science and Technology published by Universitas Syiahkuala, Aceh and Jurnal Reaktor published by Universitas Diponegoro, Semarang.

We are very grateful to Rector of Universitas Trisakti-Jakarta, Dean of Faculty of Landscape Architecture and Environmental Technology, Head of Environmental Engineering Department, Head of Landscape Architecture Department, Head of Urban and Regional Planning Department, the members of Advisory board, the members of Steering committee, Peer Reviewers, and Organizing committee that very supported and helpful within the preparations and conduction this seminar. Our grateful deliver as well to Association of Experts in Sanitary Engineering (TP) and Environmental Engineering (TL), IATPI (Ikatan Ahli Teknik Penyehatan dan Teknik Lingkungan), for their support to the seminar. Our appreciation is also for all the participants who have actively written excellent research papers. It is desired to have a sustainable seminar to be continuously held in the future times, as we are challenged to make a sustainable building and environment for a sophisticated life. Last but not least, we would like to express our gratitude to all the session chairs, reviewers, participants, institutions for their contribution to 4th ISoSUD 2017.

We believe that this event will be able to facilitate a good networking among researchers, scientists, engineers and practitioners of with common interest especially in sharing the latest research results, ideas, development and applications in the Sustainable Urban Development. We hope you enjoy the seminar and find this experience inspiring and helpful in your professional field. We look forward to seeing you at our upcoming seminar in the next 3 years on 5th ISoSUD 2020.

The Editors
Dr. Astri Rinanti
Dr. Rositayanti Hadisoebroto
Dr. Ade Gafar Abdullah
Dr. Eng. Asep Bayu Dani Nandiyanto

IOP Conf. Series: Earth and Environmental Science **106** (2017) 011001 doi:10.1088/1755-1315/106/1/011001

LIST OF COMMITTEE

Conference Chair:
Dr. Astri Rinanti

International Advisory Boards:

Prof. Dr. Ir. Dato' Zainal Mohamed, Universiti Teknologi Malaysia, Malaysia
Prof. Hadi Nur, Ph.D., Universiti Teknologi Malaysia, Malaysia
Prof. Ir. Joni Hermana, M.Sc., ES, Ph.D., Institut Teknologi Sepuluh Nopember, Indonesia
Dr. Nada Khaleefah Mohammed Ali Alrikabi, University of Baghdad, Iraq
Prof. Ir. Iwan Kridasantausa Hadihardaja, M.Sc., Ph.D., Institut Teknologi Bandung, Indonesia
Prof. Dr. Ing. Ir. Prayatni Soewondo, MS., Institut Teknologi Bandung, Indonesia
Prof. Ir. Suprihanto Notodarmojo, Ph.D., Institut Teknologi Bandung, Indonesia
Prof. Dr. Ir. Enri Damanhuri, Institut Teknologi Bandung, Indonesia
Prof. Haryoto Kusnoputranto, M.D., Dr. PH, Universitas Indonesia, Indonesia
Prof. Ir. Puji Lestari, Ph.D., Institut Teknologi Bandung, Indonesia
Assoc. Prof. Ir. Edwan Kardena, Ph.D., Institut Teknologi Bandung, Indonesia
Dr. Ade Gafar Abdullah, Universitas Pendidikan Indonesa, Indonesia
Dr. Asep Bayu Nandyanto, Universitas Pendidikan Indonesa, Indonesia
Prof. Dr. Ir. Budi Tjahyati, Universitas Trisakti, Indonesia
Prof. Ir. Agus Budi Purnomo, M.Sc., Ph.D., Universitas Trisakti, Indonesia
Prof. Dr. Ir. Dadan Umar Daihani, DEA, Universitas Trisakti, Indonesia
Prof. Dr. Ir. Indra Surjati, MT., Universitas Trisakti, Indonesia
Prof. Ir. Asri Nugrahanti Adjidarmo, MS., Ph.D., Universitas Trisakti, Indonesia
Prof. Dr. Lambang Subagiyo, M.Si., Universitas Lambung Mangkurat, Indonesia
Assoc. Prof. Dr. Ir. Rustam Hakim, MT, Universitas Trisakti, Indonesia
Dr. Ir. Hefni Effendi, Institut Pertanian Bogor, Indonesia

Scientific Committee

Dr. Iriana Safitri Zen, Universiti Teknologi Malaysia, Malaysia
Dr. Ina Krisantia, MS., Universitas Trisakti, Indonesia
Dr. Ir. Rustam Hakim, MT., Universitas Trisakti, Indonesia
Dr .Ir. Dwi Indrawati, MS., Universitas Trisakti, Indonesia
Dr. Ir. Ratnaningsih, MT., Universitas Trisakti, Indonesia
Dr. Melati Ferianita Fachrul, MS., Universitas Trisakti, Indonesia
Dr. Ir. Diana Hendrawan, M.Si., Universitas Trisakti, Indonesia

Organizing committee

Dr. Rositayanti Hadisoebroto, ST., MT - Universitas Trisakti, Indonesia
Dr. Ir. Diana Hendrawan, M.Si - Universitas Trisakti, Indonesia
Dr. rer. Nat. Widyatmoko - Universitas Trisakti, Indonesia
Dr. Yayat Supriatna, MS - Universitas Trisakti, Indonesia
Ir. Winarni, M.Sc - Universitas Trisakti, Indonesia
Pramiati Purwaningrum, ST., MT - Universitas Trisakti, Indonesia
Ir. Etty Indrawati, MT - Universitas Trisakti, Indonesia
Ir. Titiek Purwanti Debora - Universitas Trisakti, Indonesia
Ir. Bambang Iswanto, MT - Universitas Trisakti, Indonesia
Drs. M. Lindu, MT - Universitas Trisakti, Indonesia
Tazkiaturrizki, ST., MT - Universitas Trisakti, Indonesia
Lailatus Siami, ST., MT - Universitas Trisakti, Indonesia

Ir. Qurrotu Aini Besila, M.Si - Universitas Trisakti, Indonesia
Olivia Seanders, ST - Universitas Trisakti, Indonesia
Irmawati, SE - Universitas Trisakti, Indonesia
Arie Yulia Prasasti, A.Md - Universitas Trisakti, Indonesia

CONFERENCE PHOTOGRAPH

IOP Conf. Series: Earth and Environmental Science **106** (2017) 011001 doi:10.1088/1755-1315/106/1/011001

IOP Conf. Series: Earth and Environmental Science **106** (2017) 011001 doi:10.1088/1755-1315/106/1/011001

KEYNOTE SPEAKER
INVITED SPEAKER
CHAIRPERSON
TOSHIBA
MODERATOR
PRESENTER

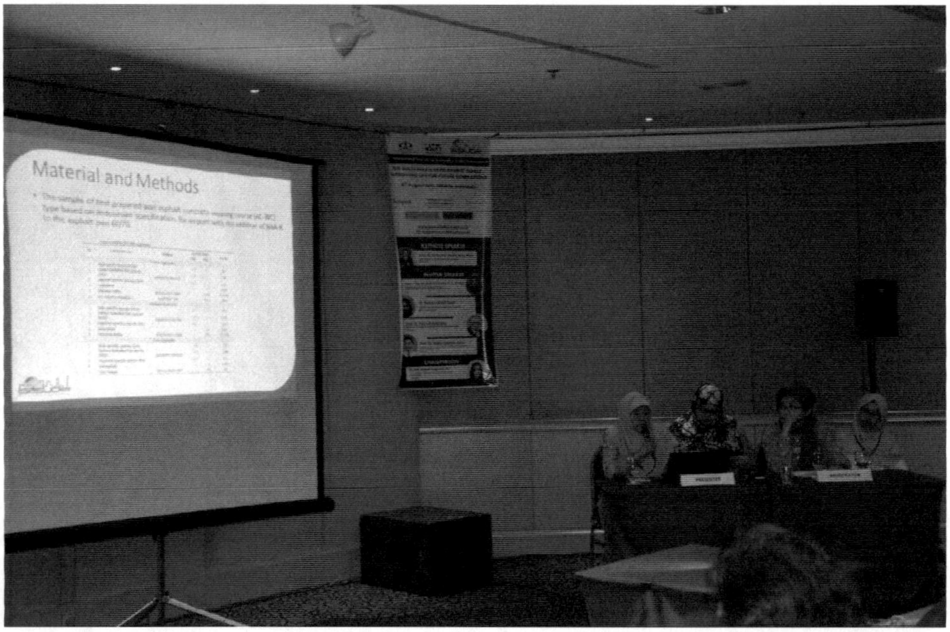
Material and Methods

IOP Conf. Series: Earth and Environmental Science **106** (2017) 011001 doi:10.1088/1755-1315/106/1/011001

4th International Seminar on Sustainable Urban Developme

Peer review statement

All papers published in this volume of *IOP Conference Series: Earth and Environmental Science* have been peer reviewed through processes administered by the proceedings Editors. Reviews were conducted by expert referees to the professional and scientific standards expected of a proceedings journal published by IOP Publishing.

Georadar and geoelectricity method to identify the determine zone of sliding landslide

Y K Dalimunthe, A Hamid*

Faculty of Earth Technology and Energy, Universitas Trisakti, Jakarta, Indonesia

*Corresponding Author: yusraida@trisakti.ac.id

Abstract. The aim of this research is to determine the contrast between the sliding plane by observing the parameters of rock types, fractures, and faults that could potentially land slides in Bandar Baru, Lampung Barat, Indonesia by both methods of georadar and geoelectricity. This research uses radar reflection profiling configuration for georadar and dipole-dipole configuration for geoelectricity. For georadar data processing has been done with Reflexwave software and for geoelectricity, data processing has been done with Earthimager 2DINV software to interpret subsurface section. Results of research by both methods of georadar and geoelectricity shows the area of contact between the sand stone with resistivity value of 200–1449 Ωm and clay stone with a resistivity value of 32–100 Ωm at the limit depth of 9 m as a potential zone of sliding landslides where the physical properties of clay stone easily derail massive material on it.
Keywords: fault, fracture, geoelectricity, georadar, landslide

1. Introduction

Studies of geophysical exploration in recent years have been progressing in electronics technology, producing new methods or electronic devices and affecting the development of radar equipment used in explorations. GPR method (Ground Penetrating Radar) is a new approach that is applied in this study. Tantamount to the seismic method in exploration using wave sources—either the natural wave coming from the earth itself or the artificial ones that are intentionally generated and inserted through the earth's surface to be captured by itself—the method of GPR employs radar pulses that are transmitted, reflected, and scattered by subsurface structures and anomalies. Physical parameters including magnetic permeability, electrical permittivity, and conductivity determine the characteristics of the electromagnetic wave radiation in the earth structure which will be forwarded, scattered, and reflected. The excellence of georadar method, among others, is the accuracy in detecting or mapping the subsurface structures such as ground water exploration, pipeline, fossils of ancient archaeology, exploration of minerals, and so on [4].

Geoelectric method is also a commonly used approach in geophysical exploration to investigate the subsurface structures based on the differences of rock resistivity [8]. Both of these methods are known as environmentally friendly approaches and quite good in imaging shallow layers.

West Lampung Province is an area that has a high level of mass movement hazard, traversed by Sumatra active fault lines of Semangko segments which make Sumatra island vulnerable areas of disasters, such as earthquakes and landslides) [5].

Based on these descriptions, this study takes advantage of georadar and geoelectric method to see the contrast between the sliding plane by observing the parameters of rock types, fractures, and faults that could potentially cause landslides in Bandar Baru, Lampung Barat, Indonesia.

1.1. Landslide

Landslides are formed by movements or destruction of rocks, soil, or mixture materials, shifting down or out of the slopes. The process of landslides begins with the water soaking into the soil that will add weight to the ground. If the water penetrating the soil acts as a watertight sliding plane, the ground becomes slippery, and the soil weathering on it will divert to follow the slope and eventually out of the slopes.

The soil movements that could potentially cause landslides are classified into five categories based on their mechanism and material types, i.e., falling, toppling, sliding (slump/rotational sliding and lateral sliding), spreading, and flowing [7].

1.2. Georadar

In principle, GPR method works by using signal reflection as shown in Figure 1. GPR system has a transmitter circuit connected to the pulse source and a receiver circuit linked to a signal processing unit. The transmitter circuit will generate an electrical pulse to form PRF (pulse repetition frequency), energy, and certain duration. The antenna will emit pulses into the ground, where the pulse will experience attenuation and signal defect during their propagation [1]. Homogeneous soil will reflect weak signals. If the pulse has struck inhomogeneities on the ground, then there will be reflected signals to the receiving antenna. Furthermore, these signals will be processed by the receiver circuit. By measuring the time interval between the pulse transmission and acceptance, the depth of the object can be determined. In this time interval, pulses go back and forth from the antenna to the object and back to the antenna. If t describes the interval, v represents the speed of electromagnetic wave propagation in the soil, and h denotes the depth of objects, then the equation can be written as follows:

$$h = \frac{1}{2}tv \quad (1)$$

The velocity of electromagnetic wave propagation must be known in order to determine the depth of the detected object. The propagation speed depends on the speed of light in the air and the relative dielectric constant of the propagation medium,written as follows:

$$v = \frac{c}{\sqrt{\varepsilon_r}} \quad (2)$$

1.3. Geoelectricity

Geoelectricity is one of the geophysical methods that utilize the resistivity properties to investigate the circumstances under the earth's surface. This approach is carried out by injecting electric currents through two electrodes, penetrating into the earth. Subsequently, the potential difference formed between them and two additional potential electrodes in another place is observed to reflect the circumstances beneath the earth's surface. The concept of electrical current propagation in a homogeneous isotropic medium is used in this method, where equal values of electric currents move in all directions. Based on this assumption, if there is any distinguished anomaly meeting the amount of the flowing current, it is probably caused by the difference of the irregular resistivity. This anomaly will be used then to reconstruct the state of the subsurface geology.

When the two electrodes with the current I are injected into the homogeneous isotropic soil/rock medium with a certain (finite) distance, then the potential at points near the surface will be influenced by the two electrodes of the stream, so that the equipotential generated from two point sources is the sum of the potential caused by the two electrodes.

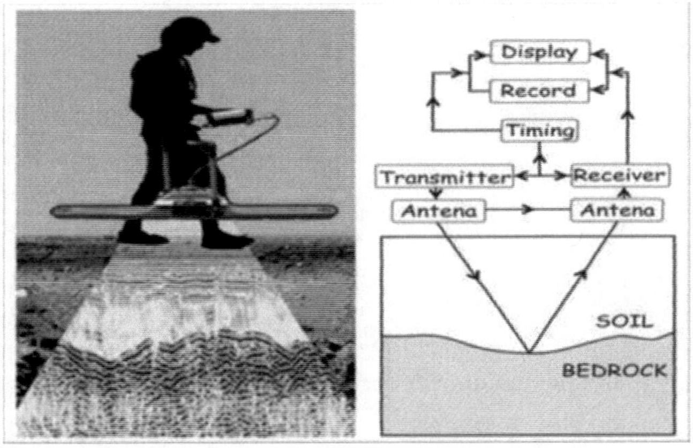

Figure 1. Working principles of GPR method [3].

If a cross section through the source C1 and C2 were made, the visible pattern of the equipotential field distribution would be as shown in Figure 2. To determine the potential difference between two points generated by two current electrodes C1 and C2, other two potential electrodes of P1 and P2 are placed near the sources as depicted in Figure 2.

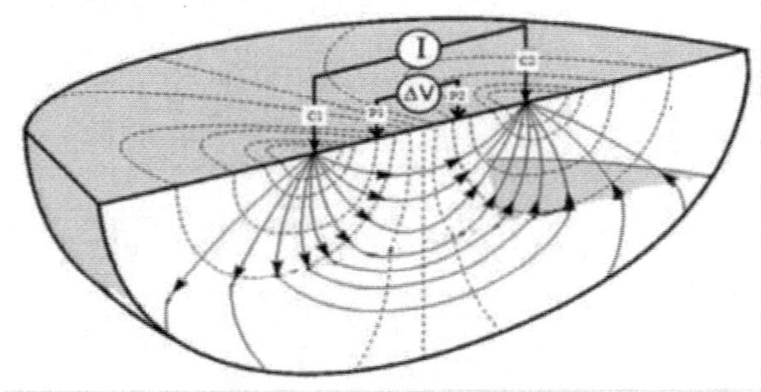

Figure 2. The configuration of the current electrodes and the potential electrodes on the surface of a homogeneous isotropic medium [6].

2. Research Method

This research was conducted in Bandar Baru, Lampung Barat, Indonesia (04°57'34.1"–04°57'36.3"S latitude and 104°03'23.6"–104°03'26.4"E longitude). A set of GSSI (*Geophysical Survey Systems Inc.*) SIR-20 georadar equipment and Geoelectricity Multichannel Supersting R8 were used to attain the objectives of this study.

Data for georadar method were acquired using an antenna with a frequency of 200 MHz along the trajectory measurements, while the geoelectricity approach was performed by dipole-dipole configuration with an interelectrode distance of six meters. The maximum line length measurement was adapted to the topography of the study area. Measurements of line and area for georadar and geoelectricity methods are the same (Figure 3).

Reflexwave software was used for georadar data processing, whereas *Earthimager 2DINV* module was used for geoelectricity data processing. For georadar, data based on the radargram analysis of subsurface rock types were interpreted through each radar facies as shown in Figure 4. Furthermore, based on Table 1, the resistivity value was adjusted for geoelectricity data.

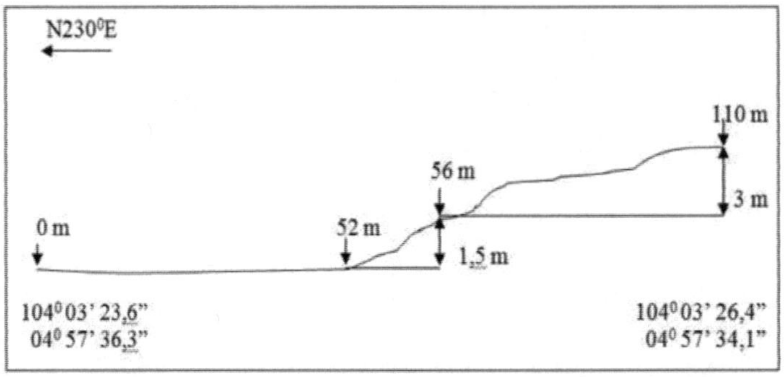

Figure 3. Line and area of georadar and geoelectricity measurements.

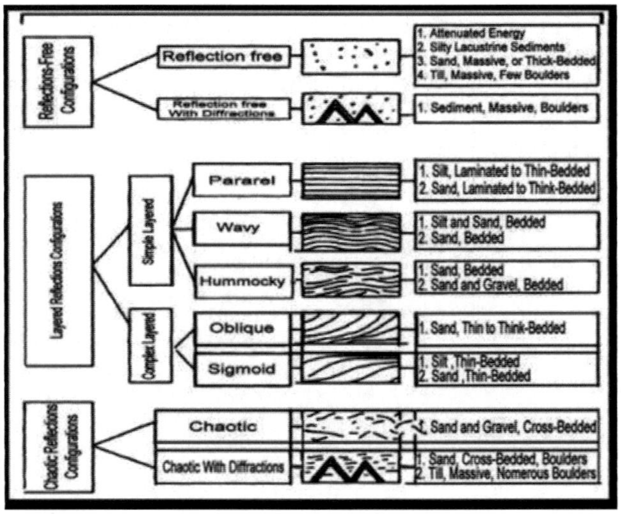

Figure 4. Determination of subsurface sediment types [2].

3. Results and Discussion

Georadar data measurements with 200 MHz frequency transducer were carried out by passing the Semangko segment of Sumatra fault, and the measurement position was determined using GPS Garmin 76 CSX. The results are shown in Figure 5. Measurements with geoelectric method were performed on the same trajectory with the georadar approach using dipole-dipole configuration type on an interelectrode of upto six meters. The results are depicted in Figure 6.

3.1. Georadar

Topographic measurements on track 2 generate the form of low undulating hills with a lithological condition of sandy loam surface. This trajectory is the main fault lineament of Semangko segment. At varying depths of 0–9 m, layered reflections in the form of simple layered configurations of hummocky/subparallel were found, instead of constantly intermediate to high possibilities such as sand and gravel. At a depth of 9–12 m, reflection-free configurations were obtained in the form of low-amplitude possibilities of clay, sand, fine to coarse-grained sediments (chewy to solid level).

IOP Conf. Series: Earth and Environmental Science **106** (2017) 012001 doi:10.1088/1755-1315/106/1/012001

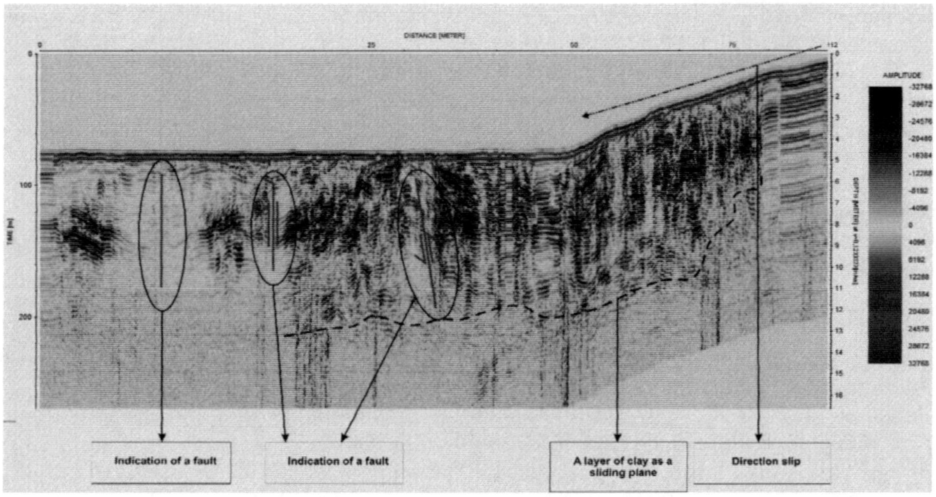

Figure 5. The results of georadar data processing on line and their interpretation.

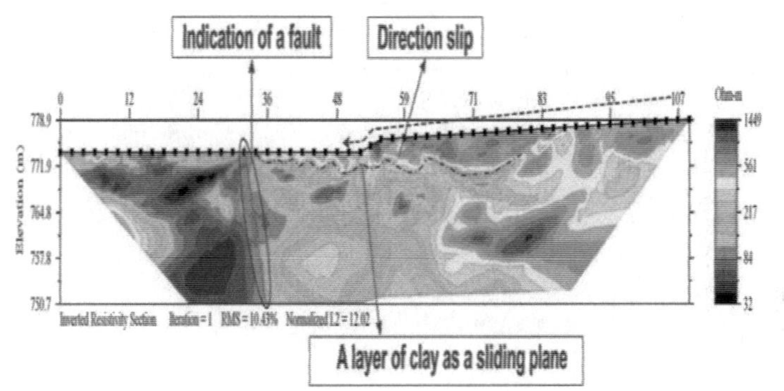

Figure 6. The results of geoelectricity data processing on line and their interpretation.

3.2. Geoelectricity

In this layer, the dark–light blue range with a resistivity value of 32–100 Ωm were estimated to consist of clay, while green–red line with a resistivity value of 200–1449 Ωm were estimated to consist of sandy and igneous rocks. Data processing of geoelectric and georadar indicate a boundary between the sandstone and clay. This clay layer under the sand structure can function as a sliding plane where the sandstone rock absorbs water so that the weight increases, especially during the rainy season. Besides, the trajectory profile sketch shows the angle of 4.45^0, indicating that the soil easilyshifts from its original position. Georadar and geoelectric data processing also indicate the presence of a fault, and in case of rain, the rocks and sand can be slipped into the area by cesarean formed clays.

Based on the results, discussion, as well as the geological maps, this area is prone to ground movements. On account of the ground motion trajectory profile, this region can be included on the type of soil translational movement with a characteristic of flatly or gently sloping corrugated ramps.

4. Conclusion

Georadar measurement results state that at varying depths of 0–9 m, layered reflections in the form of simple layered configurations of hummocky/subparallel were found, instead of constantly intermediate to high possibilities such as sand and gravel. Mean while, at a depth of 9–12 m, reflection-free configurations were obtained in the form of low-amplitude possibilities of clay, sand, fine to coarse-

IOP Conf. Series: Earth and Environmental Science **106** (2017) 012001 doi:10.1088/1755-1315/106/1/012001

grained sediments (chewy to solid level). The regional subsurface structure of Bandar Baru, Lampung Barat, Indonesia, is composed of clay stone with resistivity values of 32–100 Ωm, as well as sandy rocks and igneous rocks (200–1449 Ωm), according to the measurement by geoelectricity method. Some indications of a weak zone or fault were encountered at the red line on the data interpretation. The depth limit of 9 m is considered as a potential zone of sliding landslides where the physical properties of clay stone easily derail massive materials on it.

Acknowledgment

Geology Survey Center of The Ministry of Energy and Mineral Resources is acknowledged for the support to carry out the present study.

References

[1] Annan A P. 2003. Ground Penetrating Radar Principles, Procedures & Applications, Sensor & Software Inc., 1040 Stacey Court, Mississauga, ON L4W 2x8 (Canada).

[2] Beres M, and Haeni F P. 1991. Application of Ground Penetrating Radar Methods in Hydrogeologic Studies, *Ground Water.* 3(29):375–386.

[3] Ervan M. 2013. The Basic Theory of Ground Penetrating Radar, Puslitbang (Bandung).

[4] Deniyatno. 2014. Sliding Field Identification Zone Landslide with Georadar Method. *Journal of Physics Applications*. 7(2).

[5] Khairani Y, Sismanto, Ervan M, and Januar. 2014. Identification of Mass Movement Area in Liwa, Lampung Barat by Using Georadar and Geoelectricity Method. *International Conference on Physics* (ICP 2014).

[6] Knödel K, Krummel, H, and Lange G. 1997. Handbuch zur Erkundung des Untergrundes von Deponien und Altlasten, Band 3—Geophysik, Bundesanstalt fur Geowissenschaften und Rohstoffe (Springer Verlag, 1063 S).

[7] Vernes DJ. 1978. Slope Movement Types and Processes, in Schuster, R L , and Krizek, R J, eds., Landslides. Analysis and Control: National Research Council, Washington, D.C., Transportation Research Board, Special Report. 176:11–33.

[8] Waluyo, 2012. Free Workshop Exploration Geophysics. Theory and Applications. The Geophysical Laboratory, Faculty of Mathematics and Natural Sciences UGM, Yogyakarta.

[9] http://www.subsurfacesurveys.com/pdf/Methods.pdf.

Analysis of slope stabillity and controlling factor on residual soil of folded breccia formation

S Rachman*[1,2], D Muslim[2], N Sulaksana[2], M Burhannuddinnur[1], H Pramudito[1]

[1]Universitas Trisakti, Jakarta, Indonesia
[2]Universitas Padjadjaran, Bandung, Indonesia

*Corresponding Author: sofyan@trisakti.ac.id

Abstract. This research aims to obtain a potential landslide zonation.Theresearch area is located in Depok Village and surroundings, Jatigede District, Sumedang regency, West Java province. Geographically located at the point of coordinates 06°50'33 "- 06°51'00" South Latitude and 108°05'37 "- 108°06'17" East Longitude. This research is intended to mapping the identification of landslide and soil properties data. The mapping and soil sampling were conducted only in the research area. The methodology used was mapping and finding the safety factor with Bishop Analysis. The morphological condition of the study area indicates moderate conditions undulating hilly area with slopes between15° - 40°, with a tick soil layer was covering the slope. This condition is greatly affected by rainfall. This research is to know the type of ground movement along with the value of the safety factor of the slope so that can provide suggestions for overcoming instability in the study area.
Keywords: bishop, landslide, safety factor

1. Introduction

The process of moving large material from one place to another lower due to the influence of gravity both fast and slow [1]. The slope material may consist of soil or rocks [2].

Slope stability is one of factor in occupation that related with mining and human activities of construction. Anthropogenic activity such as excavation and ground levelling, can modify the existing environmental condition, including soil composition [3].

This stability of slope is related with availability of various works, i.e. at construction site, dam construction, road making process and irrigation flow etc. In doing the activities it was necessary to make anticipation about the slope problem. Each slopes have differ stability, so it is necessary to get some information related to slope stability. The information about the zoning of landslide potential areas is very important to be used as a basis for consideration of regional development. This matter is done to reduce victim that caused by landslide. This research is intended to mapping the identification of landslide and soil properties data [4]. This research aims to obtain a potential landslide zonation in a downstream of Dam. These natural phenomena included in one of most economic-destructive disaster, either wealth loss or victim especially if it occurred in crowded area where people do daily activities. Even though land slide cannot be prevented, however there is a possibility to reduce the cause of this disaster.

IOP Conf. Series: Earth and Environmental Science **106** (2017) 012002 doi:10.1088/1755-1315/106/1/012002

2. Research Method

In this research, the method is using basic geological mapping, lab analysis and studio analysis, therefore in this research can be divided into several steps (Figure 1).

2.1. Geological setting and analysis of slope stability

2.1.1. Geomorphology of Area

According to researchers [5] in the study area is divided into 3 forms of morphology based on relief, i.e. :

1) Geomorphological Unit of Fluvial Plain
2) Geomorphological Unit of Wavy Hills
3) Geomorphological Unit of Very Steep Hill.

2.1.2. Geomorphological Unit of Fluvial Plain

This fluvial plain geomorphological unit is located in the southern and south western of research area. The spread of this morphological unit occupies about 20% of the total research area. In this plain fluvial area, there are river stream. In this unit also found a high erosion river.

This geomorphological unit is located in southwestern and eastern researched area. Distribution of this unit cover in about 50% of the total researched area. With the percentage of degree of slope 14%-20%, and has a difference elevation of 180 m -27 m, with parallel contour pattern and corrugated contour pattern.

This very steep hill geomorphological unit is located alongside the western area. Distributions from this geomorphological unit cover about 30% from the total of researched area. With percentage of slope degree 21% – 50%, and have altitude 210 m – 300 m, with rounded contour pattern and wavy contour pattern. Lithology in this unit is claystone and breccia tuff.

2.1.3. Stratigraphy of Area

According to researchers [5], researched area can be categorized in Bogor Zone. And based on Physiographic Zone, this area has age comparison according to researcher [6]. Stratigraphy of researched area is described as from older age to the younger one: Volcanic Breccia Unit, Claystone Unit, Breccia Tuff Unit and Alluvial.

2.1.4. Volcanic Breccia Unit

Outcrops are clearly visible to be observed as they appear along the cliff. However, at some place this unit quite weathered. This breccia has characteristic like fragmentation with pebble size and categorized as andesitic. The andesitic categorized due to its dark grey colored outcrop with sandstone matrix.

2.1.5. Claystone Unit

Outcrop looks quite fresh as exposed in the cliff. This unit has properties that are easily expandable and in the field will look like inserted with volcanic breccia and breccia tuff. This unit consists of claystone that have clayey property. The appearance of this unit has gray color, soft and cracked.

2.1.6. Alluvial

This unit named with river deposition because in this area, occurred river depositions formed pebble and cobble stone. Deposition type that visible looks vary from fresh cobble to weathered cobble. This unit consists of river deposition from clay size, sand size, to cobble deposited mainly in river base and at the base of a cliff.

2.1.7. Rainfall

In research area, rainfall is effected by topographical of surrounding area, and weather condition of research area. The rainfall intensity on this area is about 2500 – 3000 mm/year. Rainfall condition on this area as one factor of landslide occurrence, if those areas have high intensity of rainfall and poor-resistant lithology so it is possible to occur a landslide.

2.1.8. Land Use

The land use of research area divided into 5 (five) cover of land use

1) Paddy field
   This land use cover about 10% from the total of land use. The paddy field located at eastern of research area with wavy hill-shaped geomorphology.
2) Bushes
   This area covers about 40% from area the total of research area. The bushy area located in northern and southwestern area with steep hill-shaped geomorphology.
3) Settlements
   The settlement area covers about 20% area from the total of researched area. This settlement located in eastern and southeastern of research area with wavy hill-shaped geomorphology.
4) Field
   The field of this area covers about 10% area from the total of research area. The field is located in eastern of researched area with very steep hill-shaped geomorphology.
5) Farm
   Farm in this area cover about 20% area from the total of research area. The farm area is located in northern and southwestern of researched area with steep hill morphology.

2.2. Slope stability analisys

The research area divided to a several segment with some section. This division purposed to acquire the Safety Factor Score (SF Score) to know a land movement on that cell area (Figure 1).

Figure 1. Safety Factor Score (SF Score) to know a ground movement on that CELL area.

Table 1. Safety factor analisys in the Cell area.

No	Cell	Slope angle (degree)	Safety factor	Remark
1	2	22	1.520	Stable
2	3	16	1.337	Stable
3	10	18	1.397	Stable
4	9	19	1.347	Stable
5	12	19	1.388	Stable
6	6	26	1.820	Stable
7	7	19	0.540	Landslidealreadyoccurredbefore

No	Cell	Slope angle (degree)	Safety factor	Remark
8	4	12	1.047	Critical
9	8	19	1.156	Critical
10	11	-	-	Rock
11	1	-	-	Rock
12	5	-	-	Rock

Based on safety analysis factor in the cell area, slopestability and also relatively unstable slope with sometime or frequent chance of landslide occurrences was obtained (table.1).

3. Results and Discussion

3.1. Geomorphological Effect

Geomorphological effect to ground movement is caused by some factors that caused landslide; morphology shape, slope tilt, rock condition of an area, rainfall intensity and land use [7]. In research are a geomorphological unit divided into low tilted geomorphological unit, wavy hill geomorphological unit, and steep hill geomorphological unit. Indication of landslide occurrence caused in very steep geomorphological unit where the slope on that area is quite steep.

3.2. Rainfall Effect

Rainfall can trigger a landslide trough accumulation of slope weighting and increasing pore in land or rock. Rainfall is a main factor of a landslide occurrence if the intensity of rainfall is very high so that speed up the erosion and weathering process in rock. Also in this area, the rainfall is categorized as a normal one 2500 – 3000 mm/year.

3.3. Land Use Effect

In research area, land use including paddy fields, bushes, settlements, field, and farm. The vulnerable area from a landslide are located in paddy field and field area. Paddy field area are more vulnerable than bushes, settlements, field, and farm area. This is happen because water is saturated in paddy field area so that creating a sliding plane especially in steep slope. If the intensity of rainfall in research area is high, it will cause a water accumulation so that potentially to occur a landslide

3.4. Lithological Effect

In the research area, lithological unit is divided by 3 (three) units [2]. Breccia tuff unit, claystone unit, and volcanic breccia unit. From that three unit, breccia tuff are very vulnerable of landslide because the physical properties of this unit is very easily weathered.

3.5. Vulnerable Zone of Ground Movement

Vulnerable Zone of Low Ground Movement

1) Cover plain fluvial geomorphological unit (0 – 2 %)
2) Lithology: breccia tuff
3) Land use: farm, bushes
4) Have safety factor score (SF) between 1,397 – 2,190 and included in relatively stable slope with rare chance of landslide

Vulnerable Zone of Middle Ground Movement

1) Cover steep hills geomorphological unit (14 –20%)
2) Lithology: breccia tuff, claystone, and volcanic breccia
3) Land use: bushes and farm
4) Have safety factor score (SF) between 1,330 – 1,049 and included in critical slope with single chance of landslide

Vulnerable Zone of High Ground Movement

1) Cover very steep hills geomorphological unit (21–55%)
2) Lithology: breccia tuff, volcanic breccia
3) Land use: bushes
4) Have safety factor score (SF) between 0.540 – 0.525 and included in slightly unstable slope with normal or frequent chance of landslide

3.6. Landslide Mitigation

There are several ways to overcome a disaster especially landslide. There are reducing water intake on the surface, big-root tree planting to reduce water flow on the surface, make a steep slope to a more plain, reducing pore pressure on slope, installing wall like *'bronjong'*.

4. Conclusion

Morphology of research area according to geomorphological analysis can be divided by 3 (three) geomorphological unit there are, plain fluvial geomorphological unit 0 – 2 %, and very steep hills geomorphological unit 21–55%. River flow pattern in research area are dendritic with adult stage. Lithological unit in researched area divided into 4 (four) units from old to young there are. Volcanicbreccia unit, claystone unit, brecciatuff unit, andalluvial. Rainfall intensity at researched area is about 2500 – 3000 mm/year. Land use of researched area divided into 5 (five), there are rice fields, bushes, settlement, field, and farm. From slope stability analysis of research area, slope stability divided into 3 (three) there are vulnerable zone of low ground movement, vulnerable zone of middle ground movement, and vulnerable zone of high ground movement. The mitigation of the area which have a moderate to high ground movement should make a plain slope, and also make some drainage control, make a holding wall and installment of wall like *'bronjong'*.

Acknowledgment

Acknowledgments to the Dirjen DIKTI for grant competed with the organization. And support in geology engineering programs.

References

[1] Van Zuidam RA, and FI Van Zuidam Cancelado, 1979. Terrain Analysis and Classifacation Using Aerial.
[2] Karnawati, D., 2005, Bencana Alam Gerakan Massa Tanah di Indonesia dan Upaya Penanggulangannya. Jurusan Teknik Geologi, Universitas Gadjah Mada, Indonesia. ISBN 979-95811-3-3.
[3] Pradhan, B. and Lee, S. 2010: Landslide susceptibility assessment and factor effect analysis: backpropagation artificial neural networks and their comparison with frequency ratio and bivariate logis-tic regression modeling, Environ. Modell. Softw., 25:747–759,
[4] Varnes, D J. 1984: Landslide hazard zonation: a review ofprinciples and practice, Commission of Landslide of IAEG, UNESCO, Natural Hazads, 3, UNESCO, Paris, 63 pp.
[5] Bishop, A. W., Morgenstren, N. R., 1960. Stability Coefficient for Earth Slopes. Geotechnic,Vol.10
[6] Carrara, A.: A multivariate model for landslide hazard evaluation, Math. Geol., 15, 403–426, 1983
[7] Djuri, 1995. Peta Geologi Lembar Arjawinangun, Jawa Barat, Skala1 : 100.000. Direktorat Geologi Bandung, Bandung.

Intelligent seismic risk mitigation system on structure building

R Suryanita[1*], H Maizir[2], E Yuniorto[1], H Jingga[1]

[1]Civil Engineering Department, Faculty of Engineering, Universitas Riau, Pekanbaru, Indonesia
[2]Civil Engineering Department, Sekolah Tinggi Teknologi Pekanbaru, Indonesia

*Corresponding Author: reni.suryanita@eng.unri.ac.id

Abstract. Indonesia located on the Pacific Ring of Fire, is one of the highest-risk seismic zone in the world. The strong ground motion might cause catastrophic collapse of the building which leads to casualties and property damages. Therefore, it is imperative to properly design the structural response of building against seismic hazard. Seismic-resistant building design process requires structural analysis to be performed to obtain the necessary building responses. However, the structural analysis could be very difficult and time consuming. This study aims to predict the structural response includes displacement, velocity, and acceleration of multi-storey building with the fixed floor plan using Artificial Neural Network (ANN) method based on the 2010 Indonesian seismic hazard map. By varying the building height, soil condition, and seismic location in 47 cities in Indonesia, 6345 data sets were obtained and fed into the ANN model for the learning process. The trained ANN can predict the displacement, velocity, and acceleration responses with up to 96% of predicted rate. The trained ANN architecture and weight factors were later used to build a simple tool in Visual Basic program which possesses the features for prediction of structural response as mentioned previously.
Keywords: artificial neural network, mitigation, seismic, structural analysis

1. Introduction

Indonesia is one of the high-risk seismic-zone in the world, which refers to the geographical region with the most active tectonic plate and volcanic activities on earth as known as the Pacific Ring of Fire. This results in a high tendency of strong ground motion to occur due to earthquake in the Pacific Ring of Fire region. In 2004, a whopping 9.3 Richter-scale mega quake struck Aceh on the Western Coast of Sumatera Island, which then followed by a tsunami that travelled several kilometers inland. In 2009, a devastating 7.9 Richter-scale earthquake hit Padang City, also the Western Coast of Sumatera Island. Recently in March 2016, another 7.8 Richter-scale earthquake hit Southwest of Sumatra and the ground motion was felt on the whole Sumatera Island [1]. In all the aforementioned cases, the property damage was severe and the casualty was huge.

The spectral hazard maps for Sumatra and Java islands was developed by [2]. Two hazard levels for representing 10% and 2% probability of exceedance (PE) in 50 years ground motions were analysed for Sumatra and Java. The analysis implemented some improvements in seismic hazard by considering the latest seismic activities around Java and Sumatra. The authors also proposed a revision of the seismic hazard map in Indonesian Seismic Code SNI 03-1726-2002 which partially adopts the concept of UBC 1997 [3].

The new revision Indonesian seismic code as known as SNI 03-1726-2012. According to [4], prior to 2012, the seismic design criteria for buildings in Indonesia is based on map with ground motion spectral accelerations of 10% probability of being exceeded (PE) in 50 years. The seismic design criteria and map was hazard-based without considering uncertainty from collapse capacity of building structures. Otherwise, in the new seismic design criteria included 2% PE in 50 years, defined as Maximum Considered Earthquake (MCE). In the new code has adopted the most recent data and current state of knowledge in probabilistic and deterministic seismic hazard assessment methodologies, and using the most recent ground motion predictive equations.

The new MCE ground motion parameter for 1.0 second spectral acceleration, site class B with 5% of critical damping as shown in Figure 1.

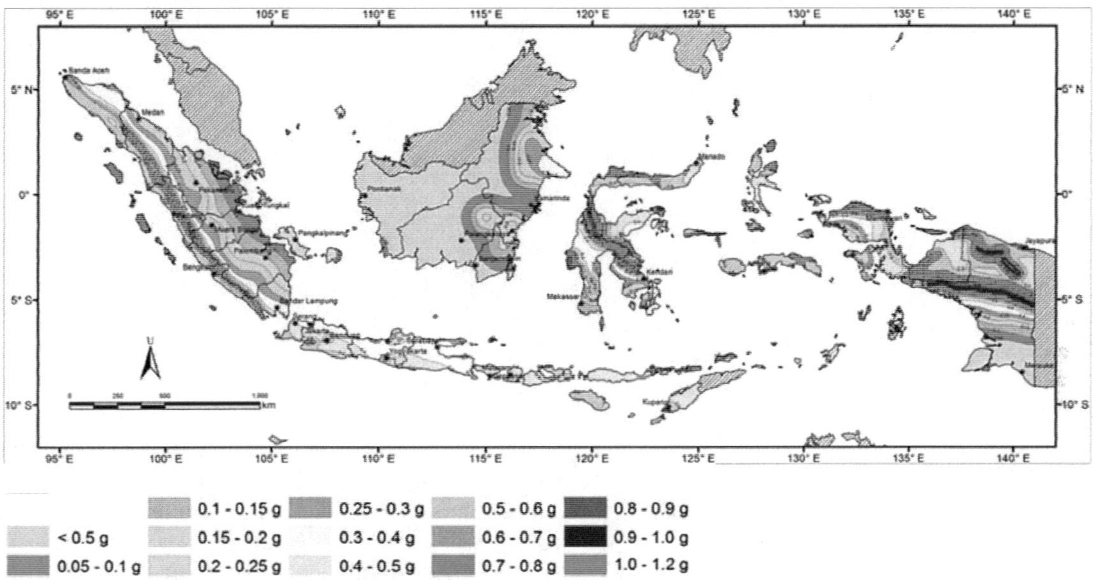

Figure1. Spectra design map of Indonesia for 1 second spectral acceleration with 5% [5].

One of the many factors that affect the aftermath of earthquake disaster is the resilience of the infrastructure building against the strong ground motion [6]. Critical infrastructure-buildings such as hospital, school, power plant office, and governmental building are most likely multi-storey buildings which are very prone to seismic loading. During strong ground motion, multi-storey building might collapse in brittle-way that endangers its occupant's due to the massive dead weight, especially for Reinforced Cement Concrete (RCC) building. Furthermore, in case of tall building is not designed properly will experience excessive displacement (storey-drift) that cause discomfort and might damage non-structural components such partition wall, window, and door which blocks evacuation passage. Due to these facts, multi-storey building shall be designed properly to exhibit ductile behaviour and controlled deformations during strong ground motion.

Seismic-resistant building design requires structural analysis to be performed first, to obtain some building response characteristics, such as storey displacement (drift), velocity, and acceleration. However, such structural analysis could be very difficult, especially for 3D building structural models. For complex building structure, the structural analysis will require the involvement of finite element structural analysis program which is very costly and time consuming to learn and operate.

Artificial Neural Network (ANN) is a mathematical model inspired by its biological neural network counterpart. The ANN system comprises of several processing layers and neurons. Just like the biological neural network, the connection and signal transfer between neurons and layers enable the ANN system to process the given input signal into appropriate outputs, which is later called prediction. ANN possesses the capability to predict output based on any given input in which the mathematical relationship between the input and output parameter is nonlinear, complex, and often vague.

IOP Conf. Series: Earth and Environmental Science **106** (2017) 012003 doi:10.1088/1755-1315/106/1/012003

The output layer of the ANN consists of output neurons that represent the output parameters to be predicted. The difference between the predicted output value and the target value (the true value according to learning data set) is the error of the ANN system. ANN neuron's functionality is analogue to the biological neuron. The synapse strength in biological neural network is represented by the weight factor in the ANN system. The initial values of the weight factors are usually random, which later modified through a process called ANN training, iteration, or learning process. The ANN learning process requires a set of data to 'train' the ANN before it is ready for testing. The trained ANN system is expected to possess the capability to predict outputs based on any given inputs at decent accuracy. The commonly adopted criteria to evaluate the performance of the ANN system are Mean-Squared-Error (MSE) and Coefficient of Correlation (R) are computed using (1) and (2), respectively.

$$MSE = 0.5\,(T_i - Y_i)^2 \tag{1}$$

$$R = \frac{n\sum T_i Y_i - (\sum T_i)(\sum Y_i)}{\sqrt{n(\sum T_i^{\,2}) - (\sum T_i)^2}\,\sqrt{n(\sum Y_i^{\,2}) - (\sum Y_i)^2}} \tag{2}$$

Where: T_i = target value based on learning data set;
Y_i = predicted output value; and
n = the number of data sets.

This research aims to predict the structural response includes displacement, velocity, and acceleration of multi-storey building in the region of seismic hazard maps of Indonesia using an Artificial Neural Network (ANN) tool. The final software was designed to be user friendly, simple-to-use, lightweight, and performs faster. The previous study [7] discussed prediction of structural response based on the seismic hazard maps of Sumatera. The study was successful to predict the story drift of multi-storey building in all the capital cities of the provinces in Sumatera. The prediction capability of the ANN-based system was achieved through a vigorous learning process with over 4000 of data sets. Meanwhile, other researchers have applied an Artificial Neural Networks to predict response spectra such as [8], and to generate the artificial earthquake such as [9], [10] and [11]

2. Research Method

The prediction system based on an ANN analysis, which requires an amount of learning data sets to perform the training, validation, and testing process. In this research, the ANN data sets were generated by performing structural analysis on several varieties of building the structure model, soil condition, and seismic location. In the following sub-sections, the methodology used in this research will be described in detail.

The multi-storey building structure models are reinforced cement concrete (RCC) moment frames combined with shear walls. In this research, 3 variations of building height are adopted: 10 storeys (Model 1), 15 storeys (Model 2), and 20 storeys (Model 3). The inter-storey height is 4.5 meters at base and 4 meters at other storeys.

Modal response spectrum analysis was performed to obtain the responses of the building structure models (storey displacement, velocity, and acceleration). The seismic load was included as seismic response spectrum plot which shows the relationship between the design structure acceleration (S_a) and the structure's period of free vibration (T). The S_a vs. T plot varies with soil condition and seismic location. In this research, 34 capital cities and 13 other cities in Indonesia were selected as seismic location with 3 soil conditions (soft, medium, and hard soil). By adopting 47 cities in Indonesia with 3 possible soil conditions, 141 seismic response spectrum plots were obtained (e.g Banda Aceh City is shown in Figure 2). For each seismic load, 10 building response data were generated from modal response spectrum analysis from Model 1, 15 data from Model 2, and 20 data from Model 3, which sums up to 45 data. Therefore, as many as 6345 data sets (141 x 45) were generated from the whole structural analysis process.

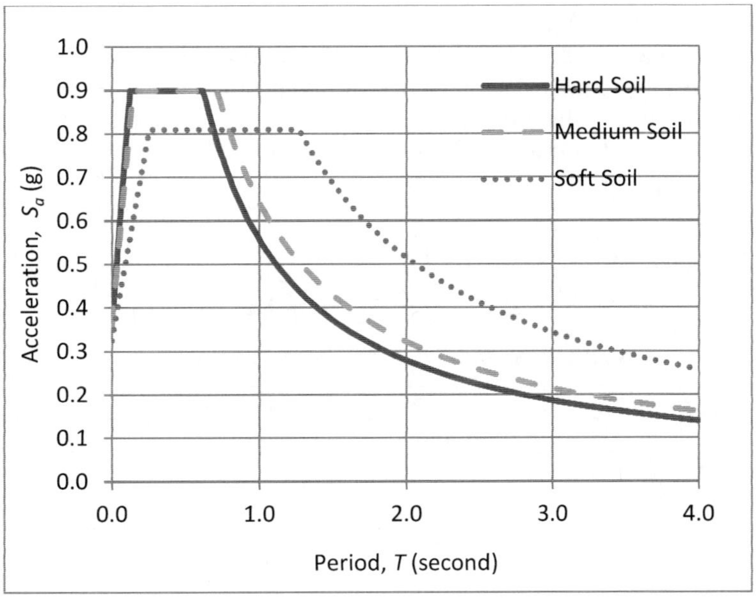

Figure 2. Seismic response spectrum plot for Banda Aceh City [5].

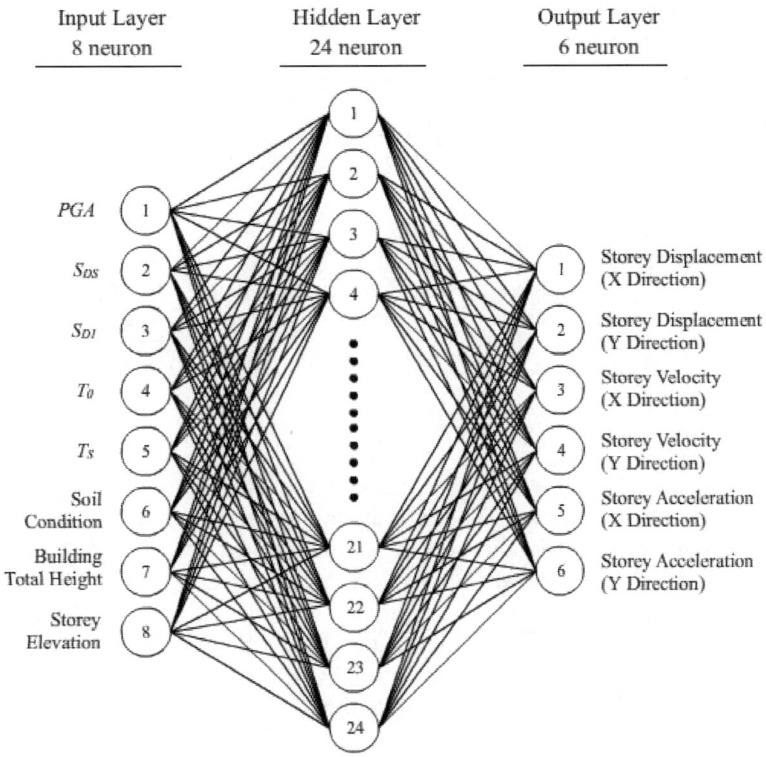

Figure 3. Proposed backpropagation ANN architecture.

The proposed backpropagation ANN architecture on the prediction of building structure response in this research as shown in Figure 3. The ANN architecture consists of 3 layers: input layer with 8 neurons, hidden layer with 24 neurons, and output layer with 6 neurons. The input parameters are peak ground acceleration (PGA), design spectral acceleration at brief period (S_{DS}), design spectral acceleration at 1 second of the period (S_{D1}), the lower limit of period that results in maximum acceleration (T_0), the upper limit of period that results in maximum acceleration (T_S), soil condition, building total height, and storey elevation (base level was not included). Whereas, the output parameters are storey displacement, velocity, and acceleration in both orthogonal horizontal directions (X and Y).

3. Results and Discussion

The details on the MSE and *R* values obtained through the ANN learning process is tabulated in Table 1 and Table 2 after 1000 epochs during the ANN learning process.

Table 1. MSE after ANN learning process.

Parameters	Mean-Squared-Error (MSE)		
	Training	Validation	Testing
Displacement X	1.09×10^{-4}	1.01×10^{-4}	1.00×10^{-4}
Displacement Y	1.05×10^{-4}	0.96×10^{-4}	0.96×10^{-4}
Velocity X	2.05×10^{-4}	2.14×10^{-4}	1.96×10^{-4}
Velocity Y	1.99×10^{-4}	1.99×10^{-4}	1.88×10^{-4}
Acceleration X	4.04×10^{-4}	4.13×10^{-4}	3.80×10^{-4}
Acceleration Y	3.80×10^{-4}	3.93×10^{-4}	3.43×10^{-4}
Average	2.34×10^{-4}	2.36×10^{-4}	2.17×10^{-4}

Table 2. *R* Details after ANN learning process.

Parameters	Coefficient of Correlation (*R*)		
	Training	Validation	Testing
Displacement X	0.982	0.981	0.988
Displacement Y	0.982	0.981	0.988
Velocity X	0.972	0.964	0.982
Velocity Y	0.972	0.965	0.983
Acceleration X	0.928	0.901	0.957
Acceleration Y	0.928	0.899	0.959
Average	0.961	0.949	0.976

4. Conclusion

The comparison of displacement, velocity and acceleration data have been concluded based on MSE mean value and regression mean value of the network model. Both comparisons show the MSE mean value decreases since the epoch increases. Whereas regression value increases close to 1 since the epoch increases. According to the results, the neural networks' method based on the displacement data has the best performance rather than velocity and acceleration data. The reason describes the displacement is derived from second time to generate the acceleration. The displacement has simpler physic quantity rather than acceleration so the convergent is approached faster. Furthermore, both calculated MSE and R value indicate that the prediction performance of the trained ANN is sufficiently accurate. The ANN is a very promising tool to provide an early prediction on structural response such as story drift (displacement, velocity and acceleration at multi-story building in the region of Indonesia to assist further Finite Element Method analysis.

References

[1] BMKG (Meteorological Climatological and Geophysical Agency), Earthquake Database, 2016. [Online]. Available: http://repogempa.bmkg.go.id/.

[2] M Irsyam, D T Dangkua, D Hoedajanto, B M Hutapea, E K. Kertapati, T Boen, and M D Petersen. 2008. Proposed seismic hazard maps of Sumatra and Java Islands and microzonation study of Jakarta city, Indonesia. *J. earth Syst. Sci.*, 117(2):865–878.

[3] M Irsyam, M Asrurifak, Hendriyawan, B Budiono, W Triyoso, and A Firmanti. 2010. Development of spectral hazard maps for a proposed revision of the Indonesian Seismic Building Code. Geomech. Geoengin. *An Int. J.* 5(1):35–47.

[4] I Wayansengara, and D Hutabarat, Development of Earthquake Risk-Targeted Ground Motions for Indonesian Earthquake Resistance Building Code SNI 1726-2012, pp. 1–6, 2015.

[5] Puskim PU, Desain Spektra Indonesia Pusat Penelitian dan Pengembangan Permukiman, 2011. [Online]. Available: http://puskim.pu.go.id/Aplikasi/desain_spektra_indonesia_2011/. [Accessed: 01-Jan-2015].

[6] U. N. G. Assembly, The Sendai Framework for Disaster Risk Reduction 2015–2030, 2015.

[7] R Suryanita, H Maizir, and H Jingga. 2016. Prediction of Structural Response due to Earthquake Load using Artificial Neural Networks, in *International conference on Engineering & Technology, Computer, Basic & Applied Sciences ECBA*, Osaka, Japan, 182(4).

[8] E Bojórquez, J Bojórquez, S E Ruiz, and A Reyes-Salazar, Prediction of inelastic response spectra using artificial neural networks. Math. Probl. Eng., vol. 2012, 2012.

[9] S. Rajasekaran. 2006. Generation of artificial earthquake motion records using wavelets and principal component analysis *J. Earthq. Eng.* 10(5):665–691.

[10] S C Lee, and S W Han. 2002. Neural-network-based models for generating artificial earthquakes and response spectra. *Comput. Struct.*. 80(20):1627–1638.

[11] F Azam, M Sharif, M Yasmin, and S Mohsin. 2014. Artificial intelligence based techniques for earthquake prediction: a review. *Sci Int.* 26(4):1495–1502.

Microclimate landscape design at southern integrated terminal Bandar Tasik Selatan, Kuala Lumpur

L H Phin[1], I Krisantia[2*]

[1]Landscape Architecture Department, Faculty of Design and Architecture, Universiti Putra Malaysia, Malaysia
[2]Landscape Architecture Department, Faculty of Landscape Architecture and Environmental Technology, Universitas Trisakti, Jakarta, Indonesia.

*Corresponding Author : inakrisantia@trisakti.ac.id

Abstract. Bandar Tasik Selatan is the integrated transport terminal has high energy consuming, high carbon emission and poor linkage. However, microclimate can be reduced through landscape design. This paper is a study to achieve energy efficiency and improve microclimate in the urban area. The research area is at Southern integrated terminal Bandar Tasik Selatan Kuala Lumpur Malaysia. It is carried out through a case study and microclimate analyzed using System Modeling method. System modelling using in this research is system energy budget of the microclimate at a site is a balance between the radiant energy supplied and the energy removed by all consumers. The finding indicated the microclimatic components that can be modified through landscape design are solar radiation, wind and precipitation can create thermal comfort, energy efficiency and others benefits.Through this research, provide more green space to achieve energy efficiency and improve microclimate of the site, introducing vertical landscape and proper planting selection to improve air quality, introducing green energy as part of the source of power supply and to promote integration of terminal building and rail systems by unify them using softscape
Keywords: energy budget, microclimatic, urban landscape

1. Introduction

Microclimatic landscape design requires knowledge of prevailing climate conditions, understanding of the ways in which objects in the landscape affect climate to create microclimates. The microclimatic components that can be modified through landscape design are wind, radiation and precipitation. Besides, air temperature and humidity can be altered and can have an effect on human comfort or to reduce energy consumption of buildings [1].

Studies by the Lawrence Berkeley National Laboratory estimate a 25%–50% reduction in annual cooling energy consumption through well-designed landscape design. Refer to researcher [2] since climatic factors affect energy use, therefore, energy usage in cooling of building can be control by controlling the microclimate. Natural element such as landform, vegetation, wind flow, water bodies and more can be make use to modify the surrounding microclimate of buildings. Thus, the energy consumption of the building will be reduced.

Other elements in landscape planning for energy conservation include building orientation, selection of materials and more. It is clear that microclimatic landscape design can help create energy efficiency and it will be possible to improve the urban fabric whilst minimizing our environmental impact include minimizing energy cost and reducing air pollution [3].

Therefore, it is vital to implement the design to all new planning. The other benefits of microclimatic design include ensuring outdoor human thermal comfort, to ensure environmental quality, to reduce pollution, and more.

Microclimate is the condition of the solar and terrestrial radiation, wind, air temperature, humidity and precipitation in a small outdoor space [1]. Microclimatic design requires a conceptual understanding on how microclimatic components such as wind and solar radiation, can be significantly affected from landscape elements [1]. Plants have a strong effect on microclimate. Trees and green spaces can help to cool our cities and save energy. Trees can provide solar protection to individual houses during summer and evapotranspiration from trees can reduce urban temperatures. Trees also help mitigate the greenhouse effect, filter pollutants, mask noise, prevent soil erosion, and calm their human observers. Shading from trees is an effective way to significantly reduce energy for cooling purposes.

System energy budget of the microclimate at a site is a balance between the radiant energy supplied and the energy removed by all consumers. Supplied Energy removed by consumer are account of terrestrial radiation, conduction, evaporation, and connection consumers. The four major available consumers are terrestrial radiation by objects at the site, conduction of heat into objects, evaporation of water, and convection of heat from objects into the air by wind.

Southern Integrated Terminal Bandar Tasik Selatan is high energy consuming since it is equipped with a 6-storey terminal building and three rail systems which are KTM, LRT and KLIA Transit. The transit zone is definitely the gathering point of buses, cars and peoples, thus, it is high carbon emissions. Besides, there is poor linkage between terminal building and rail stations. However, microclimate can be reduced through landscape design. This paper is on the study to achieve energy efficiency and improve microclimate in the urban area.

2. Research Method

2.1. Location
Southern Integrated Terminal Bandar Tasik Selatan Kuala Lumpur

2.2. Data collection
Data collected in the form of primary data and secondary data, in the form of physical data and non physical site

2.3. Method
A case study and microclimate analyzed using System Modeling method. System modelling using in this research is system energy budget of the microclimate at a site is a balance between the radiant energy supplied and the energy removed by all consumers [1]. Radiant energy supplied is a radiant energy Supplied. Energy removed by consumers is account of terrestrial radiation, conduction, evaporation, and connection.

The four major available consumers are terrestrial radiation by objects at the site, conduction of heat into objects, evaporation of water, and convection of heat from objects into the air by wind [1].

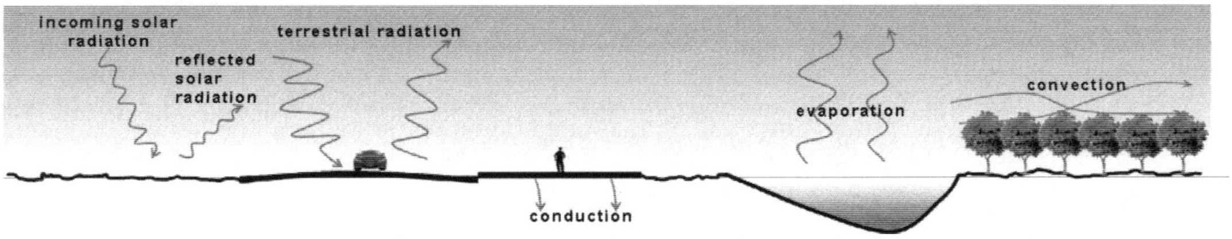

Figure 1. Illustration of Energy Budget

IOP Conf. Series: Earth and Environmental Science **106** (2017) 012004 doi:10.1088/1755-1315/106/1/012004

3. Results and Discussion

3.1. Climate

Kuala Lumpur has a tropical rainforest climate which is warm and sunny, along with abundant rain fall. The average temperature in Kuala Lumpur is 27.5°C. Kuala Lumpur's climate receives an average of 2,409 mm of rainfall per year or 201 mm per month. The average annual relative humidity is 62.6% and average monthly relative humidity ranges from 58% in March to 66% in May & November. There is an average of 2,228 hours of sunlight per year with an average of 6.1 hours of sunlight per day.

Solar energy and surface at Kuala Lumpur insolation is a measure of solar radiation energy received on a given surface area in a given time. The average insolation at Kuala Lumpur is 4.8 kWh/m2per day (Table 1).

Table 1. Solar energy and surface meteorology at Kuala Lumpur [4].

Variable	I	II	III	IV	V	VI	VII	VIII	IX	X	XI	XII
Insolation, kWh/m²/day	4.29	5.06	5.22	5.33	5.08	4.91	4.87	4.99	5.04	4.83	4.21	3.77
Clearness, 0 - 1	0.45	0.50	0.50	0.52	0.51	0.51	0.50	0.50	0.49	0.48	0.43	0.40
Temperature, °C	24.35	24.98	25.62	26.07	26.10	25.73	25.34	25.39	25.50	25.73	25.26	24.58
Wind speed, m/s	3.56	2.97	2.61	1.61	1.58	2.58	2.58	2.78	2.17	1.72	2.58	3.56
Precipitation, mm	147	137	218	264	210	130	141	154	190	268	278	232
Wet days, d	14.8	14.6	17.2	19.6	16.2	12.3	14.2	15.4	18.2	22.1	24.8	21.1

The shadow of terminal building is most obvious because the height and size of terminal building. During morning, the terminal building is only able to provide shade to west side of the building while the other areas are exposed to the sun. On the other hand, afternoon sun can shade the east side of the building. Prevailing wind at the site is from North West direction. Therefore, the potential area to develop for windy cooling purpose is area facing North West direction. Figure 2 shows the sun direction and the compare is on of shadow terminal building at 8am, 2 pm and 5pm.

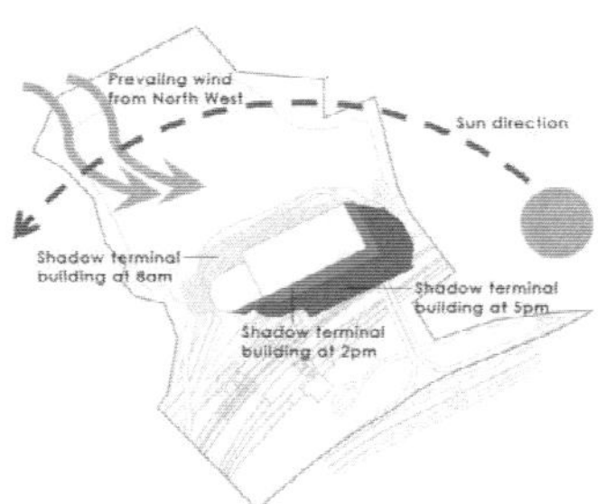

Figure 2. Analysis of microclimate.

3.2. Vegetation

The plants around Southern Integrated Terminal at Bandar Tasik Selatan are mainly the introduced landscape trees. Most of them are not mature trees since they are just newly planted at the site. The plants selections are lacking in provide shade and trap air particulates.

3.3. Topograf, Hidrology

The general topography of the site is almost flat, where it has elevation are within 40 to 50 meter above sea level. Therefore, the site is very suitable for development. The lowest elevation on site is at retention pond, which is below 45 meter above sea level. The elevation is become higher with the distance from retention pond. The retention pond is called 'Kolam Pemadam'. It is an ex-mining pond Kampung Malaysia Raya. The function of the pond is to cater the large flow of storm water to avoid from flooding of this area. There is water inlet and outlet that allow water flow in and out. The retention pond has been modified through reclamation while constructing the Southern Integrated Terminal Bandar Tasik Selatan.

3.4. Visual
As a summary of visual study, the categories of views at the site are mostly ordinary views with visual amenities of retention pond and main terminal building. Some views that have portraits the sense of place as transportation hub are under category of exceptional view.

3.5. Building
The location of terminal building, LRT, KTM and KLIA Transit station and the pedestrian bridge. Terminal building have total of 6 levels with height of 46 m. The terminal building is fully air conditioning and there is excessive lighting to light up the building during night, herefore the energy consumption of the building is very high. The existing and new pedestrian bridge is 1 level different. There is no proper ramp for users. This will trouble the users if the lift is not functioning. The width of the pedestrian bridge is notable to cater large flow of users during peak hour.

3.6. Land use

Tabel 2. The percentage of land use.

Land use	Area (m^2)	Percentage (%)
Retention pond	108,000	32.5
Public open space	46,030	13.8
Terminal, station, parking and road system	178,730	53.7
Total	332,760	100.0

Generally, the site has developed and there is limited space to be developed except for the public open space around retention pond. The land use for terminal, station, parking and road system is highest, which is 53.7%. It is apparently that road system and railway have monopoly large area of the site since it is a multi- modal integrated public transportation terminal. Therefore, large paved area has contributed to urban heat island effect and road system is highest no clear orientation can contributed to urban heat island effect because street orientation is a fundamental factor in providing shading [5]. The main green space at the site is the public open space around retention pond, which is 13.8%. It is definitely ot enough green space at the site to balance with the paved area (Table 2). Therefore, all available open space should be utilized to increase green space.

3.7. Implementation Microclimatic in Landscape Design

3.7.1. Programing
The programmes in the relation with Southern Integrated Transport Terminal include: Pick up area, Drop-off area, Parking, Waiting area and Meeting area.

While programmes in the relation with objectives include : Green area, Vertical landscape, Linkage, Green energy, Green roof according to programmes a conceptual plan of space has developed. Available open spaces are mostly proposed as green space to achieve the design objective. These supported that greenspace measures contribute to a decrease in local temperatures of up to 2°C for summer conditions [6]. Pick up area and drop-off area are separate but near to each other to avoid congestion. Solar panel is potentially to install on the rooftop of arrival platform terminal. Rooftop of car park building is utilized for green area. Elevated plaza and extended platform are function as waiting and meeting area.

3.7.2. Design Concept

A good microclimate can also be one that is energy efficient, or water efficient, or in some way makes a more positive environment [7]. Using concept of 'Manipulate the microclimate' to utilize and control the microclimatic elements i.e. solar, radiation, wind, temperature and precipitation at Southern Integrated Terminal Bandar Tasik Selatan which are renewable and sustainable natural resources in landscape design to benefit both peoples and environment. By using microclimatic landscape design, it will be possible to improve the whilst minimizing our environmental impact include minimizing energy cost and reducing air pollution.

The key consumers that influence the heat that feel by the human and objects of site is terrestrial radiation. Since Malaysia possesses a humid tropical climate with warm and uniform temperature, the result that we want to get is to reduce the terrestrial radiation. Therefore, if we would like to reduce terrestrial radiation, we should reduce radiant energy supplied, increase conduction, evaporation and convection see Figure 3. The factors how to reduce terrestrial radiation is listed on Table 3.

Table 3. Conceptual of energy budget.

Energy supplied	Energy removed by consumer			
Radiant energy supplied	Terrestrial radiation	Conduction	Evaporation	Convection
- Increase shading - Higher albedo (higher solar reflectivity)	- Try to achieve lower surface temperature	- Higher admittance	- Exist of water - More plants	- Do not block the wind

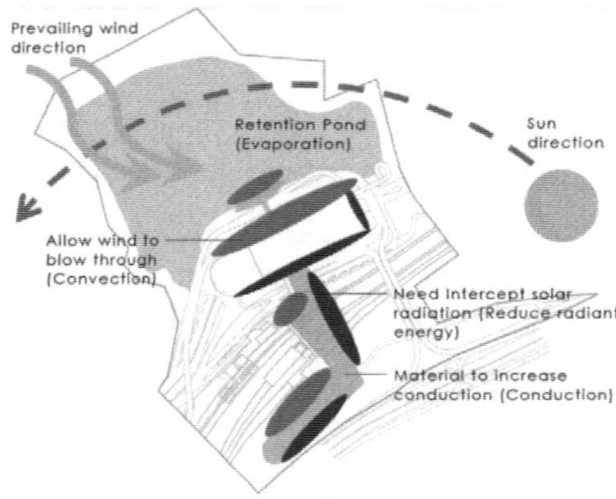

Figure 3. Conceptual Plan of energy budget.

Microclimatic landscape design is important to achieve energy efficiency, human thermal comfort and soon. Therefore, energy budget is taken into consideration such as low radiant energy supplied, high conduction, high evaporation and high convection need to achieve. Vertical landscape and shade tree with multiple canopy layers are proposed to intercept solar radiation to reduce radiant energy supplied especially shade tree with multiple canopy layers [8] and higher tree [9] and types of trees among other landscape elements strongly ameliorate the microclimate [10]. For the area that facing prevailing wind from North West direction tall trees are proposed to allow wind to blow through. This will increase convection. The retention pond and plants will contribute to increase evaporation. Material selection should be high admittance to increase conduction. To achieve the objectives, green space is an important factor. Since there is limited space for green space on ground, so vegetated wall, green roof and roof top garden are proposed to increase green space of the site. According to programmes, a conceptual plan of space has developed. Available open spaces are mostly proposed as green space to achieve the design objective. Pick up area and drop-off area are separate but near to each other to avoid congestion. Solar panel is potentially to install on the rooftop of arrival platform terminal. Rooftop of car park building is utilized for green area. Elevated plaza and extended plat form are function as waiting and meeting area.

4. Conclusion

Southern Integrated Terminal Bandar Tasik Selatan is an essential infrastructure which is needed to lead our country towards a developed country. There will be more similar infrastructure to be build all over Malaysia to upgrade the transportation system of Malaysia and thus provide convenient public transport service to residents Malaysia. But nowadays environmental issues are becoming more serious and global awareness towards the issue is high. Therefore, the built environment must be integrated with the existing green environment to reduce the negative impacts towards our Earth. Microclimate element such as solar radiation, wind and precipitation are the renewable and sustainable natural resources. We should utilize them to benefit both peoples and environment. Solar radiation can be collected as solar energy and convert to electricity to achieve energy efficiency, wind can be utilized to provide cooling effect to the surrounding and rainfall can be collected and reuse for other purpose. By using microclimatic landscape design, it will be possible to improve the urban fabric whilst minimizing our environmental impact include minimizing energy cost and reducing air pollution.

References

[1] Brown R D, and Gillespie T J, Microclimatic Landscape Design Creating Thermal Comfort and Energy Efficiency, John Wiley & Sons, Inc, 2015.

[2] El Bardisy W M, Fahmy M, El Gohary G F. 2016. Climatic sensitive landscape design: towards a better microclimate through plantation in public schools, Cairo, Egypt. *Procedia-Social and Behavioral Sciences*. 216:206–216 UPADSD 14-16 October 2015

[3] Sahidan M F, Jones P, Plant Canopy Design in Modify Urban Thermal Environment: Theory and Guidelines, PLEA 2008 – 25 Conference on Passive and Low Energy Architecture, Dublin, 22 to 24 October 2008. http://www.plea-arch.org/ARCHIVE/2008/content/papers/oral/PLEA_FinalPaper_ref_179.pdf 4.00

[4] Gaisma. Kuala Lumpur, Malaysia. Retrieve don 5 March 2011 from http://www.gaima.com/en/location/kuala-lumpur.html

[5] http://csfs.bue.edu.eg/files/Library/Papers/Sustainability%20and%20the%20Future/66.pdf, November 17 at 3.45 am

[6] Meggers F, Aschwanden G, Teitelbaum E, Guo H, Bruelisauer M . Urban Cooling Potential: System Losses from Microclimates, *The International Building Physics Conference, IBPC* 2015, Energy Procedia 00 (2015) 000–000

[7] Robinette G O, Plants and their Environmental Functions. Washington DC, US Department of the Interior. 1983.

[8] Gaitani, N, Spanou A, Saliari M, Synnefa A, Vassilakopoulou K, Papadopoulou K Pavlou K, Santamouris M, Papaioannou M, and Lagoudaki A. 2011. Improving the microclimate in urban areas: a case study in the centre of Athens. *Building Serv. Eng. Res. Technol.* 32 (1):53–71

[9] Kenawy, I M, Afifi M M, Mahmoud A. H. The effect of planting design on thermal comfort in outdoor spaces.

[10] Robert D B, Design with Microclimate, The Secret to Comfortable Outdoor Space, Printed using Berkeley and Avant Garde © 2010 Robert D. Brown Island Press, Suite 300, 1718 Connecticut Ave., NW, Washington, DC 20009

The slope and incision length of affected local cross abrasion and accretion using ASTER GDEM image analysis

A Anugrahadi

Faculty of Earth Technology and Energy, Universitas Trisakti, Jakarta, Indonesia

Corresponding Author: afiat.a@trisakti.ac.id

Abstract. Remote sensing technology is to support the identification and assessment of resources and disasters in coastal areas and oceans, because it has the advantage of covering large areas and the highest of the spatial and temporal resolution. Aster GDEM image is used to determine the slope and the length of cross the incision on exposed area abrasion and accretion. Western coastal of Banten Province has experienced abrasion with the furthest distance of 125.05 m to 274.73 m. and experienced accretion with the furthest distance of 31.65 m to 111, 58 m. ASTER GDEM results of image analysis in areas of abrasion has a slope about 1.4° to 3.3° and cross the incision length is approximately 350.52 meters to 506.57 meters. At the accretion region has a slope about 2.0° to 3.1° and cross the incision length about 306.62 m to 562.05 m.
Keywords: ASTER GDEM, coastal, abrasion, accretion, slope

1. Introduction

Indonesia in general has three coastlines with one very different natural physical conditions, especially geologic settings. For example in Banten Province, the geological setting behind the seaside coastal area (Java Sea), coastal areas between islands (Sunda Strait), and open coastal areas (Indian Ocean), have different characteristics and can be utilized to support development and simultaneously Control it [1].

Remote sensing technology is very supportive in the identification and assessment of resources in coastal areas and oceans, because it has advantages that include large areas and high temporal resolution, many choices of remote sensing satellite types that have a pretty good accuracy in identifying objects on the surface of the earth [2], [3], [4], [5], [6], [7].

In relation to the unique and highly interesting coastal deposition environments to be observed above and the advantages of remote sensing technology, it is necessary to research the relevance of abrasion and accretion processes with slope slope slopes on the coast of Banten Province.

The aim of this research is to know the slope in coastal area affected by abrasion and accretion by using Advanced Space borne Thermal Emission and Reflection Image Radiometer Global Digital Elevation Model (ASTER GDEM) with the calculation of the length of the cross-sectional line so that it gets big slope, in the west coast of Banten Province. Research Sites in coastal areas of the sea, from Anyer to Tanjung Lesung, Pandeglang District, Banten Province with coordinates 6°17'30"LS-6°32'30"LS and 105°40'30"BT-105°50 '00"BT.Map of study sites affected by abrasion, accretion and calculation of slope/slope on the coast of Banten can be seen in Figure 1.

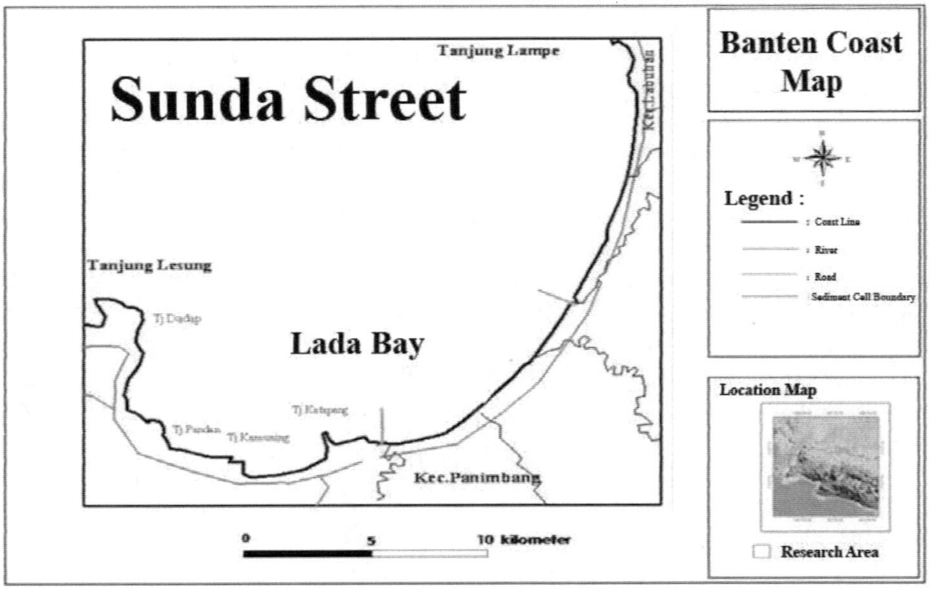

Figure 1. Map of research area.

2. Research Method

The research method used begins with literature study from the publication of some previous researchers, collecting secondary data in the form of Landsat TM and ETM + images, ASTER GDEM images of wind direction, currents, waves and bathymetry, retrieving primary data by conducting field survey to collect oceanographic data; Currents, wind direction, and bathymetry. Furthermore, the processing, analysis and discussion.

Many types of remote sensing satellites have good accuracy in identifying objects on the Earth's surface [8], [9]. Landsat Satellite (Land Satellite) is one of satellite resources that produce multispectral image. This satellite belongs to the United States, which was launched on July 23, 1972. Landsat 1, 2, 3 has MSS (Multi Spectral Scanner) and RBV (Return Beam Videocon) sensors, Landsat 4 and 5 have MSS and TM (Thematic Mapper) sensors. Landsat 6 and 7 have ETM and ETM + (Enhanced Thematic Mapper Plus) sensors. Landsat 7 ETM + has specifications such as high spectral resolution, which has 8 channels so that the ability to distinguish objects is relatively high. Landsat 7 ETM + has a temporal resolution or can record the same area every 16 days, it is very useful to obtain the latest data about the research area. The 7th generation Landsat has improved its spatial resolution with ETM + sensor in addition to generating 7-channel images such as the TM sensor, plus the 8th channel with a spatial resolution of 15 meters (panchromatic).

Method determination of abrasion and accretion area is to process Landsat image classification in 1982 and 2007 unsupervised with ER Mapper program. Enter the cropped image of the cropped image, enter the filename of the unsupervised classification result. Enter the number of iterations 100, unchanged 98.5 percent and the number of class 2 (only 2 classes because only want to do separation between land and sea).

After exiting the classification result, the next process is done with ArcView program. So the image data that has been classified needs to be exported into a format that can be read by the program ArcView. Enable layer 1982 and 2007 by ticking both layers. After the process will be generated new union results between 1982 and 2007. In view double click the union result layer and select the legend type: Unique Value and select the values field: ID. Change the color to: red = abrasion region, green = region of accretion, brown = fixed area [10].In this Landsat image it can show the West coast of Banten Province where there has been abrasion and accretion by comparing Landsat MSS and ETM +

IOP Conf. Series: Earth and Environmental Science **106** (2017) 012005 doi:10.1088/1755-1315/106/1/012005

in 1982 and 2007. Based on Unsupervised Methods of both imagery can be calculated the distance of abrasion and accretion, at 3 selected locations that have wide-shifted coastline changes seen from Landsat image in 1982 and 2007 (see Figure 2). That the research areas are in:

- Location 1 (Bay of Lada - Tanjung Lampe): has experienced Abrasion with the furthest distance = 274.73 m. Some regions experience Accretion with the furthest distance = 31.65 m
- In location 2 (Bay of Lada – Tanjung Dadap) has experienced Abrasion with the furthest distance = 206.69 m. Most beaches experience Accretion with the furthest distance = 111.58 m
- In the location of 3 coastal areas of Tanjung Lesung has experienced Abrasion with the farthest distance = 125.05 m, most beaches experience Accretion with the furthest distance = 68.71 m

Area that experienced Abrasion and Accretion by Unsupervised method that is in Location 1 Abrasion area = 615,022.77 m^2 and wide of Accretion = 9,943.14 m^2, Location 2 area of Abrasion = 848,492.98 m^2 and wide of Accretion = 24491.14 m^2, Location 3 Abrasion area = 106,966.98 m^2 and wide of Accretion = 65,841,54 m^2.Based on GDEM Aster image data can be calculated slope measurement (slope of the seafloor) and length of transverse incision, at 3 selected locations experiencing abrasion and accretion (see Figure 3, 4 and 5). The results of measurements in the research areas are:

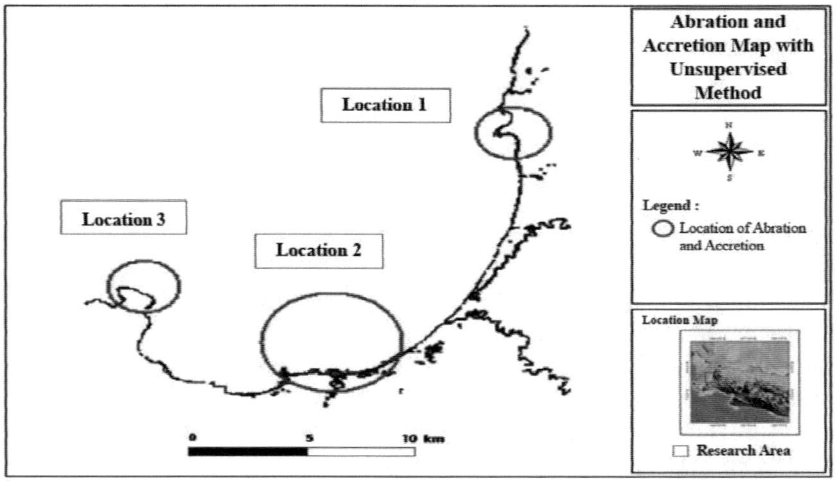

Figure 2. Results of abrasion and accretion by unsupervised method.

Location 1: in the affected area Abrasion = Slope = 2.6°, the length of Transverse Incision = 1662 feet = 506.57 meters, in the affected area Accretion = Slope = 3.1 °, Transverse Cross = 1006 feet = 306.62 meters (Figure 3a and 3b).

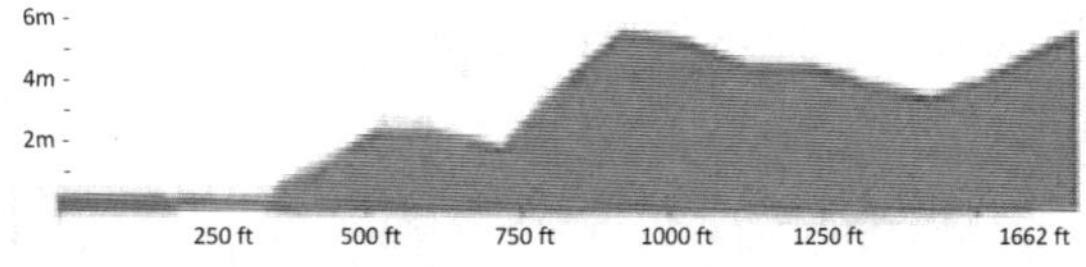

Slope = 2.6°, the length of Transverse Incision = 1662 feet = 506.57 meters

Figure 3a. Slope measurement in the abrasion location 1

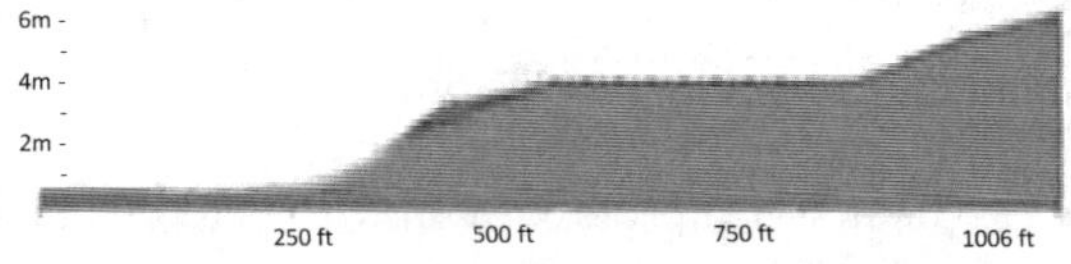

Slope = 3.1 °, Transverse Cross = 1006 feet = 306.62 meters

Figure 3b. Measurement of slope in accretion location 1

Location 2: in the affected area Abrasion = Slope = 3.3 °, the length of the Transverse Incision = 1639 feet = 499.56 meters, in the affected area Accretion = Slope = 3.1 °, Cross Crossover = 1231 feet = 375.20 meters (Figure 4a and 4b).

IOP Conf. Series: Earth and Environmental Science **106** (2017) 012005 doi:10.1088/1755-1315/106/1/012005

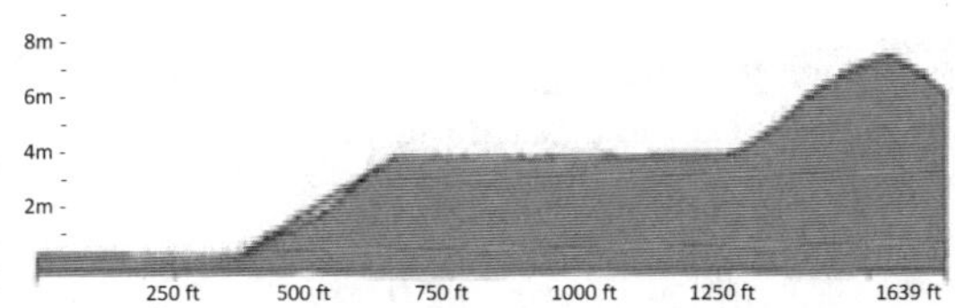

Slope = 3.3°, the length of the Transverse Incision = 1639 feet = 499.56 meters

Figure 4a. Measurement of slope in abrasion location 2.

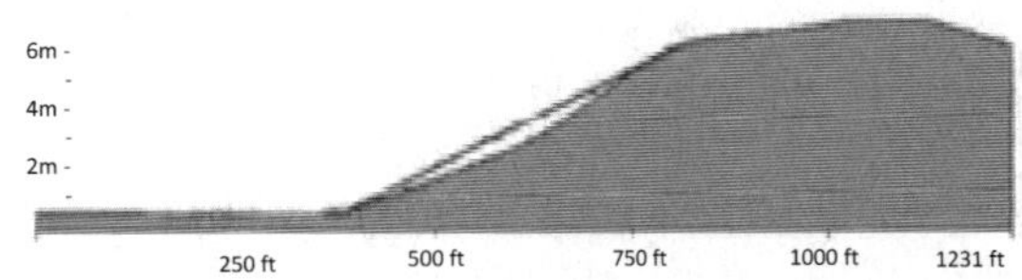

Slope = 3.1°, Cross Crossover = 1231 feet = 375.20 meters

Figure 4b. Measurement of slope in accretion location 2.

Location 3: in the affected area Abrasion = Slope = 2.0°, Transverse Cross-width = 1844 feet = 562.05 meters, in the affected area Accretion = Slope = 1.4°, Transverse Cross = 1150 feet = 350.52 meters (Figures 5a and 5b).

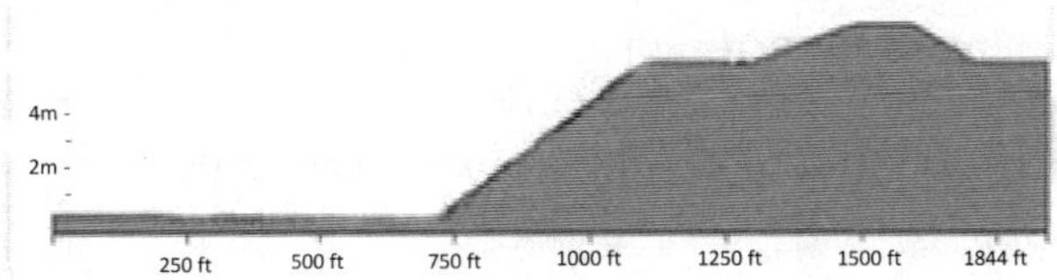

Slope = 2.0°, Transverse Cross-width = 1844 feet = 562.05 meters

Figure 5a. Slope measurement in abrasion location 3.

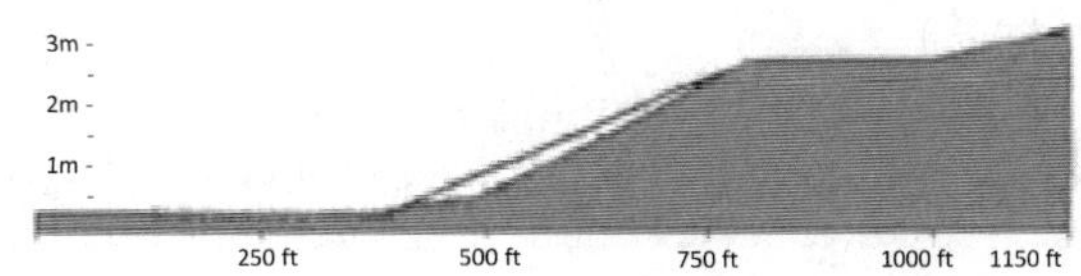

Slope = 1.4°, Transverse Cross = 1150 feet = 350.52 meters

Figure 5b. Measurement of slope in accretion location 3.

3. Results and Discussion

Based on research in the field, from 18 locations of observation (LP) of insitu data of abrasion and accretion, the location of observation is geographically grouped into 3: Tanjung Lampe, TelukLada, Tanjung Dadap and Tanjung Lesung with the following results [11].

Data obtained in the field can be concluded that in the study area is a sediment cell belonging to the low energy environment (Low Energy environment) [12], [13]. On field observations evidenced by the presence of fine grains of fine sand - containing rough fragments (rocks and shells), has a spreading from the west in November - February because the wind is blowing west wind which gives the impact of a wave/wave is greater than the impact given by the east wind. The sediment composition identifies the source, grain sediment, slope and coastal width as well as the boundary of a transport sediment (cell), whereas secondary data such as geological maps, topographic maps and Landsat images on a certain scale are effective to indicate the process Occurs in sediment transport [14].

The cell group is also called the sediment transport system as a unit in coastline management that can ultimately manage coastal resources sustainably. In this study area there are grain-sized sediments of different sizes, so the study area belongs to a group of low-energy cells. The limits of sedimentary cells in the research area are generally a natural boundary that is the convergent boundary which is the boundary of the dynamic cell, as the meeting of the direction of the sediment load transport. Convergent boundaries can be found on the beach of Tanjung Lampe, Pesisir Pantai Teluk Lada, Tanjung Dadap and Tanjung Lesung. In general, this limit occurs because of the diffraction of the wave by the growth of the Delta and by the plains that jut into the sea.

Landsat image analysis results show that the west coast of Banten Province has changed coastline, compared to Landsat TM and ETM + in 1982 and 2007, that is the occurrence of abrasion and accretion. Symptoms of abrasion at Tanjung Lampe Beach are indicated by evidence of the remains of fallen vegetation and damaged buildings caused by waves crashing and evidence of inundation of seawater in large coastal areas, suggesting that the coast is retreating towards the mainland. While the accretion area showed the reefs and sand deposits extending to the sea.

Symptoms of abrasion in the Bay of Lada region are indicated by narrow coastal skeletons and the remnants of wavelengths and break-ups that are submerged in sea water. Accretion symptoms are indicated by widening beaches, reefs and sediments towards the sea. Some areas on the coast of

IOP Conf. Series: Earth and Environmental Science **106** (2017) 012005 doi:10.1088/1755-1315/106/1/012005

Tanjung Dadap and Tanjung Lesung show the beach backward towards the land (abrasion) that is the breaking of waves that have been destroyed and submerged in sea water. Symptoms of accretion are seen on a wide beach face.

Based on Unsupervised Methods from Landsat TM and ETM + images in 1982 and 2007 it was known for 25 years to obtain the farthest distance and wide abrasion and accretion, in 3 selected locations. Areas that have experienced Abrasion in Location 1 (Bay of Lada - Tanjung Lampe): obtained the furthest distance = 274.73 m and Area of Abrasion = 615,022.77 m^2. In location 2 (Bay of Lada – Tanjung Dadap) obtained the furthest distance = 206.69 m and Area of Abrasion = 848,492.98 m2. In the location of 3 coastal areas of Cape Lesung obtained the farthest distance = 125.05 m and the Area of Abrasion = 106966,98 m^2. Areas that have experienced Accretion Location 1 obtained the furthest distance = 31.65 m and area of Accretion = 9,943.14 m^2. In location 2 obtained the furthest distance = 111.58 m and wide of Accretion = 24,491.14 m^2. At location 3 got furthest distance = 68.71 m and wide of Accretion = 65,841.54 m^2

Based on the calculation of the slope measurement and the length of the transverse incision, at 3 selected locations experiencing abrasion and accretion, measured on the image of GDEM Aster, the results obtained from the measurement in the research area located in the affected area of Abrasion at Location 1 Slope = 2.6 ^{o}, length of cross section = 1662 feet = 506.57 meters, Location 2 Slope = 3.3 ^{o}, length of cross section = 1,639 feet = 499.56 meters, in Location 3 Slope = 2.0 ^{o}, transverse incision = 1844 feet = 562.05 meters,

Affected Area at Location 1 Slope = 3.1^o, length of cross section = 1,006 feet = 306.62 meters, in Location 2: Slope = 3.1^o, transverse incision = 1,231 feet = 375.20 meters, Location 3 Slope = 1.4^o, length Cross section = 1,150 feet = 350.52 meters

The abrasion process occurs because the wind that blows in the west season (November - February) produces large waves, resulting in sediment along the coast affected by abrasion. As for the accretion process is likely to occur when the wind is blowing in the east wind season (March - October).

The occurrence of abrasion in some places in the study area because the lithology of the compilers is dominated by claystone. Slope is relatively steep, changes in wind direction, season and rainfall that can cause damage to the beach due to the large sea waves and sand mining activities.

The cause of accretion in the research area because this area has a relatively flat slope, and the presence of sediment transport from the surrounding river flow, lithology dominated by hard rock, at some point there are still areas of cliffs and reefs in coastal beaches. Much of the dumping is done on the coast to be a building that makes the beach forward and stable.

4. Conclusion

Abrasion is caused by the lithologic factor of the constituent dominated by claystone. Slope is relatively steep, changes in wind direction, seasons and rainfall sea waves and human activities in the form of sand mining. The accretion is caused by the lithologic factor of its constituent dominated by hard rock, the beach has a relatively flat slope and the sediment carrying capacity of the surrounding river flow and the backfill activity for the building. Aster Images GDEMs can calculate slopes in areas affected by abrasion, ie a steeper slope (2.0 ^{o} - 3.3 ^{o}) and the accretion area has a relatively flat slope (1.4^o - 3.1^o).

References

[1] Darsoprajitno S H. 2000. Geological Data Analysis to support coastal area management, presented in One Earth Day Seminar, HMTG, Universitas Trisakti, Jakarta.

[2] Purwadhi S H. 2001. Digital Image Interpretation. PT Grasindo. Jakarta.

[3] Trisakti B, B Hasyim, R Dewanti, M Hartuti, and G Winarso. 2003. Remote Sensing Technology in Coastal and Marine Area Management. Development Center for Remote Sensing and Technology.

[4] Anugrahadi A, and N Hendiarti. 2004. Mapping Coastal-Sea Phenomena from Satellite Images SeaWiFS and NOAA-AVHRR. *MINDAGI* 7 (1):71-75. FTM USAKTI.

[5] Hendiarti N, M Sadly, M C G Frederik, R Andiastuti A, A Sulaiman. 2006. Satelite Oceanography in Research and Technology for Monitoring Indonesian Ocean Dynamics INDOO Project. Page 20-46.

[6] Anugrahadi A, F S Purwadhi, N S Haryani. 2016. Remote Sensing Application and Geographic Information System for Hazard Mitigation in Geology Aspect. First printed, December 2016. ISBN: 978-602-9463-80-4. UniversitasTrisakti Press. Jakarta.

[7] Anugrahadi A, F S Purwadhi, N S Haryani. 2017. Remote Sensing Application and Geographic Information System for Hazard Mitigation in Geomorphology and Hydrometeorology Aspect. First printed, January 2017. ISBN 978-602-9463-81-1. Universitas Trisakti Press. Jakarta

[8] Lillesand T M, and R W Kiefer. 1979. Remote Sensing and Image Interpretation, John Wiley and Sons. New York.

[9] Sabins F F. 1997. Remote Sensing, Principles and Interpretation. W.H. Freeman and Co. San Francisco.

[10] Anugrahadi A, D S Prabowo, AI Wardhana, B M Sukojo, Y S Djajadiharja, F S Purwadhi, 2011. Software Applications Er Mapper and Arc View GIS to Measure Area and Distance of Coastline Changes during 25 Years in Coastal Pandeglang, Banten Province. *International Conference Intercarto-Intergis 17 in Bali*, December 1, 2011

[11] Anugrahadi A, and V Anjarsari. 2008. Study on Coastline Change in Landsat and Formosat Image: Case Study in Pandeglang District, Banten Province, *Proceedings of SAKE-3 Workshop*, Jakarta, October 15 - 16, 2008.

[12] Komar Paul D. 1998. Beach Processes and Sedimentation. 2nd Ed. College of Oceanic & Atmospheric Sciences, Oregon State University. Prentice Hall, Upper Saddle River, New Jersey.

[13] Sulaiman A, and I Soehardi. 2008. Introduction Coastal Quantum Geomorphology. E-book. LIPI

[14] Anugrahadi A, B M Sukojo, Y S Djajadiharja, F S Purwadhi. 2014. Coastline Change variation identification affected by abrasion and accretion. *Journal SEGARA*. 10(1) August 2014. Accreditation Number: 559/AU2/P2MI-LIPI/09/2013. Coastal and marine resources research and development centre, Marine Affairs and Fisheries.

Influence the condition land subsidence and groundwater impact of Jakarta coastal area

S Rahman[1,2*], U Sumotarto[1], H Pramudito[1]

[1]Mining Department, Faculty of Earth Technology and Energy, Universitas Trisakti, Jakarta, Indonesia
[2]Universitas Padjadjaran, Bandung, Indonesia

*Corresponding Author: sofyan@trisakti.ac.id

Abstract. Jakarta has been experiencing land subsidence for ten years due toerecting weight building and intensive extraction of groundwater for society drink water through ground water wells. Many groundwater extraction for drinkingwater has caused intensive scouring of land rock and further triggering land subsidence developed widely in coastal area of Jakarta. Measurement of the land subsidence has been performed by various experts and institutes. Between 1974 to 2010 subsidence has happened between 3 to 4.1 meters especially in Jakarta coastal area. Two major causes of the subsidence are identified. The first major cause is a result of erecting weight building such as hotels, appartments, and various human activities buildings. The second major cause is extracting ground water from aquifers bellow Jakarta land due to water deep wells down to the aquifer and traditional shallow water well of shallow or subsurface uncovered ground water. Weighter building and higher debit of water flow from deep water wells has fastened and deepened the land subsidence. Continuous measurement of land subsidence by means of geodetic as well as geophysical earth behaviour measurements need to be performed to monitor the rate, location as well as mapping of the land subsidence.
Keywords: costal area, groundwater, land subsidence

1. Introduction

Jakarta, the capital of Indonesia, has long been thesecond-largest megacity in the world, home of 10.5 million people. The city is having land subsidence problem. This problem has been worsened by the fact that the changes of climate and melting of polar ice has caused the rise of sea level. The two changes of groundwater level declining and sea level rising has caused worse land subsidence. The use of groundwater by people continuously has caused depletion of groundwater level. The land subsidence and sea water rise has been the attention to many experts.

Ground water withdrawal increased by 24 percent between 2011 and 2014, and is expected to continue to rise with increasing demand. The extraction of water has left large caverns under the city that are collapsing, a phenomenon that can be seen through the cracking of buildings, infrastructure and roads. Land subsidence is also increasing inland sea water intrusion and is causing a wider expansion of inland and coastal flooding areas.

Many experts have performed researches to solve the land subsidence problem. However, this is not a simple problem to be solved. Although it has been known that effect of human behaviour could be analysed, however, the solution is not simple. A lot of people who are using groundwater is identified to be the big problem. The depletion of ground water level had caused a more intensive land subsidence. From the research it has been recommended to the people to decrease using groundwater,

IOP Conf. Series: Earth and Environmental Science **106** (2017) 012006 doi:10.1088/1755-1315/106/1/012006

but this solution is not respected positively. It is now become the question how to reduce the rate of subsidence.

Currently informed that the head of Jakarta's environment agency has advised the city needs to "stop using groundwater" to reduce the rapid rate of subsidence. It is also informed that Jakarta is currently sinking at a rate of nine centimeters per year, with some areas experiencing subsidence rates up to 17 centimeters per year [2]. It is anticipated that this rate will increase due to the growing exploitation of groundwater and great levels of urbanisation. Water containment dams, the treatment of river waters, land reclamation, a sea wall and extended water pipelines are some of the proposed solution to solve the problem. However, the greatest investment to put into the contentious land reclamation project is now suspended due to Indonesian law problems.

2. Research Method

2.1. Groundwater Extraction

There are many researches that have been performed in connection to the land subsidence problem around Jakarta district and surrounding area. Groundwater level has been searched since long time ago. The groundwater level or piezometric level has been measured in various time period. The following information is the fact. Several wells have been searched with respect to the change of water level.

The groundwater extraction in Jakarta could be categorized into shallow (< 40 m) and deep (> 40 m) extraction. Shallow extraction is through dug wells or driven wells, operated with buckets, hand pumps or small electric pumps; whereas the deep extraction is mostly from drilled wells. Shallow extraction is mostly done by the population. It is well spread over the area, but its extraction rate per well is relatively low. Deep extraction is usually conducted by industry. It is usually more concentrated, and has a relatively high extraction rate per well [4], [5].

The subsidence rate is closely related to the rate of piezometric water level (head) deepening in the middle and lower aquifers. In the case of Jakarta, the increases in both population and industry, which require a lot of groundwater, could explain the above declining trend of piezometric heads, as shown in Figure 1. According to [5] , the piezometric level in North Jakarta has changed from 12.5 m above sea level in 1910 to about sea level in 1970's, and then deepened significantly to 30-50 m below sea level in 1990's [4], [5].

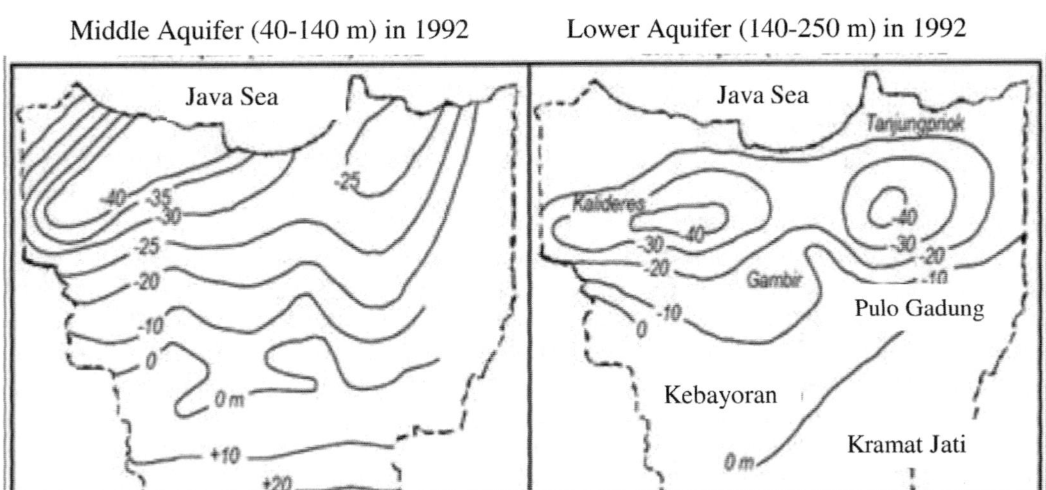

Figure1. Piezometric water level contours (in metres) inside Middle and Lower Aquifers of Jakarta in 1992 [4].

2.2. Land Subsidence Research

Experts also perform survey and research in land subsidence probelm of Jakarta area. The corelation between land subsidence and excessive groundwater taking in Jakarta can be illustrated using the subsidence results obtained from leveling surveys. Figure 2 shows the observed land subsidence during the period of 1982-1991 and 1991-97. Maximum subsidence during the period of 1982-1991 is about 80 cm, while for the period of 1991-97 is about 160 cm. In general the subsidence rates in Jakarta area during this period is about 1-5 cm/year and can reach 26 cm/year at several locations. During the period between 1982 and 1991, the maximum rate of subsidence is about -8 cm/year, while during the period between 1991 and 1997 it is about -26 cm/year.

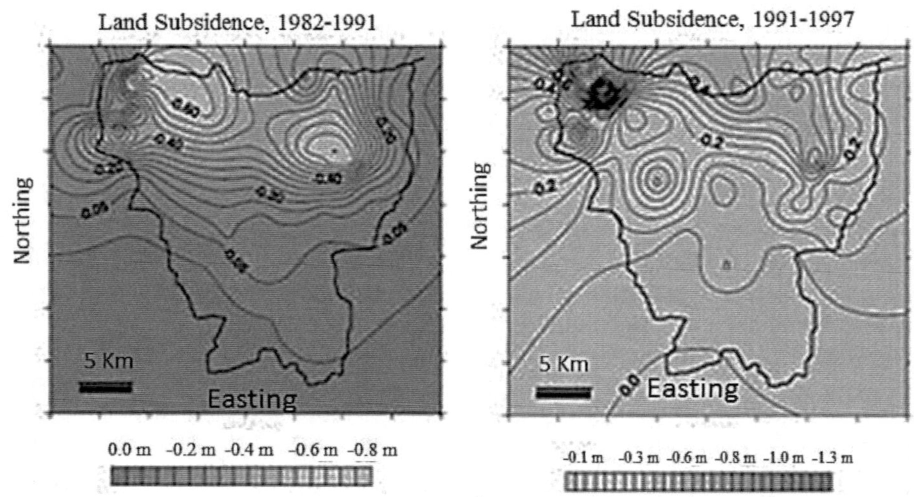

Figure 2. Land subsidence in Jakarta measured from leveling surveys (in metres), over the periods of 1982 – 1991 (left) and 1991 – 1997 (right) [6].

3. Results and Discussion

Other experts also performed survey and research about land subsidence in Jakarta area. Figure 3 shows the research output of Deltares research team in the periode between1974 and 2010. It is shown that the deepest land subsidence has happened at Ancol Beach and surrounding area. From this picture it is shown that the area has been subsided as deep as 4.1 meter during the period.

Figure 3 it can also be recognized that to the southern part of Jakarta region, the land subsidence is lessened, however, this problem cannot be neglected. The subsidence gives difficulties of people to erect building, houses, etc, because the land surface is continuously subsidized. The problem is effected by groundwater depletion has long been happened and continuously rating the subsidence because the water containing rocks undergoing decrease vertically that has caused subsidence up to the ground surface.The people population around Jakarta area also gives heavy impact to the subsidence problem, because the increase of population will need living area and other building needs (shops, markets, station, bus terminals). With the increase buildings and other needs it will increase the land surface weight and thus deepening the subsidence. Figure 4 shows the increase of living area and prerequisites causing by population increase in Jakarta region.

IOP Conf. Series: Earth and Environmental Science **106** (2017) 012006 doi:10.1088/1755-1315/106/1/012006

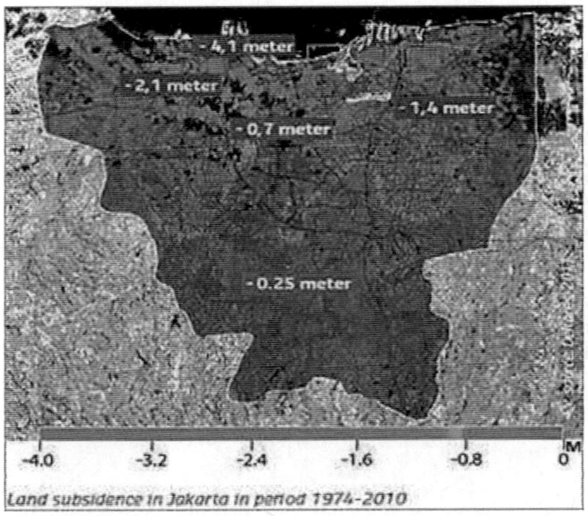

Figure 3. Output of land subsidence research around Jakarta area in a period of 1974-2010 [3].

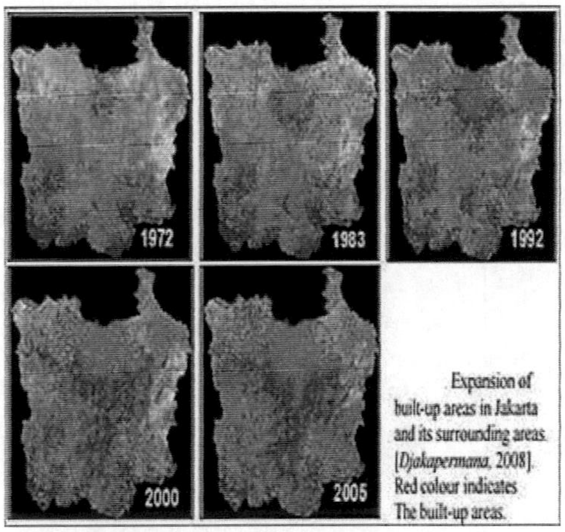

Figure 4. The development of people living area and its required facilities in Jakarta region and surrounding areas [4].

With the above explanation, it is clear that Jakarta people and local government face a number of problems in connection with land subsidence. Variety of researches can be performed to identify the land subsidence or any other land problems such as land slide, slump, etc. As an example, a land slide or subduction has been surveyed by geophysical survey and reasearch.

A group of geologists of Landslide Research and Risk Management Division (LRRMD), NBRO has performed to conduct an investigation on a land subsidence at Achchuweli area, Jaffna, suddenly occurred in the morning of 23rd January, 2016. It was observed clearly that cracks (1-5cm) have developed on the ground in a circular area with an approximate diameter of 70m. The affected area is a farming land. A house on the perimeter of said circle was severely cracked and tilted slightly. It was also noted that the cracks in the house have been developing. LRRMD geologists explored the area on 27th January, 2016 and also conducted a geophysical survey to understand the present subsurface ground condition of the affected area [8].

Geologically the area is composed of limestone which is a weak sedimentary rock and hence, easily dissolvable. As the formation of dissolution cavities and caverns are quite common in this kind of terrains, underground tunnels and cavities are found in many parts of the Jaffna Peninsula. According to the geophysical studies conducted, some underground discontinuities of the limestone terrain of the affected area were revealed indicating underground cavities possibly filled with water or moisture soil. Due to activation of such dissolution cavities, ground settling can be expected with time, leading to further subsidence in the affected area. Figure 5 shows geophysical measurement line especially resistivity method in the survey area.

IOP Conf. Series: Earth and Environmental Science **106** (2017) 012006 doi:10.1088/1755-1315/106/1/012006

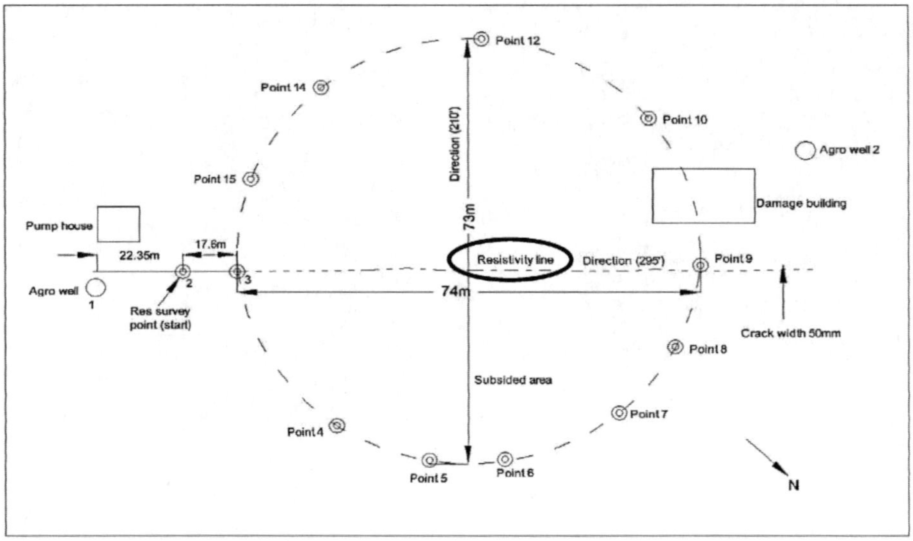

Figure 5. Geophysical measurement line in the land subsidence area of Achchuweli, Jaffna (Srilanka). Red line mark shows gephysical measurement line.

Land subsidence, land slide, slumps and other land problems has happened in Indonesia big cities including Jakarta, Semarang, Surabaya, and others. Many big cities in the world are positioned on the above of sedimentary rock layers that undergo land sliding or sunsidence. The world big cities other than Jakarta are Bangkok (Thailand), Tokyo (Japan), Osaka (Japan), Niigata (Japan), Taipei (Taiwan), Shanghai (China), Mexico, Venice (Italy), London (England), and other states in the United States of America also have land subsidence problem. The land subsidence that happens in non-coastal area cities may not give serious impact, however, for the coastal cities they may undergo trouble problems by rising sea water level into the land.

With the land subsidence causing from heavy load above the land surface such as big buildings, houses, and other infrastructures has been added the problem by increasing extraction of groundwater. Research may be performed into detail such as analysing the micro aspect of rock lithology. Figure 6 shows rock pores condition filled with groundwater. The pore pressure is continuously getting higher due to overburden pressure caused by load of heavy surface human buildings and land subsidence. When the water is extracted by deep water wells the rock grains will be compressed and crushed due to the load and land subsidence. Detail reasearch of petrology could be performed continuously to recognize the impact of land subsidence and groundwater extraction.

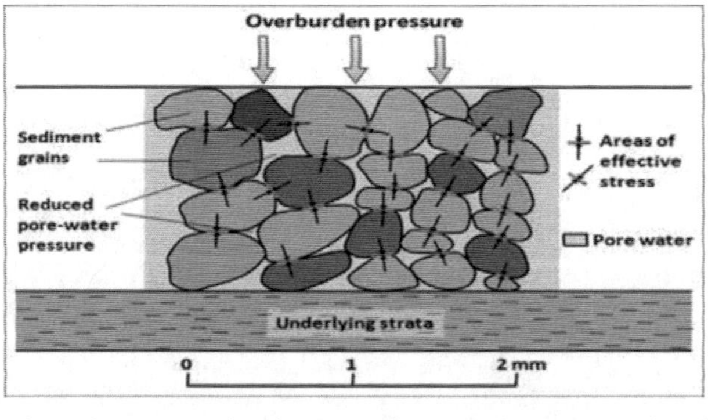

Figure 6. Detail investigation of compressing impact of a rock due to over burden pressure caused by overload and land subsidence problems.

Land subsidence, groundwater extraction and sea level rise have caused possible declining and dropping the city land below sea level. Although endeavors to solve the problem have been performed various researches could be performed. The following are possible new research area:

a) Measurement of the rate of land subsidence arround specific area such as Ancol Beach (North Jakarta) using GPS. This measurement is planned to ensure and acurateness of land subsidence rate in the area.
b) Geophysical measurement to analyse the geological structures causing by overload and land subsidence. Resistivity and refraction seismic could be performed to detect subsurface subsidence during a period of time. Geophysical method can be applied pasively to compare in a period of time or between different measuring locations. The geophysical methods can also operated actively by observing the land surface change in a period of time.
c) Rock and mineral sampling for geochemical analysis both on rock character changes and groundwater extraction impact. This can be performed using shallow well.

4. Conclusion

Land subsidence, groundwaterextractionand sea level rise are hardly faced by people in the cities. Measurements and reasearch with respect to solve the problem have been conducted. The subsidence, however, could continuously developed to be hardly overcome. Researches of different basic, method and purpose need to be performed to find possible solution. Geophysical and geochemical research and measurement using certain methods may be done to find data and information with respect to physical, chemical behaviour and geological structures that can be used to support data in finding the solution of the problem. Detail reasearch such as petrological and mineralogical research, experiments, tests and other ways to get important data and information could be performed. To perform further research, mesurement, and monitoring the land subsidence continuously, it is expected to receive sponsor and support to supply requirement including financial cost in order to solve the land subsidence, groundwater exctraction, and impact of sea water level rise problem particularly in Jakarta city.

Acknowledgments

The author would like to bethanking to the Universitas Trisakti for their moral and material support for the realization of this paper.

References

[1] de Lange G, Bucx T, Stuurman R, & Erkens G, Land subsidence, sea level rise and urban flooding, Deltares Research Institute, Aug 10, 2015.
[2] Haweya Ismail, Indonesia: Groundwater Decline is Causing Jakarta to Sink, Global Food and Water Crises Research Programme, Future Directions International Pty Ltd., 2016.
[3] Abidin H Z, Presentasi Asahi Research Finding 2014, LPPM-Institute Teknologi Bandung, 2014.
[4] Abidin H Z, Andreas H, Gumilar I, Gamal M, Fukuda Y, and Deguchi T, 2009. Land Subsidence and Urban Development in Jakarta (Indonesia), *7th FIG Regional Conference Spatial Data Serving People: Land Governance and the Environment – Building the Capacity*, Hanoi, Vietnam, 19-22 October 2009.
[5] Djakapermana R D, 2008, Kebijakan Penataan Ruang Jabodetabekjur, Paper presented at *One day Seminar celebrating World Water Day*, Environmental Management Technology Expertise Group, FTSL ITB, Bandung, 31 Maret 2008.
[6] Soetrisno S, Satrio H, and Haryadi T, 1997, To Anticipate Impacts of Reclamation of Jakarta Bay, A Groundwater Conservation's Perspective. Paper presented at Workshop on Coastal and Nearshore Geological/Oceanographical Assessment of Jakarta Bay : A Basis for Coastal

Zone Management and Development, Jakarta, 25 - 28 June.
[7] Abidin H Z, R Djaja, D Darmawan, S Hadi, A Akbar, H Rajiyowiryono, Y Sudibyo, I Meilano, M A Kusuma, J Kahar, C Subarya. March 2001. Land Subsidence of Jakarta (Indonesia) and its Geodetic-Based Monitoring System. Natural Hazards. *Journal of the International Society of the Prevention and Mitigation of Natural Hazards*. 23(2/3):365 - 387.
[8] Landslide Research and Risk Management Division, Recent Land Subsidence I Investigation at Achchuweli, Jaffna, National Building Research Organisation, 2016.

Potential collapse due to geological structures influence in Seropan Cave, Gunung Kidul, Yogyakarta, Indonesia

B Nugroho*, P A Pranantya, R Witjahjati, Rofinus

Mining Department, Faculty of Earth Technology and Energy, Universitas Trisakti, Jakarta, Indonesia

*Corresponding Author: bani@trisakti.ac.id

Abstract. This study aims to estimate the potential collapse in the Seropan cave, based on the existing geological structure conditions in the cave. This is very necessary because in the Seropan cave will be built Microhydro installation for power plants. The electricity will be used to raise the underground river water in the cave to a barren soil surface, which can be used for surface irrigation. The method used is analysis the quality of rock mass along the cave. Analysis of rock mass quality using Geomechanical Classification or Rock Mass Rating (RMR), to determine the magnitude of the effect of geological structure on rock mass stability. The research path is divided into several sections and quality analysis is performed on each section. The results show that the influence of geological structure is very large and along the cave where the research there are several places that have the potential to collapse, so need to get serious attention in handling it. Nevertheless, the construction of this Microhydro installation can still be carried out by making a reinforcement on potentially collapsing parts
Keywords: cave, collapse, geological structure, geomechanics classification, microhydro installation

1. Introduction

Karst area is an area consist of limestone that is soluble Carbonate as the main constituent of the limestone is dissolved by the surface water entering through the crack or joint of rock formed by the geological structures that work in the area. As a result of the dissolution, the landscape of the karst area has many surface holes and cavities beneath the earth's surface. The landscape on the surface of many holes is called Eksokarst, while the cavities in subsurface surface are referred to as Endokarst [1]. The landscape is as shown in Figure 1.

In karst areas, the landscape of dry season and rainy season is very different. In the dry season, the conditions are very dry, not many plants are alive, the water is very difficult, so make the soil become barren. While in rainy season, water on the surface so much that many plants that live with fertile. Although in the dry season the surface water is very difficult and the soil is barren, but in subsurface there are many water and become underground river [2]. Seropan underground river is located at a depth of about 60 - 75 m. To utilize this river water, then the river water will be pumped by using microhydro installation in cave [3], [4]. In the framework of making these installations, there should be research on rock stability and the potential collapse due to the influence of geological structures in the area.

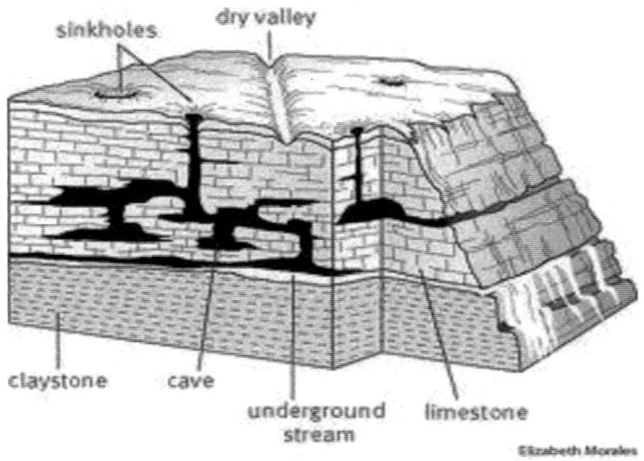

Figure 1. Illustration shows some morphology on the surface (*eksokarst*) and subsurface (*endokarst*) in karst area [1].

Figure 2. Landscape condition in part of Gunung Kidul karst area.
Condition in dry season (left) and condition in rainy season (right)

Seropan Cave is one of the big cave in karst area in Gunung Kidul, Yogyakarta (Figure 2). Inside in this cave is an underground river (Figure 3) with a fairly large debit, about 800 liters/sec, so it can be used to move the turbine that produces electricity. Therefore, in the underground river will be built microhydro installation to pump subsurface water. Although in surface are barren conditions, but in subsurface water is quite a lot and even a river, such underground river Seropan. However, due to the geological structure in the form of cracks or faults or stocky it will cause rock conditions around the cave to become unstable [5], [6]. This is potentially the occurrence of ruins on the walls and roof of the cave [2]. For that it is necessary to research to determine the potential for such collapse, for the safety and security of microhydro installation.

To know the potential of collapse, conducted research on the techniques quality of limestone in Seropan cave. The method used is to make the Rock Mass Classification with Rock Mass Rating system or known as Geomechanics classification [6]. With the classification of this Geomechanics will be known the techniques quality of limestone in the cave Seropan and how big the influence of geological structures on the rock mass stability in the cave [7]. Based on the description, the main objective of this research is to know how big the potential collapse of rock mass in this Seropan cave.

2. Research Method

The method used in this study is the analysis based on observation and calculation of existing data. Research area have divided into 7 (seven) segments, and the analysis was performed on each segment along the research path using Geomechanics Classification. To create a Geomechanical Classification, six parameters are required: Uniaxial Compessive Strength (UCS), Rock Quality Designation (RQD), Spacing of Discontinuous, Discontinuous plane conditions, groundwater and Orientation from the discontinuous field (Figure 4). Based on the analysis of these parameters will be obtained class and rock mass quality. The influence of geological structure can be identified from the quality and class of rock mass [8]. From the magnitude of the influence of these geological structures will be able to know the stability of the rock and its potential collapse, and at the same time how to overcome them.

Figure 3. Main entrance to Seropan cave and Seropan subsurface river

Figure 4. Measure discontinue plane and other datas to make geomechanic classification, in Seropan subsurface river.

3. Results and Discussion

After obtaining the data, both data for Geomechanical classification parameter and other required data, the next step is to analyze the existing geological structure, condition analysis and location of discontinuous fields then analyze the potential of collapse in each segment. The geological structure was analyzed by Geomechanical classification and the making of Roset Diagram, while Potential collapse was analyzed from the position, condition and orientation of the discontinuous plane to the cave position.

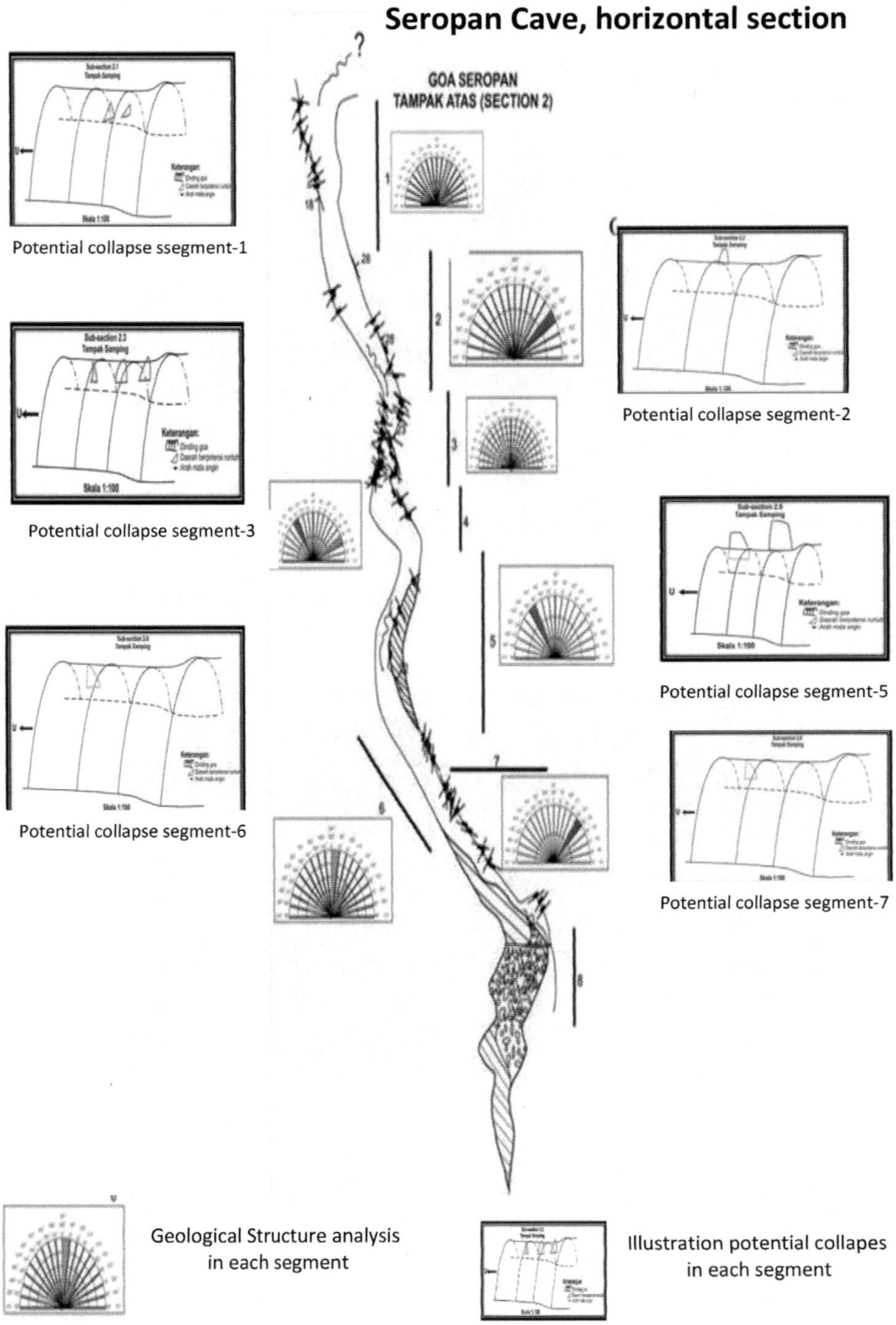

Figure 5. Correlation of geological structure condition and potential collape in each segment along Seropan cave research.

The results obtained shown in Figure 5, as follows:

1. In segment-1, there are some discontinuous planes whose existence is very unfavorable orientation, such as the N78°E, N100°E, N152°E, and N151°E directed intersections will intersect above the right cave, where this condition has the potential to collapse if there is movement along the discontinuous plane.
2. In segment-2, there are several discontinuous planes whose existence is very unfavorable orientation, such as N78°E, N100°E, N152°E, and N151°E directed intersections over the right cave, where this condition has the potential to collapse if there is movement along the discontinuous plane.
3. In segment-3, there are some discontinuous planes whose existence of orientation is very unfavorable, such as N179°E, N194°E, N183°E, N165°E, N331°E and N334°E will intersect each other which produces three (3) crash-prone locations when there is movement along the discontinuous plane.
4. In segment-4, there are some discontinuous fields whose orientation direction is very unfavorable, but when it is reconstructed all the fields, there are no pieces of discontinuous plane that can collapse.
5. In segment-5, there is a discontinuous plane whose existence of orientation is very unfavorable, such as N146°E oriented braces, where this field may intersect with other discontinuous fields which produce two (2) crash-prone locations where there is movement along the discontinuous plane .
6. In segment-6, there are some discontinuous planes whose existence is very unfavorable orientation, such as N7°E, N103°E braced, and N114°E will intersect above the cave, where this piece has the potential to collapse if there is movement along the discontinuous plane.
7. In segment-7, there is no discontinuous field in which the orientation direction is very unfavorable, and when reconstructed all the fields, there is no discontinuous pieces of fallible pieces, but with rocky conditions that are quite fragile on the roof, then in sub section 7 This remains to be warranted because with rock conditions that are quite fragile this can be a potential collapse in this area.

4. Conclusion

Based on the results of the analysis on each segment, it can be concluded that in each segment there is potential for collapse, especially on the roof of the cave. Although each segment has the potential to collapse, but the potential for each segment varies. Discontinuous plane conditions such as strike direction, slope direction and magnitude of slope, discontinuous plane orientation as well as the presence or absence of discontinuity at the top or roof of the cave have great effect on the potential for collapse. Potential collapse of the roof of the cave can be overcome or reduced the level of risk by using some reinforcement techniques, for example with rock bolt, supporting system, shotcreate, or a combination of several methods. Although there is a possibility of collapse, but with some methods such reinforcement techniques, Microhydro Installation can still be built.

Acknowledgment

Acknowledgements to the Center for Water Resources Research and Development, Ministry of Public Works and People's Housing, which has granted permission to conduct research in the Seropan cave, as well as to the Acintyacunyata Speological Club who has assisted and provided security during the research in the Seropan cave, Parties directly or indirectly providing assistance until the completion of this research.

References

[1] Ford D & Williams P. Karst Geomorphology and Hydrology. Unwin Hyman Ltd, London, Fetter. 1989.

IOP Conf. Series: Earth and Environmental Science **106** (2017) 012007 doi:10.1088/1755-1315/106/1/012007

[2] Acintyacunyata Spleological Club (ASC). Map of Seropan Cave, Gunung Kidul, Yogyakarta, Indonesia. 1988.
[3] Wibawa, Yanto. 2005. Study of the Potential of Microhydro Application for the Provision of Raw Water and Electricity in Irrigation Tumiyang Channel, Grumbul, Jurangmangu, Tumiyang Village, Pekuncen, Banyumas.
[4] Pusat Litbang Sumber Daya Air. 2009. Assessment and Application of Hydro-Pumps for Provision of Raw Water and Electricity in Mountain Areas. Final report.
[5] Nugroho Bani, and Arya Pranantya, P. 2014. Underground Dams in karst region, case study Bribin – Seropan, Yogyakarta, Indonesia. *International Symposium on Dams in Global Environmental Challenge.* Bali. Indonesia. June 1st – 6 th.
[6] Beiniawski, Z T. Engineering Rock Mass Classification, a complete manual for engineers and geologist in mining, civil and petroleum engineering. Willey & Sons, New York. 1989.
[7] Nugroho Bani, and Arya Pranantya P. 2012. Geotechnical Classification of Seropan Cave, Gunung Kidul, Yogyakarta. *Symposium and Seminar of Geomechanics-1*, Yogyakarta.
[8] Karlsruhe Institute of Technology (KIT). 2010. Pre-Design of the Hydropower Plat with Wood Stave Pipeline in Gua Seropan. Joint Project Integrated Water Rescources Management (IWRM) in Gunung Kidul, Yogyakarta, Indonesia.

IOP Conf. Series: Earth and Environmental Science **106** (2017) 012008 doi:10.1088/1755-1315/106/1/012008

The implementation of local wisdom in reducing natural disaster risk: a case study from West Sumatera

Z Zulfadrim*, **Y Toyoda**, **H Kanegae**

Graduate School of Policy Science, Ritsumeikan University, Iwakura-cho 2-150, Ibaraki, Osaka, Japan.

*Corresponding Author: ps0335hh@ed.ritsumei.ac.jp

Abstract. The purpose of this paper is to introduce some local wisdoms in West Sumatra and propose their challenges that modern values have degraded its knowledge. In a contemporary context, traditional stories (written and oral stories) still relevant to be used and internalized in disaster risk reduction. Traditional knowledge or local wisdom is a system of knowledge derived from long experienced process in the past, adopted and handed over to next generation through evolutionary process. Indigenous or traditional knowledge can be practiced in understanding the nature of natural disaster, to propose the best action in mitigation, to respond in emergency phase, and to suggest more option for recovery process based on previous experience. The paper based on four weeks field research in west Sumatra which is known with their natural hazards due to its geographical location. In the beginning, this paper discusses the nature of local wisdom and how it can be matched in disaster management, then continues to the specific case how the traditional stories in West Sumatera can be internalized and integrated with contemporary disaster risk reduction. This paper proves that local wisdom can be useful as an effective instrument to deal with natural disaster or natural hazard.
Keywords: disaster risk reduction, local wisdom, natural hazard, West Sumatra

1. Introduction

When the success story of Simeuluean in Aceh and Moken society in Andaman Sea widely publicized aftermath of 2004 Indian Ocean earthquake and tsunami, the studies of local wisdom substantially increase. Knowledge that helped those communities survives show the importance of local wisdom (written and oral stories) to disaster risk reduction. Even though the documentation and report about the application of local wisdom related to disaster risk reduction can be tracked back long before this disaster, as Dekens [1] stated that there was an evidence since 1970's that local knowledge and practice could improve preparedness in facing natural disaster, however, only in recent years practitioners have paid more attention to this issue and integrated it in policy making and planning [2].

Even though there are many definitions of local wisdom; depending on the field of study, the characteristic of local wisdom can be generalized as follow: (a) locally bound (b) context specific (c) non-formal and (d) dynamic [3]. Local wisdom is a result of experiment and continues process that makes them a dynamic evolution and non-formal form of knowledge. Local wisdom usually handed down from one generation to the next generation through oral histories, myth, song, and legends. Unfortunately, in some region such knowledge handed down through oral and written stories to a certain key member of the society only and not to be shared widely. This information may be considered as a sacred and taboo and not disseminated among family or other community members. Only key person in community such as clan leaders, or elders can keep the information with their own memories. This practice makes this knowledge less accessible.

According to Rajid [4], local wisdom can be divided into two aspects, which are technology and belief system. Below, we will discuss more about technology and belief system from local wisdom in West Sumatra. Traditional housing system, food storing system as technology aspect are considered as preparedness measure in disaster risk reduction while in the other hand during the post disaster and recovery process the appliance of *Tuddukat* and *gotong royong* is an example of local wisdom practices that meet local and socio-economic context.

2. Research Method

The research was conducted in two phases, from January – March 2017 and from Mei – June 2017. West Sumatra was chosen for their high vulnerability to the impact of natural disaster as well as their cultural diversity in that area. Community and groups' leaders, traditional and religious leaders, local government, Non-Governmental Organization (NGOs), and experts were involved in the action research that included interviews, field observation and focus discussion group (FGD). Local wisdom and practices related to natural hazard were identified and documented. The first phase is a collecting data process while the second phase is validating the data through FGD to establish scientific base for local wisdom.

According to UNESCO [5] Local wisdom refers to the understanding, skill, and philosophies developed by communities in their interaction with natural surroundings. From that definition, it implies that this type of knowledge dynamic evolved in day-to-day activities [6]. Moreover, researcher agreed that term local wisdom is analogous to local knowledge, peasant knowledge, community knowledge, traditional knowledge and indigenous technical knowledge.

3. Result and Discussion

West Sumatra is the homeland of *Minangkabau* people; one of the largest ethnic groups in Indonesia situated in the middle of the western coast of Sumatra. It has a distinct culture for its matriarchal system which makes them different from the rest of the Islands; it is influenced by Islam culture and adapted it into their own way of life. Geographical feature of west Sumatra varies from plains, mountainous volcanic, highland and an offshore archipelago called Mentawai Islands. From the following discussion, West Sumatra inhabitant can be distinguished into two main groups, first *Minangkabau,* those who live in the main island, second *Mentawaians* those who inhabit in Mentawai Islands, that are also part of the province, however their culture and language are intelligible with neither *Minangkabau* nor Indonesians. Thus, they have different local wisdom among others.

Based on Indonesian Disaster risk index, West Sumatra is one of the highest disaster risk area in Indonesia, due to its geographical location which is located between the confluence two major continental plates, (The Eurasian plate and Indo-Australian Plate) and Great Sumatera Faults [7] and based on 2003 Indonesian Seismic Zoning Map [8] belongs to Zone 5 -6 of zoning type (1=lowest, 6=highest). It is also vulnerable to other natural disasters such as cyclone, drought, flood, landslide and combine with environmental degradation such as pollution and deforestation. This vulnerable condition exacerbated by social problem like poverty and inequality.

West Sumatra province has experienced many kinds of natural disaster in the past. Based on historical document, since 1822 till nowadays, approximately there were more than 15 destructive earthquakes occurred in west Sumatra, some of them followed by tsunami. Another natural disaster that affects west Sumatra people is flood that occurs almost every year in rainy season. The heavy rainfall is not only resulting floods but also followed by landslides. In coastal area, they have a different problem. Other hydro meteorological hazards that threaten their community are tropical cyclone, coastal erosion and sea level rise.

3.1. Local Wisdom in Minangkabau Community

Communities in urban and rural area in Minangkabau have developed ways to prevent or to reduce disaster risk by adaptation and well preparation. Their ancestor has experienced recurrent of hydro meteorological disaster and utilize their knowledge to deal with them. The most interesting features about *Minangkabau* communities are their traditional house. Most people know about traditional house of West Sumatra, which has a unique triangular roof. This traditional house is called *Rumah Gadang* (Big House) because of its function as the main venue for several activities and festivities. In addition to its function as a house for daily life activities, it is also a place for clan meetings and marriages [9]. This house built on stilts with high ground space to protect them from flood, animal attack and moisture, sometimes it is also being used as a place to keep their cattle stock. This traditional house also has an ability to withstand an earthquake, since the column of this building is normally laid on stone pedestal foundation rather than inserted into the soil [10].

Generally hydro meteorological hazards often lead to food shortage; however, in West Sumatra mostly on rural area this phenomenon is not existed. Next to the traditional house or *Rumah Gadang* (Figure 1, left) there is another structure called *Rangkiang/Lumbuang* (Figure 1, right). This structure used as granary or rice barn use in times of scarcity and unpredicted events. *Rangkiang* can be found in courtyard of *Rumah Gadang*. Every clan (family) in *Minangkabau* has their own *Rumah Gadang* and *Rangkiang*. When the disaster occurs, they use their stockpile to fulfill their needs.

Figure 1. *Rumah gadang* (left) and *rangkiang* (right).

Another local wisdom in coping disaster effect that can be identified in West Sumatra is mutual assistance i.e people helping each other. This is a common practice in many rural areas in Indonesia; the term for this practice is *GotongRoyong*. According to Mardiasmo [11] this value suggests a principle that all activities are performed in cooperative and equal manner, and promote efficiency in order to quickly re-build the community. The recovery process from 2009 earthquake in Padang showed that this practice was preferred method of disaster recovery rather than aid and initiatives from Indonesian government. In West Sumatra, Gotong Royong not only takes place in disaster event but almost in every occasion, they keep working together in good time as well as in bad time, from building a house to marriage parties. This activity not only exists in the disaster moment, moreover this "togetherness" may occur and carry on in many other aspect of lives.

3.2. Local Wisdom in Mentawai Community

In Mentawai Island, M*entawaians* also have their own local wisdom for natural disaster. As one of indigenous tribe in Indonesia, they still preserve many of their original values. As Hilshorst [12] stated that the foundation of indigenous people's resilience in the face of natural disaster provided by the intimate relationship between indigenous community and nature. Mentawai Islands had experienced some small, medium and big disasters such as earthquake, tsunami, flood and landslide. The last disaster took place in this area was 2010 tsunami that caused hundreds of people lost their lives. In this case, as told by Jimi Richard disaster mitigation activist in Mentawai, quoted that local wisdom that belongs to experience knowledge have saved many lives in Tumale, Silabu village. All villagers who inhabited in Tumale survived because they applied small *Tuddukat* to give warning signal to people for incoming tsunami. *Tuddukat* itself is the traditional instrument use to communicate with clan members (Figure 2, left). It is a drum made from wood, which is struck, to produce sound. In their original land, deep in the Siberut Island, *Tuddukat* is an accessory every *Uma* should have.

Uma is traditional house for *Mentawaians* (Figure 2, right). This kind of building has existed for century and may have experienced their toughest test by nature for any kind of natural disaster and phenomenon [13]. Similar to Minangkabau, this construction plays a significant role in mitigation process, mostly for flood and earthquake disaster. From field observation, there were evidences that this construction could survive from severe earthquake and flood because of the earthquake resistant technology in these houses. Some of key points which might have contributed for their superior performance in earthquake are; the use of light structure using wood as building structure frame, the use of knock down and tied joint in joint system and the use of lightweight material for roof in this house.

For storing food, they buried *sago* (sago is the main intake for *Mentawaians*) underground, where it will last for months without putrefying. This method already practiced by their ancestors and used for their food stock in their agricultural field or can be used for emergency. When disaster occurs, the community relies on the storage in every *Uma.*

Figure 2. A set of *Tuddukat* (left) and *Uma* (right).

Both technology aspect and belief system from of local wisdom practices recognized in Minangkabau Community are questionable soon. Fuadi [14] blamed social and economic changes as reasons for a decline of function and number of traditional houses in Minangkabau. A discussion with Z. Dt. Tumanggung, one of *tukangtuo* (head carpenter) in Batusangkar, revealed the factors behind the declining (interview held in June 5th 2017). Underlying causes that were mentioned was lacking*tukangtuo* existing nowadays, while the young generation more interested on modern technology. Secondly, social values changes that caused clan members to build a house for their nuclear family. And thirdly, modern values simply replace traditional values and practices where using new designs and materials are considered prestigious and modern.

Most of interviewees in all research sites felt that *gotong royong* has been declining in this modern era. In everyday practices, this activity is limited to formal use only. *Gotong royong* is held by request from the government officers. They believe that the integration of modern technology make people are less dependent on others, until the late of 1980, *gotong royong* still being used in farming, from land cultivating until the end of process which is harvest time, always carried on by these methods. When new technology replaced human labor, people felt they were less dependent to their neighbor. However, they still believed that *gotong royong* revived in emergency time.

In Mentawai, the *Tuddukat* which designated to communicate among the clan members; to tell about death, born and their successful in capturing the prey; was modified to alarming villagers that a disaster might occur in their area. The *Tuddukat* itself may not look like the original one, yet the spirit that brought by that sacred tool remains in the community minds. As an early warning system, this type of knowledge may be found in another area, however the *Mentawaians* case is quite unique, since the tool itself may represent their original technical knowledge. Food storing practice may help *Mentawaians* survive in their hard time, when their land struck by severe disaster such as flood, landslide and tsunami. Their food supply is still available till the time they can cultivate their agricultural land.

Both practices still can be found in many traditional villages in Mentawai islands. Nevertheless, for another reason, which is national policy implementation that urges indigenous people to be integrated with national community, the knowledge and understanding of their native culture and environment is now considerably low. The policy itself intends to increase the quality of life for indigenous people through PKMT (Pembinaan Kesejahteraan Masyarakat Terasing) or resettlement for Indigenous Community, where they are relocated into area alongside the main river and coastline [15]. This policy, for native Mentawai means the eradication of their cultural practices. They are not allowed to practice their religion and ritual behavior since their practices considerably do not match with basic principle of Indonesia.

4. Conclusion

Local wisdom can be applied as an effective instrument to undertake hazard and disaster in many communities. The usefulness of this knowledge in disaster risk reduction has been demonstrated in many case studies conducted with indigenous culture. Despite the limited record of samples, it is likely that there are more similar phenomena remain "un-touched" that exist to support this finding. In this paper, we present the local wisdom that can be found in West Sumatra, however in the modern era, technologies and social value changes made this knowledge lost their meaning. Such action to promote local wisdom and its integration with science may improve and increase their resilience to face natural hazard and disaster. It is important to emphasize that implementing local wisdom for disaster risk reduction is not a panacea. However, some local wisdom can be classified as transferable local wisdom, where this knowledge having potential to be applied to other area with the same risk and context.

References

[1] Dekens Julie, Local Knowledge for Disaster Preparedness : A Literature Review, Kathmandu : International Centre for Integrated Mountain Development (ICIMOD), 2007.

[2] McAdoo BG, Moore A, Baulmwoll J, Indiegenous knowlegde and the near field population response during 2007 Solomon Islands tsunami, Nat Hazards, 2009, 48 : 73-82.

[3] Shaw Rajib, Sharma Anshu and Takeuchi Yukiko, Local wisdom and disaster risk reduction : from practice to policy, New York, 2009, Nova Science Publisher.

[4] Shaw Rajib, Uy N, Braumwoll J, editor. Local wisdom for disaster risk reduction: good practise and lesson learnt from Asia-Pacific region, Bangkok, 2008, UNISDR Asia and Pacific.

[5] UNESCO, What is local and local wisdom?, [http://www.unesco.org/new/en/natural-science/priority-areas/links/related-information/what-is-local-and-indgenous-knowledge/];undated.

[6] Chruickshank J, Do galsier, Listen? Local knowledge, colonial encounters, and social imagination, Vancouver, 2005, UBC Press.

[7] Natawijaya, D H, Sieh, K, Chlieh, M. Source parameters of the great Sumatran Megathrust Earthquake of 1797 and 1833 inferred from coral microatolls, *J. Geophs. Res*, 111, B06403. 2006.

[8] Departemen Pekerjaan Umum, SNI 1726-2003, Tata Cara Perencanaan Ketahanan Gempa untuk Bangunan Gedung (Indonesia Seismic Code for Building), 2003.

[9] Elfida Agus. Kajian Topologi. Morfologid antipologi pada Rumah Gadang Minangkabau, Universitas Bung Hatta, available from (www.fab.utm.my/download/Conference Seminar/ICCI12006S5PP21.pdf), 2001.

[10] Boen Teddy, Pribadi Krisnha, 2007, Engineering and non engineered houses for better earthquake resistance in Indonesia. *Proceeding of the DRH* contents meeting-EDMNIED, Kobe Japan. www.edm.bosai.go.jp/old/Phase2/1Documents/8_proceeding/7_PT8_P.pdf.

[11] Mardiasmo Diaswati, Barnes Paul H, 2015, Community response to disasters in Indonesia :

Gotong royong a double edged-sword, *Proceeding of the 9th Annual International Conference of International Institute for Infrastructure Renewal and Reconstruction*, Queensland University of Technology, Brisbane, Australia; pp 301-307.

[12] Hilhorst D, 2003, Responding to disasters : diversity of bureaucrats, technocrats, and local people, *International Journal of Mass Emergencies and Disasters*, 21(3):37-55.

[13] Hernawati S Tarida, Uma : Fenomena keterkaitan manusia dengan alam, Padang, Yayasan Citra Mandiri, 2007.

[14] Fuadi Al Busra, 2001, Transformasi ruang dalam pada rumah gadang studi kasus: nagari koto baru, Kecamatan Sungai Pagu, Kabupaten Solok Selatan, *Seminar Arsitektur Universitas Bung Hatta*, (www.fab.utm.my/download/ConferenceSeminar/ICCI2006S5PP21.pdf).

[15] Persoon G, Local leaders on Siberut : a creation not yet completed, P. Quarles van Ufford (Ed), Local Leadership and Programme Implementation in Indonesia, Amsterdam: Free University Press, 1987.

A review on disaster risk mitigation in the oil and gas project

N N Rodhi*, **N Anwar, I P A Wiguna**

Institut Teknologi Sepuluh Nopember, Surabaya, Indonesia

*Corresponding Author : nova.nevila@gmail.com

Abstract. In addition to the very complex risks, hazards potentially lead to disasters in the oil and gas projects. These risks can certainly be anticipated with the application of risk management, but an unsystematic and ineffective implementation of risk management will still bring adverse impacts. According to the eleven risk management principles in ISO 31000:2009, the application of risk management must pay attention to all aspects, both internal and external factors. Thus, this paper aims to identify variables that could affect the disaster mitigation efforts of oil and gas projects. This research began with literature study to determine the problems of risk management in oil and gas projects, so the affecting variables as the study objectives can be specified subsequently based on the literature review as well. The variables that must be considered in the efforts of disaster risk mitigation of oil and gas project are the risk factors and sustainability aspect.
Keywords: oil and gas project, risk management, sustainable development

1. Introduction

Oil and gas projects are at risk due to the massive capital investment, the involvement of many parties, the use of complex technology, as well as the high environmental and social impacts [1]. The risk is the possibility of the occurrence of some uncertain, unexpected, and even unwanted events, which will change the probability prospect of a given investment [2]. There have been social conflicts and casualties due to the gas leakage that always occurs in the area of oil and gas industry [3]. In general, risk factors and hazards could be induced in the oil and gas sector by the parameters contained in the industrial system [4]. Disasters can happen at any time and certainly cannot be prevented, but their impacts can be minimized by the adoption of a strategy for dealing with disasters. Risks should be mitigated since the beginning to prevent disasters. Therefore, any risk analysis must be integrated with the disaster risk assessment.

All kinds of inherent risks in the project life cycle must be identified so a framework for measuring risks can be prepared to achieve the success of the project. It should also be noted that related to the abatement of social, environmental, and economic challenges in the construction industry; the constructors should give priority to the concept of sustainable development in their activities [5].Based on the description above, this paper aims to discuss variables that could affect the disaster mitigation efforts of oil and gas projects.

IOP Conf. Series: Earth and Environmental Science **106** (2017) 012009 doi:10.1088/1755-1315/106/1/012009

2. Research Method

2.1. Project risk in oil and gas industry

A risk is an internal or external condition or situation that can influence and change the initial state of a project, as well as affect its time and cost [6]. A risk is an uncertain event that could have an impact on project objectives including the scope, schedule, cost, and quality parameters [7]. The risk can be defined from various viewpoints. Based on the process point of view, the risks are factors that may affect the achievement of objectives, leading to unintended consequences. On the other hand,from the result standpoint, the risk is an unpredictable, uncertain, and undesirable outcome because it is considered counterproductive. Therefore, the risk plays an important role in the decision-making and can affect the performance of the project [8].

There are two (2) types of project risks, i.e., the risk of a single project and multi-project risk combination [9]. There are two (2) sources of risk factors, namely the internal and external aspects [10]. Oil and gas projects are in jeopardy due to the large capital investment, the involvement of many parties, the use of complex technology, as well as the high environmental and social impacts [1]. Every activity in the oil and gas industry poses risks that not only could obstruct the project but also threaten the environment, society, and economy [11], [12].

2.2. Disaster in oil and gas industry

Oil and gas projects also have the potential of becoming dangerous to life, property, and the environment if the activity is not controlled and regulated appropriately [13]. Moreover, the gas leakage that frequently occurs in this industry leads to social conflicts and casualties [3]. In addition to risks, hazards entail the potency of turning into disasters [4]. Disasters are certainly not possible to be avoided and can occur at any time, but their circumstances can be anticipated by considering the risks and the level of toughness [14].

Hazards are classified into three (3) categories:

1. Natural hazard is the harm caused by factors related to the climatic and geographical aspects.
2. The technological hazard is the harm caused by technological or industrial accidents, also the presence of dangerous procedures, infrastructure failures, or other events that may cause loss of life or injuries, property damage, social and economic disruptions, as well as environmental degradation.
3. Environmental degradation is a deterioration process of natural resources or an adverse alteration of natural ecosystems caused by human activities and behaviors, sometimes even combined with natural disasters.

The application of a strategy for dealing with disasters can help to minimize their impacts since one the disaster causes is the absence or the lack of risk mitigation efforts. Hence, disaster risk assessments must be performed along with risk analysis.

3. Results and Discussion

3.1. Identification of risk factors

Identification results obtained by previous research mentioned that both internal and external risk factors affect oil and gas projects. The risk factors are summarized in Table 1.

IOP Conf. Series: Earth and Environmental Science **106** (2017) 012009 doi:10.1088/1755-1315/106/1/012009

Table 1. Risk factors of oil and gas projects.

Risk factors	References
Internal factors	
Design	16, 1, 17, 18
Construction	16, 1, 12
Procurement	16, 17, 12
Finance	16, 1, 17, 18
Operational factor	16, 1, 17, 19
Contract	1, 17, 20, 18
Stakeholders	1, 17
Logistics	1, 17
Technology	20, 4, 15, 21, 18
Accidents	19, 18
Employee distribution and lost	1, 17, 20, 18
Management	17, 12, 14, 18
External factors	
Social factor	11, 16, 1, 22, 12, 13, 14
Political factor	16, 12
Law	1, 17
Production competition	23, 24
Economy	16, 24, 25, 12, 13, 14
Environment	11, 22, 16, 1, 12, 13, 25, 26, 14, 18
Property	13

Disaster factors of oil and gas projects are listed in Table 2.

Table 2. Disaster factors of oil and gas projects.

Disaster Factors	References
Third party	
Water pollution	27, 28, 29, 30, 31, 32
Noise pollution	27, 28, 29, 30, 31, 32
Soil pollution	27, 28, 29, 30, 31, 32
Operations with a removal of vegetation	27, 28, 29, 30, 31, 32
Operations with a high potential of erosion	27, 28, 29, 30, 31, 32
Chemical pollution	27, 28, 29, 30, 31, 32
Public health effects	27, 28, 29, 30, 31, 32
Social disruption	27, 28, 29, 30, 31, 32
Theft/embezzlement	27, 28, 29, 30, 31, 32
Technological failures	27, 28, 29, 30, 31, 32
Explosion	27, 28, 29, 30, 31, 32
Fire	27, 28, 29, 30, 31, 32
Economic vulnerability	27, 28, 29, 30, 31, 32
Physical vulnerability	27, 28, 29, 30, 31, 32
Vulnerability of life support infrastructure	27, 28, 29, 30, 31, 32
Chemical exposure	27, 28, 29, 30, 31, 32
Social exposure	27, 28, 29, 30, 31, 32
Economic exposure	27, 28, 29, 30, 31, 32
Physical exposure	27, 28, 29, 30, 31, 32
Exposure to life support infrastructure	27, 28, 29, 30, 31, 32
Ecotoxicity	27, 28, 29, 30, 31, 32
Earthquake	27, 28, 29, 30, 31, 32

3.2. Risk mitigation

Currently, the leading oil and gas companies must have applied the risk management and complied with ISO 31000:2009 as a framework that can integrate various management processes—including the management of HSE (Health, Safety, and Environment) risks—to achieve sustainable development. It cannot be denied nonetheless that the activities that continue to this day still cause negative impacts, particularly for the surrounding environment. In addition, international oil and gas companies nowadays place more emphasis on preventive measures than coping methods [33], indicating that the concept of sustainable development has not been fully considered in risk management applications.

The risk management of construction projects must be carried out with respect to the objectives of Sustainable Development Goals which strive to achieve sustainability and eradicate poverty. The efforts to alleviate poverty must be accompanied by a progressive economic growth so that all sorts of risk that could adversely affect the economy can be well managed, including those related to disaster risks and vulnerabilities in the development plan. It is indeed on account that disaster can be a significant threat to achieving and sustaining development plans and objectives [34].

The importance of risk management in the pursuit of sustainable development objectives encourages researchers to examine them with different methods. Researchers [27] developed the concept of risk management by the AHP (Analytic Hierarchy Process) approach that can support the decision-making system on the maintenance aspects of oil and gas pipelines so that the pipeline project can survive by taking the quality of pipes and environmental conditions into account.

Reseacher [28] presented the concept of risk management in pipeline projects, referring not only to political and economic benefits but also the needs of society and environmental conditions to achieve sustainable projects. The results of research conducted using theoretical statistical techniques show that risk management can be used comprehensively and effectively in project management to reduce the impact of emerging risks. The concept presented in the outcome of the study only analyzes the risks in general and does not present how to manage the risks in the context of attaining sustainable projects.

The oil and gas pipeline project entails the potential of a dangerous failure so that methods to predict failures are highly required. Reseachers [30] proposed a Neurofuzzy approach that has been proven to be able to predict leak-related failures and explosions that impact financial and environmental conditions. The method can be applied in the planning and maintenance phase. Failures in the pipeline project can also be caused by the theft of oil and gas through pipelines. Therefore, Reseacher [31] performed a risk analysis using AHP (Analytic Hierarchy Process) which can support the decision-making system on the improvement of pipeline network security and optimization of security cost.

4. Conclusion

The results of this study state that the risk factors of projects in the oil and gas industry are truly complex that the risk analysis must be carried out to address various aspects of the project itself, the environment, as well as disaster occurrence. Also, it should be integrated with sustainable development aspects.

References

[1] Thuyet, Nguyen Van, et al. 2007. Risk management in oil and gas construction projects in Vietnam. *International Journal of Energy Sector Management*. 1(2):175-194.

[2] A Enhassi, J A Mosa. 2008. Risk Management in Building Projects: Owners' Perspective. *The Islamic University Journal (Series of Natural Studies and Engineering)*: 95 – 123.

[3] Ambarsari, DC et al. 2009. Minyak Kami Tanggung Jawab Kami – Transparansi Migas untuk Pembangunan Berkelanjutan: Belajar dari Blora dan Bojonegoro.

[4] Bolado, et al. 2012. Best practices and methodological guidelines for conducting gas risk assessments. European Union.

[5] Zabihi, Hossein et al. 2012. Sustainability in Building and Construction: Revising Definitions and Concepts. *International Journal of Emerging Sciences*. 2(4): 570 – 578.

[6] Choudhry R, and Iqbal, K. 2013. Identification of risk management systems in construction industry in Pakistan. *J. Manage. Eng*., 10.1061/(ASCE) ME.1943-5479.0000122: 42–49.

[7] Anonimus. 2008. A Guide to the Project Management Body of Knowledge (PMBOK GUIDE). *Fourth Edition an American National Standard*. ANSI / PMI 99–001–2008.

[8] Wiguna, I Putu Artama, and Scott Stephen. 2005. Nature of The Critical Risk Factors Affecting Project Performance In Indonesian Building Contracts. *Association of Researchers in Construction Management*. 1:225-235.

[9] Corvellec H. 2009. The practice of risk management: Silence is not absence. Risk Manage. 11(3–4):285–304.

[10] El-Sayegh Sameh M, and Mansour Mahmoud H. 2015. Risk Assessment and Allocation in Highway Construction Projects in the UAE. *Journal of Management Engineering*. DOI:10.1061/(ASCE)ME.1943-5479.0000365. American Society of Civil Engineers. Pp. 04015004-1 - 04015004-11.

[11] Ogwu, Friday Adejoh. 2011. Challenges of Oil and Gas Pipeline Network and the role of Physical Planners in Nigeria. *FORUM Ejournal*:41-51.

[12] Rodhi, Nova Nevila et al. 2012. Kajian Risiko Penggunaan Sumber Daya Air Berbasis Paradigma Bottom-up Approach (Studi kasus wilayah eksplorasi Banyu Urip Bojonegoro). *Thesis*. Magister Teknik Sipil. Universitas Diponegoro. Semarang.

[13] Achaw, Osei-Wusu, and Boateng. 2012. Safety practices in the oil and gas industries in Ghana. *International Journal of Development and Sustainability*. Online ISSN: 2168-8662. 1(2):456-465.

[14] Voogd Henk. 2004. Disaster Prevention in Urban Environments. *European Journal of Spatial Development*:1–20.

[15] Usman, Raheem et al. 2013. Disaster Risk Management and Social Impact Assessment: Understanding Preparedness, Response and Recovery in Community Projects. Environmental Change and Sustainability:259 – 274.

[16] Lubab, Irbabaul. 2005. Manajemen Risiko Proyek Pemboran dalam Upaya Meminimalkan Biaya Investasi Pemboran. Paper WOP VI – Manajemen Risiko.

[17] Mojtahedi, et al. 2008. Risk Identification and Analysis Concurrently: Group Decision Making Approach. *Proceedings of the IEEE ICMIT*.

[18] Namian, et al. 2016. Improving Hazard-Recognition Performance and Safety Training Outcomes: Integrating Strategies for Training Transfer. *J. Constr. Eng. Manage*. 04016048. ASCE, ISSN.

[19] Campidelli M, et al. 2015. Blast Design-Basis Threat Uncertainty and Its Effects on Probabilistic Risk Assessment. *American Society of Civil Engineers:* 04015012-1 - 04015012-15.

[20] Zhendong, Wu et al. 2011. Risk Assessment of Oil and Gas Drilling Engineering Cooperation Project Based on Fuzzy Comprehensive Evaluation. *Proceedings of the 7th International Conference on Innovation and Management*:1662 – 1667.

[21] Collins, Dave, and Junghans, Antje. 2015. Sustainable facilities management and green leasing: The company strategic approach. *Procedia Economics and Finance*. 21:128 – 136.

[22] Zolfagharian, et al. 2012. Environmental Impacts Assessment on Construction Sites. *Construction Research Congress*. ASCE:1750 – 1759.

[23] Razaque, Abdul, et al. 2012. Fostering Project Scheduling and Controlling Risk Management. *International Journal of Business and Social Science*. 3(14):118 – 127.

[24] Mursitama, Tirta N, and Yudono Maisa. 2010. Strategi Tiga Naga Ekonomi Politik Industri Minyak Cina di Indonesia. Kepik Ungu, Depok, Indonesia.

[25] Bonstrom, Holly. Corotis, Ross B, and Porter, Keith. 2012. Overcoming Public and Political Challenges for Natural Hazard Risk Investment Decisions. *Journal of Integrated Disaster Risk Management*. ISSN:2185-8322:1 – 23.

[26] El Attar Ibrahim, and Khattab Wael. 2013. Risks Assessment of Oil and Gas Transportation in the Suez Canal Sector. *The International Maritime Transport & Logistics Conference* (MARLOG 2). Sustainable Development of Suez Canal Region:1- 11.

[27] Al-Khalil M, Assaf S, and Al-Anazi F. 2005. Risk-Based Maintenance Planning of Cross-Country Pipelines. *Journal of Performance of Constructed Facilities*. 19(2).
[28] Nielsen KR. 2006. Risk Management: Lessons from Six Continents. *Journal of Management in Engineering*. 22(2)
[29] El-Abbasy M, Senouci A, Zayed T, Mirahadi F, and Parvizsedghy L. 2014. Condition Prediction Models for Oil and Gas Pipelines Using Regression Analysis. *Journal of Construction Engineering and Management*.
[30] Parvizsedghy L, and Zayed T. 2015. Consequence of Failure: Neurofuzzy-Based Prediction Model for Gas Pipelines. *Journal of Performance of Constructed Facilities*.
[31] Hasan A. 2016. Security of Cross-Country Oil and Gas Pipelines: A Risk-Based Model. *Journal of Pipeline Systems Engineering and Practice*.
[32] Conner H. 2015. Managing Environmental Risk in the Oil and Gas Industry. CMC Senior Theses.CMC Student Scholarship. Claremont Colleges.
[33] Bakhtiari S. 2014. Risk Management: A Powerful Instrument For Sustainable Development. *OIDA International Journal of Sustainable Development*:95 – 104.

IOP Conf. Series: Earth and Environmental Science **106** (2017) 012010 doi:10.1088/1755-1315/106/1/012010

Settlement preferences in the disaster-prone areas of Brantas River

S Hariyani

Urban and Regional Planning Department, Faculty of Engineering, Universitas Brawijaya, Malang, East Java, Indonesia

*Corresponding Author: septianahariyani@ub.ac.id

Abstract. Kota Lama is one of the urban villages in Malang city that has settlements along the Brantas River. Kota Lama experienced three landslides and flooding in 2015 and one in 2016. Those disasters caused the community to take action of post-disaster recovery, yet the people still choose to remain living in Kota Lama. Therefore, the study aims at determining the preferences of the citizens living in disaster-prone areas in Brantas River. The research used a factor analysis of 12 variables: 1) neighbourhood situation, 2) air condition, 3) relations between neighbours, 4) security, 5) location, 6) customs, 7) ethnic diversity, 8) the presence of social groups, 9) the community's customs and habits, 10) proximity to the economic facilities, 11) adequate educational facilities, and 12) adequate medical/health facilities. The results show that two factors have been formed, namely Factor 1 (access) comprising variables of neighbourhood situation, air condition, relations between neighbours, location, ethnic diversity, the presence of a social group, supporting positive habits at home, close to the economic facilities, educational facilities, as well as medical facilities, and Factor 2 (assurance) consisting of customs and security.
Keywords: analysis factor, disaster-prone area, settlement preferences

1. Introduction

Settlement preference is the environmental quality reflecting the quality of human life, in this case, comprising of: 1) situational variables of distance to service centres, climate, and topography; 2) physical variables of space organization, clean air, and calm ambience; 3) psychological variables of population density and luxury; as well as 4) socioeconomic variables of tribe, social status, crime rate, and education [1]. The choice of living is influenced by the attractiveness of settlements and conditions that are considered ideal [2]. Preference is an attitude of choice to a thing influenced by internal and external factors. It can be based on several circumstances including social scope (such as occupation, income, number of family members, dependents, and children education), housing (condition, type, and price of the house), community aspect, location scope, as well as physical aspect [3]. Assessing the housing location of one individual and another is not the same because the background of the level of needs and interests are different [4]. Kota Lama is one of the areas prone to floods and landslides in Malang City. Data of disaster occurrence from BPBD Malang has recorded three landslide events in 2015 and one in 2016 taking place in this village [5, 6]. As a result of the flood disaster, residents had to perform post-flood recovery actions. Although they are facing the risks of flood and landslides, they still choose to stay in Kota Lama. The tendency of the people who remain to live in this location is a big question to be studied.

IOP Conf. Series: Earth and Environmental Science **106** (2017) 012010 doi:10.1088/1755-1315/106/1/012010

2. Research Method

The study took place in a neighborhood (RT) in Kota Lama. The basic consideration is the previous research that all RWs in Kota Lama fall into a high category of hazard and vulnerability levels, except for those located very far from Brantas River [7]. This statement is supported further by the actual condition of the existing study area (results of interview surveys in 2016) showing that only houses nearby or directly adjacent to Brantas River were impacted by the flood disaster at Kota Lama in 2003, 2005, 2009, 2013. Similarly, those areas were struck by landslides, three times in 2015 and once in June 2016 [6]. This research used a sampling size technique of proportionate stratified random sampling where all individuals or elements in the population have the same opportunity to be a respondent [8]. The total sample size was 279 households.

Factor analysis is a statistical method used to describe variability among observed, correlated variables regarding the potentially lower number of unobserved variables called factors. The observed variables were modelled as a linear combination of the potential factors, plus 'error' terms. Factor analysis aims at finding independent latent variables. Followers of factor analytic methods believe that the information gained concerning the interdependencies between observed variables can be used later to reduce the set of variables in the dataset [9]. There are 12 variables analyzed in this research as shown in Table 1.

Table 1. Variables used in this study.

	Variable
Conditions in the neighbourhood (neighbourhood situation)	X_1
Healthy air in the area (air condition)	X_2
Good relations between neighbours (relations between neighbours)	X_3
Adequate security (security)	X_4
Strategic location to stay (location)	X_5
Positive behaviour and habits of the community (customs)	X_6
People diversity is not an issue in the area (ethnic diversity)	X_7
The presence of a social groups	X_8
The community's customs and habits	X_9
Progressive economic development in the area (close to the economic facilities)	X_{10}
Adequate educational facilities	X_{11}
Adequate medical/health facilities	X_{12}

The factor analysis in this study employed data taken through questionnaires using the Likert scale regarding people's preferences of the residential environment. Table 2 lists the Likert scale used in the survey.

Table 2. Likert scale.

Measurement Scales	Descriptions
5	Strongly agree
4	Agree
3	Simply agree
2	Disagree
1	Strongly disagree

The order of factor analysis in this study is as followed in Figure 1.

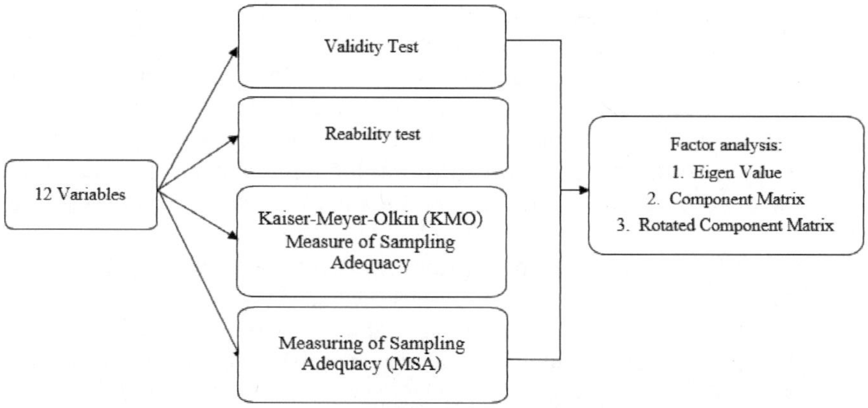

Figure 1. Procedural scheme of the methods applied in this study.

3. Results and Discussion

3.1. Feasibility testing of variables

The feasibility testing of variables was carried out through validity and reliability tests. There were 279 samples (n) in this research, then the degree of freedom (df) = n – 2 = 277. Using a significance level of 5%, r table value was obtained of 0.1175. The next step was to conduct a 12 variables test by: 1) validity test, 2) reliability test, 3) Kaiser-Meyer-Olkin (KMO) and Bartlett's Test, as well as 4) Measures of Sampling Adequacy (MSA). The results of those tests showed that the 12 variables were suitable for further processing which was the factor analysis.

3.2. Factor formation

Once the variables were determined, selected, and their correlation calculations have met the requirements for further analysis, the next step was to establish the factors to find the underlying structure of the relationship between initial variables. The method used in the factor formation was the principal component analysis. The two primary measures in the factor formation were the determination of the number of factors and the formed rotation of the factors.

Determination of the number of factors

The number of factors was determined by combining several criteria to obtain the most appropriate number of factors for the research data (Table 3).

Table 3. Total variance explained.

Factor	Initial Eigenvalues			Extraction Sums of Squared Loadings		
	Total	% of Variance	Cumulative %	Total	% of Variance	Cumulative %
1	5.852	48.766	48.766	5.852	48.766	48.766
2	1.179	9.826	58.593	1.179	9.826	58.593
3	0.865	7.205	65.798			
4	0.853	7.108	72.906			
5	0.695	5.793	78.699			
6	0.591	4.927	83.626			
7	0.507	4.226	87.852			
8	0.439	3.658	91.510			
9	0.346	2.880	94.391			
10	0.331	2.758	97.148			
11	0.227	1.894	99.043			
12	0.115	0.957	100.000			

IOP Conf. Series: Earth and Environmental Science **106** (2017) 012010 doi:10.1088/1755-1315/106/1/012010

The first criterion used was the eigenvalues. Factors having more than one eigenvalue would be maintained, and factors with less than one eigenvalue would not be included in the model. As seen in the table above, eigenvalues greater than one were obtained on one factor and two factors. Thus, the number of factors was two factors. The second criterion was the determination based on the percentage value of the total variance that could be explained by the number of factors to be formed. From the table above, interpretations could be made related to the total cumulative variance of the sample. If those variables were summarized into several factors, then the total value of the explanatory variance was as follows:

- If the eight variables were extracted into one factor, the total unexplained variance would be 5.852 / 12 x 100% = 48.766%.
- If the eight variables were extracted into two factors, the total explained variance would be 1.179 / 12 x 100% = 9.826%, and the cumulative total variance for two factors would be 48.766% + 9.826% = 58.593%.

By extracting the initial parameters into two factors, a considerable cumulative total variance of 58.593% was generated, implying that the two factors formed can already represent the 12 parameters of urban residency preference in Kota Lama. Hence, the extraction of two factors has met the second criterion. The third criterion was the determination based on the scree plot which is a plot of eigenvalues against the number of extracted factors. The point at which the scree begins to occur shows the number of appropriate factors. This point occurs when the scree starts to look flat. As seen in Figure 2, the scree plot starts horizontally on the extraction of the initial variables into two factors.

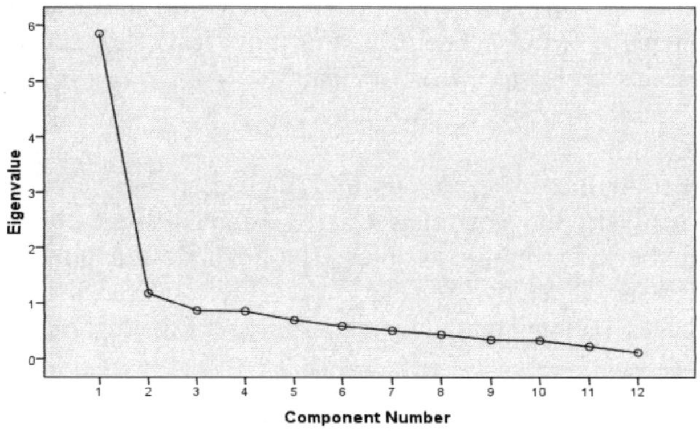

Figure 2. Scree plot.

From the combination of the three criteria, it could be concluded that the most appropriate factor extraction was two factors.

Component matrix

The component matrix shows the distribution of the 12 parameters on two factors formed, while the figures in the table are factor loadings indicating the correlation between a parameter with Factor 1 and Factor 2. The process of determining which parameters would be entered into which factor is done by a comprehensive comparison of the correlation of each row. More details could be seen in the following Table 4.

Table 4. Component matrix.

Variable	Factor	
	1	2
X_6	0.833	-0.172
X_8	0.828	0.152
X_3	0.822	-0.219
X_1	0.785	-0.149

Variable	Factor	
	1	2
X_5	0.784	-0.169
X_{11}	0.737	0.164
X_7	0.717	-0.275
X_{12}	0.664	0.106
X_{10}	0.622	0.099
X_2	0.582	-0.039
X_4	0.545	0.366
X_9	0.241	0.877

Rotation

The rotation in this study aims to obtain factors with loading values that are clear enough for interpretation. Rotated component matrix is a correlation matrix that shows a more distinct and apparent distribution of variables compared with the component matrix. More details is seen in the following Table 5.

Table 5. Rotated component matrix.

Variabel	Factor	
	1	2
X_3	0.846	0.093
X_6	0.839	0.140
X_5	0.792	0.125
X_1	0.786	0.144
X_7	0.758	-0.002
X_8	0.718	0.440
X_{11}	0.629	0.419
X_{12}	0.581	0.338
X_2	0.557	0.173
X_{10}	0.545	0.316
X_9	-0.091	0.905
X_4	0.377	0.538

The results indicate that the loading values of the variables for the two factors have been sufficiently differentiated and are ready for interpretation. All variables have a high loading on one factor and a small loading for the other factor.

From the results of factor analysis, two primary factors affecting the settlement preference of the residents in disaster-prone areas of Kota Lama, Kedung Kandang district, Malang City, have been obtained. The factor grouping is shown in Table 6.

Table 6. Results of classifying variables into factors.

Factor	Component Variables
Factor 1(access)	X_3, X_6, X_5, X_1, X_7, X_8, X_{11}, X_{12}, X_2, X_{10}
Factor 2 (security)	X_9, X_4

Factor 1, the biggest variation contributor, has close relationships with ten component variables. Factor 1 consists of relationships between neighbors X_3 (0.846), behavior and habits of the society X_6 (0.839), suitable location for residence X_5 (0.792), and neighborhood conditions X_1 (0.786), healthy air X_2 (0.758), community diversity X_7(0.718), the presence of a social group X_8 (0.629), adequate educational facilities X_{11} (0,581), adequate health facilities X_{12} (0.557), and progressive economic development X_{10} (0.545).

Factor 2, the second largest contributor, has close relations with two component variables, i.e., positive customs and habits X_9 (0.905) and adequate security X_4 (0.538). The factors' names were then determined based on the characteristics of their members by examining the variables constituting

Factor 1 and 2. Giving a new name for each factor is subjective; it could also be derived from the variable having the highest factor loading value [10]. The interpretation of factors that have been formed in particular gives a name that is considered to represent the variables of the factors members [11]. Considering both theories, by looking at the variables making up the factors, Factor 1 could be determined as Access Factor and Factor 2 could be named as Security Factor.

The two factors formed are in line with the statement of [1], [2], [3], [4]. Settlement preference is the need for access to resources, especially economic resources that imply the social needs as well as the convenience to reach the workplace to obtain a source of income. Settlements must also provide security guarantee, in the sense of ensuring family circumstances. Both factors have proven that the accounts related to the community preferences to remain living in Kota Lama were houses or shelters providing access to resources and security guarantee.

4. Conclusion

Based on the factor analysis, it can be concluded that two primary factors are affecting the settlement preferences in disaster-prone areas of Kota Lama, Malang City, namely access and security aspects. Access Factor is the new label obtained from the interpretation of factor analysis, considered as representative for the variables constituting Factor 1, i.e., relationships between neighbors, behavior and customs of the community, residential location, environmental condition, community diversity, the presence of social groups, educational facilities, health facilities, healthy air, and economic development. The second factor, named 'security', represents the variables of customs and habits of the community as well as the adequate security.

Acknowledgment

The author would like to express gratitude to Ardan Alif Ramadlan, an undergraduate student in Urban and Regional Planning, Faculty of Engineering, Brawijaya University (PWK FT UB), for the survey and data compilation in the interest of this paper.

References

[1] Rapoport, Amos, Human Aspects of Urban Form: Towards a Man-Environment Approach to Urban Form and Design. 1st Edition, Pergamon. E-Book ISBN: 9781483156828, Oxford, 1977.

[2] Sinulingga B D, Pembangunan Kota: Tinjauan Regional dan Lokal, Pustaka Sinar Harapan, Jakarta, 1999.

[3] Nurhadi Imam, Preferensi Masyarakat dalam Memilih Perumahan Perkotaan di kota Tangerang (unpublished research), Universitas Diponegoro, Semarang, 2004.

[4] Knox Paul, Urban Social Geography 6th edition, Pearson Education Limited, England, 2010.

[5] Badan Penanggulangan Bencana Daerah Kota Malang (BPBD). 2015.

[6] Badan Penanggulangan Bencana Daerah Kota Malang (BPBD). 2016.

[7] Anggraeni, Mustika, Ari, Ismu Rini Dwi, Santosa Endratno Budi, and Widayanti Reny. Climate Change & Home Location Preferences in Flood Prone Areas of Bojonegoro Regency, *The 4 th International Conference on Sustainable Future for Human Security*. SustaiN 2013. *Procedia Environmental Sciences*, 2014, p. 703 – 711.

[8] Isaac S, & Michael W B, Handbook in Research and Evaluation, Edits Publisher, San Diego, 1981.

[9] Child, D, The Essentials of Factor Analysis, 3rd edition, Bloomsbury Academic Press, 2006.

[10] Ghozali Imam, Aplikasi Analisis Multivariate dengan Program IBM SPSS 21, Badan Penerbit Universitas Diponegoro, Semarang, 2013.

[11] Santoso Singgih, Buku Latihan SPSS Statistik Multivariat, PT. Elex Media Komputindo, Jakarta, 2006.

Acceptance and Utilization of Technology (UTAUT) as a Method of Technology Acceptance Model of Mitigation Disaster Website

T Siswanto[1*], R Shofiati[2], H Hartini[3]

[1]Information System Department, Faculty of Industrial Technology, Universitas Trisakti, Jakarta, Indonesia
[2]Informatics Department, Faculty of Industrial Technology, Universitas Trisakti, Jakarta, Indonesia
[3]Management Departement, Faculty of Economic and Business, Universitas Trisakti, Jakarta, Indonesia

*Corresponding Author: teddysiswanto@yahoo.com

Abstract. www.mitigasi-bencana.com as a knowledge management website created based on survey results in April-July 2014 in East Java and Central Java provinces, indicates a gap between the expectations and reality that exist in the services provided by the regional disaster management agency. Based on condition analysis, the gaps that occur can be reduced if the community has the understanding and knowledge of adequate disaster mitigation. The problem that arises later is whether the chosen technology solution is appropriate and acceptable to the public? The methodology used in this study using the Technology Acceptance Model development is the Unified Theory of Acceptance and Utilization of Technology (UTAUT). Feedback obtained from respondents KarangTaruna youth SelogedongBantul, www.mitigasi-bencana.com can be accepted by the respondents, but from processed data is obtained only UTAUT hypotheses on the relationship dimension eligible for Social Expectancy on the Attitude toward technology , which means the higher the perception of the Social Expectancy, the higher the perception of the Attitude toward technology. Because www.mitigasi-bencana.com is new socialized so that society still need time to explore content information and knowledge contained therein. To be accepted by user, a knowledge management application must prepare various aspects of Performance Expectancy, Effort Expectancy, Social Factors, Facilitating Conditions and Attitude.
Keywords: attitude, effort expectancy, performance expectancy, social factors, UTAUT

1. Introduction

Implementation of the knowledge management website www.mitigasi-bencana.com, needs to be measured to find out how far the application can be used and accepted by the community. The application was made because of the previous survey there was a significant difference between the expectations of the community regarding the quality of services provided by the authorities in the provinces of East Java and Central Java in the handling of disasters. Some of the things that trigger the lack of services from authorized agencies in addition to geographical factors Indonesia is very broad and limited access to transportation, there is also a problem of lack of awareness of the importance of disaster mitigation that is disaster risk reduction efforts and also from officers who sometimes confused how the situation in the field. This is due to the lack or difficulty of accessing the operating

IOP Conf. Series: Earth and Environmental Science **106** (2017) 012011 doi:10.1088/1755-1315/106/1/012011

system documentation of disaster management procedures. With the knowledge management website www.mitigasi-bencana.com expected quality of service to the community will be better and the community also aware of the importance of disaster mitigation so as to increase knowledge of disaster mitigation. To perform the measurement of technology acceptance then used method of Unified Theory of Acceptance and Use of Technology (UTAUT)

Utilization of UTAUT as a method of acceptance and use of technology by researchers has been extended to integrate for various settings such as differences in users, organizations, technology types, tasks, time and differences in location. First, the categorization of technology users into different group such as employee, consumer, and society. As by [10] using employee respondents at all levels of the organization ie boards of directors, senior managers, middle managers, and staff of operational personnel). [18] studied use the consumer service user respondents. [15] studied the use of e-government services by the community. There are also other researchers using specific responders such as teachers conducted by [13] and doctor respondents by [6]. Secondly, categorization is grouped into organizations by industry sectors such as manufacturing and services, or public and private organizations. Researchers are conducting at various organizations, such as schools [13], hospitals [6] and government organizations [9]. Thirdly, categorization is grouped into different types of technologies such as the internet in society by [9], mobile banking [18], and e-government services [15]. Fourth, categorization is grouped into various types that include idea-making and decision-making in technology design [4], income tax filings [5], and medical diagnostics [6]. The fifth, the categorization is grouped in different time ie the initial use or subsequent use), as done by [18] focusing on the adoption of mobile banking users, while [14] adoption, early use, and use of post adoption. The sixth, categorization is grouped in different locations ie the State or economic sector. The study was conducted in locations other than West countries, such as India [9], China [16] and Korea [11]. Other research focuses on certain economic sectors, such as services by [10], education by [7], food services by [17], medical and health services by [12] and the public sector by [8].

2. Research Method

Sampling was done to young man Karang Taruna Selogedong, Sedayu Bantul, Yogjakarta with number of respondents as 186 young people on October 22-23, 2016.

2.1. Research Design

The UTAUT model used in this study was previously undertaken by Guyana University researchers[1]. Regardless of the inconsistencies of effects observed in the literature, it is expectedto find confirmation of the basic form of the UTAUT model coupled with the Attitude variable.

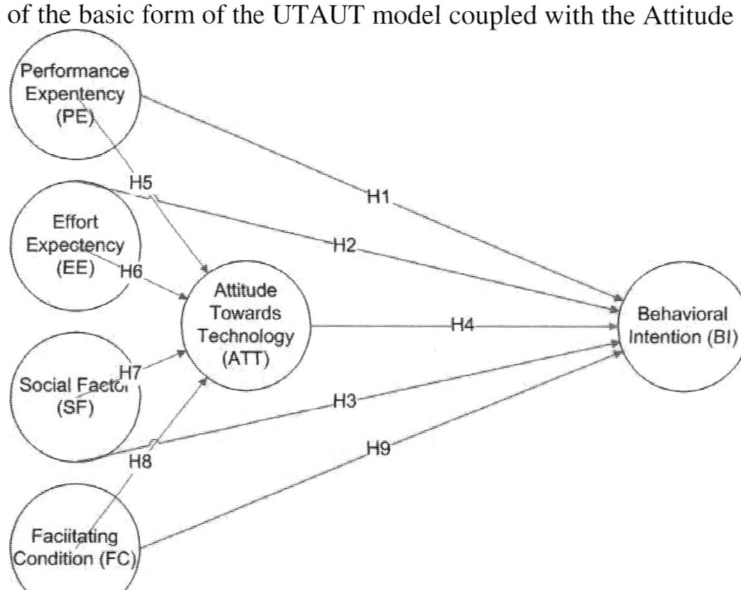

Figure 1 : UTAUT model with effect of attitude.

2.2. Hypothesis
Hypothesis 1 : Performance expectancy has a positive effect on behavioral intention.
Hypothesis 2 : Effort expectancy has a positive effect on behavioral intention.
Hypothesis 3 : Social factor has a positive effect on behavioral intention.

One UTAUT hypothesis is Attitude to Use Behavior has no effect on Behavioural Intention even if the effect of Performance Expectancy and Effort Expectancy can be controlled. As a result, Attitude is not explicitly included in the UTAUT model. However, research conducted by [2], [3] found a positive impact on Behavioural Intention. Both studies were conducted in non-western countries. The next hypothesis is:
Hypothesis 4 : Attitude to Use of Technology for learning is positively associated with Behavioral Intention.
Hypothesis 5 : Performance Expectancy is positively associated with Attitude.
Hypothesis 6 : Effort Expectancy is positively associated with Attitude.
Hypothesis 7 : Social Factors that are positively related to Attitude
Hypothesis 8 : Facilitating Condition is positively associated with Attitude.
Hypothesis 9 : Facilitating Condition is positively associated with Behavioral Intention

2.3. Data Collection Techniques and Instruments
Using Random Sampling Data Collection technique with questionnaire scale instrument of Linkert 1-5 and analysis tool: SEM (Structural Equation Modelling).

3. Results and Discussion
To perform data processing into information that can be concluded, need to be tested on the data to check there is not invalid data. In testing performed using Instrument Test and Goodness-of-Fits Test, while to determine the relationship between variables used SEM Analysis.

3.1. Testing
The tests are Instrument Test and Goodness-of-Fits Test. The Table 1 shows the results of validity and reliability testing. With a sample of 186 respondents, the Hair's Factor Loading value is 0.40. From the table above shows that in addition to the variable attitude towards technology, all indicators are able to form a variable construct because it has a factor loading value greater than 0.40. Therefore ATT1 is not valid form attitude towards technology variable, then the indicator is issued so that form attitud toward technology only 2 indicator that is ATT2 and ATT3. The table above also shows that the Cronbach's alpha value is higher than 0.6 so that the indicators are reliable in forming the construct variable.

Table 1. Instruments test.

Indicator	Factor Loadings	Cronbach's Alpha	Result
Variable: Effort Expectancy			
EE1	.805		
EE2	.830	0,734	Valid and Reliable
EE3	.789		
Variable: Attitude Towards the Use of Technologies			
ATT1	.310		
ATT2	.805		ATT1 Not Valid
ATT3	.855		

Based on the test results on the feasibility of the model, it is found that at the value of RMSEA smaller than 0.10 so concluded the goodness-of-fit model (Table 2). As for the criteria IFI, TLI, and CFI obtained the value of the default model is smaller than 0.9 so it is concluded Marginal-of-Fits model. Due to the RMSEA-based Goodness-of-Fits model, the model can be used to test the hypothesis

Table 2. Goodness-of-fitstest.

Indicator	Criteria	Score	Result
Chi-Square	Prob> 0.05	0.000	Poor-of-Fits
RMSEA	< 0.10	0.072	Goodness-of-Fits
NFI	≥ 0.90	0.730	Poor-of-Fits
RFI	≥ 0.90	0.669	Poor-of-Fits
IFI	≥ 0.90	0.847	Marginal-of-Fits
TLI	≥ 0.90	0.805	Marginal-of-Fits
CFI	≥ 0.90	0.841	Marginal-of-Fits

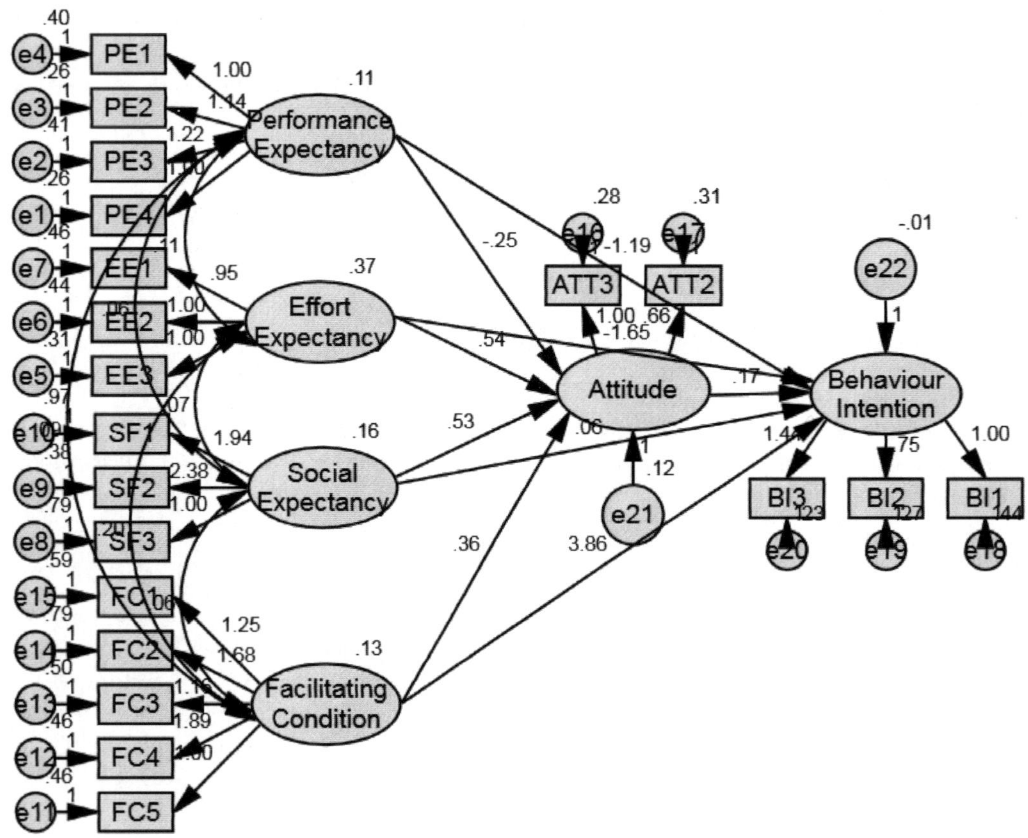

Figure 2. UTAUT measurement parameters.

3.2. SEM Analysis

1. Performance Expectancy and Behavior Intention
   Performance Expectancy has a coefficient of -1.195 with probability value of 0.483 greater than 0.05 so Ho fails to be rejected or no relationship Performance Expectancy and Behavior Intention
2. Effort Expectancy and Behavior intention
   Effort Expectancy has a coefficient value of 0.973 with probability value of 0.395 is greater than 0.05 so that Ho fails to be rejected or no relationship Effort Expectancy and Behavior intention.
3. Social Expectancy and Behavior Intention
   Social Expectancy has a coefficient value of 0.063 with a probability value of 0.891 greater than 0.05, so Ho fails to be rejected or no relationship Social Expectancy and Behavior Intention.
4. Attitude Toward Technology and Behavior intention

Attitude Toward Technology has a coefficient value of 0.171 with probability value of 0.702 greater than 0.05, so Ho fails to be rejected or no relationship Attitude Toward Technology and Behavioral Intention.

5. Performance Expectancy and Attitude Toward Technology
   Performance Expectancy has a coefficient value of -0.284 with probability value of 0.706 greater than 0.05 so Ho fails to be rejected or no relationship Performance Expectancy and Attitude Toward Technology.
6. Effort Expectancy and Attitude Toward Technology
   Effort Expectancy has a coefficient of -1.654 with probability value of 0.446 which is greater than 0.05 so that Ho fails to be rejected or no relation Effort Expectancy and Attitude Toward Technology.
7. Social Expectancy and Attitude Toward Technology
   Social Expectancy has a coefficient value of 0.532 with a probability value of 0.009 which is smaller than 0.05, so Ho is rejected or there is a relationship Social Expectancy and Attitude Toward Technology. The higher the perception of Social Expectancy then the higher the perception of Attitude Toward Technology.
8. Facilitating Conditions and Attitude Toward Technology
   Facilitating Conditions has a coefficient value of 0.363 with a probability value of 0.823 which is greater than 0.05, so Ho fails to be rejected or there is no relationship between Facilitating Conditions and Attitude Toward Technology.
9. Facilitating Conditions and Behavior Intention
   Facilitating Conditions has a coefficient value of 3.865 with probability value of 0.383 which is greater than 0.05, so Ho failed to be rejected or no relationship of Facilitating Conditions and Behavior Intention.

Table 3. Hypotesis.

Variable	Coefficient	Prob	Result
Performance Expectancy → Behavior Intention	-1.195	0.483	Ho failed to reject
Effort Expectancy → Behavior Intention	3.865	0.395	Ho failed to reject
Social Expectancy → Behavior Intention	0.063	0.891	Ho failed to reject
Attitude toward technology → Behavior Intention	0.171	0.702	Ho failed to reject
Performance Expectancy → Attitude Toward Technology	-0.248	0.706	Ho failed to reject
Effort Expectancy → Attitude Toward Technology	-1.654	0.446	Ho failed to reject
Social Expectancy → Attitude Toward Technology	0.532	0.009	Ho rejected
Facilitating Conditions → Attitude Toward Technology	0.363	0.823	Ho failed to reject
Facilitating Conditions → Behavior Intentention	3.865	0.383	Ho failed to reject

4. Conclusion

- Feed back obtained from youth respondents KarangTaruna Selogedong, Bantul Yogjakarta shows website knowledge management www.mitigasi-bencana.com can be accepted by respondents, but from processed data UTAUT only obtained hypotheses that qualify in dimension relationship for Social Expectancy to Attitude Toward Technology, which means that the higher the perception of Social Expectancy then the higher the perception of Attitude Toward Technology.
- For the acceptance of an information technology by the public should have attention to factors Performance Expectancy, Effort Expectancy, Social Factor, Facilitating Condition and Attitude.
- It is necessary to increase cooperation with local governments, disaster-affected communities for the continued dissemination of disaster mitigation information and to add disaster information and knowledge content from community experiences.
- The website will be optimized for its use by adding sharing content and animation sharing.

IOP Conf. Series: Earth and Environmental Science **106** (2017) 012011 doi:10.1088/1755-1315/106/1/012011

References

[1] Thomas T D, Singh L, & Gaffar K. (2013). The utility of the UTAUT model in explaining mobile learning adoption in higher education in Guyana. *International Journal of Education and Development using Information and Communication Technology (IJEDICT).* 9(3):71-85

[2] Jairak K, Praneetpolgrang P, &Mekhabunchakij K. (2009). An Acceptance of Mobile Learning for Higher Education Students in Thailand. *Special Issue of the International Journal of the Computer, the Internet and Management.* 17(SP3):36.1–36.8.

[3] Nassuora A B. (2012). Student Acceptance of Mobile Learning for Higher Education. *American Academic & Scholarly Research Journal.* 4:0–5.

[4] Brown, S. A., Dennis, A. R., & Venkatesh, V. (2010). Predicting collaboration technology use: Integrating technology adoption and collaboration research. Journal of Management Information Systems, 27(2), 9-53.

[5] Carter, L., & Schaupp, L. C. (2008). Efficacy and acceptance in e-file adoption. In Proceedings of the 14th Americas Conference on Information Systems (pp. 1-11).

[6] Chang, I. C., Hwang, H. G., Hung, W. F., & Li, Y. C. (2007). Physicians' acceptance of pharmacokinetics based clinical decision support systems. Expert Systems with Applications, 33(2), 296-303.

[7] Chiu, C. M., & Wang, E. T. G. (2008). Understanding web-based learning continuance intention: The role of subjective task value. Information & Management, 45(3), 194-201.

[8] Dasgupta, S., & Gupta, B. (2011). Impact of organizational culture on technology use in a developing country. In Proceedings of the 17th Americas Conference on Information Systems (pp. 1-10). Detroit, Michigan.

[9] Gupta, B., Dasgupta S., & Gupta, A. (2008). Adoption of ICT in a government organization in a developing country: An empirical study. Journal of Strategic Information Systems, 17(2), 140-154.

[10] Hong, W., Thong, J. Y. L., Chasalow, L. C., & Dhillon, G. (2011). User acceptance of agile information systems: A model and empirical test. Journal of Management Information Systems, 28(1), 235-272.

[11] Im, I., Hong, S., & Kang, M. S. (2011). An international comparison of technology adoption: Testing the UTAUT model. Information & Management, 48(1), 1-8.

[12] Liang, H. G., Xue, Y. J., Ke, W. L., & Wei, K. K. (2010). Understanding the influence of team climate on IT use. Journal of the Association for Information Systems, 11(8), 414-432.

[13] Pynoo, B., Devolder, P., Tondeur, J., van Braak, J., Duyck, W., & Duyck, P. (2011). Predicting secondary school teachers' acceptance and use of a digital learning environment: A cross-sectional study. Computers in Human Behavior, 27(1), 568-575.

[14] Venkatesh, V., Brown, S. A., Maruping, L. M., & Bala, H. (2008). Predicting different conceptualizations of system use: The competing roles of behavioral intention, facilitating conditions, and behavioral expectation. MIS Quarterly, 32(3), 483-502.

[15] Venkatesh, V., Thong, J. Y. L., Chan, F. K. Y., Hu, P. J. H., & Brown, S. A. (2011). Extending the two-stage information systems continuance model: Incorporating UTAUT predictors and the role of context. Information Systems Journal, 21(6), 527-555.

[16] Venkatesh, V., & Zhang, X. (2010). Unified theory of acceptance and use of technology: US vs. China. Journal of Global Information Technology Management, 13(1), 5-27.

[17] Yoo, S. J., Han, S. H., & Huang, W. H. (2012). The roles of intrinsic motivators and extrinsic motivators in promoting e-Learning in the workplace: A case from South Korea. Computers in Human Behavior, 28(3), 942-950.

[18] Zhou, T., Lu, Y. B., & Wang, B. (2010). Integrating TTF and UTAUT to explain mobile banking user adoption. Computers in Human Behavior, 26(4), 760-767.

Landslide early warning system prototype with GIS analysis indicates by soil movement and rainfall

Y Artha*, E S Julian

Electrical Engineering Department, Faculty of Industrial Technology, Universitas Trisakti, Jakarta, Indonesia

*Corresponding Author: yudhiarta@gmail.com

Abstract. The aim of this paper is developing and testing of landslide early warning system. The early warning system uses accelerometersas ground movement and tilt-sensing device and a water flow sensor. A microcentroller is used to process the input signal and activate the alarm. An LCD is used to display the acceleration in x,y and z axis. When the soil moved or shifted and rainfall reached 100 mm/day, the alarm rang and signal were sentto the monitoring center via a telemetry system.Data logging information and GIS spatial data can be monitored remotely as tables and graphics as well as in the form of geographical map with the help of web-GIS interface. The system were tested at Kampung Gerendong, Desa Putat Nutug, Kecamatan Ciseeng, Kabupaten Bogor. This area has 3.15 cumulative score, which mean vulnerable to landslide. The results show that the early warning system worked as planned.
Keywords: early warning system; ground movement; landslide; tilt-sensing

1. Introduction

A landslide is the downslope movement of soil, rock, and organic materials under the effects of gravity and also the landform that results from such movement [1]. Landslides are known also as landslips, slumps or slope failure. In the beginning of 2017 several countries hit by landslide such as Colombia, Italy, Indonesia, New Zealand, India, Bosnia, Kyrgyztan. In March 31, 2017 a massive landslide hit Mocoa Columbia and at least 254 people had been killed, more than 200 were missing and at least 400 were injured [2]. An early warning system can be used to minimize the impact imposed by landslide on human, damage to property and loss of live. The ability to monitor slope movements in timely fashion will inform the people of possible slope failures – giving them adequate lead time to relocate to a safer place.

This landslide early warning system integrates the rainfall sensor, accelerometer, and inclinometer with visual web-GIS (geographic information system) interface so that the factors causing the landslide and its effects (movement of the soil) can be monitor. The system is equipped with wireless telemetry so that the data obtained from accelerometers and rainfall gauges can be send to a remote monitoring or observation center and recorded [3]. Initial analysis on the rock structure/texture, classification on the soil slope, permeability, and weathering in landslide prone area were done in order to make it easier to obtain map and to determine the cost. Mapping, investigation, inspection, and monitoring [4], [5] are parts of this research process. The hardware and software design along with GIS analysis was done carefully to obtain efficient and feasible and slide monitoring system. The GIS analysis is presented in the form of spatial data and geographic maps in the web-GIS software interface in real time [6].

Landslide can be caused by two main factors, namely controlling factor and triggering factor. Both main factors will provide mutual synergy to form landslide if there is an excessive ability threshold. The controlling factor is the factors affecting the material condition such as geology, slope, lithology, shear (fault) and invariance on rocks. The triggering factor is the factor causing the movement of materials such as rainfall, earthquake, erosion of slope and human activities [7].

Slopes with inclination more than 20° have potential to move or for landslide, but slopes with inclination or sloping land has always prone to landslide depend on the geological conditions of the slopes [8]. According to type of movement there are 6 types of landslides i.e. translational, rotational, creep, topple, fall and flow as shown in Figure 1 [9].

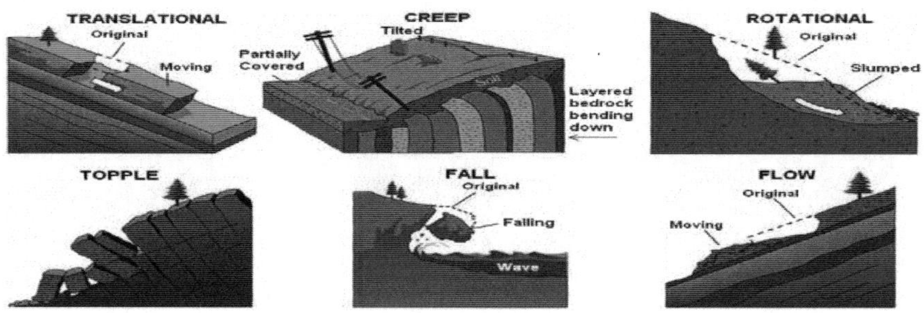

Figure 1. Types of landslides.

2. Research method

The landslide early warning system was developed in three stages: 1) GIS analysis and field study, 2) landslide detection and rainfall monitoring system development, and 3) web-GIS interface development.

2.1. GIS analysis and field study

Data such as slope inclination, soil lithology and texture, local geological data, local rainfall data, road map and land use were collected, analyzed, measured and mapped. Data processing stages are shown in Figure 2.

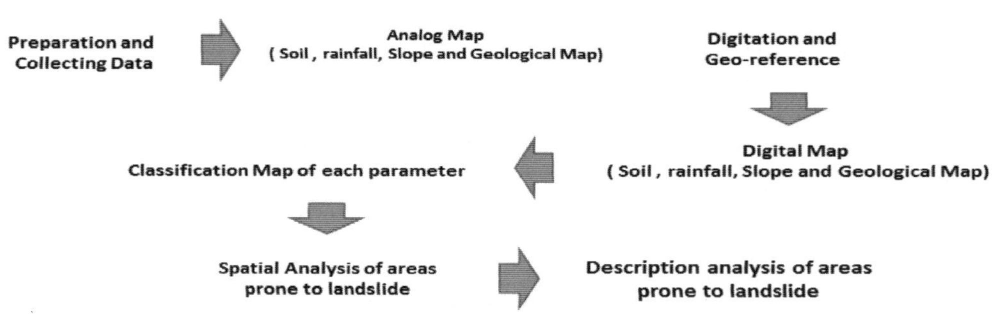

Figure 2. Flow chart of spatial data processing stages.

All the spatial data were inputted into the computer and converted into digital map,subsequently data attribute and each weighting parameter were also inputted into the computer. The parameters used to determine the vulnerability level are rainfall, type of soil, geology (rocks), land-cover, and topography. The scoring model is shown in equation 1.

$$\text{Cummulative Score} = 30\% \text{ x Rainfall Factor}) + (20\% \text{ x Soil Factor}) + (20\% \text{ x Geological Factor}) + (15\% \text{ x Land Use Factor}) + (15\% \text{ x Slope Factor}) \quad (1)$$

Rainfall is the dominant factor causing landslide disaster therefore the weight is higher than other parameters [10]. Rainfall has 30% weight, while soil and geology have the same weight that is 20%, land use and slope each has 15% weight.

Rainfall classification is shown in Table 1 and slope classification is shown in Table 2 [11].

Table 1. Rainfall classification [11].

No	Rainfall Intensity (mm/year)	Parameter	Permeability
1	2000 – 2500	Moist	1
2	2500 – 3000	Wet	2
3	> 3000	Very wet	3

Table 2. Slope classification [11].

No	Class	Slope (%)	Description	Score
1	I	0 – 8	Flat	1
2	II	8-15	Sloping	2
3	III	15-25	Rather Steep	3
4	IV	25-40	Steep	4
5	V	> 40	Very Steep	5

The soil classifications are: alluvial, score=1; association of brown and yellow latosol, reddish brown latosol association, reddish yellow latosol reddish complex and reddish brown latosol association, score=2; association of latosol brown and regosol, score = 3; andosol, podzolic yellowish, association andosol regosol, yellowish podzolic and red podzolic, score = 4; and regosol, score = 5 [11].

While the land cover/use classifications are: forest or dense vegetation and water bodies, score=1; shrubs, score=2; plantation or irrigated rice fields, score=3; industrial and residential area or village, score=4; and empty land, score=5 [11]. The classification as well for rock distribution types are: alluvial (Qav, Qa, a), score=1; Volcanic-1 (Qvsl, Qvu, Qvcp, Qvl, Qvpo, Qvk, Qvba), score=2; Sediment-1 (Tmn, Tmj), score=3, Volcanic-2 (Qvsb, Qvst, Qvb, Qvt) and Sediment-2 (Tmb, Tmbl, Tmtb), score=4 [11].

Based on the cumulative score results, the vulnerable (prone) landslide areas are grouped into three classes, namely (i) very vulnerable, (ii) vulnerable, and (iii) less vulnerable. The vulnerability grade scores are less vulnerable ($\leq$ 2,5), vulnerable ($\geq$ 2.6 – $\leq$ 3.6), and very vulnerable ($\geq$ 3.7).

2.2. *Landslide detection and rainfall monitoring system hardware development*

This landslide early warning system is divided into two parts, namely the transmitter located at the research location, and the receiver located on the remote monitor. ADXL335 accelerometers were use as ground movement and tilt sensor.

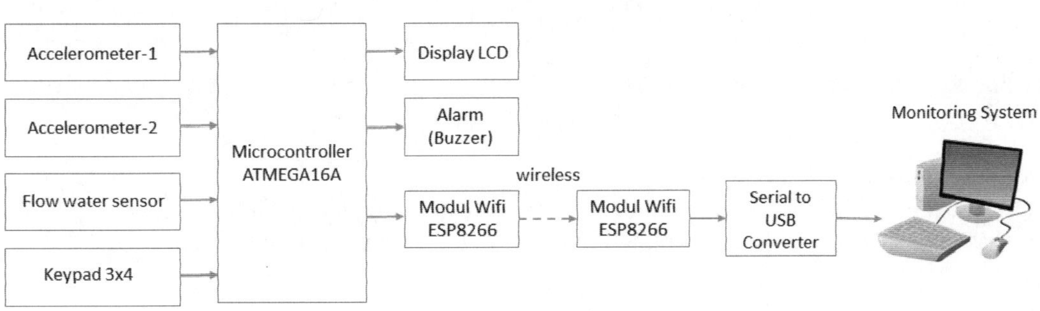

Figure 3. The landslide early warning system block diagram.

A water flow sensor is used as rainfall sensor. An ATMega 16A microcontroller is used as the processor. The ESP8266 (telemetry system) transmit the data to the receivers located in remote location through wireless networking. The receiver receives the data transmitted by the transmitter and displays the data on the computer for remote monitoring purposes. The transmitter and receiver have point-to-point connections. The whole system is shown in Figure 3.

2.3. Web-GIS interface development

Web-GIS is web-based geographical information system of some related components. The web-GIS is combination of digital map and geographical analysis, computer programming and interconnected database.

3. Results and Discussion

Each part of the landslide early warning system i.e. accelerometer, rainfall sensors, LCD (liquid crystal display), wireless telemetry, topographic and hill model analysis, risk/landslide risk analysis, I/O software, web-GIS display was tested individually followed by overall system testing.

The landslide parameters used are rainfall, geology, soil type, slope inclination, and land use.

3.1. Rainfall

Rainfall is one of the climate elements that have big roles on the landslide. Infiltrations of water into soil layer filled the soil and weaken the slope-forming material to trigger a landslide. Heavy and high intensity rain can trigger land movement. The rainfall map of the research area is shown in Figure 4a. The rainfall is classified as high intensity.

3.2. Geology

The geology map of the research area is shown in Figure 4b. The rocks composed from volcanic materials such as glass fragments, fine dust and sand. This area is very water-resistant and can be very slippery in water-saturated conditions resulting from high rainfall.

3.3. Type of soil

The type of soil of the research area is shown Figure 4c. The type of soil is latosol with high permeability. It means that it has ability to pass water at rain and add the soil weight. Therefore, in certain inclination, it is prone to landslide.

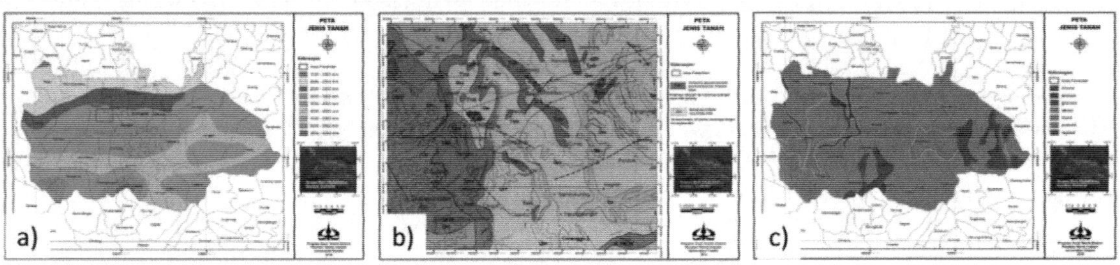

Figure 4.a) Rainfall map of Kabupaten Bogor, b) geological map of research area, c) type of soil map of Kabupaten Bogor.

3.4. Slope inclination

In May 2016, topography data were collected in the research area using GPS from the edge of main access roads until to the edge of Cisadane River. The topography model can be seen in Figure 5.

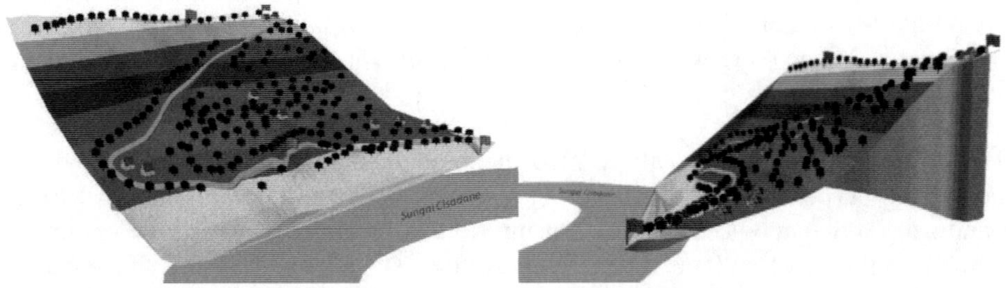

Figure 5. The 3D topography map (a) view form west and (b) view from east.

3.5. *Land cover*

In the research area, most of land cover type is housing and mixed plantation. This housing area is distributed evenly in overall research area. The housing is located in steep to very steep inclination, which mean very vulnerable to landslide.

3.6. *Landslide area analysis*

The cumulative score shows that the research area can be categorized as vulnerable to landslide.

$$\begin{aligned} Cumulative\ Score &= (30\%\ x\ 3) + (20\%\ x\ 2) + (20\%\ x\ 4) + (15\%\ x\ 4) + (15\%\ x\ 3) \\ &= 0.9 + 0.4 + 0.8 + 0.6 + 0.45 = \mathbf{3.15} \end{aligned} \quad (2)$$

Based on ground check at several landslide points in RT 01/01KampungGerendong, Desa Putat Nutug, Kecamatan Ciseeng, Kabupaten Bogor which is located ± 2km from the research area, there was a landslide on the night of April 7, 2016 caused by heavy rain, so the main road body collapsed.

The landslide early warning system was tested in the research area for three hours in sunny weather. In this test, the sensor was placed on the 1 m length rod, planted in the ground with 0.5 m depth. The rain monitor showed no changes.

The tests were conducted at 4 locations to obtain pattern of land movement. The best results can be seen in Figure 6.

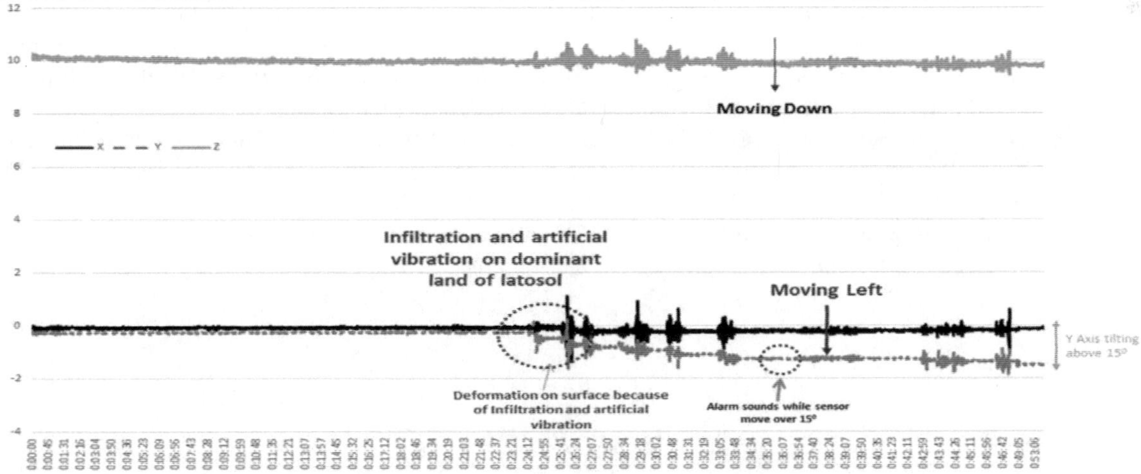

Figure 6. Measurement result and analysis.

From the data in figure 6, it can be seen that there is a spike on each axis indicating the presence of vibration generated from heavy vehicles passing through the roads. To stimulate soil movement, an artificial infiltration in the form of water spray was given towards the rod. Vibration/pounding to deform the ground surface in order to move/tilt the sensor. This is shown on baseline curve Y and X moving downwards. When there was movement that caused15° angle the alarm was activated.

4. Conclusion

A landslide early warning system with accelerometers and a rainfall sensor equipped with Wi-Fi for remote monitoring purposes was developed. The advantages of this system are real time monitoring, remote monitoring, and can be used to monitor multiple points in observation area at once. The system was tested in Kampung Gerendong, Desa Putat Nutug, Kecamatan Ciseeng, Kabupaten Bogor. The coordinate of point study area is 6°28'14.16" South and 106°40'8.40" East. The results show that the system can display and store slope and acceleration data caused by soil movement and deformation, as well as display and print graph of acceleration, angle and rainfall during monitoring. The web-GIS interface can help interpretation of landslide reconstruction pattern through the data received from the system and it can correlate with other reference data such as geological map, structure, rainfall and other information in spatial form (map).

References

[1] Highland Lynn, and Bobrowsky P A, Landslide Handbook – A Guide to Understanding Landslide". Reston, Virginia, U.S. Geological Survey Circular 1325, p 4. 2008.
[2] TW April 2. Massive mudslides leave hundreds dead and missing in Mocoa, Colombia. https://watchers.news/2017/04/02/massive-mudslides-leave-hundreds-dead-and-missing-in-mocoa-colombia/. 2017.
[3] Rahmaniar W. Prototype of Volcano Eruption Monitoring Robot Based on Temperature Detection using Amplitude Shifting Keying Modulation Telemetry. Yogyakarta, Department of Physics FMIPA UGM. 2009.
[4] Zuidam V. Terrain Analysis and Classification Using Aerial Photograph A Geomorphological Approach. 1979.
[5] Rachmansyah A. 2006. Moisture effect on Compaction at Strength of Clay Land. *Journal of Engineering XIV* p 29-35.
[6] Rusyana N, The Practical Side of Knowing the WEB-GIS, Bogor, LojiGIS, 2012.
[7] Naryanto H S, Application of Science and Technology in Landslide Risk Reduction (Soil Movement) in Indonesia, Jakarta, BPPT PRESS, p 3-6, 2016.
[8] Khadiyanto P, Landslide. http://parfikh.blogspot.com/2008/12/gerakan-tanah-longsoran.html. 2008.
[9] NN. Introduction of Ground Movement. http://www.esdm.go.id/publikasi /lainlain/doc_download/ 489-pengenalan-gerakantanah.html, 2008.
[10] Directorate of Volcanology and Geological Hazard Mitigation, Landslide Disaster Management, 2005.
[11] Sugiarto D P, Protected Area and Scoring Method (Slope, Soil, Rain) Forest Area Function, https://tnrawku.wordpress.com/2013/06/26/kawasan-lindung-dan-metode-skoring-kelerengan-tanah-hujan-fungsi-kawasan-hutan/, 2013.

IOP Conf. Series: Earth and Environmental Science **106** (2017) 012013 doi:10.1088/1755-1315/106/1/012013

Identification of rice supply chain risk to DKI Jakarta through Cipinang primary rice market

D Sugiarto[1*], A Ariwibowo[1], I Mardianto[2] and D Surjasa[3]

[1]Information System Program, Faculty of Industrial Technology, Universitas Trisakti, Jakarta, Indonesia
[2]Informatics Program, Faculty of Industrial Technology, Universitas Trisakti, Jakarta, Indonesia
[3]Industrial Engineering Master Program, Faculty of Industrial Technology, Universitas Trisakti, Jakarta, Indonesia

*Corresponding Author : dedy@trisakti.ac.id

Abstract. This paper identifies several sources of risks in DKI Jakarta rice supply chain that through Cipinang Primary Rice Market (CPRM). Secondary data from several sources were collected and analysed using pareto chart and time series analysis. Based on the pareto analysis, it was known that there was a change in the order of suppliers whereas in 2011, 80% of the supply came only from Cirebon, Karawang and Bandung (West Java Province). While in 2015 the main source of supply changed to Cirebon, Central Java and Karawang. Linear trend equation using decomposition model for Cirebon and Karawang showed trend of decreasing monthly supply while Central Java had a positive trend. Harvest area of wetland paddy in Cirebon and Karawang showed a negative trend in the last 6 years. The data also showed that West Java Province was the province with the largest rice crop area affected by plant organism attack and drought disaster in 2015. DKI Jakarta had several potential supply chain risks from rice supply, drought risk and pests risk where the province of West Java, which previously could become a major supplier began to require supply assistance from other provinces, especially Central Java.
Keyword: pareto analysed, pest risk, rice market, rice supply, supply chain risk

1. Introduction

DKI Jakarta is the area of rice consumption due to the very limited amount of rice production in this city. So it requires supply from various producer areas. This big city needs about 876912 tons of rice to meet the food needs of its people in 2010 and it has been assured that DKI Jakarta can only fulfill about less than 1 percent [1]. The supply and distribution of rice in DKI Jakarta has been regulated by the Regional Government by assigning Cipinang Primary Rice Market (CPRM) managed by PT. Food Station Tjipinang Jaya as a regional-owned enterprise that maintains food security especially rice in DKI Jakarta. Rice entering through CPRM comes from three rice production centers namely Cirebon, Karawang and Bandung. The supply volume of Karawang and Cirebon tends to be high in March-April (first harvest season) and September-October (second harvest season) and low volume around December-February [2].

Rice supply chain as well as general supply chain face various risk factors. Several type of risks are supply risk, operational risk, demand risk, security risk, macro risk, policy risk, competitive risk and resource risk [3]. Another classification of risks are market related risks, logistical and infrastructure related risk, political related risk, policy and institutional related risk, financial related risk and operational managerial related risks [4]. Specifically for agricultural supply chain, risk mainly comes from weather/natural disasters related as well as biological and environmental related [5], [6].

Global supply chain risk management and mitigation consists of five steps namely risk identification, risk assessment and evaluation, selection of appropriate risk management, implementation of supply chain risk management strategy and mitigation of supply chain risk [3]. This paper identifies risks related to market related risk (supply), natural disaster risk, and biology/environmental risk in DKI Jakarta supply chain that through CPRM using Pareto and time series analysis. Previous research related to the use of time series analysis for risk analysis was done by Leila to analyse risk and vulnerabilities in supply chain networks using ARIMA model [7].

2. Research Method

This study used secondary data coming from various institutions. Secondary data comes from various institutions namely PT. Food Station, statistical center bureau [8], [9] and Food Protection Directorate [10], [11]. The stage began with the collection of monthly rice supply data from various regions entering the rice Cipinang rice market for 5 years (2011 - 2015). The largest source of supply was analysed using a pareto diagram. Time series data are then analysed using decomposition method that can be used for trends and seasonal conditions [12], [13]. Time series could have trend, seasonal, or cyclical pattern. A trend is a gradual upward or downward movement of a time series over time and a seasonal effectis one that repeats at fixed intervals of time [14]. . Decomposition computes time series decomposition in trend, seasonal, cycle and error components. The equations of this modelis:

$$Yt = f\ (St\,; Tt\,; Et) \quad (1)$$

where Y_t= data at period t
S_t= seasonal component at period t
T_t= trend component at period t
E_t= remainder (or irregular or error) component at period t

For additive decomposition :$Y_t = S_t + T_t + E_t$ and for multiplicative decomposition : $Y_t = S_t * T_t * E_t$
The additive model is most appropriate if the magnitude of the seasonal fluctuations or the variation around the trend-cycle does not vary with the level of the time series [12].

The analysis was then continued by looking at various factors that are likely to affect the pattern changes in the pare to diagram and time series analysis. The least squares regression method is used for wetland availability and paddy production trend analysis. Wetland area and paddy production data were from data Attack of plant pest organisms and drought disasters especially in West Java Province were analysed using bar chart.

3. Result and Discussion

The results showed that CPRM get supplies from various rice production centers mainly Cirebon (including Indramayu), Karawang, and Bandung in 2011 (Figure 1). Total supply increased from 706308 tonnes in 2011 to 815402 tonnes in 2015. Based on the pareto analysis, it was known that there was a change in the order of suppliers, which about 85% supply previously in 2011 came from three regions namely Cirebon, Karawang and Bandung (part of West Java Province) has changed to Cirebon, Central Java and Karawang (Figure 2). Bandung regency has turned into the fourth largest region that supplies rice to Jakarta.

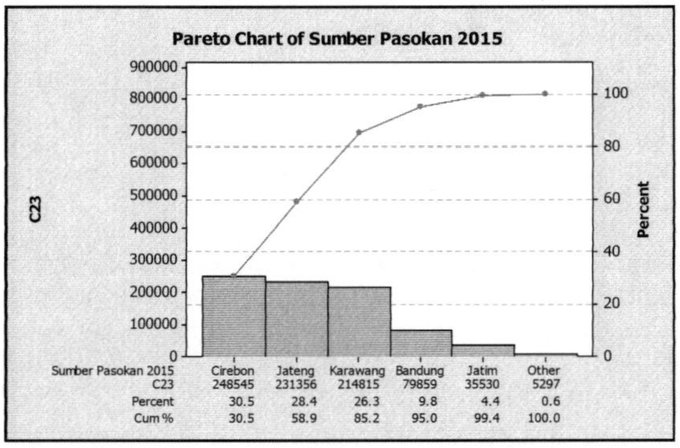

Figure 1. Pareto chart of supply source area in 2011

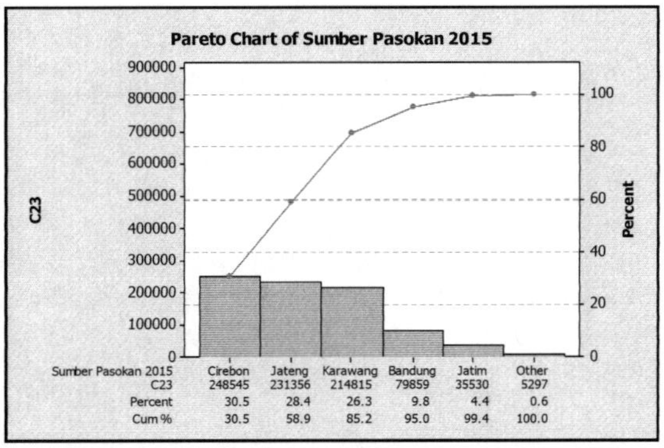

Figure 2. Pareto chart of supply source area in 2015

Time series analysis using Minitab software for rice supply from Cirebon can be seen in Figure 3. Linear trend model for Cirebon showed trend of decreasing supply amount with negative gradient. The negative slope coefficient indicates a potential risk factor on the supply side from Cirebon and surrounding areas (including Indramayu). Indramayu and Karawang are the two largest rice production centers in West Java [8]. R statistical programming software was used to visualize seasonal, trend and randomness separately (Figure 4). The trend patterns also show a declining pattern over the last 5 years. Seasonal patterns show a recurring pattern every year. Supply amount of rice will be large at the time of the first harvest (during March and April) and second harvest (during September and October) and will be very low at around the end and the beginning of the year.

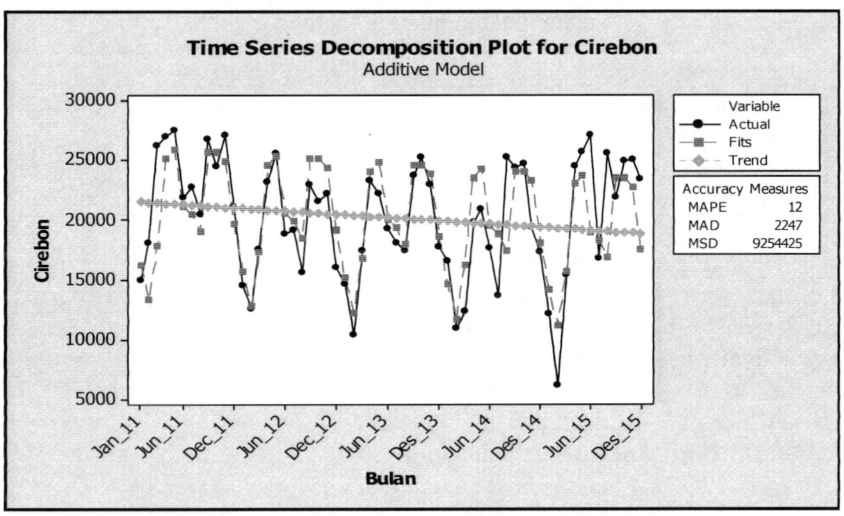

Figure 3.Time Series Decomposition for supply from Cirebon (Additive Model).

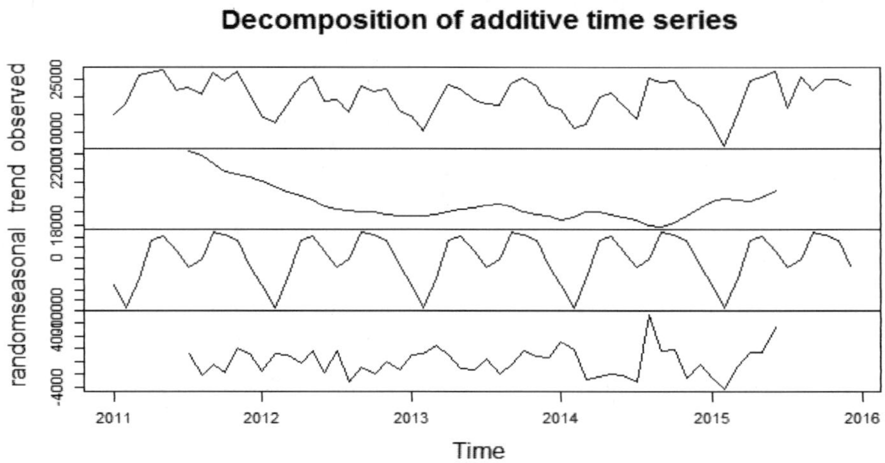

Figure 4. Decomposition for supply from Cirebon (Additive Model) using R.

Fitted trend equation supply from Cirebon, Karawang and Bandung showed trend of decreasing supply amount with gradient coefficient respectively -44.6986, -25.9552 and -21.614 (Table 1). The negative slope coefficient indicates a risk factor on the supply side from West Java Province. Unlike the supply from Cirebon, Karawang and Bandung, the supply from Central Java Province shows a positive trend (Table 1). Gradient coefficient was 207.

Table 1. Fitted trend equation for monthly rice supply to cprm 2011-2015 using decomposition method.

Region of Supplier	Fitted Trend Equation
Cirebon	$Y_t = 21557 - 44.6986*t$
Karawang	$Y_t = 17005 - 25.9552*t$
Bandung	$Y_t = 7916 - 21.6314*t$
Central Java	$Y_t = 7274 + 207*t.$

Harvest area of wetland paddy in Cirebon, Karawang and Indramayu Regency showed a negative trend in the last 6 years where based of gradient coefficient, the biggest decline occurred in Indramayu district (Table 2).The yearly data comes from [8]. Based on [15] for 13 years in the period 2000-2013, the decrease of rice fields in Karawang regency amounted to 12,402 ha or an average of 954 ha per year or 0.9% per year. In the year 2000 the area of rice field is 116,268 ha or 60.6% of the total area of Karawang regency, while in 2013 decreased the area to 103,866 ha (54.1%). Part of the area where rice fields are reduced is mainly near the urban and toll roads that are in the central part of Karawang regency [15].

Table 2. Linier trend model for wetland paddy harvested area and production 2010-2015 using least square regression method.

Region	Harvested Area (ha)	Production (tons)
Cirebon Regency	Y_t=86771 – 1236.00*t	Y_t=522247 – 6299.39*t
Karawang Regency	Y_t=185194 – 106.914*t	Y_t=1079174 + 12118*t
Indramayu Regency	Y_t=230238 – 6113.54*t	Y_t=1342109 – 13349.1*t
Bandung Regency	Y_t=68408 + 1607*t	Y_t=432542 + 9533*t
West Java Province	Y_t=1907441 – 18836.9*t	Y_t=112716656 – 44959.7*t

However, it is not always a decrease in the area of wetland rice is proportional to the level of rice production. Although the area of paddy field in Karawang regency decreased, the data sourced from

Central Statistics Agency of West Java showed that rice production still showed positive trend (Table 2). This is different from other rice supplier areas to Jakarta, namely Cirebon and Indramayu (areas adjacent to Cirebon). In both areas, rice production showed a negative trend. In total, this condition is similar to West Java province which originally became the main supplier of rice to DKI Jakarta. Rice production in West Java province shows a negative trend which indicates a supply risk (market related risk).

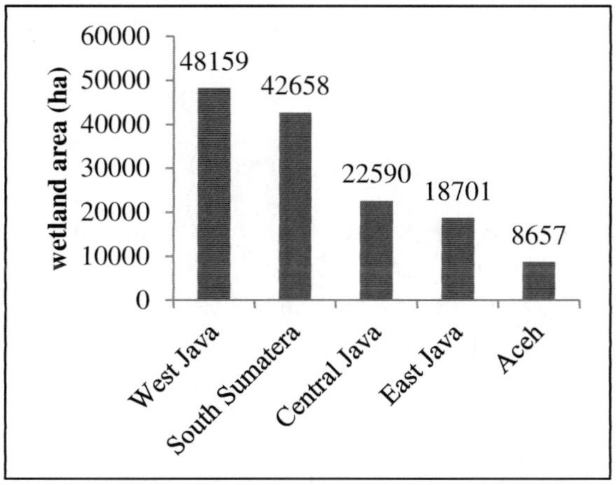

Figure 5. Wetland paddy area affected by pest (ha) in 2015.

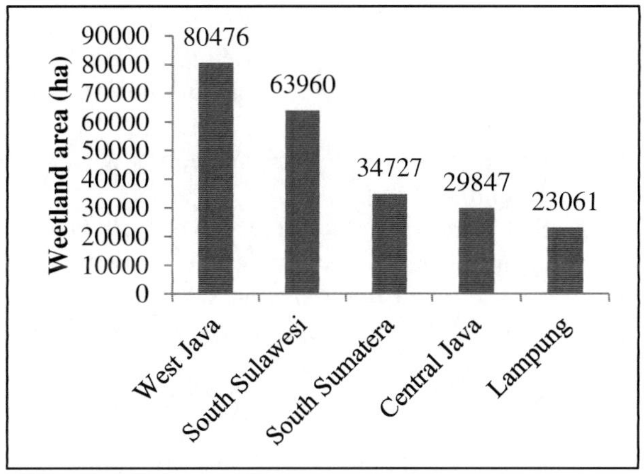

Figure 6. Wetland paddy area exposed to drought (ha) in 2015.

In addition to supply-side risk factors, this paper also reviews the risk factors in terms of natural disasters and environmental risks.The data also showed that West Java Province was the province with the largest rice crop area affected by plant organism attack and drought disaster in 2015 that is equal to respectively 48159 ha (Figure 5) and 80476 ha (Figure 6). Pest and disease attacks are a risk that must be faced by rice supply chain from West Java. This risk management is not always based on the use of chemical pesticides but can be done by utilizing various other control technologies. Such as the utilization of biological agents and vegetable pesticides [10]. The best practice on biological and environmental risk mitigation in Bandung Regency of West Java Province has been done by an exemplary farmer group named Sarinah Organik through the use of vegetable pesticides [16].

4. Conclusion

The results showed that DKI Jakarta rice supply chain had several potential risks from supply risk, drought risk and pests risk where the province of West Java, which previously could become a major supplier began to require supply assistance from other provinces, especially Central Java.

References

[1] Mahardika TKS. Study of Rice Distribution in Regional Jakarta through Cipinang Rice Market Center. Thesis (Bogor : Bogor Agricultural Institute). 2013.

[2] Sugiarto D, Surjasa D, Rahmawati N, and Solihah B. 2013. Statistical analysis of rice supply chain through Cipinang Rice Market. *Proc. Indonesia Statistical Analysis Conf.*

[3] Manuj I, and Mentzer JT. 2008. Global supply chain risk management. *J. of Business Logistics*, 29(1), p.133.

[4] Jaffee S, Siegel P and Andrews C. Rapid agricultural supply chain risk assessment: a conceptual framework. Washington. D.C : The World Ban). 2010.

[5] Antonaci L, Demeke M, and Vezzani, A. The challenges of managing agricultural price and production risks in sub-Saharan Africa. Rome : FAO. 2014.

[6] Yeboah M, Edmond N, Feng Y, Daniel OS and Joseph NB. 2014. Agricultural Supply Chain Risk Identification. A Case Finding from Ghana. *J. of Management and Strategy*. 5(2).

[7] Sakli L, Hennet JC and Mercantini JM. 2014. An analysis of risks and vulnerabilities in supply networks *Proc. of the 19th World Congress The International Federation of Automatic Control.*
[8] Central Bureau of Statistics West Java Province. Production of Paddy and Secondary Food Crops in Wes Java 2011-2015. 2016.
[9] Central Bureau of Statistics West Java Province. Production of Paddy and Secondary Food Crops in Wes Java 2010-2014. 2015
[10] Food Plant Protection Directorate. Performance Report 2014. Directorate General of Food Crops. Ministry of Agriculture. 2015.
[11] Food Plant Protection Directorate. Performance Report 2015. DirectorateGeneral of Food Crops. Ministry of Agriculture. 2016
[12] Hyndman R.J and Athanasopoulos G. 2013. Forecasting: principles and practice. O Texts - Online Textbooks.
[13] Sen J and Chaudhuri TD. 2016. Decomposition of Time Series Data of Stock Markets and its Implications for Prediction: An Application for the Indian Auto Sector. *Proc. of the 2nd National Conference on Advances in Business Research and Management Practices*.
[14] Evans, J. R. Business Analytics: Methods, Models, and Decisions. (Boston : Pearson Education, Inc). 2013.
[15] Rafiuddin, A. The Spatial Model of Land Use Change to Maintain The Karawang Regency As a Rice Contributor. Thesis (Bogor : Bogor Agricultural Institute). 2016.
[16] Surjasa D, Sugiarto D, Solihah B, Ratnawati N and Jingga BA. 2014. Knowledge management implementation using SECI model. *Proc. Int. Sem. on Industrial Engineering and Management.*

The empowerment of sustainable design in food packaging as designer responsibilities

V Setiadi

Visual Communication Design Department, Faculty of Art and Design, Universitas Trisakti, Jakarta

Corresponding Author: nia@trisakti.ac.id

Abstract. The purpose of this paper is emphasized on the empowerment of sustainable design in providing the dual function of a food packaging. Which can extend the life of paper, cardboard, plastic, aluminum foil so as to reduce the contribution of waste on earth. The methodology used in this research is using qualitative research. With the main approach taken on the layout of the packaging design, the approach that relies heavily on the data in the form of packaging design. For the process of observation, the authors should compare with the forms of food packaging designs that are contained in the diversity of food packaging types from products outside Indonesia. The purpose of this study is also intended as a recommendation through observation of data interviews and survey related products. Conclusion through material exploration, packaging structure exploration, efficient exploration of ink usage and packaging usage patterns.
Keywords: lifestyle, packaging, ink usage, packaging usage patterns, sustainable design

1. Introduction

Today with the entry of the free market era, the competitiveness of foreign products increases in the nation's economic arena. In Indonesia, especially in the city of Jakarta as the capital of the country has an acceleration lifestyle and the velocity of money is so great. The purchasing power and consumption of both individuals and families is so dynamic. Seen from the urban lifestyle is very fond of buying all the daily needs of both urgent and optional nature is no longer a special attention to the function of the things he bought. Starting from the needs of clothing and food, as well as things related to hobbies and lifestyle. The author's attention is focused on the amount of wasted packaging after the content of the product is no longer in use. The most common thing is seen from the purchase of food products everyday. The period of the packaging function is lost after the product is consumed. In the case study this time is on the packaging of food such as on Boneto Milk Packaging, Frisian Flag Milky Packaging, Honey Stars Cereal Packaging, Gulaku Sugar Stick Boxes Packaging, Quick Melt Kraft Cheese Packaging, Gong Pudding Packaging and Marga Mulia Misua Packaging. Other functions of packaging in addition to protecting the products therein may also serve as another useful thing as a sustainable design. So that the period of disposal of packaging that is no longer used. It is hoped that the form of responsibility of a Visual Communication Designer and Product Designer can be implemented to the dual function of packaging design. At least we can support the reduction of waste in our family first for a better environment in the future base on sustainable design theory. The purpose of this paper is emphasized on the empowerment of sustainable design in providing the dual function of a food packaging.

2. Research Method

The methodology used in this research is using qualitative research. With the main approach taken on the layout of the packaging design, the approach that relies heavily on the data in the form of packaging design. Kristina Niedder and Seymour Roworth-Stokes clarified the use of practice in research into three major groups, ie research involving practice; Critical inquiry; And practice through reflection and/or research. Research involving practice includes practice-based research, studio-based research, practice-centered research, art-based research, and design-based research. In this case the research process is based on or rooted in practice or practice playing a steering role in the inquiry process. Interventions / experiments are an investigative 'framework' of how practices can be improved or improved [1]. The purpose of this study is to generate a direct contribution to or relevant to the progress of practice. Another goal is that the practice informs the theory of research in order to gain new insights, knowledge or understanding. There are several supporting theories in the observation of the author, which is based on Theory of Victor Papanek in his book entitled 'The Green Imperative Ecology and Ethics in design and Architecture', which states as follows: “The relationship between design and ecology is very closed one, and makes for some unexpected complexities. The designed product goes, as shown, through at least six potentially ecologically dangerous phases. Product Life Cycle Assesment is the evaluation incorporating all of them, from the original acquisition of raw materials, through the manufacturing process and assembly, the purchase of the complete product (which also includes shipping, packaging, advertising and the printing of instruction manuals), the use, the collection of the product after use, and finally the re-use or recycling and final disposal. It can best be understood through the hexagonal diagram, the six-sided ‘Funtion Matrix’ (Figure 1). At the moment Life Cycle Assesment is very new, and can be profoundly complicated, demanding a great deal of study, testing and experimentation.”

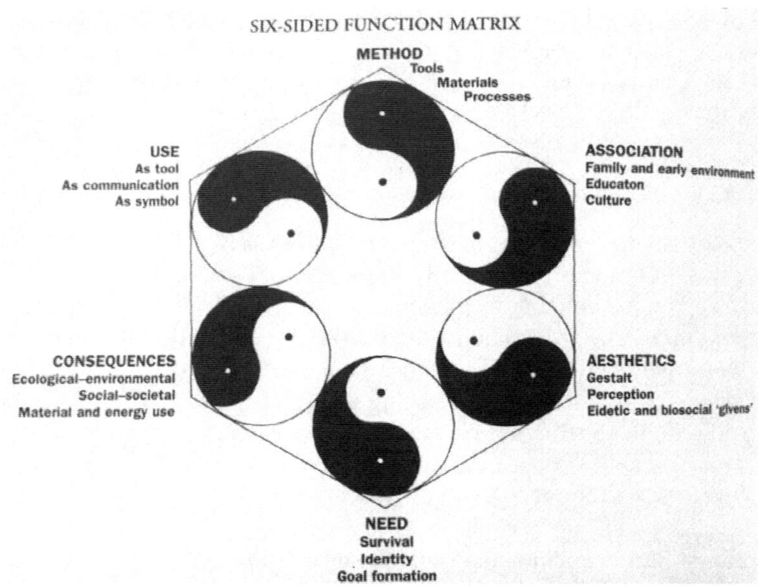

Figure 1. Six function matrix [2].

In his book Chuck Groth entitled 'Exploring Package Design' he gave the example of the packaging process design with the case studies '(Baking Mix) Quick & Easy to make Blueberry Muffins', describes how:

1) Understand the problem and determine the solution. For example visual packaging should show quality, clear, and organized.
2) Define the hierarchy, create a list of elements and determine the location of elements on the panel [3].

3. Results and Discussion

For the process of observation, the authors should compare with the forms of food packaging designs that are contained in the diversity of food packaging types from products outside Indonesia. Why is that? Because not many domestic food packaging products that have awareness of sustainable design. Here are some of the results of the author's observations in collecting some of the food packaging contained in major cities in Indonesia, especially in hypermarkets in Jakarta. One of them has taken advantage of other functions of a package.

Among these we can observe from several products, including: Boneto Milk Packaging Products (Figure 2a & 2b), Frisian Flag Milky Packaging (Figure 3a & 3b), Honey Stars Cereal Packaging (Figure 4a & 4b), Sugar Stick Boxes (Figure 5a & 5b), Quick Melt Kraft Cheese Packaging (Figure 6a & 6b), Gong Pudding Instan Package (Figure 7a & 7b), and Mulia Misua Marga Packaging (Figure 8a & 8b).

Figure 2a & 2b. Boneto milk packaging front side & back side [4].

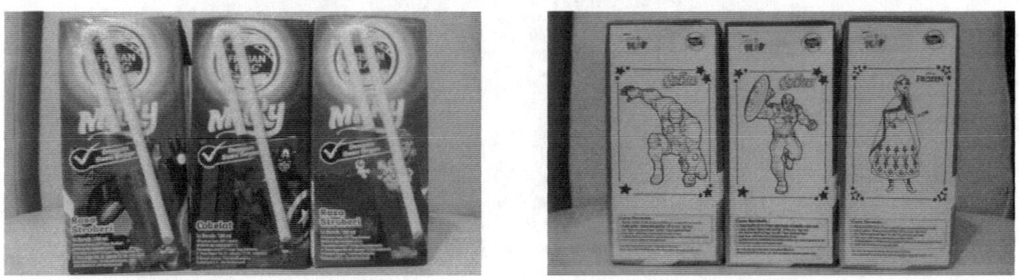

Figure 3a & 3b. Frisian Flag milky packaging front side & back side [4].

Figure 4a & 4b. Honey Stars cereal packaging front side & back side [4].

IOP Conf. Series: Earth and Environmental Science **106** (2017) 012014 doi:10.1088/1755-1315/106/1/012014

Figure 5a & 5b. Gulaku Sugar stick boxes front side & side view [4].

Figure 6a & 6b. Quick Melt Kraft Cheese packaging front side & back side [4].

Figure 7a & 7b. Gong pudding instan packaging front side & back side [4].

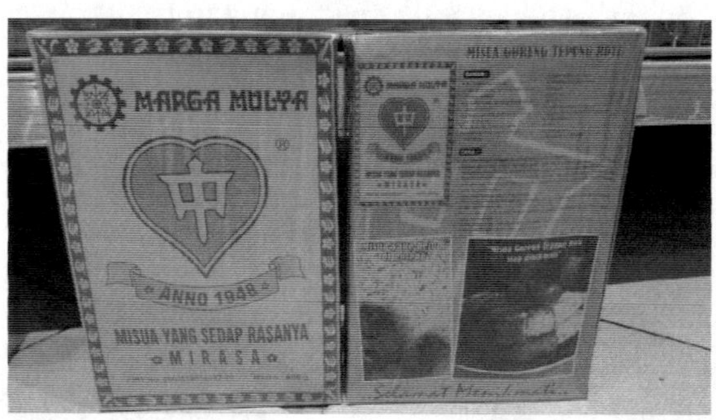

Figure 8a & 8b. Marga Mulia Misua packaging front side & back side [4].

From the overall case study observed, it can be concluded that from seven packs, six of them newly optimally use the rear side of the packaging as another function. While only one pack, that is on Sugar Stick Box Gulaku product that can be reused crib as container with other uses. For Boneto Milk Packaging Products and Honey Stars Cereal Packaging, Frisian Flag Milky Packaging uses the back side of the packaging for children's board games. On the back of Frisian Flag Milky Packaging is used as a coloring area. While on the Quick Melt Kraft Cheese Packaging use optimally both sides of the rear area to put pictures of recipes related to cheese. While in Gong Pudding Packaging and Marga Mulia Misua Packaging only use one side on the back of the packaging to put recipes related to their products. Consider also the current weight of distribution, content security, cost and the possibility of realizing it. Here are examples of forms in material exploration (Figure 9a, 9b, 9c, 9d, 9e, 9f, 9g, & 9h):

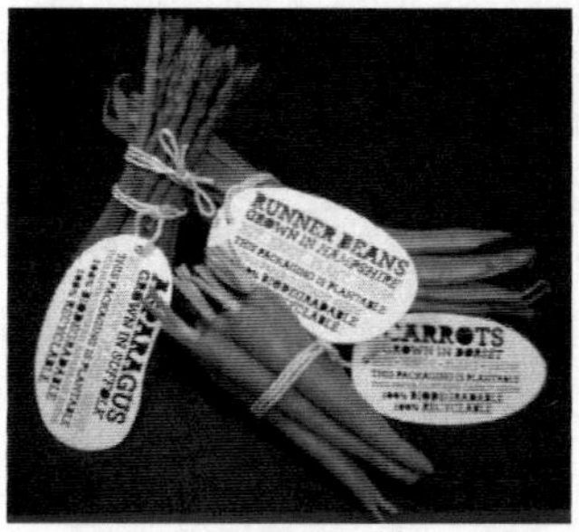

Figure 9a. Vegetable tag - seed paper [5].

Figure 9b. Solar mill - seed paper [6].

Figure 9c. Pan - non plastic bread packaging [7].

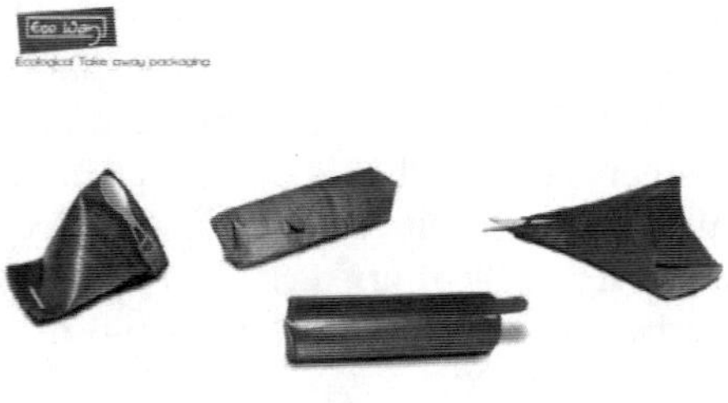

Figure 9d. Eco way-take away packaging [8].

Figure 9e. Harvest hills [9]

Figure 9f. Premium thai pomelo [10].

Figure 9g. Besek (alternative packaging) [11]

Figure 9h. Folded box [12]

In the exploration of creative packaging design structures and can be a packaging waste solution by reducing the carbon footprint. Things to note include: 1) Size, piece, waste; 2) Stage in production process; 3) The amount of weight at the time of distribution; 4) Product content security issues

Here is an example of the forms in the exploration of the packaging design structure. Exploration of ink usage, among others, in the following ways: 1) Taking into account the use of the number of colors; 2) Pay attention to the color area. Here are examples of forms in the exploration of ink usage. Therefore, the full responsibility of visual communication designers and product designers is required to participate in using the recommendations of the exploration forms described above. The conception of sustainable design should be applied as soon as possible and socialized significantly into society. As a form of responsibility towards world civilization and the survival of mankind. Exploration of packaging usage, among others, as follows: 1) Multiple Functions; 2) Disposable; 3) Biodegradable; 4) Return the packaging to the manufacturer.

4. Conclusion

Conclusion through material exploration, exploration of packaging structures, efficient exploration of ink usage and packaging usage patterns. Material exploration is expected to find more sustainable materials without compromising economic and social functions. As for some recommendations that can be done, among others: 1) by using biodegradable paper; 2) using environmentally friendly plastics; 3) using glass material; 4) using aluminum; 5) using recycled corrugated; 6) using other alternative materials.

References

[1] Guntur, Metode Penelitian Artistik, Surakarta : ISI Press. 2016. p27-28.
[2] Papanek V. 1983. The Green Imperative Ecology and Ethics in Design and Architecture, Singapore : Thames and Hudson. p32-34.
[3] Chuck G. 2006. Exploring Package Design (New York: Thomson Delmar Learning. p72-73.
[4] Virginia S. Setiadi. 2015. Daya Tarik Warna pada Rancangan-rancangan Kemasan Makanan Lokal. *Jurnal Dimensi Seni Rupa dan Desain FSRD Usakti*. 12(2). ISSN : 1693-6337, September 2015, single writer.
[5] http://lovelypackage.com/student-work-ben-huttly/, Accessed on October 2017.
[6] http://www.botanicalpaperworks.com/blog/read,article/273/seed-paper-packaging-for-solarmill|, Accessed on October 2017. https://www.facebook.com/SolarMill, Accessed on October 2017.
[7] http://lovelypackage.com/student-work-leslie-ortiz/#more-20234, Accessed on October 2017.
[8] http://www.designboom.com/project/eco-way-ecological-take-away-packages/, Accessed on October 2017.
[9] http://lovelypackage.com/student-work-jose-rivas/, Accessed on October 2017.

[10] http://www.thedieline.com/blog/2012/6/18/eco-sustainable-premium-thai-pomelo-packaging.html, Accessed on October 2017.
[11] http://grosirbesek.blogspot.co.id/2016/05/grosir-besek-bambu-langsung-dari.html, Accessed on October 2017.
[12] http://wallpart.com/poster/desain-kemasan-ukm-kemasan-produk-kemasan-makanan-desain-kemasan-927446613164, Accessed on October 2017.

Model of cyclist accident characteristics in the city of Malang and Blitar

M Z Arifin[1], I W Agustin[2]*

[1]Civil Engineering Department, Engineering Faculty, Universitas Brawijaya, Malang, Indonesia
[2]Urban and Regional Planning Department, Engineering Faculty, Universitas Brawijaya, Malang, Indonesia

*Corresponding Author: immasaitama@ub.ac.id

Abstract. Utilization of bicycles as an environmentally friendly mode of transportation is re-concerned as the development of sustainable transportation programs. The use of bicycles in some developed countries such as the Netherlands is 27 per cent of total travel, while for developing countries such as Indonesia, cyclists are less than 1% of total travel with low educated characteristics (65 per cent) and low income (48 per cent). Cyclist reduces dependencies on petroleum and environmental pollution as well as lessen the occurrence of congestion and traffic accidents involving motor vehicles. It was necessary to know the behavior and interaction of bicycle riders with other vehicle users in a heterogeneous traffic flow. The main purpose of the research is to create a model of bicycle accidents to increase the road traffic safety in Malang city and Blitar city. The research used analyses of frequency, the road's level of service, and multiple linear regression. The results showed that there was a need for a development basis of a special lane for bicycle. It aims to reduce the level and number of cyclist accidents and to achieve transportation safety as well as to raise public awareness in traffic safety.
Keywords: bicycle, road traffic safety, traffic accidents, sustainable transportation

1. Introduction

The development of transport is currently very rapid since the needs of the community are increasing, directly proportional to the needs of transportation modes to support their activities. According to [1], the development of vehicles as a means of transportation brings a positive impact on the fulfilment and improvement of human welfare, especially as a means of mobilization in order to facilitate daily activities but it was also accompanied by the emergence of some undesirable negative effects such as congestion and the increasing number of traffic accidents. Based on [2] regulates non-motorized vehicles, stating that the government must provide facilities for cyclists in traffic. Cyclists are entitled to have the right to facilities supporting security, orderliness, fluency, and safety in traffic. Based on the regulation of the Local Government in Malang and Blitar City, bike paths have been constructed to provide comfort and flexibility for cyclist.

Accidents are triggered by several aspects, including human factors, vehicle features, road factors, and environmental aspects. Accidents due to human factors are influenced by the individual characteristics of the riders. Vehicle factors are affected by the reliability of the components in the vehicle to maneuver when it is moving. Road factors are influenced by the condition of road infrastructure that ought to support comfort, safety, and security in driving, while environmental factors are affected by the bad weather around the crash site that sways the driver's visibility in driving [3] ,[4].

Malang is the second largest city in East Java Province which has transportation problems such as cengestion and a high rate of accidents which is increasing annually. There were 330 cases took place on highways in 2015, increased by 66 per cent compared with the year of 2014 where there were only 199 cases of accidents. This number of accidents continued to rise by 8 per cent in 2016 of 356 cases causing 82 deaths and 428 minor injuries. The total loss caused by accidents in Malang City in 2016 reached Rp 245,905,000.00. As for accidents involving cyclists in Malang, there are 4 cases in 2014–2015, and it declined to 3 accidents in 2016. In general, accidents involving bicycles in the City of Malang occurred on the road that does not have a particular bike lane. Based on the statistics, bicycle accidents have happened at almost all road functions such as arterial, collector, and local roads [5].

2. Research Method

2.1. Location

The study locations for this research were Malang and Blitar City. Questionnaires were distributed in both cities to cyclist respondents who had been cycling during the last six months at the minimum and had passed the path along arterial, collector, and local/neighborhood roads in Malang and Blitar with or without particular bike lanes. Road segments used as the location to collect data on road geometric characteristics in this research were those with frequent bike accidents based on the police data from 2014–2016. Consideration was made that each road function is represented by two different road locations (Table 1, Table 2, and Figure 1).

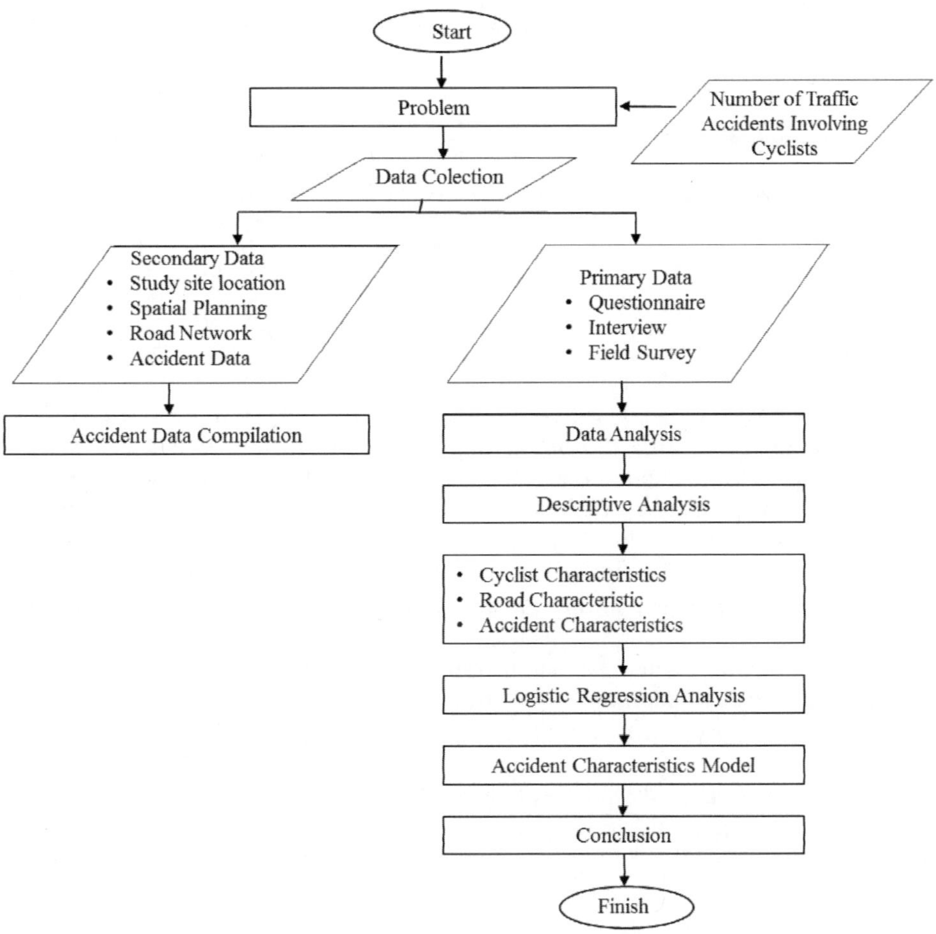

Figure 1. Research framework.

2.2. Types and sources of data

The required data are statistics related to the main purpose of this study, i.e., the characteristics of the drivers, road geometric, and accidents as well as the model of cyclist accidents probabilities in both cities. The study used primary data (obtained directly from the survey) and secondary data.

Table 1. Survey locations of road geometry in Malang City.

No.	Location	Road Function
1	Ahmad Yani	Arterial
2	Borobudur	Arterial
3	Pattimura	Collector
4	Jenderal Basuki Rahmat	Collector
5	Sawojajar	Local
6	IkanTombro	Local

Table 2. Survey locations of road geometry in Blitar City.

No.	Location	Road Function
1	Tanjung	Arterial
2	Raya Kediri – Blitar	Arterial
3	Bengawan Solo, Pakunden Sub-district, Sukorejo Sub-district	Collector
4	Raya Pakisrejo Village, Srengat Sub-district	Collector
5	Tawangrejo Village, Wonodadi	Local
6	Simpang 4 Selokajang Village	Local

The data obtained were then processed and analyzed by SPSS program to create a model of accident characteristics that can occur to cyclists based on the road function. Secondary data were gathered indirectly by the researchers from some relevant agencies to support the primary data analysis.

2.3. Data Collection

The data collected in this study were in the form of interviews and questionnaires as well as direct surveys in the field:

- An interview is a conversation conducted by two parties the interviewer and the interviewee with a specific intention. In this research, the interviewee is a cyclist who had been in an accident. Interviews with cyclists are intended to explore more information about the experienced accidents.
- A questionnaire is an attempt to collect certain information by submitting a number of written questions to be answered in writing by the respondent. Respondents from this questionnaire are cyclists who had been active for at least six months. The purpose of collecting questionnaires in this study is to obtain information regarding the research variables such as the rider characteristics and accident details.
- Field survey is an attempt to gather further information about the object to be assessed through direct action to see and measure conditions in the field. In this study, a field survey was used to gather information about the road geometric characteristics, the volume of traffic, and the speed of the vehicle.

The sampling technique used the proportionate stratified random sampling. According to [6], this approach is applied when the population has members/elements that are not homogeneous and stratified professionally. Determination of the number of samples was calculated using the method of time linear function because the population of cyclists is very dynamic, especially those who use not only a bike but also a motor vehicle as a mode of transportation. Here is the sample calculation formula with time linear function method [7]:

$$n = \frac{T - t_0}{t_1} \tag{1}$$

N = minimum sample size; T = time available for research; T0 = sampling time; T1 = time spent by respondents to fill out the questionnaire/interview

$$n = \frac{225-180}{0.25} = \frac{45}{0.25} = 180\ respondents \tag{2}$$

2.4. Descriptive analysis
Questionnaires data in the form of cyclists characteristics, road geometric characteristics, and accident characteristics in the study area were then analyzed using the frequency analysis. This method was chosen because compared with other analysis, it has advantages of being easily and quickly comprehended through its data arrangement in the form of diagrams [8], [9].

2.5. Statistics logistic regression
In this study, accident model was developed to determine the probability of cyclists experiencing accidents by predicting the rider's behavior, accident characteristics, and road geometric characteristics. The logistic regression method was employed because a multivariate normal distribution could not be fulfilled and the explanatory variables were a mix between continuous and categorical variables. Another account of this choice is that the probability values were in the range 0–1, distinguishing this approach from the usual linear regression where the value of the response variable (dependent) is less than 0 or more than 1.

According to [10,11], the formation of a log it model is based on a cumulative logistics opportunity function which is specified as follows:

$$P_i = F(\beta_0 + \beta_1 X_{1i}) = \frac{1}{1+e^{-z}} = \frac{1}{1+e^{-(\beta_0+\beta_1 X_{1i})}} \tag{3}$$

P (BA) = chance of a bicycle accident occurrence; e = natural number (2.71828); β = explanatory variable coefficient (predictor); X = explanatory variable (predictor)

3. Results and Discussion

3.1. Location characteristics
The continuously increasing number of motor vehicles, uncompensated by the infrastructure development, leads to the aggravating transportation problems including the number of accidents. There were at least 356 cases of accidents in Malang City in 2016 which only about 1 per cent or 3 cases of them involving bicycle users as victims and 165 accidents in Blitar with 3 cases resulting in cyclists as victims [5].

Figure 2. Mixed traffic condition between motorized and non-motorized vehicles.

Figure 3. Bicycle path condition used for other activities.

Figure 2 shows the mixed traffic condition between motorized and non-motorized vehicles where all types of vehicles move together in the same lane (not separated). Figure 3 shows that many of currently operating trails do not conform to the hierarchy of road functions, especially for bike lanes that are often used by motor vehicles and used for other activities.

The study locations are Malang and Blitar. Both were chosen considering that Malang is the second largest city in East Java Province representing a medium town and Blitar portrays a small town. The selected road segments were those with arterial, collector, and local roads function in Malang and Blitar, with and without the existence of particular bike lanes.

3.2. General Conditions

Malang is one of the cities in East Java Province located on the plateau and is + 90 km south of Surabaya. Based on BPS data of Malang City, its population in 2015 is 851,298 inhabitants comprising 419,713 males and 431,585 females (Figure 4).

Blitar City is one of the small towns in East Java Province. It lies on the coordinates of 112.014°–112.028° East Longitude and 8.02°–8.08°South Latitude with an altitude of +156 m above the sea level. According to Blitar BPS data, its male and female ratio of population is 99.15 per cent. Its total population is 146,155 inhabitants consisting of 54,193 people in Sananwetan District, 49,783 residents in Sukorejo District, and 42,179 people in Kepanjenkidul District (Figure 5).

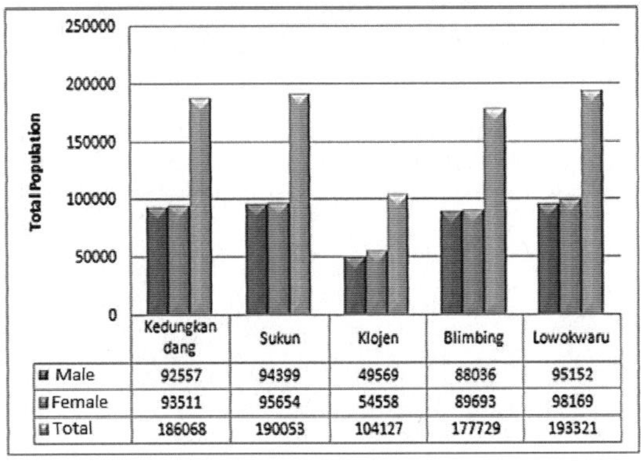

Figure 4. Total population of Malang City in 2015.

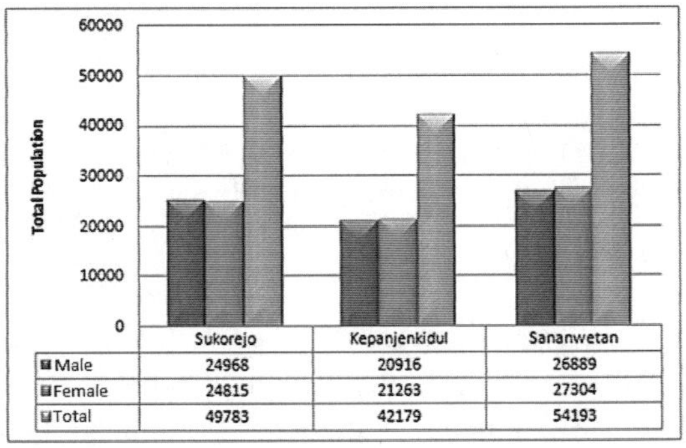

Figure 5. Total population of Blitar City in 2015.

3.3. Cyclist characteristics

3.3.1. Socioeconomic aspect

The majority of bicycle users Malang and Blitar city is the male gender of 73 per cent and 68 per cent, respectively. This gender domination is related to the age of respondents in the study area, proven by the value of chi-square test between those two variables of 0.004 (Table 3).

Table 3. Socioeconomic aspect recapitulation of cyclists.

Variables	Malang	Blitar
Gender	Male	Male
Age	15–25 years old	15–25 years old
Education	Senior high school	Senior high school
Job	Student	Entrepreneur
Income	IDR 1,000,000 – IDR2,000,000	< IDR 1,000,000

3.3.2. Movement characteristics

The majority of cyclists in Malang City use bikes with the intention to exercise of as much as 28 per cent since the bicycling movement there is only crowded on the weekend, indicated by the uncertain intensity. On the other hand, the majority of cyclists in Blitar use bicycle with the purpose to go to work of as much as 27 per cent because they are mostly self-employed and routinely do this.

3.3.3. Behavioral characteristics

Cyclists in Malang tend to never check the condition of the bicycle lights before driving, but always check the brakes, tires, and sometimes check the bike chain. They have the habit of not using safety tools such as helmets, gloves, as well as knee and elbow protectors while driving. Their other behavioural traits are that they sometimes wear light-coloured clothing, never ride or walk in groups, never joke on the street, never go through red lights, never carry overloaded items, occasionally be prepared for vehicles from the right direction, always give a sign when they will turn right, and sometimes keep on going under the rain.

Similarly, the cyclists in Blitar tend to never check the condition of bicycle lights before driving, but always check the brakes, tires, and sometimes check the bike chain. They also do not wear safety equipment such as helmets, gloves, and knee and elbow protector while driving. As for the other behaviors, they sometimes wear light-colored clothing, never ride but sometimes walk in groups, never joke on the street, never go through red lights, never carry overloaded goods, sometimes be prepared for vehicles from the right direction, always give a sign when they will turn right, and sometimes keep on going under the rain.

3.3.4. Accident characteristics

The percentages of respondents having experienced an accident in Malang and Blitar are 61 per cent and 44 per cent, respectively. Accidents involving bicyclists in Malang are caused by slippery or damaged road conditions of 36.07 per cent, while the account of bike accidents in Blitar based on the respondents' experiences is their own mistakes such as drowsiness, daydreaming, and joking of as much as 29.55 per cent.

4. Conclusion

The majority of bicycle accidents in Malang and Blitar occurred in the afternoon with a percentage of 39.34 per cent and 36.36 per cent, respectively. Accidents involving bicycle users in Malang are caused by slippery or damaged road conditions, while the account of bike accidents in Blitar is the bicyclists' own mistakes such as drowsiness, daydreaming, and joking. The respondents of both Malang and Blitar cities have also expressed the need for bike lanes on the road that is usually passed by cyclists. Hence, the development of special lanes for bicycles is necessary to reduce the level and number of bike accidents and to achieve the transportation safety as well as to raise public awareness regarding traffic safety.

References

[1] Marsaid, Hidayat M, Ahsan. 2013. Factors Associated with Traffic Accidents Event on Motorcycle Riders in Polres District of Malang Regency. *Journal of Nursing Science*. 1(2).
[2] Republic of Indonesia, Law of the Republic of Indonesia No.22 of 2009 on Traffic and Transportation, 2009.
[3] Warpani, Management of Traffic and Road Transport, ITB, Bandung, 1999.
[4] Machsus, Sulstio H, Wicaksono A, Djakfar L. 2013. The Prediction Models of Motorcycle Accidents on Surabaya Arterial Roads Using Generalized Linear Models, Middle-East Journal of Scientific Research. 18(12):1859-1866.
[5] Polda East Java. 2017. Traffic Accident Data. http://korlantas.info.
[6] I W Agustin, Combination Model for Rural urban Fringe, 埼玉大学大学院理工学研究科, Dissertation, 2013.

[7] Sari Puspita, et al. Analysis of Factors Affecting Consumer Satisfaction against Wingko (Case Study of Wingko UKM in Babat Lamongan Subdistrict). Unpublished Thesis. Universitas Brawijaya. Malang, Indonesia, 2013.
[8] I W Agustin. 2015. *Jurnal Sosial Ekonomi Pekerjaan Umum*, Vol. 7 No. 2 pp. 143-158.
[9] I W Agustin, Y Sumantri. 2017. The Effect of Industrial Vehicles on the Road's Level of Services of Industrial Area in Malang City. *Journal IOP Conference Series: Earth and Environmental Science*. 70(1), 012015.
[10] Indriastuti, Amelia K and Sulistio H. 2010. Motorcycle Accident Model for Severity Level and Collision Type. *International Journal of Academic Research.* 2(5), September.
[11] Machsus, Sulstio H, Wicaksono A, Djakfar L. 2014. The Effect of Access Points on Motorcycle Accident Rates on Surabaya Arterial Roads. *Australian Journal of Basic and Applied Sciences.* 8(10): 38-43.

Measuring the scale of sustainability of new town development based on the assessment of the residents of the native settlement around the new town area of Gading Serpong Tangerang

M Ischak[1*], B Setioko[2], D Nurgandarum[3]

[1,3]Architecture Department, Faculty of Civil Engineering and Planning Universitas Trisakti, Jakarta, Indonesia
[2]Architecture Department, Universitas Diponegoro, Semarang, Indonesia

*Corresponding Author : m.ischak@trisakti.ac.id

Abstract. The growth trend of Jakarta city as a Metropolitan city nowadays is the construction of large-scale planned settlement that is often referred to as a new town and is carried out by major developers. The process of land tenure and the process of constructing the new town are directly tangent to the original pre-existing settlements and shape the pattern or types of original settlements in the context of their relationship with the new town. This research was intended to measure the scale of sustainability due to land expansion by new town developers and was measured from the side of the original settlers who still exist. The research method used was descriptive explorative that is by formulating sustainability criteria that match best with research context and using the criteria as a tool to measure the sustainability level of new city development at research site that is new town of Gading Serpong Tangerang. The research concludes that despite the apparent displacement and restriction of original settlement' lands, it indicates, overall, that new town development meets sustainability criteria when viewed from the residents of three types of the original settlements.
Keywords: development sustainability, new town, original settlement

1. Introduction

The United Nations reported that in 2008 the population of the urban population was larger than the rural population. Furthermore in 2014, World Urbanization Prospects indicated that 3.9 billion people or 54% (and predicted to rise to 66% in 2050) of the world's population live in urban areas. The growth of cities towards rural causes increased density in sub-urban areas in developed country cities [1] and especially cities in developing countries including Jakarta [2].

It reveals that one of the important issues in the design of cities in Indonesia is the shrinking of centers and the mist of suburban borders [3]. This is due to the growth of new centers in the core suburbs and is driven by the provision of transportation network built and spread in various directions, and connecting the city center with the new growth centers. The urbanization in Indonesia as well as in Southeast Asia is also characterized by the blurred distinction between 'rural' and 'urban' [4]. Both agricultural and non-agricultural activities take place side by side in the adjacent areas of the urban centers, while the urban physical development extends beyond city administrative boundaries. [1] saw it as positive thing by analyzing it as an anabolic and catabolic process in the metabolism of the human body, and the keyword of that metabolism is mobility. The phenomenon of urban growth towards the

IOP Conf. Series: Earth and Environmental Science **106** (2017) 012016 doi:10.1088/1755-1315/106/1/012016

suburb creates a very distinctive area due to the meeting of two cantonal characters, which are the character of the city and the rural character. Some municipal experts refer to this area as peri-urban or semi-urban defined as "a combination of suburbs as a combined process of population and economic growth with the expansion of urban space [5]. The existence of planned settlements that are mushrooming around the suburbs and wrapped in the term new town is basically an indication of the development and dynamics of the city's growth including within the context of the Metropolitan city of Jakarta. Thus, the existence of a new town, whether in the context of urban sprawl or urban expansion is basically a necessity associated with the growth trend of cities around the world [1].

In the context of regional development, the most important thing is the achievement of quality improvement on all sides, either physically, regionally or for the society, both native local people and immigrants as new residents of planned housing. There have been many studies linking the development of planned settlements (new towns) to the concept of sustainability [6], [7], [8] , but the material used as benchmarks in the implementation of the sustainability concept is more limited to its physical development only. The question that arises is which community is the benchmark for the implementation of the concept of sustainable development? In the context of the growth of the new town, it is actually very important to study from the side of the affected society, which is the people who are currently living in the original settlements adjacent to the newly developed town. This is the intention and the underlying question of research; is the development of a new town already in the context of sustainability to affected society? This is based on the idea that the development of a region by itself will have an impact on the lives of the surrounding society while the development of new town by many developers take up existing settlements land.

2. Research Method

To examine problems in the field and answer research questions, the method used was descriptive-explorative. The first step was to formulate the most appropriate sustainable criteria to be implemented based on the conditions in the field. This is due to the many concepts and theories about "sustainable". The criteria formulated were then used to analyze and evaluate the case of the development of the new town of Gading Serpong in Tangerang. Overall, the research steps are in the scheme:

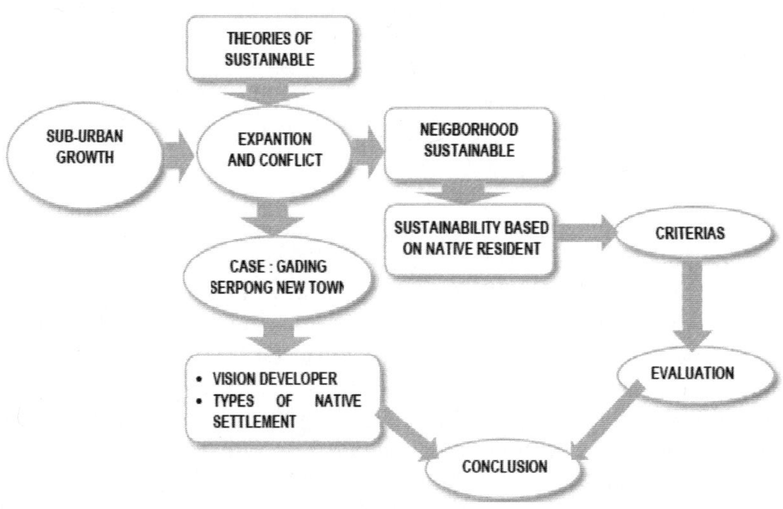

Figure 1. Research steps.

The research location was conducted in the new town area of Gading Serpong that developed by two developers, PT. Summarecon Agung, and PT. Paramount Land located in Kelapa Dua and Pagedangan District, Tangerang Regency

3. Results and Discussion

3.1. Review of theories

Sustainable development is one word popularized by the Word Commission on Environment and Development in the 1987 report, "Our Common Future" and by the Earth Summit in Rio de Janeiro in 1992. It is simply stated the concept of sustainability is an ethic, set of principles and views that are oriented towards the future. Sustainability means protecting and continuing local potential (genius loci) and the way it works is by always linking it to a specific area [8] with the following details:

a. Environmental Sustainability is the ability to manage the environment or the quality of the environment so that it is more meaningful to the surrounding which means managing the factors and actions that directly contribute to the quality of the environment
b. Economical Sustainability means that sustainability on economic factors is broader than the concept of sustainability in terms of welfare, local sources, or profit margins. Basically, sustainable economy is about providing what society needs with what kind of supply is needed. It is about how we select and use the right local resources for the future.
c. Social Sustainability requires social relation and its potential that is directed to the common goal of achieving and fulfilling the individual's basic needs regarding health, safety, shelter, education and cultural expression.

The sustainable concept in a settlement and urban environment is closely related to the subject of ecology, which focuses on the relationship between humans and the environment [10]. Wherein the planning and design of settlements should be persuasive to the analysis of ecological systems that can provide a decent place to live in. One of the characters of society in the sub-urban settlement is a strong sense of neighborliness, with the thickness of human scale among the buildings of their houses [11] yet cities cannot be considered sustainable if their component parts, such as neighbor hoods, do not meet sustainability criteria [12]. The term neighborhood has been in the urban dweller's vocabulary for a very long time, with many of the same characteristics regardless of the culture to which one is referring.

3.2. Types of native settlements due to land expansion by developers

The pattern of land acquisition and development carried out by the developer produces an effect on the formation of the native settlement type into at least three forms: small enclave settlements [11], large enclave settlements [3], and settlements that are cut off and directly adjacent to the development land [4] (Figure 2).

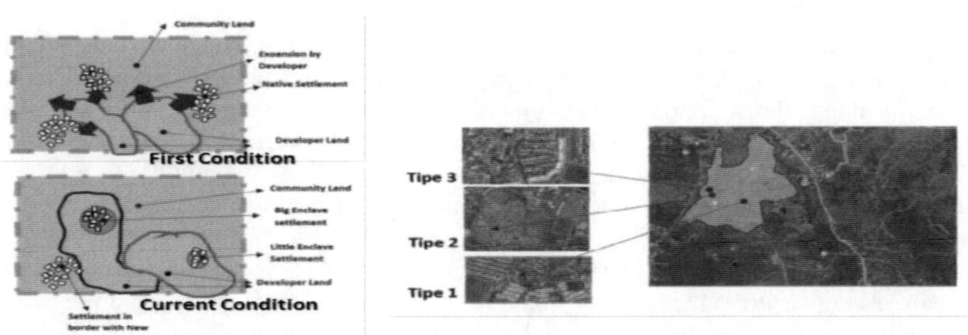

Figure 2. The pattern of land acquisition and development.

3.3. Vision and Mission of the Developers

To measure the level of development sustainability undertaken by developers, the first measuring tool used is to measure the concepts and desires of the developer companies summarized in the vision and mission of the company. As seen in the Table 1, it can be observed that the vision of the two development companies is very ambitious to be the most advanced company in the field of property business development. This is actually not a strange thing, just like any other company's vision. It is interesting to observe the mission of both companies. Neither company explicitly discloses the application of a comprehensive concept of sustainability, including the environment and affected society around the new town development site.

Table 1. The vision of the two development companies.

Developers	Vision	Mision
Summarecon Agung Grup	To be the "Crown Jewel" of property developers in Indonesia that continuously delivers optimal economic value to its customers, employees, shareholders and other stakeholders, as well as fulfilling its social and environmental responsibilities.	1. Focus on maximizing the benefits of an integrated development model, including townships development, by utilizing competitive advantages in each sector: residential, shopping centers and other commercials. 2. Developing high-value products for the consumers, encourage entrepreneurship, implementing the principles of good corporate governance and continuous improvement. 3. Developing high quality Human Resources, responsible, prosperous, according to the companys values and culture.
Paramount Land	To be a leading property company in Indonesia, by building homes and people across the nation with heart	To faithfully serve with the highest commitment to our customers and communities ,so they can live holistically and sustainably by developing townships, real estate, large mixed-use development, infrastructure and high quality living environments, whilst continuously also observing and supporting education both quantitatively and qualitatively, as Indonesia becomes firmly grounded as a globally recognized nation of strength, moving into its golden era of development over the next 20-30 years.

3.4. Scope of Sustainability

From the study of sustainable development, there is a scheme of relevance between sustainability elements and variables influencing the context of sustainable new town development which is seen from the side of the resident of native settlement, inside and around the developed area as the new town (Figure 3).

3.5. Sustainability

3.5.1. Social Sustainability

Social sustainability studies are measured by four criteria. The first is community accessibility to social facilities and public facilities. Only Type 1 of native settlements have constraints due to access to education facilities (basic and secondary), health (clinic, general practitioner), economy (environmental market), and religion (mosque) that must first exit from the boundaries of their settlements. In the type 2 of native settlements almost all social facilities and public facilities are in the settlement. While in type 3, access to social facilities and public facilities are not influenced by the existence of new town.

The second criterion is the guarantee of the continuity of social and religious events. Until now residents in all types of native settlements can still do social and religious events in any environment level without obstacles. The third criterion is the bonding and family territory and social harmony. For this criterion the residents' type 1 and type 3 are those who feel it the most. Due to the expansion of land by the developers most of the neighbors and relatives moved out of the village area. The release of the neighboring ties and territories is handled by still inviting relatives and neighbors who have moved out to social and religious events within the family or village. The fourth criterion is convenience in carrying out activities and social interaction. For the fourth criterion, the residents of all types can do activities and social interactions smoothly without any constraints.

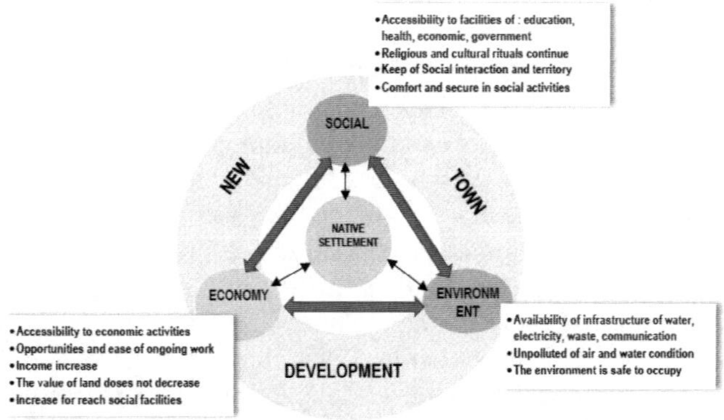

Figure 3. Sustainable new town development.

3.5.2. Economic Sustainability
Economic sustainability is seen from five criteria: 1) accessibility to a new economic activity center, 2) availability and employment opportunities, 3) increase in family income, 4) increase in land value, 5) increase in family expenditure to meet basic needs. In terms of economic sustainability, almost all criteria are applied to all types of native settlements. The presence and development of new town by developers brings a huge economic value to the surrounding society. The criteria felt the most by the society in all types of native settlements are the availability and job opportunities such as shopkeepers, office employees, security personnel, parking, park management, and as housekeepers. Another economic opportunity is the emergence of opportunities due to the large number of workers in the new town coming from outside the region in need of rental housing and the need for cheap eating places.

This condition results in the emergence of many boarding houses, rented houses and cheap food stalls in all types of native settlements. Thus, if there is an initiative from the inhabitants to capture the opportunity, the criteria for increasing family income can be met. Another economic benefit felt by the society in the three types of native settlements due to the construction of new town of Gading Serpong is the higher price of land occupied.

3.5.3. Environment Sustainability
Environment sustainability is seen from three criteria: 1) availability and affordability of road, electricity, clean water, disposal, 2) air and water quality, and 3) environmental conditions in support of habitable activities. On the whole, what stands out in environmental sustainability is the physical constraint of a massive wall that divides developer lands to what does not belong to them. The developers only restrict land with walls, but socially and infrastructure component is not restrictive at all; by giving access to the entrance and exit so that the residents of the native settlements in the three types still feel the supply of clean water, electricity, roads, and very good sewer. Likewise, the air and groundwater conditions have no significant constraints. These causes people in three types of native settlements to feel the comfort of the environment.

4. Conclusion
The construction of new town implemented by developers results in development that is oriented only on commercial values, so it does not meet the concept of sustainability when viewed comprehensively with the surrounding environment and society. Whereas if the new town development is seen from the people living in the native settlements around, it can be concluded that the concept of sustainability can run except in the residential of native settlement of type 1 which is small enclave settlements located within the developer's land whether the new town is not built, is being built or has been built around it. Society in all types of native settlements take the advantage from the existence of new towns, and this is greatly influenced by the treatment of developers who generally open access to all infrastructure and supporting facilities built to be traversed and used by surrounding society.

References

[1] Sieverts T. 2013. Cities Without Cities, Spon Press, London,Stan, Angelica, Morphological Patterns of Urban Sprawl Territories Urbanism. *Arhitectură. Construcţii.* 4(4).
[2] Pribadi O D, & Pauleit S, 2015. The dynamics of peri-urban agriculture during rapid urbanization of Jabodetabek Metropolitan Area. *Land Use Policy*. 48:13-24
[3] Wirtomartono A, Bagus P, Seni\ Bangunan dan Seni Binakota di Indonesia, Penerbit PT. Gramedia Pustaka Utama, Jakarta, 1995.
[4] Mulyana W. 2014. Rural-Urban Linkages: Indonesia Case Study. Working Paper Series N° 126. Working Group: Development with Territorial Cohesion. Territorial Cohesion for Development Program. Rimisp, Santiago, Chile.
[5] Ricci L. 2016. Reinterpreting Sub-Saharan Cities through the Concept of Adaptive Capacity, An Analysis of Autonomous Adaptation in Response to Environmental Changes in Peri-Urban Areas. Jo*urnal Springer Briefs in Environment, Security, Development and Peace*. Vol. 26
[6] Firman T. 2004. New town development in Jakarta Metropolitan Region: a perspective of spatial segregation. *Habitat International*. 28:349–368
[7] Setioko B, Pandelaki E E, Murtini W, Titien. 2013. towards sustainable urban growth: The unaffected fisherman settlement setting (with case study Semarang coastal area). *Procedia Environmental Sciences*. 17: 401 – 407
[8] Diningrat A R. 2015. Segregasi Spasial Perumahan Skala Besar: Studi Kasus Kota Baru Kota Harapan Indah (KHI) Bekasi. *Jurnal Perencanaan Wilayah dan Kota*. 26(2): 111-129
[9] Hosseini S. 2010. Studi of Sustainable Settlement of Hawraman, Kurdistan and Its Sustainable Approach to Design of a Residential Complex in Mousa Tourist Village, Iran. *Journal of Sustainable Development.* 3(4).
[10] Burton H. 2005. Towards a conceptual model for healthy, sustainable settlements. *Built Environment.* 31(4).
[11] Neal Peter (Ed.), Urban villages and the making of communities, Spon Press. 2003.
[12] Choguill L, Charles. 2008. Developing sustainable neighborhoods. *Habitat International.* 32:41–48
[13] Firman, T. 2007. The Patterns of Indonesia's Urbanization, 1980-2007, unpublished paper, Bandung.
[14] Agustina H, Ina. Kajian Tentang Konsep Keberlanjutan pada Beberapa Kota Baru dan Permukiman Berskala Besar, *Jurnal PWK Unisba.*
[15] Rachman, F Hamzah. Kajian Pola Spasial Pertumbuhan Kawasan Perumahan dan Permukiman di Kecamatan Limboto Kabupaten Gorontalo. Tesis. Program Studi Magister Teknik Pembangunan Wilayah dan Kota Universitas Diponegoro. Semarang, Indonesia, 2010.

Study of water quality and carbon absorbtion in West Sunter Lake using phytoplankton

C M Rusly*, D Hendrawan, A Rinanti

Environmental Engineering Department, Faculty of Landscape Architecture and Environmental Technology, Universitas Trisakti, Jakarta, Indonesia

*Corresponding Author : dianahendrawan@trisakti.ac.id

Abstract. The purposes of this research are to analyze the water quality and measure the carbon absorbtion in water of West Sunter Lake using the phytoplankton community. The sampling and analysis of water quality and phytoplankton using the APHA method. The result show that DO is 1.1-1.4 mg/L),BOD is 10.34 to 27.35 mg/L, COD is 23-130 mg/L, and phosphate is 0.38- 0.57 mg/L. The range of index values were at 0,128 to 2,516 for the Diversity Index (H'), 0.003 to 0.924 for the Evenness Index (E), that is waters this medium polluted. The study indicated that the water quality and productivity of West Sunter Lake is impacted by the activities around West Sunter Lake, especially the household activities. The value of chlorophyll-a in the Reservoir was ranging from 3.2-386.9 mg/m^3 or 0.89 s/d 105.52 mg C/m^3 and carbon which absorbed by phytoplankton are 2x10^9 to 14x10^9 ton/year or 9x10^3 to 50x10^3 ton/m^3. The amount of carbon absorbtion by phytoplankton per year proves that phytoplankton have an important role in reducing CO_2 emissions.
Keywords: carbon absorbtion, phytoplankton, water quality, West Sunter Lake

1. Introduction

Changes of water quality depend on the physical and chemical conditions can also be reviewed from the abundance and composition of phytoplankton [1]. The presence of phytoplankton in the waters can provide information about the state of the water. As aquatic organisms, phytoplankton has many advantages as biological benchmarks that are able to show the level of ecological instability and evaluate various forms of pollution [2].

Phytoplankton is widespread in the waters, while sunlight can still penetrate the depth of the waters where phytoplankton are located. Therefore, the activity of photosynthesis is a photochemical reaction that becomes the basis of primary productivity by phytoplankton. The basic controlling factors of primary production are light and nutrients, varying their relative importance in the ecosystems according to their dynamics [3]. The monitoring of water quality using a biological approach is a relatively recent practice. The purposes of this research are to analyze the water quality, the structure of phytoplankton community such as the diversity, uniformity, also to investigate the relation between the distribution of phytoplankton with the abundance of nitrate and phosphate. These research involving phytoplankton organisms in reservoirs represent a significant advance in the determination of water quality in these aquatic systems.

2. Research Method

The research was conducted at West Sunter Lake located on Sunter Jaya Sub-district, North Jakarta. West Sunter Lake with an area of less than 19 Ha and a depth of 12 m. The sampling and analysis of water quality and phytoplankton using the APHA method [4]. The water sample was taken by using a water sampler, then inserted into a plastic jerry and then the sample bottle inserted into an ice box to preserve the sample unchanged due to the temperature effect. Other factors were measured, in the water column, such as temperature and dissolved oxygen, pH, conductivity at 25°C, water transparency (30 cm-white Secchi disk), alkalinity [5], and chlorophyll-*a* [6]. Dissolved N-NO_3^- [7], N-NH_4^+ and P-PO_4^{3-}[8]. The carbon absorbtion rate was calculated as the ratio between primary productivity rate and chlorophyll-*a* concentration, in order to evaluate the photosynthetic efficiency.

Phytoplankton samples were taken with plankton net number 25 by filtering 100 liters of water. Composite water collection. Subsequently, the phytoplankton collected on a plankton net concentrate bottle (bucket) was transferred into a sample or sample bottle and preserved ± 10 drops of lugol.

2.1. *Phytoplankton Abundance (N) [4].*

$$N = z\,x \frac{X}{Y}\,x\,\frac{1}{V} \qquad (1)$$

N = Abundance Phytoplankton (sel/L), X = The volume of water filtered (100 liter), Y = Volume 1 drops (0,05 ml), V = The volume of water filtered (250 liter), Z = The number of individual that found (cell)

2.2. *Evenness Index (E)[3].*

$$E = \frac{H'}{H'maks} \qquad (2)$$

E = Evenness Index , H' = Diversity Index, H' maks = The maximum value of diversity

2.3. *Chlorophyll-a [6].*

$$\text{Chlorophyll} - mg/\,m^3 = \frac{\{(11{,}85(E664)-1.54(E647)-0.08(E630)\ x\ Volume\ Extraksi\ L}{\text{Volume Sampel } m^3} \qquad (3)$$

E_{664} = absorbance 664 – absorbance 750 nm, E_{647} = absorbance 647 - absorbance 750 nm, E_{630}= absorbance 630 – absorbance 750 nm, V_e = The volume of extract acetone (ml), V_s = The volume sampel of strain that filtrered (250 m)

2.4. *Carbon Absorbtion [8]*

$$Chlorophyll - a\ value\ (mg/m3)\ x\ mol\ CO2\ =\ mg\ CO2/m3 \qquad (4)$$
$$mg\ CO2/m3\ x\ 12/44\ (Ar\ C/\ CO2)\ \rightarrow mg\ C/m3\ *$$

*Conversion C/m^3 as ton/ m^3 and ton/year to determine the ability of carbon uptake in a body of water and the amount of carbon uptake in a certain period of time.

3. Results and Discussion

3.1. *Water Quality*

The result of physical water measurement are temperature, turbidity and brightness. Water temperature ranges from 28–30°C. This high difference in low temperatures occurs at the time of sampling. Turbidity ranges from 24.3 to 2337 mg/L. The high turbidity due to the runoff water carried from open land and other factors that incoming wastewater contains many suspended solids. The results of brightness measurements were 0.10 to 0.22 m. In general, very shallow brightness measured less than 1 m is affected by the water condition which is very turbid as seen from the turbidity data all sampling points obtained high results. This low brightness is cause by high suspended particles in the water (indicated by turbidity) as the turbidity has highest value of 237 NTU. The relationship between brightness and turbidity is inversely proportional. The high turbidity is caused by the presence of suspended organic matter and causing less sunlight to enter the waters, so that the brightness value in these waters is also quite shallow and the brightness value is low.

The pH values range from 7.15-8.1. The DO values were 0.22-1.9 mg/L. The low concentration of DO due to the level of oxygen consumption to decomposition of organic materials is higher than the level of oxygen production. The more organic content in the wastewater, the more the dissolved oxygen required for decomposition activities as well. BOD concentration were 2.18-65.60 mg/L, where it is decreased at the next point up to the outlet showing the relationship with the value of DO which shows the increase to the outlet. COD concentration were 13.76-247.68 mg/L, while the highest concentration is at the inlet point coming from the elite housing and surrounding mosques, which contain lots of organic and organic waste persistently causing high COD values.

The highest concentration of Nitrate at the inlet point. According to [5] the concentration of nitrate will increase as it gets closer from the point of disposal, but will decrease further from the point of discharge caused by the activity of microorganisms that will oxidize ammonium to nitrite which eventually becomes nitrate. Presumably this is in accordance with the natural rule that the tendency of high nitrate value is to occur in waters close to the land.

The highest concentration of Phosphate is at the point of inlet, the water entering the inlet is derived from the slums so that much of the organic waste is residual human activity and which affects the high concentration of phosphate from detergent. High phosphate values can be utilized by phytoplankton as a nutritional requirement. As it is known that phosphorus is a form of phosphorus that can be utilized by plants [5].

3.2. Phytoplankton Abundance

The composition and abundance of phytoplankton will change at various levels in response to changes in environmental conditions physically, chemically and biologically. The abundance of phytoplankton in West Sunter Lake were 500 – 1,202,850 cells/L, which is in the dominance type of phytoplankton *Oscillatoria* sp 1. The abundance of phytoplankton is related to the enrichment of a waters. The results of this trophic status of the waters according to [2] included in the category of eutrophic waters that have rich nutrients and high productivity. Associated with the level of water fertility, this condition is influenced by nutrients inside the water.

Nutrient content that affects the presence of phytoplankton in waters is nitrate and phosphate [3], [5]. Nutrients in waters do not describe momentary conditions but are a process of overhauling of incoming nutrients into compounds absorbed by phytoplankton, therefore conditions in the West Sunter Lake where relatively has low nitrate, while phosphates was in high abundance indicate that N in the form of nitrate has been absorbed by phytoplankton.

3.3. Diversity Index

The value of Diversity Index (H') average is 1.61 showing the stability of the high biodiversity of Phytoplankton community (Figure 1). Based on the criteria of water quality in [1], the quality of the water categorized medium polluted.

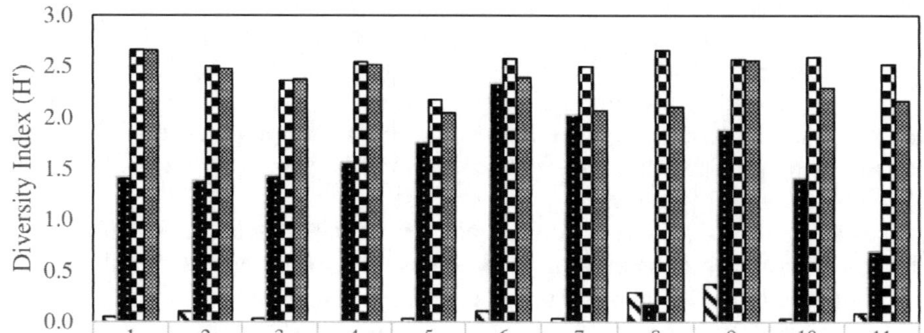

	1	2	3	4	5	6	7	8	9	10	11
1st	0.052	0.106	0.035		0.034	0.108	0.035	0.292	0.374	0.034	0.087
2nd	1.420	1.383	1.432	1.563	1.761	2.340	2.030	0.186	1.884	1.412	0.693
3rd	2.663	2.502	2.361	2.545	2.175	2.579	2.500	2.659	2.571	2.594	2.523
4th	2.66	2.48	2.37	2.52	2.05	2.39	2.07	2.10	2.56	2.29	2.16

Determination of the sampling point

1st 2nd 3rd 4th

Figure 1. Diversity Index (H').

3.4. Evenness Index (E)

The highest Evenness Index is 0.924 shows in June (Figure 2). The Evenness Index (E) closes to 1 means that the species is relatively similar and the difference is not so striking, other words the distribution is evenly distributed. The value of this similarity is directly proportional to the value of diversity due to the highest value of diversity that occurs in June. Besides, it is also influenced by the quantity value of water based on physical and chemical, which at this time of June is at neutral pH, the optimum temperature, the average BOD value does not exceed the quality standard so it is suitable as biota habitat.

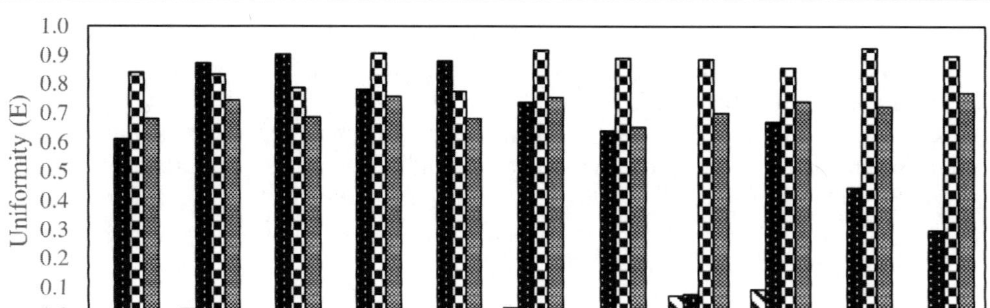

	1	2	3	4	5	6	7	8	9	10	11
1st	0.014	0.025	0.010		0.009	0.031	0.008	0.073	0.096	0.009	0.021
2nd	0.611	0.873	0.903	0.781	0.880	0.738	0.640	0.080	0.671	0.445	0.299
3rd	0.840	0.834	0.787	0.906	0.775	0.919	0.890	0.886	0.857	0.924	0.899
4th	0.68	0.75	0.69	0.76	0.68	0.75	0.65	0.70	0.74	0.72	0.77

Determination of the sampling point

1st 2nd 3rd 4th

Figure 2. Evenness Index (E).

3.5. The Carbon Absorbtion

The value of chlorophyll-a were 32.2-386.9 mg/m^3 as seen at Table 1. There is a lot of factors that influence one of the value of abundance. The trophic status, this West Sunter Lake belongs to the eutrophic category which according to [3], [7] this chlorophyll-a value on trophic status of eutrophic category are 10-500 mg/m^3. The magnitude of this chlorophyll-a value will cause the waters to be highly productive and the growth of algae will increase which causes blooming algae.

Table 1. The value of chlorophyll-a and carbon absorbtion.

Sampling Points	**Valuechlorophyll -a**		**Carbon absorption**	
	mg/m^3	**mgC/m^3**	**Ton/Tahun** (10^9)	**Ton/m^3**
1	139.58	38.07	12	1652828.96
2	125.52	34.23	11	1486371.49
3	84.29	22.99	7.6	998157.45
4	114.02	31.10	10	1350175.26
5	42.44	11.58	3,8	503006.58
6	**386.9**	42.75	**14**	1856317.50
7	97.31	26.54	8.8	1152249.91
8	**32.02**	8.73	**2.9**	379194.30
9	36.94	10.07	3.3	437429.66
10	52.77	14.39	4.7	624869.74
11	27.16	7.41	2.4	321608.56

Carbon absorption capacity ranges from 2x10^9 to 14x10^9 ton/year (Table 1). The highest value is at the point of the inlet point of 14x10^9 ton/year and the lowest at point 11 (the center near the outlet) of 2.4x10^9 ton/year. That was happened because at point 1 the value of chlorophyll-a is high and influenced by the phytoplankton composition. The higher the value of chlorophyll-a, the higher the carbon absorption. The amount of carbon absorption by phytoplankton per year, proving that phytoplankton have an important role in reducing CO_2 emissions, therefore West Sunter Lake as an ecosystem that must be protected.

4. Conclusion

There are four parameters that not comply with the quality standard of the Governor Regulation Number 582 of 1995 on Water Quality Management and Pollution Control Class III, i.e DO (1.1-1.4 mg/L), BOD (10.34-27.35 mg/L), COD (23-130 mg/L), and phosphate (0.38-0.57 mg/L). The abundance relationship with the concentration of nitrate and phosphate is weak, where the nutrients in not depend on instantaneous condition, but on a whole process of nutrients entering into compounds absorbed by phytoplankton. The abundance of phytoplankton in West Sunter Lake were 500-1,202,850 cells/L, which is in the dominance type of phytoplankton *Oscillatoria* sp 1. The value of chlorophyll-a in the West Sunter Lake were 3.2-386.9 mg / m^3 which categorized as eutrophic condition. The value of carbon uptake between 2 x10^9 to 14 x 10^9 Ton/year or 9x10^3 to 50x10^3 Ton/m^3. The amount of carbon absorption by phytoplankton per year proves that phytoplankton have an important role in reducing CO_2 emissions. The source of contaminant in West Sunter Lake were from the organic and inorganic exiles produced by the activities around West SunterLake, especially the household activities.

References

[1] Simone Jaqueline Cardoso, Fábio Roland, Simoni Maria Loverde-Oliveira, Vera Lúcia de Moraes Huszar. September 2012. Phytoplankton abundance, biomass and diversity within

and between Pantanal wetland habitats. *Limnologica - Ecology and Management of Inland Waters*. 42(3):235-24.

[2] Aleksandra M, Lewandowska, Petra Breithaupt, Helmut Hillebrand, Hans-Georg Hoppe, Klaus Jürgens, Ulrich Somme. August 2012. Responses of primary productivity to increased temperature and phytoplankton diversity. *Journal of Sea Research*. 72:87-93.

[3] Eliana Garcia Feresin, Marlene Sofia Arcifa, Lúcia Helena Sampaio da Silva, André Luiz Henríques Esguícero. 2010. Primary productivity of the phytoplankton in a tropical Brazilian shallow lake: experiments in the lake and in mesocosms. *Acta Limnol*. Bras. (Online). 22(4). Rio Claro Oct./Dec. 2010.

[4] APHA. 1995. Standar Methods for The Examination of Water and Wastewater. (19th Ed). Washington D.C: American Water Works Association, Water Pollution Control Federation.

[5] Stumm W, and Morgan J P. Aquatic chemistry. 3rd ed. New York: Wiley-Interscience. 1022 p. 1996.

[6] Jeffrey S W, and Humphrey G F. 1975. New spectrophotometric equations for determining chlorophylls a, b, $c_1$and c_2 in higher plants, algae and natural phytoplankton. *Biochemical and Physiology Pflanzen*, 167:191-194.

[7] Mackereth F J H, Heron J, and Talling J F. 1978. Water analysis: some revised methods for limnologists. Cumbria and Dorser: Freshwater Biological Association. *Scientific Publication*. Vol. 36.

[8] Golterman H L, Clymo R S, and Ohnstad M A M. 1978. Methods for physical and chemical analysis of freshwaters. 2nd Ed. Oxford: Blackwell Scientific Publications. 213 p. IBP Handbook, no. 8.

Knowledge management system for risk mitigation in supply chain uncertainty: case from automotive battery supply chain

I A Marie[1*], D Sugiarto[2], D Surjasa[1], A Witonohadi[1]

[1]Industrial Engineering Department, Faculty of Industrial Technology, Universitas Trisakti, Jakarta, Indonesia
[2]Information System Department, Faculty of Industrial Technology, Universitas Trisakti, Jakarta, Indonesia

*Corresponding Author : iveline.annemarie@trisakti.ac.id

Abstract. Automotive battery supply chain include battery manufacturer, sulphuric acid suppliers, polypropylene suppliers, lead suppliers, transportation service providers, warehouses, retailers and even customers. Due to the increasingly dynamic condition of the environment, supply chain actors were required to improve their ability to overcome various uncertainty issues in the environment. This paper aims to describe the process of designing a knowledge management system for risk mitigation in supply chain uncertainty. The design methodology began with the identification of the knowledge needed to solve the problems associated with uncertainty and analysis of system requirements. The design of the knowledge management system was described in the form of a data flow diagram. The results of the study indicated that key knowledge area that needs to be managed were the knowledge to maintain the stability of process in sulphuric acid process and knowledge to overcome the wastes in battery manufacturing process. The system was expected to be a media acquisition, dissemination and storage of knowledge associated with the uncertainty in the battery supply chain and increase the supply chain performance.
Keywords : automotive battery, risk mitigation, supply chain, uncertainty

1. Introduction

The application of knowledge management to the industry has been widely studied by researchers. Its utilization then evolves into a wider range of industries such as clusters [1] and supply chains [2], [3]. At the supply chain level, knowledge management can also occur due to a common interest in buyer and seller relationships that both want to improve their quality and efficiency. Articles by [4], [5] conducted a more specific study of simulating the application of specific knowledge management regarding knowledge of uncertainty to anticipate uncertainty in the supply chain. High market dynamics and market uncertainty have created a high risk for automotive industry companies. The article written by [6] explained that there are 14 categories of uncertainty in the supply chain. Some of them are supply uncertainty, demand and manufacturing process.

Automotive battery supply chain include battery manufacturer (PT. XYZ), sulfuric acid suppliers (PT. IAI), polypropylene suppliers, lead suppliers, transportation service providers, warehouses, retailers and even customers. Due to the increasingly dynamic condition of the environment, supply chain actors were required to improve their ability to overcome various uncertainty issues in the environment.

The aim of this paper was describe the process of designing a knowledge management system to support knowledge codification and knowledge sharing in order to mitigate risk of the uncertainty in the battery supply chain. Knowledge codification converts tacit knowledge to explicit knowledge in a usable form for the organizational members. Tacit knowledge is identified and leveraged through a form that is able to produce highest return for the business. Explicit knowledge is organized, categorized, indexed and accessed. Explicit knowledge through knowledge codification often includes decision trees, decision tables, production rule, concept map etc. [7], [8]. Codification will support training and decision-making areas.

2. Research Method

The design methodology begins with the identification of the key knowledge area based on the dominant source of uncertainty in the supply chain and the uncertainty handling strategy that needs to be done. Knowledge strategy is aligned with uncertainty and risk management strategy. Clear process steps of risk management was explained by [9] The next step were knowledge capture by interviewing the expert in the related industry and knowledge codification using decision tree and concept map and finalized by system analysis and design process. Stages of knowledge management system design began with the analysis of the needs of knowledge management system implemented by interviewing the user and then proceed with the design of data flow diagrams (DFD) level 0, level 1 and knowledge portal using Drupal content management system. Stages of research can be seen in Figure 1.

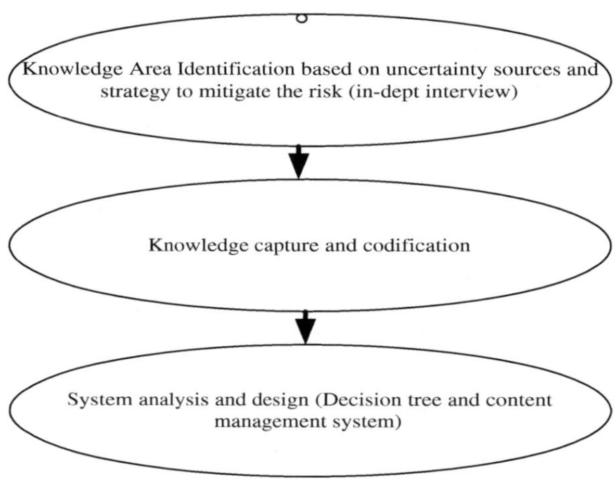

Figure 1. Research steps.

3. Results and Discussion

3.1. PT. XYZ Supply C.hain

PT. XYZ supply chain (Figure 2) started from the customer (for example PT. TY) that provides information in the form of orders to Marketing. Then Marketing will do input data process order and do forecast. Process orders and forecasts are then given to Production Planning Control (PPC). Production Planning Control (PPC) will calculate the capacity and plan production. The data is then supplied to Material Planning Control (MPC) to input the Material Requirement Planning (MRP) data and make the order planning of raw materials to be given to Purchasing. Purchasing will purchase raw materials to suppliers according to the data provided by Material Requirement Planning (MRP). Supplier will deliver the goods to the raw material warehouse and delivered to the Plant to be processed into finished goods. Then the battery packed will be sent to the finished goods warehouse to be shipped to the customer.

3.2. Knowledge Area Identification

Based on in-dept interviews, it is known that the uncertainty in the manufacturing process was prioritized to be captured, codified and shared. Therefore, strategies related to the sources of uncertainty are lean strategy (including waste elimination) and process stabilization [10]. Elimination of waste was a priority for PT. XYZ that require risk mitigation. While for one of its suppliers that is PT. IAI, the priority was in stabilization of the sulfuric acid process due to failure of the production machine that can cause delays in supply and air pollution.

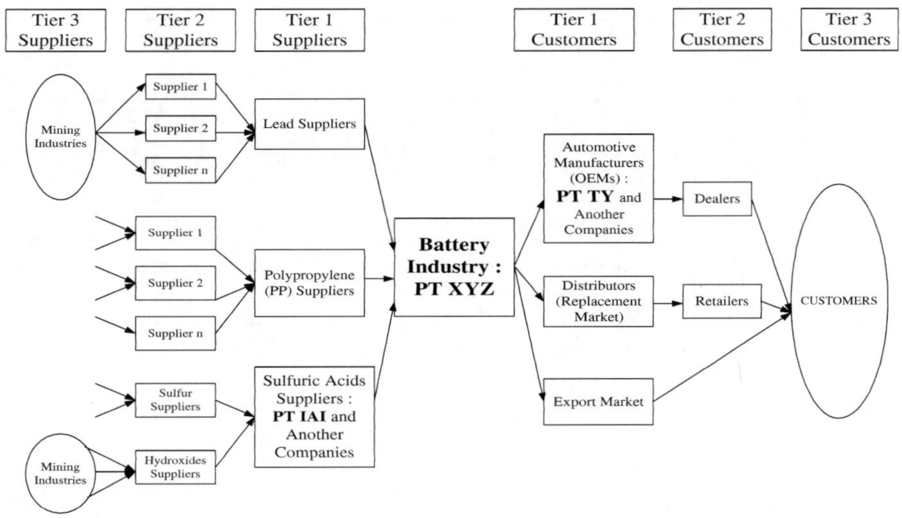

Figure 2. Supply chain network structure for battery industry.

3.3. Knowledge capture and codification

Related to the design of knowledge management system to mitigate the risk of uncertainty that occurs in the battery industry supply chain, codification process is based on the knowledge that has been obtained. The following is an example of a decision tree to mitigate the occurrence of machine breakdowns in the sulfuric acid production process associated with the smoke characteristics produced.

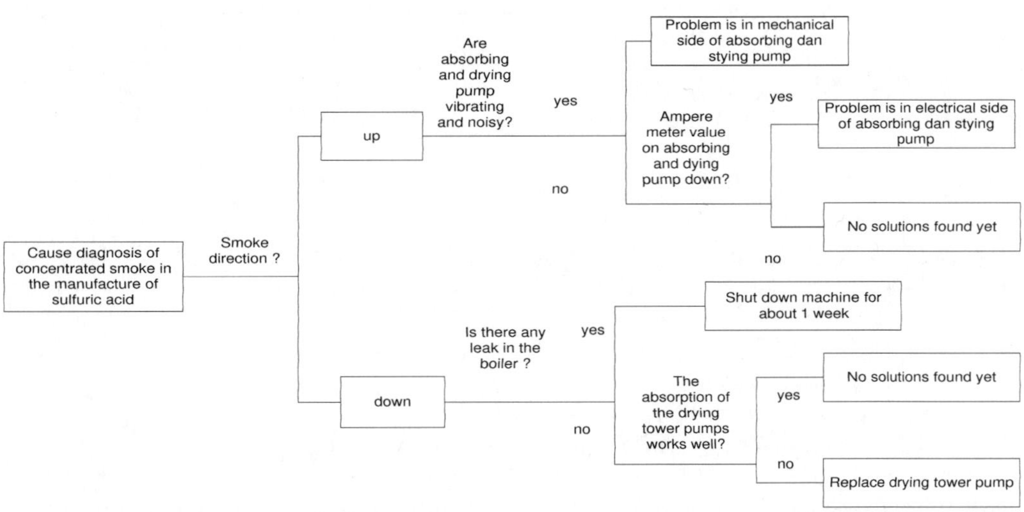

Figure 3. Decision tree for cause diagnosis of concentrat.ed smoke in sulphuric acid manufacturing process.

Uncertainty that occurred in PT. XYZ due to waste in the supply chain flow. Here is a table that contains a summary of several wastes that occurs in the supply chain PT. XYZ (Table 1).

Table 1. Example of waste at PT. XYZ.

Type	Location	Example at PT. XYZ
Defects	Casting	Defect occurs in the form of a fragile grid, broken, tilted and excess material.
	Assembling	Defect occurs in the checking process performed by the machine that makes the half-finished battery out of the process if there is a disability.
	Supplier	Defect occurs upon receipt of defective raw materials.
	Cutting Plate	Defect occurs during plate inspection at the end of the cutting plate process.
	Finishing	Defect occurs after checking done before packing
	Store Warehouse	Defect occurs when the battery is too long kept in warehouse so it requires rework.
Unnecessary Inventory	Cutting Plate	Inventory occurs when the plate is generated more than the plate used in the assembling process resulting in work in process of plate
	Store Warehouse	Inventory occurs when the battery is produced exceeds the demand causing potential damage to the battery.

3.4. Knowledge Management System Design

The result of the identification of the need for development of knowledge management portal in the battery supply chain can be seen in Table 2. The acquisition and knowledge sharing process can be mediated through the page management module, decision tree module and the concept map module (mind map). The sharing process can be mediated through forums.

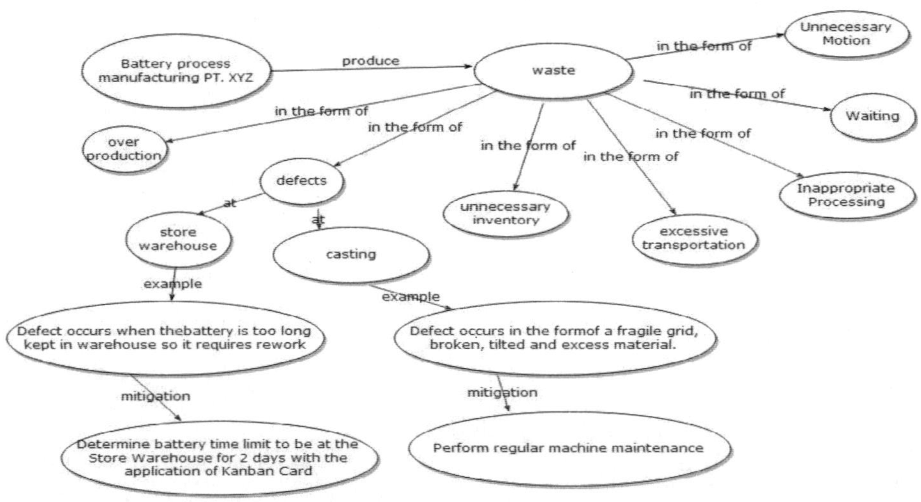

Figure 4. Risk mitigation knowledge concept diagram.

System functional requirements are the system can be used by administrators to enter and manage various sources of information and knowledge for its users, provides facilities for users to search and view various sources of knowledge, provides facilities for discussions and provides facilities for managing users. Non-functional system requirements are system should be secure for the exchange of knowledge among members within only in PT. XYZ supply chain.

Table 2. Need of knowledge management system and technology support needed.

No	Knowledge Management System Need	Web tool
1	Media that can be used to write articles about best practice handling uncertainty and risk mitigation in the supply chain	Page management
2	Media that can be used for discussion among supply chain actors	Discussion forum
3	Media that can be used for knowledge codification and application	Decision tree and knowledge map module

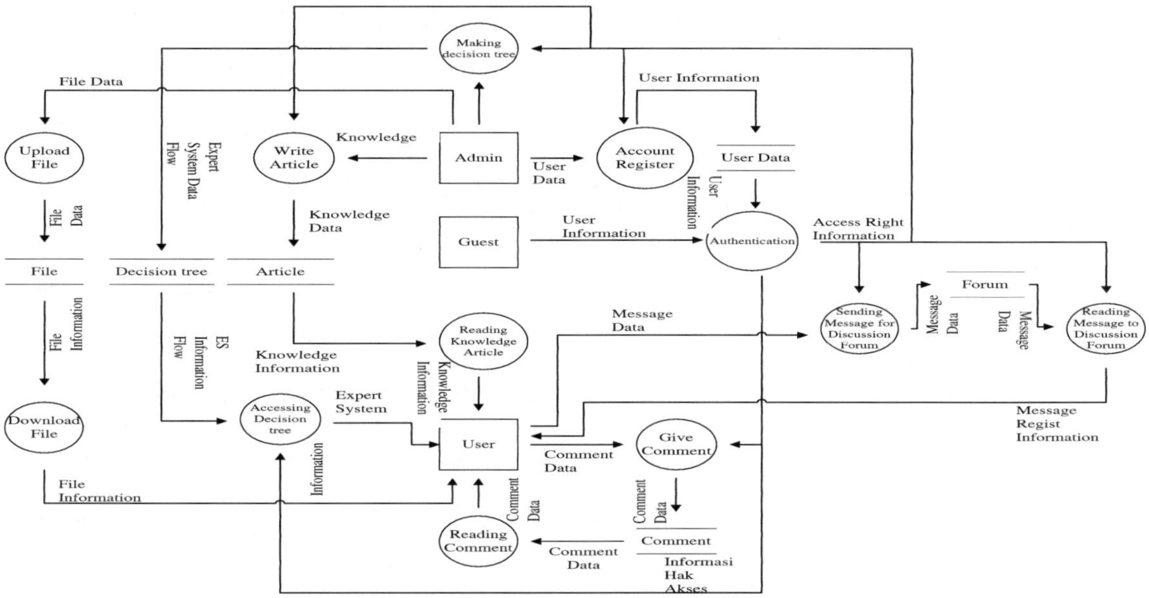

Figure 5. Data flow diagram level 1.

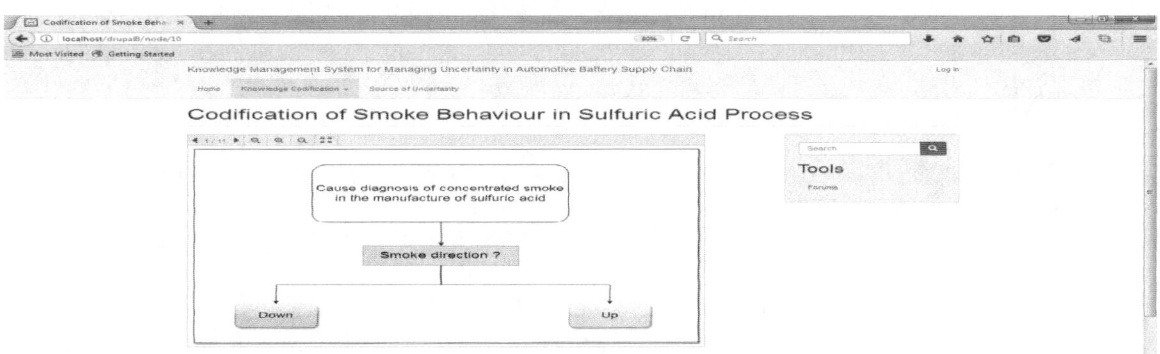

Figure 6. Example of knowledge management system menu display.

Based on interview, we collect information needs from the users of the knowledge management system. The knowledge management system has three types of users, i.e. Admin, User, and Guest. Admin is responsible for managing the system as a whole, including managing usernames and passwords, managing the knowledge stored in the system. Admins also have access to add and change knowledge codification modul and knowledge articles. Users are those who utilize knowledge management systems actively, and can access knowledge articles and knowledge codification modules. Guest must have a username and password to be a user. The process details of the context diagram are presented in DFD level 1 in figure 5. Data Flow Diagram is a description of the flow of data flow and information flowing in the system. Decision tree for the knowledge management system can be seen in Figure 6. The categories of accessible knowledge can be articles, decision tree and concept map.

4. Conclusion

Key knowledge area that needs to be managed were the knowledge to maintain the stability of process in sulphuric acid process and knowledge to eliminate wastes in battery manufacturing process. Knowledge was captured through an expert interview process and codified in the form of a decision tree and concept map. Knowledge management system based on Drupal content management system was expected to be a media acquisition, dissemination and storage of knowledge associated with the uncertainty in the battery supply chain and increase the supply chain performance.

Acknowledgement

This research was supported by the Directorate General of Higher Education, Ministry of Education and Culture of the Republic of Indonesia. Contract Research : Superior Research of Higher Education Fiscal Year 2017 with contract No.135/A/LPT/USAKTI/IV/2017.

References

[1] Sureephong P, Chakpitak N, Ouzrout Y, Neubert G, and Bouras A. 2007. Knowledge management system architecture for the industry cluster. *Proc. Int. Conference on Industrial Engineering and Engineering Management.*

[2] Sureephong P, Chakpitak N, Buzon L, and Bouras A. 2008.Cluster development and knowledge exchange in supply chain. *Proc. Int. C onf. on Software Knowledge Information Management and Applications.*

[3] Surjasa D, Sugiarto D, Solihah B, Ratnawati N. 2013. Knowledge management system model in DKI Jakarta rice supply chain. *Proc. 6th Int. Seminar on Industrial Engineering and Management.*

[4] Koh S C L, and Gunasekaran A. 2006. A Knowledge management approach for managing uncertainty in manufacturing. *Industrial Management & Data Systems*. 106(4):439-459

[5] Koh SCL, Tan KH 2006 Translating knowledge of supply chain uncertainty into business strategy and actions. *J. Of Manufacturing Technology.* 17(4):472-486

[6] Simangunsong E, Hendry L C and Stevenson M. 2012. Supply-chain uncertainty: a review and theoretical foundation for future research. *J. of Production Research.* 50(16):4493–4523

[7] Ramanigopal C S, Palaniappan G, and Mani A. 2012. Mind mapping and knowledge management : coding and implementation of KM system. *Int. J. of Management.* 3(2):250-259

[8] Chrysler E. 2006. Using decision tree analysis to develop an expert system. *Information Systems Education Journal.* 4:55

[9] Manuj I, and Mentzer J T. 2008. Global supply chain risk management. *J. of Business Logistics.* 29(1):133

[10] Jones R M, Towill D R. 2000. Coping with Uncertainty: Reducing "Bullwhip" Behaviour in Global Supply Chains. *Supply Chain Forum An International Journal.* p40-45

Urban community empowerment: context on supply chain collaboration in the SMEs

Y Masnita*, H Triyowati, M Rasyawal

Faculty of Economic and Business, Universitas Trisakti, Jakarta, Indonesia

*Corresponding Author: yolandamasnita@trisakti.ac.id

Abstract. The purpose of this research is to analyze the effect of EB application to operational performance through buyer-supplier collaboration, as well as the influence of buyer-supplier collaboration on operational performance. The context of small industries that ultimately by empowering the community will increase business competitiveness. Community empowerment is a process of creating a community and control over its environment. One concept that is considered as a new paradigm today is e-business (EB), which continues to grow. Data analysis method used is Structural Equation Model (SEM) by the use of PLS program. Based on the analysis from various industries such small garment manufacture, furniture and food, showed that all companies have a supply chain patterns are almost the same. Each company has associated premises supplier, manufacturing and enterprise users of the product. One way to empower the environment is to improve business competitiveness. The development of information and communication technology has been developing very fast and it has brought a significant impact for many aspects of life, including in the business world. EB existence of a significant impact on business practices, at least in terms of the improvement of direct marketing, and organizational transformation.
Keywords : buyer-supplier collaboration, e-business, SMEs, technology context, urban community empowerment

1. Introduction

Community empowerment is a process of creating a community and of much greater control over its environment. The process demands organizational means in order to develop. The organization enables the people in the community to manage their lives by themselves [1]. In this era of globalization and modern, information technology needs become very important for the community. The development of information and communication technology has been developing very fast and it has brought a significant impact for many aspects of life, including in the business aspects. One concept that is considered as a new paradigm today is e-business (EB), which continues to grow. EB existence of a significant impact on business practices, at least in terms of the improvement of direct marketing, organizational transformation, etc. EB refers to the definition of e-commerce more broadly, not just the buying and selling of goods and services but also serving customers, collaborating with business partners and conduct electronic transactions within an organization. Others looked at EB as a collaborative activity and intra business performed by using the internet [2].

Talking about EB there will be a lot of definitions that covers various fields, if interpreted EB as an abbreviation of electronic business, namely all forms of business activities conducted using electronic media such as computers, mobile phones, and the Internet. On the whole, EB can be defined as the use of information and communication technologies by organizations, individuals, or parties related to running and managing a business so that it can provide benefits, either in the form of security, flexibility, integration, optimization, efficient, and/or an increase in productivity and profit [3].In large enterprises, IT and e-business is used to facilitate business processes. The existence of internet applications business to business (B2B) and has provided new and innovative applications for supply chain management [2], [4]. Many companies are using e-business applications such as electronic procurement, electronic catalogs, customer relationship management applications to simplify its business processes along the supply chain [3], [5]. E-business can help ease the company's business processes, especially in managing the supply chain will be a plus for the company, because business processes are automated these companies can reduce the cost, time, and can control the activities of their supply chain, and the most important thing a company can do a collaboration with the parties that are members of the supply chain. Basically the company is unable to perform any activity the production process itself and certainly need others who can help him so that by the establishment of e-business in the supply chain can be a way to alleviate it.

EB has three main dimensions, namely interactive application, application coordination and integrative application. The implementation of three dimensions on a company engaged in services (especially banking) can perform using the interactive web-based internet technologies such as EDI (electronic data interchange), provides a private portal for suppliers, conduct audio conference, etc. In addition, the company can monitor and analyse the company's expenses, monitor supplier performance, there are records and electronic booking status, and employment contracts with suppliers organized electronic. Lastly, the system used by the company is easy to cast, integrated and processed, the changes can be seen ordering the e-procurement system, and EB applications used by the company together with applications used by the supplier and the buyer [6].

Application of EB maximum application and function properly then the collaboration that took place between buyers with suppliers will be well maintained so that the operational performance of the company will be considered to be good and has a positive value and can make it ahead of the competition. However, instead when EB application using is not running with the maximum collaboration will not be achieved as expected which in turn creates a split between the two parties and the operational performance of the company will not be achieved maximal.

Basically, the concept of EB applications in affecting operational performance through buyer-supplier collaboration was to apply the context of technology (technology context). Technology context is the level of technology intensity or sophistication of some of the information necessary to determine the attributes of a product, as a manifestation of the complexity of the products that affect supply chain strategy and performance. In this case, the core production allowing the change of logistics, the increasing number of emerging information, and makes the company relatively sophisticated in running their business activities. Thus, the influence of the buyer-supplier collaboration the operational performance of companies will increase, due to the influence of the use of technology. There are other journals [7] which states that the establishment of operational performance is influenced by technology context. In the main journal stated that for future research should consider further research suggestions technology. Thus, technology is added as a moderating variable context. This needs to be communicated to look for reasons for the importance of this research study.

One strategy to win the competition is to strengthen supply chain management. Implementation of a series of activities in the supply chain starting from the upstream suppliers to downstream customers. [8], [9]. Each of the activities in the supply chain can be trusted to give effect to the performance of the company, as an example of collaborative activities between supplier and purchaser (buyer-supplier partnership) that can either be assessors on the performance of suppliers and enterprises, cooperative relationship will either be established if their confidence (trust) between suppliers with buyers [10]. According to the theory of social exchange (social exchange theory), people will do the exchange in this case the purchase because of the relationship of trust.

2. Research Method

Referring to the research objectives, then this kind of research is causal research that aims to determine the causal relationship by manipulating one or more independent variables or to exercise control over the other connecting variables [11]. The research design used is Hypothesis Testing.

This study focused on a small industry in diverse kind, such as companies convection, furniture and food. Third attempt the same type of business - at industrial scale is still small, thus making the data bias due to the diversity of the industry can be avoided. Methods of data analysis used in this research is Structural Equation Model (SEM) by the use of PLS program [12].

3. Results and Discussion

The significance of estimated parameters provides very useful information about the relationship between research variables. Hypothesis test results can be seen in Table 1.

Table 1. Path coefficients (Mean, STDEV, T-Values).

	Original Sample (O)	Sample Mean (M)	Standard Deviation (STDEV)	Standard Error (STERR)	T Statistics (|O/STERR|)
BSC -> OP	0.325934	0.338618	0.078415	0.078415	4.156522
EB -> BSC	0.528423	0.565366	0.064521	0.064521	8.189987
EB -> OP	0.339700	0.380615	0.400883	0.400883	8.473807
EB * TC -> OP	0.715474	0.767101	0.617235	0.617235	1.159160
TC -> EB	0.481377	0.507839	0.055585	0.055585	8.660187
TC -> OP	0.098618	0.084366	0.303136	0.303136	0.325327

In the first hypothesis, to be observed is whether there are positive influence between EB Application of the Operational Performance seen. Positive effect of EB Application of the Operational Performance indicated by the variable relationship to the Operational Performance Application EB with the value t stat = 8.4738> 1.96 and with the effect of 0.3397 so that Ho rejected.

Electronic Business Application is an information system that is focused on supporting organizations and/or activate the activity of electronic processes with key suppliers. The use of multiple application demonstrates the use of an electronic application rate is good. The application is measured in this study were (i) interactive application, (ii) Coordination application, (iii) Integrative application.

Use of EB can improve operating performance indicated by many aspects, including a decrease in bookings cost, time of message, inventory turnover and the cost of raw materials preparation, and improving the quality, good quality product components as well as the durability of the product. EB using have characteristics that is more flexible, resilient to the poor performance of suppliers, flexibly to changes in demand, flexible to changes in the quantity and variety of products and constantly striving to innovate as well.

Table 2. Indirect effect.

Track	**Indirect effects**	
		Sub Total
EB *Application* BS Collaboration *Operational Performance*	(0.528423)(0.325934)	0.1722

The second hypothesis testing results show the value t stat = 8.1899 <1.96 and the magnitude of the effect of 0.5284. While the variable relationship Buyer-Supplier Collaboration on Operational Performance indicates the value t stat = 4.1565 <1.96 and with the effect of 0.3259 so that H_2 is accepted. H2 test result is supported by hypothesis mediation test with product of coefficient strategy seen in Table 2.

It can be concluded that the existence of EB application will affect the operational performance in case of collaboration between the buyer and the supplier (buyer-supplier collaboration). The results of this study support the results of previous research that has been done by [13]. Where the results of his research stating that the EB application has facilitated the company to conduct information sharing and in making the right decisions, so that the collaboration that exists between the two sides will continue to run, which in turn will improve the company's performance in terms of operating either as cost, quality, flexibility, and innovation.

The third hypothesis tested positive influence between EB Application of the Operational Performance moderated by Technology Context. Mechanisms is to look one by one relationship between EB Application of the Operational Performance and Application EB moderated by Technology Context. The results to be obtained is the direct effect and influence with moderation showed a nonsignificant value of t stat = 1.159 and the magnitude of the effect of 0.7154. It can be concluded that the H3 is rejected, which means variable Technology Context can not moderate EB Application of the Operational Performance.

Technology context is the level of technology intensity or sophistication of some of the information necessary to determine the attributes of a product, as a manifestation of the complexity of the products that affect supply chain strategy and performance. Companies that run the technology context, the core production process of being changed from manual to automatic, allowing the change of logistics, the increasing number of emerging information, and make the company a relatively sophisticated in running their business activities. Thus, the influence of the buyer-supplier collaborationi the operational performance of companies will increase, due to the influence of the use of technology.

The results of this study contrast with research that has been done by [7] which states that the technology context has associated with operational performance. Small scale industry today still have low technology intensity, so that technological turbulence is still slow. This means that the company is still rarely make changes logistic processes, products become obsolete, changing the core production processes, new products are rarely introduced. In addition, product complexity is quite high, which has a tendency technically complex, requiring a lot of information, has a large number of sub-components, necessary efforts and significant engineering expertise and skills, less sophisticated.

In small companies, the use of electronic applications is still regarded as a very big investment for the company. Most of the companies studied using only electronic applications business is still modest, so that the ordering process is not maximized supply chain. In addition, the number of companies involved in the supply chain either as a supplier or as a user company still small in number, so that the advantages - advantages that can be obtained because the use of electronic applications business is not too felt the company. Variations reservations were not complex, delivery scheduling, and shipping costs can still be handled by the company's electronics business applications simple.

The coefficient of determination R Square of buyer-suplier collaboration, EB application, Operational performance, and technology contest respectively are 0.3558, 0.3178, 0.6451 and R square value equation EB Application to the Buyer-Supplier Collaboration is 0.3558 indicating that 35.58% of the variance Buyer-Supplier Collaboration can be explained by changes in the EB Application. While the R-square for equality Buyer-Supplier Collaboration on Operational Performance is 0.6451 indicating that 64, 51% of the variance of customer integration can be explained by changes in information technology.

Based on the analysis that has been conducted on 289 respondents from various industries, namely small companies convection, furniture and food, showed that all companies have a supply chain patterns are almost the same. Each company has associated premises supplier, manufacturing and enterprise users of the product.

Links between companies look of the fabric of cooperation between both parties that the buyer and supplier, we need a container or a system that is able to unite the two so as to build and improve its performance. The test results showed that the EB application to have a positive relationship to operational performance. So also in the hypothesis testing indirectly, EB application will be a positive influence on operational performance through buyer-supplier collaboration. The results of this study support the results of previous research that has been done [13]. Where the results of his research stating that the EB application has facilitated the company to conduct information sharing and in making the right decisions, so that the collaboration that exists between the two sides will continue to run, which in turn will improve the company's performance in terms of operating either as cost, quality, flexibility, and innovation.

In the later hypothesis, it is known that the technology variable context can not moderate EB application to operational performance. The results of this study according to research conducted by [7] which states that the technology context has no connection with the operational performance. Difficulty use of technology in the context of small businesses mainly caused by the large investment that the technology used is still simple.

Based on the results of research that has been conducted found that EB application and significant positive effect on operational performance. In addition, the EB application also affects indirectly through buyer-supplier collaboration. From these results it can be said that the collaboration is an important factor in influencing the performance, especially the performance in terms of cost. In addition, the managerial implications of this collaboration is considered as a concept that building activity - activity interactions and processes, such as information sharing, joined in making the right decisions, and the sharing of cost and risk.

4. Conclusion

Management needs to ensure also that the system can transmit, and process to integrate internal and external data and ensure that the EB application is applied successfully integrated with EB system suppliers. Technology can improve the competitiveness of companies but the limitations of technology in SMEs become a certain challenge for the company, so SMEs are required to be more creative in doing business by strengthening supply chain management through community empowerment. As for suggestions for future research that research results more perfect is as follows: First, research on medium-scale industry on the variety of products and high technology. Second, further research is expected to increase the sample size, so as to better show the representation of the population. Third, further research is expected to use the measurement of operational performance indicators - real and measurable indicators, such as counting the cost, quality, innovation and flexibility with quantitative data. Fourth, increase the dependent variable is financial performance, so that the effect of EB application for performance not only in terms of its operations but it can be seen in terms of finances as well. It would be interesting to be identified, whether the collaboration can be a source of competitive advantage or a sustainable competitive advantage.

References

[1] Simon B L. 1990. Rethinking Empowerment. *Journal of Progressive Human Services*. 1(1):27-39.

[2] Da silveira G J C, and Cagliano R. 2006. The relationship between inter-organizational information system and operations performance. *International Journal of Operations & Production Management*. 26(3):232-53.

[3] Bakker E, Zheng J, Knight L, and Harland C. 2008. Putting e-commerce adoption in supply chain context. *International Journal of Operations & Production Management*. 28(4):313-30.

[4] Carter C R, Rogers D S, & Choi T Y 2015b. Toward the theory of the supply chain. *Journal of Supply Chain Management*. 51(2):89–97.

[5] Wong C W Y, Lai K H, & Benroidey E W N. 2015. The performance of contingencies of supply chain integration: The roles of product and market complexity. *International Journal of Production Economics*. 165:1–11

[6] Burt D N, Dobler D W, and Starling S L, World Class Supply Management: The Key to Supply Chain Management. 7th Edition, McGraw-Hill Irwin, New York. 2003.

[7] Iyer N S, Karthik. 2013. Operational impact of collaboration and resource specificity: The moderating role of technology context. *Journal of Business & Industrial Marketing*. 29(5): 387-399.

[8] Golicic Susan L, & Smith Carlo D A. 2013. Meta-Analysis of Environmentally Sustainable Supply Chain Management Practices and Firm Performance. *Journal of Supply Chain Management*. 49(2):78.

[9] Flynn Barbara B, Koufteros X, & Guanyi Lu. 2016. On The Theory in Supply Chain

Uncertainty and Its Implications For Supply Chain Integration. *Journal of Supply Chain Management*. 52(3).

[10] Reimann F, & Ketchen David JR. 2017. Power In Supply Chain Management. *Journal of Supply Chain Management*. 53(2):3–9

[11] Malhotra Naresh K. Marketing Research : An Applied Orientation. Pearson education, Inc., Fifth Edition, New Jersey, USA. 2004.

[12] Hair Joseph F, William C, Black Barry, J Black, and Rolph E Anderson. Multivariat data analysis: A global perspective. 7th ed. United States of America: Pearson Prentice Hall. 2010.

[13] Wiengarten F, Humphreys P, Cao G, and Fynes B. 2013. Investigating the impact of e-business applications on supply chain collaboration in the German automotive industry. *Journal of Operations & Production Management*. 33(1):25-48.

IOP Conf. Series: Earth and Environmental Science **106** (2017) 012020 doi:10.1088/1755-1315/106/1/012020

Comparison of productive house spatial planning in Kampung Batik - Central Java object of observation: Pekalongan and Lasem

E R Kridarso

Architecture Department, Faculty of Civil Engineering and Planning, Universitas Trisakti, Jakarta, Indonesia

Corresponding Author: etty.k@trisakti.ac.id

Abstract. Home is a basic human need other than clothing and food. As one of the basic needs of man, it has variety of functions, for example, as a place to protect and develop themselves occupants, also as an asset that have economic and non-economic value. Houses that have economic value can be utilized as capital to earn a living by using part of room as a working space, named as productive house. Batik products become the focus of observation with the consideration that batik is a unique Indonesian cultural richness that has been recognized internationally. Pekalongan and Lasem is a coastal city located on the north coast of Java Island, where both cities become the benchmark of batik products located in the coastal area. Kampung Batik in Pekalongan and Lasem is the location used as an object of observation for comparative pattern of productive house layout with qualitative method. The data obtained in primary and secondary, in the form of visual recordings, maps and sketches of productive layout pattern of batik houses. The comparative result shows many similarities in the pattern of productive layout of batik houses in Pekalongan and Lasem; Differences exist in existing occupants. The existing equations are due to the activities undertaken and the differences that exist are due to the growing culture in both locations of observation.
Keywords: kampung batik, lasem, productive house, Pekalongan, spatial pattern

1. Introduction

Batik is an Indonesian cultural treasure that has been recognized internationally by UNESCO (United Nations Educational, Scientific and Cultural Organization) in 2009, so that the sustainability of batik products is the responsibility of the entire nation of Indonesia. Batik cloth is a culture that developed since the era of the Majapahit kingdom, the next use of limited batik keratin. Keraton, namely the palace of Yogyakarta and the palace of Solo. Further development, batik fabrics began to be produced on the coast/coast north of Java, currently the northern coast of Java island known as batik products are in the city of Cirebon, Pekalongan and Lasem (Figure 1). The comparison between Kampung Batik takes place in Pekalongan and Lasem, since both sites are located in Central Java; With a distance between the city of Pekalongan and Lasem about 230 km. The comparative pattern of productive house layout has the purpose to know the differences and similarities between two different locations, where each location has similarity in the product of batik and the two locations are on the north coast of Java Island. The benefits of comparisons are to maintain the advantage of each location as a tourist attraction; as a guide for site development and development; As well as reference for similar studies in different locations. The method used is qualitative with the following steps: see the phenomenon of batik productive house and then compile theory based on literature relating to productive house, then empirically test object observation. The final step concludes the descriptive observation.

Figure 1. Icon of Pekalongan and Lasem
Source: Jaringan Kota Pusaka Indonesia, downloaded 04072017

2. Research Method

A. Comparison, in Big Indonesian Dictionary (online) has a comparative meaning; in this case that will be compared is a pattern of productive layout of batik house in Kampung Batik Kauman - Pekalongan and in Kampong Batagan Babagan, Lasem.

B. Home is a basic human need as a concrete manifestation of a basic need called board. Various functions can be accommodated by a house, because the house consists of arrangement of spaces that form a particular pattern. One of the principles in the pattern of spatial arrangement is 'hierarchy', which means as the degree of importance of a space. In residential or residential homes, the hierarchy of space can be grouped according to user territory, where public space, private space, service space and space between public space and private space are known.

C. A productive house is a house that functions as a place to live as well as a place to earn a living or work. In addition to the two functions already mentioned, another function of a house according to [1] is: as a refuge, as a place to develop self-occupant, as an asset that has economic value and non-economic value. Productive house models can be grouped into:

- Based on the type: mixed productive house (having equal access to occupancy and work, working space
- using part of the living space); A balanced productive home (having equal access between occupation and work, working space and separate living space); Separate productive houses (access to separate shelters and workplaces, so that places occupy and work separately).
- Based on production process: productive house with complete production process and production process incomplete, special product of batik production process from plain white cloth until fabric ready to be distributed to consumer.
- Based on management: productive houses have spatial management (shared space or unused shared space between work and occupation), time (regular or flexible working time), labor (labor from family members or others), capital (Finance mixed or separated from occupational and work activities) and wastes (wastes treated by environment or self-treated waste).

3. Results and Discussion

3.1 Batik Village in Pekalongan

The batik village of Kauman in Pekalongan (Figure 3 and Figure 4) is located in the center of the city adjacent to the square. This village consists of one main road, which is called Kauman alley 1, where on the left and right side of the main road there are several other alleys ranging from alley 2 to alley 14. The inhabitants of Kauman Batik Tourism Village consist of Javanese and a small part of the population of Arab descent and the majority are Muslim, so the employee holiday is set on Friday. Another marker from Kampung wisata batik in the form of pictures of batik motifs that exist on the wall at the entrance of the entrance (Kauman alley 1). Batik productive house located in the batik village of Kauman is as follows:

Figure 2. Kauman Batik Village, Pekalongan

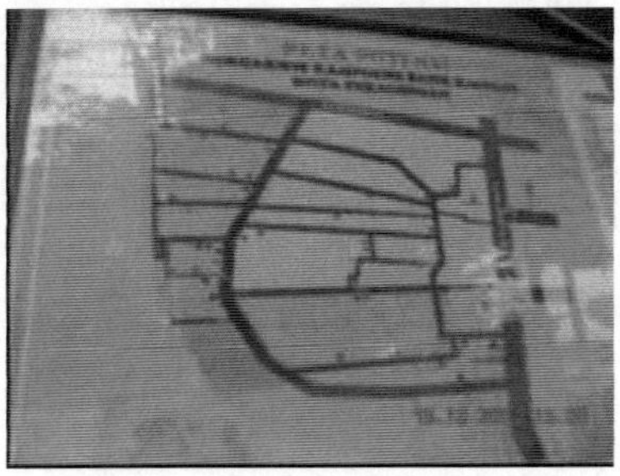

Figure 3. Pattern of Kampung batik in Pekalongan

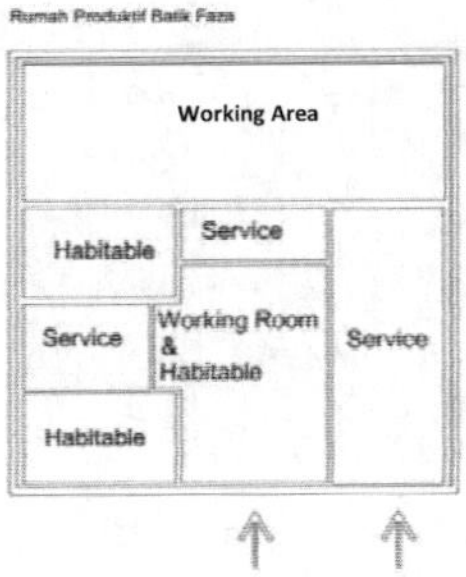

Figure 4. Productive House Faza, Pekalongan

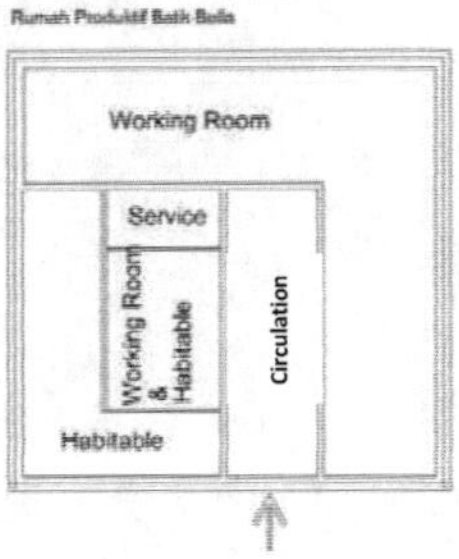

Figure 5. Productive house of Bella, Pekalongan

The productive house of Faza batik (Figure 4 and Figure 8) is a mixed-type productive house in which work space and living space have equal access and in flexible spatial management, have shared space between work and habitation. The production process is complete ranging from white *mori* cloth to batik products that are ready to be distributed to consumers. Bella's batik productive house (Figure 5 and Figure 9) is a mixed-type productive house where work space and living space have equal access and in flexible spatial management, have shared space between work and habitation. The production process is complete ranging from white *mori* cloth to batik products that are ready to be distributed to consumers.

Rizka batik productive house (Figure 6) is a mixed-type productive house where work space and living space have equal access and flexible spatial management, have shared space between work and habitation. The production process is complete ranging from white *mori* cloth to batik products that are ready to be distributed to consumers. The productive house of Mufti batik (Figure 7) is a mixed-type productive house where work space and living space have the same access and in the management of

IOP Conf. Series: Earth and Environmental Science **106** (2017) 012020 doi:10.1088/1755-1315/106/1/012020

the space has a shared / flexible space between work and habitation. The production process is complete ranging from white *mori* cloth to batik products that are ready to be distributed to consumers. In the four productive houses above, the involvement of the owners is very dominant and the relationship between workers and owners is very close.

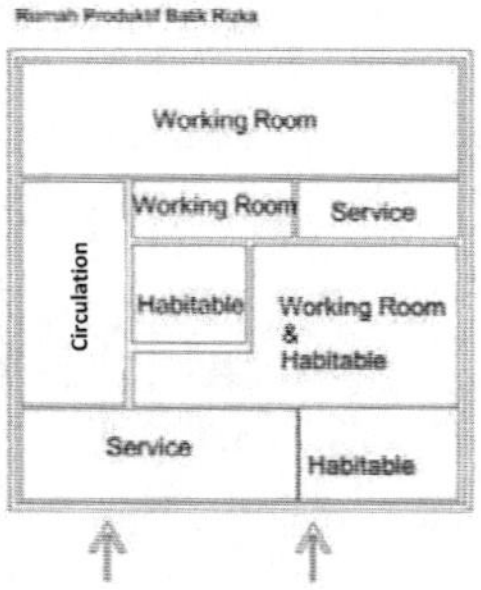

Figure 6. Productive House of Rizka, Pekalongan

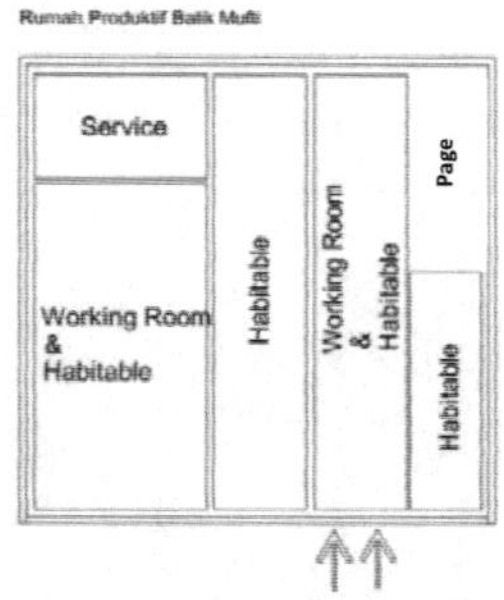

Figure 7. Productive House Mufti, Pekalongan

Figure 8. Productive house of Faza, Pekalongan

Figure 9. Productive house of Bella, Pekalongan

3.2 Kampung Batik in Lasem

The batik village of Lasem is located in Babagan Village, its location on the right side of Java's north coast road from Semarang direction to Surabaya (Figure 10 and Figure 11). Productive houses in Babagan batik kampongs are owned by indigenous people and the people of Chinese descent.

The productive house of Putri Jelita (Figure 12), productive house of Sumber Rejeki (Figure 13, Figure 15, Figure 16) and productive house of Kidang Mas (Figure 14) is a mixed productive house, which has equal access to workplace and workplace; The management of the space is flexible because it has a shared space between work and occupation activities. The production process is complete ranging from white *mori* cloth to batik cloth ready to be distributed to consumers. In the four productive houses above, the involvement of the owners is very dominant and the relationship between workers and owners is very close.

Figure 10. Batik village gate in Lasem

Figure 11. Pattern of Batik Village Babagan, Lasem

House Earning contained in Babagan, Lasem :

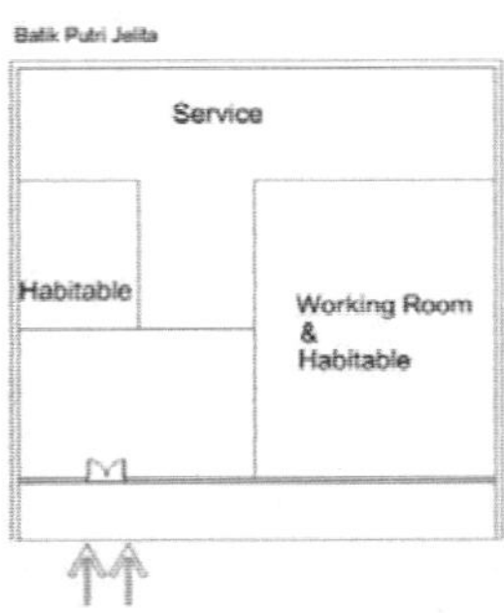

Figure 12. Productive House Putri Jelita, Lasem

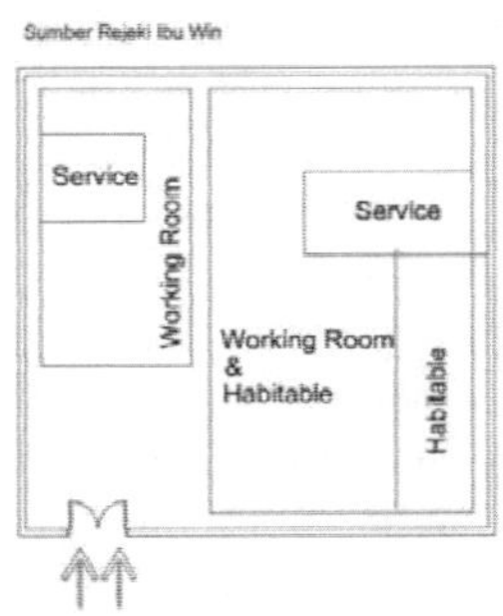

Figure 13. Productive House Sumber Rejeki, Lasem

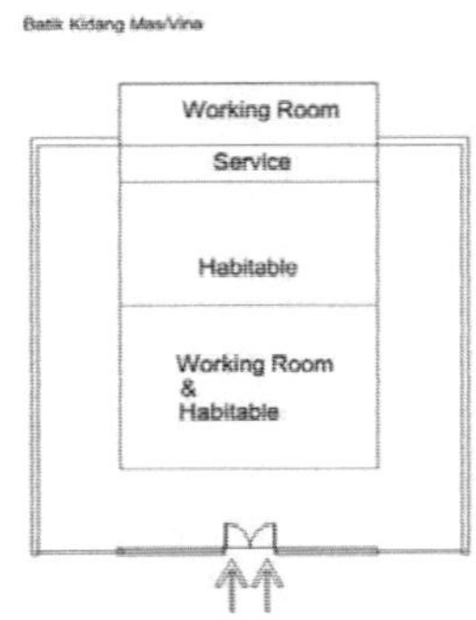

Figure 14. Productive House Kidang Mas, Lasem

Figure 15. Productive House Sumber Rejeki

Figure 16. Productive House Sumber Rejeki

3.3 Comparison

Observations on Kampung Wisata Batik/Batik Tourism Villages in Pekalongan and Lasem have similarities and differences. The equation is that productive houses in Pekalongan and Lasem are both mixed productive houses, which have equal access to occupation and work. In addition, each productive house has a shared space between occupation and work. In general, shared space is a family room located in the house or terrace on the front, side or back of the house. Based on the process done, there is

IOP Conf. Series: Earth and Environmental Science **106** (2017) 012020 doi:10.1088/1755-1315/106/1/012020

a process equation that the complete process of a batik product ranging from white mori fabric to batik cloth that is ready to be distributed / sold to consumers. Distribution / sale to the consumer is done on the front porch, side or back of the house. The use of local labor is done in productive homes of Pekalongan and Lasem. The difference exists, lies in the inhabitants of a productive home environment where in Pekalongan there is a combination of indigenous Javanese with a minority of Arab descendants; While in Lasem the combination of indigenous Javanese ethnic minority with Chinese descent.

4. Conclusion

Observations made on the comparison of batik tourism villages in Pekalongan and Lasem in three respects are location, activity and occupant / resident shows that although both have locations on the north coast of Java Island, the population growing on the site has different characteristics; In Pekalongan of Arab descent and in Lasem Chinese descent. Indigenous people and migrants alike live well side by side and both have a business/activity, especially batik products. The same activity makes the pattern of productive house layout in Pekalongan and Lasem have similarities as well.

Reference

[1] Silas J, and Partner, Productive House, in Traditional Dimensions and Empowerment, UPT ITS, Surabaya, 2000

[2] D K Ching, Architecture, Shape, Space, and Order, 2nd Edition, Erlangga, Jakarta, 2000.

[3] Groat L, Wang D, Architectural Research Methods, New York, John Wiley & Sons, Inc, 2013

[4] Haryadi, B Setiawan. Arsitektur, Lingkungan dan Perilaku; Pengantar ke Teori, Metodologi dan Aplikasi, Gajah Mada University Press, Jogyakarta, 2004

[5] Aryanti D Antariksa, San S. 2005. The Effect of Trading Activities on Spatial Patterns in Home Building - Shop in Chinatown City of Malang City. *Jurnal Dimensi.* 33(1).

[6] Kamus Besar Bahasa Indonesia (KBBI), online

[7] Kridarso E R. Relationship Between Patterns of Productive House Spatial Batik with Character Ethnicity Residents in Pekalongan City, Central Java, Disertasi, Universitas Katolik Parahyangan, Bandung, 2017.

[8] Osman W W, & Amin, S. 2012. Research Results - Home. *Prosiding*

[9] Faqih M, Purnomo H, Productive House: As a Residence and Working Place in Gold Crafts Community Settlements; Spatial Use Pattern on Household Business. National Seminar on Housing & Settlements, *Taufiqurrahman*, 2010.

[10] Changes in the Pattern of the Living Room Layout as a result of the activities of the Home Industry. Case Study: Metal Craftsmen in Ngingas Village, Waru Sub-District - Sidoarjo Regency.

[11] Sugiyono. Quantitative Research Methods, Qualitative and R & D, Alfabeta, Bandung, 2008.

[12] Wicaksono A, Tisnawati E, Interior Theory, Naga Swadaya, Jakarta. 2014.

[13] Widayati N, Settlement of Batik Entrepreneurs in Surakarta, Gajah Mada University Press, Jogjakarta, 2004.

Public open space for disaster mitigation in Tangerang housing estates

M I R Winandari

Architecture Department, Faculty of Civil Engineering and Planning, Universitas Trisakti, Jakarta, Indonesia

Corresponding Author: mi.ririk@trisakti.ac.id

Abstract. Public open space in housing estates plays an important role particularly in disaster mitigation. In some housing, there are indications of shape and use of space changes post-handover to local government. The aim of this study is to explore the relationship between public open space condition and management related to disaster mitigation in Tangerang housing estates. Multiple case study method is used to analyse of 2 housing cases. Aspects of access and boundaries were used to evaluate the cases. The results showed that gated community housing type should have more than 1 access to facilitate evacuation by considering the farthest unit to the housing gate. This is necessary to provide open spaces that are easily accessible from all units as the first evacuation site during and post disaster.
Keywords: disaster, mitigation, public open space

1. Introduction

Indonesia is a disaster-prone country. According to world risk report 2016 [1], Indonesia was at 36th rank with 10.24% world risk index which is improved 2 levels compared to 2 years ago. Referring to this report, Indonesian exposure is in the very high risk [1]. This Exposure relates to natural disaster such as earthquakes, storms, droughts, floods, and sea level rise. Based on BNPB, there are 148.4 million people living in areas prone to earthquakes, 5 million in tsunami-prone areas, 1.2 million in volcanic eruption areas, 63.7 million in flood-prone areas, and 40.9 million in areas prone to landslides [2]. Indonesian Disaster Data and Information shows that the highest mortality caused by earthquake, tsunami, and eruption [3]. These condition need to be anticipated to reduce loss of life and property by providing public open spaces in housing estates.

Several research and paper on open spaces related to disaster mitigation has been done. Based on previous researches, public open spaces in housing estates functioned as green place, protective green belt, and disaster evacuation place [4], [7]. Those spaces were the first evacuation place when disaster strikes. There were several studies about the use of open space related to flooding disaster [5], to earthquake disaster [6], to tsunami [7], to disaster resilience [8]. Related to researched by [19], during floods, streets serve as boat evacuation route while square that should be located on higher ground serving as place for rescue camps, medical facilities, and dropping of supplies. On the contrary, open spaces with limited vegetation become a refuge from fire during fire.

Researchers [6] evaluated characteristics of urban open space in Tehran as places for the relief operation and crisis management during disaster incidence. Researcher [7] evaluated green open space in Banda Aceh post disaster reconstruction. Both previous studies explored cases directly at the studies area. On the contrary, Researchers [8] explored the potential uses of public open spaces for disaster resilience through literature analysis. Based on library search, open space studies related to flood and fire in Tangerang have been no found.

Tangerang is a part of Banten Province, the fourth highest population density in Indonesia [9]. As seen in Figure 1, this area is vulnerable to disaster. This condition requires housing estates management in Tangerang to mitigate the disaster possibility. Stated by Vice Governor of Banten, Banten province is vulnerable to 14 types of disasters with fire and flooding as the most frequent disaster [10]. This study explored the role of open space for mitigation in housing estates. In this study, public open space condition in Tangerang housing estates evaluated to disaster mitigation especially flooding and fire.

Figure 1. Distribution of disaster event per district 1815 – 2017 [3].

1.1 Open Space for Mitigation

Regulation [11] stated that open space in disaster prone area functioned as evacuation place. In line with the regulation, researchers [12] stated that open space has an important role in mitigation strategies to climate change. Based on Ministerial Decree [13], the open space that must be provided for every 250 people is Neighbourhood Park with 0.3 m^2/population, parking area with 3% of the total area, and Neighbourhood Street with a width of 2 m - 5 m. In housing estate, open spaces must provide to accommodate those requirement. Both square (Neighbourhood Park or parking area) and street should be clean, accessible, and well maintained [13].

According to mitigation, open spaces network play an important role in rescuing to safer place. Regulation of the National Disaster Management Agency No 11/2008 stated that the indicators of post-disaster restoration are access availability to housing units and adequate open spaces [14]. Researcher [7] confirmed that open space network is useful for the rescue in a short time as well as for the victim evacuation. Referring to researcher [18], street as an important evacuation path requires sufficient wide for the escape route. Disaster mitigation will be more successful using community-based disaster mitigation programs. This is essential so that people are able to control the situation in the event of a disaster [18]. It should be noted that the capacity, location, and open space connections are affected by the type of open space. The space should be usable in times of emergency and in everyday life rather than special mitigation rooms [8].

2. Research Method

Multiple case study method is used to analyse open spaces related to disaster mitigation in 2 housing cases namely Bumi Serpong Damai estate (Case 1) and Gading Serpong estate (Case 2) that located in South Tangerang. Both housings were middle-high income housing estates. The observation unit is a cluster located in both housing estates. Literature review is used to explore the role and standards of open space required for disaster mitigation. The analysis includes 4 aspects consisting of type, access, location, and boundaries. These aspects were used to evaluate public open space conditions in all cases

IOP Conf. Series: Earth and Environmental Science **106** (2017) 012021 doi:10.1088/1755-1315/106/1/012021

related to disaster mitigation. The observation and interviews to the residents were conducted simultaneously. These activities have been done in sequence case after case.

2.1 Bumi Serpong Damai Estate and Gading Serpong Estate General Data

Bumi Serpong Damai Estate (Case 1) is located in Tangerang Municipality and South Tangerang Municipality. Most of these clusters areas are located in Serpong District, part of South Tangerang Municipality. Built around 1997 by Sinarmas Developer and Real Estate, this estate is developed on an area of approximately 6000 hectares [20]. This housing consists of several clusters one of which is Puspita Loka located in the middle of housing. One of Puspita Loka cluster that became the location observation is surrounded by main road and river that separate it with other clusters and located next to BSD City Park. This cluster has a grid pattern with half of units' arrangement from Northeast to Southwest and the other half from East to West.

Gading Serpong Estate (Case 2) is located in Tangerang Municipality. Built around 1993 by Summarecon and Keris Group, this housing was developed on an area of approximately 1500 hectares [21]. As in Case 1, this housing consists of several clusters. One of the clusters that became the location of this observation was the Aquamarine cluster. This cluster is part of Pondok Hijau Golf area that adjacent to the golf course and separated with other clusters by walls and street. This cluster has a grid pattern with units' arrangement from Northeast to Southwest.

3. Results and Discussion

Streets and squares in both cases are utilized by occupants in daily life and can be used during disaster. Based on researched by [8], this is in accordance with the use of space. This section describes housing and open spaces types, access, location, and boundaries. Based on the open space type, there are various types in both cases. Both cases consist of Neighbourhood Park and green housing boundaries. Besides these two types, Case 1 consists of river green barrier and street green barrier. In accordance with that researched [8], that various types of open spaces will affect the capacity and connection of space especially during a disaster, Case 1 is more responsive to disaster due to various types of open space. In addition to streets, Case 1 has river as an alternative evacuation route in flood time. The variety of open space types allows occupants to choose their most appropriate place during and after the disaster. Referring to Ministerial Decree [13], Case 1 has met the minimum standard that has not been fulfilled in Case 2. The fulfilment of open space minimum standards in Case 2 (clusters) is met through the provision of space in the housing scope.

In terms of access, both cases are gated housing type surrounded by high fences with 24-hours security guard. This condition reinforced our previous study which stated that most middle-high estates would have a restricted access entrance gate [15]. The difference between the two lies in the amount of housings access. Case 1 has 2 accesses with 24-hour guard that can be daily traversed and 2 others access that only opened at certain events. On the contrary, Case 2 has single access with 24-hour guard (Figure 2 and Figure 3). The street width in both cases already meets the minimum standards referring to researcher [18] even wider than the requirements stated in Ministerial Decree [13]. These conditions facilitate the evacuation vehicles access when disaster comes. The distance from the furthest unit to the housing gate is approximately 160 m in Case 1 whereas in Case 2 approximately 440 m. Referring to researcher's [7] statement, the occupants in Case 2 had longer time to rescue and evacuate than the Case 1 during a disaster emergency. According to Ministerial Regulation No. 8, year 2008, the distance of open space as the first evacuation place is more adequate in Case 1 than in Case 2. The distance from the furthest unit to the nearest open space in Case 1 is approximately 90 m while in Case 2 is approximately 279 m. When the fire or flood comes, the nearest open space is needed as a buffer as well as a place to put goods and tents.

Based on location, in Case 1, Neighbourhood Parks is scattered in several places with mostly triangular shapes surrounded by house units. In Case 2, there is only one Neighbourhood Park located in the middle of the housing. Scattered parks make it easier for residents to reach a refuge place from their units. In housing scope (in Case 2), the golf course that located next to the cluster provides residents an alternative place for refuge especially during fire. Provision of more and scattered open spaces in the housing serves as a refuge place during disaster. Viewed from ground level, all parks in

IOP Conf. Series: Earth and Environmental Science **106** (2017) 012021 doi:10.1088/1755-1315/106/1/012021

both cases are in a slightly higher place from the street. Referring to researched by [19], this condition has not qualified as a temporary evacuation site in case of flooding. Open space network in both cases showed that linkage among park, street, and river in Case 1 has a clear path that connects between open spaces and residential entrances.

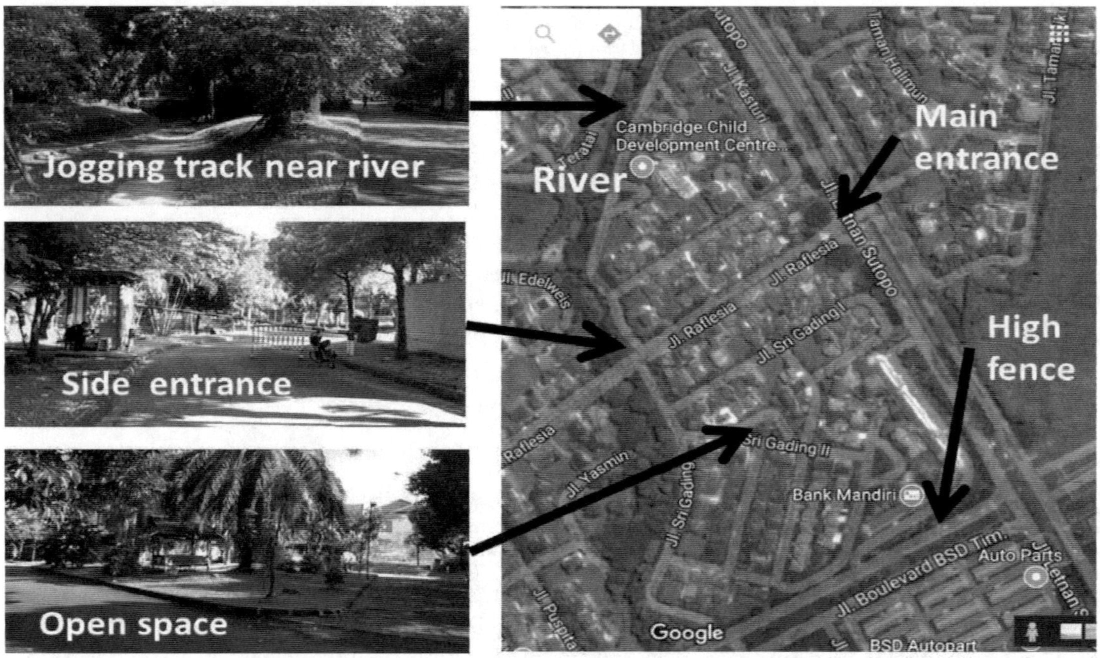

Figure 2. First case with double access and high fence [3].

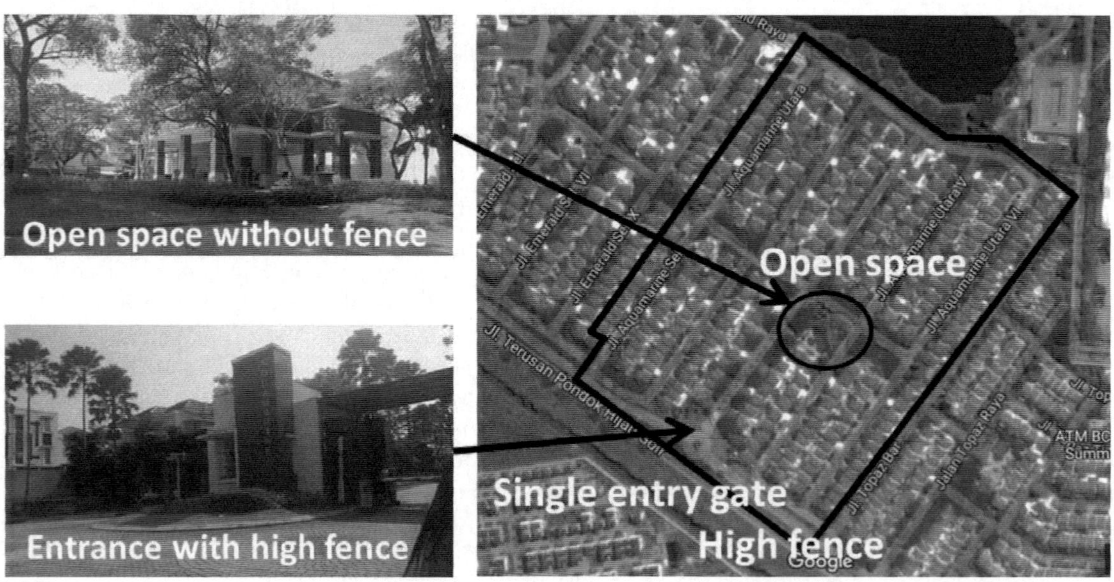

Figure 3. Second case with single access and high fence [3].

In relation to boundaries, open spaces in both cases have no divider to the street or to surrounding unit. When disaster strikes, those open spaces can easily accessible by occupants. Open spaces with no barrier in both cases facilitate resident access during disaster evacuation. Park territory is characterized by ground level that is reinforced by paving blocks around the park. These level differences are clearly seen in Case 2 with almost 50 cm height difference. Associated with fires, a more open space with few trees is much safer than the lush spaces. In Case 1, lush trees along the river may lead to a higher possible spread of fires. The more open the space, the better they reduces the possibility of fire spreading. Open spaces condition in both cases qualified Ministerial decree [13] that should be clean [16], accessible, and well maintained [17]. The occupant's involvement in both cases limited to open space management and construction related to daily life not entered disaster mitigation. Referring to researcher [18], community-based disaster mitigation programs need to be done to control the situation during disaster.

4. Conclusion

Learning from both cases that were middle-high gated housing type, there are 3 things to note related to disaster mitigation especially during floods and fires. First, gated community housing should consider the easily housing access. Housing should have more than 1 access although it can only be passed at a certain time. Secondly, the distance of the farthest unit to the housing gate should be considered to facilitate and accelerate the evacuation process. Thirdly, providing open spaces that is easily accessible from all units as the first evacuation site during and post disaster is necessary in housing estate. Scattered parks in several places gives greater possibilities for occupants as a refuge place rather than centralized park.

References

[1] UNU-EHS and Alliance Development Works. World risk report 2014 Focus: The city as a risk area (Berlin: Bündnis Entwicklung Hilft). 2014.
[2] Ika. Millions of Indonesians Live in Disaster Prone Areas https://ugm.ac.id/id/berita/13340 (Yogyakarta: Public relation UGM) in Bahasa Indonesia [accessed on February 21th. 2017].
[3] DIBI 2017. Indonesian Disaster Data and Information. Jakarta: National Disaster Management Agency. http://dibi.bnpb.go.id accessed on May 4th. 2017.
[4] Winandari M I R. Pemanfaatan Ruang Terbuka Publik di Perumahan dengan Tingkat Pendapatan yang Berbeda Studi Kasus: Perumahan Menengah Atas dan Perumahan Menengah Bawah di Yogyakarta. Dissertation in Bahasa Indonesia. Gadjah Mada University. Yogyakarta, Indonesia, 2015.
[5] Katsuhama Y. Capacity Building For Flood Management In Developing Countries Under Climate Change Dissertation. Colorado State University. Colorado, USA, 2010.
[6] Milanchi H, Rahimi M. 2014. The Role of Open Spaces in Reducing the Vulnerability of Urban Tissues Case Study: Region 17 of Tehran. *MAGNT Research Report*. 2 (7): 166-175.
[7] Mirza F. 2015. Disaster Mitigation Approach of Urban Green Structure Concept in Coastal Settlement Dimension. *Journal of Architecture and Built Environment*. 42(2): 51-58.
[8] Jayakody R R J C, Amarathunga D, Haigh R. 2016. The use of Public Open Spaces for Disaster Resilient Urban Cities. *12th International Conference of the International Institute for Infrastructure Resilience and Reconstruction*, University of Peradeniya, Sri Lanka.
[9] BPS. Statistical Yearbook of Indonesia 2016, Jakarta: BPS-Statistics Indonesia, 2016.
[10] Rosarians F, 2014, Rano Karno : Banten Terancam 14 Jenis Bencana (Jakarta: Tempo) https://m.tempo.co/read/news/2014/03/05/058559703 [accessed on May 4th 2017].
[11] Permen 05 Peraturan Menteri Pekerjaan Umum Nomor : 05/Prt/M/2008 Tentang Pedoman Penyediaan dan Pemanfaatan Ruang Terbuka Hijau di Kawasan Perkotaan (Jakarta: Direktorat Jenderal Penataan Ruang Departemen Pekerjaan Umum) in Bahasa Indonesia. 2008.
[12] Vargas J C, Meece B, Emperador S, 2014, A Framework for Using Open Green Spaces for Climate Change Adaptation and Resilience in Barranquilla, Colombia Resilient Cities – ICLEI http://www.geoadaptive.com [accessed on May 4th 2017].
[13] Kepmen "Keputusan Menteri Permukiman dan Prasarana Wilayah No. 534/KPTS/M/2001 tentang Pedoman Penentuan Standar Pelayanan Minimal Bidang Penataan Ruang, Perumahan dan Permukiman dan Pekerjaan Umum" (Jakarta: Kemen Kimpraswil). 2001.
[14] Perka "Peraturan Kepala Badan Nasional Penanggulangan Bencana Nomor 11 Tahun 2008

tentang Pedoman Rehabilitasi dan Rekonstruksi Pasca Bencana" (Jakarta: BNPB). 2008.

[15] Winandari M I R, Wibisono B H, Djunaedi A, Ahimsa-Putra H S. 2014. The Occupant's Perception Toward Open Space in Two Different Types of Housing. *Journal of Habitat Engineering and Design*. 6(1):71-82.

[16] Firman T, Surbakti I M, Idroes I C, Simarmata H A. 2011. Potential climate-change related vulnerabilities in Jakarta: Challenges and current status Habitat International. 35(2): 372-378.

[17] Mohsin A. 2015. Jakarta Under Water: The 2007 Flood and The Debate on Jakarta's Future Water Infrastructure. *Jurnal Wilayah dan Lingkungan*. 3(1): 39-57.

[18] Maskrey A. 1989. Disaster Mitigation: a Community Based Approach. (Development Guidelines No 3), Oxford: Oxfam.

[19] Saxena M R. 2016. Role of Open Spaces in Disaster Management. Conference Paper November 2016. https://www.researchgate.net/project/Role-of-Open-Spaces-in-Disaster-Mitigation [accessed on November 16th 2017]

[20] Bsdcity. 2016. BSD City. http://www.bsdcity.com [accessed on June 16th 2016].

[21] Summareconserpong. 2016. The Developer. http://www.summareconserpong.com [accessed on June 16th 2016].

Water quality study of Sunter River in Jakarta, Indonesia

Y Martinus, W Astono, D Hendrawan*

Environmental Engineering Department, Faculty of Architecture Landscape and Environmental Technology, Universitas Trisakti, Jakarta, Indonesia

*Corresponding Author: dianahendrawan@trisakti.ac.id

Abstract. Sunter River flows in the city of Jakarta with the designation of river water for agricultural purposes, and can be utilized for urban business and hydroelectric power industry. This study aims to determine the Sunter River water quality based on physical and chemical parameters. Water sampling was conducted 2 times which done in April and May with 5 sampling stations for measuring. The samples was analayzed in the laboratory according SNI methods for parameters BOD, COD, PO_4^{3-}, NO_3, Oil & Grease and Detergents. The quality status of Sunter River is determined by the Pollutant Index method. The results show that the water quality of Sunter River is influenced by organic parameter as dominant pollutant with COD concentration ranging from 48 mg/l – 182.4 mg/l and BOD concentration ranging from 14.69 mg/L – 98.91 mg/L. The Pollution Index calculation results show that the water quality status of Sunter River is moderate polluted with IP 6.47. The source of pollutants generally comes from the urban drainage channels, tributaries, and slaughtering industry. The results of this study expected to be use by the government to improve the water quality of Sunter River for better environment.
Keywords: water quality, pollution parameters, Sunter River, pollution index

1. Introduction

Rivers have been utilized by mankind for thousands of years to the extent that few of them are now in their natural condition [1]. The increase of human population and activities give impacts to the water quality. Although some water pollution may occur naturally, most water pollutants are the result of human activities [2]. Much of water pollution results from human activities such as agriculture, mining, manufacturing industries and poor management of urban waste [3]. Industrialization and urbanization can threaten river ecosystems by altering land use and generating waste, resulting in sustained problems [4]. Water quality degradation is one of the most serious of all environmental problems because it can affect human health and economic activities as well as biotic communities [5].

The Sunter River is one of 13 river rivers in Jakarta with main stream along 37 km and river basin (watershed) of 73.18 km2 [6]. The river's upstream is located in Cimpaeun Village, Depok City and downstream in Jakarta Bay, Tanjung Priok, North Jakarta. Based on [7] the Sunter River water designation is Class D, which is water that can be used for agricultural purposes, and can be utilized for urban enterprises, hydroelectric generating industries [7]. The Sunter River basin is very densely populated, and floods are frequent.

Based on description above, this study was conducted to to assess the quality of the Sunter River water by identify the type of pollutants entering the river and their impacts. The quality status of Sunter water river can be used to assess the water quality improvement program and might be considered for water resources management.

2. Reseach Method

2.1. Study Area and Sampling Stations

Sunter River under study has length of 10.4 km and located in the middle and downstream of the main river, which flows from East Jakarta to North Jakarta. To analyze the river quality, 5 sampling stations were determined for measuring. Sampling is done 2 times in April as rainy season and May as dry season. Samples were taken using Grab Sampling method (instantaneous example). Water sampling is done in the middle of the bridge above the river with bottled water sampler made of polyethylene plastic tied with rope. The water samples that have been taken are inserted into a 2L-sized plastic oxygen stored in a refrigerated foam box and later stored in the laboratory for analysis. The coordinates and elevation of the sampling point can be seen in Table 1 while Map of sampling location and Sunter River catchment area can be seen in Figure 1.

Table 1. Sampling point and coordinate.

Point	Location	Coordinate		Elevation	Distance
		E	S	(m)	(m)
1	Jl. I Gusti Ngurah Rai No.37, Klender, Duren Sawit, Kota Jakarta Timur	106°53'44.83"	6°12'48.37"	11 m	0
2	Jl. Jatinegara Kaum No.29, Jatinegara Kaum, Pulo Gadung, Kota Jakarta Timur	106°53'51.50"	6°12'9.95"	11 m	1.428
3	Jl. Palad, Pulo Gadung, Kota Jakarta Timur	106°54'18.60"	6°11'11.01"	6 m	3.915
4	Jl. Perintis Kemerdekaan No.65, Pulo Gadung, Jakarta Timur	106°53'50.47"	6°10'41.66"	6 m	5.460
5	Jl. Yos Sudarso, Klp. Gading Barat, Kota Jakarta Utara	106°53'26.34"	6° 8'44.66"	5 m	10.315

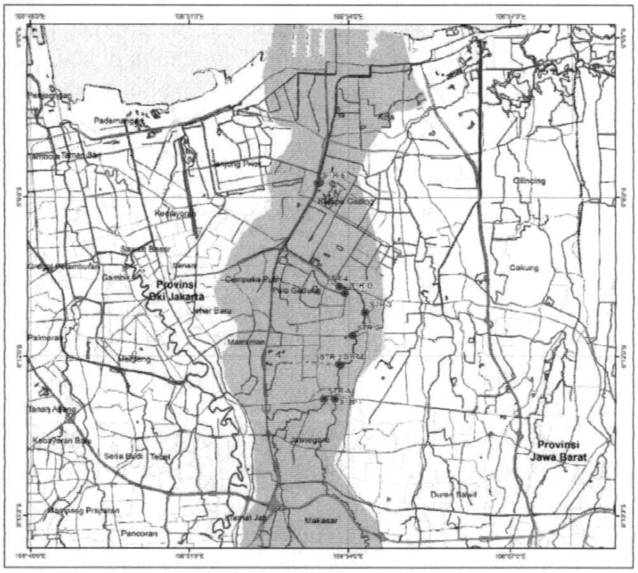

Figure 1. Map of sampling location and Sunter River Basin.

2.2. Laboratory Analysis

River water samples are analyzed in the laboratory with the parameters were TDS, BOD, DO, COD, PO_4^{3-}, NO_3, Grease Oil and Detergent. Standard analytical procedures were used in the determination of selected physical and chemical water quality parameters of the sample. Parameters analyzed and the method of analysis of the sample refers to the Indonesian National Standard which can be seen in Table 2. Results from the analysis of river pollutant parameters are then compared with river water quality standard in DKI Jakarta [7].

Table 2. Laboratorium method for surface water analysis.

No.	Parameter	Unit	Method
A	Physic		
1	Total Dissolved Solid (TDS)	mg/L	SNI. 06-6989.27-2005
2	Temperature	^oC	SNI. 06-6989.23-2004
B	Chemical		
3	Nitrate (NO_3)	mg/L	SNI. 6989.74:2009
4	pH	-	SNI. 06-6989.11-2004
5	Phosphat (PO4)	mg/L	SNI. 06-6989.31-2005
6	Grease and Oil	mg/L	SNI. 06-6989.10-2004
7	DO	mg/L	SNI. 06-6989.14-2004
8	BOD	mg/L	SNI. 06-2503-1991
9	COD	mg/L	SNI. 06-6989.2-2004
10	Detergent (MBAS)	mg/L	SNI. 06-6989.51-2005

2.3. Pollution Index

The results of laboratorium analyses were used to calculate the water quality status of the Sunter River using the pollutant index method. The Pollution Index (IP) is used to determine the level of pollution relative to the permitted water quality parameters [6]. The Pollution Index is determined for a designation, then it can be developed for some designation for all parts of the water body or part of a river. The equation for calculating the pollutant index value is as follows. The calculation results from equation below then matched to the scale in Table 3 to determine the quality status of Sunter River.

$$PIj = \left\{ \frac{\left(\frac{Ci}{Lij}\right)^2 M + \left(\frac{Ci}{Lij}\right)^2 R}{2} \right\}^{1/2} \quad (1)$$

Where Ci = Observed concentration of i parameter
Li = Permissible limit of i parameter
M = Maximum value of (Ci/Lij)
R = Average value of (Ci/Lij)

Table 3. Evaluation of Pollution Index Value.

No	Pollution Index Value	Water Criteria Status
1	**0 < IP < 1,0**	Meets Quality Standards (Good)
2	**1 < IP < 5,0**	Low Polluted
3	**5 < IP < 10**	Moderate Polluted
4	**IP > 10**	Severe Polluted

3. Results and Discussion

3.1. Sunter River Water Quality

Table 4. Sunter River water quality.

Parameter	Unit	WQS	Sampling Stations									
			1		2		3		4		5	
			April	May	April	May	April	May	April	May	April	May
Physical												
Air Temperature	°C		29	30	30	31	31	32	31	32	31	32
Water Temperature	°C		28,8	28,1	29	29,2	30,2	30,1	30,3	30,2	30,3	31
Total Dissolved Solid	mg/L	200	305	316	303	318	306	302	292	307	324	334
Conductivity	μmhos/cm	1000	606	631	603	633	613	605	585	615	649	667
Anorganic Chemical												
pH	mg/L	6 – 8.5	7.4	7.16	7.27	7.25	7.34	7.31	7.24	7.32	7.57	7.35
BOD (5 day 20°C)	mg/L	20	98.91	31.05	79.9	14.69	76.13	27.91	69.5	33.66	86.63	36.46
COD	mg/L	30	182.4	64	154	70.4	134	73.6	134	54.4	144	48
Dissolved Oxygen	mg/L	>3	5.33	2.30	6.10	4.50	5.60	3.90	4.20	4.30	2.77	3.03
Phospate (PO_4)	mg/L	0.5	5.43	8.36	12.27	7.05	3.59	4.74	13.57	5.62	15.42	5.30
Nitrate (NO_3)	mg/L	10*	0.037	0.187	0.013	0.081	0.013	0.029	0.026	0.124	0.015	0.138
Organic Chemical												
Oil and Grease	mg/L	1**	1.324	0.145	0.283	0.435	0.670	1.013	0.781	1.166	2.100	1.422
Detergent	mg/L	0.5	8.15	2.31	8.53	6.20	6.50	3.10	2.34	1.51	7.72	5.96

*) Class B Water Standard **) Class C Water Standard
WQS = Water Quality Standard (Decree of the Governor of DKI Jakarta Number 582 of 1995)

The results of laboratory analysis showed that anorganic chemical parameters, such as BOD_5, COD, and Phosphate have far exceeded the quality standard of Decree of akarta Governor Number 582 of 1995 for river water of Class D while the physical parameters such as water temperature and water conductivity meet the quality standard.

The highest BOD_5 value was recorded during rainy season (April) with concentration of 98.91 mg/l at Point 1. The mean BOD_5 concentration during dry season (May) was 28.75 mg/L. The BOD_5 value is directly proportional to the COD value, with COD always greater than BOD_5. The concentration of COD in Sunter River were 48 mg/L – 182.4 mg/L in all season. The high concentrations of BOD_5 and COD in Sunter River indicates organic waste as the dominant pollutant. Organic waste in Sunter River one of them comes from residential area that most likely directly dispose grey water into the river. COD concentrations generally indicate that levels of pollution in water bodies are high [8].

In the rainy season (April), Dissolved Oxygen (DO) of Sunter River were 2.77 mg/L - 6.10 mg/L whereas in dry season, DO values range from 2.30 mg/L to 4.50 mg/L. Overall the DO content in Sunter River still meets the quality standard (> 3 mg/L).

3.2. Sunter River Quality Status

The quality status of Sunter River water is calculated based on the results of laboratory analysis of Sunter River water. Determination of status of water quality using method of pollution index (IP) attached in Decree of Environment Minister Indonesia Number 115 Year 2003. Result of calculation status of water quality of Sunter River period of April, May, and July can be seen in following table.

Table 5. Pollution index of Sunter River at every point.

Point	Location	Polution Index		Average	Status
		April	Mei		
1	I Gusti Ngurah Rai	5.96	5.41	5.69	Moderate Polluted
2	Jl Jatinegara Kaum	6.91	5.38	6.15	Moderate Polluted
3	Jl Palad, Pulo Gadung	5.25	4.50	4.88	Moderate Polluted
4	JL Perintis Kemerdekaan	9.18	4.70	6.94	Moderate Polluted
5	Jl. Yos Sudarso, Kelapa Gading	12.44	5.01	8.72	Moderate Polluted

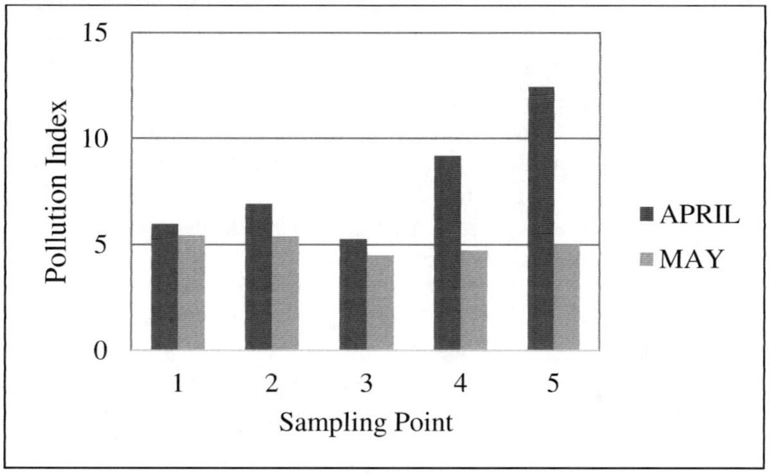

Figure 2. Pollution index graphs of Sunter River.

The Pollution Index (PI) result shows that the average status of the Sunter River is moderate polluted with average PI of all point 6.47. The calculation results may change if the water quality parameter under study increases. The Advantages of PI method is the result obtained determined based on the unit data and the water quality status at each point can be determined. However, the lack of using this PI method is the result obtained less sensitive because it depends on the number of parameters studied.

Water quality status obtained is the result of instantaneous measurement in accordance with the measurement water flow for 2 months. The discharge of water entering the Sunter River can affect the water quality and the Pollution Index. The results of calculations with the Pollution Index may change if the water quality parameters studied increase. Determination of water quality is important to assess the state of the watershed and to make the management decisions necessary to control current and future pollution of water bodies [9].

4. Conclusion

The Sunter River water quality parameters have exceeded the quality standards for BOD, COD, PO_4^{3-}, TDS, and (detergent) surfactant anionic parameters. The dominant pollutants entering the Sunter River are organic waste that may come from residential, traditional market, and chicken slaughtering house. The water quality status of the Sunter River during 2 months of monitoring was in the medium polluted category with an average IP of 6.47. The water quality status of the Sunter River in May (dry season) is better than April (rainy season). Therefore, this research can be used as a guideline for the government in improving the quality of River Sunter's water.

References

[1] Ngoye E, and Machiwa, J F. 2004. The influence of land-use patterns in the Ruvu River watershed on water quality in the River system, *Physics and Chemistry of the Earth*. 29(15-18):1161-1166.

[2] Daniels T, and Daniels K, 2003, The Environmental Planning Handbook for Sustainable Communities and Regions, Planners Press, American Planners Association Chicago.

[3] Masere T P, Munodawafa A, and Chitata T. 2012. Assessment of human impact on water quality along Manyame River. *International Journal of Development and Sustainability*. 1(3):754-765.

[4] Shen Y, Cao H, Tang M, and Deng H. 2017. The Human Threat to River Ecosystems at the Watershed Scale: An Ecological Security Assessment of the Songhua River Basin. Northeast China. *Water 2017*. 9. 219.

[5] Turpie J, and van Zyl H. 2002. Valuing the environment in water resources management, on Hirji R, Johnson P, Maro P. and Matiza-Chiuta, T. (Eds), Defining and Mainstreaming Environmental Sustainability in Water Resources Management in Southern Africa. SARDC/World Bank: Maseru/Harare/Washington DC, pp 85-110.

[6] Badan Pengelola Lingkungan Hidup Daerah Provinsi DKI Jakarta. 2015. Laporan Pelaksanaan Pemantauan Kualitas Sungai 2015. Jakarta: Pemerintah Provinsi DKI Jakarta

[7] Keputusan Gubernur DKI Jakarta Nomor : 582 Tahun 1995 Tanggal : 12 Juni 1995 tentang Penetapan Peruntukan dan Baku Mutu Air Sungai / Badan Air Serta Baku Mutu Limbah Cair di Wilayah Daerah Khusus Ibukota Jakarta

[8] Waziri M, and V O Ogugbuaja. 2010. Interrelationship between physicochemical water pollution indicators: a case study of River Yobe-Nigeria. *Journal of Scientific & Industrial Research*. 1:76–80.

[9] Khadam I M, & Kaluarachchi J J. 2006. Water quality modelling under hydrologic variability and parameter uncertainty using erosion-scaled export coefficients. *Journal of Hydrology*. 330(1-2):354-367.

Study of pollution effect on water quality of Grogol River, DKI Jakarta

S Amira, W Astono, and D Hendrawan*

Environment Engineering Department, Faculty of Landscape Architecture and Environmental Technology, Universitas Trisakti, Jakarta

*Corresponding Author : dianahendrawan@trisakti.ac.id

Abstract. A study has been conducted to identify the incoming pollutants and assess the water quality in Grogol River, DKI Jakarta, Indonesia, which has a length of 13.35 km and consists of two segments. The water quality assessment is determined by pollution index method, referring to Minister of Environment Decree No. 15/2013 on The Guidelines of Water Quality Status. The samples were taken both in rainy and dry seasons at 7 sampling points. Based on the analyses of 10 key parameters and the calculation of pollution index value, it can be concluded that Grogol River is low polluted in rainy season and moderate polluted in dry season. The information obtained from this research can be used for decision making to improve the water quality of Grogol River.
Keywords: Grogol River, pollutant index, water quality assessment

1. Introduction

Rivers in underdeveloped and developing countries are widely used as waste disposal sites for domestic and industrial waste. The waste contains various types of contaminants that contribute to the poor quality of river water [1].

Grogol River is one of the rivers that flow past DKI Jakarta (capital city of Indonesia). The upstream of this river is located in the area of Pancoran Mas, Depok City, West Java Province. The focuse of the research conducted at the downstream segment located in DKI Jakarta Province, more precisely in South Jakarta City, Central Jakarta City and West Jakarta City.

Based on the sources, there are two kinds of pollutants defiling the river, i.e., point source and non-point source. Point source pollution is water contamination that comes from a single place, usually like a pipe. Non-point source pollution is the scouring and leaching caused by rainfall-runoff processes, dissolved or solid pollutants on the ground and in the soil entering rivers, lakes, reservoirs, and the ocean, tainting the aquatic environment [2].

Based on the above description, this research needs to be done to identify pollutants entering the Grogol River and to see the quality status of the grogol river so that it can be used as a reference to improve the water quality of Grogol River.

2. Research Method

The research area of Grogol River is 13.35 km. The samplings were done at 7 points of main river point and 2 point from point source of Grogol River which are a tributary and a drainage. Sampling was conducted during the rainy and dry seasons (Table 1). The sampling point is determined by the accessibility of the location and the pollutant input location.

Table 1. Coordinates of the sampling locations of Grogol River.

Sampling Points	Types	Locations	Coordinates
1	Main Point	Jl. Kyai Moh. Syafii Hadzami	6°14'444.3"S 106°47'16.7"E
2	Main Point	Jl.Hang Lekir I	6°13'45.3"S 106°47'45.0"E
3	Main Point	Jl. Patal Senayan	6°13'7"S 106°47'49"E
4	Main Point	Jl. Palmerah Barat	6°12'28.2"S106°49'32.6"E
5	Main Point	Jl. Anggrek Rosaliana	6°11'42.9"S 106°47'37.2"E
6	Main Point	Jl. Letjen S. Parman (in front of CP Mall)	6°10'29.6"S 106°47'27.9"E
7	Main Point	Jl. Letjen S. Parman (in front of Ciputra Mall)	6°10'04.5"S 106°47'16.5"E

The methods used to analyze ten key parameters of Grogol River water samples are listed in Table 2. Two quality standards were used as references, i.e., Governor of DKI Jakarta Decree (KEPGUB) No. 582/1995 and Government Regulation (PP) No. 82/2001. The latter is used as the secondary guide if the parameter is not registered in the Governor Decree.

Table 2. Measurement methods of water quality parameters.

No.	Water Quality Parameters	Units	Analytical Methods	Standard Quality Types
A	**Physical Parameters**			
1	Total Dissolved Solids (TDS)	mg/L	SNI 06-6989.27-2005	KEPGUB DKI Class D
2	Conductivity	μmhos/cm	SNI 06-6989.11-2004	KEPGUB DKI Class D
3	Temperature	°C	SNI 06-6989.23-2004	KEPGUB DKI Class D
B	**Chemical Parameters**			
4	Nitrate (NO_3)	mg/L	SNI 6989.74:2009	PP No. 82/2012 Class 4
5	pH	-	SNI 06-6989.11-2004	KEPGUB DKI Class D
6	Phospat (PO_4)	mg/L	APHA Method 4500-P	KEPGUB DKI Class D
7	Grease and Oil	mg/L	SNI 06-6989.10-2004	PP No. 82/2001 Class 3
8	Dissolved Oxygen (DO)	mg/L	SNI 06-6989.14-2004	KEPGUB DKI Class D
9	Biochemical Oxygen Demand (BOD)	mg/L	SNI 06-2503-1991	KEPGUB DKI Class D
10	Chemical Oxygen Demand (COD)	mg/L	SNI 06-6989.2-2004	KEPGUB DKI Class D
11	Metylene Blue Active Substance (detergent)	mg/L	SNI 06-6989.51-2005	KEPGUB DKI Class D

The pollution index (PI) method was used to calculate the quality status of the river. As an index-based approach, this method is built on two quality indexes. The first is the average index (I_R) that shows the average pollution level of all parameters in one observation, while the second is the maximum index

(I_M) that indicates one type of dominant parameters causing a decrease in water quality at one observation [3]. Pollution index are calculated using equation 1.

$$PI_j = \sqrt{\frac{(Ci_i/L_{ij})_M^2 + (Ci_i/L_{ij})_R^2}{2}} \quad (1)$$

Where, Ci : Observed concentration of i parameter
Lij : Permissible limit of i parameter
M : Maximum value of (Ci/Lij)
R : average value of (Ci/Lij)

The categories of water quality assessment based on pollution index value are as follows [4]:
1) Meet quality standards: IP≤1
2) Low Polluted: 1<IP≤5
3) Moderately Polluted: 5<IP≤10
4) Heavily Polluted: IP>10

3. Results and Discussion

3.1. Results of water quality measurement of Grogol River
The result of two water quality measurements of Grogol River, i.e., in the rainy season (RS) and the dry season (DS) shown at Table 3.

Water temperature of Grogol River is ranging from 28 °C - 30 °C. It can be considered in good condition because do not have much differences compared to ambient temperature which is ranging from 27 °C - 30 °C. Water Temperature is basically important because it give effects to aquatic organism [5].

The result of TDS and conductivity of Grogol River also can be considered in good condition. TDS value ranging from 139 mg/L – 245 mg/L (RS) and 241 mg/L – 287 mg/L (DS). Conductivity value ranging from 278 µhos/cm – 490 µhos/cm (RS) and 481 µhos/cm – 575 µhos/cm (DS). Governor of DKI Jakarta Decree No. 582/1995 for Class D requires TDS <200 mg/L and conductivity <1000 µhos/cm and the result from 7 sample points both rainy season and dry season still not exceed the standard.

pH value of Grogol River showed no significant diference between each sampling point (both rainy season and dry season) and the value still fulfill the water quality standard from Governor of DKI Jakarta Decree No. 582/1995 for Class D. NO_3 value of Grogol River also shown no significant diference between each sampling point (both rainy season and dry season). NO_3 concentration is ranging from 0 mg/L – 0.02 mg/L (RS) and 0.01 mg/L – 0.03 mg/L (DS). Those concentration stil not exceed water quality standard by Government Regulation (PP) No. 82/2001, Class 4.

Dissolved oxygen of Grogol River ranged from 2.2 mg/L – 4.9 mg/L (RS) and 2.23 mg/L -3.35 mg/L (DS). In Rainy Season, 2 sampling point has DO concentration below the standard (>3 mg/L) and increased in dry season to 4 sampling point. Low DO concentration can be caused by the organic materials that pollute the river. Oxygen will be consumed to decompose organic material.

Oil and grease concentration of Grogol River ranging from 0.24 mg/L – 0.78 mg/L (RS) and 0.13 mg/L -0.9 mg/L (DS). Oil which forms a surface film on the river can coat plants and animals reducing oxygenation from the atmos-phere above. The film of oil that floats over the water body affects the transmission of light through the water body there by disturbing the process of photosynthesis in the aquatic plants [6]. Governor of DKI Jakarta Decree No. 582/1995 for Class D requires oil and grease <1 mg/L and the result both rainy season and dry season still not exceed the standard.

IOP Conf. Series: Earth and Environmental Science **106** (2017) 012023 doi:10.1088/1755-1315/106/1/012023

Table 3. Measurement results of water quality parameters in the rainy season and dry season.

Parameter	Unit	Standard	Point													
			1		2		3		4		5		6		7	
			RS	DS	RS	DS	RS	DS	RS	DS	RS	DS	RS	DS	RS	DS
Physic																
Temperature	°C	±3	29	28	29	28	29	29	29	29	29	29	29	29	30	29
TDS	mg/L	200	139	241	179	260	191	245	211	259	238	247	231	287	245	269
Conductivity	µhos/cm	1000	278	481	358	520	381	489	422	499	476	495	462	575	490	545
Anorganic Chemical																
pH		6 - 9	7.01	7.26	7.19	6.50	6.98	7.10	7.04	7.00	7.05	7.00	7.09	7.30	7.14	7.40
DO	mg/L	>3	3.40	3.25	4.90	2.23	4.70	2.30	3.95	4.50	2.57	2.80	4.85	3.55	2.20	2.70
BOD_5	mg/L	20	26.54	9.30	15.81	26.26	54.16	52.64	41.16	28.14	53.20	13.07	54.06	24.37	41.98	10.89
COD	mg/L	30	61.34	43.2	68.16	73.60	74.98	72.00	78.38	64.80	102.24	64.80	88.61	60.00	95.42	52.80
Total Phosphate	mg/L	0.5	1.18	3.76	8.85	10.45	6.04	8.76	10.05	6.72	2.59	13.16	1.92	12.82	2.52	11.61
NO_3	mg/L	10*	0.02	0.01	0.02	0.03	0.00	0.03	0.00	0.02	0.01	0.01	0.00	0.01	0.00	0.01
Organic Chemical																
Oil and Grease	mg/L	1**	0.79	0.13	0.78	0.43	0.64	0.85	0.37	0.85	0.24	0.54	0.61	0.67	0.41	0.90
Detergent (MBAS)	mg/L	0.5	1.33	1.04	1.06	2.25	1.38	2.50	1.03	2.58	1.63	3.90	4.23	2.93	3.19	2.70

Note: * Government Regulation (PP) No. 82/2001, Class 4

** Governor of DKI Jakarta Decree (KEPGUB) No. 582/1995, Class C

Pollutants that entering the Grogol River are from activities near Grogol River. The pollutants dominated by domestic waste. It is proven from the high concentration of phospate and detergent parameter. Phospate ranging from 1.18 mg/L – 10.05 mg/L (RS) and 3.76 mg/L -13.16 mg/L (DS) and detergent ranging from 1.03 mg/L – 4.23 mg/L (RS) and 1.204 mg/L -3.90 mg/L (DS). Domestic and industrial wastewater such as from laundry and dry cleaning services release high concentration of phosphates through detergents used into the land and water environment [7].

BOD and COD concentration have a straightforward value. Grogol River BOD concentration ranged from 15.81 mg/L – 54.16 mg/L (RS) and 9.3 mg/L – 52.64 mg/L (DS). COD ranged from 61.34 mg/L – 102.24 mg/L (RS) and 43.2 mg/L – 72 mg/L (DS). In the dry season, almost all parameters have increased concentration (except BOD and COD). The BOD value in rivers often increases during periods of heavy rain and high river flows as organic matter is washed in from the land [8].

3.2. Grogol river quality status by pollution index method

Based on the result of calculation of quality status, Grogol River has more moderate polluted status at its monitoring points (can be seen at Figure 1 and Table 4). Constantly, it appears that the quality status of point 1 is always in low polluted state and points 2 and 4 are constant in moderate polluted conditions. The highest Pollution Index value are at the point 7 in Dry Season in the amount of 5.889 and the lowest PI value are at point 1 in Rainy season in the amount of 2.369.

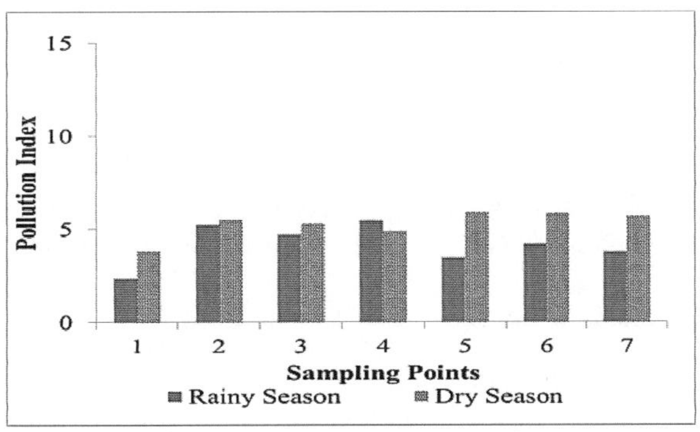

Figure 1. The pollution index value of Grogol River.

Table 4. The quality status of Grogol River by pollution index method in RS and DS.

Points	Rainy Season	Dry Season	Mean	Quality Status
1	2.369	3.896	4.098	**Low polluted**
2	5.234	5.537	5.617	**Moderate polluted**
3	4.702	5.310	5.108	**Moderate polluted**
4	5.467	4.881	5.462	**Moderate polluted**
5	3.465	5.900	5.099	**Moderate polluted**
6	4.203	5.861	5.375	**Moderate polluted**
7	3,785	5,690	5,200	**Moderate polluted**
Mean	4.175	5.295		
Quality Status	**Low polluted**	**Moderate polluted**		

Based on the data listed in tables above, it can be seen that during Rainy season with the pollution index mean value 4.175, Grogol River is polluted and the dry season with the pollution index mean value 5.295, Grogol River is moderate polluted.

4. Conclusion

The primary pollutants of Grogol River are COD, BOD_5, phospate and detergent. Those four parameters almost in all sampling points pass the quality standard both in the rainy and dry season. The pollutant index (IP) value of Grogol River range are 2.369—5.890 from both rainy season and dry season and Grogol River quality status are dominated by moderate polluted status at dry season and dominated by low polluted status at rainy season.

References

[1] Aisien F A, Aisien E T, and Shaka F. 2003. The effects of rubber factory effluent on Ikpoba river. Nigeria. *Journal of Biomedical Engineering.* 2(1):32-35.

[2] Long-xi HAN, Fei HUO, Juan SUN. 2011. Method for calculating non point source pollution distribution in plain rivers. *Water Science and Engineering.* 4(1):83-91. http://www.sciencedirect.com/science/article/pii/S1674237015301447.

[3] Mraganingrum D, Roosmin D, Pradono, and Sabar A, Diferensiasi Suber Pencemar Sungai Menggunakan Pendekatan Metode Indeks Pencemar (IP) (Studi Kasus: Hulu DAS Citarum), 2013.

[4] Keputusan Menteri Negara Lingkungan Hidup No. 115 Tahun 2003 tentang Pedoman Penentuan Status Mutu Air (Lampiran II). Ris.Geo.Tam Vol. 23, No.1, Juni 2013 (37-48)

[5] Saksena DN, Garg RK, Rao RJ. 2008. Water quality and pollution status of Chambal River in National Chambal Sanctuary, Madhya Pradesh. *Journal of Environmental Biology.* 29(5): 701-10.

[6] Lokhande, Ram S, Pravin U S. 2011. Study on Physico-Chemical Parameters of Waste Water Effluents from Taloja Industrial Area of Mumbai, India. *International Journal of Ecosystem.* 1(1):1-9

[7] Adesoye A M, Olayinka K, Olukomaiya O O, Iwuchukwu P O. 2014. The removal of phosphates from laundry wastewater using alum and ferrous sulphate as coagulants. *International Journal of Innovation and Scientific Resarch.* 8(2):256 – 260

[8] EPA. 2015. Integrated Water Quality Report 2012 - Monaghan & Louth. https://www.epa.ie/pubs/reports/water/waterqua/iwqmolou/

Study of noise level at Raja Haji Fisabilillah airport in Tanjung Pinang, Riau Islands

H Nofriandi, A Wijayanti, M F Fachrul*

Environmental Engineering Department, Faculty of Landscape Architecture and Environmental Technology, Universitas Trisakti, Jakarta, Indonesia

*Corresponding Author: melati@trisakti.ac.id

Abstract. Raja Haji Fisabilillah International Airport is a central airport located in Kampung Mekarsari, Pinang Kencana District, Tanjung Pinang City, Riau Islands Province. The aims of this study are to determine noise level at the airport and to calculate noise index using WECPNL (Weighted Equivalent Continuous Perceived Noise Level) method. The method using recommendations from the International Civil Aviation Organization (ICAO), the measurement point is based on at a distance of 300 meters parallel to the runway, as well as 1000 meters, 2000 meters, 3000 meters and 4000 meters from the runway end. The results at point 3 was 75.30 $dB_{(A)}$. Based on the noise intensity result, Boeing aircraft 737-500 was considered as the highest in the airport surrounding area, which is 95.24 dB(A) and the lowest was at point 12 with a value of 37,24 dB(A). Mapping contour shows that 3 areas of noise and point 3 with 75,30 $dB_{(A)}$ were considered as second level area and were complied to the standard required.
Keywords: noise level airport, noise contour WECPNL method

1. Introduction

The plane is one of the most popular modes of transportation today. The shortest reachable distance and the facilities offered also make the plane into one of the alternatives transportation at this time. The airport is a gateway to connect economic centres, tourist areas, and government centres [1]. Activities in the airport area, among others, is the administrative activities of the airport management headquarters Raja Haji Fisabilillah PT. Angkasa Pura II, activities around the runway in the form of fire stations, shuttle passengers from the terminal to the plane, delivery facilities passenger plane and take off and the landing of the aircraft.

The noise is the physical environmental factors that can disrupt the work or activity in both in terms of health with high noise levels for workers who are exposed to such noise exposure. Humans who received the noise in lasts for long time, it should be controlled or prevented [2], [3], [4]. Noise with a high enough level above 70 dB can cause anxiety, discomfort, hearing problems and narrowing of blood vessels. If, the noise level above 80 dB can lead to serious deterioration in a person's health in general and if it lasts long may cause temporary or permanent loss of hearing [5], [6].

2. Research Method

2.1. Time and Location

The study was conducted in October–November 2016, sampling data was conducted every 14 hours/day in runway and settlements around the airport.

2.2. Data collection

The primary data obtained in the field by measured the noise using the Sound Level Meter for 14 hours during 07.00 am - 21.00 pm. Data was collected when activities the aircraft as take off and landing. Meanwhile, the secondary data collected include frequency data of aircraft movement, type of aircraft, layout of airport and total data of passengers in 2015-2016 coming from Raja Haji Fisabilillah Airport management agency, Tanjung Pinang. The flight schedule show in Table 1.

Table 1. Flight schedule.

No	Name of Airlines	Departure Schedule	Arrival Schedule
1	Garuda Indonesia	12:10	13:20
2	Lion Air	10:00	10:35
3	Nam Air	19:00	07:00

2.3. Noise measuring point

The measurement point is based on recommendations from the International Civil Aviation Organization (ICAO) [8] at a distance of 300 meters parallel to the runway, as well as 1000 meters, 2000 meters, 3000 meters and 4000 meters from the runway end. In detail the location of the measurement point show in Table 2.

Table 2. Noise measuring point.

No.	Sampling Points Location	Distance from Runway (m)	Coordinate
1	Tower ATC Raja Haji Fisabilillah Airport	300	0°55'12.74"N; 104°31'39.74"E
2	Airport surveillance radar AIR NAV Raja Haji Fisabilillah airport	300	0°55'28.08"N 104°31'53.28"E
3	Runway-22 at Raja Haji Fisabilillah Airport	300	0°55'40.87"N 104°32'05.26"E
4	at Rajawali street, Kampung Sumber Rejo, Kelurahan Pinang Kencana	300	0°55'31.90"N 104°32'15.33"E
5	at Rajawali street, Kampung SumberRejo Kelurahan Pinang Kencana	300	0°55'17.48"N 104°32'02.86"E
6	at Rajawali, Kampung Sumber Rejo Kelurahan Pinang Kencana	300	0°55'01.88"N 104°31'50.05"E
7	at Merpati street, Kampung Sidojadi, Kelurahan Batu IX	1000	0°54'31.24"N 104°31'10.29"E
8	at Radar street, Kampung Bangun Rejo, Kelurahan Batu IX	2000	0°54'12.94"N 104°30'52.55"E
9	at Karya street, Kampung Sido Jasa, Kelurahan Batu IX	3000	0°53'48.38"N 104°30'31.19"E
10	at Garuda street, Kelurahan Batu IX	4000	0°53'31.86"N 104°30'12.99"E
11	*Obstacle* Runway-22	750	0°56'05.28"N 104°32'38.74"E
12	at Tirta Madu street, Kelurahan Batu XIV	2500	0°56'25.39"N 104°32'55.46"E

2.4. Noise measurement method

The noise level at Raja Haji Fisabilillah Airport was measured using Sound Level Meter (SLM). Measurements were conducted using a tripod with a height of 1.20 meters from ground. Data were taken when the plane movement that landing and take off. The instrument was used show in Table 3.

Table 3. Measuring equipment.

No	Equipment	Figure	Function
1	*Sound Level Meter* (SLM)		Measure the level of sound violence or to search for a sound source
2	Tripod		to lay a sound level meter
3	GPS		To determine the coordinates of the sampling point

2.5. Method of calculating the noise index at the airport

Noise calculation used by Weightened Continuous Perceived Noise Level (WECPNL) method. This method is based on the standards of the International Civil Aviation Organization (ICAO).

$$\overline{dB(A)} = 10\,log\left[\frac{10^{\frac{L1}{10}} + 10^{\frac{L2}{10}} + 10^{\frac{L3}{10}} + \ldots\ldots + 10^{\frac{Ln}{10}}}{n}\right] \qquad (1)$$

Description:

$\overline{dB(A)}$: Average Noise Intensity
L : The Noise Value at the Time of plane activities
N : Number of Aircraft

After obtained the noise intensity value of the plane from each point, then the calculation to measure the noise index at the airport is formulated in the following equation:

$$WECPNL = \overline{dB(A)} + 10logN - 27 \qquad (2)$$

Description:

N : Number of arrivals and departures in 24 hours
$\overline{dB(A)}$: Average Noise Intensity

3. Results and Discussion

3.1. The Noise Intensity

In Table 4 shows that the highest noise intensity is at point 3 and 4. This is due to the 3 and 4 points are the starting point of the aircraft take off.

Tabel 4. Maximum and minimum noise measurement results by distance.

No	Sampling Points Location	Distance from Runway (m)	Noise Level dB(A)	
			Min	Max
1	Tower ATC Raja Haji Fisabilillah Airport	300	89.42	90.60
2	Airport surveillance radar AIR NAV Raja Haji Fisabilillah airport	300	91.97	93.20
3	Runway-22 at Raja Haji Fisabilillah Airport	300	93.80	95.24
4	at Rajawali street, Kampung Sumber Rejo, Kelurahan Pinang Kencana	300	82.90	88.50
5	at Rajawali street, Kampung SumberRejo Kelurahan Pinang Kencana	300	81.50	86.20
6	at Rajawali, Kampung Sumber Rejo Kelurahan Pinang Kencana	300	78.39	81.61
7	at Merpati street, Kampung Sidojadi, Kelurahan Batu IX	1000	88.16	89.20
8	at Radar street, Kampung Bangun Rejo, Kelurahan Batu IX	2000	88.76	93.43
9	at Karya street, Kampung Sido Jasa, Kelurahan Batu IX	3000	79.39	86.37
10	at Garuda street, Kelurahan Batu IX	4000	78.12	84.65
11	*Obstacle* Runway-22	750	43.06	47.77
12	at Tirta Madu street, Kelurahan Batu XIV	2500	37.24	41.82

Refer to Table 4, then can determined the intensity of each plane noise at all sampling point. Following the noise intensity of the plane at a distance can be seen in Table 5.

Tabel 5. Table aircraft noise intensity of each point.

No	Aircraft Noise intensity of each point dB(A)											
	1	2	3	4	5	6	7	8	9	10	11	12
1	90.60	93.19	95.24	88.50	86.20	81.61	88.56	93.43	79.39	79.16	47.77	41.82
2	90.10	92.16	94.35	82.90	84.11	80.30	89.20	88.76	84.53	84.65	42.91	38.76
3	89.42	91.97	93.80	86.73	81.50	78.39	88.16	90.83	86.37	78.12	43.06	37.24

3.2. Average noise intensity

Raja Haji Fisabilillah Tanjungpinang Airport can serve the arrival and departure using Boeing Aircraft approximately 6 times a day. The intensity of noise generated by 3 aircraft units has an average intensity value at all sampling points. So, at each point it has the average noise intensity generated by all types of aircraft. The following Table 6 shows the intensity of the aircraft noise, the highest average noise intensity is at point 3 and 4 which is the initial position of the aircraft to take off. More over, the lowess noise intensity obtained at obstacle runway-22 area and Tirta Madu Street where no activity take off and landing around there. The take off and landing only at one way of the runway.

3.3. The WECPNL index (ICAO)

The frequency of flights in one day occurred between the hours of 07:00 to 21:00. At the time of the study obtained the number of aircraft movements (N) in one day (14 hours). Total N that uses runways is 6 movements within 14 hours, be it commercial aircraft or cargo planes. The assumption of any aircraft movement exposes the noise for 60 seconds based on each aircraft's movement per day at each point. The following Table 7 show the The highest WECPNL index [9] where the highest WECPNL index obtained at Obstacle runway and Tirta Madu street that the area not influence the aircraft activity.

Tabel 6. Average noise intensity.

No.	Sampling Points Location	Distance from Runway (m)	dB(A)
1	Tower ATC Raja Haji Fisabilillah Airport	300	90.10
2	Airport surveillance radar AIR NAV Raja Haji Fisabilillah Airport	300	92.47
3	Runway-22 at Raja Haji Fisabilillah Airport	300	94.50
4	at Rajawali street, Kampung Sumber Rejo, Kelurahan Pinang Kencana	300	86.61
5	at Rajawali street, Kampung Sumber Rejo Kelurahan Pinang Kencana	300	84.34
6	at Rajawali, Kampung Sumber Rejo Kelurahan Pinang Kencana	300	80.30
7	at Merpati street, Kampung Sidojadi, Kelurahan Batu IX	1000	88.66
8	at Radar street, Kampung Bangun Rejo, Kelurahan Batu IX	2000	91.43
9	at Karya street, Kampung Sido Jasa, Kelurahan Batu IX	3000	84.28
10	at Garuda street, Kelurahan Batu IX	4000	81.65
11	*Obstacle* Runway-22	750	45.21
12	at Tirta Madu street, Kelurahan Batu XIV	2500	39.70

Tabel 7. WECPNL index (ICAO) at each location

No	Sampling Points Location	Distance from Runway (m)	Noise dB(A)		WECPNL
			Min	Max	
1	Tower ATC Raja Haji Fisabilillah Airport	300	89.42	90.60	70.90
2	Airport surveillance radar AIR NAV Raja Haji Fisabilillah Airport	300	91.97	93.19	73.25
3	Runway-22 at Raja Haji Fisabilillah Airport	300	93.80	95.24	75.30
4	at Rajawali street, Kampung Sumber Rejo, Kelurahan Pinang Kencana	300	82.90	88.50	67.42
5	at Rajawali street, Kampung Sumber Rejo Kelurahan Pinang Kencana	300	81.50	86.20	65.12
6	at Rajawali, Kampung Sumber Rejo Kelurahan Pinang Kencana	300	78.39	81.61	61.10
7	at Merpati street, Kampung Sidojadi, Kelurahan Batu IX	1000	88.16	89.20	69.44
8	at Radar street, Kampung Bangun Rejo, Kelurahan Batu IX	2000	88.76	93.43	72.21
9	at Karya street, Kampung Sido Jasa, Kelurahan Batu IX	3000	79.39	86.37	65.06
10	at Garuda street, Kelurahan Batu IX	4000	78.12	84.65	62.43
11	*Obstacle* Runway-22	750	43.06	47.77	26.00
12	at Tirta Madu street, Kelurahan Batu XIV	2500	37.24	41.82	20.48

4. Conclusion

The highest noise intensity value is at point 3 which at this point is the start of the aircraft to take off. At this point the noise intensity is 90.6 dB (A). In the calculation of the average noise intensity and WECPNL index calculation the highest noise intensity is at point 3 where point 3 is the initial position of the aircraft to take off, but for residential areas outside the airport the noise intensity is not comply with the ministers of No. KM 48 2002.

References

[1] Button, K. 2008. The Impacts of Globalisation on International Air Transport Activity. Past trends and future perspectives. Global Forum on Transport and Environment in a Globalising World 10-12 November 2008, Guadalajara, Mexico. School of George Mason University, USA.

[2] Anizar. Teknik Keselamatan dan Kesehatan Kerja di Industri Yogyakarta. Graha Ilmu, 2009.

[3] Riyangga K, Purwaningsih R, Rani Rumita R. 2014. Analisis Kebisingan Untuk Peningkatan Kesehatan dan Keselamatan Pekerja pada Area Kerja Apron Ground Handling (Studi Kasus di Bandar Udara Ahmad Yani, Semarang). *Industrial Enginerering Online Journal.* 3(1).

[4] Primanda F B, Pemetaan Kebisingan Akibat Aktivitas Pesawat Dengan Software Integrated Niose Model (INM) di Sekitar Bandar Udara Internasional Soekarno-Hatta. Skripsi Sarjana Teknik. Fakultas Teknik program Studi Teknik Lingkungan, Universitas Indonesia. 2012.

[5] Suma'mur P K, Higiene Perusahaan dan Kesehatan Kerja (HIPERKES) (2nd ed), Sagung Seto. Jakarta, 2013.

[6] Correia A W, Junenette L, Peters J L, Levy J I, Melly S, Dominici F. Residential exposure to aircraft noise and hospital admissions for cardiovascular diseases: multi-airport retrospective study. BMJ 2013; 347 doi: https://doi.org/10.1136/bmj.f5561

[7] ICAO, Annex 16 to the Convection on International Civil Aviation, Enviromental Protection Volume I: Aircraft Noise 4th Edition, APP 2-1-2-23 (2005)

[8] Soo Jeong Kim, Sang Kug Chai, Keou Won Lee, Jae-Beom Park, Kyoung-Bok Min, Hyun Gwon Kil, Chan Lee, and Kyung Jong Leeb. Exposure–Response Relationship Between Aircraft Noise and Sleep Quality: A Community-based Cross-sectional Study Osong Public Health Res Perspect. 2014 Apr; 5(2): 108–114. Published online 2014 Apr 16. doi: 10.1016/j.phrp.2014.03.004

[9] Changwoo Lim, Jaehwan Kim, Jiyoung Hong, and Soogab Lee. Effect of background noise levels on community annoyance from aircraft noise. *J. Acoust. Soc. Am.* 123(2), February 2008 0001-4966/2008/123 2 /766/6/$23.00 © 2008 Acoustical Society of America.

IOP Conf. Series: Earth and Environmental Science **106** (2017) 012025 doi:10.1088/1755-1315/106/1/012025

Makrozoobentos as bioindicator water quality of Krukut River, Depok, West Java, Indonesia

A D Jayanti, M F Fachrul*, D Hendrawan

Environmental Engineering Department, Faculty of Landscape Architecture and Environmental Engineering, Universitas Trisakti, Jakarta, Indonesia

*Corresponding Author: melati@trisakti.ac.id

Abstract. The purpose of this study was to determine the Diversity Index, Similarity Index, and Dominance Index. The research location in Krukut River along 12 km. The research was conducted in March - June 2017 in 7 sampling points. Macrozoobentos obtained as many as 23 species belonging to the 9 classes are 3 types of Bivalves, 7 types of Arthopoda, 2 types of Molusca, 4 types of insects, 4 types of Oligochaeta, 2 Malacostraca, 1 Crustacea, 1 Diptera, and 1 Nematodes. The highest abundance of macrozoobentos is 10550 Ind/m^2, whereas the lowest macrozoobenthic abundance is 850 Ind/m^2. The range of Diversity Index (H') 0.32 - 1.80, Similarity Index (E) 0.58-1.07 and Dominance Index (C) 0.40-0.63. Therefore the result shows that Krukut River is in category medium pollution.
Keywords: biology indeks, Krukut river, macrozoobentos, Shannon Wiener

1. Introduction

River as one natural resource has a versatile function for human life. The function and usefulness of the river may be lost from a variety of anthropogenic activities into its capacity. Changes in agricultural land use patterns, moorings and settlements and increased industrial activity will have an impact on hydrological conditions within the watershed. Krukut River is one of the rivers that flow through in the middle development area of Depok City which has been defined as drinking water category B as drinking water intake.

Krukut river flows from Citayam to Glodok along ± 40 kilometers. Watershed Area of Krukut is Ciangke - Pesanggrahan Watershed located in District of Sawangan with a length of 12 kilometers. The river basin Krukut has an area of ± 84.9 km2. The flow of Krukut River passes Depok City, Tangerang Regency, Parung district, Bogor district, Limo subdistrict, Pancoran Mas, Cipayung, and Bojongsari.

Increased population and rapid development in Depok City in 2 decades where the population in 2000 to 2016 increased from 1,160,791 to 2,106,100 inhabitants [1]. Krukut River currently only has a width of 1.5 meters. Changes in water quality and substrates where life greatly affects the composition and abundance of macrozoobenthic invertebrates can be used as a indicator of overall biodiversity in aquatic ecosystems. So, it needs to be done a study on the quality of waters in terms of composition and abundance of macrozoobentos waters of the middle segment of the Krukut River.

IOP Conf. Series: Earth and Environmental Science **106** (2017) 012025 doi:10.1088/1755-1315/106/1/012025

2. Research Method

The sampling of water in Krukut River conducted 4 times in May, June, March and April 2017. Taking macrozoobentos refers to SNI 06-3963-1995. Taking samples of macrozoobentos was done on 7 sampling points 2 times repetition. Taking a sample of macrozoobentos was done using Eckman Grab placed at the bottom of the water for about 5 minutes to capture the biota that drifted into the net and the substrate then filtered using a multilevel filter. Further identification by using stereomicroscope and macrozoobenthic identification books can use [11], [12], [13].

Tabel 1. Sampling point position.

Sampling Point	Coordinat Point	Location	Space Distance (km)
1	06°23'40,57"LS 106°48'09,10"BT	Raya Sawangan Street No. 3, Mampang, Pancoran MAS, Depok City, West Java.	-
2	06°23'26,02"LS 106°48'07,74"BT	Swadaya VIII Street, Depok City, West Java.	2,1
3	06°23'05,82"LS 106°48'05,87"BT	Arsip Raya Street, Depok City, West Java.	1,5
4	06°22'58,59"LS 106°48'02,27"BT	Kp. Utan Jaya Street, Depok City, West Java.	2,3
5	06°22'19,79"LS 106°47'45,34"BT	Tanah Baru Beji Street, Depok City, West Java.	2,3
6	06°21'58,21"LS 106°47'40,03"BT	Jalan Curug Agung Tanah Baru, Beji Street, Depok City, West Java.	2,6
7	06°21'14,64"LS 106°47'37,93"BT	Grand Matoa Street, Krukut, Limo , Depok City, West Java.	1,2

2.1. Level of Pollution Based on Makrozoobentos Structure

2.1.1. Shannon Wiener's (H') Diversity Index

$$H' = -\sum pi\, Log_2 pi \tag{1}$$

Whereas : H' = Shannon Wiener's Diversity Index
ni = Number of individuals in one species
N = Total number of individual species found
pi = ni/N

Tabel 2. Water quality criteria.

No.	Uniformity Index	Quality	Reference
I	> 3 1 – 3 < 1	Clean water Medium Polluted Heavily Polluted	Wilha (1975)
II	3.0 – 4.0 2.0 – 3.0 1.0 – 2.0	Very Lightly Polluted Contaminated Lightly Medium Polluted	Wilha (1975)
III	2.0 2.0 – 1.0 1.5 – 1.0 < 1.0	Not Polluted Contaminated Lightly Medium Polluted Heavily Polluted	Lee, dkk (1975)

IOP Conf. Series: Earth and Environmental Science **106** (2017) 012025 doi:10.1088/1755-1315/106/1/012025

2.1.2. Uniformity Index (E)

$$E = \frac{H'}{H'max} \tag{2}$$

E = Uniformity Indeks, H' = Diversity Indeks, H'max = Maximum Diversity Values (Log_2S), S = Number of species

Criteria : E ~ 0 = There is Dominance of Species
E ~ 1 = The number of Individuals of same species
Noted : The smaller the E value, the smaller the Similarity of the Species Population

2.1.3. Dominance Index (C)

$$C = \Sigma(\frac{ni}{N})^2 \tag{3}$$

C = Dominance Index, Ni = Number of Individuals of Each Type, N = Number of Individuals All Types.
Simpson Dominance Index value ranges from 0-1, if the Index value close to 1 means there is dominance of certain species in the waters.

3. Results and Discussion

3.1. The Shanon Wiener Diversity Index (H') in the Middle Segment Krukut River
This index provides information on regularity and irregularity in an index of diversity.

Table 3. Diversity Index (H') *Macrozoobentos* of middle segment.

Sampling Point		Diversity Index (H')	Average H'	Water Quality
1	March 2017	2.40	1.79	Moderate Polluted
	April 2017	2.76		
	May 2017	1.37		
	June 2017	0.64		
2	March 2017	1.29	1.33	Moderate Polluted
	April 2017	1.42		
	May 2017	0.81		
	June 2017	1.81		
3	March 2017	1.92	1.62	Moderate Polluted
	April 2017	2.06		
	May 2017	1.57		
	June 2017	0.94		
4	March 2017	1.92	1.60	Moderate Polluted
	April 2017	1.60		
	May 2017	1.91		
	June 2017	0.95		
5	March 2017	1.58	1.14	Moderate Polluted
	April 2017	0.71		
	May 2017	1.30		
	June 2017	0.95		
6	March 2017	2.35	1.32	Moderate Polluted
	April 2017	1.07		
	May 2017	1.54		
	June 2017	0.31		

Sampling Point		Diversity Index (H')	Average H'	Water Quality
7	March 2017	2.43	1.80	Moderate Polluted
	April 2017	2.95		
	May 2017	1.51		
	June 2017	0.31		

The index value of diversity in all sampling points ranging from 1.14 to 1.79 is the category of polluted waters being. The diversity index of sampling points 7 and 1 belongs to medium-polluted category (1.80 and 1.79), since pollutants entering the middle segment flow of Krukut River come from household waste around Krukut River which is a source of organic and inorganic waste, and B3.

3.2. Uniformity Index Mcrozoobentos (E)

The middle segment of can be seen the uniformity index of type macrozoobentos (E) which describes the abundance of species. The uniformity index can be seen in Table 4.

Table 4. Uniformity Index *Mcrozoobentos* (E) Krukut River.

Sampling Points		Uniformity Index (E)	Uniformity Average
1	March 2017	0.72	0.63
	April 2017	0.92	
	May 2017	0.59	
	June 2017	0.28	
2	March 2017	0.64	0.78
	April 2017	0.89	
	May 2017	0.81	
	June 2017	0.78	
3	March 2017	0.83	1.07
	April 2017	0.89	
	May 2017	0.99	
	June 2017	1.56	
4	March 2017	0.83	0.74
	April 2017	0.69	
	May 2017	0.96	
	June 2017	0.48	
5	March 2017	0.68	0.58
	April 2017	0.36	
	May 2017	0.82	
	June 2017	0.48	
6	March 2017	0.91	0.68
	April 2017	0.53	
	May 2017	0.97	
	June 2017	0.31	
7	March 2017	0.68	0.67
	April 2017	0.93	
	May 2017	0.75	
	June 2017	0.31	

Marphisa sp is considered a widespread species in temperate and tropical regions [2]. *Viviparus* sp scattered in the waters allegedly because these locations contain small river estuaries [3]. Overall sampling points were dominated by *Viviparus* sp in May and June. That if a location is dominated by Bentos or *Viviparus* sp macroinvertebrate then the water quality in this location is good [4].

Almost all macrozoobenthic classes are always present in lotic habitats such as Gastropoda Class, Insekta, Pelecypoda and Oligochaeta which have a high attendance rate compared to other macrozoobentos classes [5]. Macrozoobentos *Melanoides* sp. is a species indicator of the presence of low dissolved oxygen (DO) and high suspended particles in river water ecosystems. There are macrozoobentos *Melanoides* sp. at sampling points 7 and 1 in March and April. Changing physical conditions - aquatic chemistry can lead to the disappearance of certain types of macrozoobentos and spur the growth of other species.

3.3. Macrozoobentos Dominance Index (C)

Dominance Index (C) is used to determine *the* quality of waters with large number of species or with a high diversity of species [6]. This situation is shown in the form of numbers that is the Dominance Index.

Table 5. Dominant Index (C) Central Segment Krukut River.

Sampling Point		Dominance Index (C)	Dominance Index (C) Average
1	March 2017	0.28	0.45
	April 2017	0.17	
	May 2017	0.55	
	June 2017	0.82	
2	March 2017	0.38	0.44
	April 2017	0.41	
	May 2017	0.63	
	June 2017	0.33	
3	March 2017	0.30	0.40
	April 2017	0.29	
	May 2017	0.34	
	June 2017	1.69	
4	March 2017	0.32	0.42
	April 2017	0.46	
	May 2017	0.28	
	June 2017	0.61	
5	March 2017	0.43	0.63
	April 2017	0.78	
	May 2017	0.47	
	June 2017	0.83	
6	March 2017	0.22	0.52
	April 2017	0.62	
	May 2017	0.35	
	June 2017	0.89	
7	March 2017	0.33	0.45
	April 2017	0.15	
	May 2017	0.42	
	June 2017	0.90	

The Dominance Index value is inversely proportional to Diversity, if the Dominance Index is high then the Diversity Index is low and vice versa. The highest dominance index is owned by the sampling point 5, with the average value of the Dominance Index being 0.63. At the sampling point 5 is dominated by macrozoobentos Viviparus sp.

3.4. Relation of Diversity Index (H'), Uniformity Index (E), and Dominance Index (C)

Table 6. Average Relationship Diversity (H'), Average Uniformity (E), and Average of Dominance Index (C).

Sampling Point	Average Diversity Index (H')	Average Uniformity Index (E)	Average of Dominance Index (C)
1	1.79	0.63	0.45
2	0.33	0.78	0.44
3	0.62	1.07	0.40
4	0.60	0.74	0.42
5	1.14	0.58	0.63
6	0.32	0.68	0.52
7	1.80	0.67	0.45

Within a community type diversity, uniformity and dominance can describe the disruption of aquatic ecosystems [7]. In a dominant community will control the community, so that if the dominant organism is lost it will cause important changes in the community, not only the biotic community but also in the physical environment. Dominance index is inversely related to diversity and uniformity whereas diversity and uniformity have a positive relationship [8]. The high diversity index shows the condition of the aquatic environment in the station is quite good and supports the life of biota in it. Crustacean group has a wide range of life from the muddy habitat to the clean waters and the Gastropoda group has a high adaptability to the environment and the type of feeder depositor on the mud surface [9]. Then from the group of Bivalves with *Corbicula javanica* type found with low abundance due to its nature cannot move actively and settled somewhere. Many studies which conclude that the Mollusca group can function as a bioindicator of freshwater pollution [10].

4. Conclusion

Macrozoobentos obtained as many as 23 species belonging to the 9 classes are 3 types of Bivalves, 7 types of Arthopoda, 2 types of Molusca, 4 types of Insects, 4 types of Oligochaeta, 2 types of Malacostraca, 1 Crustacea, 1 Diptera, and 1 Nematodes. The range of values of Diversity Index (H ') 0.32-1.80, Uniformity Index (E) 0.58-1.07 and Dominant Index (C) 0.40-0.63. The results show the waters of the middle segment of Krukut River in medium contaminated state.

References

[1] Badan Pusat Statistik Indonesia. 2016. Depok dalam Angka
[2] Molina C, Isabel Carrera, Parra Luis F. 2015. Reinstatement of three species of the Marphysa sanguinea complex (Polychaeta: Eunicidae) from the Grand Carribean Region. Zootaxa 3925 (1): 037-055. Mexico.
[3] Satria, Hendra Sunarno, Mas, Tri Joko. 2009. Karakteristik Habitat Calon Suaka Perikanan di Danau Sentani. Bawal. 2(4). Balai Riset Pemulihan Sumberdaya Ikan. Indonesia.
[4] Purwanti A A, Sunarto, & Setyaningsih, R. 2015. Kualitas Air Tanah Sekitar Aliran Sungai Pepe. Biosmart, 66-71.
[5] Oktarina, Syamsudin, Suryati. 2015. Keanekaragaman dan Distribusi Makrozoobentos di Perairan Lotik dan Lentik Kawasan Kampus Institut Teknologi Bandung, Jatinangor. (1)2. Jurusan Biologi Institut Teknologi Bandung. Sumedang.
[6] Fachrul M F. 2007. Metode Sampling Bioekologi. Jakarta: Penerbit Bumi Aksara

IOP Conf. Series: Earth and Environmental Science **106** (2017) 012025 doi:10.1088/1755-1315/106/1/012025

[7] Sinaga T. 2009. Keanekaragaman Makrozoobentos Sebagai Indikator Kualitas Perairan Danau Toba Balige Kabupaten Toba Samosir. Sekolah Pascasarjana Universitas Sumatera Utara. Medan.
[8] Arbi U C. 2012. Komunitas Moluska di Padang Lamun Pantai Wori, Sulawesi Utara, *Jurnal UPT Loka Konservasi Boita Laut* – LIPI Bitung.
[9] Fitriana Y R. 2006. Keanekaragaman dan Kemelimpahan Makrozoobentos di Hutan Mangrove Hasil Rehabilitasi Taman Hutan Raya Ngurah Rai Bali. *Biodiversitas.* 7(1):67-72.
[10] Roy S, & Gupta, A. 2010. Molluscan Diversity in River Barak and Tributaries, Assam. *India. J. Scie & Tech. Biological and Environmental Sciences*. 5(1):109-113
[11] Pennak R. 1978. Freshwater Intervertebrates of the United States. 2nd ed. New York. John Wiley & Sons
[12] Edmondson W T,ed 1959. Freshwater Biology. 2nd ed. New York. John Wiley & Sons
[13] Needham J G, and P R Needham. 1962. A Giude to the study of freshwater Biology. 5th ed. San Franscisco. Holden-Day

The role of social message using norm abstraction level and ecological value orientation to achieve sustainable consumption

A Ekasari

Faculty of Economic and Business, Universitas Trisakti, Jakarta, Indonesia

Coresponding Author: ayu.ekasari@trisakti.ac.id

Abstract. Pro-environmental behavior is one of human activities to achieve sustainability. In order to encourage people to do so, it needs contribution from marketing discipline using social message. The research aims to investigate the effect of social message framed by norm abstraction level and ecological value orientation on attitude and intention to act pro-environmental behavior in the context of littering. This study implemented a 3 (message framing: biospheric/altruistic/egoistic) x 2 (norm abstraction level : abstract/concrete) between subject experimental design to collect the data. An independent sample t test was used to analyze the data. The results indicate that a social message using concrete norm combined with the three ecological value orientation gains more positive response than the use of abstract norm with the same ecological value orientations. Findings of the research are expected to help government or other institutions to create an appropriate social message in anti littering campaign and motivates people to change their behavior in practicing sustainable consumption.
Keywords: ecological value orientation, norm abstraction level, pro-environmental behavior, sustainable consumption, social message

1. Introduction

During these past decades, there has been an enormously increasing in economic development as well as consumption that contributes to environmental degradation caused by natural resources exploitation [1]. This leads to increased awareness of people, government and environmentalists to produce, consume and live in sustainable ways. The concept of sustainability has been adopted by many countries and big corporations as national policy to alter consumption pattern since the Rio Earth Summit in 1992 with the publication of Brundtland Report "Our Common Future" created by World Commission on Environment and Development [2], [3]. The report proposed that it requires interdependence and balance between nature, well-being and economic activity [2], [4].

The publication of Brundtland Report brought up the debate about the role of marketing as a contributor to environmental problem by introducing sustainable consumption or a trigger for the development of global consumption [5]. The term sustainable consumption defined by The United Nations Environment Programme [6] is "the use of services and related products that responds to basic needs and bring a better quality of life while minimizing the use of natural resources and toxic materials as well as the emissions of waste and pollutants over the life cycle so as not to jeopardize the needs of future generations". Meanwhile, marketing is often referred to as the main trigger of consumption while sustainability is a concept that encourages people to satisfy their needs without sacrificing future generations' quality of life, both are in conflict with each other [7].

Social marketing is a form of marketing that has been widely used and considered to support campaigns of consumption reduction, one type of sustainable consumption [5]. From its first emergence in the early seventies and the debate about its identity crisis, the term social marketing finally reached its position as [8] defined social marketing as "the use of marketing principles and techniques to influence a target audience to voluntarily accept, reject, modify or abandon a behavior for the benefit for individuals, groups and society as a whole".

Researchers [5] were amongst the first to mention the use of social marketing to promote sustainability through health and environmental issue campaign that focused on delivering quality of life and well-being that are the main objectives in the concept of sustainability. One aspect of scholarly studies about sustainable consumption found that in order to change people's behavior, it needs psychological understanding of consumers which are their thinking, feeling, attitude and belief about sustainable consumption [4]. Therefore, sustainable consumption starts from an individual through the value he or she holds about the environment that will be implemented in his or her day-to-day lives, such as consuming organic foods, recycling and not littering. The three ecological value orientations introduced by [9] are egoistic, altruistic and biospheric.

Previous research indicated that three ecological value orientations, namely egoistic, altruistic and biospheric contribute to pro-environmental behavior although the results varied. People with egoistic orientation that protect the environment for their own sake and those who have highly altruistic orientation focusing on others tend to practicing pro-environmental behavior, but to some extent they do not do it [10], [11], [12], [13], [14], [15], [16]. However, people with biospheric orientation believe that bad environmental condition can bring adverse consequences for ecological systems and they show consistent attitude and predisposition to protect the environment by behaving in pro environmental ways [11], [12], [18], [19], [20].

Based on the facts above, it is important to analyze the use of social message as one of the social marketing concepts consists of information to encourage people especially with egoistic orientation to behave pro environmentally as well as to highlight biospheric and altruistic orientation in the message. Researchers [20], [21] proposed goal-framing theory derived from sociology to guide pro-environmental behavior that out of the three overarching goals (hedonic, gain and normative), the normative goal-frame condition is the best to encourage pro-environmental behavior. According to the two authors, when a goal is activated or become focal, it is a combination of motives and activated knowledge structure. A goal-frame is a focal goal with its framing effects, that is the effect of focal goals on cognitive processes. Then the focal goal will guide people's behavior. The main goal in a normative goal-frame is 'to act appropriately' [20].

In their earlier papers, researchers [20], [22] as well as [23] mentioned that in the context of sociology there are social norms held by any society that once it is getting larger, the social norms become more abstract because they have to cover diversity of people that need to be translated to a given situation. The problem is the more abstract social norms are, the more difficult for people to know what behaviors are or are not appropriate.

It is argued that to understand how social norms guide behavior, people can be put in a 'frame' that combines cognitive and motivational processes. Thus, motivational processes (goals) will guide cognitive processes that leads to behavior. As it is discussed in sociology literature, concrete norms are presumed more effectively in conveying social message to guide people to behave in a specific situation because they are more assertive and clear rather than abstract norms [20], [21], [22].

Related to the use of goal-framing theory to encourage people to behave pro-environmentally, researchers [20] stated that the normative goal-frame aims to boost people's willingness to act appropriately by protecting the environment and it can be activated using advertisement consists of social message [20]. On the other hand, few studies found that social message framed by ecological value orientation are effective in gaining people's response and intention to act in pro-environment ways [15], [16].

This study is pioneer in examining the effectiveness of social message framed by three ecological value orientations and norm abstraction level in pro-environmental behavior campaign as well as testing the concept of normative goal-frame condition introduced by [20], [21] to foster pro-environmental behavior.

2. Research Method

The design of this study was a 3 (message framing: biospheric/altruistic/egoistic) x 2 (norm abstraction level: abstract/concrete) between subject experimental design to examine the effect of social advertisement on attitude towards the social message and intention to do pro-environmental behavior.

A pilot study was conducted to determine the messages that represent the norm abstraction level combined with three ecological value orientations and the design of the advertisement in the context of anti-littering campaign. The abstract norm message was: 'Preserve your environment' and the concrete norm message was: 'Dispose your waste by type'. The biospheric message framing was: "Soil fertility and ecosystem life will be preserved'. The altruistic message framing was: "People will feel the freshness, cleanliness and become healthier". The egoistic message framing was: "You will feel the freshness, cleanliness and become healthier". The six social advertisements were manipulated using three levels of ecological value orientation message framing: biospheric, altruistic, egoistic and two levels of norm abstraction level: abstract and concrete.

There were 260 students from Universitas Trisakti participated in the experiment and they were placed in each of the six experiment cells using random assignment method. Next they were presented with a description of the new local regulation related with waste management. Subjects were then given the social advertisements and instructed to answer each question using seven-point Likert scale about their attitude towards the advertisement and their intention to act pro-environment behavior. Attitude towards the advertisement was measured using the same items from [24], while intention to behave pro-environmentally was measured using items adopted from [25], and [26].

Manipulation check was done to test whether the constructs manipulated, message framing and norm abstraction level, were well understood by the participants. There were 214 participants that passed the manipulation test and the hypotheses testing were then performed.

3. Results and Discussion

The main effect of norm abstraction level on attitude towards social advertisement is statistically significant (F: 449.02, sig: 0.000) as well as its effect on intention to behave pro-environmentally (F: 338.933, sig: 0.000). The message framing statistically influences attitude towards social advertisement (F: 44.691, sig: 0.000) and intention to behave pro-environmentally (F:28.300, sig: 0.000).

The interaction effect of norm abstraction level and message framing on attitude towards social advertisement (F:26.752, sig : 0.000) and intention to behave pro-environmentally (F: 38.188, sig : 0.000) are significant.

An independent sample t test was used to test the hypotheses proposed and the results can be seen on Figure 1 and 2. Hypotheses 1a and 1b are supported since participants in concrete-biospheric condition respond more positively on attitude towards social advertisement and intention to behave pro-environmentally compared to those that were shown a abstract-biospheric message. The mean difference is significant (ρ: 0.000). Hypotheses 2a dan 2b are also supported, because the mean difference between participants in concrete-egoistic condition and abstract-egoistic condition is statistically significant (ρ: 0.000). The mean difference is statistically significant (ρ: 0.000) for attitude towards social advertisement and intention to act pro-environmental behavior both in groups in concrete-altruistic and concrete-biospheric conditions compared to participants that were shown an abstract-altruistic and abstract-biospheric advertisements. Thus, hypotheses 3a, 3b, 4a and 4b are supported.

Hypotheses 5a is not supported because there is no significant difference (ρ: 0.121) between participants in concrete-altruistic condition with those who are in concrete-egoistic condition. This finding confirms the results found by [16] that social advertisement framed by egoistic and altruistic value orientation in the context of pro-environmental product did not produce different response from the participants. However, hypothesis 5b is supported (ρ: 0.000) because participants in concrete-altruistic condition respond differently compared to participants in the concrete-egoistic condition.

IOP Conf. Series: Earth and Environmental Science **106** (2017) 012026 doi:10.1088/1755-1315/106/1/012026

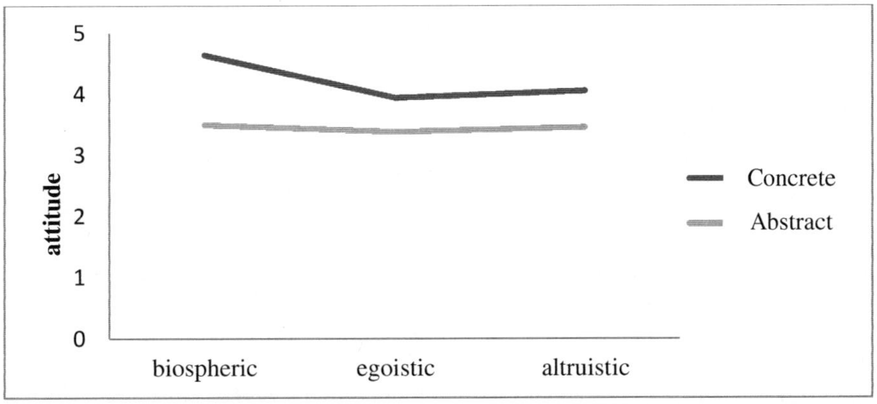

Figure 1. Mean plot attitude concrete vs abstract on message framing (biospheric, egoistic, altruistic).

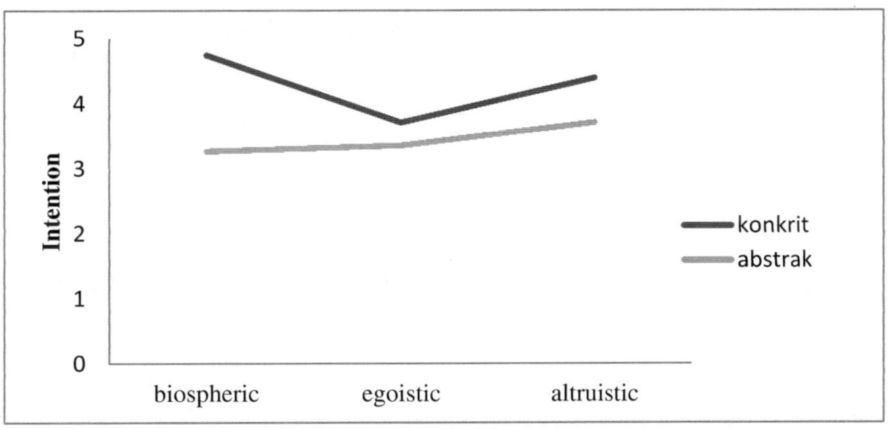

Figure 2. Mean plot attitude concrete vs abstract on message framing (biospheric, egoistic, altruistic).

Overall, results show that people will respond more positively to social advertisement using concrete norm ('disposed your waste by type')framed by the three ecological value orientations. This finding proves Lindenberg and Steg's goal-framing theory that normative goal-frame is the appropriate goal that can be placed as a focal goal in social advertisement to encourage pe Concrete ge in pro-environmental behavior [20]. Moreover, the study also found that the use of c Abstract ı or smart norm rather than the abstract norm is more easy to be triggered and ɔ specific environmental behavior, because it represents the main goal in normative goal-frame : to act appropriately, in this case to behave pro-environmentally.

4. Conclusion

Social marketing campaign in pro-environmental behavior requires good social advertisement and this study found that message framing using concrete norm and three ecological orientations (biospheric, egoistic and altruistic) can trigger people to behave pro-environmentally. Theoretically, this study enriches the social marketing domain within the context of sustainable consumption. This finding also strengthen the role of the fourth P in marketing mix, which is promotion or marketing communication especially in conveying information about the importance of pro-environmental behavior. This study helps social marketer/government to design social marketing campaign consists of a very clear information and instruction about what people have to do. The social advertisement used could also be framed by three ecological value orientations (biospheric, altruistic and biospheric) because naturally each individual holds an ecological value that can be highlighted through an appropriate social message.

References

[1] Tan Booi Chen, and Lau Teck Chai. 2010. Attitudes toward the environment and green product. *Management Science and Engineering*. 4(2):27-29.
[2] Peattie Ken, Crane Andrew. 2005. Green Marketing : legend, myth or prophecy? *Qualitative Market Research.* 8(4):357-370.
[3] Schaefer Anja, and Crane Andrew. 2005. Addressing sustainability and consumption. *Journal of Macromarketing*. 25(76):1-18.
[4] Belz Frank-Martin, and Peattie Ken. *Sustainability Marketing A Global Perspective.* Second Edition. John Wiley & Sons Ltd, 2012.
[5] Peattie Ken, and Peattie Sue. 2008. Social Marketing: a Pathway to Consumption Reduction? *Journal of Business Research*, xx, p. 1-9.
[6] *Sustainable Cosumption*. Available at:http//www.unep.fr/scp/sc.
[7] Her Majesty's Government. 2005. *Securing the future*, Vol. 6467, cm. 6467. Availabe at: http//www.sustainable-development.gov.uk/publications/uk-strategy/index/htm.
[8] Kotler Philip, Roberto Eduardo, and Lee Nancy, *Social Marketing: Strategies for Changing Public Behavior*, Thousand Oaks, CA, Sage Publications, 2002.
[9] Stern Paul, and Dietz Thomas. 1994. The Value Basis of Environmental Concern. *Journal of Social Issues*. 50(3):65-84.
[10] Milfont Taciano L, Duckitt John, Cameron, Linda. 2006. A Cross Cultural Study of Environmental Motive Concerns and Their Implications for Pro environmental Behavior. *Environment and Behavior*. 38(6):745-767.
[11] De Groot, Judith I M. et.al. 2012. Environmental Values in Post-socialist Hungary: Is It Useful to Distinguish Egoistic, Altruistic and Biospheric Values? *Czech Sociological Review,* 48(3):421–440.
[12] De Groot, Judith I M, Steg Linda. 2010. Relationships between valueorientations, self-determined motivational types and pro-environmental behavioural intentions. *Journal of Environmental Psychology*. 30:368- 378 .
[13] Khacatryan Haik, Joireman Jeff, and Casavant Kent. 2013. Relating Values and Consideration of Future and Immediate Conssequences to Consumer Preference for biofuels: a three-dimensional social dilemma analysis. *Journal of Environmental Psychology.* (34):97-108.
[14] Gatersleben B, White E, Abrahams W, Uzzell D. 2010. Values and Sustainable Lifestyle. *Architectural Science Review*, 53:37-50.
[15] Bastaman Aam. 2011. Pengaruh Pembingkaian Pesan Orientasi Nilai Ekologis, Bentuk Tampilan Iklan dan Sumber Pesan Terhadap Respons Konsumen: Suatu Perspektif Demarketing Energi Menggunakan Iklan Sosial. *Dissertation*, Universitas Indonesia.
[16] Hansla Andre. 2011. Value Orientation, Awareness of Consequences, and Environmental Concern. *Dissertation*. Department of Psychology, University of Gothenburg.
[17] De Groot, Judith I M, and Steg Linda. 2007. General Beliefs and the Theory of Planned Behavior: The Role of Environmental Concerns in the TPB. *Journal of Applied Social Psychology,* 37(8):1817–1836.
[18] Midori Aoygi-Usui,Vinken Henk, and Atsuko Kuribayashi. 2003. *Human Ecology Review.* 10(1):23-31.
[19] Steg Linda, Dreijerink Lieke, and Abrahamse Wokje. 2005. Factors influencing the acceptability of energy policies: A test of VBN Theory. *Journal of Environmental Psychology.* 25(2005):415-425.
[20] Lindenberg, Siegwart, and Steg Linda. 2007. Normative, Gain and Hedonic Goal Frames Guiding Enviornmental Behavior. *Journal of Social Issues*. 63(1):117-137.

[21] Lindenberg, Siegwart, and Steg Linda, Goal-Framing Theory and Norm-Guided Behavior. In H. van Trijp *(eds) Encouraging Sustainable Behavior,* pp. 37-54. New York : Psychology Press, 2013.

[22] Lindenberg Siegwart. Smart norms. How do they work and does the school have an important function for making them work? in: Veugelers W, Bosman M H. (Eds.) *De strijd om het curriculum* (pp.85-107). Antwerpen/Apeldoorn: Garant. 2005.

[23] Lindenberg Siegwart, Why framing should be all about the impact of goals on cognitions and evaluations, In : Hartmut Esser Erklarende Soziologie, pp. 53-77.

[24] Homer, Pamela A. 1990. The mediating role of attitude toward the ad: some additional evidence. *Journal of Marketing Research*. Vol XXVII, p. 78-86.

[25] Schifman Leon G, and Kanuk Leslie Lazar. *Consumer Behavior*. Global Edition. Pearson Education, Inc, 2010.

[26] Fielding Kelly L, McDonald Rachel, and Louis, Winnifred. 2008. Theory of planned behavior, identity and intentions to engage in environmental activism. *Journal of Environmental Pscycholgy*, 28:318-326.

Speed–volume relationship and headway distribution analysis of motorcycle (case study: Teuku Nyak Arief Road)

E Prahara*, **R A Prasetya**

Civil Engineering Department, Faculty of Engineering, Universitas Bina Nusantara, Jakarta, Indonesia

*Corresponding Author: eduardi@gmail.com

Abstract. In many developing countries, transportation modes are more varied than the other country. For example, in Jakarta, Indonesia, in some roadway, motorcycle is the most dominant vehicle, with total volume is four times higher than a passenger car. Thus, the traffic characteristic in motorcycle-dominated traffic differs from a common traffic situation. The purpose of this study is to apply the concept and theory developed to analyze motorcycle behaviour under motorcycle-dominated traffic condition. The survey is applied by recording the traffic flow movement of research location at specified time period. The macroscopic characteristic analyzed in this research is a speed-flow relationship based on motorcycle equivalent unit (MCU). Furthermore, a detail microscopic characteristic analyzed that is motorcycle time headway regarding traffic flow. MCU values computed were consists of motorcycle (MC), light vehicle (LV) and heavy vehicle (HV). Those values were calculated 1.00, 6.13 and 10.71 respectively. The speed and volume relationship result is showing a linear regression model with R^2 value is 0.58, it can be explained that the correlation between two variables is intermediate. The headway distribution of motorcycle is compatible with the negative exponential distribution which fitted with the proposed theory for a small vehicle such as a motorcycle.
Keywords: flow, headway distribution, motorcycle equivalent unit, transportation modes

1. Introduction

Congestion is the main problem occurred in developed country such as Indonesia, particularly in their capital city Jakarta. In Jakarta, for example, two-wheelers are the most dominant vehicle with the total volume is four times higher than a passenger car. Due to a high number, Jakarta has been suffering from a congestion problem, which is mostly caused by two-wheel vehicles. The high proportion of motorcycle in traffic will affect the traffic characteristic. Characteristic in motorcycle-dominated traffic is differ from a common traffic situation due to its mixed proportion of vehicles, it will convert the characteristic from a homogeneous traffic flow to heterogeneous traffic flow. Regarding heterogeneous characteristic, the homogeneous traffic concept has a limitation in the application for heterogeneous traffic [1].

There was a study conducted to identify traffic characteristic that has significant motorcycle proportion. Minh, Sano and Matsumoto [2] were developed a motorcycle unit equivalent known as MCU in order to estimate volume under heterogeneous traffic flow condition. MCU is developed with a consideration of the dynamic characteristic of moving vehicle [2].

IOP Conf. Series: Earth and Environmental Science **106** (2017) 012027 doi:10.1088/1755-1315/106/1/012027

The present study is intended to adopt and applied the concept of motorcycle equivalent unit (MCU) as a proper method to analyze a characteristic of heterogeneous traffic flow at Indonesian roadways, specifically in Jakarta, which have a significant motorcycle proportion in traffic flow. MCU is considered due to the limitation of passenger car unit (PCU) as a common method used to analyze a characteristic of heterogeneous traffic flow. In order to conduct this study, site location used must have a sufficient motorcycle proportion. The objectives of the study are to obtain several parameters, such as MCU value for each type of vehicle in traffic, speed and flow relationship based on the MCU value obtained, and motorcycle headway analysis.

2. Research Method

There are two types of data used in this research, primary and secondary data. Primary consists of speed and volume of vehicles in traffic streams, motorcycle headway, and roadway geometry. Speed, volume, and headway were collected by using video camera and tripod located at the pedestrian bridge above site location. The video camera was recorded all the traffic flow at specified time period. Roadway geometry data was collected by measuring roller to measure length and width of the site location. To avoid any unusual behaviour by the driver, all of the data were collected under clear weather.

Secondary data consists of vehicle categories and vehicle dimension. In this study, vehicles are classified into three categories according to Indonesian Highway Capacity Manual (also known as MKJI 1997), there are light vehicle (LV), heavy vehicle (HV), and motorcycle (MC). Dimensions for each category were determined from domestic auto market data on 2016 in order to represent the real condition in Indonesia by showing the highest market for each category of the vehicles. Thus, the data obtained were used as a guideline for computing MCU. Domestic market data of light vehicle (LV) and heavy vehicle (HV) were obtained from the data that published by Association of Indonesia Automotive Industries (GAIKINDO). Meanwhile, the domestic market data of motorcycle (MC) were obtained from the data that published by Association of Indonesia Motorcycle Industries (AISI).

3. Results and Discussion

The site location chosen was located at Teuku Nyak Arief Road, Jakarta, Indonesia. It's a two-lane roadway with 3 m width for each lane. The observed length was measured 40 m. It can be noted that the distance measured should have a clear sight from camera perspective in order to obtain a good data. Furthermore, the percentage of motorcycle (MC), light vehicle (LV) and heavy vehicle (HV) are 74.29%, 24.55% and 1.16% respectively.

3.1. Motorcycle unit

Motorcycle equivalent unit (MCU) was calculated based on a dynamic characteristic of moving vehicle. This characteristic represents the correlation between speed and occupied space of vehicles in the lane. The maneuver is one of the main characteristic of motorcycle behaviour in traffic, due to its small size of the vehicle. Regarding this characteristic, the motorcycle does not move in the lane. Meanwhile, in traffic with lane discipline, the occupancy is controlled by the length of the vehicle. However, in the condition where vehicles do not follow lanes strictly, the occupancy is better reflected by area [3]. The equation for determining MCU is described below:

$$MCU_i = \frac{V_{mc}/V_i}{A_{mc}/A_i} \tag{1}$$

Where

MCU_i = Motorcycle equivalent unit of type i vehicle;
V_{mc} = Mean speed of motorcycles (km/hr);
V_i = Mean speed of type i vehicle (km/hr);
A_{mc} = Projected rectangular area of motorcycles (m^2);
A_i = Projected rectangular area of type i vehicle (m^2).

Table 1. Vehicle categories and sizes.

No.	**Category**	**Dimension (m)**		**Area (m²)**
		Length	**Width**	
1	Motorcycle (MC)	1.856	0.666	1.236
2	Light Vehicle (LV)	4.190	1.660	6.955
3	Heavy Vehicle (LV)	6.026	1.945	11.721

Table 1 is showing vehicle dimension for each type respectively. The length and width of all vehicles were obtained from domestic auto market data that published by Association of Indonesia Automotive Industries (GAIKINDO) and Association of Indonesia Motorcycle Industries (AISI) in 2016. It represents the highest market for each type of vehicles, then the length and width of those vehicles were used as a guideline for computing MCU.

Mean speed of all vehicles was determined by using spot speed technique. To obtain the speed of the vehicle, the time required for travelling observation distance was counted by stopwatch. Stopwatch begins counting when vehicle entry an observed point and stop counting when the vehicle passes the end-point. Then, all of the speed obtained was converted into kilometer per hour. Furthermore, MCU value can be calculated by using the equation above.

Table 2. MCU value for each type of vehicle.

Motorcycle (MC)	**Light Vehicle (LV)**	**Heavy Vehicle (LV)**
1.00	6.13	10.71

The computed results of MCU from equation (1) are shown in Table 2. The value of MCU for each category of vehicle is affected by the ratio between mean speed and the projected area of the motorcycle and another type of vehicle. Therefore, if the motorcycle has a significant difference mean speed than another type (i.e. light vehicle and heavy vehicle), MCU value will be increased as well.

3.2. Speed – volume relationship

There are three macroscopic characteristics of traffic flow; speed, volume, and density. In order to calculate the capacity of the carriageway, speed – volume relationship is usually applied due to the flexibility of collecting the data needed. The data should be collected at non-congested traffic and no traffic flow data was collected at or over the capacity [2].

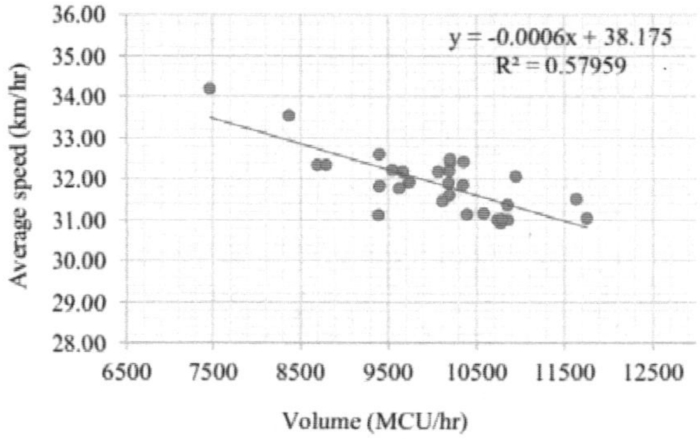

Figure 1. Speed – Volume relationship.

Figure 1 is showing speed and volume relationship at Teuku Nyak Arief Road. The lowest traffic flow at site location is 7468 MCU/hr, and the highest flow is 11756 MCU/hr. The average stream speed of traffic on minimum flow and peak condition at site location does not have a significant difference. The speed varies approximately 4 km/hr. The average stream speed is affected by the driver behaviour. According to the data collected, the motorcycle has the highest average stream speed with 34.13 km/hr, while light vehicle and heavy vehicle are 31.31 km/hr and 30.21 km/hr respectively. Light vehicle and heavy vehicle tends to reduce their speed on traffic in order to reduce the risk of accidents due to motorcycle behaviour of maneuvering.

3.3. Motorcycle headway distribution

Researcher [4] described that time headway distribution varied considerably as the traffic flow rate increased due to the increasing interactions between vehicle and traffic streams. He proposed three classifications of time headway distribution based on traffic flow rate; random headway state for low flow condition, intermediate headway state for moderate flow condition, and constant headway state for high flow condition. The negative exponential distribution represents the random headway state flow. The normal distribution can be used when time headways are all constant. Meanwhile, intermediate headway state is a state lies between those two previous headway states (random and constant), and the mathematical distribution model approach proposed is Pearson type III distribution [4].

However, researcher [5] suggested negative exponential distribution represents its compatibility to a traffic flow that consists of a significant percentage of smaller vehicles such as a motorcycle. From the study conducted, the motorcycles arrivals are in the random state even during medium and high flow condition. Therefore, the negative exponential distribution is applied in order to obtain the best-fitted distribution for motorcycles. Slovin's method with 5 percent margin of error was used to determine the observation sample number. Thus, from the total flow 3600 veh/hr, 360 motorcycles were observed.

Table 3. Statistical parameters of motorcycle headway.

Min (s)	Max (s)	Mean (s)	St. Dev (s)	Median (s)	Mode (s)	Range	Number of classes	Interval
0.22	8.84	1.15	1.07	0.81	0.53	8.62	9	1

Table 3 is showing statistical parameters of motorcycle headway taken from the observed sample. The negative exponential distribution can be derived from Poisson distribution. If there are no vehicle arrivals in an interval t, there must be a headway greater than or equal to t [6]. The equation is described below:

$$Probability\,(headway > t) = exp\,(-qt) \tag{2}$$

Where,

q = Mean time headway (veh/sec);
t = Time interval (sec).

Table 4. Observed and theoretical headway data of motorcycle.

Class	Time headway classes (s)	Observed frequency (fo)	Cumulative observed frequency (fo c)	Observed frequency percentage (%)	Theoretical frequency percentage (%)	Cumulative theoretical frequency (ft c)	Theoretical frequency (ft)
0	0.0 – 0.9	230	360	100.00	100.00	360.00	227.56
1	1.0 – 1.9	77	130	36.11	36.79	132.44	83.72
2	2.0 – 2.9	35	53	14.72	13.53	48.72	30.80
3	3.0 – 3.9	7	18	5.00	4.98	17.92	11.33
4	4.0 – 4.9	7	11	3.06	1.83	6.59	4.17

Class	Time headway classes (s)	Observed frequency (fo)	Cumulative observed frequency (fo c)	Observed frequency percentage (%)	Theoretical frequency percentage (%)	Cumulative theoretical frequency (ft c)	Theoretical frequency (ft)
5	5.0 – 5.9	1	4	1.11	0.67	2.43	1.53
6	6.0 – 6.9	1	3	0.83	0.25	0.89	0.56
7	7.0 – 7.9	1	2	0.56	0.09	0.33	0.21
8	8.0 – 8.9	1	1	0.28	0.03	0.12	0.12

The computed results between observed and theoretical frequencies of motorcycle headway are shown in Table 4. The calculation of theoretical frequency percentage is obtained from equation (2) and multiplied by 100, where q = 3600/3600 = 1 veh/sec and t = 0, 1, 2, ..., 8. Then, the value multiplied by a total amount of headway observed in order to obtain cumulative theoretical frequency (f_t c). Theoretical frequency headway in class (f_t) can be obtained from the difference between successive values of cumulative theoretical frequency.

To compare the observed distribution with the probability distribution, the non-parametric goodness of fit Chi-Square test was applied. The equation of the test is described below:

$$\chi^2 = \sum \frac{(observed\ frequency - theoretical\ frequency)^2}{theoritical\ frequency} \quad (3)$$

Table 5. Chi-Square test.

Observed frequency (f_o)	Theoretical frequency (f_t)	$(f_o - f_t)^2 / f_t$
230	227.56	0.026
77	83.72	0.539
35	30.80	0.573
7	11.33	1.655
11	6.59	2.945
	Total	5.738

Table 5 is showing the Chi-Square calculation. The last five classes of a theoretical frequency have a value less than 5, therefore, the remaining classes should be combined. According to the Chi-Square distribution tables, with the 5 percent significant level (α) and 3 degrees of freedom (df), the value is 7.815. Thus, χ^2 calculated < χ^2 table, so there is no significant difference between observed and theoretical frequency. Consequently, the headway distribution of motorcycle at Teuku Nyak Arief roadway is negative exponential distribution. The goodness of fit of the negative exponential distribution to the observed values is shown in the graphic below.

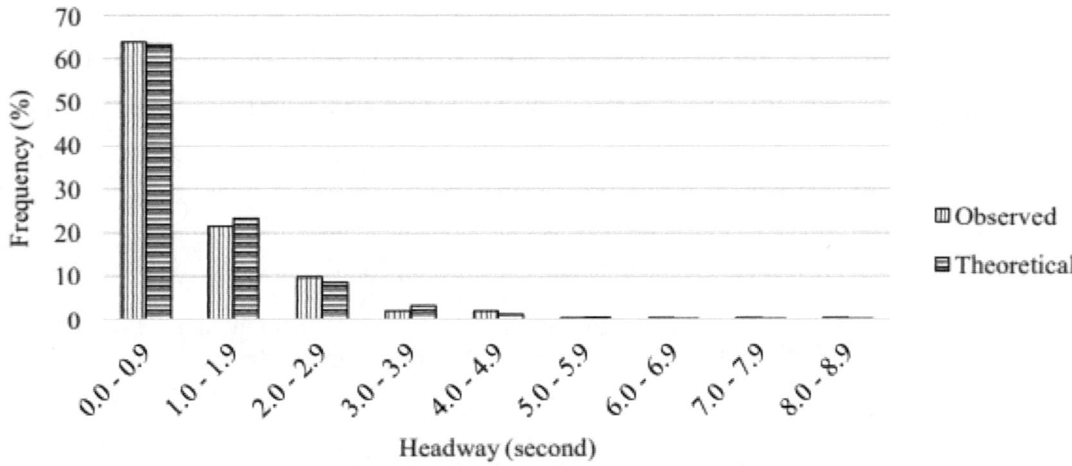

Figure 2. A comparison of observed and theoretical headway frequencies.

4. Conclusion

MCU values computed were consists of motorcycle (MC), light vehicle (LV) and heavy vehicle (HV). Those values were calculated 1.00, 6.13 and 10.71 respectively. The MCU values considerably affected by the dynamic characteristic of moving vehicle. In this case, the parameters are speed (km/hr) and projected rectangular area (m^2). The speed and volume relationship result is showing a linear regression model with R^2 value is 0.58, it can be explained that the correlation between two variables is intermediate. The lowest traffic flow at site location is 7468 MCU/hr, and the highest flow is 11,756 MCU/hr. Meanwhile, the average stream speed of the motorcycle, light vehicle and heavy vehicle are 34.13, 31.31 km/hr and 30.21 km/hr respectively The headway distribution of motorcycle is compatible with the negative exponential distribution which fitted with the proposed theory for a small vehicle such as a motorcycle. The average time headway of traffic flow is 1.15 sec, while the headway ranges from 0.22 to 8.84 sec, and the standard deviation is 1.07 sec. The present study providing an information that can be used in the future as a simulation model for developing the traffic operation, particularly in the developed country like Indonesia, due to a significant proportion of motorcycle in traffic flow.

References

[1] Minh C C, Sano K, Mai T T, and Matsumoto S. 2010. Development of motorcycle unit (MCU) for motorcycle-dominated traffic. *J. East Asia Soc. Transp. Stud.* 8:1596-1608.

[2] Minh C C, Sano K, and Matsumoto S. 2005. The speed, flow, and headway analyses of motorcycle traffic. *J. East Asia Soc. Transp. Stud.* 6:1496-1508

[3] Chandra S, and Kumar U. 2003. Effect on lane width on capacity under mixed traffic conditions in India. *J. Transp. Eng.*129:155-60.

[4] May A D, Traffic Flow Fundamental, Ed. V L McCarthy, New Jersey: Prentice-Hall, Inc.,1990.

[5] Arasan V T, and Koshy R Z. 2003. Headway distribution of heterogeneous traffic on urban arterials. *J. Inst. Eng.* 84:210-15.

[6] Salter R J, Highway Traffic Analysis and Design, London: The Macmillan Press Ltd., 1974.

The dynamic of urban and protected areas at Balai Raja Wildlife Reserve, Riau, Indonesia: a social ecology approach

Suwondo[1,2*], Darmadi[1,2], M Yunus[1]

[1]The Center for Environmental Studies, Research and Community Services Institution, Universitas Riau, Kampus Bina Widya, Pekanbaru, Indonesia
[2]Departement of Biology Education, Faculty of Education and Teacher Training, Universitas Riau, Kampus Bina Widya, Pekanbaru, Indonesia

*Corresponding author: suwondo@lecturer.unri.ac.id

Abstract. The development process has resulted in deforestation. A comprehensive study is needed to obtain an objective solution by integrating the ecological dimension and human dimension. This study was conducted within Balai Raja Wildlife Reserve (BRWR), Bengkalis Regency, Riau Province, Indonesia. We used the social-ecological systems (SES) approach based on local characteristics, categorized into ecological status, social status and actors. Each factoris ranked using Multi-Dimensional Scaling (MDS).BRWR sustainability levels are in moderate condition. The ecological dimension is in a less sustainable state, with leverage: (1) forest conversion; (2) local ecological knowledge; (3) high conservation value. The social dimension is in a less sustainable state, with leverage: (1) community empowerment; (2) social conflict; (3) participation in landscape management. Dimensions actors are on a fairly sustainable status, with leverage: (1) institutional interaction; (2) stakeholder's commitment; (3) law enforcement. We recommend strengthening community empowerment, local ecological knowledge, interaction, and stakeholder commitment
Keywords: protected area, social ecology, sustainability, urban

1. Introduction

Sumatra Island has lost 6.6 million ha of forest or 557,000 ha per year since the early 1970s. The largest area, about 30% was lost from Riau Province, Indonesia [1]. This condition occurs in areas managed by the public as well as by the state, one of which is Balai Raja Wildlife Reserve (BRWR). Forest areas are subsequently converted to commercial activities, such as settlements, plantations, agriculture, mining, and infrastructure.

Deforestation is closely related to the human dimension, on the one hand, humans have made the greatest contribution to today's landscape changes [2], and on the other hand, human beings are the beneficiaries of the negative environmental impacts [3]. Population growth and urbanization [4] became one of the factors driving natural resource degradation [5], [6]. Therefore, in the context of sustainable area management, we need examining the interrelationships between humans and natural resources, complex and involves identifying feedback relationships among people, resources and institutions [7], [8], [9].

Efforts to improve the knowledge of the multi-dimensional and multi-scale complex interactions between human development and the natural environment have led to novel ideas of social-ecological systems (SES) for studying these processes [10]. Fundamentally, a "social-ecological system" is composed of an ecosystem and the people that interact with it. Such systems are complex and adaptive

IOP Conf. Series: Earth and Environmental Science **106** (2017) 012028 doi:10.1088/1755-1315/106/1/012028

society and environment are considered to "co-evolve" [11]. SESs are systems that include social, economic, and ecological elements as well as the interactions between them. The concept of an SES is useful for protected area (PA) management because it explicitly implies that the manager, other stakeholders, and related institutions are part of a cohesive whole, the system. This, in turn, suggests that approaches that incorporate these elements into dynamic models of system interactions, rather than treating them as immutable external influences on ecosystems, may identify opportunities to enhance the resilience of systems that would otherwise be overlooked. Moreover, PAs do not exist in a vacuum and interact with, contain, and/or are nested within other SESs [12].

2. Research Method

This study was conducted within BRWR (0°24' N-01°25' S, 101°50' E-101°55' E), Bengkalis Regency, Riau Province, Indonesia from November 2016 to February 2017. This research is conducted through the following stages: (a) desk study; (b) determination of key factors; (c) surveys; (d) data analysis. The type of data used in this study consists of primary data and secondary data. Primary data comes from survey results and focus group discussions (FGDs). Secondary data is obtained from relevant agencies at the national and local levels. We use the modification of SES framework [13], [14] to analyze the ecological status, social status, and actors. The framework can comprise multiple, concurrent sets of each subsystem. Analysis of the socio-ecological status is done with the approach Multi-Dimensional Scaling (MDS) using RAPFISH [15]. As for visualization actors, we use GEPHI [16].

3. Results and Discussion

The results of the analysis show that the BRWR sustainability index for the ecological dimension is 47.06 (less sustainable), social dimension is 47.96 (less sustainable), and the dimension's actors are 59.13 (sustainable enough). The results of Monte Carlo analysis show that 95% confidence level for each dimension is not much difference (relatively small difference). This condition shows that this simulation has a high level of trust. The differences or differences between the sustainability indexes of MDS and Monte Carlo are relatively small, with a range of values of 0.30-0.77. This indicates that the simulation is quite good as one of the BRWR sustainability evaluation tools. The sustainability index has a stress value ranging from 0.13-0.14 and the determination value (R2) between 0.94-0.95. Based on the value of stress smaller than 0.25 and the value of determination (R2) close to 1.0 indicates that the results obtained are quite accurate and can be accounted for. In general, BRWR sustainability levels are in moderate condition. However, this condition still has a chance to turn bad if not anticipated by concrete action and joint action.

Table 1. Result of MDS, Monte Carlo, and statistic analysis.

Dimensions	Sustainability Index		Difference	Statistic	
	MDS	Monte Carlo		Stress	R^2
Ecological	47.06	46.76	0.30	0.13	0.95
Social	47.96	47.56	0.40	0.13	0.95
Actors	59.13	5836	0.77	0.14	0.94

We used 10 attributes to analyze the sustainability of ecological dimensions, namely: (1) ecosystem services; (2) pressure on forests; (3) important conservation values; (4) carbon potential; (5) local wisdom; (6) wildlife; (7) biodiversity; (8) land cover; (9) land use suitability; (10) forest conversion. The ecological dimension is in a less sustainable state, with leverage: (1) forest conversion; (2) local ecological knowledge; (3) high conservation value (Figure 2). BRWR has component complexity and interaction, a mix of PA and urban (Duri City). BRWR has an area of 15,343.95 Ha in accordance with the Decree of the Minister of Forestry Number: 3978/Menhut-VII/KUH/2014.

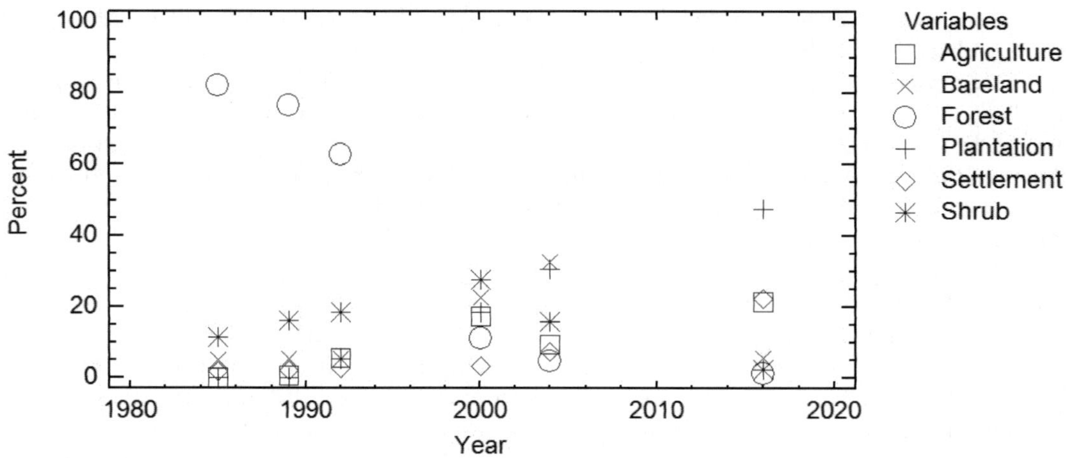

Figure 1. Land cover changes in BRWR (1985-2015).

The result of time series analysis in 1985, 1989, 1992, 2000, 2004 (PPLH IPB & CPI, 2005) and 2015, indicates that there has been a decrease of forest cover area in BRWR as follows 82% (13,705.19 ha), 76.44 % (12,767.78 ha), 62.89% (10,504.84 ha), 11.17% (1,865.88 ha), 4.64% (705.85 ha) and 0.98% (150.66 ha). Thus there is a loss of forests of 4.99% per annum on average. While there was an increase in land use for settlements and bare land from 1985 of 280.51 ha (settlement) and 825.78 ha (bare land) increased in 2004 to 1,103.36 ha (settlement) and 4,947.59 ha (bare land).

The forest conversion occurring at BRWR creates a negative reciprocal relationship. The local wisdom that once was able to maintain harmony between nature and human now loses its existence. In addition, this conversion also poses a threat to high conservation value, such as human-elephant conflict [17] and Tiger [18], [19].

We used 7 attributes to analyse the sustainability of social dimensions, namely: (1) forest area management; (2) community empowerment; (3) public awareness; (4) social conflict; (5) the role of indigenous peoples; (6) participation in landscape management; (7) community education. The social dimension is in a less sustainable state, with leverage: (1) community empowerment; (2) social conflict; (3) participation in landscape management (Figure 2).

Community empowerment is very important in the protection and management of BRWR, as they are the main actors of any policy implementation. Empowerment implemented so far is not optimal, especially in terms of empowerment program, where the role of government is very dominant. So there is a very high dependence on funding from the state, while the state has limited funding. In the end, the implementation of the program cannot be implemented as planned.

Community empowerment is related to social conflict, both as cause and effect. Weak development of partnerships leads to inappropriate relationships. As a result, there is a gap in access to resources that impact on the emergence of social conflict. The issue of social conflict at BRWR shows an increasing intensity, leading to society versus government. On the one hand, communities are prohibited from engaging in conservation areas. On the other hand, the government also issued permits to various activities undertaken by the community. The issue of migrants is very important in this region, as population growth and migrants increase rapidly.

Population growth would have an impact on the increased needs of life. The most significant implication is the conversion of agricultural land into settlements, according to [20], the conversion of agricultural land into settlements in Duri City is 669.68 Ha in the period 1995-2005, and 378.60 Ha in the period of 2005-2015. Not only agricultural land, the dominant change is plantation land of 57.42 Ha in the period 1995-2005 and 64.85 Ha in the period of 2005-2015. Surely this poses a special threat to BRWR's sustainability, due to its adjacent location. In the context of the relationship between urban and protected areas, these conditions can be both opportunities and challenges. The current global urbanization implies two things of importance. First, that urban environment is getting closer to the large area of global protection [21]. And second, that protection of urban nature is increasingly important for sustainable development of cities as well as safeguarding urban biodiversity as part of global biodiversity [22].

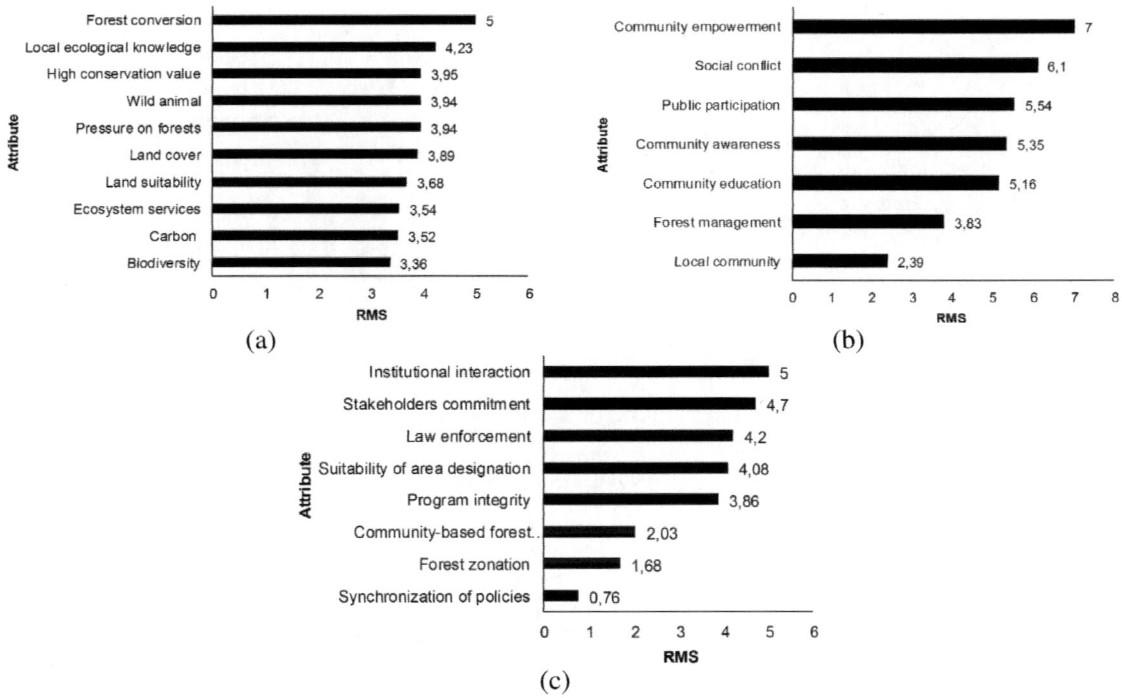

Figure 2. Leverage of attributes on sustainability score: (a) ecological; (b) social; (c) actors.

We used 8 attributes to analyse the sustainability of dimensions' actors, namely: (1) zoning of the region; (2) forest area management; (3) interaction between institutions; (4) the suitability of regional designation; (5) stakeholder's commitment; (6) program integrity; (7) law enforcement; (8) policy synchronization. Dimensions actors are on a fairly sustainable status, with leverage: (1) institutional interaction; (2) stakeholder's commitment; (3) law enforcement.

We analyse stakeholder interactions and identify their contribution to BRWR management. We integrate between nodes and edges, where nodes represent interconnected parties. In Figure 3, the size and colour of the node is proportional to the number of contributions each side: the more they interact, the larger the edge size. The most active parties are Natural Resources Conservation Agency, Environment and Forestry Agency, NGOs, local people, and University. We also consider the intensity of the interaction of stakeholders, represented by the width of each edge, which is proportional to the number of interactions held together. We observed several strong ties between the stakeholders: Natural Resources Conservation Agency-local people, Natural Resources Conservation Agency-law enforcer, NGOs-local people, private company-local people.

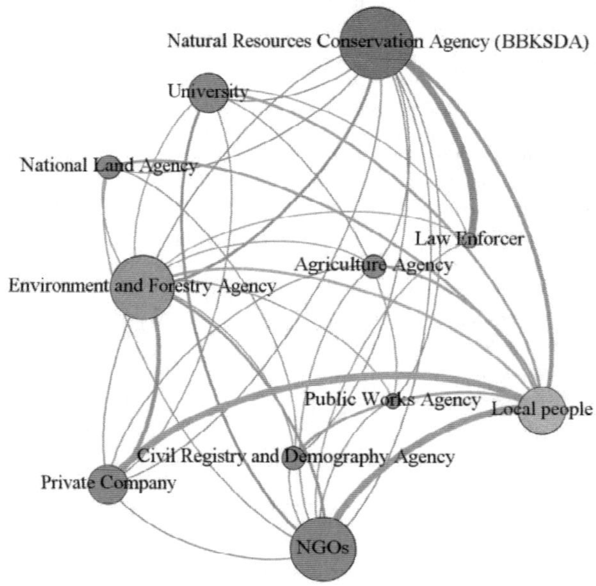

Figure 3. Interaction between actors.

4. Conclusion

BRWR sustainability levels are in moderate condition. The ecological dimension is in a less sustainable state, with leverage: (1) forest conversion; (2) local ecological knowledge; (3) high conservation value. The social dimension is in a less sustainable state, with leverage: (1) community empowerment; (2) social conflict; (3) participation in landscape management. Dimensions actors are on a fairly sustainable status, with leverage: (1) institutional interaction; (2) stakeholders commitment; (3) law enforcement. We recommend strengthening community empowerment, local ecological knowledge, interaction, and stakeholder commitment.

Acknowledgments

This project is an initiative of Conservation International Indonesia and was funded by them. The project is supported by Universitas Riau. We wish to thank the participants in this study who gave generously of their time and intellectual resources in interviews, workshops, meetings and focus groups discussion.

References

[1] Margono B A, Turubanova S, Zhuravleva I, Potapov P, Tyukavina A, Baccini A, Goetz S and Hansen M C. 2012. Mapping and monitoring deforestation and forest degradation in Sumatra (Indonesia) using Landsat time series data sets from 1990 to 2010. *Environ. Res. Lett.*7 34010.

[2] Peterson C H. 1993. Improvement of environmental impact analysis by application of principles derived from manipulative ecology: Lessons from coastal marine case histories. *Aust. J. Ecol.* 18:21–52

[3] Stoltman J P. 21st Century Geography: A Reference Handbook, California: SAGE Publications, Inc, 2012.

[4] Firman T, Kombaitan B and Pradono P. 2007. The Dynamics of Indonesia's Urbanisation, 1980–2006. *Urban Policy Res.* 25:433–454

[5] Hunter P. 2007. The human impact on biological diversity. How species adapt to urban challenges sheds light on evolution and provides clues about conservation. *EMBO Rep.* 8:316–8.

[6] McDonald R I, Marcotullio P J, and Güneralp B, Urbanization and Global Trends in Biodiversity and Ecosystem Services. *Urbanization, Biodiversity and Ecosystem Services: Challenges and Opportunities: A Global Assessment* Ed, 2013.
[7] T Elmqvist, M Fragkias, J Goodness, B Güneralp, P J Marcotullio, R I McDonald, S Parnell, M Schewenius, M Sendstad, K C Seto and C Wilkinson (Dordrecht: Springer Netherlands) pp 31–52
[8] Liu J, Dietz T, Carpenter S R, Alberti M, Folke C, Moran E, Pell A N, Deadman P, Kratz T, Lubchenco J, Ostrom E, Ouyang Z, Provencher W, Redman C L, Schneider S H, and Taylor W W. 2007. Complexity of Coupled Human and Natural Systems. *Science (80-).*3171513 LP-1516
[9] Schlüter M, McAllister R R J, Arlinghaus R, Bunnefeld N, Eisenack K, Hölker F, Milner-Gulland E J, Müller B, Nicholson E, Quaas M and Stöven M. 2012. New horizons for managing the environment: A review of coupled social-ecological systems modelling. *Nat. Resour. Model.* 25:219–72.
[10] Nagendra H, and Ostrom E. 2014. Applying the social-ecological system framework to the diagnosis of urban lake commons in Bangalore, India. *Ecol. Soc.*19.
[11] Folke C. 2006. Resilience: The emergence of a perspective for social–ecological systems analyses. *Glob. Environ. Chang.*16:253–67.
[12] Tretter F, and Halliday A. 2012. Modelling Social-Ecological Systems Bridging the Gap Between Natural and Social Sciences Human-Nature Interactions in the Anthropocene Potentials of Social-Ecological Systems Analysis Ed, M Glaser, G Krause, B Ratter and M Welp (New York: Routledge) pp 60–89.
[13] Cumming G S, Allen C R, Ban N C, Biggs D, Biggs H C, Cumming D H M, De Vos A, Epstein G, Etienne M, Maciejewski K, Mathevet R, Moore C, Nenadovic M and Schoon M. 2015 Understanding protected area resilience: a multi-scale, social-ecological approach. *Ecol. Appl.* 25:299–319.
[14] Ostrom E. 2009. A General Framework for Analyzing Sustainability of Social-Ecological Systems *Science.* (80):325419 LP-422.
[15] McGinnis M D, and Ostrom E. Social-ecological system framework: initial changes and continuing challenges. *Ecol. Soc.*19
[16] Pitcher T J. 1999. *Rapfish, a rapid appraisal technique for fisheries, and its application to the code of conduct for responsible fisheries* (Rome).
[17] Bastian M, Heymann S and Jacomy M. 2009. Gephi: An Open Source Software for Exploring and Manipulating Networks. *Third International AAAI Conference on Weblogs and Social Media* (California: The Association for the Advancement of Artificial Intelligence). pp 361–2.
[18] Desai A A, and Riddle H S. 2005. *Human-Elephant Conflict in Asia* (Virginia: U.S. Fish and Wildlife Service: Asian Elephant Support).
[19] Greenpeace. 2013. *Licence to kill: How deforestation for palm oil is driving Sumatran tigers toward extinction.* (Amsterdam).
[20] Soehartono T, Wibisono H T, Sunarto, Martyr D, Susilo H D, Maddox, T and Priatna D. 2007. *Conservation Strategy and Action Plan for the Sumatran Tiger (Panthera tigris sumatrae) Indonesia 2007-2017.* Jakarta.
[21] Sandri D, and Rudiarto I. 2016. Pola Perkembangan Penggunaan Lahan dan Struktur Ruang di Sekitar Wilayah Eksploitasi Minyak Bumi di Kota Duri. *J. Pembang. Wil. Kota.* 12:361–72.
[22] Mcdonald R I, Kareiva P, and Forman R T T. 2008. The implications of current and future urbanization for global protected areas and biodiversity conservation. *Biol. Conserv.* 141: 1695–703.

IOP Conf. Series: Earth and Environmental Science **106** (2017) 012029 doi:10.1088/1755-1315/106/1/012029

Enhancing the open space of Jabodetabek area, Indonesia

A S Wartaman*, R Situmorang, B B Suharto

Urban and Regional Planning Department, Faculty of Landscape Architecture and Environmental Technology, Universitas Trisakti, Indonesia.

*Corresponding Author: asitawati@trisakti.ac.id

Abstract. Jabodetabek area is planned to be developed as a sustainable development region. The Government of Indonesia (GoI) committed in Paris Conference that the target of GHG emissions would be reduced by 29% up to 2030 with her own efforts and by 41% with supported by international assistance. This research attempts to apply specific methods such as stratified planting method to improve the quality of green open space by increasing the absorbed carbon level. The study showed that with the planting vegetation of stratified model, the total capacity of the mainland to absorb CO_2 gas is quite significant increase with approximately 20% compared to that of homogeneous vegetation. The stratified planting model is worthwhile to achieve the commitment of the GoI mentioned above.
Keywords: CO_2, land absorption, open space

1. Introduction

Indonesia is designated National Spatial Plan (RTRWN) through Government Regulation No. 26 year 2008. One of them is the Jakarta Metropolitan Area (JMA), known as Jabodetabek as a National Strategic Area or Kawasan Stratejik Nasional (KSN). The development of KSN Jabodetabek influences widespread of the built-up areas as a place of economic activity. The expansion of the built-up areas may impact on decreasing the carrying capacity of the environment, as for limited ability the land area to absorb CO_2. Such a condition would result in the increase of emissions of greenhouse gasses (GHG). GHG emissions have been related closely to climate change. Carbon dioxide (CO_2) emissions from land-use change and deforestation account for about 13% of total anthropogenic CO_2 emissions [1]. Such issue would lead to the disaster in all aspects of life, ranging from the physical, social to economic aspects. To mitigate the impact of climate change, one would argue that emissions reduction is one of the priorities for the world's target. One approach to achieving the national GHG emissions reduction targets needs a specific policy related to land-based fields, i.e. a direct regulation of land use restrictions, known as a Spatial Planning Policy. *Conference of the Parties* (COP) 21/*Conferences of the Parties serving as meeting of parties to the Kyoto Protocol* (CMP) 11 *United Nations Framework Convention on Climate Change* (UNFCCC) in Paris, the Government of Indonesian (GoI) committed to reduce GHG emissions by 29% up to 2030 with her own efforts and by 41% with support of international assistance [2]. In the last few decades, Indonesia was a frontrunner in deforestation [3]. Based on such a commitment, this study attempts to apply specific methods to improve the quality of green open space to reach the sustainable development of the Jakarta Metropolitan Area. Some researchers have examined the relationship between different types of land cover and CO_2 gas absorption, among others land-use change and CO_2 emissions in Indonesia by 2020 [4], the potential of agroforestry to reduce atmospheric CO_2 concentrations [5]. Previous research similarly focused on land use absorption emission. But, they did not utilize stratified vegetation planting method in calculating total gas absorption. Considering

IOP Conf. Series: Earth and Environmental Science **106** (2017) 012029 doi:10.1088/1755-1315/106/1/012029

variety of various vegetation in terms of CO2 capacity absorption, this paper argues to apply such methodology. Jabodetabek area is located on the island of Java, Indonesia, covering three provinces, Banten, Jakarta and West Java. Located between 106°19'BT-107°30' BT and 5°18'LS-6°48' LS. The range of height for JMA is 0-700 meters above sea level. The region chosen is based on the consideration that the JMA is the largest metropolitan area in Indonesia, Southeast Asia, and the second-largest in the world after Keihin metropolitan areas in Japan consisting of Tokyo, Kawasaki, and Yokohama. The total population of Jabodetabek is 26,746 million people in the year 2013, compared to that of Tokyo-Yokohama which is estimated to 37,239 million people [6].

2. Research Method

According to [7], one of the best approaches to mitigate the land-based National Action Plan for GHG RAN-GRK field is using the prospective method. This method combines the information of trends land's changes in the past and the anticipation of future behavior regarding the land use changes to predict the rate of change and location. In addition, one of the efforts to reduce GHG emissions is induced by the management of land resources, namely the maintenance of carbon stocks (carbon stock) and GHG. The level of CO_2 gas absorption is inversely proportional to the level of GHG emission reduction. In other words, if the rate of absorption of CO_2 gas decreases, the level of GHG emissions will be increased. By using the above approach, previous studies carried out of [8]. This research attempts to discover methods of improving the quality of open spaces in order to increase the stock of carbon in an effort to realize the sustainable development of the Greater Jakarta area. Analysis using Regulation of the Minister of Public Works on guidelines for the provision and utilization of open space in urban areas [9]. The data used in this study are mostly secondary data and cross-section (Table 1).

Table 1. Data needs.

Aspects	Type of Data	Data Forms
Physical	Size of Green Open Space Land Use Prediction in 2020	Secondary
	Topographic	Secondary
Other Supporting Aspects	Related Policies	Secondary
	GHG reduction target	Secondary
	The relationship between vegetated land cover and the ability to absorb CO_2	Secondary

Source of data is obtained from various institutions. Additionally, direct field observation was conducted to see the justification for the existing secondary data. Furthermore, based on the above approach, the method of data analysis includes:

1. Identification of the area of open space based on the slope in each region in the JMA. Slope classification based on criteria and procedure for determination the forest protection, forest conservation, and forest production. Identification conducted by using Geographic Information System (GIS) (overlay between land use map open space and slope).
2. Prediction distribution of open space in 2020. Prediction is conducted by using a linear projection formula approach and the assumption that there is land's conversion of open space into a built area on a slope of 0-15%. The basic consideration of using linear projection method is due to the pattern of development of land use 5 years previously in the form of a straight line
3. Identify the composition of the plants will be planted with planting vegetation method in open space. Type of plant vegetation on an open space consists of trees, shrubs, and agricultures
4. Estimation of the mainland's ability to absorb CO_2 Gas, calculations are made by applying the model relationship between the type of vegetation and land to cover CO_2 absorption [10] (see Table 2).

IOP Conf. Series: Earth and Environmental Science **106** (2017) 012029 doi:10.1088/1755-1315/106/1/012029

Table 2. The relationship between different types of Land Cover and CO_2 Gas Absorption [6].

Types of Land Cover	CO_2 Gas Absorption (kg/ha/hour)	CO_2 Gas Absorption (ton/ha/year)
Tree	129,92	569,07
Bush	12,56	55,00
Prairie/grass	2,74	12,00
Agricultural land	2,74	12,00

2.1. Descriptive Analysis
The objective of this part is to formulate and interpret the existing phenomena. Components of analysis are processed to cover a theoretical study, empirical results, the standard technical criteria and related policies. One of the relevant policies is the Regulation of the Minister of Public Works on guidelines for the provision and utilization of green open space in urban areas [9].

3. Results and Discussion
This discussion consisted of (i) the prediction of area of green open space land use in 2020 according to class slope, and (ii) the calculation of predictive capability of mainland land Jabodetabek absorb CO2 in 2020.

3.1. Prediction of the Size of Open Space Areas in 2020 according to the Class Slope
Prediction of open space areas will be focused on the land use for forests, agricultural areas, and parks.

3.2. Prediction Forest Areas in Jabodetabek in 2020.
Expansion of the built-up areas has consequences on the decrease of open space areas. In general, lands that are in the slope of 0-15% are lands that are vulnerable to land conversion. Based on these assumptions, the estimation of the size of forest green space areas is as follows:

Table 3. Prediction of size of forest areas in 2020 (Ha).

Region	Area based on Slope (Ha)					Total
	0- 8%	8-15%	15-25%	25-40 %	>40 %	
Bekasi Regency	0	0	0	0	0	0
Depok City	0	0	0	0	0	0
Jakarta City	0	0	0	0	0	0
Bekasi City	0	0	0	0	0	0
Bogor Regency	0	11,974.70	36,848.03	21,195.34	15,693.13	85,711.21
Bogor City	87.00	0	0	0	0	87.00
Tangerang Regency	0	0	0	0	0	0
Tangerang City	0	0	0	0	0	0
Tangsel City	0	0	0	0	0	0
Total	87.00	11,974.70	36,848.03	21,195.34	15,693.13	85,798.21

Sources: Calculated Result, Authors, 2016

Table 3 shows that in 2020, predicted use of forest land mostly in Bogor Regency. Nearly 90% of the forest area located in Bogor Regency

3.3. Prediction Agricultural Areas in 2020
Agricultural land is the land use type that is sensitive to be developed into a built-up area. Considering topography, the land area located within the slope of 0-15% is vulnerable for conversion. Based on such assumptions, the prediction area of agricultural land locations spread in any class of slope (Table 4).

Table 4. Predicted size of agricultural areas in 2020 (Ha).

Region	Area based on Slope (Ha)					Total
	0- 8%	8-15%	15-25%	25-40 %	>40 %	
Bekasi Regency	0	152.09	0	0	0	152.09
Depok City	795.30	0	0	0	0	795.30
Jakarta City	0	0	0	0	0	0
Bekasi City	1,008.56	0	0	0	0	1,008.56
Bogor Regency	66,607.43	31,639.47	18,735.38	5,146.20	8,336.41	130,464.89
Bogor City	1,647.22	1,642.58	162.36	0	0	3,452.15
Tangerang Regency	0	136.87	0	0	0	136.87
Tangerang City	617.32	0	0	0	0	617.32
Bekasi Regency	2,963.84	0	0	0	0	2,963.84
Total	73,639.67	33,571.01	18,897.74	5,146.20	8,336.41	139,591.02

Sources: Calculated Result, Authors, 2016

3.4. Prediction Park Area in 2020

Estimation size of park areas in 2020 is 1.178.99 Ha. The locations are spread throughout the region of Jabodetabek and located on the slope of 0-8%.

3.5. Calculation of Carbon Absorbed in 2020

Stratified planting vegetation is one of the methods to improve the quality of green open space. Stratified vegetation is a type of vegetation by planting not only trees and grasses but also shrubs.

3.6. Park Area Capacity in Absorbing CO_2 Gas with Planting Stratified Vegetation

The estimated Park area in 2020 is 1,178.99 Ha (Table 5). The ability of the parks area to absorb CO_2 Gas with stratified planting vegetation, consisting of 50% trees, 30% shrubs, and 20% grasses, is as follows:

Table 5. Park capacity in absorbing CO_2 gas with planting stratified vegetation.

Type of Vegetation	CO_2 Gas Absorption (ton/ha/year)	Area (Ha)	Total CO_2 Gas Absorption (ton/ year)
Tree 50%	569.07	589.50	335,463.92
Bush 30%	55.00	353.70	19,453.34
Grass 20%	12.00	235.80	2,829.58
Total		1,178.99	357,746.83

Sources: Calculated Result, Authors, 2016

Table 5 shows that the amount of carbon absorbed in park areas with the stratified planting vegetation method, with 50% trees, 30% shrubs and 20% grass, reaches up to approximately 357,746.83 tons/year. Compared with the ability to absorb CO_2 gas without planting vegetation method, there is an increase in the capability in absorbing CO_2 gas of 25 times.

3.7. Agricultural Area Capacity in Absorbing CO_2 Gas with Planting Stratified Vegetation

The estimation of the amount of agricultural area spread in 2020 reaches up to 139,591.02 Ha. The ability of the agricultural land use to absorb CO_2 Gas with stratified planting vegetation, consists of 20% trees, 30% shrubs and 50% agricultural land is as follows:

Table 6. Agricultural capacity in absorbing CO_2 Gas with planting stratified vegetation.

Type of Vegetation	CO_2 Gas Absorption (ton/ha/year)	Area (Ha)	Total CO_2 Gas Absorption (ton/year)
Trees (20%)	539.07	27,918.20	15,049,866.23
Bush (30%)	55.00	41,877.31	2,303,251.83
Grass	12.00	-	-
Agricultural-land (50%)	12.00	69,795.51	837,546.12
Total		139,591.02	18,190,664.18

Sources: Calculated Result, Authors, 2016

Table 6 shows that planting with the stratified vegetation, consists of 20% trees, 30% shrubs, and 50% paddy fields on agricultural land use, the rate of CO_2 gas absorption reaches up to 18,190,664.18 ton/year. Compared with the ability to absorb CO_2 gas without stratified vegetation planting (1,675,092.20 tons/year), there is an increase in ability to absorb CO_2 gas is about 10%.

3.8. Forest Area Capacity in Absorbing CO_2 Gas with Planting Stratified Vegetation
In detail, with consideration to the canopy width and spacing of trees, the maximum total of trees that can be planted on forest land is approximately 90%. By estimating the size of the forest areas in 2020 will reach up to 85,798.21 Ha. With the composition of trees and bushes of 90% to 10%, the ability of the land area to absorb CO_2 as follows (Table 7).

Table 7. Forest capacity in absorbing CO_2 Gas with planting stratified vegetation.

Type of Vegetation	CO_2 Gas Absorption (ton/ha/year)	Area (Ha)	Total CO_2 Gas Absorption (ton/ year)
Trees 90%	539.07	77,218.39	41,626,117.50
Bush10%	55.00	8,579.82	471,890.16
Grass	12.00		-
Agricultural land	12.00		
Total		85,798.21	42,098,007.65

Sources: Calculated Result, Authors, 2016

Table 8. Capability to absorb CO_2 between Homogeneous and Stratified Vegetation.

Green Open Space	The Ability to Absorb CO_2 Gas (Ton/Year)	
	Homogeneous Vegetation	Stratified Vegetation
Parks	14,147.82	357,746.83
Agricultural Land	1,675,092.20	18,190,664.18
Forests	48,825,187.36	42,098,007.65
Total	50,514,427.38	60,646,418.66

Sources: Calculated Result, Authors, 2016

Table 8 shows that with the stratified planting vegetation which consists of 90% trees and 10% shrubs in agricultural land use, the rate of CO_2 gas absorption is 42,098,007.65 ton/year. Compared with the ability to absorb CO_2 gas without stratified planting vegetation (48,825,187.36 tons/year), there is a decrease in ability to absorb CO_2 gas for about 14%. Based on the review above, with the stratified planting vegetation, the total ability of the mainland to absorb CO_2 gas is 60,646,418.66 tons/year. The amount is increased by approximately 20% compared to the ability of the land to absorb CO_2 gas regions with homogeneous vegetation (Table 8). An increase that happens is not in accordance with the expected increase to comply the commitments the Government of Indonesia in the G-20 meeting in Pittsburgh. However, when the increase is accumulated with the amount of CO_2 gas absorption of CO_2 gas that comes from the yard of the house, green belt, roads railways green open space, etc., the capability of CO_2 absorption.

4. Conclusion

The study results showed that with the stratified planting vegetation method, there is an increase in the rate of CO_2 gas absorption by about 20%. Unfortunately, such improvements are still not enough to meet the commitment of the GoI to increase the absorption of CO_2 gas by 29% in 2030. However, when accumulated with the rate of CO_2 gas absorption that comes from the yard of the house, green belt, roads, and railways green space, and etc., the ability to absorb CO_2 gas is expected to achieve the commitment of the GoI mentioned above. Based on this research, we may conclude that stratified planting vegetation is the most suitable planting method for lands with the height of 0-700 m above sea level.

Acknowledgment

This research was supported by Ministry of Research, Technology and Higher Education of the Republic of Indonesia through the mechanism of research grants namely "Hibah Bersaing". We gratefully acknowledge the support and generosity of Ministry of Research, Technology and Higher Education of the Republic of Indonesia mentioned above.

References

[1] Friedlingstein P, Houghton RA, Marland G, Hackler J, Boden TA, Conway TJ, Canadell JG, Raupach MR, Ciais P, Le Quere C. 2010. Update on CO2 emissions. *Nature Geoscience.* 3(12):811–812.

[2] Kementerian Lingkungan Hidup dan Kehutanan. Perubahan Iklim, Perjanjian Paris, dan Nationally Determined Contribution. ISBN: 978-602-74011-1-2; 2016.

[3] Margono B A, Potapov P V, Turubanova S, Stolle F, Hansen M C. 2014. Primary forest cover loss in Indonesia over 2000–2012. *Nature Climate Change.* 4(8):730–735.

[4] Liselotte Schebek, Jan T. Mizgajski, Rüdiger Schaldach and Florian Wimmer. 2018. Land-Use Change and CO2 Emissions Associated with Oil Palm Expansion in Indonesia by 2020, Springer International Publishing AG. B. Otjacques et al. (eds.), From Science to Society.

[5] Abbas F, Hammad HM, Fahad S, Cerdà A, Rizwan M, Farhad W, Ehsan S. Bakhat HF. 2017. Agroforestry: a sustainable environmental practice for carbon sequestration under the climate change scenarios- a review. *Environmental Science Pollution Research.*24:11177–11191.

[6] Amri A B. Jabodetabek, Megacity World's Second Largest. Available from URL: http://dunia.news.viva.co.id/news /read/404299-jabodetabek---i-megacity--i--terbesar-kedua-di-dunia. Accessed August 10, 2015.

[7] Ministry of National Development Planning. Guidelines for the Implementation Action Plan for Greenhouse Gas Emission Reduction; 2011. pp: 37-8.

[8] Wartaman A S. 2015. Sustainable Development of The Jabodetabek Area, West Java Province, Indonesia: The Second Largest Mega City Area in The World. *African Journal of Science and Research.* 4(5):09-12 ISSN: 2306-5877

[9] Ministry of Public Works. Guidelines for The Provision and Utilization of Green Open Space in Urban Areas. No. 05/ PRT/M/2008 the year 2008.

[10] Velayati LH et al. 2013. Green Open Space Needs Analysis Based Gas Absorption of CO2 in Pontianak. *Journal of Environmental Engineering of Tanjungpura University.*

Accessibility of low-income family flats in North Jakarta city

T A Feminin[*], H W Wiranegara, Y Supriatna

Urban and Regional Planning Department, Faculty of Landscape Architecture and Environmental Technology, Universitas Trisakti, Jakarta, Indonesia

*Corresponding Author: trieayufeminin@gmail.com

Abstract. The majority of relocated, low-income families in North Jakarta city who residing the flats, complained at decreasing their accessibility to the workplaces and to the social facilities. The aim of this research was to identify the changing of their accessibility before and after relocated, viewed from three dimensions: distance, travel time, and travel cost to the workplaces, educational facilities, and shopping areas. The research design was questionnaire survey containing the degree of accessibility before and after resided the flats. Five flats were chosen as cases. Their inhabitants were chosen as respondents which used simple random sampling. The result showed that their flats accessibility to the workplaces in all three dimensions was lower than when they resided in the slum area. Also, in distance and travel time accessibility to shopping areas was lower. Only accessibility to educational facilities measured in those three dimensions was higher after they moved. Supply for affordable public transport from their flats to reach their workplaces is needed to raise their accessibility. Also, they need subsidizeto rent of their flats so the burden to their income lesser.Using the ground space of their flats for retail activities was to make more accessible for their shopping activities.
Keywords: accessibility, low-income families, slum area

1. Introduction

There are three phenomena related to poverty for getting worse, i.e. the high level of transportation cost, the high cost of rental housing, and the expensive price of basic need goods. One problem faced by the low-income families and related to transportation cost is the accessibility of their housing. Accessibility for low-income families has the potential to reduce poverty and therefore contribute to sustainability [1]. Low-income families are people with limited purchasing power who need support from the government to get the affordable and livable house. Based on Public Housing Ministerial Regulation [2], low-income families are people who have income between 3.5-5.5 million rupiahs per month.

The new low-income families flats location in North Jakarta city was faced the accessibility problem to their workplaces and social facilities. Development of low-income family flats here was to accommodate the inhabitants of slums areas who resided on the riverbanks and whom the target of relocation. But the majority of them who moved to the flats claimed that their locations were not accessible. Their travel cost raised their living cost. To reach educational facilities also burden their children who still continued their study in the old schools when they resided slums areas. Meanwhile moving to the new school which nearer from their flats was not easy, especially for them who took the national final test.

"Accessibility expresses the ability of an environment to guarantee everyone an independent life regardless of their age, gender, cultural background and physical, sensory or cognitive abilities" [3].

In this research, a phenomenon of accessibility was focused on housing (i.e. the flats), low-income families, three major activities which usually do by the member of family (work, shop, education). Flats development goal for low-income families is how to facilitate them for meeting their social and economic needs. The most important factor for low-income families is accessibility to facilities location where they do their activities. To represent all the activity of family member, this research chose three kinds of activities and travel destinations namely the workplaces which represent activity and travel of fathers or husband, shopping facilities which represent the mother or housewife activity, and educational facilities which represent their children activity.

The degree of accessibility of low-income family flats to reach the workplaces, educational facilities, and shopping areas was viewed in three dimensions: distance, travel time, and travel cost. It was compared before they moved and after they resided the flats. Flats location affects the pattern of spending of family income, especially travel cost to working areas and educational facilities which are daily activities. The research aim was to compare their degree of accessibility of their residential area, before and after they moved to the flats.

2. Research Method

Design research was a survey questionnaire with open questions. Questions directed to record two conditions namely a degree of accessibility before and after low-income families moved to the flats. From their response, the analysis was focused to compare their degree of accessibility in three dimensions (distance, travel time, travel cost) to the workplaces, educational facilities, and shopping areas before and after moving to the flats. Five public flats in North Jakarta city were chosen as samples. They were Marunda, Kapuk Muara, Sukapura, Muara Baru, and Penjaringan. Respondents were chosen using simple random sampling method. A questionnaire survey was conducted from 18th November 2016 until 18th December 2016.

3. Results and Discussion

Low-income families that were studied had the characteristics as follows: 3-4 persons in family size, a level of education of majority is from elementary to high school. Their occupations are construction laborers, have own business, or private employees. The degree of accessibility of their flats was measured by physical distance, time distance, and travel cost to the workplaces, educational facilities, and shopping areas.

Changing the degree of accessibility of the low-income families before and after they moved to the flats can be described below.

3.1. Trend of degree of accessibility to the workplaces before and after moving to the flats

In general, in all five cases, a trend of the accessibility to the work places from their flats in all three dimensions was decreasing or getting lower than before they moved, see Table 1. It was the big problem for low-income families caused by flats location chosen. The more distance or time travel, the more cost of public transport, the lesser their quality of life.

Break down into each case, it could be found that Marunda and Kapuk Muara were decreasing in all dimension of accessibility. In these locations, there were no kinds of employment available suitable for their background and unavailability of affordable public transport for them. Those were the ultimate caused for decreasing in accessibility. They faced more distance to the workplaces so were their travel time and travel cost which could affect their quality of life.

Penjaringan and Muara Baru had two dimensions of accessibility which its trend was increasing. In both cases, there was available employment suitable with the kinds of their jobs. Surrounding of Penjaringan there was the center of Jakarta city containing offices which employment availability. Shorter travel time and lower travel cost to working areas here were caused by the availability of public transport. Location of Muara Baru is near seaport with available jobs fit with some of their kinds of jobs. The trend of accessibility in distance and travel time was increasing, but in travel-cost was decreasing. Sukapura is also near industry and offices areas which could be their workplaces. Only accessibility in travel time was an increase in its trend here.

Table 1. Changing their accessibility to the work places before and after moving to the flats.

Name of flats	Accessibility dimensions to working areas		
	Distance	Travel time	Travel cost
Marunda	decrease	decrease	decrease
Sukapura	decrease	increase	decrease
Muara Baru	increase	increase	decrease
Penjaringan	decrease	increase	increase
Kapuk Muara	decrease	decrease	decrease
All 5	decrease	decrease	decrease

Table 1 showed clearly that trend of accessibility in distance and travel cost was decreasing more in a number of cases. Only accessibility in travel time was more in a number of cases which was increasing. For low-income families, travel cost is the most important dimension in accessibility has to be considered when choosing the location for their flats. Lack of affordable public transport made the travel cost to the workplaces was rising and push their quality of life or their welfare lower. For low-income families to pay rent for their flats had already burdened them. Hence, local government besides providing subsidize for public transport also need to give subsidize the cost of the rental flats. It was just to sustain their level of welfare.

3.2. Trend of degree of accessibility to educational facilities before and after moving to the flats

The trend of a degree of accessibility to educational facilities in all dimension and all cases was increasing, see Table 2. This result was contrary to the trend of accessibility to the work places. Increasing accessibility was caused by the availability of educational facilities surrounding their flats and availability of free school buses for the students.

Marunda and Kapuk Muara showed increasing accessibility in all dimensions. Decreasing of accessibility to educational facilities in certain dimensions found in Sukapura and Muara Baru as well as Penjaringan. In these three flats showed the accessibility in distance was decreasing. The distance accessibility to educational facilities here was not fit with the Indonesian National Standard.

Table 2. Changing their accessibility to educational facilities before and after moving to the flats.

Name of flats	Accessibility dimensions to educational facilities		
	Distance	Travel time	Travel cost
Marunda	increase	increase	increase
Sukapura	decrease	increase	decrease
Muara Baru	decrease	increase	decrease
Penjaringan	decrease	increase	increase
Kapuk Muara	increase	increase	increase
All 5	increase	increase	increase

Although in some flats their distance accessibility to educational facilities decreased, the availability of school buses services made the accessibility in travel time in all cases were increasing. Decreasing in distance accessibility was the locational failure in developing new flats for low-income families which depend on the availability of land owned by local government. To overcome this problem they chose using a motorcycle that could be destructed to the quality of the environment. In other cases found that the increasing accessibility in travel cost caused by the availability of free school buses. It could be found in Marunda, Penjaringan, and Kapuk Muara. It showed that subsidize for public transportation was very helpful for low-income families to raise the accessibility of their flats location.

3.3. Trend of degree of accessibility to shopping area before and after moving to the flats

The trend of accessibility in distance and travel time to shopping facilities in all cases together was decreasing, only in travel cost was increasing, see Table 3. Marunda was the only one decreasing in all dimensions. It caused by incompleteness of goods offered by the nearest shopping place, so they chose to shop further facilities. It made the accessibility of all dimensions getting lower. Sukapura was also decreasing, but in two dimensions: distance and travel time. In these two flats mentioned, a trend of accessibility in distance was decreasing. It was caused by the distance were longer, so was the travel time.

Table 3. Changing their accessibility to shopping facilities before and after moving to the flats .

Name of flats	Accessibility dimensions to shopping facilities		
	Distance	Travel time	Travel cost
Marunda	decrease	decrease	decrease
Sukapura	decrease	decrease	increase
Muara Baru	increase	increase	increase
Penjaringan	increase	fixed	increase
Kapuk Muara	decrease	increase	increase
All 5	decrease	decrease	increase

On the contrary, Muara Baru was the only one showed increasing in the trend of accessibility to shopping facilities in all dimensions. In Muara Baru, its shopping facilities could fulfill their needs. A trend of accessibility to shopping facilities in Penjaringan and Kapuk Muara was increasing too, with the same reason as in Muara Baru. There was no changing in travel time between before and after they moved to the flats in Penjaringan. In Kapuk Muara, the accessibility in travel time and travel cost were increasing caused by using motorcycle which made faster and cheaper to reach its destination.

In general, the trend of accessibility in distance and travel time to shopping facilities was decreasing. From this result could be concluded that local government had to be considered providing shopping facilities needed near and/or supplying the subsidized public transportation to support them. Using the ground space in the tower of flats to accommodate retail activities was the best solution to increase the accessibility in all dimensions. Almost all cases showed increasing accessibility in travel cost except Marunda. In Marunda inhabitant found more distance and travel time to reach shopping facilities.

3.4. Discussion

There are several factors to be noticed in making choice for low-income housing location i.e. accessibility to the workplaces as well as commercial and educational facilities [4]. As a criterion, the walking distance to working areas is 20-30 minutes or 1.5-2.75 km [5]. Also, the radius and the walking distance criteria for educational and shopping facilities that can be used are as follow: elementary and secondary schools are 1 km or 15 minutes; high school is 3 km or 30 minutes; shopping areas is 2 km or 20 minutes [6]. The best radius for housing to the market, bank, hospital, working areas, school, etc.is 1.5 km, it is called walker's paradise [7]. These criteria had to be considered when local government providing flats for low-income families if they want to sustain or even raise their quality of life and keep in the pace of sustainable urban development.

Comparing the changing of accessibility among destinations, it could be ordered from the lowest to the highest: the workplaces, shopping facilities, and educational facilities. For low-income families, accessibility to the work places is the most important to sustain their level of welfare. In the case of development of the new flats for them in the North Jakarta City, this factor likely was abandoned. The trend of accessibility to the work places in all dimensions was decreasing. This was not an appropriate location for the low-income families flats that built by the local government. In building the new flats for low-income families, local government depended on the availability of land owned. So, to increase the accessibility to the work places is needed providing subsidized public transportation. In the long run, needs land banking mechanism which can support the availability of suitable location for low-income.

The trend of accessibility to shopping facilities was also decreasing, but there was travel cost dimension rising. In general, low-income families chose the complete shopping facilities even it further. Why this happened can be the next research to answer.

Accessibility to educational facilities was increasing in three dimensions. It sounds good. Local government provided dispersed free school buses for students from low-income families.

Table 4. Comparing changing accessibility among all destination in all dimensions.

No	Destinations	Accessibility dimensions		
		Distance	Travel time	Travel cost
1	Working areas	decrease	decrease	decrease
2	Shopping facilities	decrease	decrease	Increase
3	Educational facilities	increase	increase	Increase
	All destinations	decrease	decrease	increase

If compared to dimensions, a trend of accessibility in distance and travel time had the same pattern, they were decreasing, see Table 4. The more distance between the flats and destination, the more decreasing accessibility in travel time and travel cost. Only to educational facilities, the distance more accessible because of the availability of educational facilities themselves. So were accessibility in travel time and travel cost was increasing because of availability of free school buses.

To increase the accessibility in all dimensions for all destinations for low-income families flats, the local government needs providing the subsidize for public transportation and subsidize the cost of renting the flats. Transport cost will get lower with subsidize, the rent cost will too. These two subsidize will lessen the burden of their living cost. Not only the subsidize for public transportation but also the spatial distribution of bus networks that reachable must be provided. As mentioned earlier, lesser transport cost and rent cost were the contributors to sustaining their welfare.

4. Conclusion

After moving to the flats, low income families in North Jakarta city had decreasing accessibility to the work places in all three dimensions (distance, travel time, travel cost), decreasing accessibility to shopping areas in two dimensions (distance, travel time), and increasing accessibility to educational facilities in all three dimensions (distance, travel time, travel cost). The good supply of affordable public transportation through subsidize and/or providing free public transport like school buses had helped increasing accessibility of low-income families. Fast and affordable public transport has to be supplied to overcome decreasing accessibility especially accessibility to the workplaces. For a longer period, to prevent decreasing accessibility, the local government had to be considered land banking mechanism to set the location for flats development which suitable with the characteristics of low-income families. But, immediate effort to raise the accessibility in shopping activity is using the ground space in the tower of their flats for retail.

References

[1] Chapin FS. Urban Land Use Planning. Urbana: Univ. of Illinois Press.1972.
[2] Edelman A, and Pudenzi A C M. Accessibility of Housing: A Handbook of Inclusive Affordable Housing Solutions for Persons with Disabilities and Older Persons. 978-92-1-132635-26. 2014.
[3] Florence accessibility lab. Accessibility to cultural heritage & human development. England: Universita Degli Studi Firenze. p.3. 2013.
[4] National Standard of Indonesia number 03-1733. Procedures for Urban Housing Planning.

Jakarta: National Standardization Institution. 2004.
[5] Public Housing Ministerial Regulation Number 27 the year 2012. October 8, 2012. Housing Procurement through Credit/Financing for Prosperous Home Ownership with the Support of Housing Financing Liquidity Facility. Jakarta
[6] Schmitz A. 3rd Residential Development Handbook. Washington DC: The Urban Land Institute. January, 2004.
[7] Yin J. Urban Planning for Dummies. Chapter 6 p. 115-116. Canada: John Wiley & Sons Ltd. 2012

Simulating and evaluating an adaptive and integrated traffic lights control system for smart city application

E Djuana*, K Rahardjo, F Gozali, S Tan, R Rambung, D Adrian

Computer Systems Laboratory, Electrical Engineering Department, Faculty of Industrial Technology, Universitas Trisakti, Jakarta, Indonesia

*Corresponding Author : edjuana@trisakti.ac.id

Abstract. A city could be categorized as a smart city when the information technology has been developed to the point that the administration could sense, understand, and control every resource to serve its people and sustain the development of the city. One of the smart city aspects is transportation and traffic management. This paper presents a research project to design an adaptive traffic lights control system as a part of the smart system for optimizing road utilization and reducing congestion. Research problems presented include: (1) Congestion in one direction toward an intersection due to dynamic traffic condition from time to time during the day, while the timing cycles in traffic lights system are mostly static; (2) No timing synchronization among traffic lights in adjacent intersections that is causing unsteady flows; (3) Difficulties in traffic condition monitoring on the intersection and the lack of facility for remotely controlling traffic lights. In this research, a simulator has been built to model the adaptivity and integration among different traffic lights controllers in adjacent intersections, and a case study consisting of three sets of intersections along Jalan K. H. Hasyim Ashari has been simulated. It can be concluded that timing slots synchronization among traffic lights is crucial for maintaining a steady traffic flow.
Keywords: economic cost of congestion, smart city, smart traffic light control

1. Introduction

The introduction of Indonesia Smart City index in 2015 has created a national standard to measure how major cities in Indonesia are developing, in terms of smart city achievement [1]. A city could be categorized as a smart city when Information Technology (IT) has been developed to the point that the administration could sense, understand, and control every resource to serve its people and sustain the development of the city [1]. Several cities in Indonesia have geared up toward the smart city concept, such as Jakarta Smart City [2], and 98 cities have been surveyed in a national competition including Depok, Bandung, Semarang, Surabaya, and Tangerang as the champions [3].

One of the smart city aspects is transportation and traffic management. As the economic cost of congestion in major cities (Jakarta and Bandung) is estimated as high as IDR 35 trillion (28 trillion in Jakarta and 7 trillion in Bandung) every year [5], a coordinated effort which may alleviate a city's congestion is deemed necessary, especially in tackling busy and complex intersections at the center of the problem to reduce losses.

One stream of four focuses in Trisakti University's Research Strategic Plan [4] is "Eco-region" with one of its sub focuses of "City Transportation" covering "infrastructure and transportation as well as transportation integration and economic cost of congestion". Based on this focus, a research project has been initiated to design an adaptive and integrated traffic lights control system as a part of the smart system for optimizing road utilization and reducing congestion. Therefore, the purpose of this research is to design a simulator for the traffic lights control system with the adaptivity and integration as its core characteristics.

In each intersection/crossroad, traffic lights are installed to direct all the vehicles going through that intersection from all directions. The purposes of those traffic lights are (1) to prevent accidents from happening, caused by uncoordinated traffics from all directions, and (2) to prevent congestions that might happen because of the substantial number of vehicles or traffic density. Although adequate traffic lights have been installed, heavy congestion may still occur due to the incorrect timing setup compared to the traffic volume at different time of the day, e.g., the duration of green light is too short at busy time yet the traffic is heavy, therefore causing a long queue in one direction, while at vacant time the duration of green light is too long, causing non-optimal utilization of lanes/direction.

Nowadays, to monitor and coordinate traffic in multiple sites of the city, a traffic management center (TMC) has usually been established. The purpose of this TMC is to monitor the controller status in each intersection and is equipped to overwrite the timing cycle in each local controller. Also, to improve the effectiveness and coordination among multiple controllers, TMC can synchronize the timing of nearby controllers in the adjacent intersections [6], [7].

According to the design framework described above, the literature reviews for different controller implementation regarding adaptivity and integration features have been conducted. Several local controller designs using sensors to provide adaptivity [8], [9], [10], [11] equipped with a programmable logic controller (PLC) for a convenient change timing [8], [9], [10] have been implemented. However, they did not provide flexibility for testing out a different configuration of intersections. Several monitoring systems for remote controllers in intersections have been implemented [12], [13]. However, they did not provide synchronization among different local controllers in the adjacent intersections.

2. Research Method

2.1. Research problems, aim, and scope

This study focuses on three (3) research problems as follows: (1) Congestion in one direction toward an intersection due to dynamic traffic conditions from time to time during the day, while the timing cycles in traffic lights system are mostly static; (2) No timing synchronization among traffic lights in adjacent intersections, causing unsteady flows; and (3) Difficulties in traffic condition monitoring on one intersection and the lack of facility for remote controlling of traffic lights.

Specifically, this research aims to build a simulator, combining hardware and software as a tool to assist in the analysis and design process of the traffic lights control system. Thus, all kinds of traffic lights configurations and timings can be simulated before the actual implementation.

This simulator design process follows the system engineering methodology consisting of the following tasks: requirements analysis, design and modelling, implementation, testing and verification, and finally, system deployment. All these tasks can be carried out in multiple cycles (iterations) which produce a gradual progress of completed system. In this paper, the discussion is limited to the first two tasks which are the two dominant stages in producing the simulator: (1) requirement analysis and (2) design and modelling.

2.2. Case study selection

As a case study for simulating adaptivity and coordination among traffic light controllers in adjacent intersections, a series of intersections along Jalan K. H. Hasyim Ashari nearby Trisakti campus in West Jakarta was selected, started from Jalan Gajah Mada to Grogol intersection. A modeled map is presented in Figure 1. This location was chosen based on certain criteria as follows:

1. One intersection along this road has enough complexity for modeling adaptivity in traffic management, i.e. the Cideng intersection which has four directions of traffic.
2. Alongside this road, there is a busway lane that adds one extra traffic to manage which contributes to merging traffic on the flyover bridge which may slow down traffic leaving Roxy intersection.
3. The distance between adjacent intersections (Cideng and Petojo) is relatively short which may need

IOP Conf. Series: Earth and Environmental Science **106** (2017) 012031 doi:10.1088/1755-1315/106/1/012031

coordination in case of priority traffic request, e.g., an important government official convoy or emergency vehicles (ambulance, fire department, or police) demanding to travel along without disruption.

4. With this condition, a total deadlock has frequently happened in the intersections along this road especially at Cideng and Petojo during busy hour, caused by static timing (no adaptivity to traffic changes on different time of the day) and uncoordinated traffic light timing (see Point 3) as well as undisciplined drivers who did not want to wait before entering the intersection while traffic was heavy and still stuck in the middle of intersection.

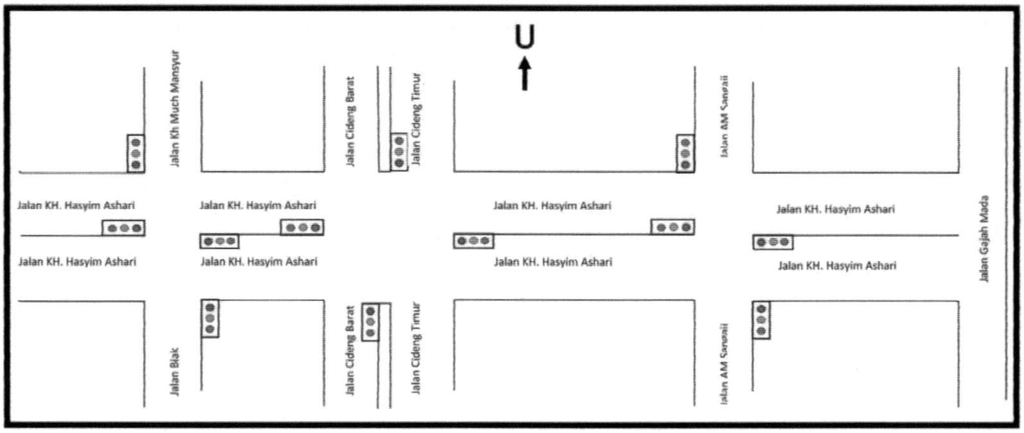

Figure 1. Modeled map of Jl. KH. Hasyim Ashari from Jl. Gajah Mada to Grogol for simulation purpose.

3. Results and Discussion

3.1. Analysis phase

In the analysis phase, timing sequence analyses of current condition for all traffic lights in each intersection alongside Jl. K. H. Hasyim Ashari were conducted as a basis for local traffic controller design. A sample timing sequence for one intersection is provided in Figure 2.

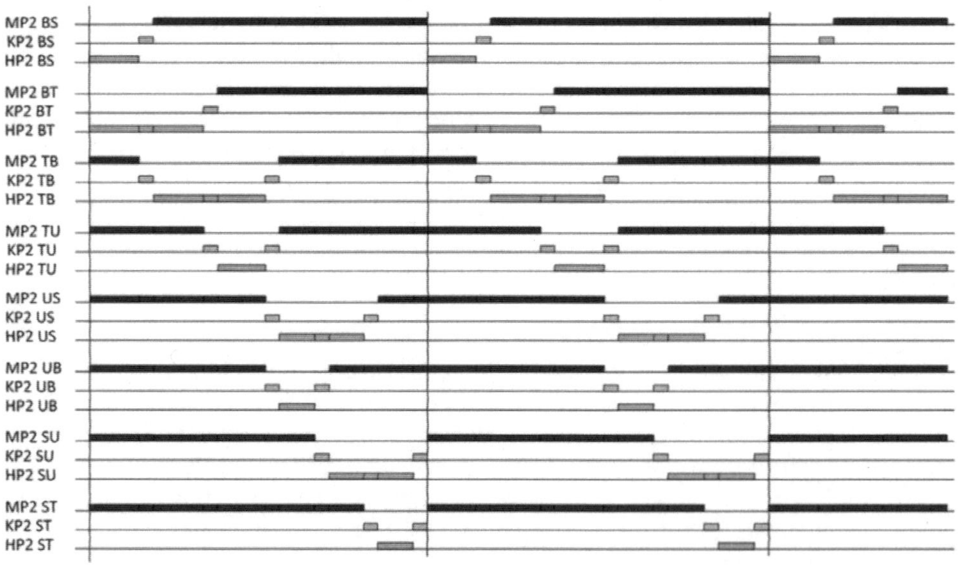

Figure 2. Traffic light timing sequence at Cideng intersection

IOP Conf. Series: Earth and Environmental Science **106** (2017) 012031 doi:10.1088/1755-1315/106/1/012031

Legend for Figure 2:
MP2 BS, KP2 BS, HP2 BS: Traffic light timing sequence alongside Jl. K. H. Hasyim Ashari from Roxy direction turning to Jl. Cideng Timur. MP2 BT, KP2 BT, HP2 BT: Traffic light timing sequence alongside Jl. K. H. Hasyim Ashari from Roxy direction straight to Jl. Gajah Mada. MP1 TB, KP1 TB, HP1 TB: Traffic light timing sequence alongside Jl. K. H. Hasyim Ashari from Jl. Gajah Mada direction straight to Roxy. MP2 TU, KP2 TU, HP2 TU: Traffic light timing sequence alongside Jl. K. H. Hasyim Ashari from Jl. Gajah Mada turning to Jl. Cideng Barat. MP2 US, KP2 US, HP2 US: Traffic light timing sequence alongside Jl. Cideng Timur in a direction straight to Tanah Abang MP2 UB, KP2 UB, HP2 UB: Traffic light timing sequence alongside Jl. Cideng Timur turning to Jl. K. H. Hasyim Ashari to Roxy direction. MP2 SU, KP2 SU, HP2 SU: Traffic light timing sequence alongside Jl. Cideng Barat in a direction to Jl. Cideng Barat. MP2 ST, KP2 ST, HP2 ST: Traffic light timing sequence alongside Jl. Cideng Barat turning to K. H. Hasyim Ashari to Jl. Gajah Mada direction.

For each intersection, currently, there is a local controller in which the sequence can be set and can be over-written manually. The sequence is synchronized to prevent race condition which can cause an accident. Once it is set, the sequence will run automatically, but if a timing change is needed for one lane, for example, it needs to be arranged manually on the spot. There is an indication of timer use for establishing discrepant timing sequence at a different time of the day. However, a manual set up of the operation is still needed whenever there is a new sequence. There are also no functions of synchronizing adjacent controllers, sending new sequence remotely, or automatically detecting current traffic.

Based on this analysis regarding the adaptivity and integration, means to address all these limitations are needed. To be able to provide adaptivity to cater the dynamic traffic change, a traffic sensor is required for detecting traffic changes and determining at what threshold level the sequence needs to be altered. Therefore, a statistical evaluation is necessary to serve as a basis for the dynamic sequence change calculation.

To be able to provide integration among multiple intersections, a centralized controller is required to monitor the status of each intersection and to define which can remotely synchronize and overwrite the sequence in each intersection. An analysis and rearrangement of timing slots in each intersection need to be conducted to attain synchronization among controllers. Several steps of rearrangement that have been carried out in this study are as follows:

1) Combining the timing diagrams of multiple intersections in one place for the ease of re-arrangement
2) Identifying continuous timing slots which can be forwarded from an intersection to the next one
3) Rearranging timing slots and relabelling timing variables to facilitate the synchronization
4) Performing the adaptive and synchronized change just by changing one timing variable (the rest of timing slots will alter accordingly, and synchronization could be maintained) once the timing slots are continuous (correspond to counterparts from other intersections or other directions)

The results of this rearrangement are depicted in Figure 3, showing that all corresponding time-slots are put together in the same order and with the same label. Changing one timing sequence, e.g., extending a green light duration in one direction in case of priority traffic, requires only a change of the corresponding variable.

3.2. Design phase

Based on all factors described in the analysis phase, a high-level System Block Diagram with an indication of technological components have been defined as guidance for detailed design as depicted in Figure 4. It shows the role of the central traffic controller which may reside in TMC which can synchronize multiple intersections by sending out synchronizing command based on the synchronized timing diagrams discussed in Section 3.1. It is also capable of sending out those commands remotely through the utilization of Internet of Things (IoT) platform in which every controller is linked to each other in a loosely coupled manner through the Internet.

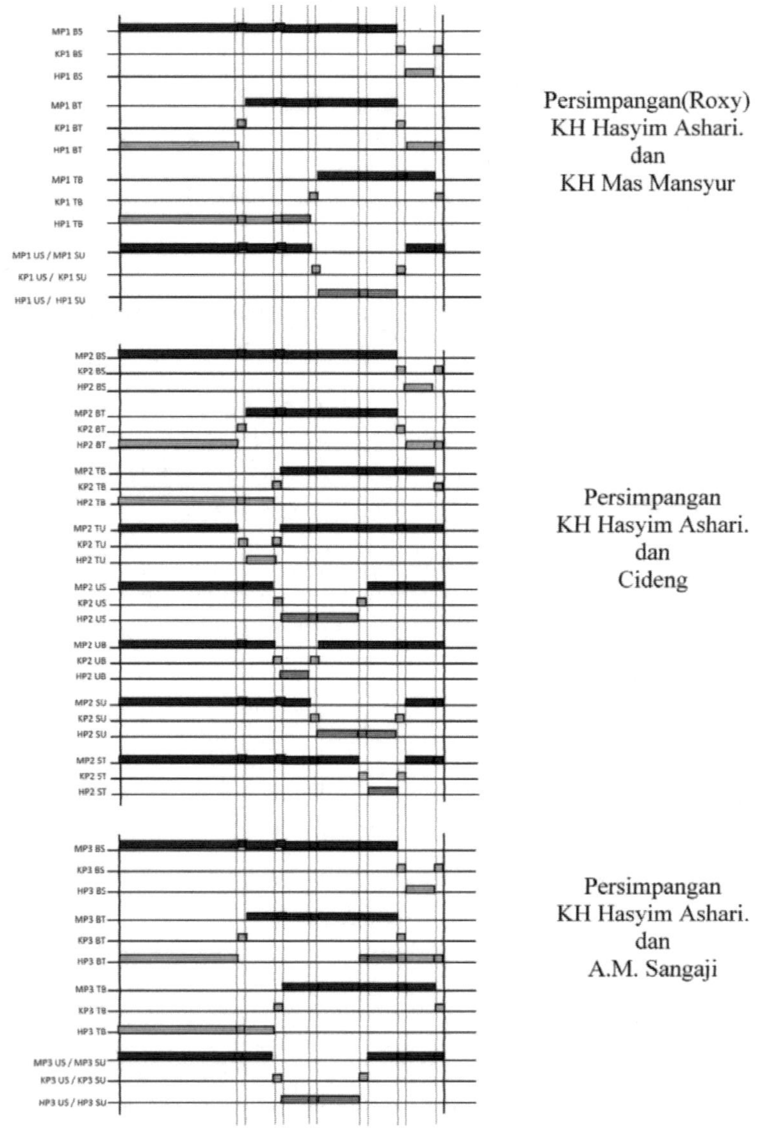

Figure 3. Integrated and synchronized timing sequence diagram to facilitate synchronization and flexibility of change in the sequence.

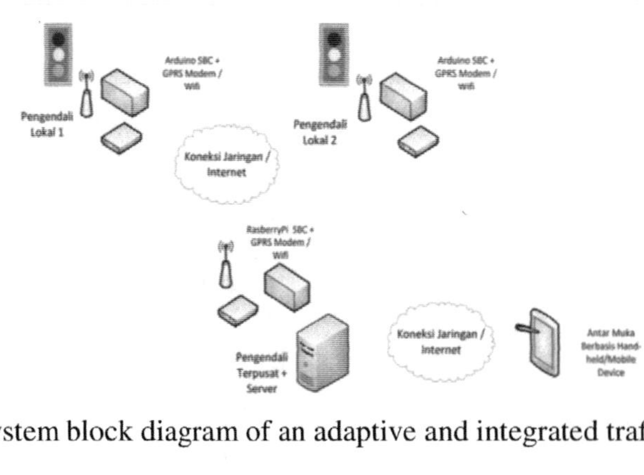

Figure 4. High-level system block diagram of an adaptive and integrated traffic lights control system.

IOP Conf. Series: Earth and Environmental Science **106** (2017) 012031 doi:10.1088/1755-1315/106/1/012031

4. Conclusion

The high economic cost to solve congestion in critical intersections will lead to the losses in productivity. A case study has been carried out regarding how to tackle frequent congestion due to the complexity of the intersections and uncoordinated traffic lights controllers. Integration and adaptivity among intersections can be facilitated by performing an analysis of the timing diagrams of related intersections as well as a rearrangement of timing slots. This synchronized timing has been simulated in a simulator in this study and can be implemented and deployed in real world scenario.

Acknowledgment

The authors thank the generous funding, support, and guidance provided by Faculty of Industrial Technology, Trisakti University, in conducting this research.

References

[1] Budiari I. 2015. Smart city index coming to Indonesia. *The Jakarta Post* (url:http://www.thejakartapost.com/news/2015/03/25/smart-city-index-coming-indonesia.html, accessed: June 25th, 2017)

[2] Jakarta Smart City. 2014. *Portal Jakarta smart city* (url: http://smartcity.jakarta.go.id/, accessed: June 25th, 2017).

[3] The Jakarta Post. 2015. Indonesia's 15 smart cities announced. *The Jakarta Post* (url: http://www.thejakartapost.com/news/2015/08/13/indonesia-s-15-smart-cities-announced.html, accessed: June 25th, 2017).

[4] Trisakti Research Institute (Lemlit). 2016. *Universitas Trisakti strategic research plan (rencana induk penelitian) 2016-2020.* Jakarta : Trisakti Research Institute.

[5] Primanita A. 2016. Chronic congestion costs big cities Rp35t a year. *The Jakarta Post* (url: http://www.thejakartapost.com/news/2016/03/29/chronic-congestion-costs-big-cities-rp-35t-a-year.html, accessed: 25th June 2017).

[6] Djuana E. 2005. Analysis and design of traffic lights control system by design and analysis of real-time systems (DARTS) methodology. Unpublished.

[7] Djuana E, Rahardjo T S K, Gozali F, Tan S, and Rambung R. 2016. Perancangan simulator sistem pengendalian lampu lalu lintas adaptif dan terintegrasi. *Prosiding Seminar Nasional Teknologi Industri V 2016.* Jakarta: Fakultas Teknologi Industri, Universitas Trisakti. 373–378.

[8] Swamardika A. 2005. Simulasi kontrol lampu lalu lintas sistem detektor dengan menggunakan PLC untuk persimpangan Jalan Waribang-W. R. Supratman Denpasar. *Jurnal Teknologi Elektro.* 4:17–21. Universitas Udayana, Juli–Desember 2005.

[9] Pissesti A. 2012. Sistem simulasi kontrol lampu lalu lintas berdasarkan kepadatan jalan berbasis programmable logic controller. S1 Final Project Universitas Negeri Yogyakarta.

[10] Cesardarmantya H, Triyanto D, and Brianorman Y. 2013. Prototype lampu lalu lintas berbasis PLC berdasarkan panjang antrian kendaraan pada perempatan jalan. *Jurnal Coding Sistem Komputer Universitas Tanjungpura.* 01(2):28–39.

[11] Riyadi K P, Wahyunggoro O, and Prabowo H. 2014. Simulasi lampu lalu lintas dengan sensor di simpang empat menggunakan software Automation Studio 5.0. *Jurnal Penelitian Teknik Elektro dan Teknologi Informasi.* 1(1): 24–28. Universitas Gajah Mada. April 2014.

[12] Asr M, Zainuddin Z, and Ilham A A. 2012. Pengembangan sistem kontrol dan monitoring lampu lalu lintas. Universitas Hasannudin Research Report.

[13] Budiman J M. 2012. Sistem monitoring dan kontrol lalu-lintas perkotaan. *Jurnal Pasca Sarjana Univ tas Hasannudin*. Makassar.

Management of Hazardous Waste in Indonesia

H Widyatmoko

Environmental Engineering Department, Faculty of Landscape Architecture and Environmental Technology, Universitas Trisakti, Jakarta, Indonesia

Corresponding Author : widyatmoko@trisakti.ac.id

Abstract. Indonesia needs to build four Treatment Centrals for 229,907 tons per year produced hazardous waste. But almost all hazardous waste treatment is managed by just one company at present, namely PT. PPLI (Prasada Pamunah Limbah Industri). This research is based on collected data which identifies payback period of 0.69 years and rate of return 85 %. PT PPLI is located within the Cileungsi District of the Bogor Regency of West Java Province. Records from nearest rainfall station at Cibinong indicate that annual average rainfall for the site is about 3,600 mm. It is situated on hilly terrain and is characterized by steep slopes as well as has a very complex geological structure. The Tertiary sequence was folded to form an assymetric anticline with axis trend in an East–West direction. Three major faults cut the middle of the site in a North–South direction with a vertical displacement of about 1.5 meters and a zone width of 1 meter. The high concentration of Chemical Oxygen Demand (COD) 2500 ppm in Secondary Leachate Collection System (SLCS) indicate a possible failure of the Primary Leachate Clection System (PLCS), which need correct action to prevent groundwater contamination.
Keywords: groundwater contamination, hazardous waste, leachate collection system, treatment centrals

1. Introduction
Industries in Indonesia generate hazardous wastes about 229,907 tons per year. Unfortunately, Indonesia has only one treatment plant with international standard now, namely PT. PPLI. It is obviously foreseen that there is not enough plant to receive an annual amount of about 68,000 tons of hazardous wastes from West Java. Therefore, it is impossible to treat, store and dispose these wastes safely. As a result, there is seriously potential environmental contamination to occur which causes health hazards.

The plant is located between two cement plants, PT. Indo Cement is 3 Km west of the site and PT. Semen Cibinong is 2.5 Km northeast of the site, in Cileungsi District of the Bogor Regency of West Java Province, about 35 Km south of central Jakarta. The Surrounding area is largely rural land with villages forming the population centres. The area has monsoonal rains during the months of October to April and a drier season between May and September. The nearest rainfall station at Cibinong records that annual average rainfall for the site is about 3,600 mm. It is on hilly terrain and characterized by steep slopes and has complex geological structures. Topographic elevations vary between about 80 m and 140 m above sea level.

This research aim to:
- asses the probability risk of ground water contamination from site plant waste
- asses the feasibility of hazardous waste deposal facility according to geological structures
- analyse the finance and management of hazardous waste deposal plant

- evaluate the current efforts and future requirements

2. Research Method

Below are the data obtained from field research and questionnaire:

1) Types and amounts of hazardous wastes were obtained from the 215 PT. PPLI costumers.
2) Facilities, instrumentations, laboratory, processing of the waste in the rotary kiln for the organic chemical, netralization plant, detoxification plant, precipitation plant and dewatering plant.
3) Each instrument price taken from Catalog.
4) Cost-benefit analysis based on report of hazardous waste treatment from Independent Auditors [1], [2], [3]
5) Landfill design category
6) The influence of geologic structures and lithology on groundwater

3. Results and Discussion

Hazardous waste landfill in Indonesia is classified in three types. Type I has two geomembranes, type II has one geomembrane, and type III has no geomembrane. PT. PPLI has a licence to manage landfill type I (Figure 1).

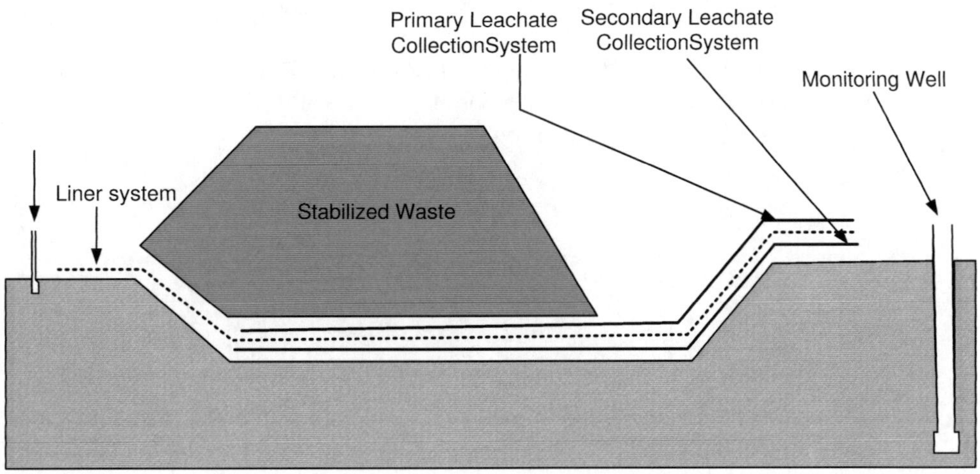

Figure 1. Liner system of landfill type I with Primary Leachate Collection System (PLCS) and Secundary Leachate Cillection System (SLCS).

On 6 January 2003, a group of residents who live near PT. PPLI complained about the odour and filed a class-action lawsuit over the release of various hazardous substances that they claimed had created a severe health risk and had contaminated their properties. Five employees who were medically examined indicated no personal exposure (Table 1).

Table 1. Air and personal monitoring analysis at PT. PPLI .

Parameters (mg/m^3)	Main	Cecep	Anim	Dian	Syafi'i	NAB (mg/m^3)
Metyl Etyl Keton	4,1566	3,0966	22,2700	3,8500	4,4433	590
Aceton	3,8353	4,0423	5,2105	4,7205	3,5771	1780
Dichloro Etilin	79,8230	67,6659	91,2376	72,8650	68,2132	793
Total Hydrocarbon	0,0849	0,0390	0,1730	0,0235	0,0695	1
Xylene	7,3657	1,7442	58,7142	3,6371	16,3326	434
Toluene	8,1549	4,6375	5,3926	1,6197	4,0528	188
Ether	6,1769	ttd	4,7511	8,1180	ttd	49
TCE	11,1200	ttd	8,5533	14,6146	ttd	269
Ethyl Acrylate	6,8467	5,9273	6,9124	5,6847	5,6761	20
Isoprophyl Alcohol	6,2397	5,6109	8,1763	6,3325	5,3615	983
Ethyl Acetate	6,4781	16,2785	97,9948	27,1668	8,6015	1440
Styrene	5,2254	2,206	1,9306	1,3833	2,2042	85
Buthyl Acetate	3,5608	4,4755	5,2105	4,7205	3,5771	713
Methylene Chloride	68,4588	57,8341	77,9809	62,2778	58,3019	174
Buthyl Acrylate	8,7687	7,5866	8,8528	7,2805	7,2695	52

The negative correlations between COD concentrations in the PLCS and SLCS, and the high concentration of COD 2500 ppm in SLCS indicate a possible failure of the PLCS (Figure 2), which need correct action to prevent groundwater contamination.

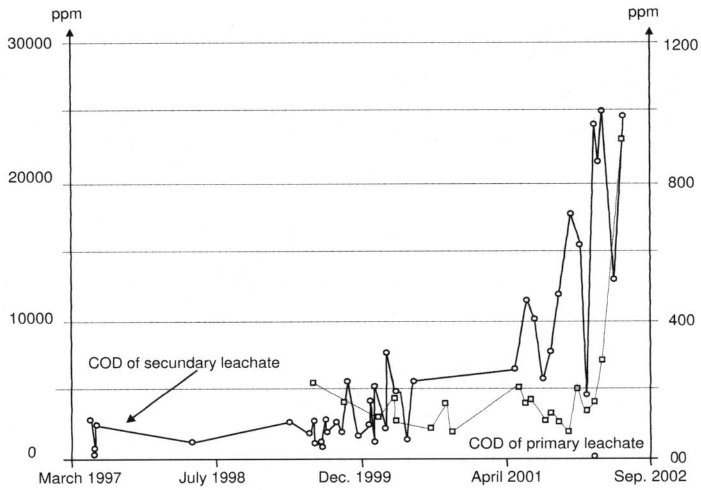

Figure 2. COD concentrations in primary and secondary leachates.

3.1. Geology

The geological mapping and drilling provided useful information on the lithology and structure of the sedimentary sequence beneath the site. The site is underlain by folded and faulted sedimentary rocks of Tertiary (Miocene) age. Tertiary rocks are covered by less than 5 m thick unconsolidated deposits of quaternary age. Tertiary sequence forms an asymmetric anticline with a fold axis which trends in an east-west direction (Figure 3).

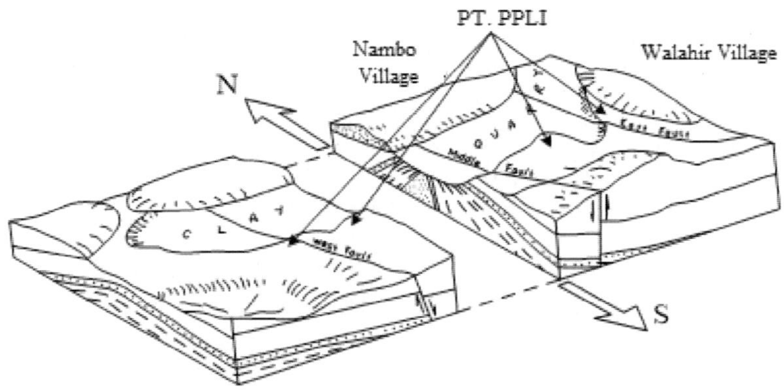

Figure 3. Evidence of faulting involving throws of a few metres was found on site indicating the presence of fault blocks.

The southern limb of the anticline dips southwards at a low angle of about 20°. The northern limb dips northward at a low angle in west and at high angle, approaching 90°, in east. The tertiary sedimentary sequence underlying the site forms a complex multi-layered aquifer system comprising mudstones, sandstones and limetones. The tertiary aquifer system is recharged by rainfall. Ground water discharges in lower lying areas in the north-eastern part of the site occurs from springs in the depth of water table increases with rising topographic elevation (less than 5 m in the lower areas). Groundwater flows from the southwest and southeast along the valleys towards the northeast. Groundwater generally has pH 6.2-6.8.

3.2. Economic Analysis

Hazardous waste trading is one of the best ways to increase regional revenue. Indonesia produces 229,907 tons of wastes yearly [4]. Based on this figure, Indonesia Government decided to construct four hazardous waste treatment centres: Jakarta and surrounding areas (Jabotabek) for 75,741 tons per year, Surabaya and its vicinity (Gerbang Kertosusilo) for 147,760, East Kalimantan for 46,256 tons per year, and Lhoksumawe 29,150 tons per year. Unfortunately, only one of the four treatments centres (Jabotabek, owned by PT. PPLI), could be realized. That means almost all hazardous waste in Indonesia should be processed by PT PPLI.

The largest amount of waste processed in 1997 could be interpreted as better law enforcement of government control program than more awareness of the industrialists to environmental issues. To support this assumption a comparative study was made in East Java.

Table 2. Estimation of volume, price and treatment of hazardous waste.

No.	Volume (%)	Servisces	Rate /ton/m^3 (USD)	Average (USD)
1	70%	Stabilization +LF	\$ 380/ton	0.7 × 380 = 266
2	20%	Direct Landfill/LF	\$ 175/ton	0.2 × 175 = 35
3	5%	Fuel Blending *	\$ 425/m^3	0.05 × 425 = 21.25*)
4	5%	Storage	\$ 150/ton	0.05 × 150 = 7.50
	100%			**Average = 329.75**

*)USD.85 per 200 liters.

Feasibility study for a treatment plant in East Java described payback period of 0.69 years and rate of return 85 % as shown on revenue and cash flow analysis (Table 3). The pricing based upon US$296.7

IOP Conf. Series: Earth and Environmental Science **106** (2017) 012032 doi:10.1088/1755-1315/106/1/012032

/ton hazardous waste or Rp 2,670,975/ton (1 USD = R 9,000) or 10 % lower than the average price of processing services in PPLI (US $329.75/ton or m^3; see Table 2).

Estimation of product
1st year = 5000 ton = Rp 13,354,875,000.00
2nd year = 10000 ton = Rp 26,709,750,000.00
3rd year – 25th year = 30,000 ton = Rp. 80,129,250,000.00

Annual revenue
= Rp {13,354,875,000.00 + 26,709,750,000.00 + (80,129,250,000.00× **22**)}**:** 25
= Rp 72,116,325,000.00

Average operation cost
= {(4.9 + 18.7 + 22.2 + 19.4 + 16.4 + 12.01 + (11.2 × **19**)}**:** 25
= 12.242 billion

a) Average rate of Return

Revenue	72.11
OPR cost	12.24
Depreciation	3
	--------- -
Profit BT	56.87
Company tax	14.79
	--------- -
Profit AT	42.87
Rate of return/year	(42.87 : 50) × 100 % = 85.74 %

b) Payback Period **50 : 72.11 = 0.69 years**

Table 3. Financial analysis.

Year	**Product capacity (ton)**	**Revenue 2,37/ ton**	**Cost**	**Profit AT 30%**	**Outstanding loan**	**Repayment loan**	**Interest 20%**	Cash / available	Capital accumulation
1	5000	13.35	4.90	5.91	50	0	0	5.91	5.91
2	10000	26.70	18.70	5.60	60	5	10	0.60	6.51
3	30000	80.13	22.20	40.55	66	25	11	15.51	22.06
4	30000	80.13	19.40	42.51	49.2	25	8.2	17.51	39.57
5	30000	80.13	16.04	44.86	29.04	25	4.84	19.86	59.43
6	30000	80.13	12.01	47.71	4.85	4.85	0.81	43.67	102.29
7	30000	80.13			0				

Based on this financial analysis, it is feasible to develop the hazardous treatment plant in East Java with the capacity of 5,000 tons in the 1st year, 10,000 tons in the 2nd year, and 30,000 tons in the next 3rd year – 25th year. After six years all the outstanding loan to be zero and the Capital accumulation Rp. 102.29 billion (see Table 3). The investation of Rp. 50 billion is spent for feasibility study, funding

IOP Conf. Series: Earth and Environmental Science **106** (2017) 012032 doi:10.1088/1755-1315/106/1/012032

for 30 hectares of land compensation, construction, equipment, training and operation (salary, overhead, maintenace, insurance, socialisazion program, and marketing)

4. Conclusion

This study reveals that groundwater and rainfall are potential to transport contaminants leached from the facility offsite to agricutural area downstream and create potential health hazards. Therefore, PPLI needs special engineering barriers to isolate waste stored in the landfill.

The negative correlations of the COD concentrations of primary and secondary leachates indicate a possible failure of primary leachate liner system, which requires correct action to prevent groundwater contamination. Negative impacts of waste treatment plants to surrounding areas comes from over capacity, inssufficient knowledge of geological structure, risk of existing faults, groundwater flow system to quifers.

Hazardous business simulation identifies Payback Period of 0.69 years and Rate of Return 85 %. High profit and monopoly practice quickly resulted in the new development of treatment plant in Cileungsi by PPLI.

References

[1] Siswanto Sutojo, Studi Kelayakan Proyek, PT. Damar Mulya Pustaka, Jakarta, 2010

[2] Siswanto Sutoyo, Pembiayaan Investasi Proyek, PT. Damar Mulya Pustaka, Jakarta, 2009.

[3] Vincent Teo Hup Ee. Operation of Off-Shore Landfill and Incineration Plants in Singapore. Joint Workshop on Solid Waste Management. Jakarta, 2002

[4] BAPEDAL, Head of BAPEDAL Decree Number Kep-06/BAPEDAL/02/1999, 26 February 20-2011 on Permit for Treatment of PCB Waste as Syntetic Fuel Oil for PT. PPLI in Year I, II PT Semen Cibinong, BAPEDAL, Permit for PT. PPLI, 1993.

IOP Conf. Series: Earth and Environmental Science **106** (2017) 012033 doi:10.1088/1755-1315/106/1/012033

The correlation between temperature and humidity with the population density of *Aedes aegypti* as dengue fever's vector

M M Sintorini

Environmental Engineering Departement, Faculty of Architecture Landscape and Environmental Technology, Universitas Trisakti, Jakarta, Indonesia

Corresponding author: sintorini@trisakti.ac.id

Abstract. The weather change in South East Asia have triggered the increase of dengue fever illness in Indonesia. Jakarta has been declared as one of dengue fever endemic region. This research aim to gain the dynamic of dengue fever incidents related to temperature, humidity and the population density of Aedes aegypti. This research implementated Design of Ecology Study. The samples were collected from April 2015 to March 2016, from houses located in the suburbs i.e. Pasar Minggu, Ciracas, Sunter Agung, Palmerah and Bendungan Hilir. The sampling based on Sampling Design Cluster and each suburb represents 153 samples. The research shows correlation between temperature (p value 0.000) and humidity (p value 0,000) with Aedes aegypti as dengue fever's Vector. Therefore, an early warning system should be developed based on environmental factors to anticipate the spread of dengue fever.
Keywords: aedes aegypti, dengue fever, humidity, temperature

1. Introduction

As a DHF vector, Aedes aegypti can adapt to urban environments, contact with humans and transmit disease [1]. Climate change in a short time is a complex system because it is influenced by many factors that interact with each other. Dynamic models and stochastic models can be used to analyze dengue cases and climatic factors, mapping strategies, prevention and monitoring of DHF. The dynamic character of this method comes from a unique structural model of a simulation model designed to predict future cases [2], [3]. Dynamics analysis is developed into a dynamic structure pattern in which each structure has a different pattern that is expressed through a loop or feedback. There are four more influence subsystems to contribute to disease: climatic subsystem, vector, human disease and dengue. These subsystems are interrelated, mutually influence and have separated the dynamics. This research was aimed at gaining the dynamic of dengue fever incidents that was related to temperature and neighbourhood's humidity as well as the density population of Aedes aegypti.

2. Research Method

2.1. Research Design

This research used the cross sectional and ecologic study design [4].

2.2. Place and Time of Research

This research is performed in dry and wet season in one year period, since April 2015 until March 2016. All data was collected from South Jakarta (district Pasar Minggu), East Jakarta (district Ciracas), North Jakarta (distict Sunter Agung), West Jakarta (district Palmerah), and Central Jakarta (district

IOP Conf. Series: Earth and Environmental Science **106** (2017) 012033 doi:10.1088/1755-1315/106/1/012033

Bendungan Hilir). These locations includes as region, which stated as endemic region of DHF by Government of Jakarta based on DHF outbreak.

2.3. Population and Sample

Vector of Aedes mosquito: calculation of the maturity Aedes is performed one time per week by calculate Man Landing Rate (MLR) by people feedback and calculating of resting habit as follow [4][5].

$$\mathbf{MLR} = \frac{\textbf{total of captured mosquito per species}}{\textbf{total of the captured hour} * \textbf{total of collector}} \quad (1)$$

3. Results and Discussion

The observation result showed that there was a significant correlation between temperature, humidity and Aedes mosquito population. Aedes population increase is an indicator of increase in DHF cases. The lowest average environmental temperature was 32.37 ° C in East Jakarta and the highest average was 33.77 ° C in North Jakarta (Table 1).

Table 1. Temperature (°C) in Jakarta, April 2015-March 2016.

Month	South Jakarta	East Jakarta	North Jakarta	West Jakarta	Central Jakarta	Mean
April	32.17	33.41	34.28	33.23	32.73	33.16
May	34.22	33.57	34.47	33.29	34.22	33.95
June	32.83	34.65	35.26	35.17	34.58	34.50
July	32.26	30.87	32.74	32.45	33.37	32.34
August	32.75	31.18	32.58	32.49	32.85	32.37
September	33.20	31.77	33.23	33.51	35.19	33.38
October	34.65	32.25	33.86	32.72	33.64	33.42
November	33.75	32.63	33.59	33.45	34.27	33.54
December	30.12	32.14	34.42	33.83	32.25	32.55
January	31.64	31.51	33.25	33.72	32.36	32.50
February	32.15	31.87	33.67	31.73	32.29	32.34
March	32.10	32.55	33.92	32.59	33.74	32.98
Mean	32.65	32.37	33.77	33.18	33.46	33.09

The lowest average humidity per region during one year of observation was 64.51% in East Jakarta and the highest of 67.12% in North Jakarta (Table 2).

3.1. Vector and DHF cases

MLR is calculated each two weeks once time, it shows total of the perched adult mosquito per person per hours are 1.46 (1.46 + 0.63). The lower of MLR: occur in July in Central Jakarta: 0.50, and higher 2.75 in September (South Jakarta). Average of MLR in West Jakarta is low, maximum occur in July-September and February. Although the positive correlation is low which occur in July, September and February Differ in Central Jakarta. From April until July, MLR feel decrease step by step and reach the first its peak in September when the rainfall curve feels decrease.

Table 2. Relative humidity (%) in Jakarta, April 2015–Maret 2016.

Month	**South Jakarta**	**East Jakarta**	**North Jakarta**	**West Jakarta**	**Central Jakarta**	**Mean**
April	68.33	69.13	69.25	71.65	72.33	70.14
May	58.75	65.87	67.95	62.33	62.75	63.53
June	58.92	64.25	64.17	58.27	59.28	60.69
July	59.77	58.73	64.21	62.00	62.33	61.41
August	60.40	59.16	62.05	56.25	56.75	58.92
September	60.65	60.28	62.87	58.20	56.83	59.77
October	58.65	56.05	65.35	64.25	63.90	61.64
November	60.78	61.29	64.73	61.72	62.15	62.13
December	62.25	62.83	63.81	60.25	65.95	63.02
January	78.50	70.65	72.65	75.15	78.00	74.99
February	77.60	74.15	73.32	77.30	78.70	76.21
March	72.45	71.76	75.07	78.25	78.55	75.22
Mean	64.75	64.51	67.12	65.47	66.46	65.66

Multivariate analysis was used to calculate the p value of temperature (p: 0.000) and humidity (p: 0.000) as independent variable. The man landing rate (MLR) as dependent variable (Table 3). The correlation between temperature, humidity and man landing rate (MLR) is shown on the graph of Figure 1.

Table 3. Environmental factors with MLR in Jakarta 2015–2016.

Variable	□	Std. Error	Beta	P value
Temperature	- 0.002	0.014	-0.004	0.000
Humidity	- 0.009	0.002	-0.137	0.000

Dep.Var.: MLR

Tabel 4. MLR of *Aedes* in Jakarta, April 2015–March 2016.

Month	South Jakarta	East Jakarta	North Jakarta	West Jakarta	Central Jakarta	Mean
April	1.48	1.12	1.93	1.05	1.72	1.46
May	1.03	1.20	1.70	0.84	1.51	1.26
June	1.21	0.74	1.58	1.57	0.89	1.20
July	1.68	0.65	1.00	1.85	0.50	1.14
August	2.17	1.15	0.73	1.52	0.81	1.28
September	2.75	1.30	1.34	1.57	1.59	1.71
October	1.82	0.87	1.02	1.22	1.40	1.27
November	0.61	1.20	0.73	0.56	1.25	0.87
December	0.72	1.82	0.50	1.10	1.92	1.21
January	1.55	2.15	1.36	1.64	2.33	1.81
February	2.21	2.53	2.12	1.89	2.72	2.29
March	1.72	2.41	1.98	1.22	2.53	1.97
Mean	1.58	1.43	1,33	1,34	1,60	1.46

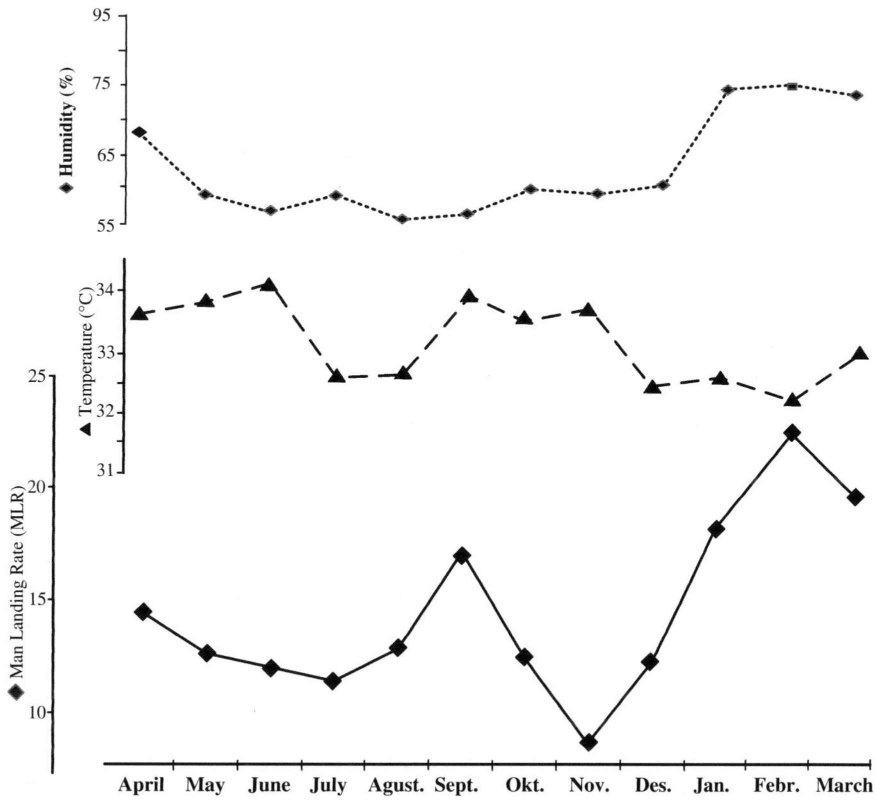

Figure 1. Correlation of MLR, temperature and humidity in Jakarta, April 2015–March 2016.

The MLR is exception in September, when occur increasing of MLR, can explain that higher temperature and lower humidity, MLR will increase but its transmission rate is low. High transmission is occur when higher rainfall, lower temperature and higher humidity [6], as occur in December until February.

3.2. Prediction DHF Cases Based on Model of Environmental Factors

Based on Figure 2, there are for related subsystem by occurrence process of DHF that is climate subsystem, Aedes subsystem, human subsystem, and DHF subsystem. Four subsystems are interrelated and inter influencing.

First subsystem is climate. In this subsystem, there is rainfall, temperature, and relative humidity. Second subsystem is the life cycle of Aedes mosquito. Connecting factor to climate and mosquito subsystem is location of breeding places (BP). BP is more influenced by rainfall [7]. Other connecting factor from climate subsystem is environment, which influence Extrinsic Incubation Period (EIP). The EIP is influenced by environment temperature, humidity, viremia rate in human, and virus strain. Increasingly in temperature can minimized EIP and increase the transmission process [8]. The temperature can increase until 34°C will influence water temperature of BP and also can influence-hatching egg become the larva. DHF occurrence is determined by contact between mosquito and human[6]. Human activities are indicated by how much using fosil fuel. Impact from this is emission of CO_2. Based on statistic analysis is known that there is significant correlation between MLR, temperature and humidity, by p-value :0.000.

IOP Conf. Series: Earth and Environmental Science **106** (2017) 012033 doi:10.1088/1755-1315/106/1/012033

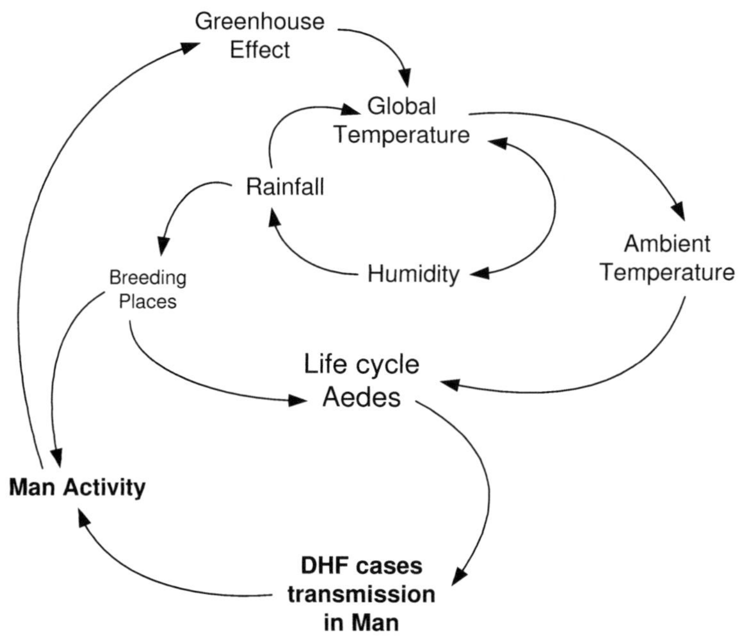

Figure 2. Diagram of DHF cases and environmental factors in Jakarta.

4. Conclusion

From the finding and discussion above can be concluded as follows:

1. Mean of man landing rate 1.46, temperature 33.09°C and relative humidity 65.66% in Jakarta.
2. There is significant correlation among man landing rate, temperature (p: 0.000), and relative humidity (p:0.000).
3. The climate factor, which influence man landing rate, is temperature, outdoor humidity, and which influence the man landing rate of Aedes mosquitoes.
4. To anticipate the spread of Dengue Hemorrhagic Fever cases, there should be early warning system through temperature and humidity as the environmental factors.

References

[1] Gubler Duane J, Kuno G, 2007, *Dengue and Dengue Hemorrhagic Fever*, CABI Publishing, 10 E 40th Street Suite 3203 NY 10016, USA.

[2] Karina A, Sari, Sri Y I, Sumardi Uun, Setiawati, Elsa Pudji. 2015. Incidence of Dengue Hemorrhagic Fever Related to Annual Rainfall, Population Density, Larval Free Index and Prevention Program in Bandung 2008 to 2011. *Althea Medical Journal*, 262-267

[3] Zhang Y, Bi P, Hiller J E. 2008, *Climate change and the transmission of vector-borne diseases: a review. Asia Pac J Public Health.* 20(1):64-76.

[4] World Health Organization, 2012, *Report on Insect Vectors and Human Health, Scientific Working Group*, Geneva, Switzerland.

[5] Health Department, 2013, *Prevention and Control to Dengue and Dengue Hemorraghic Fever, The Complete Manual Translated from WHO Regional Publication SEARO no 29, Prevention and Control of Dengue and Dengue Haemorrhagic Fever*, Cooperation with Health Department, Jakarta.

[6] Xiaoxu Wu, Yongmei Lu, Sen Zhou, Lifan Chen, Bing Xu. 2016. *Impact of climate change on human infectious diseases: Empirical evidence and human adaptation, Environment International Journal.* 86:14–23. homepage:www.elsevier.com/locate/envint

[7] Chee-Fu Yung, Kim-Sung Lee, 2015, Dengue Serotype-Specific Differences in Clinical Manifestation, Laboratory Parameters and Risk of Severe Disease in Adults, Singapore, The American Society of Tropical Medicine and Hygiene, 999 - 1005
[8] Kadam D B, Sonali Salvi, Ajay Chandanwale. 2016. Expanded Dengue. *Journal of The Association of Physicians of India*, 64:59-63

The impact of total suspended particulate concentration on workers' health at ceramic industry

M M Sintorini

Environmental Engineering Department, Faculty of Landscape Architecture and Environmental Technology, Universitas Trisakti, Jakarta, Indonesia

Corresponding Author: sintorini@trisakti.ac.id

Abstract. Ceramic production process pollutes the air with particulate matter at high concentration and has negative impact on the workers. The objective of this research was to determine the particulate concentration in the air and to analyse its impact on the workers. This research used cross sectional method to correlate the particulate concentration, temperature, humidity, smoke level and level of workers' compliance with safety regulations. Sampling was conducted from April to May 2012 in three locations, i.e. exposure area (Mass Preparation I, II) and non-exposure area (Forming area). In the exposure area (Mass Preparation I and II) where the particulate concentrations were 22.3673 $mg/m^3$and 14.8277 mg/m^3, and 58.33%, the workers had bad health status. In the non-exposure area, where the particulate concentration was 3.2185 mg/m^3 and 25% the workers had bad health status. The Odds Ratio among the workers in exposure area was 4.2 times higher than the workers in the non-exposure area.
Keywords: ceramic industry, health risk ratio, particulate exposure

1. Introduction

Indoor air quality gets serious attention because of its impact on employee performance.High concentration of Particulate Matter (PM) in air endangers workers' health in the ceramic industry. Respiratory disorders are common in workers who do not use protective equipment while working [1]. Symptoms that occur are headache, loss of concentration, dry throat, respiratory problems and eye irritation.

PM exposure can causes lung and systemic inflammation, which in turn leads to vascular endothelial dysfunction, a key step in the initiation and progression of atherosclerosis. Endothelial dysfunction was assessed by vascular response to acetylcholine (ACh) and sodium nitroprusside (SNP). PM_{10} exposure increased lung macrophages ($P < 0.02$), macrophages containing particles ($P < 0.001$), and activated macrophages ($P < 0.006$) [2].

Relatively low levels of fine particulate exposure from either air pollution or second-hand cigarette smoke are sufficient to induce adverse biological responses increasing the risk of cardiovascular disease mortality. The exposure-response relationship between cardiovascular disease mortality and fine particulate matter is relatively steep at low levels of exposure and flattens out at higher exposures [3]. The meta-relative risk for lung cancer associated with $PM_{2.5}$ was 1.09 (95% CI: 1.04, 1.14). The meta-relative risk of lung cancer associated with PM_{10} was similar, but less precise: 1.08 (95% CI: 1.00, 1.17) [4].

IOP Conf. Series: Earth and Environmental Science **106** (2017) 012034 doi:10.1088/1755-1315/106/1/012034

Objective of this study is to determine the concentration of total suspended particulate (TSP) and exposure of TSP to respiratory problems of workers in the tile production unit Indonesian Ceramic Association.

2. Research Method

The research design used is Cross-sectional [5]. The research was conducted on April - May 2012 at Indonesian Ceramic Associations, Bogor.

The conceptual framework used in the study is respiratory distress as dependent variable, total suspended particulate as independent variable, while confounding variable is temperature, humidity, age, length of service, smoking and the use of personal protective equipment (Figure 1).

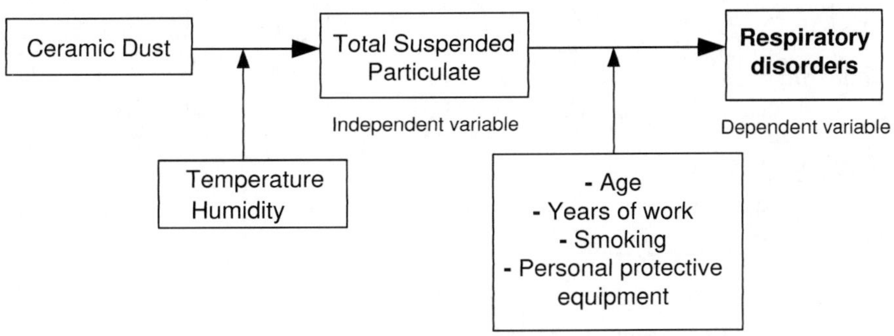

Figure 1. Conceptual framework.

2.1. Data collection

Data collected were total suspended particulate (TSP), temperature, relative humidity, and factory workers interview data of 48 people, ie 20 people in Mass Preparation unit I, 16 people in Mass Preparation unit II, and 12 people in forming unit [6]. The determination of respiratory symptoms is based on the standard Respiratory Symptoms and Patterns of Pulmonary Dysfunction [7].

Respondents were classified with respiratory distress if they complained of more than one cough symptoms, shortness of breath, bronchitis, pneumonia, pulmonary tuberculosis, emphysema and asthma. If the respondent complains about one or none of the above disorders then be classified as workers with good health status. The calculation of probability for a person suffering from respiratory distress, using the following provisions:

$$p = \frac{1}{1+exp^{-(a+bX)}} \tag{1}$$

The formula for measuring dust is:

$$X = \frac{A}{V} \tag{2}$$

X = TSP concentration (mg/m^3)
A = Weight of TSP on filter paper (mg)
V = Volume of air flow through filter x sampling time (L)

3. Results and Discussion

The result shows that all workers (68.75%) are men with the majority of age between 21-30 years. The most of 48 workers (79.17%) worked less than 10 years, but 42.11% of them have already suffered respiratory problems. The majority of workers (68.75%) between 21-30 years of age with respiratory disorders (51.52%). 52.08% of workers were smokers, with respiratory disorders proportions of 64%. All the workers were dutifully using personal protective equipment (100%) but apparently did not guarantee they were free of health problems [8].

Table 1. Frequency distribution of respondent characteristics in Indonesian Ceramic.

No	Variable	Total	%	Suffer	Proportion	%
1.	Experience :					
	a. < 10years	38	79,17	16	16/38	42,11
	b. > 10years	10	20,83	8	8/10	80
2.	Age :					
	a. <21 years	3	6,25	-	-	-
	b. 21–30years	33	68,75	17	17/33	51,52
	c. > 30 years	12	25	7	7/12	58,33
3.	Smoking					
	a. Yes	25	52,08	16	16/25	64
	b. No	23	47,92	8	8/23	34,78
4.	Using PPE					
	a. Yes	48	100	24	24/48	50
	b. No	-	-	-	-	-

The statistical analysis shows a significant correlation between works place temperature and works place TSP concentration, with p value: 0.000 (Table 2.). Figure 2 shows the concentration of TSP, temperature and humidity in all workspaces where sampling takes place. The highest TSP concentration comes from Mass Preparation room I (MP I). This space is an early stage of soil processing before being processed into tile, i.e. box feeder, crushing, milling and screening. While the space Mass Preparation II (MP II) where the final processing of earth, including crushing and screening, then it put into the Space Forming where the tile model is formed.

Table 2. Correlation between TSP and working room temperature in Indonesian Ceramic.

Model	B	R	R^2	Sig
Constant	-136,484			0,000
Temp. (°C)	4,568	0,869	0,755	0,000

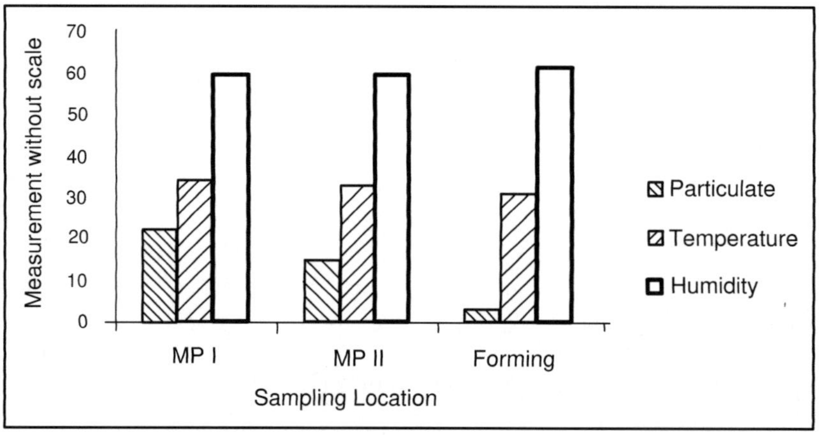

Figure 2. Total suspended particulate concentration, temperature and humidity in Indonesian Ceramic.

From the survey it was found that workers with the most health problems were from MP I unit (Table 3). While in Forming units are unexposed, workers (unexposed).

Table 3. Frequency distribution of worker health status in Indonesian Ceramic.

No.	Health Status	Location			Total	%
		MP I	MP II	Forming		
1.	Good	8	7	9	24	50
2.	Bad	12	9	3	24	50
	Total	20	16	12	48	100

From the on site observation, it is known that the highest respiratory unit is in the unit of MP I, ie 87.04%. This unit has the highest TSP concentration of 22.3673 mg / m^3, while the industry standard is 10,000 mg / m^3 (Table 4).

Table 4. Probability of respiratory disorders in Indonesian Ceramic workers.

Location	TSP (mg/m^3)	Probability (%)
MP I	22,3673	87,04
MP II	14,8277	70,97
Forming	3,2185	34,04

Statistically there is a significant correlation between relative humidity and TSP concentration in the workspace (Table 5.).

Table 5. Correlation between TSP and humidity of workspace in Indonesian Ceramic.

Model	B	R	R^2	Sig
Constant	48,543			0,006
Humidity (%)	-0,536	0,354	0,125	0,055

The development of long-term degenerative diseases such as bronchitis and emphysema is a cumulative response to poor environmental conditions. Total suspended particulate (TSP) may also cause infection in the lower respiratory tract. It is known that chronic bronchitis is suffered by workers exposed to continuous dust in high concentrations, such as workers in the cement industry, ceramics, coal and so on. This chronic bronchitis is diagnosed as a slimy cough throughout the day in 3 months for 1 year or more, without any underlying disease [9].

Respiratory disorders in workers who become dependent variables have a statistically significant relationship with TSP concentration in the workplace with a value of p: 0.004 (Table 6.).

Table 6. Correlation between TSP concentration and respiratory symptoms in workers Indonesian Ceramic.

Model	B	SE	Wald	df	Sig
TSP (mg/m^3)	0,134	0,046	8,332	1	0,004
Constant	-1,093	0,817	6,567	1	0,010

Y = respiratory symptoms

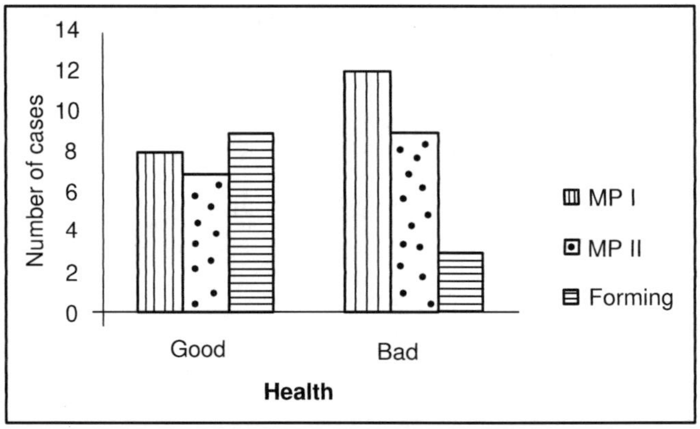

Figure 3. Health problems for workers in Indonesian Ceramic.

Workers who had better health were more in the Forming room, whereas those with the worst health were mostly in MP I room (Figure 3). Because the TSP exposure in the forming space is at least.

Based on multivariate statistic analysis it can be seen that the highest risk of health problems is particulate exposure [10]. The risk is 1,942 times, greater than the exposure that is caused by other variables (Table 7).

Table 7. Correlation between TSP concentration and respiratory disorders in workers Indonesian Ceramic.

Variable	B	SE	Wald	df	Sig	Exp (B)
TSP concentration	1,008	1,389	5,135	1	0,023	1,942
Temperature	0,021	1,003	5,545	1	0,019	1,001
Humidity	0,216	1,088	5,278	1	0,022	1,083
Age	0,179	0,823	0,761	1	0,675	1,078
Years of work	0,077	0,753	6,089	1	0,008	1,010
Smoker	0,601	1,475	0,126	1	0,722	1,295
Constant	-30,9415	3,217	5,885	1		

Compared to all production spaces of MP I, MP II and Forming, the greatest risk exists in MP I and II spaces that are 4.2 times larger than Forming rooms (Odds ratio: 4.2) (Table 8).

Table 8. Health status of workers Indonesian Ceramic.

No.	Health Status	Location MP I & MP II (exposure)	Forming (non exposure)	Total
1.	Good < 1 symptom	15	9	24
2.	Bad > 1 symptom	21	3	24
	Total	36	12	48

4. Conclusion

In the exposure area (Mass Preparation I and II) indicated that the particulate concentration was 22.3673 mg/m^3 and 14.8277 mg/m^3, and 58.33% the workers had bad health status. In the non-exposure area, the particulate concentration was 3.2185 mg/m^3 and 25% the workers had bad health status. The Odds Ratio among the workers in exposure area was 4.2 times higher than the workers in the non-exposure area.

References

[1] Cho Y, Min Choi, Jun-Pyo Myong, Hyoung-Ryoul Kim, Hye Eun Lee, Tae-Won Jang, Jung Wan Koo. 2015. The association between bronchial anthracofibrosis and pneumoconiosis: A retrospective cross-sectional study. *Journal of Occupational Health*. 57:110–117.

[2] Tamagawa E, Ni Bai, Morimoto K, Gray C, Mui T, Yatera K, Zhang X, Xing L, Li Y, Laher I, Sin D D, Man S F P, Van Eeden S F. 2008. Particulate matter exposure induces persistent lung inflammation and endothelial dysfunction. *American Journal of Physiology - Lung Cellular and Molecular Physiology*. Published 3 July 2008, 295 (1) : 79-85, DOI:10.1152/ajplung.00048.2007

[3] Pope C A, Burnett R T, Krewski D, Jerrett M, Shi Y, Calle E E, Thun M J. 2009. Cardiovascular Mortality and Exposure to Airborne Fine Particulate Matter and Cigarette Smoke Shape of the Exposure-*Response Relationship*. https:/ /doi.org/ 10.1161/ CIRCULATIONAHA. 109.857888. 120(11):941 – 948.

[4] Hamra G B, Guha N, Cohen A, Laden F, Nielsen O R, Samet J M, Vineis P, Forastiere F, Saldiva P, Yorifuji T, Loomis D. 2014. Outdoor Particulate Matter Exposure and Lung Cancer: A Systematic Review and Meta-Analysis. *Environmental Health Perspectives*.122(9), http://nrs.harvard.edu/urn-3:HUL.InstRepos:dash.current.termsof-use#LAA, DOI: 10.1289/ ehp /1408092.

[5] Mori K, Nagata M, Hiraoka M, Nagata T, Kajiki S. 2015. Surveys on the competencies of specialist occupational physicians and effective methods for acquisition of competencies in Japan. *Journal of Occupational Health*. 57:126–141.

[6] Albert A, Matthew R H, Kleiner B M. 2014. Emerging Strategies for Construction Safety and Health Hazard Recognition. *Journal of Safety Health and Environmental Research*, 10(2):152–161.

[7] Thepaksorn P, Pongpanich S, Siriwong W, Chapman R S, Taneepanichskul S. 2013. Respiratory Symptoms and Patterns of Pulmonary Dysfunction among Roofing Fiber Cement Workers in the South of Thailand. *Journal of Occupational Health*. 55:21 – 28.

[8] Noordin F, S J Lewis, P N S O'Donnell, P A J Crosbie. 2014. Manual Work and Lung Cancer Risk in High Risk Populations. *Journal of Occupational Safety and Health*. National Institute of Occupational Safety and Health, Ministry of Human Resources Malaysia. 11(1):5–14.

[9] Boiani J M, Steege A L, Sweeney M H. 2014. Adherence to Safe Handling Guidelines by Health Care Workers Who Administer Antineoplastic Drugs. *Journal of Occupational and Environmental Hygienes*, 11(11):728–740.

[10] Tan Chin Keng, Razak NA. 2014. Case Study on the Safety Management at Construction Site. *Journal of Sustainability Science and Management*. 9(2):90–108.

Public spaces and urban sustainability in the tropical built environment

Y M Yusof[1*], M Kozlowski[2]

[1]Centre for Advanced Research (CARe), FASS Building, Universiti Brunei Darussalam, Jalan Tungku Link, Gadong BE 1410, Brunei
[2]Department of Architecture, Faculty of Design and Architecture, University Putra Malaysia, 43400 Serdang, Selangor, Malaysia

*Corresponding Author : myyusnani@gmail.com

Abstract. Sustainability is an overarching sense of responsibility towards the future. On a city-wide level, urban sustainability incorporates a wide body of changes especially as they relate to the built environment, all of which intended at creating a livable place. This paper discusses existing public spaces in view of their achievement against a set of criteria for the built environment. The paper introduces performance design criteria for the tropical built environment. The key findings indicate that long-term strategies, guidance and directions for the city and region can achieve development which corresponds to local climate, synergies and provide a higher proportion of public spaces that offer something for everyone.
Keyword: built environment, public space, sustainable planning, landscape design, permeable urban form

1. Introduction

Conventional wisdom says that the cities undergo complex spatial organization and reorganization over time and it's the interrelated spaces where people move in and out every day and every time that essentially contribute to sustainable spaces. Even though there is still no consensus on how to comprehensively define the concept of sustainability as it pertains to the built environment, nor is there consensus on what aspects of the built environment should be considered in evaluating the sustainability of a built facility [1], the tropical built environment is not all different to other built environments. The built environment of cities in the tropics seems to suggest micro planning of public spaces with consideration given to temperature variations as to where the city warms and cools more slowly; experience increased surface runoff during heavy rainfall, less evaporation; public spaces more likely having a surge in usage at certain hours of the day; with potential for vernacular solutions to public spaces problem. All of which eventually impacts on how people consume the public spaces.

Deriving from the dialogues [2], [3], [4], [5], it can be said that to understand the synergies between public spaces and the tropical built environment and its contribution to sustainability, requires an analytical framework which is geared for a specific environment, in this case, the tropics, and a strategic set of performance design criteria. Thus following Weiss [6] Theory of Change, the purpose of this study is to propose a conceptual road map that becomes the basis of performance measurement criteria for the tropical environments. The desired outcome can be used to inform planning authorities, managers, and the community, on achievements of a public space and how well it did it. This method will then be illustrated in a case study of Bandar Seri Begawan (BSB), Brunei on the island of Borneo, by a critical evaluation of the state of the existing public space.

1.1 What about urban public spaces in the tropical built environment: how to measure its performance?
Flagging off Carmona's [7] research on new and regenerated public space, the author reconceptualises the nature of public space to conceive of a set of normative principles that are evolving, balanced, diverse, delineated, social, free, engaging, meaningful, comfortable and robust. Performance measurement of the space is then an on-going and responsive process and the design criteria measure an aspect or various aspects of a place and/or site in order to reflect on how best the area fits within a certain standard.

Table 1. Measurement of urban tropical public spaces.

Public-street Spaces	Landscape Design	Building-street Interface	Building Design	Sustainable Planning and Design
Intangibles Character, Form, and Function: size and shape, scale, bulk and character of surrounding buildings	Connected, walkable, convenient access, continuous, perceptible urban green	Building/street edge, building orientation, setback, height, character and facade	Integration of built form with the natural environment, permeability at ground level, horizontal and vertical landscapes	Increase water supply, reliability, waste management
Measurement Street width (building height to street width ratio not greater than 3:1), urban enclosure, access and connectivity, robustness, pedestrian, cycling friendly environments, outdoor activities, security and surveillance, view corridors, connectivity to water elements.	Canopy trees, soft landscaping, floor surface treatment, street furniture, lighting, way-finding and signage, shaded shelters, water fountains	Treatment and the location of service and car parking areas and utility installations	Building articulation, openings, and sun-shading devices, verandas on the external façade, use of climate responsive building materials	Capturing rain and stormwater: swales, infiltration, rooftop rainwater harvesting, energy efficiency, sustainable waste management

Source: [8-25]

2. Research Method

In this study, it is then possible to envisage performance design criteria (PDC) for tropical urban environments founded on the knowledge arising from scholarly and evidence-based research, and the most commonly used criteria or indicators in planning and designing of subtropical and tropical urban environments. 40 PDCs have been identified [7 – 23]. Each criterion is assigned one unit (1) making a total of 40 units. Visual analysis of each criterion ensures currency of the selected sites. The range of recurring criteria manifests consideration of local character, respect of topography, permeable urban and built form, optimizing solar access and air movement, use of traditional elements and vernacular structures, avoiding high fences and blank walls, good visual and physical connectivity, presence of water, promoting active building/street interface, extended pedestrian sidewalks, quality floors capes and street furniture, dense tree planting, and access for pedestrians and cyclists. These criteria fully summarized and broadly categorized as Public-street Spaces, Landscape Design, Building-Street Interface, Building Design, and Sustainable Planning and Design as shown in Table 1.

The major elements of the urban environment include streets, open spaces (squares, plazas, urban parks) and buildings [7]. For this research, the public-street spaces address diverse character, form, and function, size, shape and scale, economically vibrant space. Significant tropical climate urban design may include street width, urban enclosure, access and connectivity, robustness and provision of pedestrian and cycling friendly environments, outdoor activities, security and surveillance, important view corridors and connectivity to water elements. Permeable and active building/ street interface, outdoor activities, planting of vegetation and canopy shady trees, climate responsive surface materials and vertical landscapes in new buildings influence the outcome of the public-street spaces [24].

The perceptible landscape design requires connectedness, walkable, convenient access, and fluidity which emphasizes on the provision of canopy trees and soft landscaping, floor surface treatment, street furniture, lighting, way finding and signage, provision of shaded shelters and water fountains. Soft landscaping and canopy shady trees significantly contribute in mitigating the urban heat island effect. Meaningful building street interface addresses building/street edge, building orientation, setback, height, character and façade treatment and the location of service and car parking areas and utility installations. Evolving building design aims at integrating the built form with the natural environment, permeability at ground level, horizontal and vertical landscapes. The focus is on building articulation, provision of openings and sun-shading devices and semi-open verandas on the external façade, and the use of climate responsive building materials [18], [19], [20]. Sustainable planning and design address issues related to capturing rain and storm water, energy efficiency, and sustainable waste management. All these criteria are prerequisites for the achievement of a fully ecological sustainable urban environment [22], [23].

3. Results and Discussion

3.1 Case Study of Bandar Seri Begawan, Brunei

Brunei developed a network of water villages and in the 16th century known as the 'Venice of the East'. During the British administration in the early 20th-century, urban development commenced on dry land opposite Kampung Ayer (the water village). In the 1970s, the Pusat Bandar or the Central Business District of Bandar Seri Begawan seated major national institutions, commercial offices, and retail establishments. However, with the expansion of the city, many government institutions and commercial establishments moved to the new suburbs and the traditional city centre commenced on a trajectory of gradual decline [24]. Attempts to revive public spaces and promote new development have only partially revitalised the centre. Much of new developments are economic-led growth which often neglects the local conditions, natural settings, topography and the local tropical climate. Subsequently, delineating of spaces are unclear and architecture less meaningful thus affecting the quality of public space. Under the Bandar Seri Begawan Master Plan, the Central Business District (CBD) is designated as the hub of finance, business, education, and multinational headquarters, underpinning a diversification of the economy. Following those recommendations, the intent of the local authority is to transform the center into a thriving waterfront precinct which is characterized by strong cultural context featuring MIB architecture and re-adaptation of existing buildings to mixed-uses.

3.2 Correlations between public spaces and tropical built environment

The urban enclosure is generally not suitable for tropical climates, with the height of some buildings significantly less than the width of the street. The street has practically no defined enclosure along the southern side which contributes to the heat impact during the daytime. There are clear view corridors from the street, southwards towards Sungai Brunei and Kampung Ayer and eastwards featuring Bukit Subok. The majority of surface treatment materials (mainly concrete pavers) are not appropriate for tropical conditions, and the street lacks soft landscape, canopy native trees which have an impact on the thermal comfort of the end-users. Many of the modern buildings along Jalan Kianggeh are devoid of tropical design elements. Specifically, the correlations to build form, landscape design and built environment is shown in Table 2.

Table 2. Typology of correlations between public spaces and tropical built environment.

Correlation to built form	Correlation to landscape design	Correlation to tropical built environment
• Loss of visual connectivity • Defined five-foot footways create sheltered approach for pedestrian • Incidental spaces have a less coherent form than other spaces and are generally less clean and comfortable • Building articulation inconsistent with tropical environment with few buildings designed with semi-open verandas on the external façade	• Consumption spaces shows active day-time frontages • Insignificant greenery is a common feature of all space types • Public spaces tend to lack sun-shading devices • Private spaces generally rundown	• Public urban spaces are the most active, they are less enclosed and more open • Lack of fine urban grain with few network of small and detailed streetscapes • Spaces are usually contemporary in style generally showing less regard to historical context • Public buildings and spaces vary significantly in their architectural treatments • Feel like very safe places.

3.3 Sustainability compromised

The analysis of the streets and public spaces in the Bandar Seri Begawan CBD identified a few distinct stereotypes (see Figure 1). The analyzed streets are physically and visually well connected and provide a good sense of enclosed and contained urban space but lack tropical landscaping and buildings with tropical design features. There are pedestrian sidewalks however the absence of shading devices, sitting and resting areas, canopy tropical trees, and other soft landscaping elements significantly reduces the comfort for pedestrians. Dermaga di-Raja, the principal waterfront area in Bandar Seri Begawan, representing a historical sense of place for all Bruneians, was redeveloped along the typical model of 'a western style waterfront' which looks out of context with its predominant hard floor-scaping suitable for temperate climatic zones. The redevelopment was not in keeping with the eco-corridor [26] proposed a few meters away from the site. As a result, the waterfront area is mainly used as the gateway to Kampung Ayer.

Visual sight analysis	Location	Findings
	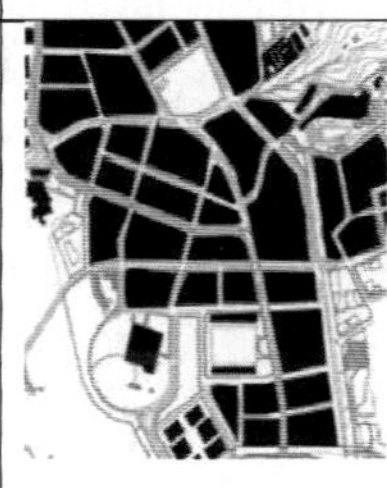	Waterfront 1. Detachment of visual connectivity leading to the loss of active frontage 2. The old Customs Building blocks the view of Kg Ayer from users on Jalan McArthur 3. Deficiency in softscaping elements and street furniture, surface materials not suitable for tropical conditions, absence of continuous pedestrian connectivity, lacking activities along Brunei River

IOP Conf. Series: Earth and Environmental Science **106** (2017) 012035 doi:10.1088/1755-1315/106/1/012035

	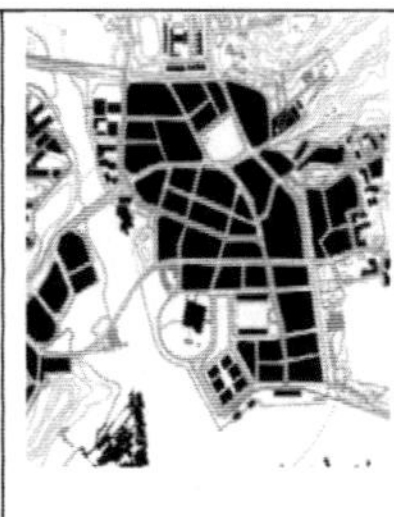	McArthur Road 1. Buildings yielding an unappealing sight 2. Lack of shade and greenery 3. Large sections of the street exposed to direct sunlight, lack of canopy trees and softscaping elements, many buildings devoid of tropical character and local motifs, potential to be converted to city boulevard
		Kianggeh Road 1. Flanked by Brunei River and Kianggeh River, limited diversity in activities 2. Potential for kinetic use of land 3. Street softened by water edge (Sungai Kianggeh), shortage of tropical design elements on majority of building facades, lack of sufficient pedestrian sidewalk on the western side 4. Absence of canopy trees and softscaping elements, and active frontages on the western side. Potential to be converted to city boulevard
		Alleyway 1. Tunnel effect 2. Not forming a part of pedestrian network 3. Lacking good lighting in the evening 4. Lack of active frontages and landscaping elements

Figure 1. Stereotypes of public spaces.

3.4 Homogenous public-streets

Homogenous existing street typology of Jalan Macarthur, Jalan Sultan, Jalan Roberts and Jalan Kianggeh impedes vibrancy of the city center (see Figure 2). Even though the streets are important east-west connectors of the CBD and provide connectivity but hollowed out buildings flanking the road create a deserted oasis from early evening onward. The pedestrian sidewalk is narrow on one side while on-street parking is rampant, to top it all, there is lack of natural shading and shading devices.

Figure 2. Existing street typology.

In all the selected public space, the average full compliance with the performance design criteria for tropical urban environments is 5.7 out of 40. The average non-compliance with the PDC is 6.7 out of 40. This implies major deficiencies in addressing tropical design elements in the urban environment of Bandar Seri Begawan's CBD. Shortcomings are observed in the design of modern buildings and in landscape design where local tropical climate themes are often non-existent. The study revealed the strong presence of vernacular styles in the 1990s which is outweighed by a modern, austere approach to design, typical of the late global modernist period. There is a visible pedestrian environment in the study area however its attractiveness for the end-users is hindered by low thermal comfort and leisure activities. The city needs canopy trees, permeable building edge and provision of active frontages which is rare here. Today, despite catalyst rejuvenation projects, the CBD is still characterized by a large number of decayed buildings, struggling businesses, vacant premises, and of significance, local Bruneians do not frequent the area.

4. Conclusion

The analysis unveils an urgent need for tropical urban design guidelines and green interventions for public spaces and building forms to address the local tropical climate at all scales and levels: the regional, city-wide district, neighborhood and site levels. Such guidelines provide a series of performance measures necessary for the tropical revitalization of streets and public spaces and promote the sustainable tropical urban environment. Tropical design and planning should be a leading theme in the national, regional, city- wide local and neighborhood planning documents. Developing tropical design guidelines for the Bandar Seri Begawan CBD in due course fills-up a noticeable gap in the field of urban studies in Brunei. It will identify measures that could soften the impact of global economy on the changing urban environment. This research offers an innovative contribution in the disciplines of urban planning and design by opening a Bruneian tropical design chapter. Responding to the local climate and enforcing them at regional and city-wide level with urban design/planning tools is important as it provides long-term strategies, guidance and directions for the region and the city. Arguably, whilst Bandar Seri Begawan is a medium-sized city with a less diverse range of public spaces, the city will need a higher proportion of spaces that offer something for everyone. In turn, the city center needs to put a break on decantation of the city center. Local community and stakeholders would be at the core of the planning stages ensuring design led by and needed by the community.

References

[1] Pearce A R, Vanegas J A. 2002. A parametric review of the built environment sustainability literature. *Int. J. of Environmental Technology and Management*. 2(1/2/3):54-93.

[2] Cowley R. 2015. Reframing the Problem of Public Space in the Sustainable City. Paper presented at *Int. Conf. on The Ideal City: between myth and reality*. Representations, policies, contradictions and challenges for tomorrow's urban life (Urbino Italy) pp 27-29.

[3] Banerjee T, and Southworth M, 1990, City Sense and City Design: Writings and Projects of Kevin Lynch (Cambridge: MIT Press)

[4] Devereux M, and Littlefield D, 2017, A Literature Review on the Privatisation of Public Space. Project Report (UWE) Available from: http://eprints.uwe.ac.uk/31529

[5] Tonnelat S, 2010, Sociology of Urban Spaces, In Wang Hongyang, Savy Michel and ZhaiI Guofang (eds.), Territorial Evolution and Planning Solution: Experiences from China, France. (Paris: Atlantis Press)

[6] Weiss C H, 1995, Nothing as Practical as Good Theory: Exploring theory-based Evaluation for Comprehensive Community Initiatives for Children and Families. (Washington DC: Aspen Institute)

[7] Carmona M, Heath T, Taner O C, and Tiesdell S, 2010, Public Spaces-Urban Spaces: The Dimensions of Urban Design (Oxford: Architectural Press)

[8] Kennedy R J, 2010, Subtropical Design in South East Queensland: A Handbook for Planners, Developers and Decision Makers, Centre for Subtropical Design, Queensland University of Technology, Brisbane.

[9] Gold Coast City Council, 2011, Gold Coast Rapid Transit Corridor Study Phase 2: Urban Design Framework and Guidelines, Accessed on February 2015 at www.goldcoast.qld.gov.au
[10] Gold Coast City Council, 2010, Central Southport Master Plan Accessed at www.goldcoast.qld.gov.au
[11] Brisbane City Council, 2011, Brisbane Streetscape Design Guidelines Brisbane.
[12] Urban Redevelopment Authority, 2008, Architecture and Urban Design Excellence www.ura.gov.sg
[13] Lehmann S, 2010, The Principles of Green Urbanism: Transforming the City for Sustainability (Washington: Earthscan)
[14] Carmona M, Wunderlich F M, 2012, Capital Spaces: The Multiple Complex Public Spaces of a Global City (London: Routledge)
[15] Farr D, 2008, Sustainable Urbanism: Urban Design with Nature (London: Wiley and Sons)
[16] Gehl J, and Melbourne City Council (2004) Res. Places for People-Melbourne Australia EDRA Awards
[17] Salleh E, 2006, Tropical Urban Street Canyons Tropical Sustainable Architecture: Social and Environmental Dimensions, Ed Bay P J H and Ong B L, pp 202-217 (Oxford: Architectural Press)
[18] Yeang K, 2006, Green Design in the Hot Humid Tropical Zone Tropical Sustainable Architecture: Social and Environmental Dimensions, Ed Bay P J H and Ong B L, pp 45-57 (Oxford: Architectural Press)
[19] Tei Lin H, 2006, Policy and Evaluation System for Green Building in Subtropical Taiwan Tropical Sustainable Architecture: Social and Environmental Dimensions, Ed Bay P J H and Ong B L, pp. 101-125 (Oxford: Architectural Press)
[20] Bay P J H, Wang N, Liang Q, Kong P, 2006, Socio-Environmental Dimensions In Tropical Semi Open Spaces of High Rise Housing in Singapore, pp 59-83 (Oxford: Architectural Press)
[21] Nyuk Hien W, and Yu C, 2006, The Urban Heat Island Effect in Singapore Tropical Sustainable Architecture: Social and Environmental Dimensions, Ed Bay P J H and Ong B L, pp 181-201 (Oxford: Architectural Press)
[22] Engblom S, and Bonham-Carter C, 2010, The Era of Ecological Metropolis in (ORO editions) Climate Design: Design and Planning for Age of Climate Change, pp 187-198
[23] Asselin S, 2010, Water Urban Landscapes and Water Management in (ORO editions) Climate Design: Design and Planning for Age of Climate Change, pp 54-62
[24] Dewan Bandaraya Kuala Lumpur, 2014, Urban Design Guidelines for Kuala Lumpur City Centre, DBKL.
[25] Yunos R, 2011, Our Brunei Heritage: A Collection of Brunei's Historical Accounts (Brunei Publications)
[26] Negara Brunei Darussalam, Ministry of Development, Department of Town and Country Planning 2008 The National Land Use Master Plan 2006-2025.

Carbon emission disclosure: does it matter

Y A Sudibyo

Faculty of Economic and Business, Universitas Trisakti, Jakarta, Indonesia

Corresponding author: yvonneags@gmail.com

Abstract. The purpose of this research were to test empirically the relationship of Volume of Carbon emission, Carbon Management Practice disclosure and Carbon disclosure emission with firm value, especially in Indonesia as developing Country. This research using data from Indonesian sustainability Award in 2013-2015. The instrument of this research was adapted from CDP Questionnaires to score the disclosure of Carbon Management Practice. While the carbon emission disclosure instrument was dummy variable. For volume of carbon emission, this research used the quantity or volume of carbon reported in sustainability reporting. We find that Volume of carbon emission was not related to Firm value. Also Carbon disclosure Emission does not have relationship with Firm value. Both hypotheses were not consistent with [8] which was doing their research in Developed Country. While Carbon Management Practice Disclosure, using CDP Questionnaires, has positive relationship with Firm value. The conclusion is developing country as resource constraint need to be motivated to report and disclose carbon emission from voluntary reporting to mandatory by regulation from government, not just only for high sensitive industry but also low sensitive industry. Then developing country which has resource constraint need to have more proactive strategy to prevent carbon emission instead of reducing carbon emission.

Keywords: carbon management practice, CDP questionnaires, developing country, volume of carbon emissions

1. Introduction

The issues of Climate change has been growing for a decades in developed country, but only few about Climate change in Developing countries. We need to encourage and bring private and public sectors to move immediately and put into action on both mitigation and adaptation. This action will get impact and stimulate the market for generating more investment to achieve sustainable energy in the future. According to [1] the effective tool for combating the climate change is carbon pricing, which translates carbon pollution into a price that governments, business or company and customers can factor into their reinvestment decisions. This is also considered in Sustainable finance.

Researcher [2] said that there is a need of strong public pressure globally to mitigate GHG emission even though regulatory requirements for GHG mitigation are vary greatly among countries. It means that there might be differences in regulation and policy between developed and developing countries.

This is also discussed in [3], that sustainability reporting becomes an integral to global action which need to considers environmental and social problems. As a result , there is a need also to consider the policies, regulations, standards and other instruments and should have encourage organizations to report it.

The consumption of household in Indonesia is much increasing in accordance with the increasing of the income. There is also shift the consumer social life percentage in a decade from 2005 until 2015. The number of households with high annual disposable income of over IDR 120 million (on average of US$ 10,000) or IDR 10,000,0000 per month. The estimation of Indonesia's middle class which was consisting of income between IDR 36-120 millions, will also expand to achieve 32 million households by 2020. This means represent a growth in consumption. As a result, the energy consumption for household product will also be increasing and also the carbon emission. This will be a big problem for carbon emission reporting and disclosure. At the end will influence the cost of using energy. Will it be reduced? Or we need to have strategy for carbon emission. That is a matter of fact we need to solve.

While the regulation for motivate and drive to make the environment activities or carbon emission is not set properly and procedure or guidance for Indonesia especially in those activities only did by some few of company. Almost company with enough funding will make the sustainability report. From 525 company only 100 more make and produce the report. So it means that is such voluntary reporting. If there is mandatory and there is penalty and forced by government, there will be most company prefer to report and disclose it. This is the second matter we need to consider, funding constraint. After we explain about the problem of disclosing and reporting. We discuss why to report.

What is the contain or the company need to disclose and reporting? It is also a matter. GRI 4, extended to GRI standard is just talking how to report environmentally, but not especially carbon emission. In addition, there is no information about Indonesia's company included in CDP. It means there is no official staff concerning Indonesia carbon emission.

Researchers [4] doing research on Determinants of Environmental Disclosure which considers industry type, size of the firm and profitability. The samples are Dutch listed company as developed country. They find out that industry type and firm size have significant influenced on Environmental disclosure, but profitability does not have relationship to environmental disclosure. Their argument is the research using data in financial crisis condition, which is some company have debt and financial distress.

Researchers [5] discussed about mix reporting using mandatory and voluntary for carbon emissions. Some of the literature notes the significance and the important of reporting instead of detailed investigation or reporting assessment of different type of carbon emissions, carbon intensity, volume of carbon and carbon reporting.

Why carbon emission disclosure and reporting is very important? The reasons are there is a need from investor to take a look at the environmental risk of the company. It means that investor needs information to assess the sustainable development and how the environmental protection and activities being done by the company. So that the investor will make more better in decision making for their portfolio.

Since research topics about CSR and sustainability has spread and done by some researchers in Indonesia, but only little concern in doing research on especially in Environmental activities and also in Carbon emission. That is motivation of doing research in topic of carbon. They did their research with the sample of Australia, Dustch company as developed country, so in this research using Indonesia as developing country.

Researchers [6] discuss the carbon emission by comparing the developed and developing country. From their findings, developing country as resource scarcity or shortage will have little commitment to carbon mitigation and disclosure. Their samples of developing country were China, India and Russia at CDP global Report.

They find out that carbon disclosure ahs direct and strong relationship with the resource availability, especially in developing country. It means that shortage of financial resources will be the important reason for the lack of commitment to carbon mitigation and disclosure in developing country. This is related to resource constraint theory. Their results also indicated that company tend to disclose carbon information if their shares are owned by CDP signatories. This is because their company will be viewed as more powerful stakeholders.

The other findings was the environmental reporting which was detailed in content in developed country (Deegan, 2002), but developing country only stated in sentences, not detailed, few supported with data in physically and or monetary

Reserachers [7] doing research with title of carbon emission and financial Performance with the mediation of carbon disclosure in the UK. They extended the instrument used as CDP Questionnaires for 18 items into 42 items. Their Findings were volume of carbon emission has negative relationship with financial performance, but carbon emission disclosures have positive and significant relationship with financial performance. Also, the level of implementation of carbon emissions reduction strategies within the UK industrial sectors is fairly "low" and varies significantly across the four sectors; with relatively high uptake in the energy and utilities sector, and low uptake in the construction sector. The level of implementation of change management initiatives to deal with carbon emissions reduction initiatives is also relatively "low".

The significance of doing this research is first, there is increasing awareness regarding environmental issues, so that company have the consideration for disclosing carbon emission. This increasing awareness disclosure of carbon emission is happened in developed country such America, Europe and other Asian developed Country such as Japan and Singapore. Indonesia is included as developing country which is disclosing carbon is still new and hot issue and only few companies reported and disclose the carbon emission. Developing country have less funding to disclose compare to developed country [6]. The empirical evidence and research findings also supported that developing countries are falling behind in carbon disclosure compare to developed country. Second, for carbon disclosures and reporting the company need funding for that activities. In general, developing countries has less funding compare to developed country. The question is that even though reporting and disclose carbon emission is important, if the company does not have enough funding, how can they disclose? So this need to be checked or researched. Third, [6] argued that carbon disclosure in developing countries was optional or voluntary and not as mandatory. Because of those reasons, this research will be conducted in Indonesia as developing country.

The result of this research will be important as information for suggesting OJK to consider about investment opportunity related to Carbon disclosure emission. Also OJK as Government need to consider for drive the mandatory reporting of carbon emission disclosure, so that it will support transparency or governance, not just only as positive information that will increase reputation, but also if there is negative information which might be signaling of company image and reputation. Then, as CEO of the company, this negative information will give faster information for decision making for reducing risk in the future.

2. Research Method

2.1. Data Collection Procedure

The population of this research was all company participating in the ISRA (Indonesian Sustainability Reporting Award) for the year of 2013-2015. The criteria for the sample that will be used are: companies participating in the ISRA, publish the annual report and sustainability report in 2013-2015, and as well as companies that implicitly or explicitly disclose carbon emissions.

2.2. Research Variables

Volume of Carbon Emission is measured by content analysis to find the amount of volume of carbon emission which was disclosed in Sustainability Reports. This is adapted from [8].

This research was adapted the CDP Questionnaires which has 18 items. Disclosure of Carbon Management Practice is measured by scoring, by giving maximum score is 18 and the minimum score is 0. Each item is worth 1 if the company discloses all of the information in the report so that mean company score is 18. Score on each company then totaled and divided by 18. The Carbon Emission Disclosure checklist, adapted from [9] can be seen in Table 1.

Tabel 1. Carbon emission disclosure checklist.

Climate Change: risks and opportunities	CC1- Assessment/description of the risks (regulatory, physical or general) relating to climate change and actions taken or to be taken to manage the risks
	CC2- Assessment/description of current (and future) financial implications, business implications and opportunities of climate change
GHG Emission	GHG1- Description of the methodology used to calculate GHG emissions (e.g. GHG protocol or ISO)
	GHG2- Existence external verification of quantity of GHG emission- if so by whom and on what basis
	GHG3- Total GHG Emissions – metric tons CO_2-e emitted
	GHG4- Disclosure of scopes 1 and 2, or scope direct GHG emissions
	GHG5- Disclosure of GHG emissions by sources (e.g. coal, electricity, etc.)
	GHG6- Disclosure of GHG emissions by facility or segment level
	GHG7- Comparison of GHG emissions with previous years
Energy Consumption	EC1- Total energy consumed (e.g. tera-joules or peta-joules)
	EC2- Quantification of energy used from renewable sources
	EC3- Disclosure by type, facility or segment
GHG Reduction and Cost	RC1- Detail of plans or strategies to reduce GHG emissions
	RC2- Specification of GHG emissions reduction target level and target year
	RC3- Emissions reductions and associated costs or savings
	RC4- Cost of future emissions factored into capital expenditure planning
Carbon Emission Accountability	AEC1- Indication of which board committee (or other executive body) has overall responsibility for actions related to climate change
	AEC2- Description of the mechanism by which the board (or other executive body) reviews the company's progress regarding climate change

Carbon emission disclosure was measured using a dummy variable, where the value of 1 to companies that disclose Carbon Management in the annual report and sustainability report, while a value of 0 is the opposite where the company does not disclose Carbon Management in the annual report and sustainability report [8].

3. Results and Discussion

3.1. Overall Significance of Regression Test Samples (F Statistic Test)

For testing the research framework goodness of fit, we used F test and find out that all the hypotheses are accepted, because the significance result is 0.016 which is lower than 0.05. It means that Volume of Carbon Emission, Disclosure of Carbon Management Practice and Carbon Emission disclosure will influence and have relationship with Firm value.

According to the research [8], the estimated coefficient on Carbon Emission per unit is given by the negative net assets, profit before extraordinary things and profit forecasts. The greater the volume of Carbon Emission, means that the cost of the company is increasing. Also, it will increase the risk of the company. As a result the profit of the company will decreasing and it will influence the price of stock. Instead, the company expect more greater market value of the firm's. It means that the companyneed to reduce the carbon emission. While [10]. supported that the volume of carbon emission have a negative association with firm value. It means that every additional thousand metric tons of carbon emissions will decrease the Firm value.

Table 2. Simultaneous test results in regression analysis (F Test).

The Regression Model	F	Sig.	Conclusion
Multiple Regression Model	3.575	.016	Ha accepted

IOP Conf. Series: Earth and Environmental Science **106** (2017) 012036 doi:10.1088/1755-1315/106/1/012036

From the Table 2, the F test supported the hypotheses that Volume of carbon, Carbon Management Practice and Carbon disclosure Emission have relationship to Firm Value. This finding is supported the research of [8].

3.2. Significant Individual Test Parameters (T Statistic Test)

Table 3. Individual parameter test results (T Test).

Variable	Regression Coeficient (β)	T-Stat	Sig.	Conclusion
Firm Value	11.468	1.564	0.127	-
Volume of Carbon Emission	-1.383	-0.688	0.496	H_1 rejected
Disclosure of Carbon Management Practice	33.836	3.084	0.004	H_2 accepted
Carbon Management Disclosure	-15.009	-1.486	0.147	H_3 rejected

3.3. Findings and Discussion

Hypotheses testing h1 was rejected because of the significance value, but the coefficient regression was negative which was supported the hypothesis area. It means that if volume of carbon emission was increasing, it will reduce the firm value. The reason was this research used only 14 companies, while the population of ISRA was 168. Data was reduced due to incomplete or not reporting the volume of carbon. In general Indonesian Company only explain there is carbon emission, but not recording the volume. This is also consider as lack of management practice and knowledge of reporting [8].

Hypothesis 2 is supported and consistent with [8], even though we used different way of calculation. It means that Carbon Management Practice Disclosure have positive relationship to firm value. The higher score of carbon Management Practice disclosure it will increase the Firm Value.

Hypotheses 3 is rejected, means that Carbon disclosure emissions has not relationship to Firm Value. The reason was the used of dummy variable which not really explained the motives of disclosure. Only 3 companies doesn't disclose the carbon emission. The logical thinking that it should have relationship, but the statistical result is not supported.

While Firm Value was measured using Tobin's Q which included short term and long term debt of company, the company still have burden to pay financially in the future. That is the reason hypotheses 1 and 3 were not supported. If the company just only using Market Value of Equity, it will be better findings. So future research we need to use Market Value of Equity for measuring Firm Value.

4. Conclusion

The research model was accepted, means there were relationship of volume of carbon emission, carbon management practice disclosure and carbon emission disclosure with firm value. While t test for individual hypotheses only carbon management practice disclosure have relationship with firm performance. The hypnoses which were not accepted or rejected were the volume of carbon emission has not relationship with firm value and carbon emission disclosure has not relationship with firm value. The evidence suggests that resource shortages may constrain a firm management's carbon decisions. As the regulatory environment becomes more stringent, firms, particularly those in developing countries need to take a more proactive strategy to tackle global warming challenges and balance the need to achieve financial goals and prevent carbon pollution with their limited resources. The use of Tobins Q as measurement of Firm Value is not a proper or match instrument. In Tobins Q contains of short term and Long term Debt which company need to pay in the future. Therefore there is burden to pay and contain also with risk. So how come with the company to legitimate with disclosing Carbon emission while there is still burden financially First, there is possibility to use Market value of Equity instead of Tobins Q for better instrument of Firm Value. Second, the carbon disclosure score is measured directly from individual companies' annual reports and sustainability reports. A checklist is established to determine the breadth and depth of the information on related to climate change and carbon emissions incorporated in these publicly available reports [2].

References

[1] Kim, Yong Jim, Carbon Pricing Leadership Report, 2016-2017
[2] Aggarwal R, & Dow S. 2011. Greenhouse Gas Emissions Mitigation and Firm Value : A Study of Large North-American and European Firms.
[3] King, Fogelberg, Hoballah, Malan,2016, Carrot and Sticks Report: Global trends in sustainability reporting regulation and policy
[4] Burgwal and Vieira, 2014, Environmental Disclosure Determinants in Dutch Listed Companies, R. Cont. Fin. – USP, São Paulo, v. 25, n. 64, p. 60-78, jan./fev./mar./abr.
[5] Haigh M, & Shapiro M A. 2012. *Auditing & Accountability Journal.* 25(1):105-125.
[6] Luo L, Tang Q, & Lan Y C. 2013. Comparison of propensity for carbon disclosure between developing and developed countries : A resource constraint perspective. *Accounting Research Journal*. 26(1):6-34
[7] Liu S, Zhou X, Yang J, & Hoepner A. Corporate carbon emission and financial performance: does carbon disclosure mediate the relationship in the UK? *In: European Accounting Association Annual Congress 2017*, 10-12, May, 2017, Valencia, Spain.
[8] Saka C, & Oshika T. 2014. Disclosure Effects, Carbon Emission and Coporate Value. Sustainability Accounting. *Management and Policy Journal.* 5(1).
[9] Choi B B, Lee D, & Psaros J. 2013. An analysis of Australian company carbon emission disclosure. *Pacif Accounting Review*. 25(1):58-79.
[10] Prakash MatsumuraR, & Vera-Muñoz S. Carbon Emissions and Firm Value. 2010
[11] Fani E, Pengaruh Pengungkapan Emisi Karbon dan Corporate Social Responsibility Terhadap Nilai Perusahaan (Studi Empiris Pada Perusahaan Manufaktur yang terdaftar di BEI Periode (2010-2013), 2015.

Evaluation of axial pile bearing capacity based on pile driving analyzer (PDA) test using Neural Network

H Maizir*, R Suryanita

*Civil Engineering Department, Sekolah Tinggi Teknologi Pekanbaru, Pekanbaru, Indonesia
Civil Engineering Department, Faculty of Engineering, Universitas Riau, Pekanbaru, Indonesia

* Corresponding Author: harnedi@sttp-yds.ac.id

Abstract. A few decades, many methods have been developed to predict and evaluate the bearing capacity of driven piles. The problem of the predicting and assessing the bearing capacity of the pile is very complicated and not yet established, different soil testing and evaluation produce a widely different solution. However, the most important thing is to determine methods used to predict and evaluate the bearing capacity of the pile to the required degree of accuracy and consistency value. Accurate prediction and evaluation of axial bearing capacity depend on some variables, such as the type of soil, diameter, and length of pile, etc. The aims of the study of Artificial Neural Networks (ANNs) are utilized to obtain more accurate and consistent axial bearing capacity of a driven pile. ANNs can be described as mapping an input to the target output data. The method using the ANN model developed to predict and evaluate the axial bearing capacity of the pile based on the pile driving analyzer (PDA) test data for more than 200 selected data. The results of the predictions obtained by the ANN model and the PDA test were then compared. This research as the neural network models give a right prediction and evaluation of the axial bearing capacity of piles using neural networks.
Keywords: axial bearing capacity, driven pile, neural networks, pile driving analyzer

1. Introduction

Analytical and empirical estimation of the axial bearing capacity of piles based on properties of soil has improved considerably over the years. The results are still not capable subject to many factors such as inherent soil variability. Furthermore, interpretation of site investigation data, and different assumptions used in the prediction methods as well as disturbance during pile penetration. Dynamic formulas were introduced for the driven pile, but the stages are inaccurate due to the simplicity in modeling the driving system. Therefore, most design codes require some Static Load Test (SLT), (ASTM-D5780-10, 2010) to be performed at construction sites to make sure that the actual capacity of the piles.

The SLT is not favored by practicing engineers because it is costly and time-consuming. In some cases such as very soft near surface soil and marine environment, the test is unmanageable due to the massive setup. Some researchers [1], [2] and [3] developed a High Strain Dynamic Pile Test (HSDPT) to study the strain, internal loading and acceleration of a pile under an push loading. The data is used to estimate the bearing capacity and the structural integrity of the pile. Moreover, the hammer performance, pile stresses and soil characteristics such as damping coefficients and quake values are also evaluating by the engineer. The procedure is standardized in ASTM-D4945-08, 2008. The HSDPT system is most widely employed using a combination of Pile Driving Analyzer (PDA) and Case Pile Wave Analysis Program (CAPWAP) software, because of the data more simplicity and quick handling.

The capability of the HSDPT to accurately predict static capacity from dynamic pile testing has been the subject of many studies. Based on the previous research concluded that, if performed and interpreted correctly, the high strain dynamic re-strike testing with CAPWAP analysis can provide reasonable agreement with the results of SLT regarding design capacity [4]. However, it indicated that the accuracy of HSDPT output relies mainly on the input parameters such as hammer efficiency and damping factor [5]. Therefore, many researchers still unwilling to adopt the estimation of pile bearing capacity according to the PDA test. There are some cases due to the uncertainty parameters such as hammer system and estimate of the soil damping factor and wave stress propagation theory. The use of the reduction factor was proposed in design codes [6] (e.g., Australian Standard AS2159) when using CAPWAP simulated static load test.

2. Research Method

2.1. Concept of ANN

ANN are scientific inventions inspired by the function of the human brain and nervous system. The ANN must be trained by inputting data repeatedly together with target output. The ANN learned to know the patterns of the data, hence; creating an internal model of the governing data process. The ANN uses the internal model to make estimations for new input. The variables of the input as known as neurons, and neurons send signals to other neurons. Some inputs to the neuron could have greater importance than the other, and this is modeled by weighting the input to the neuron Figure 1. Thus, the neuron can be thought of as a small computing engine that takes in inputs, processes them, and then transmits an output. Equation.1 gives the output of the neuron.

$$z_j = f\left(\sum_{i=0}^{n} w_{ji} x_i\right) + \theta_j \quad (1)$$

$$y_j = f(z_j)$$

Whereas,

- z : output prediction
- wji : connection weight ke-i
- xi : input data ke-i
- f() : a non linear transfer function
- n : number of iteration
- θ : bias
- y : target output

In this study, a three-layer feedforward backpropagation network was investigated. Some hidden neurons is an essential thing in backpropagation networks. Furthermore, there is no exact method for determining the number of hidden layer neurons. Using too many will make the training time longer, but using fewer hidden layer will make learning algorithm becoming trapped in local minimum [4]. Thus, it is trial and error that we use an absolute minimum number of hidden neurons, which will perform adequately. It suggested that the nodes on the hidden layer should be between average and the sum of the nodes on the input and output layers [5]. There are rough estimations to find the best-hidden layer size.

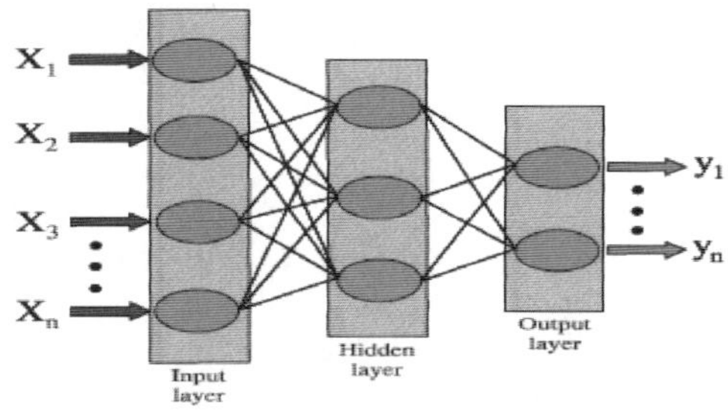

Figure 1. Artificial Neural Networks with inputs and outputs.

A variety of different transfer functions were investigated to achieve the best results in training. The optimal performance was obtained from tan sigmoid-log sigmoid-log sigmoid-linear activation functions in the first, second, third and output layers respectively.

2.2. Data Collection

The dynamic load test data were performed using a PDA test and based on ASTM D 4945, and the analysis of bearing capacity was evaluated by CAPWAP (Case Pile Wave Analysis Program). The PDA test data used in this study were collected from the two areas:

(a) 119 load test records got from Malaysia area

(b) 97 load test records got from Indonesia area

The database consists of a full spectrum of variety in a pile diameter, D. The embedment length is L, compression stress is CS, tension stress is TS, vertical displacement is DFN, ram weight is W, drop height is H, and energy is EMX. The axial bearing capacity of the single pile, Q is the only target output variable. All of the pile load tests are spun piles and made of the concrete piles.

The data of the factors recorded in the test are ordered to emphasize individual effect on the axial bearing capacity of the pile. This spun pile used in this study are the circular shape of which the diameter ranges from 300 to 500 mm and rectangular way of which the side length from 200 to 400 mm. The measures of the piles are varied from 10 to 35 m, and bearing capacity of the single piles is about 50 to 250 ton.

2.3. Calculation and Data Analysis

All databases are prepared for the training and testing procedure of the neural network. The data were divided into training subject constitutes 70% of database whereas the testing 30%. All of the input and target output variables were normalized in the values range between 0 - 1 before training. The rescaling is often accomplished by using a linear interpolation formula [6]. The following Equation.2 gives the method :

$$x_i' = (max_{target} - min_{target})\left[\frac{(x_i - min_{value})}{(max_{value} - min_{value})}\right] + min_{target} \tag{2}$$

Whereas,

$(max_{target} - min_{target}) \neq 0$, when $(max_{target} - min_{target}) = 0$or a feature, it indicates a constant value for the feature in the data. . When a feature value is found in the data with a constant value, it should be removed because it does not provide any information to the neural network. The network was trained with the backpropagation algorithm. Back-propagation neural networks can provide accurate predictions to any continuous function with sufficient neurons. In ANN model, we are not looking for the best training data, but we looked for the best responds of the training data that used for testing data in a better manner. However, to get the good reliable model, the number of hidden nodes and the value minimum sum-squared error goal was varied based on the coefficient of determination R2 of the testing result [7]. The R2 value described the contribution of input value in predicting the target output value, which means by how many errors are happened when predicting the axial bearing capacity of a driven pile by using the input information.

In this study, the model neural network uses eight nodes in the input layer, eight nodes in the hidden layer, and a node in the target output, as shown in Figure 2.

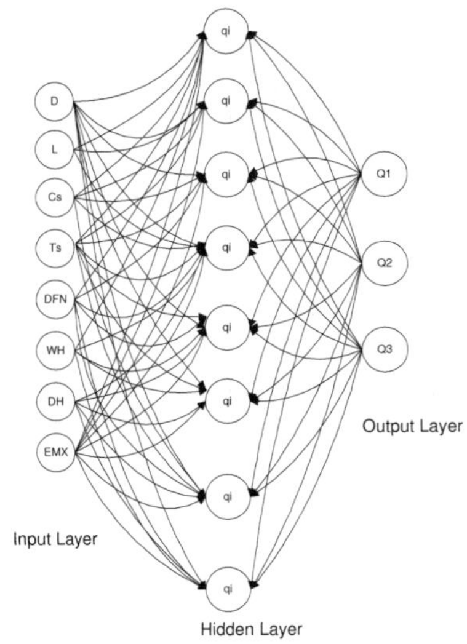

Figure 2. The architecture of the ANN Model.

3. Results and Discussion

Once the ANN model is evaluated in the training phase, the measured axial pile bearing capacities are compared with the capabilities obtained by BPNN. Then, the neural network models are capable of reproducing the target output values with minimal error. Furthermore, the result of the trained model is used for a new set of data is examined. Once the training, testing, and validation phases are accomplished, the neural network obtained can be used as a tool for predicting the axial bearing capacity of the single driven pile.

An ANN model (named HM) was developed to predict the axial bearing capacity of the driven pile. Figure 3 shows the comparison of Measured axial pile bearing Capacity and the prediction by HM model during the training Phase. From the 207 test piles used in the training set, the network HM was able to learn all the pattern presented with the high coefficients of correlation (R2) of 0.936. The mean square error of the neural network model was 0.0094, which is acceptable. The training process was determined after 15 (fifteen) iterations and then continued with testing phase. As shown in Figure 4, the model gives slightly lower coefficients of correlation (R2) of 0.916 after the completion of the testing phase

The model was used to predict the bearing capacity of three groups of piles (small, medium and large) and the results are shown in Figure 5, 6, 7. The figures indicate that the ultimate pile bearing capacity for spun piles predicted by HM model is in best agreement with the results of CAPWAP from PDA tests. Statistical analysis of the axial bearing capacity gives the R2 value of 0.811 for small piles, 0.911 for medium piles and 0.913 for large piles. The study indicates that the number of data is not always related to the quality of the prediction.

Artificial neural networks are used as a tool for predictive models on various geotechnical problems such as the piles bearing capacity. The primary idea behind this application is to develop optimal models using simple data. For example, some researchers [8] and [9] simulated data obtained from model pile load tests using in-situ pile load test. The output of the model for prediction of the ultimate bearing capacity while the input includes penetration depth of the pile, the diameter of the pile and the mean normal-stress. The result of the study shows that the ANN model gives a maximum error not more than 25%. Similarly, researcher [10] introduced a general regression neural network for

IOP Conf. Series: Earth and Environmental Science **106** (2017) 012037 doi:10.1088/1755-1315/106/1/012037

predicting the capacity of a driven pile in cohesionless soil. The variables were selected as input data: angle of friction of soil, effective overburden pressure, pile length and cross-sectional pile area. The results show that the ANN model gives a better prediction when compared with analytical and empirical methods, given by researchers [11], [12] and [13]. Researchers [14] also studied a method for design and analysis of deep-foundation using artificial intelligence techniques. The inputs of the network for training and testing correspond to the N-SPT value and pile dimension. Back-propagation Neural Network (BPNN) models are used for predictions. During the training phase, the measured axial pile capacities were compared with the capabilities obtained by BPNN. The developed neural network models were capable of reproducing the target output values with minimal error.

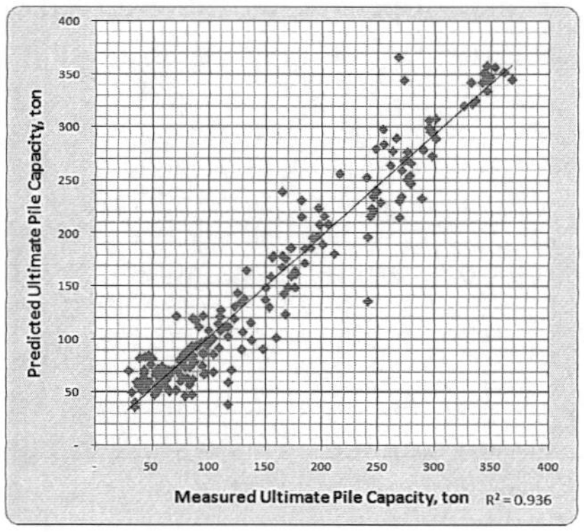

Figure 3. Predicting of axial bearing capacity. (training phase).

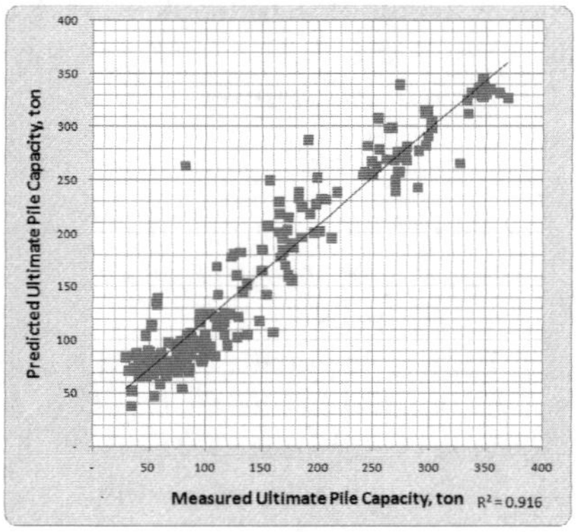

Figure 4. Predicting of axial bearing capacity. (testing phase).

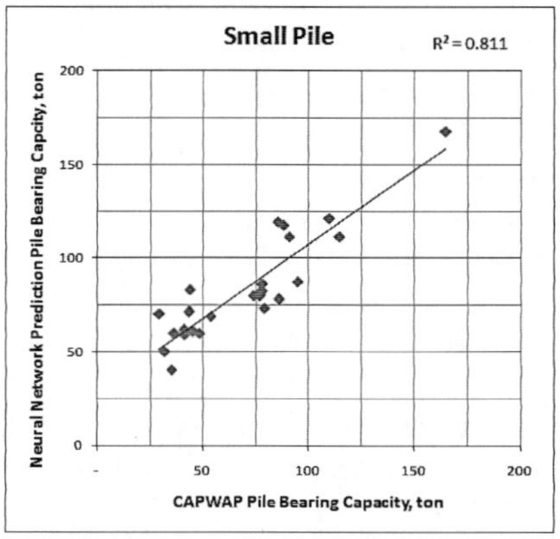

Figure 5. Predicted small pile capacity.

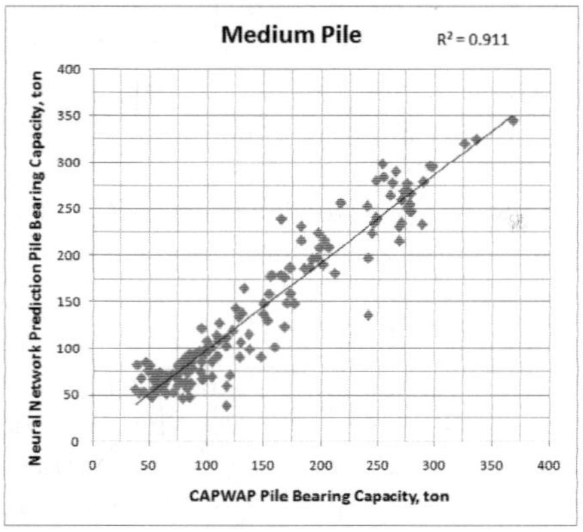

Figure 6. Predicted medium pile capacity.

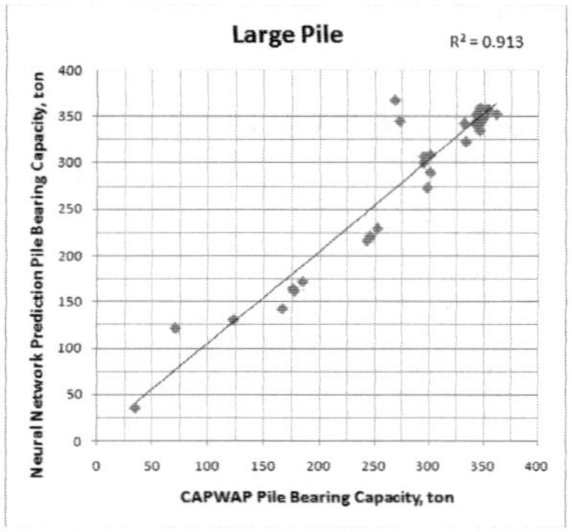

Figure 7. Predicted large pile capacity.

4. Conclusion

According to the simulation results, the technique that using artificial neural network can be applied to predict the ultimate pile capacity of the single driven pile. The ANN model gives a better prediction of the measured axial pile capacities. This model can be compared with the capabilities obtained by Back-propagation Neural Network (BPNN) models which are used for predictions model. It is easy to use, and the algorithm is easy to implement. However, it is hard to know how many neurons and layers are necessary, because of that factor the learning time may take longer.

Acknowledgment

The authors would like to thank PT. Delta Systech Indonesia for permitting the use of their field data for this paper and CV. Harista Karsa for sharing the quality of the site investigation and pile test data using Pile Dynamic Analyzer.

References

[1] D.-M. Cui, W. Yan, X.-Q. Wang, and L.-M. Lu, Towards Intelligent Interpretation of Low Strain Pile Integrity Testing Results Using Machine Learning Techniques, *Sensors*, vol. 17, no. 11, p. 2443, 2017

[2] M. Ehsan, Application of Artificial Neural Network for Predicting Shaft and Tip Resistances of Concrete Piles,*Turkish J. Eng. Environ. Sci. ISSN 1300-0160*, vol. 40, no. 1, 2016

[3] E. A. Sellountou, P. H. Duffy, and T. P. Holman, Optimization of drilled shaft design using high strain dynamic monitoring, in *Geotechnical and Structural Engineering Congress 2016*, 2016, pp. 724–734

[4] Likins G, Piscsalko G, Rausche F, & Hussein M. Monitoring quality assurance for deep foundation. ASCE Geotechnical Eng, 222-238, 2004

[5] Lok T M H, and W F Che, Axial capacity prediction for driven piles using ANN: Model comparison, American Society of Civil Engineers, 2004.

[6] Lee I M, & Lee J H. 1996. Prediction of Pile Bearing Capacity Using Artificial Neural Networks. *Computers and Geotechnics*. 18:189-200.

[7] Abu Kiefa M A. 1998. General Regression Neural Networks for Driven Piles in Cohessionless Soils. *Journal Geotech. & Geoenv. Eng.* 124:1177-1185.

[8] Meyerhoff G G. 1976. Bearing capacity and settlement of pile foundations. *Journal Geotech. Engrg. ASCE*. 102:196-228.

[9] Coyle H M, & Castello R R. 1981. New Design Correlations for Piles in Sand. *Journal of the Geotechnical Engineering Division*, American Society of Civil Engineers. 107:965-986.

[10] Randolph F M. 1985. Capacity of Piles driven into dense sand. Rep. Soils [Online], TR 171.

[11] Nawari N O, R Liang, and J Nusairat. 2005. Artificial Intelligence Techniques for the Design and Analysis of Deep Foundations. *Engineering Journal of Geotechnical Engineering*.

[12] Berke L, and P Hajela, Application of neural nets in structural optimisation. NATO/AGARD Avanced study Institute. Germany. 23(I-II): 731-745, 1991.

[13] Priddy K L, and P E Keller, Artificial Neural Network, An Introduction. New Delhi, Prentice-Hall of India, 2007.

[14] Park H I, and C W Cho. 2010. Neural Network Model for Predicting the Resistance of Driven Pile. *Marine Georesources and Geotechnology*. 28:324-344.

Water accounting implementation: water footprint and water efficiency of the coffee shop in Indonesia

S P Hendratno*, Y Agustine

Accounting Department, Doctoral Program in Economy, Universitas Trisakti, Jakarta, Indonesia

*Corresponding Author: setianihendratno@gmail.com

Abstract. The purpose of this paper is for understand the water accounting practice in the company, especially beverage industry in Indonesia. The sample in this study is one coffee shop near Jakarta. Case study has been choosen as the method in this study. We collect data with semi-structured interview, observation, and survey about the water efficiency in the coffee shop. The operational officers such as barista, cashier, supervisor, and store manager are the respondents in this study. Operational management already understand about the importance of water efficiency in the coffee shop operation, but it can't be implemented because their standard operation haven't use the water efficiency as part of their procedures. The coffee shop's operational standard in cleaning always takes much time and use so much water. The cleaning itself takes one until two hours each day only for cleaning bar and all operational equipment. This paper is for understand the water efficiency in the coffee shop with the focus is in their water footprint, operational standard that used every day in the coffee shop, and the connection between operational standard and the water efficiency.
Keywords: coffee shop, water accounting, water footprint, water efficiency

1. Introduction

Water, the element in humans' life that can't be separated in their daily life as an important aspect. Not only human who consume water in their life, but also the companies generally also consume water as part of their production. The companies always consume the natural resources, but there are still many of them don't pay attention about what happenned in our planet, a climate change. Actually, water is a part of natural resources that can be limited and scarcity when the extreme weather in the place that need water and using biomass as new product. Development of biomass products as one of renewable energy products is in countries in Africa, Europe, Southeast Asia, and Australia [1]. Therefore, Indonesia as one of the various countries in the part of Southeast Asia, should have started to focus on the scarcity of water occurring in Indonesia, especially urban areas and industries that use water as their production.

Industries that use water are not just large industries like manufacturing industry, but also food and beverage industries, for example in coffee shops. The availability of clean water is also a concern for coffee shop entrepreneurs and baristas as well as the coffee industry itself because water itself is the core in every coffee creations that can affect the quality of coffee, both from coffee beans until the coffee is served [5]. Water contributes to the presentation of coffee, not just for brewing, but also for other activities. Water can be used for hand washing, mopping floors, cleaning pitchers, ice machines, dipper wells, and for rest room [4]. This means that enough water is used by the cafe or coffee shop inside serving one cup of coffee due to the large amount of water used for supporting activities.

IOP Conf. Series: Earth and Environmental Science **106** (2017) 012038 doi:10.1088/1755-1315/106/1/012038

Indonesian coffee consumption continues to grow, growing faster than existing production at the moment [12]. Researchers [12] explained that the industry is reporting coffee growth is currently being led by the consumption of coffee dissolves. The retail coffee shop also continues to grow, growing in the business of shopping centers such as malls and transportation hubs such as the Starbucks coffee shop on Sudiman station. Judging from the phenomenon and statistical data, the development of people's interest in coffee continues to grow and allows it to become a market that is enough targeted by entrepreneurs. The phenomenon that occurs in Indonesia about the consumption of coffee that continues to increase from year to year, is depicted in the Figure 1.

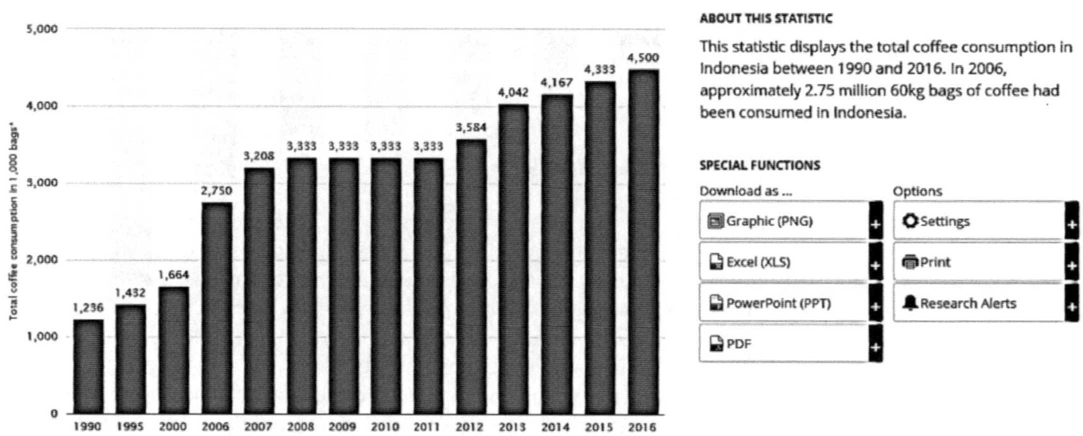

Figure 1. Growth data on coffee consumption in Indonesia.

Stakeholder theory further explains how managers or stakeholders should act and should look at organizational goals, based on some ethical principles. This theory suggests that managers treat stakeholders according to stakeholder concepts that will be more successful organizations in the long term [6]. In this study, the management sees the interests of the coffee shop in the long term regarding environmental conservation. With the influence directly or indirectly, coffee shops are expected to be more environmentally conserving in accordance with the desired by the other party.

Water footprint is a geographically explicit indicator, not only showing the volume of water use and pollution, but also the location [2]. Compared to other water accounting tools, the water footprint provides the most extensive data and complete water counting methods, as it includes both direct and indirect use of water and considers water consumption and pollution. Water footprint is the volume of freshwater used to produce the product, measured through various stages of the production chain. The use of water is measured in terms of the volume of water consumed or polluted. Water consumption refers to the water used or produced into the manufacture of the product [7].

Further water footprint concepts are specified for specific processes or products, and for every well-defined consumer group (eg individual, family, village, city, province, state, state) or producer (eg public organization, private company, economy sector). From the perspective of producers and consumers, WF is an indicator of both direct and indirect use of WF water is geographic and temporal indicators [3]. WF not only shows the volume of water use and pollution, but also the location of the presence of water.

Efficiency can be defined generally, the ratio between the desired output and the input, ie the quantity of resources consumed in the process of increasing efficiency means creating more value with less resource consumption [1]. Water efficiency is a multi-faceted concept. This means doing more and better with less by getting more value with available resources, by reducing resource consumption and reducing pollution and the environmental impact of water use for the production of goods and services at each stage of the value chain. Improving water efficiency means increasing water productivity, reducing the intensity of water use and pollution from socio-economic activities by maximizing the value of water use, improving water allocation among competing water uses so as to obtain socioeconomic value that saves water flow, and improves technical efficiency Water services and management of their provision over the complete life cycle [11].

This study discusses the water footprint and water efficiency of water use in the coffee shop. Water footprint is an indicator of water use that is seen both directly and indirectly from consumers or producers [2]. Water footprint for an individual, a community or a business is defined as the total volume of freshwater used to produce goods and services consumed by an individual or society or produced by a business [7]. If the water is produced by the business, then the need for efficient use of clean water, considering the water is one of the natural resources that are already quite rare nowadays. The purpose of measuring the efficiency of water use is to be able to complete the desired task by using the minimum amount of water without harming the existing system [10]. This study will try to find out how the implementation of water footprint and water efficiency in a coffee shop near the capital Jakarta is in Tangerang City because Tangerang is a city buffer city of Jakarta and many shopping centers and other public places in Tangerang as a place selected by coffee shops for selling their products. With the large or increasing number of coffee shops opened, the consumption of water used in coffee shops is also more and more. This is because every coffee shop uses water to produce or perform support activities such as washing equipment, mopping, and other things related to the coffee shop business.

2. Research Method

The method used in this research is using case study method. The sample in this research is a coffee shop located in Tangerang, Indonesia. It needs to be observed and field survey in advance about the use of water path or water footprint in the coffee shop and allotment of existing water flow. In addition to the observation and field survey of water footprint, the water efficiency in the coffee shop can be known by conducting interviews or interviews to employees of the coffee shop. The interview was done to the store manager, supervisor, and barista of the coffee shop. From the results of observations, surveys, and interviews conducted, it will be analyzed on the efficiency of water that has been done by the coffee shop based on water footprint and information on awareness of water efficiency.

The first step in the study is to conduct semi-structured interviews or interviews based on the interview list [8], [9] along with observations and surveys of the coffee shop. Interviews were conducted with store managers, supervisors, baristas, and cashiers who were part of the organizational structure of the coffee shop concerned with awareness of the importance of water and water efficiency of employees.

3. Results and Discussion

The problem of water efficiency in a coffee shop lies in the efficiency in consuming water. SOPs in the coffee shop that are related to water usage is one of them by washing all the existing equipment on the bar (where the drink in the coffee shop) and also the supporting equipment. The support equipment in the bar is not just for making drinks, but also as a storage place, presentation of raw materials, base to put glass, and also equipment to present other complementary materials such as green tea powder, caramel sauce, and others.

Preparation of water footprint was first performed to determine the flow of water used by the coffee shop for production activities, buying and selling, and also supporters (see Figure 2). This production activity means making drinks and serving pastry or food. The buying and selling activities are of course done on the cashier side and also the presentation of the finished drinks on the right side of the bar (where the drinks are taken). Support activities that exist within the coffee shop is to include cleaning the table, floor, windows, and more to the maintenance store.

IOP Conf. Series: Earth and Environmental Science **106** (2017) 012038 doi:10.1088/1755-1315/106/1/012038

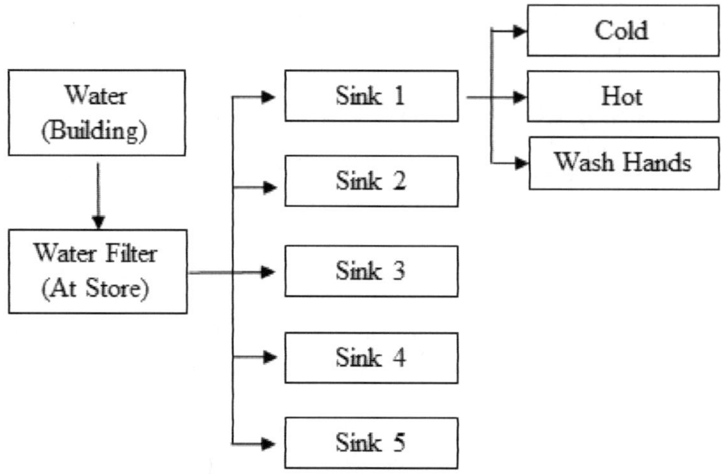

Figure 2. Water footprint in the coffee shop.

Based on the results of the interview, the workers in the coffee shop, ranging from the line store manager to the barista and cashier convey that espresso machine used by the coffee shop using a special tablet washing machine espresso that stay put into the machine and processed by turning the engine in time Every two days, so do not use water directly washed. The machine does not use much water in its washing and is only wiped by a warm cloth for its outer side.

The machine we use is the latest espresso coffee machine with the standard that the coffee shop has prepared. Cleaning it by removing all the coffee beans from the espresso machine, the outside of the machine to muddy the coffee beans divakum, then every two days using a special machine washing tablet to clean the inside of the machine.

All sinks in the coffee shop use filtered water and can be consumed for drinking. All water sinks use cold and hot water for each tap in the sink. Water Sink in the coffee shop there are at least two types of water temperature ie hot and cold temperatures. Hot water temperatures help the baristas to clean the equipment in the bar, such as pitchers, plates, spoon wash, and also to wash the blender. All employees in the coffee shop implement SOP cleaning of production equipment with the standard coffee shop. Average washing after closing the shop for bar equipment, pastry, and also cleaning the shop area, takes 1-2 hours.

Not just baristas and store managers who feel responsible for the coffee shop, but also other parts, such as cashier and pastry section. Although the section is less related to water use than the rest of the coffee shop, the pastry section also knows that efficiency is important in its role in the coffee shop. As in the interview results, one employee, the cashier who also doubles as a pastry section knows the importance of water use that is related to the efficiency of water consumption and affects the cost efficiency of the coffee shop.

For interview results at the supervisor level, supervisors are aware that the water consumed by the coffee shop is quite a lot per day, as it requires laundering many of the equipment in the coffee shop. For example, on the matt or base for to make a drink that is on the bar by washing it every day. This is because the number of droplets of raw materials at the time of making drinks ordered by customers, especially syrup or sugar. With that, it needs constant washing and the supervisor realizes that it is faster to wash with hot water than regular water. At the supervisor level it is more to the awareness that there is a lot of water consumption, but realize that efficiency can not be done with the will of within the operational lines and management of one store of the coffee shop. This is because the coffee shop is very strict in providing standards or SOPs on every operational undertaken, including on cleanliness.

From the results of interviews and observations made, it can be seen that the coffee shop is actually trying to do the water efficiency by reducing the use of water to wash equipment that is not smudged or dirty. However, back again with a running coffee shop SOP, would be a boomerang if the operational line management does water efficiency outside of the existing SOP inside the coffee shop. Therefore, representing all employees, the cashier section has disclosed that coffee shops should change SOP related to water efficiency or in other words is to incorporate water efficiency variables into company SOP.

As mentioned earlier on the definition of water footprint, water footprint is the volume of freshwater used to produce the product, measured through various stages of the production chain. Water efficiency is a multi-faceted concept of doing more and better with less by getting more value, by reducing resource consumption and reducing pollution and the environmental impact of water use for the production of goods and services at each stage of the value chain. These two sections relate to the value chain in which there is a production chain used to produce products and interact with each other. Therefore, it is important in a company to know the water footprint in its business and how to reduce the use of excess water in order to reduce pollution and the poor environment of the production.

Limitations in this study are the first, the lack of data on water use and the cost of using water to be obtained at middle level management. Secondly, additional research is needed by asking for opinions on middle-level management lines to better understand their views on water efficiency. This is because middle management also has an important role to provide value related to the water efficiency that is already mandated by upper management to the operational line management. In addition, middle management also intermediates the constraints, problems, solutions, innovation, strategy implementation, and operational coordination from the operational line management to upper management. With the delivery of suggestions from the operational line management, it is expected that top management can know and make decisions on those mentioned through middle level management. For further research it can broaden the scope of the research by interviewing middle to upper level management and can request data in the form of quantitative data on the efficiency of water use in the coffee shop.

4. Conclusion

Water footprint and water efficiency are important in a single unit of value chain and can be applied to companies that directly use water in their production or indirectly. By knowing water footprint first, will be easier for the company or management to be able to create water efficiency in the production process. For line of operational management of course very base with SOP in company or effort, so that need a policy taken by level strategic related to existence of water efficiency. Although the SOP does not currently include any visible water efficiency, the operational line management of the coffee shop has realized that water efficiency is necessary. This is known during the dialogue with supervisors, baristas, and cashiers, they are quite surprised because it turns out a lot of water used to clean a coffee shop and takes a long time to wash their equipment. With the awareness of the operational line, it is expected to be key in decision making at the upper level of management level related to water efficiency.

References

[1] Donoso G, Aldaya M M, Cabral de Sousa W J, Cai X, Chico D, de Miguel A, Zarate E. 2014. Water efficiency: Status and trends. *Water for Food and Wellbeing in Latin America and the Caribbean. Social and Environmental Implications for a Globalized Economy.* 261–283. Retrieved from http://new-test.fundacionbotin.org/89dguuytdfr276ed_uploads/Observatorio Tendencias/PUBLICACIONES/LIBROS SEM INTERN/water for food security/capitulo10.pdf

[2] Ercin A E, Aldaya M M, & Hoekstra A Y. 2011. Corporate Water Footprint Accounting and Impact Assessment: The Case of the Water Footprint of a Sugar-Containing Carbonated Beverage. *Water Resources Management.* 25(2):721–741. https://doi.org/10.1007/s11269-010-9723-8.

[3] Ercin A E, & Hoekstra A Y. 2012. UNITED NATIONS WORLD WATER ASSESSMENT PROGRAMME Carbon and Water Footprints Concepts, Methodologies and Policy Responses.

[4] Fasman D, Water Usage in the Cafe, at the Farm, and in the Future: Episode 1 – The Cafe. http://www.baristaguildofamerica.net, 2015.

[5] Fasman D, Water Usage in the Cafe, at the Farm, and in the Future: Episode 3 – The Future. http://www.baristaguildofamerica.net, 2016.
[6] Fontaine C, Haarman A, & Schmid S. 2006. The stakeholder theory. *Available in Http://www. Edalys. Fr/documents/Stakeholders% 20theory.pdf.* (December).
[7] Hoekstra A Y. 2015. The Water Footprint: The Relation Between Human Consumption and Water Use, 1–256. https://doi.org/10.1007/978-3-319-16393-2
[8] Kraeutler T, Krowiak A, Mercurio D, & Oriol L. 2015. Coffee Energy Audit: An Analysis of Coffee Energy Consumption and Efficiency at Boston College. (April). 1–31.
[9] PWC, Best practices for managing water risks and opportunities. *PwC Global Best Practices® Focus Paper*, 2011.
[10] Section D, Harmon J, Schneider N, Rose D, Rodriguez J, & Garrido N. 2013. Water Efficiency and Self-Conducted Water Audits at Commercial and Institutional Facilities, 2.
[11] United Nations. Water and Energy Efficiency, 1–5. Retrieved from http://www.un.org/waterforlifedecade/pdf/01_2014_water_energy_efficiency.pdf. 2014.
[12] Wright T, & Rahmanulloh A, Indonesia Coffee Annual Indonesia Coffee Annual Report 2016, 2016.

Implementation green and low cost on landscape design of Manggarai Integrated Station, Jakarta

T Suryanti*, H Meilianti

Landscape Architecture Department, Faculty of Landscape Architecture and Environmental Technology, Universitas Trisakti, Jakarta, Indonesia

*Corresponding Author : titiensr@trisakti.ac.id

Abstract. The Manggarai Integrated Station is the transit of various transportation modes. The Integrated Station located in Manggarai Jakarta and managed by PT. KAI. The Manggarai station is integrated and have terminal nature of transit areas (switching mode of transportation). There are several problems in the site, such as the problem of the site conditions in the urban area, topography, soil, vegetation, space, visual, users on the site can provide ideas for the concepts. The data was analyzed using the quantitative descriptive methode. The purpose of this research is to design the integrated station atmosphere, not only can support of the activities station users, but can also accommodate the needs of the community. It will "Green, Low cost" at the Manggarai integrated transport transit station in Jakarta. The potential that exists in this area is the lowliest integrated from various areas of the mode of transportation that make the users to facilitate transit transportation to the other. The basic concept of this design refers to the "Green, Low Cost" which unite with theme "user friendly" land use on a more efficient and effective site. The result of this research is landscape design development of Manggarai integrated station. Its consists of landscape design in west and east area, transition area, parking area, solar panel area, and social interaction area.
Keywords: landscape design, open space, urban transit

1. Introduction

Jakarta's urban transportation was growing very rapidly due to the high mobility of its people who need vehicles. The increasing human population is also increasing the need for transportation causing an imbalance between roads and vehicles. Less support for road facilities provided by the government causes traffic congestion and violations by motor vehicle users. The problem occurs because many people prefer private vehicles compared to public transportation.

Data from Jakarta Central Bureau of Statistics [1] shows that the population growth rate from 2000 to 2010 in the capital reaches 1.40 percent per year. The rate of population growth should be adjusted with adequate means of transportation. To facilitate the movement or movement of people from and to various places, it is necessary a mode of transportation. For Jabodetabek, one alternative mode of transportation is the train. All the countries in the world recognize that developing rail transportation as the core of the urban public transportation system is the only way to solve the traffic problem in the city [2]. For service users who need access to Jakarta and outside Jakarta, the railway functions as an irreplaceable means of connecting [3]. This is due to the low cost and the ability to transport large numbers of passengers in a relatively short time.

Selection of Manggarai location because of the area is a strategic area which becomes the point of crossing several modes of transportation from various regions planned by the Provincial Government of

DKI Jakarta. Manggarai Station which is one of the modes of transportation in the area, is one of the stations that can be categorized as a big station in Jakarta. Based on the planning of the Provincial Government of DKI Jakarta, Manggarai Station will be the final station for long-distance trains and airport railways (Soekarno-Hatta) integrated with TransJakarta Bus Stop and Manggarai Bus Terminal, the design of the integrated station will be designed around the Manggarai area.

The scope of these researchs is to design integrated Manggarai station landscape as a transport transit intermod. The design of the transit station landscape by providing a more organized Manggarai station atmosphere to be greener and to provide green space as a conservation area. Green ransportation is aimed to improve active mobility and public transit system with more integrated and convenient service [4]. The surrounding communities that use transportation services and which influence the inhospitable life in the Manggarai station are also considered in the design of the integrated Manggarai station landscape area, to provide facilities and circulation to avoid passengers accumulation at one point.

2. Research Method

The research was done at Manggarai station in the Centre of Jakarta. The wide of research was 9.4 ha. The border area: south at Jl Raya Utara, east at neighborhood, west at traditional Manggarai market, north at Jl Raya Balai Yasa. The method used in the design of the transit area of Manggarai Integrated Station transportation was quantitative descriptive methode that consisted of five phases as folows;

2.1. Data collection

At this step the data collected in the form of primary data and secondary data, in the form of physical data and non physical site. It needs to be done to deepen and understand the physical and non physical conditions. Selection of appropriate data will help the next process required appropriate data and concerns about the potential and constraints.

2.2. Analysis

This step needs to be traversed to understand and pay attention to the potential or existing constraints of the tread, so that it can be known in terms of what needs to be maintained and improved. In-depth review also be done with literature studies and comparative studies. It is necessary to enrich references on the design of landscapes designed and facilitated area when creating a draft concept.

2.3. Programming Site Requirement

This step provides details of what needs to be applied to the site, such as the utility needs, facilities, and green requirements needed to minimize the problems at the site.

2.4. Draft Concept

Creating landscape design concepts that can minimize or reduce existing problems on the tread, already understood through the process of analysis that has been done.

3. Results and Discussion

3.1. Basic Design Concepts

For this design in accordance with the analysis and space needs of the transit area of transportation integrated Manggarai station landscape Jakarta. It has the basic concept of outer space pattern adapted to the potential and function of the area.

In accordance with the theme of "user friendly" it means that this region refers to one destination for the user. Users will feel comfortable if the security and comfort is well maintained, so that from one theme can be categorized as Green and Low cost .Plant configuration near the site should the principle of harmonization. Ussually we choose defferent kind of plans, defferent planting methodes and combine with the building to form an open or semi open area [2].

3.2. Landscape Design Development
The zoning system at the site is as follows:

3.2.1. West and East Entrance
There is 2 entrances at the site. The main entrance is the east entrance and the second entrance is the west entrance (Figure 1). At main the entrance, there is a circulation of vehicles and pedestrian for access to Manggarai station. These two entrances are distinguished by the distinction of a node in each area (east with sculpture, west part with fountain).

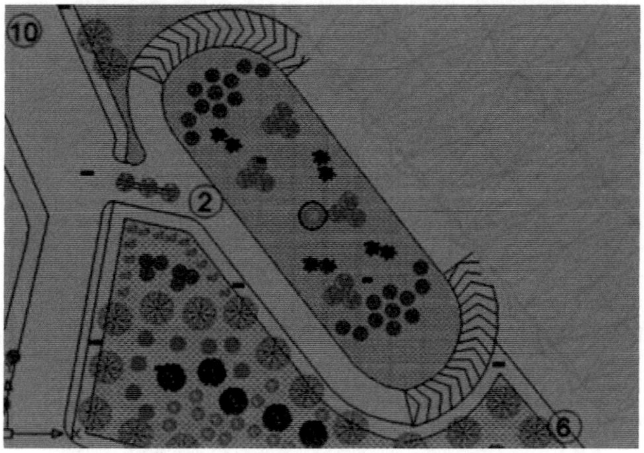

Figure 1. West entrance [4].

3.2.2. Transfer / Transition Area
The transition area is an area that serves as a transitional place for users of rail transport services. The transition area (Figure 2) is under a rail platform with a height of 12 m and an area of 13,724,832 m^2. This area contains only a few seats and a fountain pool as a point of view area. This area is designed to ease users to do transportation mode transfer.

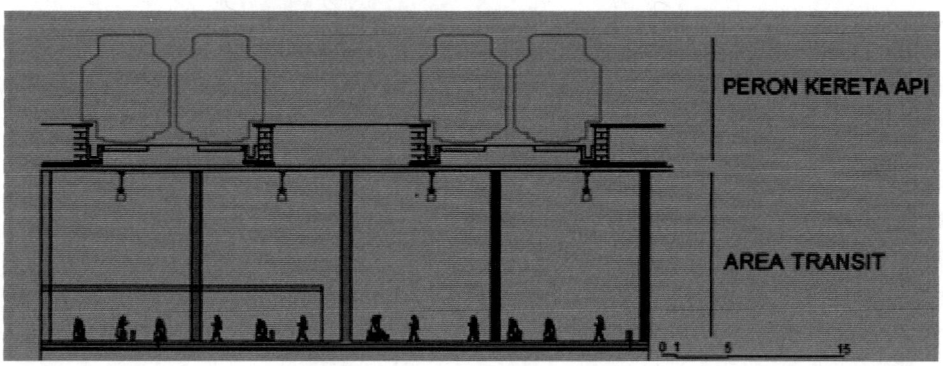

Figure 2. Transition area [4].

3.2.3. Parking area
The parking area is under a railway platform (Figure 3), with an area of 360.52 m^2 [6]. Because this parking area is shaded by a massive roof in the form of rail platforms, so there is no need to add vegetation for shade.

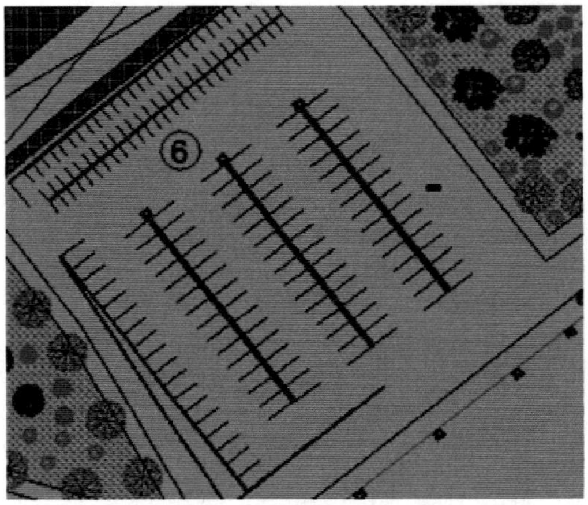

Figure 3. Parking area [4].

3.2.4. Waiting Area

The waiting area (Figure 4) is used for transportation service users to wait for city bus or public transportation for a long time. The waiting area has a seating area, a smoking area and a fountain.

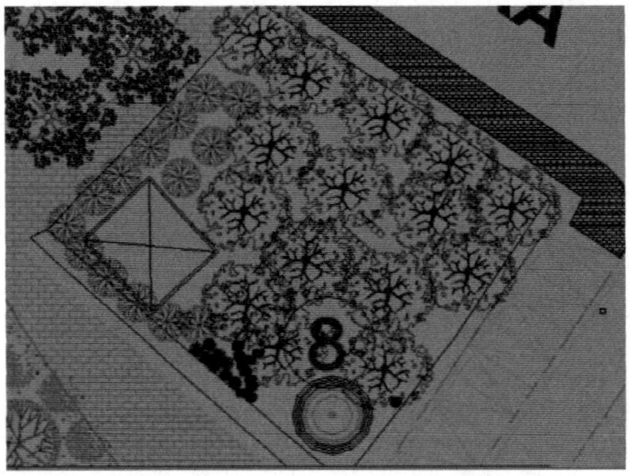

Figure 4. Waiting area [4].

3.2.5. Solar Panel Area

The solar panel area (Figure 5) consists of several solar panels that are used to capture solar thermal energy for use as electricity in the tread area. In addition to energy-efficient, using solar panels means spending on reduced station electricity (low cost concept). The function vegetation in This area are for buffer as barriers and to control climate [8].

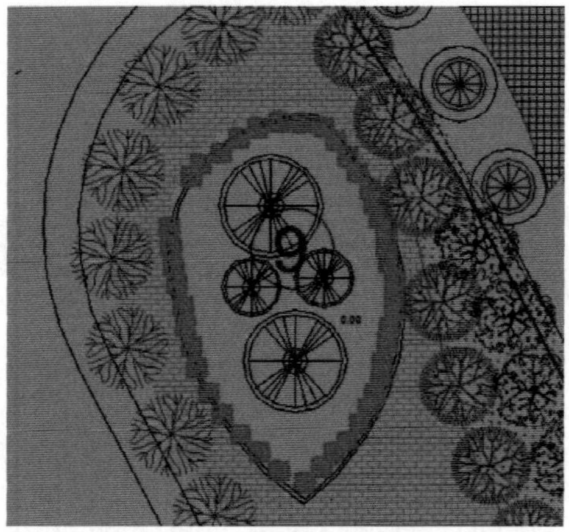

Figure 5. Solar Panel Area [4]

3.2.6. Social Interaction Area

Social interaction area (Figure 6) has 2 function i.e for conservation (soil, water, air) and for social interaction (aesthetic). As a function for vegetation conservation, it uses stratified from ground cover, shrubs, shrubs, trees. Conservation areas are needed to minimize flooding, air pollution and noise pollution. While as function for social interaction prefer to use aesthetic vegetation, both the leaf canopy and the color that makes the users relax psychologically.

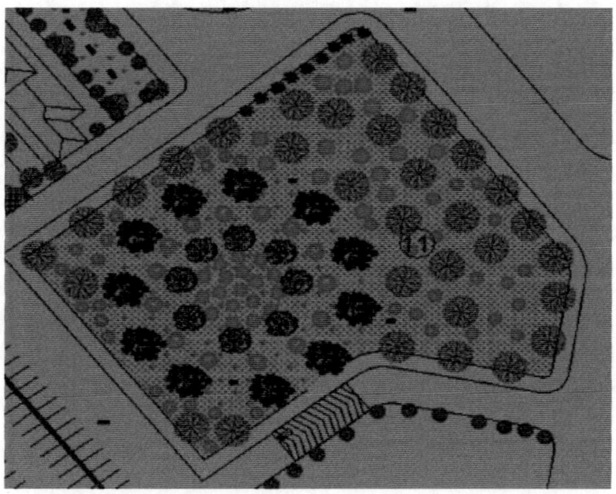

Figure 6. Social interaction area.

4. Conclusion

The basic concept of this design were the "green, low cost" which unite with theme "user friendly" land use on a more efficient and effective site. Landscape design of Manggarai integrated station transfer area consists of west and east entrance area, transition area, parking area, waiting area, solar panel area, and social interaction area. The system of outer space function on the site is generally as a transit area of a systemic nature. Other functions that exist in the tread is the function of space and function of supporting activities. The functional system is supported by facilities in each area as the main supporting subsystem. The activity of the transport mode is good from the train in the mode of transportation. Crowded atmosphere occurs during peak hours of community leave and go home to work. The unifying function of the space is accommodated by the receiving plaza and the connecting plaza. The plazas connect between one area and another.Urban rail transport plays an important role in the process of city development because of its obvious advantages. [3].so the landscape design of station Manggarai as the transit of integrated transportation is very important.

References

[1] Hasil Sensus Penduduk Provinsi DKI Jakarta, accessed from http://jakarta.bps.go.id/index.php?bWVudT0yMjAwJnBhZ2U9cHVibGlrYXNpJmlkPTIw on Oktober 2, 2012 at 00:08.
[2] Ning Zhang; Meng Yuan Liu; Suming Guo, Analysis on the sub way station landscape o the southern Mount Zijin in Nanjing. IOP. Cenference series; Earth and Environmental Bingyi. Liu, Landscape planning and design of city road. Southeast University Press. 2002
[3] Grava, Siguard. Urban Transportation System. USA: McGraw-Hill. 1999
[4] Baoping Mao, Planning and design of city rail traffic. China Communication Press.2006
[5] Chang,SK. Challenges for TOD in developing cities; Invited presentation in train the trainer SUMA. Programe sponsored by ADB and GIZ, Beijing China. 2006
[6] H. Meilianti. Landscape Design of Manggarai Integrated Station Transit Area, Jakarta, Bachelor Degree of Landscape Architecture. 2014
[7] Philip L Carpenter, Theodore D Walker, Lamphear F, Plant in the landscape.1975

Dynamics of land - use change in urban area in West Jakarta

R L Pangaribowo

Landscape Architecture Department, Faculty of Landscape Architecture and Environmental Technology, Universitas Trisakti, Jakarta, Indonesia

Corresponding Author: pangaribowo2006@yahoo.com

Abstract. This aim to research is to know how land use change in West Jakarta period 2000 - 2010. The research method used is descriptive method with a quantitative approach. Data analysis was done by using the result of research instrument to find out the driving of land change and to know the change of was analyzed using GIS (Geographic Information System) in Arc View GIS 3.3 program and Quantitative Analysis Model Location Quotient (LQ) and Shift-Share Analysis (SSA) In this study. The research instrument used in the analysis was observation and documentation. Based on the analysis conducted, the results of research on land use change in West Jakarta in the period of 10 years from 2000 until 2010 is caused by several aspects that are related to each other, namely political, economic, demographic, and cultural aspects. The land use change occurred in the area which decreased by minus 367,79 hectares (2.87%), the open space area decreased by minus 103.36 hectares (0.8%), the built up area increased by 201.13 hectares (1.57%), and the settlement area was 27.14 hectares (0.21%).
Keywords: GIS, land use, land-use change

1. Introduction

The phenomenon of need for land tends to increase, which is the resultant of economic development and population growth. This in turn will give rise to the phenomenon of land use competition, which is actually a manifestation of the law of supply and demand [1]. This can be understood, given the land is a very important natural resources [2]. Almost all aspects of life and development, whether directly or indirectly related to land issues.Along with the growth of the region including the growth of the city, the need for land resources tends to increase [3]. Meanwhile, it is seen from its availability in terms of land area within administrative boundaries is limited [4].

Therefore, with the change of economic structure, which is characterized by the development of industrial sector, the increasing of activity and the variety of specialization outside agriculture and the increasing of population which among others caused by the existence of urbanization, allegedly will cause pressure on agricultural land and trigger the shift pattern.

This phenomenon in turn will result in changes in the quality of the environment and increased land values. Increasing the value of the land if associated with urban areas will be more related to location of its location (location factor), while for rural areas (rural), the increase of the value of the land is more due to fertility factor. Land use is also a reflection of economic structure and community preference. Because the structure of the economy and the preferences of this society are dynamic whose orientations are always changing at any time in line with population growth and development dynamics, the structure of land use is dynamic, and thus changes in land use patterns are inevitable [5].

IOP Conf. Series: Earth and Environmental Science **106** (2017) 012040 doi:10.1088/1755-1315/106/1/012040

Even within a broader framework, land use and land-use phenomena will have considerable implications for regional economic diversity, resource allocation and labor and spatial structure.Land use competition in market mechanism is determined by the economic value of the land offered. The use of land that offers the economic value of high land is likely to win the competition.

The most significant change in land use is its effect on the sustainability of water resources is the change from forest area to other uses, such as agriculture, housing or industry. If the symptoms are not immediately managed properly, then the consequences can cause excess water (flood) during the rainy season and drought during the dry season. This is due to the rain falling mostly into surface runoff. Therefore, efforts to conserve water resources are needed through the arrangement of land use within a watershed.

Thus the intensity of the current movement of people (labor), goods and services in the area of DKI Jakarta is quite high. This phenomenon in turn has made this area has grown rapidly, which among others is characterized by the development of population and relatively high economic growth rate, as well as the development of increased accessibility. Thus, it becomes easy to understand if land use tends to be dynamic, not only because there are many factors that affect the utilization and use of land that occurs in a region. However, the change in land use has significant implications for the diversity of the economy, the allocation of resources and labor and spatial structure in the region concerned [6].

Similarly, the type of land use in the study sites to date is more dominated by the use of non green space. This is of course due to various causes and one of them is the physical factor of the environment. Therefore, the study of physical environmental factors on land use change is quite interesting to do.

1.1 Research Purposes

The purpose of this study is to know land use and its changes, therefore it is necessary to analyze the spread and its condition both locally and nationally which is expected to provide information or insight understanding about the relationship between land condition and its use while providing information to the related planners with an alternative choice in the development and development of a nationally optimal area.

2. Research Method

The study conducted in this research is Descriptive Research, which is supported by quantitative analysis. The main focus is the dynamics of land use in urban areas. Research data collected through documentation method of time series. The analytical model that will be used to assess the dynamics of land use change is the Quantitative Location Quotient (LQ) and Shift-Share Analysis (SSA) quantitative analysis model [7]. In this study, the LQ method is used to compare the area of land use for certain activities with the total area of aggregate. While SSA is one way or technique that can give an idea about the causes of land use change that occur in a region and Geographic analysis of system information [8].

3. Results and Discussion

In RUTR DKI Jakarta 2005 it was suggested that the development of land value was not accompanied by the development of socio-economic level of the community, causing inefficient and effective land use. In addition, the formal juridical provisions concerning the regulation, control and supervision of the land are still not solid.

Therefore some parts of urban land are used and overused, which are occupied by a large group of urbanites.Some of the urban land has been used formally and not optimally, while in some other areas have not been utilized.

3.1. Changes in Land Use from 2000 to 2010

Land use is actually a human effort in its interaction with the physical resources of the land in order to meet the needs of his life. From the data collected in this research for a decade in West Jakarta region there has been a change of land use structure. For more details can be seen in Table 1 below :

Table 1. West Jakarta land use 2000, 2005 and 2010 (hectare).

NO	Land Use	Year2000	Year2005	Year2010	Changes
1	Settlement	6,097.71	6,114.29	6,124.85	**+ 27.14**
2	Open Space	2,003.55	2,037.57	1,900.19	**-103.36**
3	Lake and River	238.46	238.46	139.81	**- 98.85**
4	Green Area	506.67	455.62	138.88	**- 367.79**
5	Build Up Area	2,251.32	2,251.77	2,452.45	**+201.13**
6	Street	1,710.24	1,710.24	2,051.77	**+ 341.53**
7	Total	**12,807.95**	**12,807.95**	**12,807.95**	

Source: Results of Data Processing

From the above table, can be seen that the reduction of green open space and open space in West Jakarta area during the 10 years period reaches minus 367,79 hectares (2.87%) and minus 103.36 hectares (0.8%) of the total area .

West Jakarta has changed the structure of land use. Lands that tend to experience the greatest reduction/depletion are the use of green open space, the use of lakes and rivers; and unused open spaces whereas other uses tend to increase, especially for use for settlements; Industry and services or built-in areas (Table 1).

Land use is actually a human effort in its interaction with the physical resources of the land in order to meet the needs of his life. The change of land use structure is not merely a physical phenomenon of reduced land area and increased use of land for other uses, but has a close relationship with changes in economic, social, cultural and political orientation of society. This orientation change is related to the transformation of economic structure characterized by decreasing of primary sector (agriculture) and increasing of secondary and tertiary sector (industry and service). Thus economic development is directed to reduce the dependence of a region's economy on the primary sector which has lower added value compared to the secondary and tertiary sectors.

3.2. Land Use Change and Regional Growth

The phenomenon of land use change in West Jakarta is undeniably influenced by population growth and economic development. For a decade, the population of Jakarta City has grown by 18.6%, with an average of 1.8% per year. Similarly, the economic growth of urban areas, for almost a decade the city of Jakarta has experienced a very significant economic development (Table 2).

Table 2. Economic growth in Jakarta.

No	Agricultural Sector	Year		growth rate
		2000	2010	
1	Agricultural	162,682	247,576	6.52
2	Mining and excavation	65,056	78,885	2.65
3	Processing industry	623,509	1,064,499	8.84
4	Electricity and clean water	37,553	39,050	0.40
5	Building	361,044	445,025	2.90
6	Trade, hotel & restaurant	803,950	1,037,250	3.62
7	Transportation & communication	536,588	890,120	8.23
8	Finance	296,643	1,159,261	36.34
9	Services	728,002	840,637	1.93
Gross Regional Domestic Bruto		3,615,027	5,802,307	7.56

Source : Data processing from the Central Bureau of Statistics Jakarta

The table above shows that for almost a decade the economy of Jakarta has grown by an average of 7% per year. While the growth rate of each sector, shows that the financial sector, rental and service

IOP Conf. Series: Earth and Environmental Science **106** (2017) 012040 doi:10.1088/1755-1315/106/1/012040

companies have the highest growth rate compared with other economic sectors, which amounted to 36.34% per year. While the smallest growth rate is in the electricity and water sector, which is 0.49% bigger for almost a decade. Competitively the economic sector contributes to regional growth. It can also be seen in the table below:

Table 3. LQ and Shift-Share Value of Economic Sector in Jakarta City

No	Agricultural Sector	Value of LQ 2000	2010	Value of shift differential
1	Agricultural	0.09	0.10	0.14
2	Mining and excavation	0.68	0.56	- 0.13
3	Processing industry	1.27	1.35	0.22
4	Electricity and clean water	3.05	1.81	- 0.57
5	Building	1.90	1.54	- 0.19
6	Trade, hotel & restaurant	1.40	1.12	- 0.20
7	Transportation & communication	2.66	2.39	- 0.05
8	Finance	2.22	2.52	0.71
9	Services	2.17	1.89	- 0.07

Source : Data processing

Table 3 shows that for LQ values that reflect comparative advantage, for almost a decade there is no significant change. That is, that the economic sector in the City of Jakarta outside the agricultural sector and the mining sector and excavation all have comparative advantages, both in 2000 and 2010. While for the value of differential shifts that reflect competitive keuanggulan, seen that the economy in this city in the long term has the advantage In the agricultural sector; Non-oil and gas processing and finance, leasing and business services relative to other regions on a regional or provincial scale.

On the basis of this, it becomes easy to understand if population growth and increased development and economic activity will lead to increased demand for land. The relatively fixed land area on the one hand and the increasing demand for land on the other, causing land use change in an area is inevitable. Therefore, the higher rate of growth of the region requires the allocation of land use that is more effective and efficient. The high rate of growth of the region that can be characterized by the more complete and varied availability of infrastructure facilities and or the number of population causes the increasing importance of land functions. These conditions resulted in less productive (extensively) lands being diverted into more productive (intensive) lands. Along with the occurrence of economic growth in a region, then the change of land use is a process that has the rate, pattern and impact.

The rate of change of land use as scale will be affected by changes in speed and time dimension. The time dimension here is defined as the development or growth of the region. That is, the region will evolve over time, assuming that the developmental component of the region develops in line with the development of the region [9].

4. Conclusion

From the above description then the conclusions can be drawn is as follows :Changes in land use in West Jakarta occur along with the growth and development of city areas that reflectthe rate; Patterns and impacts that vary in each part of the city area. Economic growth factors, population growth and community preference (which is a reflection of capital variables, information and accessibility) are factors that trigger changes in land use as reflected in changesin urban spatial use patterns. For almost a decade West Jakarta's economic growth is supported by the economic sector outside the agricultural sector and the mining & quarrying sector. While in the long run the city's economic growth competition is supported by the agricultural sector; Non-oil and gas processing industry; And financial sector, leasing & corporate services.

References

[1] Anon, Green Open Space, City Planning Department of the Special Capital City Government of Jakarta, Jakarta. 2000.

[2] Lumempouw R J, Study Factors influencing the shrinkage of water catchment area of South Jakarta city, Province Special Capital Region of Jakarta, Tesis Program Pasca SarjanaUniversitas Sam Ratulangi, Manado. 2000.

[3] Purwadhi FSH. Rational Approach in the effort of Spatial Planning of a Region, *Prosiding Seminar Sehari, Jurusan Geografi FMIPA – UI*. 1994.

[4] Sandy, I M. Land, Earth, UUPA 1960 – 1995, PT Indograph Bakti, FMIPA – UI, Jakarta. 1996.

[5] Haggett, Peter. Geography A Global Synthesis, Prentice Hall, London. 2001.

[6] Muiz, Abdul. Analysis of Land Use Change in Sukabumi District (Tesis) Sekolah Pascasarjana Institut Pertanian Bogor. 2009.

[7] Blakely, Edward J. Planning Local Economic Development: Theory and Practice Second Edition. Sage Publication. California. 1994.

[8] David J Maguire, Michael F Goodchild, David W Rhind. Geographycal Information Systems, Principles and Applications, Longman Scientific & Technical, Copublished in the United States ang Canada with John Wiley & Sons, Inc, New York. 1991.

[9] De Blij, H J & Muller, P O, Geography Regions and Concepts, John Wiley & Sons, Inc, Canada. 1992.

Study on water quality around mangrove ecosystem for coastal rehabilitation

G Guntur[1], A B Sambah[2,3], D M Arisandi[4]*, A Jauhari[2], A A Jaziri[5]

[1]Marine Science Department, Faculty of Fisheries and Marine Science, Universitas Brawijaya, Malang, Indonesia
[2]Fisheries Resources Management Department, Faculty of Fisheries and Marine Science, Universitas Brawijaya, Malang, Indonesia
[3]Marine Resources Exploration and Management (MEXMA) Research Group, Faculty of Fisheries and Marine Science, Universitas Brawijaya, Malang, Indonesia
[4]Aquaculture Department, Faculty of Fisheries and Marine Science, Universitas Brawijaya, Malang, Indonesia
[5]Technology of Fisheries Product Department, Faculty of Fisheries and Marine Science, Universitas Brawijaya, Malang, Indonesia

*Corresponding Author: defrianmarzaarisandi@gmail.com

Abstract. Coastal ecosystems are vulnerable to environmental degradation including the declining water quality in the coastal environment due to the influence of human activities where the river becomes one of the input channels. Some areas in the coastal regions of East Java directly facing the Madura Strait indicate having experienced the environmental degradation, especially regarding the water quality. This research was conducted in the coastal area of Probolinggo Regency, East Java, aiming to analyze the water quality as the basis for coastal rehabilitation planning. This study was carried out using survey and observation methods. Water quality measurement results were analyzed conforming to predetermined quality standards. The coastal area rehabilitation planning as a means to restore the degraded water quality parameters is presumably implemented through mangrove planting. Thus, the mangrove mapping was also devised in this research. Based on 40 sampling points, the results illustrate that according to the quality standard, the water quality in the study area is likely to be deteriorated. On account of the mapping analysis of mangrove distribution in the study area, the rehabilitation of the coastal zone can be done through planning the mangrove forest plantation. The recommended coastal area maintenance is a periodic water quality observation planning in the river region which is divided into three zones to monitor the impact of fluctuating changes in land use or human activities on the coastal water quality.
Keywords: coastal, mangrove, rehabilitation, water quality

1. Introduction

Marine ecosystems in the northern coastal areas of Java and Madura Strait have been declining functionally during the last ten years. Sporadic and perfunctory marine ecosystem improvements will certainly not sufficiently compensate for the deterioration, even are considered futile having spent a lot of time and cost. In general, the ecosystem condition of the sea on the island of Java has greatly declined. Therefore, ecosystem control activities must be initiated, starting from restoring the damaged ecosystem first. East Java Province has conducted coastal environmental rehabilitation activities through the empowerment of surrounding communities. Fish planting projects have been commenced in four regencies of Banyuwangi, Lamongan, Probolinggo, and Pacitan. Rehabilitation of corals has also been conducted by making artificial reefs in Situbondo and Malang. Those measures of fish stocking, artificial reefs, and coral transplants have been carried out in the last four years, yet have not demonstrated a concrete success in restoring the coastal ecosystems in East Java. These unsuccessful attempts are allegedly affected by two factors. Firstly, the rehabilitation activities are lacking, so there is no equilibrium point between the healthy and damaged environments which resulting in the shortage of environmental supporting capacity to accelerate the rehabilitation. Secondly, there is the influence of wider environmental conditions. Marine waters are runoff from the upstream areas, and if these upper waters are contaminated, the pollutants will be possibly accumulating in the coastal regions.

Rehabilitation is one of the options in increasing the carrying capacity of the coastal environment, particularly in damaged areas. In the long run, the coastal rehabilitation is expected to improve the function and benefits of coastal natural resources for the community [1]. The planning of coastal environmental rehabilitation should be followed by directional, measurable, and comprehensive studies. Technical improvements of ecosystems can be made through the initial assessment in the form of observations and water quality analysis. One manifestation of coastal rehabilitation is the cultivation of mangrove forests along with the debarment of coastal areas destruction by their own communities [2]. This study aims to analyze the water quality in the study area as a basis for coastal rehabilitation planning.

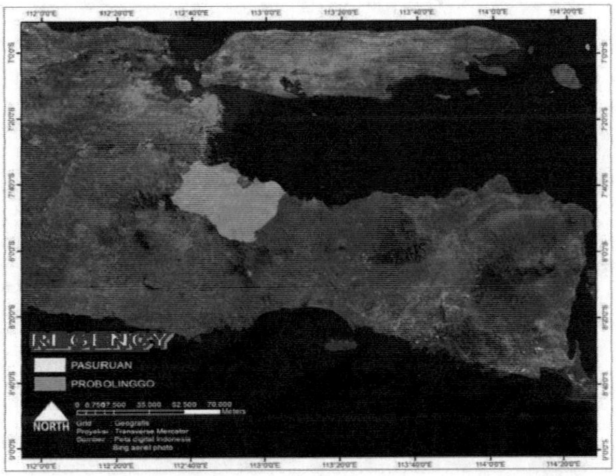

Figure 1. Research area.

2. Research Method

2.1. Research Areas
This study was conducted in coastal regions of Probolinggo Regency (Figure 1). This research area is part of the Madura Strait waters which gets a lot of groundwater input towards the ocean.

2.2. Methods
This study is a survey research—research that does not make changes or give special treatments to the variables studied. Survey research is a study conducted to obtain the facts of the symptoms that exist and seek information factually. One of the first important steps is the collection of data and information related to the study to ensure that the activities are going well and reaches the objectives. The statistics used in this research consist of water quality parameter data through samplings, land use data, spatial planning documents, mangrove area map, and profile of the research area.

IOP Conf. Series: Earth and Environmental Science **106** (2017) 012041 doi:10.1088/1755-1315/106/1/012041

2.3. Data Analysis

Data analysis was done by comparing the sampling and laboratory analysis results with the water quality standard as stated in the Decree of Minister of Environment No. 51 of 2004 on the Quality Standard of Sea Water [3]. The parameters of seawater quality observed in this study consist of physical and chemical characteristics. The information obtained is a temporary water condition that represents one season at the time of sampling. The water quality analysis comprises of (1) biological parameters of coliform Bacteria; (2) chemical parameters of DO, BOD, NH_3-N, detergent/surfactants, and phenol compounds; as well as (3) physical parameters of turbidity level and suspended solids. The spatial analysis was conducted to obtain data on land use conditions in coastal areas and coastlines, as well as to overlap various spatial data using a mapping software.

2.4. Research Flow

The research flow explains the activities undertaken from the beginning to obtain results and to deduce conclusions at the end of this study. In general, the research concepts employed in these two areas are described in Figure 2 below.

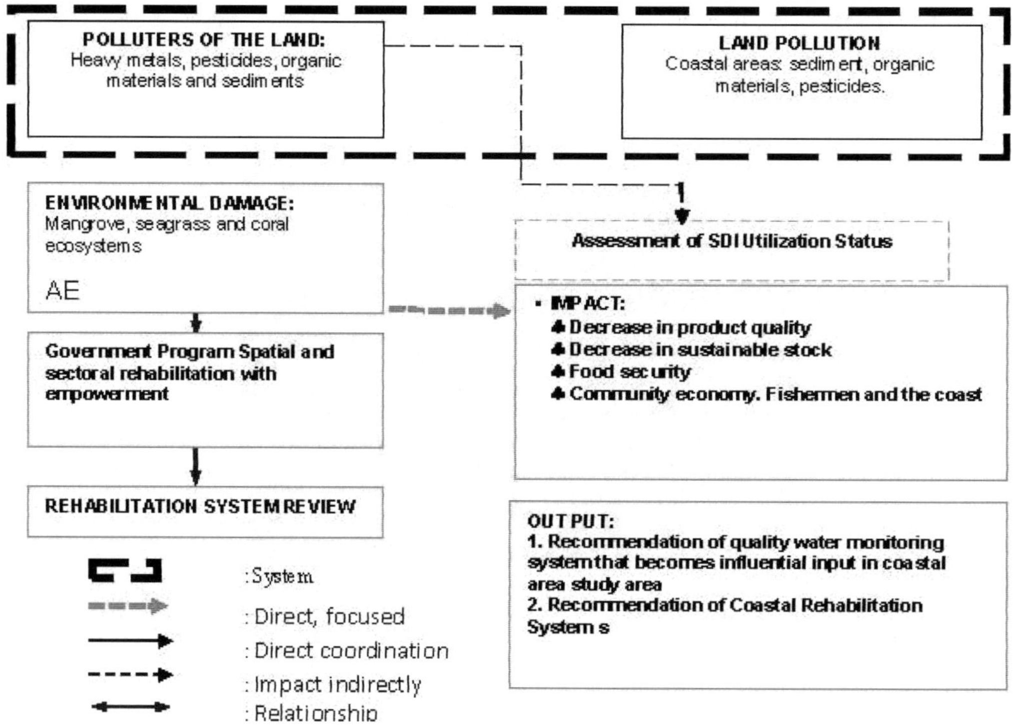

Figure 2. Research flow diagram.

3. Results and Discussion

3.1. Research Results

Water quality is a measure of water conditions seen from its physical, chemical, and biological characteristics [4]. The water quality also shows the water conditions relative to the needs of water and human biota [5]. The water conditions vary over time depending on the local ecological states so that the water quality includes complex subjects of the environmental science. Industrial activities such as manufacturing, mining, construction, and transportation are major causes of water pollution, as well as surface runoff from agriculture and urban areas. It is considered a complicated matter, reflected in the type of measurement and water indicator used. This measurement will be more accurate if it is done directly on site because the water is in equilibrium condition with the environment. Sampling in this study was conducted in some regions that are assumed to be marine areas affected by the river as illustrated in the map of Figure 3.

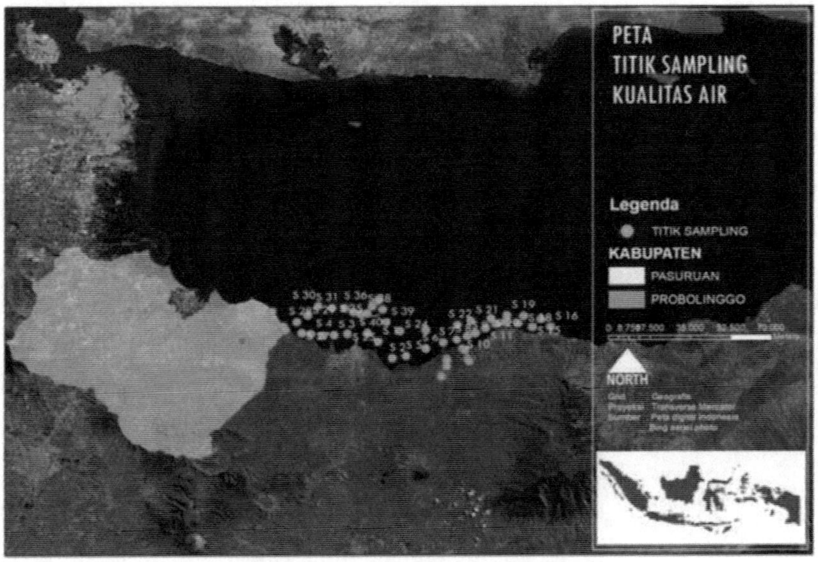

Figure 3. Sampling location map.

The location used as the sampling point in this study is a large coastal area whose input water gets influence from the mainland through the river. The distribution of the sampling points and the parameters tested are presented in Table 1.

Table 1. Sampling points and field measurement results on water quality parameters.

Station Name	Position Lat.	Lon.	Sal (ppt)	DO (mg/L)	Temperature (°C)	pH	Brightness	Phytoplankton (ind/L)	Zooplankton (ind/L)
S 1	S7 43 44.5	E113 15 30.0	30	8.4	29.4	7.19	1.50	7,458,333,336	4,166,666
S 2	S7 45 38.1	E113 17 07.3	29	9.9	28.9	7.1	2.25	6,835,186,351	1,153,765,471
S 3	S7 42 59.9	E113 10 37.8	28	9.1	29.2	7.14	1.80	20,833,333	2,083,333
S 4	S7 42 52.6	E113 08 04.1	28	7.7	30.5	7.16	1.80	20,833,333	2,083,333
S 5	S7 45 21.6	E113 18 39.2	26	7.1	29.8	7.14	-	3,356,408,643	1,223,690,651
S 6	S7 44 29.6	E113 20 47.7	28	6.4	28.8	7.16	0.50	191,666,666	2,038,333
S 7	S7 43 47.3	E113 22 43.5	27	5.2	29	7.6	1.25	175,000	6,250
S 8	S7 43 26.3	E113 24 16.4	25	7.8	29.4	7.16	2.00	1,905,461,157	2,447,381,302
S 9	S7 43 04.3	E113 25 24.5	26	6.1	30	7.12	1.00	1,922,942,452	1,573,316,551
S 10	S7 42 43.2	E113 26 20.2	23	5.8	30.1	7.14	3.50	1,940,423,747	4,166,667
S 11	S7 42 09.1	E113 27 26.3	25	7.6	30.5	7.12	1.50	14,583,334	3,671,071,953
S 12	S7 41 39.1	E113 28 35.2	28	5.9	29.9	7.12	2.00	1,083,840,291	6,992,518,006
S 13	S7 41 25.5	E113 29 36.7	26	7.2	31.7	7.12	4.20	1,485,910,076	0
S 14	S7 41 34.9	E113 31 02.7	26	7.1	30.8	7.15	6.50	29,166,667	8,740,647,507
S 15	S7 41 59.8	E113 32 33.7	29	7.8	30.2	7.12	17.0	1,011,157,438	2,272,568,352
S 16	S7 41 58.6	E113 34 32.3	26	6.7	28.4	7.25	10.20	1,730,648,206	3,146,633,103
S 17	S7 41 01.7	E113 33 28.6	28	6.8	30.9	7.24	10.0	3,915,810,083	4,195,510,803
S 18	S7 40 40.6	E113 31 34.7	27	7.4	30.1	7.22	5.00	3,199,076,988	2,097,755,402
S 19	S7 40 40.6	E113 29 45.8	28	6.9	30.3	7.23	11.0	8,006,433,117	1,573,316,551
S 20	S7 41 00.7	E113 28 01.0	28	6.8	28.6	7.24	8.80	1,241,171,946	2,097,755,402
S 21	S7 41 18.4	E113 25 47.6	29	6.7	28.5	7.27	7.00	5,226,907,209	1,922,942,452
S 22	S7 41 46.0	E113 24 17.8	28	7.1	28.1	7.28	3.50	1,206,209,356	1,048,877,701
S 23	S7 42 20.6	E113 20 47.6	30	6.4	29.9	7.15	2.30	7,272,218,726	3,146,633,103
S 24	S7 42 29.0	E113 17 55.7	28	5.1	27.9	7.12	2.00	4,510,174,114	4,370,323,754
S 25	S7 42 50.2	E113 12 07.5	29	4.3	28.8	7.11	1.00	5,226,907,209	41,666,666
S 26	S7 43 03.3	E113 09 19.3	28	6.1	28.6	7.1	0.50	5,436,682,749	6,468,079,155
S 27	S7 42 45.6	E113 07 02.9	27	6.4	28.3	7.09	0.50	8,333,333	0
S 28	S7 41 28.6	E113 06 29.0	27	5.6	28.5	7.5	1.00	1,031,396,406	1,223,690,651
S 29	S7 40 44.2	E113 07 39.7	28	4.5	29	7.13	2.00	12,499,999	8,740,647,507
S 30	S7 39 38.0	E113 08 49.4	29	4.9	29.2	7.12	2.20	5,069,575,554	41,666,663

Station Name	Position Lat.	Position Lon.	Sal (ppt)	DO (mg/L)	Temperature (°C)	pH	Brightness	Phytoplankton (ind/L)	Zooplankton (ind/L)
S 31	S7 39 50.5	E113 10 23.8	29	5.7	29.6	7.2	2.00	7,691,769,806	1,048,877,701
S 32	S7 39 52.7	E113 11 50.3	28	5.1	29.5	7.17	8.00	613,341,068	-
S 33	S7 40 20.3	E113 12 46.9	28	6.0	29.7	7.69	7.00	4,575,051,553	-
S 34	S7 40 33.7	E113 13 34.4	29	6.1	29.3	7.09	5.10	3,016,692,427	-
S 35	S7 40 28.8	E113 14 17.4	27	6.9	28.8	7.09	7.00	14,583,333	-
S 36	S7 39 39.5	E113 14 48.8	26	6	29.1	7.8	6.50	3,234,039,578	-
S 37	S7 38 57.6	E113 15 32.1	30	7	29	7.08	5.00	-	-
S 38	S7 40 01.7	E113 15 59.9	29	7.4	29.7	7.11	6.00	-	-
S 39	S7 41 28.0	E113 16 24.4	29	6.6	31.7	7.16	3.50	-	-
S 40	S7 42 44.9	E113 14 18.6	27	6.8	33.3	7.2	3.10	-	-
	Average		27.65	6.61	29.575	7.2045	4.175	-	-

The water quality in several locations of both coasts and rivers having a direct or indirect impact on the coastal ecosystem cannot be separated from the influence of the surrounding land use. The results of field identification explain some types of land use that have the opportunity to affect the river flow until the stream reaches the coastal area and adversely affect the dynamics of the existing ecosystem. Land covers around the river include the form of mangroves, ponds, settlements, industries, mooring activities, and intensive ponds.

3.2. Rehabilitation Prospect Analysis

This study identifies some of the internal strengths of the coastal area of Probolinggo Regency, i.e., the community's positive perception of rehabilitation activities, the availability of potential sites for rehabilitation, and the full support of local authorities. In addition, there are opportunities where the program can get help from the government (e.g., seedlings). However, the lack of technical capacity and the threat of coastal environmental degradation due to the input of pollutants into the river and carried down to the coast are the two limiting factors that need to be handled properly. With the dominant strengths and opportunities described above, coastal rehabilitation activities have prospects to be implemented in the coastal area of Probolinggo Regency. These strengths and opportunities must be optimized, while the limiting factors must be overcome. Based on the survey results including the measurements of the water quality, this coastal rehabilitation program is highly feasible. Thus, the rehabilitation of mangroves is recommended, starting from the area as illustrated in Figure 4. The suggested planting system is for a location of the expanse type, then planting can be done with a medium to a dense distance of (1 m x 1 m), (1 m x 2 m), or (2 m x 2 m) from the back side toward the front zone. Reaching the regions of inundation and strong wave currents, the planting should be stopped (As for the planting along drains or ditches, it can be carried out more closely with the distance Figure 4a). between plants of 20–60 cm in 1–5 rows according to the conditions in the field (Figure 4b).

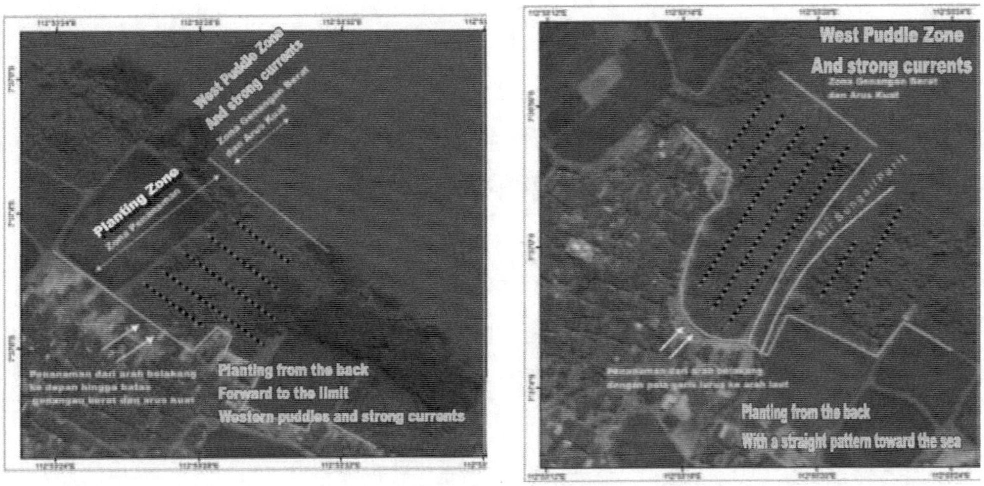

(a) (b)

Figure 4. (a) Planting patterns in sandy coastal landscapes; (b) Planting patterns in the area along the river/ditch channel.

4. Conclusion

The results of water quality measurements in the coastal area of Probolinggo Regency indicate an environmental degradation, probably due to the influence of river runoff that leads to the coastal region of Probolinggo. The outcome of the tabulation of strengths, weaknesses, opportunities, and threats of the coastal area rehabilitation planning suggests a means of mangrove rehabilitation. This study is a multi-year research that will assess the impact of water quality reduction on fisheries activities as well as the design of control systems on the quality of waters in coastal areas.

References

[1] Wibisono ITC, Kajian Kelayakan dan Pengembangan Desain Teknis Rehabilitasi Pesisir, Working paper 227, Bogor, Indonesia: World Agroforestry Centre (ICRAF) Southeast Asia Regional Program. DOI. http://dx.doi.org/10.5716/WP16037.PDF. 2016.

[2] Haryanto R, Rehabilitasi Hutan Mangrove: Pelestarian Ekosistem Pesisir Pantai dan Pemberdayaan Masyarakat Pesisir, Vol. XIV No. 2, KARSA, 2008.

[3] Keputusan Menteri Negara Lingkungan Hidup Nomor 179 Tahun 2004 Tentang Ralat Atas Keputusan Menteri Negara Lingkungan Hidup Nomor 51 Tahun 2004 Tentang Baku Mutu Air Laut.

[4] Diersing, Nancy. Water Quality: Frequently Asked Questions. Florida Brooks National Marine Sanctuary, Key West, FL, 2009.

[5] Johnson D L, S H Ambrose, T J Bassett, M L Bowen, D E Crummer, J S Isaacson, D N Johnson P Lamb, M Saul, and A E Winter-Nelson. 1997. Meanings of environmental terms. *Journal of Environmental Quality*. 26: 581-589.DOI:10.2134/jeq1997.00472425002600030002x. 1997.

Urban simulation evaluation with study case of the Singapore Management University, Singapore

O Seanders

Scuola di Architettura Urbanistica Ingegneria delle Costruzioni – MI, Politecnico di Milano, Milan, Italia

Corresponding Author: olivia.seanders@mail.polimi.it

Abstract. This paper reports and discusses about the urban simulation evaluation with a study case, The Singapore Managemant University (SMU), the first major university to be located in the city centre. It is located in Bras Basah District, with some controversy on the geographical establishment, the physical realization of the University in the original plan required some demolishes, urban historical building, a public park and in the end will impact the lose of some certain qualities of the urban space. From this case we can see that the urban design and cultural heritage principles could come into conflicts with the more practical concerns of space constraints and transportation efficiency. This SMU case reflect the problem of the developing countries that have to decide between conservation of buildings and green spaces and space demands. In this case, for Singapore, it marks a progress in the step of greater community involvement in the planning process.
Keywords: land use, urban simulation, urban evaluation.

1. Introduction

The Singapore Management University (SMU) is a new university established in year 2000, Singapore's third major university, with 6000 students enrolled currently. It currently occupies four plots of land. It will be gradually expanded to up to eight plots covering 238,000 square metres with a target student population of 15,000 [1]. The difference between this university and the other two long-established universities in Singapore is that this is the first major university to be located in the city centre. It is located in the bustling and lively Bras Basah District [1]. But its geographical establishment is not without controversy, as the physical realization of the University in the original plan required some sacrifices including the demolition of a well-loved building of the National Library, the lost of a public park which is an important green lung of the city, and the loss of certain qualities of the urban space such as the visual impact. On this research, the research method is descriptive analysis with The Singapore Management University as object study or research. The aim of this paper is to emphasize the need of community involvement (the 'skilled citizen' and 'unskilled citizen') in urban planning in this case Singaporean.

2. Research Method

The object of study is the Singapore Management University (SMU), the plots covering 238,000 square metres. It is located in Bras Basah District, Singapore [1], [2]. The method used is descriptive analysis method. Data were collected by survey and field observation as a primary data to feel the space of location and social aspect in surrounding area. In the sometimes, study of literature as a

IOP Conf. Series: Earth and Environmental Science **106** (2017) 012042 doi:10.1088/1755-1315/106/1/012042

secondary data for supporting knowledge in the application of theory in the object study or research area.

3. Results and Discussion

3.1. The Case

The original urban plan of the SMU (Figure 1) was crafted by the national planning authority: the Urban Redevelopment Authority (URA) [2], which proposed that the new university campus be located in the Bras Basah District. It was also known as the 'Museum District' because of the existence of a number of museums, as well as colonial buildings of historical value, and private educational institutions [3]. The plan of SMU falls in line with the vision of the URA to establish the Bras Basah District as a cultural and educational hub of the city [2].

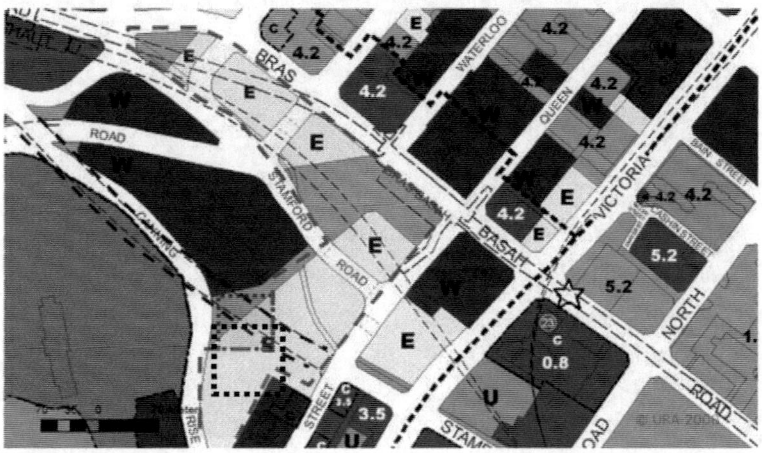

"E" (Educational) – Proposed sites of SMU, Area bounded by green lines – original size of Bras Basah Park, Area bounded by red lines – original site the demolished Old National Library, Area bounded by blue lines – new Fort Canning Tunnel

Figure 1. Master plan showing the sites of the SMU [1].

The original plan, however had several points of contention that have been brought up by urban planners, architects and local Singaporeans. Some of these have stronger social concerns (cultural and natural resources at stake), and some are more of a technical concern (such as traffic generation). These contentions prompted strong public debates and involvement in the planning process that is seldom seen in Singapore. Some urban planners and architects proposed alternative plans in the attempt to salvage the cultural and natural resources that are threatened by the new plan. These are outlined as follows:

3.1.1 The loss of the Old National Library

The Old National Library is regarded as a national icon to many Singaporeans. It was built in 1960 and was demolished in 2004 [3], despite strong public dissent, to give way to the new SMU and the new Fort Canning tunnel.

Rationale of Plan: The URA recommends for the demolition of the Old National Library because there was a strong (future) need to build the Fort Canning vehicular tunnel to facilitate heavy traffic going to and fro the central business district to the shopping belt, which crosses Bras Basah, as well as for anticipated new road users coming to the new SMU campus [2].

Also the URA deems that the Old National Library to be "not of great architectural merit and should not be conserved".

Evaluation: Failure to conserve a monument but planning process was participative and included many public opinions. Despite the 'negative' result of the demolition of the library even after a series of public debate, this issue has created substantial public awareness and generated a level of public participation that is seldom seen in Singapore. Yet on the other hand, some harsher critics note that despite a strong public dissent and even the submission of alternative plans to save the Library, the original plan has been implemented.

3.1.2 The Bras Basah Park reduces in size

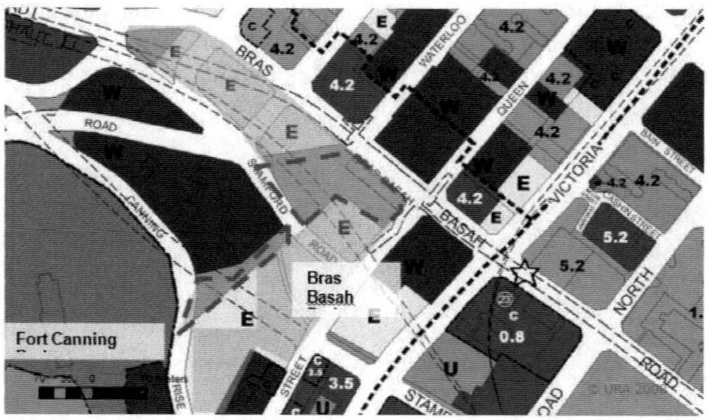

The original Bras Basah Park Within dotted lines: the planned Bras Basah Park

Figure 2. Master Plan showing the sites of the SMU [2].

The map (Figure 2) give a visual explanation of the establishment of the various buildings of the campus (indicated by "E") can result in the reduction of the size of the Bras Basah Park.

Rationale of Plan As seen by the micro urban structure here which is quite densely built-up, as well as height restrictions for this area, it offers less flexibility for the location of the buildings of the campus. Thus the proposal is to take some parts of Park, and including the Old National Library to be used for the new campus.

Evaluation: The Bras Basah Park has served as a green lung of this fairly dense area. In its original dimensions, the park can be seen as a connector from the densely built-up city to the Fort Canning Park. However with the reduction in size of the Bras Basah Park, there is left just a tenuous link to the Fort Canning Park, as represented by the little strip of green.

3.1.3 Traffic volume generation/ New Tunnel

The construction of the Fort Canning is to fulfil the aim of relieving the heavy load of anticipated (even current) traffic that comes into and passes through Bras Basah District [4]. Considerations have been made to design the road to go under the Fort Canning Hill instead of cutting through it to protect the historical and natural significance of the hill. Although the tunnel is underground, the exit of the tunnel is at the location where the Old National Library was, and thus it had to go.

3.1.4 Visual quality and "Space Syntax"

Yet the purpose of maintaining visual corridors goes beyond satisfying aesthetical requirements, but also has to do with how the various stakeholders and users of different parcels of land relate and interact with each other on the spatial dimension. Expounds on the idea of "Space Syntax" which recognises "the autonomous potential of space to form patterns was, in effect, seen as the means through which it is able to give expression to social meanings" [5], [6], [7]. The analysis of space relations has bearings on the movements and flows of people – those that use the space and pass through it. The land use in area also consists of an interesting variety. The colonial buildings and the museums, residential areas and the Fort Canning, local small scale shop houses found in the narrower streets, commercial buildings with shops and offices, educational institutions, and a great variety of

IOP Conf. Series: Earth and Environmental Science **106** (2017) 012042 doi:10.1088/1755-1315/106/1/012042

eating places (which are very important social places in Singapore). The variety of attractions generate movements – walking can be done to and fro buildings and in smaller street, while vehicular traffic uses the roads to enter or pass through the area. The idea of co-presence "We find in effect a basic partition of the city into a dual pattern, the one created by and responding to micro-economic forces, the other, the residential part, responding to cultural forces, the one more integrated the other more segregated." The study of Yang and Putra (undated) [9] noted the importance of the Bras Basah Park in maintaining "the quality of openness and visual connection" between its surrounding uses. There s also the concern of generating urban activities and promoting stronger interactions between stakeholders in the area, which includes students, local residents, business people and tourists. On a more macro scale, architect Tay Kheng Soon illustrates a mental map to show how the Bras Basah District is functionally and geographically a connector with four important "petals", or distinctive areas, in the central region (Figure 3) [10].

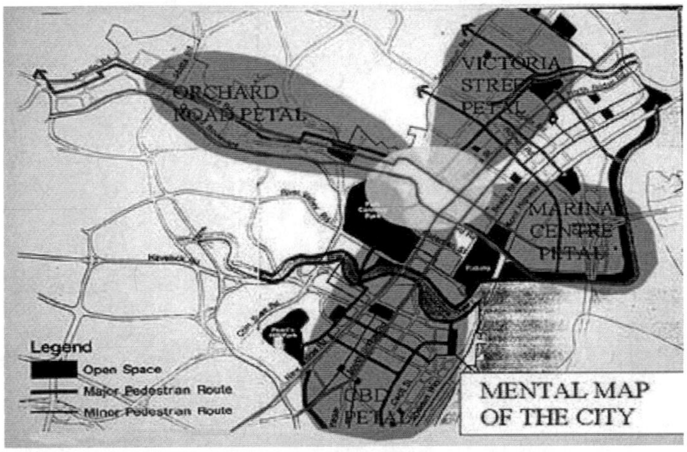

Figure 3. Mental map of Architect Tay Kheng Soon, describing how the Bras Basah District (the white oval) connects four important areas of the central region of Singapore (not to scale) [10].

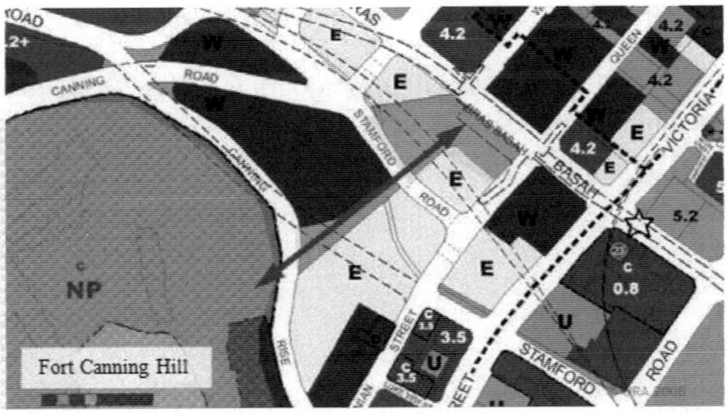

Figure 4. Included in the design guidelines, this sight-line must be preserved to maintain the vista of the Fort Canning Hill (dotted shading) from the Bras Basah Park (not to scale) (gray shading in the arrow line). [2], [4]

In the approved plan (Figure 4), it can be seen that the Bras Basah Park has been reduced in size, although a narrow line of view to the Fort Canning Hill has been maintained. The below study further elaborates the visual impacts of a number of alternative plans [9].

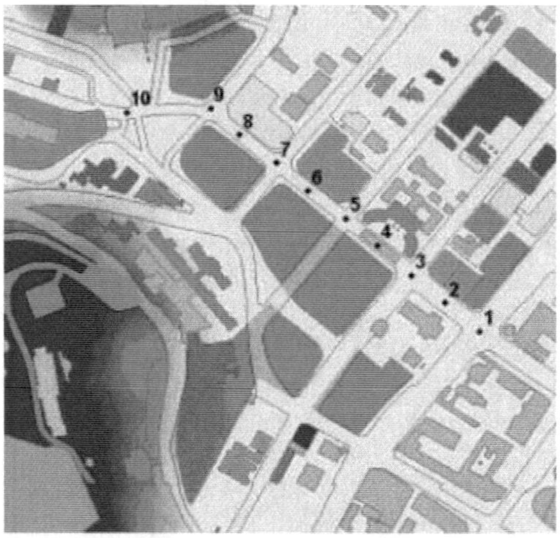

Figure 5. Map with point of observation [9].

Map with Point of Observation, numbered 1 to 10 along Bras Basah Road, looking from the road towards Fort Canning Hill. This is to establish and compare the extent of vista based on different proposals on the physical locations of the new buildings (Figure 5).

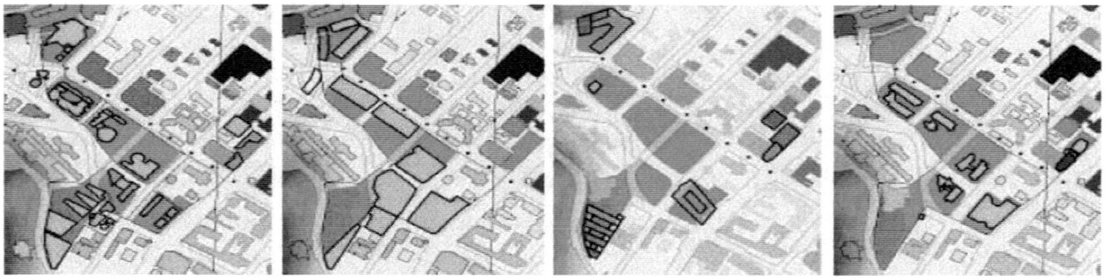

Figure 6. Map with block and point of observation [9]

Figure 6 show four separate proposals. The building outlines show how the various buildings of the new campus will be located. A *View shed Area Comparison* was made and the results are as follows.

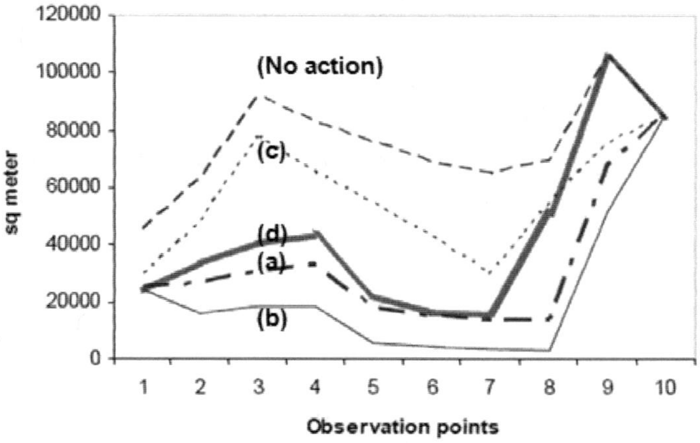

Figure 7. Result of view shed area comparison [9].

In the View shed Area Comparison, it can be seen that if "no action" was taken, that is, no plan was implemented, the open space visual quality is the highest at all ten observation points. Proposal C on Figure 7. Result of View shed Area Comparison seem to yield comparatively good result for keep the visual quality, but the buildings of the campus would have to be quite "scattered".

4. Conclusion

This case may represent how in the case of a multi-criteria evaluation, the urban design and cultural heritage principles can come into conflicts with the more practical concerns of space constraints and efficiency in transportation. This has been a consistent issue in Singapore which is very land scarce, but it also reflects the problem with many fast developing countries that have to decide between conservation of buildings and green spaces and space demands. In terms of the final results of the plan, though it did not manage to save the Old National Library and keep the Bras Basah park intact, it nevertheless had gone though some modifications to keep a fair level of urban visual quality. Yet in this case, for Singapore, it marks a progress in the step of greater community involvement in the planning process, where not just lay citizens voice their views and opinions, but also where "skilled citizens", meaning architects and planners, in their private capacities, draw up alternative proposals that could perhaps better represent the of local communities of their aspirations for the shaping of the spaces and places that they inhabit.

References

[1]. The Singapore Management university (SMU) https://www.smu.edu.sg/ accessed on 6 July 2017 h. 22.34 GMT +7

[2]. Urban Redevelopment Authority https://www.ura.gov.sg/skyline/skyline00/skyline00-03.pdf. accessed on 7 July 2017 h. 00.12 GMT +7

[3]. Bra Basah Bugis.http://brasbasahbugis.sg/about/history-of-bras-basah-bugis-precinct accessed on 7 July 2017 h. 01.55 GMT +7

[4]. Land Transport and Authority, https://www.lta.gov.sg/ accessed on 7 July 2017 h. 01.55 GMT +7

[5]. Hillier and Vaughan, The Social Logic of Space, Cambridge University Press, Cambridge, 2014 [e-book edition]

[6]. Hillier and Vaughan, The City as One Thing, Progress in Planning, 67 (3), 2007, p. 205-230

[7]. Hillier and Stonor, Space Syntax – Strategic Urban Design, City Planning Review 284, Special Issue Future of Urban Space and Humanity, City Planning Institute of Japan, 25 June 2010.

[8]. Stonor, Tim. 2014, Space Syntax : a SMART approach to urban planning, design and governance., http://www.spacesyntax.com/wp-content/uploads/2015/08/Space-Syntax-a-SMART-approach-to-urban-planning-design-and-governance.pdf [7 July 2017]

[9]. Yang P, & Putra S Y, An analysis of alternative plans for SMU using GIS. Sendai Instructiosn to Prepare a Final Paper for CUPUM'03, 2003.

[10]. Tay K. H. 2000. SMU: An Urban University Centered at Fort Canning Hill. (The Alternative [11] Plan of Architect Tay Kheng Soon), http://www.akitektenggara.com/ideas/smu/smu.htm. [8 July 2017].

Identification and analysis the illegal dumping spot of solid waste at Ciliwung segment 5 riverbanks

D Indrawati*, P Purwaningrum

Environmental Engineering Department, Faculty of Landscape Architecture and Environmental Technology, Universitas Trisakti, Indonesia.

*Corresponding Author: dindrawati@trisakti.ac.id

Abstract. Ciliwung River is the main river in the area of Jakarta that is divided into six segments across West Java and Jakarta. The study focuses on the fifth segment which is 30 km long, covering from Kelapa Dua Depok to Manggarai, South Jakarta. The survey of the river consists of 3 sub-segments: Lenteng Agung, Pejaten Timur and Manggarai. Objectives of the study are to describe the characteristics and typology of the residential surrounding the Ciliwung Segment 5 Riverbank, to identification the illegal dumping spot of solid waste, to measure the volume and composition of solid waste in the riverbank, to decide solid waste management for residential area surrounding river banks to control the river pollution. The study shows that there are 11 illegal dumping spot of solid waste consisting of 4.37 m^3 solid waste volume. The average composition of solid waste consists of 44% organic, 14% woods, 12% papers, 11% plastics, 3% rubbers, 1% metals and 2% others. To control the river pollution efforts are restoring the function of riverbanks to become green open space area, installing the trash rack into the river, to manage domestic solid waste based on 3R (Reduce, Reuse, Recycle) concept.
Keywords: illegal dumping spot, riverbank, solid waste

1. Introduction

Population increase and the increase of lifestyle are very influential to the volume of solid waste generation, especially in larger cities. The daily solid waste generation is increasing and increase in variation, which often makes solid waste management still a problem. Moreover, the space of final disposal is limited. The environmental problem due to the solid waste are among others is water pollution [1]. The tendency of the population to release the garbage into the river is basically a manifestation of the population perception [10]. The general problem in solid waste management in the cities is limited space, pollution, and social problem. Therefore it is necessary to find a solid waste management model that can benefit as effective as possible; using an easy technology, economical, environmentally friendly and acceptable to the population; and producing the product that is saleable like compost or other products. To support the idea, the solid waste management model is required having the characteristic of riverbanks within the integrated solid waste management concept.

The practice of dumping solid waste into the river by ignoring the consequences to the environment is affected by certain socioeconomic statuses such as gender, educational attainment, and household income and geographic risk factors such as residential areas and distance of residential homes to permitted waste disposal facilities [4]. Eight states of India throw their waste into the river. In Varanasi also Nagar Nigam waste accounts for 60 percent of the total volume of waste dumped into the Ganges River, and the industry accounts for about 15 percent of river pollution. Industrial waste is smaller in number but the impact is more dangerous than waste pollution in rivers [6]. Whether in NGO or government spear headed projects, community in riverbanks area will participate in restoration activities if the benefits to them involved. The level of participation and sustainability depends on the appropriateness of the approach and process used in participation [8]. The concept of riverbanks management must begin with a good understanding of the conditions surrounding riverbanks and all the limitations to achieving ideal conditions [9]. From several studies that have been done, it shows that research related to solid waste management in riverbanks area has not much done yet, especially the research related to the amount and composition of solid waste.

The objective of this study is to find the characteristic or residential typology in the Ciliwung segment 5 riverbanks; to map the illegal dumping spot of solid waste in the riverbanks; to find out the source and composition of the solid waste in the riverbanks and to formulate the integrated solid waste management model for the residential area in the riverbanks.

2. Research Method

2.1. Sampling Location

This research was conducted at three sampling locations along the Ciliwung riverbanks on Segment 5 are located in Jakarta, namely Lenteng Agung (Sub-Segment 1), Pejaten Timur (Sub-segment 2), and Manggarai (Sub-segment 3). For more details of the research, location can be seen in Figure 1. Determining sampling locations were selected based on the amount of spot solid waste were obtained from a preliminary survey, the site makes it possible to do sampling and at the Ciliwung Segment 5 Riverbanks many of the inhabitants.

2.2. Identification of Solid Waste Illegal Dumping Spot and Measurement of Solid Waste Volume and Composition

The identification of the illegal dumping spot of solid waste along the Ciliwung Segment 5 riverbank is done by mapping directly in the field, through river trekking by boat and by road. Sampling was conducted using a plastic trash bag with a capacity of 50 L. The volume of solid waste was measured at each illegal dumping spot. Determination of waste composition is done by the following formula:

$$\%\,Components\ of\ solid\ waste\ =\ [Component\ weight\ (kg){:}\ Total\ solid\ waste\ weight\ (kg)]x\ 100\% \quad (1)$$

2.3. Data analysis

Data have been obtained subsequently tabulated and analyzed descriptively.

3. Results and Discussion

3.1. Characteristic and typology of the Ciliwung Segment 5 Riverbanks

Ciliwung Segment 5 River has 30 km length which starts from Kelapa Dua, Depok until floodgate Manggarai, South Jakarta. The riverbank is the left and right of the area along the river, including the artificial river that has important benefits for maintaining the preservation of the river. In the riverbanks area is forbidden: a). disposing of garbage, solid or liquid waste; b). establishing permanent buildings for residential and business premises [5].

The area along the riverbank, in fact, is not supported by the presence of green open space (RTH) which is capable of functioning as an ecological, aesthetic and socio-cultural and economic, it happens due to the imbalance in the proportion and distribution of green open space on the riverbank area. Household waste is one of the factors which led to pollution of river water, for household waste being dumped and passed through the gutters will ultimately lead to the river. In addition to containing organic and inorganic substances, household waste also carry germs that can be transmitted to humans, which can result in extensive epidemics in the community.

The increasing human population in the metropolitan city of Douala, has led to an increase in agricultural, and industrial activities in this area. Therefore, a better distribution system, effluent discharge, and municipal waste management for the safety of public health and environment in Douala, is imperitive [2]. The results of research by [7] concluded that one of the reasons residents live on the riverbanks is the ease in discharging both liquid and solid waste directly into the river.

Figure 1. Sampling location in Ciliwung segment 5 riverbank.

Ciliwung Segment 5 riverbank covered of 8 districts and 21 urban villages in East Jakarta and South Jakarta. The land use changes so fast in the Jakarta area as well due to the construction of housing and other facilities are also driven by the weak control of land use by the local government of DKI Jakarta. In the effort of management of residential area on the Ciliwung riverbanks is the involvement of the community in spatial planning. In this case, it is demanded by the government to act as facilitator and do excellent services.

3.2. Solid Waste Volume

Ciliwung Segment 5 is 30 km long covering from Lenteng Agung District Area to Manggarai. The illegal dumping spot along the Ciliwung Segment 5 Riverbanks consist of 11 illegal spot and of various volume of the solid waste, see Table 1.

Referring to the masterplan solid waste management of DKI Jakarta Province [3], the generation rate of solid waste is 3,1 l/cap/day, then it can be calculated the volume of solid waste in the Ciliwung Segment 5 Riverbanks is 286,17 m3/day. Therefore it can be estimated that 35.17 m3/day of solid waste is not transported and 17% of solid waste is released into the river, riverbanks and burnt by the community of the riverbanks. The average volume of solid waste volume in illegal dumping spot of solid waste along the Ciliwung Segment 5 Riverbank amounting to 0.40 m3. The largest volume of solid waste is located in spot 11th is 1.43 m3 and the smallest is located in spot 6th and 10th each one consist of 0.02 m3.

The condition of the physical environment of each illegal dumping spot that on spot 11th of the Ciliwung Segment 5 Riverbank is covered with a limited number of shrubs and trees. The location is full of community residential and the illegal dumping spot is located behind their residence along 50 meters. The community living in the riverbanks are using the river water for daily activities. The area is considered as a high slum area.

Table 1. Estimates of solid waste volume at the illegal dumping spot.

Illegal dumping spot	Latitude	Longitudinal	District Area	Estimates of Solid Waste Volume (m^3)
1.	-6,3322	106,8407	Lenteng Agung	0.60
2.	-6,3237	106,8415	Lenteng Agung	0.85
3.	-6,3028	106,8546	Gedong	0.85
4.	-6,2826	106.8488	Pejaten Timur	0.02
5.	-6,2705	106,8549	Pejaten Timur	0.08
6.	-6,2937	106,8537	Bale Kambang	0.30
7.	-6,2572	106,8600	Rawajati	0.08
8.	-6,2573	106,8599	Rawajati	0.09
9.	-6,2344	106,8662	Kebon Baru	0.05
10.	-6,2385	106,8631	Kebon Baru	0.02
11.	-6,2077	106,8485	Manggarai	1.43

3.3. Solid Waste Composition

Solid waste composition in illegal dumping spot along the Ciliwung Segment 5 Riverbanks is conducted by 3 days continuous sampling approach. In three location considered representative, namely in Sub-Segment 1 (Lenteng Agung District Area), Sub-Segment 2 (Pejaten Timur District Area) and Sub-Segment 3 (Manggarai District Area). Solid waste composition based on sampling result in the location see Table 2.

Analysis of solid waste composition in the 3 sub-segment shows that organic waste is 42%, inorganic waste that consist of wood 14%, paper and plastic each one consist of 12%, glass 11%, others 4%, textile, rubber and metal each one consist of 1% at the Ciliwung Segment 5 Riverbank. The solid waste composition diagram see at Figure 2. Based on the current condition and to reduce the illegal dumping spot, in solid

IOP Conf. Series: Earth and Environmental Science **106** (2017) 012043 doi:10.1088/1755-1315/106/1/012043

waste management can be implemented in integrated solid waste management in the area along the along the Ciliwung Segment 5 Riverbank.

Table 2. Solid waste composition in Ciliwung segment 5 riverbanks.

Composition	Location		
	Sub-Segment 1 (%)	**Sub-Segment 2 (%)**	**Sub-Segment 3 (%)**
Organic	35	47	44
Inorganic			
• Wood	18	11	14
• Paper	15	11	11
• Plastic	14	11	11
• Textile	1	0	2
• Rubber	1	0	3
• Metal	3	0	1
• Glass	9	14	11
• Others	4	7	2

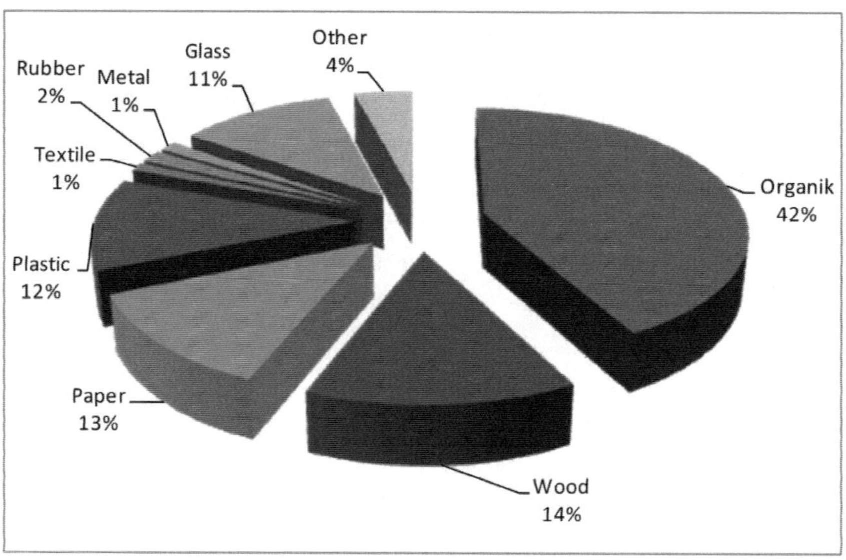

Figure 2. Solid waste composition.

4. Conclusion

The area along the Ciliwung riverbank, in fact, is not supported by the presence of green open space (RTH) which is capable of functioning as an ecological, aesthetic and socio-cultural and economic, it happens due to the imbalance in the proportion and distribution of green open space on the riverbank area.

The number of illegal dumping spot along the Ciliwung Segment 5 Riverbank are 11 spots, with the total volume of solid waste that is equal to 4.37 m^3. Average percentage of solid waste composition in Ciliwung

Segment 5 Riverbanks is in Lenteng Agung, Pejaten Timur and Manggarai consists of organic waste by 42%, followed by 14% wood, 13% of paper, and plastic respectively by 12%, glass by 11%, miscellaneous junk by 4%, rubber by 2%, textile/leather and metal respectively by 1%.

In the effort of management of residential area on the Ciliwung riverbanks is the involvement of the community in spatial planning. In this case, it is demanded by the local government to act as facilitator and do excellent services.

References

[1] Achmad, L. 2009. Material Source and Water Pollution. *Journal.* 1(I)(http://www.chem-is-try.org [accessed 1 Februari 2016].

[2] Akello, S., N. Turyahabwe, H. Sseguya, P. Okullo, J.G. Agea. 2017. Local Community Participation in Restoration of Watersheds in Uganda. American Journal of Environmental Protection, 2017, Vol. 5, No. 2, 25-32

[3] Cleanliness Services of DKI Jakarta Provinces. 2012. Masterplan and academic study of Solid Waste Management of DKI Jakarta Provinces, Jakarta.

[4] Eric, T.T., D. N Eric, K. Théophile. 2014. Trend Analysis of Waste Disposal in an Afrotropical Urban River Water Body." Journal of Environment Pollution and Human Health 2.4 (2014): 81-84.

[5] Government regulation No. 38 Year of 2011 regarding the river.

[6] Kumar, M. N. 2015. Assessment of solid waste management in Ganga basin-Varanasi: a case study. The Journal of Community Health Management, Volume : 2, Issue : 1, Print ISSN : 2394-272X. Online ISSN : 2394-2738.

[7] Mamady, K. 2016. Factors Influencing Attitude, Safety Behavior, and Knowledge regarding Household Waste Management in Guinea: A Cross-Sectional Study. Journal of Environmental and Public Health Volume 2016 (2016), Article ID 9305768, 9 pages http://dx.doi.org/10.1155/2016/9305768

[8] Mulyandari, H. 2011. Efforts to manage the building land on riverbanks with environmentally based in Sleman District, Yogyakarta. Journal of Civil Engineering and Planning, No.1 Vol. 13. January 2011, Page: 31-40.

[9] Poedjioetami Esty. 2008. Rearrangement of the Area Along the Riverbanks by Presenting Economis and Recreational Centers of The City (Case Study Dinoyo Tenun Area, Surabaya). *Journal Engineering Planning*. Architecture. Institute of Teknologi Adhi Tama Surabaya. Surabaya.

[10] Susilowati, W, et.al. 2012. Community Base Participation in Improving The Environmental Quality of The Watershed Ciliwung. *Journal Technology Polytechnic.* 11(1):55-63.

Park system concept for environmental sustainabilityin urban spatial development

Q Uniaty

Landscape Architecture Department, Faculty of Landscape Architecture and Environmental Technology, Universitas Trisakti, Jakarta, Indonesia

CorrespondingAuthor: quintarina@trisakti.ac.id

Abstract. Urban Park System is an integrated concept between nature system and urban life. The problems caused by urban population activity resulted in the need to increase the balance between two systems. Establishment of urban park system is a response to the need for resilience of urban space structures. As an ideal requirement it needs to be built integration between the ecological, social, economic, aesthetic aspects of urban landscape architecture. The methodology was developed based on an approach to issues affecting the conditions due to urban issues and its relation to the development efforts of urban park system; Observation of Jakarta problem was obtained based on published studies and data, literature, characteristic and potential analyzes, local physical, from limited field observations. Both are simple methods aimed to describe the nature of a condition as well as form characteristics of problems in controlling the development of region, to examine the causes and symptoms. This method try to assess an object study compared between the conditions before and after. The benefits of urban park system development will not only improve the urban environment, but the value of urban pride, identity and control urban growth in line with efforts to improve the balance between conservation and development. Integrated urban park system will enhance the multifunctional role, connectivity, habitability, durability, identity and investment.
Keywords: conservation and development, urban park system, urban environment

1. Introduction

Increased urban land use change in Jakarta of green areas into developed regions resulted in problems for the environmental balance of the city. The excessive carrying capacity is the main problem for Jakarta. Overlapping issues; including overcrowding, litter, water pollution, air, soil, flooding has established a network problem, signaled to the municipality to act in resolving systemically; through the urban park system. In modern landscape architecture, the park system is collaborating with the idea of planning greenways, which run through urban and rural areas. These systems can serve the landscape through ecological, recreational, social, cultural, and healthful measures, and are designed with intentions of sustainability. Carrying capacity is the maximum number of individuals that can be supported or served by the resources within an ecosystem. Or, the ability of the environment (ecosystem) to support the lives of all creatures that are in it on an ongoing basis. Urban Park System is an important part of the environmental support systems. As the main support of urban park system in urban areas, it is necessary to develop green infrastructure. Social dimensions of carrying capacity include lifestyle aspirations, epidemiological factors, and patterns of socially controlled resource distribution, the disparity between private and social costs, the difficulty in formulating rational policy in the face of uncertainty, and various other features of human sociopolitical and economic organization [7]. Urban parks and open space improve the physical and psychological health, strengthening communities, and make cities and neighborhoods more attractive places to live and work.

2. Research Method

Jakarta as the capital of the state and metropolitan cities have a complex problem. As a citizen liaison application that allows citizens to actively report the surrounding conditions to the government or related parties, it has released the first quarter data of 2017 regarding reports of incoming citizens through its application. Data problems most frequently reported in the first three months of 2017: the need for clean water, waste, parking, and rent on the street, misuse of public facilities, damaged roads, street vendors, traffic congestion, street lighting and flood [6].

Observations of Jakarta city cases were obtained based on published studies and data, literature, characteristic and potential analyzes, local physical, social, economic and cultural approaches from limited field observations. Problem description is the result of observation; Both are simple methods aimed at describing the nature of a condition while forming the characteristics of the problem, to examine the causes and symptoms. With this method try to judge a comparison between the condition of the object before and after.

3. Results and Discussion

Today, most people live in urban regions, and many try to move from rural to urban areas for jobs, education, and cultural experiences. Urbanization is the consequence of this phenomenon. High population density and built context are growing in urban regions, leaving only small green spaces next to tall buildings. This urbanization has enormous impacts on biodiversity and ecosystems in cities. Urban areas cover a small portion of land, but people in urban areas have powerful effects on the earth by using natural resources in a most profound manner [9]. Growing urban populations cause many environmental problems around the world (Figure 1) [16]. Urban population and protected areas are overlapping in the map, and these results suggest four category classification schemes which can help us understand the extent of contemporary urbanization in the global scale.

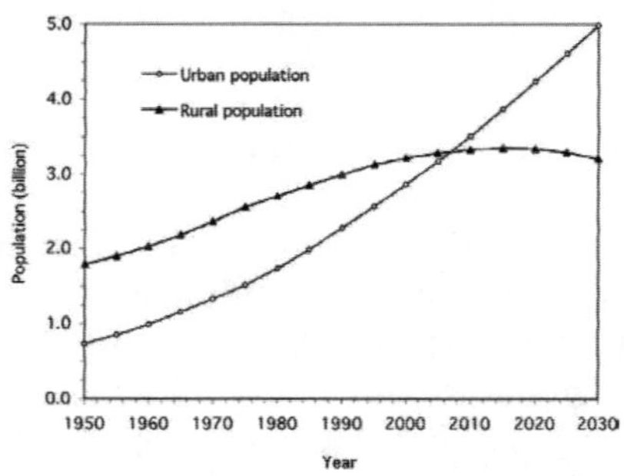

Figure 1. The growth of urban and rural populations around the world, 1950-2050 [16].

3.1. Jakarta's Spatial Problem

The rapid development of Jakarta has reduced the green area. In 1970, the green area is still 70% of the city's width, and is currently only about 10%. Within 40 years there has been a decrease in green space almost 60 percent. Based on the prevailing standards, the green space is at least 30 percent of the total area built up. The transfer of agricultural land into an area built outside Jakarta has an impact on the environmental conditions in Jakarta, especially flooding [11]. In 1992, the area awakened in the upstream is still 100 thousand ha, currently has become 225 thousand ha. Thus, the existence of the Jabodetabekjur Spatial Plan as a means of controlling the utilization of space becomes vital. Based on the scenario RPJMN 2015-2019, spatial arrangement has entered the stage of space utilization and control of space utilization. The unfinished spatial plan has impact on the weakness of spatial use control, especially in coastal areas. Meanwhile, it is fully realized that many of the problems that occurred in Jakarta started from the not optimal implementation of spatial planning. There are still many spatial violations that result in increased traffic generation, congestion, flooding and environmental pollution.

Jakarta's Environmental Problems

Environmental problems in Jakarta, especially garbage, flooding, congestion and the need for clean water is a group of classical and most fundamental problem. The number of residents in Jakarta, plus the newcomers, and commuters who live in suburbs but work in Jakarta causing waste production. Data Production of waste in Jakarta continues to rise and is now estimated at 6,000 tons per day. While the cooperation contract of DKI Jakarta Provincial Government with PT Godang Tua Jaya as the manager of Bantar Gebang Integrated Waste Disposal Site, which is 3,000 tons per day which means 3000 tons of un-transported garbage, spread in the Jakarta area, plus the lack of public awareness waste problem becomes serious [5]. Air pollution is generally considered as a major concern in urban areas, and as being among the major risk factors contributing to the global burden of disease, with for example high levels of particulate matter (PM) air pollution being associated with excess mortality and morbidity in the urban population [3].

Jakarta lies between river basins such as the Ciliwung River, Pesanggrahan River, Krukut River and Sunter River, where this area lacks good river management, waste is dumped in rivers, and illegal housing on the banks of the river also triggers flooding. Water absorption area continues to decrease due to the construction of buildings and settlements so that water from rain can not seep very well down to the ground. Jakarta's location is lower than sea level and water from Bogor is a factor that worsens flooding.

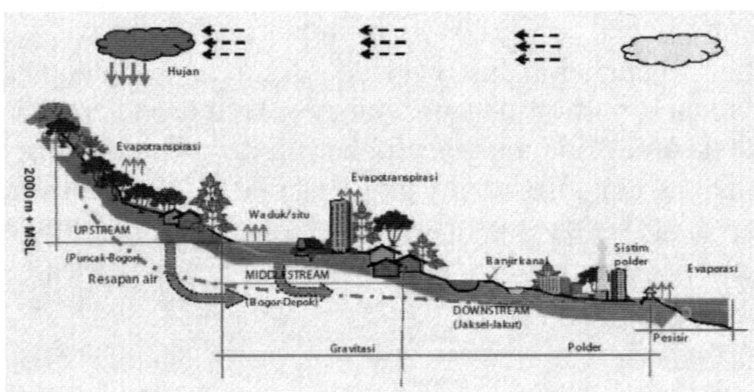

Figure 2. Pieces North-South Jakarta (Source: Pem Prov DKI Jakarta) [13].

Slum areas in Jakarta are 174 neighborhoods in five administrative municipalities, resulting from poverty. Illegal migrants with low socioeconomic conditions typically lead to the growth of illegal settlements and declining housing conditions that they occupy into slums. Congestion and air pollution problems supported by commuters from satellite towns around the capital that works in Jakarta, plus the growth of the highway in Jakarta a 0.01% per annum from 6.9 million meters wide road equivalent to 6900 kilometers. While the growth of vehicles in Jakarta per day up to 6000. Statistic Institution [2] makes congestion into serious problems in this megapolitan city, resulting in air pollution due to motor vehicle fumes, plus the extent of inadequate green areas make pollution cannot be absorbed by plants. Inadequate public transport facilities make people more likely to use private vehicles. A systematic review of the cooling effects of urban greening, including specifically the category of urban parks. This review also included a meta-analysis to statistically synthesise data on the temperature differences between urban parks and non-green urban areas to quantify the average cooling effect of a park across studies. In the context of climate change, the sequestration of carbon by urban trees and other vegetation plays an important role. Five of the articles included COx removal by urban parks, agreeing that parks act as 'carbon sinks' [8], [11], [10], [17], [5].

The Provincial Government of DKI Jakarta has not been able to adequately fulfill clean water needs for its citizens. So much ground water suction for clean water. The presence of large-scale groundwater is also done with the needs of high-rise buildings in offices and commercial areas. This resulted in significant reductions in groundwater levels and land subsidence.

IOP Conf. Series: Earth and Environmental Science **106** (2017) 012044 doi:10.1088/1755-1315/106/1/012044

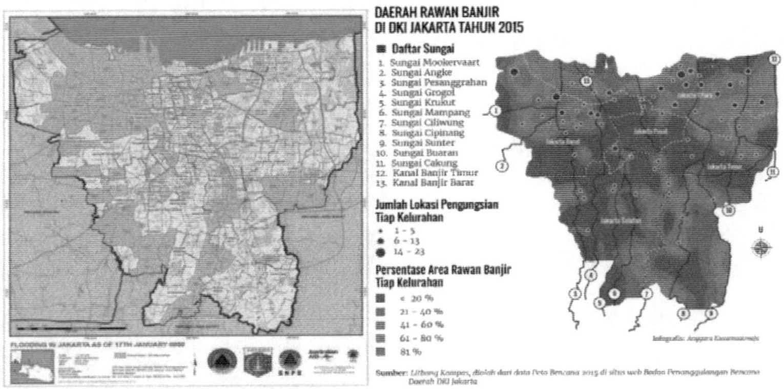

Figure 3. Map of flood area and flood hazard area in Jakarta [13].

Treat people living habits and their physical environment at odds with natural physiographic characteristics of the landscape. From the house, the manufacturing industry, offices, markets and shops, workshops, recreation centers and others, they left garbage in the drainage, causing flooding in the anal and drainage channels silted and filled with sand, plastic cans and bottles, sachets of pure water, plastic bags and others [1]; Urban green space promotes physical activity and public health; Many US minority communities lack green space access, an environmental injustice; US and Chinese cities have developed innovative ways to create new green space ;Urban greening can, however, create paradoxical effects such as gentrification; Urban green space projects need more integrative sustainability policies to protect communities [14].

Ecosystem-services-based approaches can help regulate the urban water cycle by reducing the amount of stormwater runoff and to improve water quality by removing pollutants from runoff. Bioswales, rain gardens, green roofs and other green infrastructure components can help reduce runoff e.g., due increased infiltration rates for non-paved surfaces. Increased infiltration would promote groundwater recharge and evapotranspiration from vegetated surfaces, and thus help to improve climatic conditions in the city [15], [4]. Urban landscapes with 50–90% impervious cover can lose 40–83% of rainfall to surface runoff [4]. Solving problems of the city should be developed within the concept of management at national, regional and local planning on short, medium and long term. Basically problem solving can be anticipated through; awareness raising for stakeholders; intensive increase community involvement; Enhancing partnerships with the private sector, universities, communities and NGOs; Implementation of regulations and law enforcement related to space utilization and management.

For the sustainability of urban resilience, it is necessary to implement the management of all urban areas mainly in watershed areas as the control of the water system. (i) Water catchment areas management, restoring lakes/situ functions around Jakarta that support Jakarta's water system. (ii) Water Resources Management. Development of Water Resources, Drought Prediction, Drought Relief, Water Use Permit, Water Allocation, Water Distribution. (iii) Flood Control. Flood Monitoring and Prediction, Flood Control and Prevention, Flood Mitigation, Flood Damage Improvement. (iv) River Environment Management. Land Use Planning of River Border Area, Control of River Border Use, Water Biota Preservation, Tourism Development, Sports, and Transformation of Water. (v) Community Empowerment.

Examples of Watershed Management and Development: (i) East Canal Floods and West Canal Floods were built with the aim of channeling rainwater and water from upstream to the sea, so water does not flood Jakarta; (ii) The river dredging program in Jakarta is conducted systematically and planned from 2008, the government has dredged 6 critical river rivers; 2009 with dredging 66 river segments; 2010 with 6 segments of river dredging in five areas of Jakarta

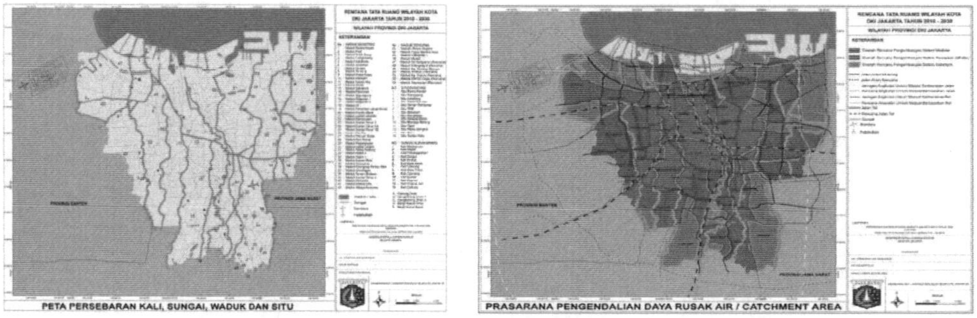

Figure 4. Map of the distribution of rivers, reservoirs and situ in Jakarta and Map of water catchment area control infrastructure [12]

The large scale river normalization program in Angke started in 2003. The 5 meter wide river width is widened by 40 meters. Angke residents who have lived in huts on the banks of Kali Angke illegally moved into flats in Muara Angke and Cengkareng.

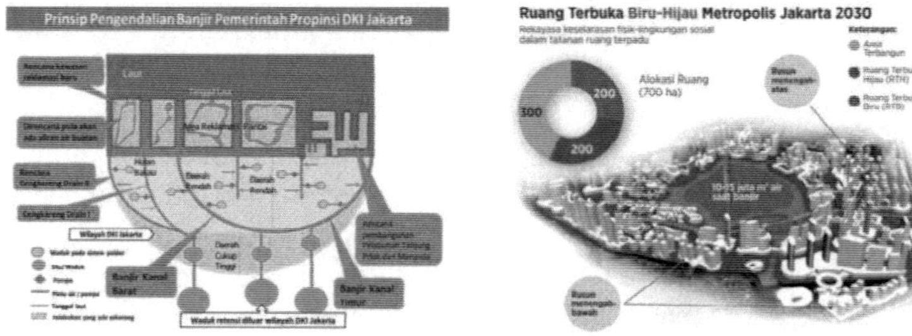

Figure 5. Flood control principles of DKI Jakarta Provincial Government [13].

Urban green space, such as parks, forests, green roofs, streams, and community gardens, provides critical ecosystem services. Green space also promotes physical activity, psychological well-being, and the general public health of urban residents [14]. A direct relationship between biodiversity to human well-being, is the establishment of ecosystem function which is a positive impact for the city park air quality improvement and carbon sequestration; related to water management: the park's contribution can cope with the effects of flooding. And contributing to the climate is cooling urban areas. As the development of the Jakarta Urban Park System and resolving disputes cities need to be built in the Green Infrastructure planning application-based urban design ecology would be a wise solution strategies; a network planning and management strategies for the sustainability of ecological processes, landscape and green open spaces of the city that will protect the values and functions of natural ecosystems and to improve the benefits for human life. The city of the future can be realized if Jakarta accommodates the realization of urban park system for the target of achievement of RTH by 30% in the City Spatial Planning (RTRW) in line with the effort to build a sustainable city as mandated by Law no. 26 Year 2007 on Spatial Planning. Building the city's resilience to climate change is a top priority for the city. In addition to mitigation, cities must now build an integrated concept in the form of Urban Park System in the adaptation and handling of conservation of environmental systems.

4. Conclusion

Jakarta Urban Park System must: (i) Being a major consideration in planning, development and maintenance of Jakarta eco-city. (ii) Designated as a network that varies, is widespread, strategically planned and interconnected. (iii) Become an element in land assessment and decision on the determination of housing density and urban structures. (iv) Accessible to local communities and provide alternative means of transportation. (v) Designed to reflect and accentuate local character, including landscape and local habitat. (vi) Required surface water flow data, such as rivers, reservoirs, situations, and waterlogged areas during rain to develop a flood control RTH pattern by determining areas that should not be built for conservation and preservation functions so that hydrological processes can still take place to support the Green Infrastructure Intensification strategy. (vii) Multifunctional; Integration and interaction of different functions on the same footprint and across the infrastructure network as a whole.(viii) Implemented through planning and management coordination through local authorities and constraints across sectors. (ix) Able to achieve physical and functional relationships between sites at each level; City or sub-region. (x) Implemented intact through the main strategies of green infrastructure intensification and spatial strategy planning systems from regional spatial strategies to localdevelopment framework that are formally adopted through provisions and planning documents established with the support of ongoing allocation of maintenance and adaptation funding. (xi) Support the legal aspect of political will in the form of a national policy set forth in a higher regulation or law.

References

[1] A, Aderogba, Kofo. 2012. Department of Geography & Environmental Sciences,Tai Solarin University of Education, Ijebu Ode, Ogun State, Nigeria; Qualitative Studies of Recent Floods and Sustainable Growth and Development of Cities and Towns in Nigeria; *International Journal of Academic Research in Economics and Management Sciences.* 1 (3). ISSN: 2226-36. 24 June 2012,

[2] BPS Jakarta, 2015

[3] Cavanagh, J.-A.E., Zawar-Reza, P., Wilson, J.G., 2009. Spatial attenuation of ambient particulate matter air pollution within an urbanised native forest patch. Urban Forestry & Urban Greening, 21-30.

[4] D.E., Pataki, Carreiro, et al 2011, Coupling biogeochemical cycles in urban environments: ecosystem services, green solutions, and misconceptions. Frontiers in Ecology and the Environment, 27-36

[5] E., Paoletti, Bardelli, T., Giovannini, G., Pecchioli, L., 2011. Air quality impact of an urban park over time. Procedia Environmental Sciences 2011, 10-16

[6] https://wiraabiyoso.wordpress.com/2017.

[7] Ibes, Dorothy C. 2014. Sustainable Urban Park Systems, Cities and The Environment (CATE), Artikel 8. 7. Loyola Marymount University.

[8] Jo, H.-K., 2002. Impacts of urban greenspace on offsetting carbon emissions for middle Korea. *Journal of Environmental Management*, 115-126.

[9] Kim Hyunja. 2012. Spatial Ecology, A Park System Model, Ball State University, Muncie, Indiana.

[10] Kordowski, K., Kuttler, W., 2010. Carbon dioxide fluxes over an urban park area. Atmospheric Environment, 2722- 2730.

[11] Makhelouf, A. 2009. The effect of green spaces on urban climate and pollution. *Iranian Journal of Environmental Health Science and Engineering*, 35-40

[12] Muadzin Mungkasa, Oswar, 2014, Masalah dan Solusi, Direktorat Tata Ruang dan Pertanahan BAPPENAS. Jakarta.

[13] Pemerintah Provinsi DKI Jakarta, Bebas Banjir 2015

[14] R. Wolcha, Jennifer, Jason Byrne b, Joshua P. Newell , Urban green space, public health, and environmental justice: The challenge of making cities 'just green enough'; Research Paper, Landscape and Urban Planning, 125:234-244, May 2014

[15] S., Pauleit, Duhme, F., 2000. Assessing the environmental performance of land cover types for urban planning. Landscape and Urban Planning, 1-20.

[16] Wu Jianguo. 2008. Making the Case for Landscape Ecology: An Effective Approach to Urban Sustainability. *Landscape Journal.* 27.

[17] Z.G., Davies, et al., 2011. Mapping an urban ecosystem service: Quantifying above-ground carbon storage at a city-wide scale. *Journal of Applied Ecology,* 1125-1134

Composing the theme of city to be diverse and sustainable

H W Wiranegara

Urban and Regional Planning Department, Faculty of Landscape Architecture and Environmental Technology, Universitas Trisakti, Jakarta, Indonesia

Corresponding Author: hanny.w@trisakti.ac.id

Abstract. To give a path for developing a city needs a theme. City's goal stated in a document of a spatial plan were too broad and insufficient detail in giving a direction. To make more detail and precise, every city has to compose a city theme. It is developed based on the potential, the uniqueness, the excellence, and the sustainability of its human resources, natural resources, and man-made resources. An integration among the three of resources which have the highest score become a theme of the city. The aim of this research was to formulate the conceptual framework to compose a city theme. The research design was the interview survey in Banda Aceh, Banjarmasin, and Kupang. Informants were the government officials, academics, figures, the private sector and public who considered related to the intended information being collected. Having set the conceptual framework, the interview directed to check the implementation in realities. The result was that the conceptual framework could accommodate the phenomenon of composing the theme of the city. Yet, it was a preliminary in nature and needed more research to get a complete result.
Keywords: city theme, spatial plan, sustainability

1. Introduction

Cities in Indonesia are tending to look alike as if they lost their identity. The city identity can be determined by sociocultural structure, socio-economical structure, physical environment and image elements [1]. All is the city resources which are interrelated. It's important to keep the city identity, it could lead to cities' diversity. Ensuring the diversity is the key [2] for sustainability [3]. Generally, the statement of vision, mission, and goals in city planning document was too broad, as we can see in master plan of Banda Aceh 2009-2029[4], Banjarmasin 2013-2032 [5], Kupang 2011-2031 [6], Jakarta 2030 [7], Bandung 2011-2031 [8], Surabaya 2014-2034 [9], etc. Most of these contained the future of city function(s). In realities, the vision of Mayor became a guidance in the development process that might not direct to the goal of the city plan directly. Also, it might be no consistency among the vision of the mayors for every five years consecutively and at the end after twenty years, the goal of city plan was a failure to be achieved. Cities' planned functions weren't enough to make cities diverse like the themes were. Diversity among the parts of city or diversity among cities was an essential to be the sustainable cities. Looking for a theme of a city is an essential effort to make city has an identity and a specific role and function which will affect its identity, growth, and sustainability. The theme is composed by considering and integrating of three kinds of city resources: human resources, natural resources, and man-made resources altogether which will make a city diverse. Now and then, every city faces the problem of urbanization and the global threat. The city can only survive if it has a thematic which directing its development process and differentiating it from the other [10]. Composing the city theme is based on the potential, uniqueness, excellence, and sustainability of its human, natural, and man-made resources altogether [11]. This research aim was to develop the conceptual framework how to compose the city theme. This research was a preliminary in nature, it needed more cases to get the better result especially in the formulation of the thematic statement.

2. Research Method

The research design was the interview survey. There were three cities chosen as cases namely Banda Aceh, Banjarmasin, and Kupang. These cities were assumed having the potential or uniqueness of one kind of its resources. The interviewees were the government officials, academics, figures, the private sector and ordinary people. The government officials chose from several institutions related to the spatial planning. The question for them related to the mayor vision and development programs. Academics were chosen who's background were a historian, sociologic, or city planning. From them were gotten about the process of formulation of the mayor's vision; the relation between the mayor vision and the development program; etc. From figures were gotten about the community activities related to the mayor vision. From the private sector (property developer) was gotten information about formulation the project theme as a comparison and to know the relation between the project theme and the official city plan. The people were everyone found randomly on the road just to check if they recognized or heard about the vision of their mayor as a signed it was socialized.

3. Results and Discussion

The process to formulate the conceptual framework: Firstly, study the empirical condition how the theme of the city was practiced by cities. The mayor vision chose to represent the theme of the city. The goal statement or planned city function in city plan document couldn't be used. It was too long in the time dimension and too broad. Secondly, it was developed the hypothetical framework based on literature review. Finally, the conceptual framework proved if it could be applied in real cases.

3.1. Empirical condition related to a city theme

In Banda Aceh city, a vision of the elected mayor was to create a *tammadun* city. *Tammadun* was the concept of civilization, which reflected an advanced state of moral, intellectual, and technological development. This vision was stressed more on human resources. The condition of human resources which related to *tammadun* Islam could be described as the people uphold the Islamic values and teaching; they used the customary law in regulating social life and as a social control; the people traditions were more lead to a bond unity; there was the community involvement in *diniyah* education at schools and in neighborhood; they still held the concept of mutual cooperation.

Banjarmasin was named as *the city of thousands of rivers.* It had naturally of 102 rivers spread out throughout the city with the total length was 185.303 km [12]. No other cities like Banjarmasin. Its rivers characteristic became its identity. Having been realized its potential and uniqueness, this natural identity became its vision and the local government wanted to review the spatial plan with this vision. To manifest its vision, there was the development of man-made and human resources which closely related to its potential and uniqueness of the rivers (natural resources).

Kupang still neglected the city resources uniqueness in the development process. Kupang is located on Nusa Tenggara Island, east of Bali Island is one of the city which is still looking for its identity. It has a long coastal region which can be its identity. But, its vision is to create as a waterfront city or a *minapolitan* (fisheries city), human and man-made resources haven't yet directed to it.

3.2. The conceptual framework of composing a city theme

The city which has a theme is called a thematic city. The thematic city is the city which has a clear character, built based on its potential, uniqueness, excellence, and sustainable resources; its space easy to recognise; rooted in economic, social, and cultural life of the people; as well as be the attraction to outsiders to come in which will guarantee its sustainability in its development [11]. The benefit for cities which has a theme in development process are: directing city more clearly and suitable to its character; building the city based on its theme will guarantee its sustainability (economic growth, equity, ecological function); supporting the city independence; manifesting the diversity of cities; aligning with spatial plans and supported by sectoral development plan; responding to the national and global issues (food/water/energy security, climate change, global warming, etc.). There are three kinds of city resources. After exploring cities in Indonesia, the elements of each resource is as follows. Natural resources include geological, hydrological, topographical, and geographical resources, climate, a beauty of natural landscape (mountain, etc.), the sea ecosystem, biodiversity, flora, and fauna being rare, crop production, etc. Human resources include local genius, religious values, cultural values or tradition need to be maintained, skill and creativity to produce the typical product, mastery of technology, the community needs, institutional capacity. Man-made resources include city structure, its morphology, infrastructures, socio-econocultural facilities, historic buildings, monument/landmark, traditional buildings, the beauty of cityscape, etc [11]. There are four criteria of resources used in building a city theme. They are 1) its potential, 2) its uniqueness, 3) its excellence, and 4) its sustainability. The potential means its resources that could be exploited because of having economic value, giving more added value, or showing the domination. The uniqueness of resources means its resource utilization which is typical or distinguishing with other cities. The excellence of resources is its resource utilization which became a city competitiveness or the competitiveness to its growth. The sustainability of resources is the integration/synergy in using of its various renewable resources or otherwise be preserved from extinction because of the threats (natural disasters, human behavior, globalization, climate change, etc.) [11]. The conceptual framework for composing the city theme can be drawn in figure 1. The city theme is determined by the potential, the uniqueness, the excellent, and the sustainability of city natural resources, man-made resources, and human resources altogether.

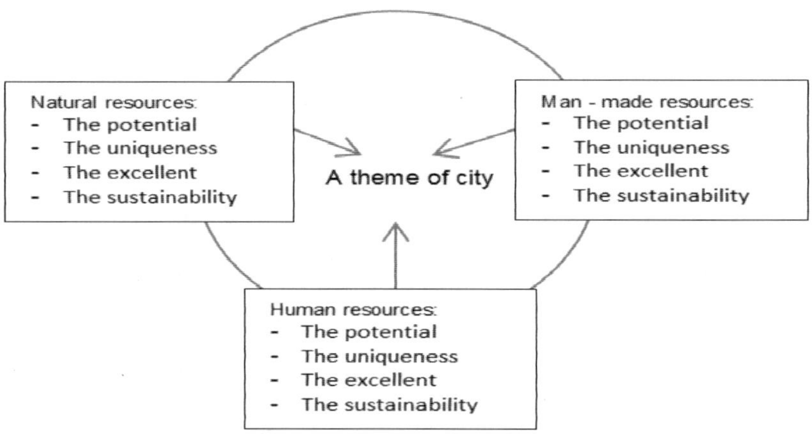

Figure 1. A Conceptual framework to compose the theme of a city [11].

For each resource and its criteria were set the indicators. Got a geological resource as an example. Its indicator for the potential criterium is 'geological resources that economically valuable or has added value or show domination'. Its indicator for the excellent criterium is 'geological resources that having competitiveness or competitiveness for city growth'. Its indicator for the uniqueness criterium is 'geological resources that are characteristic or distinguishing'. Its indicator from the sustainability criterium is 'geological resource that is occupied synergistically integrated or resources that can be renewed or preserved'. The same pattern of indicator statement was applied to other elements of all resources.These indicators used to value each of the elements from all resources.

3.3. The implementation of a conceptual framework

The conceptual framework drawn in fig 1 was checked for its implementation in three city cases. In each city conducted interview limited to these informants including the head of local government bodies or its staff who can represent; the academics from a local university; and the people. Assuming the vision of mayor as a city theme, to check the implementation of the conceptual framework was done by checking the relation between the mayor's vision and the element from the three resources.

In Banda Aceh city, the head of several local government bodies interviewed was stated that the mayor asked them to set the programs accordance with his/her vision. To create a *tammadun* city of

Banda Aceh (assumed as a theme of Banda Aceh city) was conducted by setting five years plan goal based on the conditions of human resources, man-made resources, and natural resources which were related to the concept of *tammadun*. Conditions of man-made resources of Banda Aceh city related to this concept were: mosque and *meunasah* were functioned as centers of all community activities; many relics of history was the remaining of the royal Aceh at the age of Islamic *tamaddun*. Conditions of natural resources of Banda Aceh city were related to this concept were like the rivers were functioned as an important means like in the age of the kingdom of Islamic *tamaddun*. Also, related to human resources in all programs were related to the manifestation of Islam teaching comprehensively. They called it as *kaffah*. In every aspect of human life was based on the goals to manifested a *tamaddun* completely. The people in Banda Aceh had already known the *tammadun* as the mayor vision and implemented in everyday life such as determining halal and haram food, using Islamic cloth, etc. From the case of Banda Aceh, it could be concluded temporarily how the conceptual framework was implemented by checking how to manifest a vision by counting to all of the three resources. A theme of Banjarmasin city as *the thousand rivers city* (in Bahasa: Kota Seribu Sungai), all the programs were set to manifesting it. In developing man-made and human resources was based on the rivers which were its unique character. Water transportation revived. Piers water transportation propagated. Similarly, the objects of tourism in river banks was developed. The care river community was also built. These conditions described the relationship among the three of resources in manifested the city theme. In Kupang city was found the ambiguity in choosing its theme between the waterfront city and *minapolitan* (fisheries city). The potential of its resources was not much explored and exploited. As waterfront city or *minapolitan*, it couldn't be implemented yet because of the lack of information of city's resources (its potential, uniqueness, excellent, and sustainability).

From all cases could be concluded that the conceptual framework for composing the city theme was implemented. The diagram showing how a theme was determined by the three resources altogether. The arrow was directed to the city theme from the resources. When proving the conceptual framework in realities used the reversed arrow, started from the theme to the resources. But, both were shown the same relation between a theme and its supported resources. To achieve a vision needs supporting the resources. It is implied that to compose the theme of a city depends on the city's resources.

3.4. The steps of composing a city theme

The steps for composing a city theme can be explained as follows, see Table 1.

Table 1. Scoring the elements of resources by four criteria [11].

Resources	"High" for the Potential criterium (P)	The excellent criterium (E)			The uniqueness criterium (U)			The sustainability criterium (S)			Score Total (P*E*U*S)
		High	Medium	Low	High	Medium	Low	High	Medium	Low	
Natural	N_1										
	…										
	Nn										
Human	H_1										
	…										
	H_n										
Man-made	M_1										
	…										
	M_n										
Example	Tradition	5	-	-	-	3	-	-	3	-	225

Information:
- Make a tabular for "high" potential criterium, try for spatial hierarchy in the scale of regional, within the city boundary, and part of the city (or district).
- Repeat making a tabular for "medium" potential criterium, if there is no "high" potential resource at all.

The first, conducting mapping the condition of the elements from all resources for the potential criterium first. There are three categories: high, medium, low. It gave 5 for high, 3 for medium, and 1 for low. See the definitions in Table 2 for scoring these categories. Choose the element by using a potential criterium with high category first. Medium and low potential category drop to make a list simpler. So, only the high potential category to be mapped in Table 1. To value the degree of each element is done by the local expert. Separate the table for a different scale. The scale of the potential criterium can be a regional, city, or part of the city in nature. The second, valuing the degree (high, medium, low) for three other criteria (its excellent, uniqueness, and sustainability). The third, counting the total score for each element for all criteria in the second step. The total score is the result of the multiplication of its score from all criteria. The fourth, choose one element from each kind of resources which has the highest total score. The fifth, do the analysis the relationship between those elements of resources in step four (has the highest score). The sixth, every couple of the elements from three resources which had closed relation was counted for its score. This score is the result of multiplication of the total score of each element, see Table 3. The last step is to determine the theme based on the result of the last step. High related elements from the natural, human, and man-made resources which have the highest score become the ultimate theme of the city. The score below it become the sub-theme if it has a strong relationship with the ultimate theme. It is a choice for the city to have more than one theme. To illustrate more clearly to compose a city theme see again all the tables.

Table 2. The definition of scoring the criteria [11].

Criterium	High (score 5)	Medium (score 3)	Low (score 1)
The potential	Can be used by giving high added value or have high economic value	Can be used by giving medium added value or have medium economic value	Can be used by giving low added value or have low economic value
The excellent	Having high competitive value for the city growth	Having medium competitive value for the city growth	Having low competitive value for the city growth
The uniqueness	Are the specific characteristic or clear/high differentiating with other areas	Are the specific characteristic or medium of differentiating with other areas	Are the specific characteristic or low of differentiating with other areas
The sustainability	Can be renewed or be preserved so continuously being beneficial	Can be renewed or be preserved so being beneficial in the long-term	Can be renewed or be preserved so being beneficial in the medium term

Table 3. Analysis the relationships among the elements of resources [11].

Alternative of the combination	Element of resources with high score which interrelated			Total score (X*Y*Z)
	Natural resources (score X)	Human resources (score Y)	Man-made resources (score Z)	
Alternative 1	NRp (score P)	HRq (score Q)	MRr (score R)	PQR
Alternative 2	NRm (score M)	HRn (score N)	MRo (score O)	MNO
Etc.				

3.5. Discussion

As a result of scoring, the highest score determines the theme of the city. The second best is to be the subtheme or another theme if not related to. The theme of the city can be implemented in a strategic area as a part of the city or to the city as a whole or to an urban area which boundaries based on its function. It depends on spatial hierarchy in the determination of its degree of the criteria when doing mapping such as shown in Table 1. For the city as a whole may have more than one theme and will make it more diversified. The diversified city is good for its sustainability. More diversified a city is a more productive and more sustainable consecutively.

A theme of the city is also to support the goal statement in a document of a city plan to give more detail direction. It can direct activities and functional planning than the goal statement can. In practice, the goal formulation in city planning document is too general and not the result of a deep analysis of its interrelated of three resources. In this research, it was showed how to formulate the theme of the city was a long process which involved relationships of all resources in four criteria. For an example, the theme for Banjarmasin city is a thousand of rivers. It's clear for all activities and functions to be planned related to the character of the river. Also, developing human resources and man-made resources are directed to create Banjarmasin as a unique city with thousands of rivers. This city theme is also can be used as city brand to make it more attractive to outsiders to come in. It has consequences on the city income to use in the process of development continuously. Related to the general election, push the candidate to formulate the vision derived from the theme. With the existence of the theme of a city makes the candidate easier. It will guarantee the city plan used more in the process of five years development and in direct to achieve its goal. If all the cities have a theme or themes based on the potential, uniqueness, excellent, and sustainability of its resources and to be the thematic city, then we will find the cities in diversity. It will create bounding among the parts of a city and bound among cities which will form to be the web which will guarantee the sustainability of the system of cities.

The conceptual framework for composing a city theme was stressed to internal factors namely its resources. To make more realistic, it needs more research which considers the external factors.

4. Conclusion

The theme of the city is a need for its existence and sustainability. To formulate the theme of the city needs mapping and analyzing of its potential, uniqueness, excellent, and sustainability of its resources and the synergistic relation among them. Based on its resources, every city will have a different theme. It will create city diverse and sustainable.

Acknowledgment

I would like to express my special thanks of gratitude to Directorate General of Spatial Planning, The Ministry of Agrarian and Spatial Planning which gave me the opportunity to do this wonderful project on the topic The preparation of material-technical guidelines for using the space of the thematic city, which also helped me in doing a research and I came to know about to compose a theme of city as a new things I am really thankful to them. Secondly, I would also like to thank my team who helped me a lot in finalizing this project within the limited time frame.

References

[1] Yaldız E, Aydın D, Sıramkaya S.B 2014 *Procedia - Social and Behavioral Sciences* Loss of City Identities in The Process of Change: The City of Konya-Turkey 140 221-33
[2] Fainstein S 2005. *Urban Affairs Review* Cities and Diversity: Should We Want It? Can We Plan For It? 41 3-19. 10.1177/1078087405278968.
[3] Syed, J 2006 Diversity for sustainability: An interdependent model of corporate social responsibility. Fondazione Eni Enrico Mattei Series. CSR Paper 10.2006.
[4] http://jdih.bandaacehkota.go.id/xperaturan/pdf/qanun%20nomor%204%20rtrw%202009-2029%20final%20.pdf
[5] http://www.jdih.setjen.kemendagri.go.id/files/KOTA_BANJARMASIN_5_2013.pdf
[6] http://sipd.bangda.kemendagri.go.id/dokumen/uploads/rtrw_332_2016.pdf
[7] http://pertamananpemakaman.jakarta.go.id/assets/data/data-peraturan/PERDA-01_TAHUN_2012-TENTANG-RTRW_2030.pdf
[8] https://www.scribd.com/document/189807891/Peraturan-Daerah-Kota-Bandung-Nomor-18-tahun-2011-Tentang-Rencana-Tata-Ruang-wilayah-Kota-Bandung-Tahun-2011-2031
[9] http://www.surabaya.bpk.go.id/wp-content/uploads/2015/10/PERATURAN-DAERAH-NOMOR-12-TAHUN-2014.pdf
[10] Directorate of Area Arrangement, Directorate General of Spatial Planning. 2015. A thematic city: responding to global challenges and the issue of sustainability. Ministry of Agrarian and Spatial Planning, Jakarta.
[11] Directorate General of Spatial Planning. 2016. The preparation of material-technical guidelines for using the space of the thematic city. Ministry of Agrarian and Spatial Planning, Jakarta.
[12] Directorate of Space Utilization, Directorate General of Spatial Planning. 2017. Spatial Planning Bulletin. Ministry of Agrarian and Spatial Planning, Jakarta.

IOP Conf. Series: Earth and Environmental Science **106** (2017) 012046 doi:10.1088/1755-1315/106/1/012046

Determinants of corporate dividend policy in Indonesia

H S Lestari

Faculty of Economic and Business, Universitas Trisakti, Jakarta Barat, Indonesia

Corresponding Author: henny_setyo_lestari@trisakti.ac.id

Abstract. This study aims to investigate the determinants factors that effect the dividend policy. The sample used in this research is manufacture companies listed in Indonesia Stock Exchange (IDX) and the period 2011 – 2015. There are independent variables such as earning, cash flow, free cash flow, debt, growth opportunities, investment opportunities, firm size, largest shareholder, firm risk, lagged dividend and dividend policy used as dependent variable. The study examines a total of 32 manufacture companies. After analyzing the data using the program software Eviews 9.0 by multiples regression analysis reveal that earning, cash flow, free cash flow, firm size, and lagged dividend have significant effect on dividend policy, whereas debt, growth opportunities, investment opportunities, largest shareholder, and firm risk have no significant effect on dividend policy. The results of this study are expected to be implemented by the financial managers in improving corporate profits and basic information as return on investment decisions.
Keywords: cash flow, debt, dividend per share, earning, firm risk, firm size, free cash flow

1. Introduction

Dividend policy is important to firms because it is used to show the firm stability and its growth [1]. Shareholders and potential investors will decide to invest if they see the firm's ability to generate dividends. Furthermore, dividend policy can be used by management or shareholders to reduce the agency cost. Many studies have been discussed determinants of dividend policy. Firms with high profit and stable earning are able to manage greater cash flow [2]. Firms with high earnings pay larger dividends to shareholders. A positive relationship found between earnings and dividend policy.

Cash flow is one of determinants of dividend policy. Firm with high cash flow leads to pay greater dividends [3]. Cash flow has a positive relationship with dividend payment. Free cash flow is the residual cash flow at the end of period after paying expenses. Literature stated that greater free cash flow leds to greater dividend payment [4]. A positive association is found between free cash flow and dividends.

High level of debt results in lower dividend payment as firms choose to retain its funds to pay the debt than distribute them as dividends. A negative association between debt and dividends. Growth and investment as determinants of dividend policy is in line with the agency cost theory, whereby firms with no growth of fewer investment opportunities have greater exposure to agency cost [5]. In reducing the agency cost, these firms will pay higher dividends to the shareholders [4]. Firms with high growth and investment opportunities will need the internally generate funds to finance those investments, and thus tend to pay little or no dividends [5].

Firm size positively associated with dividends. A large firm typically has better access to capital markets and finds it easier to raise funds with lower cost and fewer constraints compared to a small firm [5]. Therefore, large firms are more likely to afford paying higher dividends to shareholders.

Largest shareholders play a greater role in monitoring management as compared to small shareholders, as they have greater voting power to influence a firm's decision [6]. They found a positive relationship between largest shareholder and dividends. Next factor that affects the firm's dividend policy is firm risk. In minimizing the firm's risk, a firm will pay lower dividend [7]. A negative association found between firm risk and dividend policy.

Lagged dividend are the significant factors in determining dividend policy as current dividens affected by the firm's past dividens. This is due to belief that the shareholders favor a reasonably stable rate of dividends [3]. Based on the above argument, a study was conducted to analyze determinants of dividend policy in manufacturing companies.

2. Research Method

The research design used for this research is hypothesis testing with purposed to understand the influence of earning, cash flow, free cash flow, debt, growth opportunities, investment opportunities, firm size, largest shareholder, firm risk and lagged dividend toward dividend policy. Data collected from financial report of manufacturing company that shared dividend to the shareholders and listed in Indonesia Stock Exchange for the period of 2011-2015. The analysis tool for this research is regression using E-views software.

2.1. Dependent variable

Dependent variable in this study is dividend policy. Dividend policy in this study measured by dividend per share [7].

$$\text{DPS} = \frac{\text{Cash Dividend}}{\text{Outstanding Common Stock}} \tag{1}$$

2.2. Independent variable

Independent variables in this study are earning, cash flow, free cash flow, debt, growth opportunities, investment opportunities, firm size, largest shareholder, firm risk, and lagged dividend.

$$\text{EPS} = \frac{\text{Net Income}}{\text{Number of Shares Outstanding}} \tag{2}$$

$$\text{CF} = \frac{\text{Operation Cash Flow}}{\text{Total Asset}} \tag{3}$$

$$\text{FCF} = \frac{\text{OCF} - \Delta\text{NFA} - \Delta\text{NWC}}{\text{Total asset}} \tag{4}$$

$$\text{DEBT} = \frac{\text{Total Liabilities}}{\text{Total Asset}} \tag{5}$$

$$\text{GROWTH} = \frac{\text{Current Sales} - \text{Last Year Sales}}{\text{Last Year Sales}} \tag{6}$$

$$\text{INV} = \frac{\text{Retained Earnings}}{\text{Total Asset}} \tag{7}$$

$$\text{SIZE} = \text{LOG of Total Asset} \tag{8}$$

$$\text{LARGE} = \text{Percentage Shares Owned by Largest Shareholder} \tag{9}$$

$$\text{RISK} = \text{Standard Deviation of Monthly Share Returns} \tag{10}$$

$$\text{D(t} - 1) = \text{Last Year Dividend} \tag{11}$$

$$\text{SIZE} = \text{LOG of Total Asset} \tag{12}$$

IOP Conf. Series: Earth and Environmental Science **106** (2017) 012046 doi:10.1088/1755-1315/106/1/012046

2.3. Data analysis method

Analysis method used in this study is multiple linear regression. Purpose of this analysis method is to identify relationship of independent variables toward dependent variable. Multiple regression model in this study:

$$\mathrm{DPS} = \beta 0 + \beta 1\mathrm{E} + \beta 2\mathrm{CF} + \beta 3\mathrm{FCF} + \beta 4\mathrm{DEBT} + \beta 5\mathrm{GR} + \beta 6\mathrm{INV} + \beta 7\mathrm{SIZE} + \beta 8\mathrm{LARGE} + \beta 9\mathrm{RISK} + \beta 10\mathrm{D}(t-1) \quad (13)$$

$\beta 1$ = regression coefficient for earning; $\beta 2$ = regression coefficient for cash flow; $\beta 3$ = regression coefficient for free cash flow; $\beta 4$ = regression coefficient for debt; $\beta 5$ = regression coefficient for growth opportunities; $\beta 6$ = regression coefficient for investment opportunities; $\beta 7$ = regression coefficient for firm size; $\beta 8$ = regression coefficient for largest shareholder; $\beta 9$ = regression coefficient for firm risk; $\beta 10$ = regression coefficient for lagged dividend

T test is used to examine the influence of independent variables individually toward dependent variable. Independent variables are earning, cash flow, free cash flow, debt, growth opportunities, investment opportunities, firm size, largest shareholder, firm risk, and lagged dividend. Dependent variable is dividend policy.

3. Results and Discussion

The objects of this study is manufacturing companies listed in Indonesia Stock Exchange for the period of 2011-2015. Sampling technique used for this study is purposive sampling method, the sampling based on certain criteria, namely manufacturing companies that have gone public and listed in Indonesia Stock Exchange and distribute it dividend for the period of 2011-2015. Data gathered from 30 companies within a period of 5 years.

Table 1. Descriptive statistics.

	N	Mean	Median	Maximum	Minimum	Std. Dev
DPS	150	1.2144	0.1470	16.0000	0.0031	3.0901
EPS	150	1.6500	0.4445	17.6210	0.0090	3.5312
CF	150	0.4007	0.3343	0.8895	0.0933	0.1814
FCF	150	0.3193	0.2431	0.8671	-0.1388	0.2009
DEBT	150	0.3490	0.3239	0.8013	0.0904	0.1476
GR	150	0.1184	0.1209	0.5301	-0.2990	0.1149
INV	150	0.0851	0.0068	0.7773	5.75E-06	0.1910
SIZE	150	6.5275	6.4114	7.9630	4.9913	0.7391
LARGE	150	0.5733	0.5477	0.9818	0.1017	0.2458
RISK	150	0.1105	0.0884	0.4386	0.0282	0.0707
LAGGED	150	1.1559	0.1440	14.0000	0.0031	2.9249

Soure: Panel Data Regression Output Eviews 9

3.1. Earning

The result of the coefficient showed the influence of earning on dividend policy is 0.286027. Statistical test showed the p-value is $0.0000 < 0.05$. The result of this study indicate earning has a positive effect on dividend policy. Consistent with literature which found positive effect of earning on dividend policy [3]. Supported by literature, earning positively related with dividend policy [8]. The higher earning of the firm, the more likely the firm to pay higher dividend. Earning has positive impact on dividend policy, thus firm with high and stable earning will be able to manage it cash, therefore will leads to pay greater dividends [2].

Table 2. T test result.

Variable	Coefficient	Probability	Conclusion
DPS	13.134	0.0077	
EARNING	0.2860	0.0000	Significant
CF	-4.4463	0.0003	Significant
FCF	3.3182	0.0000	Significant
DEBT	0.7107	0.3391	Insignificant
GROWTH	-0.5897	0.2278	Inignificant
INV	-1.4268	0.2373	Insignificant
SIZE	-1.6249	0.0124	Significant
LARGE	-2.0493	0.3070	Insignificant
RISK	-1.2405	0.0827	Insignificant
LAGGED	0.1665	0.0226	Significant

Source: Output Data Eviews 9

3.2. Cash Flow
The result of the coefficient showed the influence of cash flow on dividend policy is -4.446376. Statistical test showed the p-value is 0.003 < 0.05. The result of this study reported cash flow has a negative effect on dividend policy. In contrast, literature found cash flow has no significant effect on dividend policy [3]. But it is in line with stated firm with lower cash flow will pay higher amount of dividend than the firm with high cash flow [9]. As cash flow increases, it will be difficult for investors to monitor the cash flow. Thus, the higher the cash flow, the higher agency cost. Therefore firm with higher cash flow tends to has higher agency cost and it results in reduce dividend payment. The negative effect is exists between cash flow and dividend policy.

3.3. Free Cash Flow
The result of the coefficient showed the influence of free cash flow on dividend policy is 3.318259. Statistical test showed the p-value is 0.0000 < 0.005. The result of this study reported free cash flow has a positive effect on dividend policy. In the other hand, [3] showed free cash flow has no effect on dividend policy. But this study supported by literature who found positive effect between free cash flow and dividend policy [10]. Availability of free cash flow leads the management to increase dividend payment to mitigate the agency cost. Free cash flow has positive effect on dividend policy [11]. Greater free cash flow is profitable as the remaining cash will be used to pay dividend to shareholders to maximize their wealth.

3.4. Debt
The result of the p-value is 0.3391 > 0.05 therefore it can be said it is insignificant. The result of this study reported debt has no significant effect on dividend policy. In contrast, a study by literature found a negative effect between debt and dividend policy [3]. Firm with high debt level leads to decrease in dividend payment. However this result found there is no significant effect of debt on dividend policy [12]. This means that the amount of debt does not effect firm's dividend payment. This result also found no significant effect of debt on dividend policy [13].

3.5. Growth Opportunities
The result of the p-value is 0.2278 > 0.05 therefore it can be said it is insignificant. The result of this study reported growth opportunities has no significant effect on dividend policy. It is supported by literature on their study that growth opportunities does not effect dividend policy [3]. Supported by literature growth opportunities has no impact on firm's dividend policy [13]. When growth opportunities of the firm increasing, it will not determine the amount of dividend paid by the firm. Thus there is no significant effect of growth opportunities on dividend policy.

3.6. Investment Opportunities
The result of the p-value is 0.2373 > 0.05 therefore it can be said it is insignificant. The result of this study reported investment opportunities has no significant effect on dividend policy. This result is not in line with literature which found investment opportunities has positive effect on dividend policy [3]. Firm with high investment opportunities will have a better cash flow then it can pay dividend in high amount. However result of this study supported by literature which there is no significant effect of investment opportunities on dividend policy [14]. Dividend payment depends on the investment decisions made by the firm. Thus there is no significant effect of investment opportunities on dividend policy.

3.7. Firm Size
The result of the coefficient showed the influence of firm size on dividend policy is -1.624961. Statistical test showed the p-value is 0.0124 < 0.05. The result of this study reported firm size has negative effect on dividend policy. In contrast, a significant positive association between firm size and dividend policy [3]. Large firms generally have various ownership. This increase the agency cost. Then to mitigate the agency problem between principal and agent, firm need to pay dividends. This result consistent with literature they found negative effect between firm size and dividend policy [12]. This happen as the larger size of the firm, the greater the expenses it owned, and it will leads to reduce the proportion of dividends distributed to shareholders. There is negative effect of firm size on dividend policy.

3.8. Largest Shareholder
The result of the p-value is 0.3070 > 0.05 therefore it can be said it is insignificant. The result of this study show largest shareholder has no significant effect on dividend policy. This result is in contrast to a study by literature who stated that largest shareholder has a positive effect on dividend policy [16]. The higher percentage shares owned by it largest shareholder, the dividend payment will be higher. The result of this study is supported by literature which found largest shareholder has no effect on dividend policy [15]. There is no effect of largest shareholder on dividend policy.

3.9. Firm Risk
The result of the p-value is 0.0857 > 0.05 therefore it can be said it is insignificant. The result of this study show firm risk has no significant effect on dividend policy. This supported by literature, they discovered firm risk has no impact on dividend policy [3]. This means that the amount dividend payment is not depends on the risk had by the firm. A study also stated no effect between firm risk and dividend policy [16]. Literature study found that firm risk has no significant effect on dividend policy [17].

3.10. Lagged Dividend
The result of the coefficient show the influence of lagged dividend on dividend policy is 0.166561. Statistical test show the p-value is 0.0226 < 0.05. The result of this study reported lagged dividend has positive effect on dividend policy. In contrast, literature found no significant effect between lagged dividend and dividend policy [3]. However it is supported by literature that discovered a significant positive association between lagged dividend and dividend policy [4]. Previous year's dividend has positive relation on current dividend [12]. If the dividend of the previous year was in a stable form, then current dividend payment will be in a stable level or even better, increased.

4. Conclusion
This research is done to examine the influence of earning, cash flow, free cash flow, debt, growth opportunities, investment opportunities, firm size, largest shareholder, firm risk, and lagged dividend on dividend policy (dividend per share) in companies listed in Indonesia Stock Exchange for the period of 2011-2015. Based on above analysis and discussions, it can be concluded that significant positive effect on dividend policy: *earning, free cash flow, lagged dividend*; significant negative effect

on dividend policy: *cash flow, firm size*; significant positive effect on dividend policy: *debt, growth opportunities, investment opportunities, largest shareholder, firm risk.*

References

[1] Cheng Z, Cullinan C, & Zhang J. 2014. Free Cash Flow, Growth Opportunities, and Dividends: Does Cross-Listing of Shares Matter? *The Journal of Applied Business Research.* 30(2):587-598.

[2] Manneh M, and Naser K. 2015. Determinants of Corporate Dividends Policy: Evidence from an Emerging Economy. *International Journal of Economics and Finance*. 7(7):10.

[3] D'Souza J. and Saxena A. 1999. Agency Cost, Market Risk, Investment Opportunities and Dividend Policy – An International Perspective. *Managerial Finance.* 25(6):35–43.

[4] Hussain I. 2013. The Paradox of Rising Dividend Payouts in a Recession: Evidence from Pakistan. *The Lahore Journal of Business.* 1(2):97–116.

[5] Grinstein Y, & Michaely R. 2005. Institutional Holdings and Payout Policy. *Journal of Finance.* 60:1389–1426.

[6] Al-Kuwari D. 2010. To Pay or Not to Pay: Using Emerging Panel Data to Identity Factors Influencing Corporate Dividend Payout Decisions. *International Research Journal of Finance and Economics*. 42:19-36.

[7] Ahmed H, and Javid A. 2009. The determinants of dividend policy in Pakistan, *International Research Journal of Finance and Economics.* 29:110-125.

[8] Amidu M, and Abor J. 2006. Determinants of Dividend Payout Ratios in Ghana. *The Journal of Risk Finance*. 7(2):136-145.

[9] Maladjian C, and Khoury R. 2014. Determinants of Dividend Policy: An Empirical Study on the Lebanese Listed Banks. *International Journal of Economics and Finance.* 6(4):E-SSN 1916-1978.

[10] Bradley, M, Capozza, D R. & Sequin, P J. 1998. Dividend Policy and Cash-Flow Uncertainty, *Real Estate Economics*, 26(4), 555-572.

[11] Al-Malkawi H A, 2007. Determinants of Corporate Dividend Policy in Jordan: An Application of the Tobit Model. *Journal of Economic and Administrative Sciences*. 23(2):44-70.

[12] Naser K, Nuseibeh R, and Rashed W. 2013. Managers' Pereption of Dividend Poicy: Evidende from Companies Listed on Abu Dhabi Securities Exchange. Issues in *Business Management and Economics*. 1(1):001-012

[13] Naceur S, Goaied M, and Belanes A. 2006. On the Determinants and Dynamics of Dividends Policy. *Journal of International review of finance*. P. 1-23.

[14] Truong T, and Heaney R. 2007. Largest Shareholder and Dividend Policy Around World. *The Quarterly Review of Economics and Finance.* 47(5):667 687

[15] Rozeff M. 1982. Growth, Beta and Agency Costs as Determinants of Dividend Payout Ratios. *The Journal of Financial Research*. 5(3):249-259.

[16] Yusof Y, and Ismail S. 2016. Determinants of Dividend Policy of Public Listed Companies in Malaysia. *Review of* International *Business and Strategy*. 26:88 – 99.

[17] Zameer H, Rasool S, Iqbal S, and Arshad U. 2013. Determinants of Dividend Policy: A Case of Banking Sector in Pakistan. *Middle-East Journal of Science Research.* 18(3):410-424.

Performance of warm mix asphalt with Buton natural asphalt-rubber and zeolite as an additives

N Wahjuningsih*, **S P Hadiwardoyo, R J Sumabrata, M Anis**

Civil Engineering Department, Faculty of Engineering, Universitas Indonesia

*Corresponding Author: nwahjuningsih@gmail.com

Abstract. The aim of this research is improving of asphalt industry to decrease the fuel consumption by lowering the temperature of mixing and compacting of asphalt mixture. This technology known as Warm Mix Asphalt (WMA). Buton Natural Asphalt Rubber (BNA-R) as a function of the additive has been able to improve the performance of HMA. Zeolit has a function as an additive to lowering the mixing temperature. Aggregate composition using the composition of aggregate grading specifications for airport pavement, with the composition of BNA-R 5% and 10% and Zeolite content of 2%. The mixture have produced Resilient Modulus value by using the Universal Material Testing Apparatus (UMATTA) on optimum bitumen content each of which has been obtained from the Marshall test. Furthermore, the value of permanent deformation of asphalt mixtures tested using Wheel Tracking Machine (WTM). The result shows that BNA-R modified binder for WMA can decrease the rutting potential. The additive of local materials has improved the performance of the WMA for airport pavement with certain restrictions. From this research it is known there have been changes in the characteristics of resilient modulus and permanent deformation due to the addition of BNA-R for type of aggregate composition.
Keywords: aggregate grading, natural asphalt rubber, warm mix asphalt

1. Introduction

The commonly materials used for flexible pavement is Hot Mix Asphalt (HMA) that was produced at high temperature (150 – 190°C). This temperature will influence the fuel consumption and bad for environment. Warm Mix Asphalt (WMA) is one of innovation in the manufacturing of asphalt mixture that reduce the mixing and compaction temperature about 20-40°C with involve the organic additives, chemical additives and water-based or water-containing processes [1], [2].

Many advantages of WMA as reported the researcher like reduced emission, lower energy consumption in production of asphalt mixture, better working condition because absence of harmful gasses, longer hauling distances, and extended paving window. Besides, the drawbacks of this mixture still found such are higher costs because there is still a certain reticence to using them, lack of data concerning their long-term performance, greater moisture susceptibility due to lower temperatures and coating and bonding problems [1]. The laboratory evaluation of WMA mixture for use in airport pavement rehabilitation showed that produced or mixed at 30°C lower than normal can be recommended. This mixture with a coarse gradation exhibited comparable performance compare with the control HMA [3].

Indonesia has many natural deposits such as rock asphalt from Buton Island and zeolite. Some areas in southern Buton Island, Indonesia possess natural rock asphalt resources with deposit of approximately 60,991,554.38 ton (24,352,833.07 barrel oil equivalent). Natural rock asphalt composed of approximately 30% bitumen and 70% mineral [4]. Buton natural asphalt has been applied as an additive or modifier to petroleum asphalt. Because of it has low penetration, the addition of BNA could improve the properties of virgin asphalt [5]. BNA has many forms such like aggregate (raw material), semi extraction and full extraction. BNA-R is a semi extraction of Buton rock asphalt with crumb rubber. The utilization of crumb rubber modified (CRM) bitumen lately more popular because its several benefits such as increasing the stiffness of bitumen, improving the temperature susceptibility, increasing the rutting, decreasing the noisy and maintenance cost [6], [7].

One of WMA types is produced by foaming process. The foaming process is characterized by the injection of small amounts of water directly into the mix chamber or into the hot binder; in fact, when the water is dispersed in hot asphalt, it vaporizes (from contact with the hot asphalt) and it results in a binder expansion with a consequent reduction of the mix viscosity. One of the methodology of this process is water containing technology using synthetic or natural zeolite. Zeolite is one of WMA additives that can be lowering the mixing and compaction temperature because of the foaming processes when heating the mixture [2], [8], [9].

The resilient modulus is the elastic modulus to be used with the elastic theory. It is well known that most of paving materials are not elastic but experience some permanent deformation after load repetition due to traffic. The permanent deformation is one of principal structural distress mode of asphalt mixture. Permanent deformation is an important factor in flexible pavement design. With the increase in traffic load and tire pressure, most of the permanent deformation occurs in the upper layers [10], [11]. The main objective of this research is evaluate the use of additive from local material (BNA-R and Zeolite) on the engineering properties of WMA for airport pavement especially from the point of view of the resilient modulus value and the rate deformation.

2. Research Method

This research use AC WC (Asphalt Concrete Wearing Course) types based on airport pavement standard of Indonesia with the maximum size of aggregate of 19 mm and dense gradation. The BNA-R was use as modifier for virgin asphalt Pen 60/70 from Shell with the content of 5% and 10% of weight of asphalt binder. The virgin asphalt (0% of BNA-R) is still used in this research for comparation.

In warm mix technology where mixing and compaction temperature must lower than hot mix asphalt, the natural zeolite was used in this research. The natural zeolite from Tasikmalaya area West Java that has been activated by chemical process was used in this research because of has the high water content [12]. 2% zeolite of weight of total mixture was used and replace as filler in the aggregate composition. The Marshall test was performed as a standard test for asphalt mixture to obtain the maximum binder content. 5 samples was prepared for every mixture with the content of BNA-R of 0%, 5% and 10%. Based on the optimum binder content, the samples for further test was prepared.

Wheel tracking test (WTT) was used to determine the deformation resistance of mixtures under high temperature. In this method a rubber faced tire moved back and forth in the middle of the 300 x 300 x 50 mm specimen at the speed of 42 passes/min. A high tire-specimen contact pressure of 1.38 MPa was used to fully simulate the aircraft (Boeing 747) loading condition. The specimen was compacted by a rolling compactor, and then held in an environmental chamber for a minimum of 6 h. The test was performed at temperature of 40°C for an hour.

Resilient Modulus is an important parameter to determine the performance of pavement, to analysis the pavement response to traffic loading. The test was done by measuring the indirect tensile strength in repeated loading using Universal Material Testing Apparatus (UMATTA). Specimens at their optimum bitumen content i.e. mixture with 0%, 5% and 10% BNA-R were made and loaded by diametrical force in pulse loading. This apparatus consists of Control and Data Acquisition System (CDAS), personal computer and related integrated software. The test follows the ASTM Designation D 4123-82. Data inputted were condition pulse count = 5, condition pulse period = 3,000 ms, test pulse period = 2,000 ms, rise time = 50 ms, peak loading force = 1,000 N and estimated poisons ratio = 0.4. The temperature of test was selected as room temperature of 25°C, 35°C and 45°C.

3. Results and Discussion

The optimum binder content which obtained from the Marshall test is 5.9%, 5.5% and 5.4% for warm mix asphalt with content of BNA-R 0%, 5% and 10% respectively. All mixtures have stability over 1,000 kg where this is the minimum value required of AC-WC mixtures for airport pavement. The addition of BNA-R as modifier for warm mix asphalt not give significant enhancement on stability value.

The result of resilient modulus test using UMATTA is concluded as shown in Table 1. With the addition of BNA-R to the WMA, the value of resilient modulus looked increasing in 5% of BNA-R but decrease at 10%. Federal Aviation Association (FAA) stated the minimum value resilient modulus for the surface of flexible pavement at 32°C of airport viz 1,320 MPa. All mixture (with or without BNA-R) can fulfill the requirement at the temperature of test of 25°C. While at the test temperature test of 35°C and 45°C, all mixture failed to fulfill. The highest value of resilient modulus has shown by WMA mixture with 5% of BNA-R for all test temperature.

Table 1. Characteristics of WMA with BNA-R modifier.

Type of WMA	Optimum Binder Content (%)	Temperature of Test (°C)	Resilient Modulus (MPa)
		25	3,053
0% BNA-R	5.9	35	853
		45	338
		25	4,386
5% BNA-R	5.5	35	1,016
		45	400
		25	3,156
10% BNA-R	5.4	35	745
		45	363

From the wheel tracking test using wheel tracking machine at the temperature test of 40°C, the result of WMA with asphalt binder modified by BNA-R shown in Figure 1. In deformation in mm, the addition of BNA-R as asphalt modifier give the significant influence in decreasing deformation value. It means that the rutting potential can be lowered by modify the asphalt binder with BNA-R.

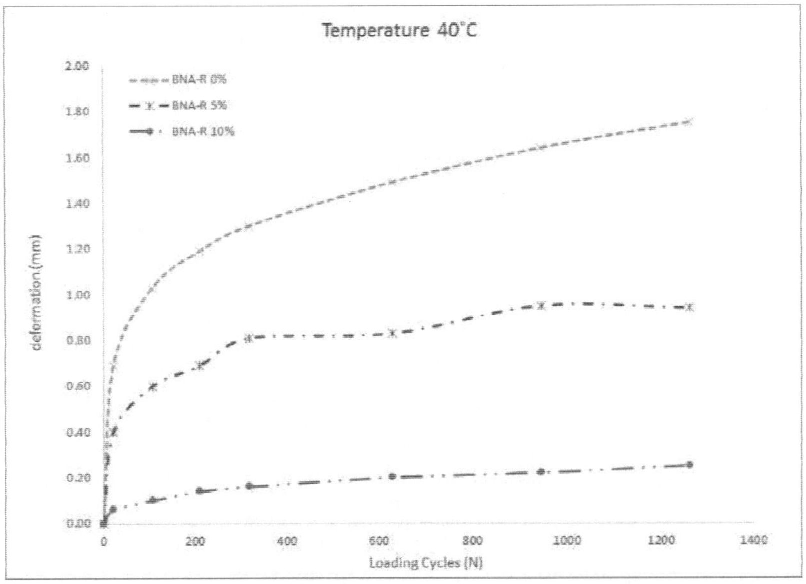

Figure 1. Deformation characteristic of WMA.

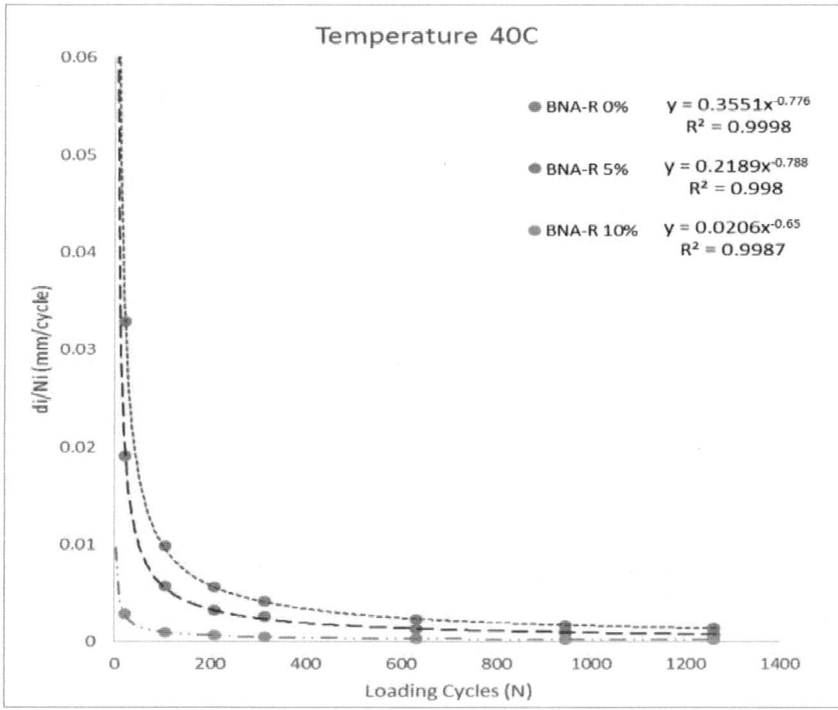

Figure 2. Rate of deformation performance of WMA.

The rate of deformation (di/Ni) versus the cycle Ni for different samples types and temperatures of 40° C is shown in Figure 2. At the beginning of the loading time, the value di/Ni is high, and the value decreases rapidly to near zero. Figure 2 show that the relationship between di/Ni and Ni is a power function that can be written as:

$$di/Ni = a.(Ni)^{-b} \tag{1}$$

or

$$di = Ni.a.(Ni)^{-b} \tag{2}$$

The parameter of ".*a*" and "*b*" can be identified by a series of wheel tracking machine test. The value *b* is almost same at the same test temperature.

4. Conclusion

The summary and conclusion based on the results of the research presented in this paper are the local material can be used as an additive or modifier on the asphalt mixture. BNA-R modifier can enhance the ability of asphalt mixture to withstand the loading that presented in stability value at the certain content. And zeolite powder can use as filler to help reducing the temperature of mixing and compaction. The warm mix asphalt (WMA) with local materials (BNA-R and Zeolite) still can used as pavement material at airport with certain restrictions. Further research is needed to evaluate the potential of warm mix asphalt as the pavement material.

Acknowledgment

The authors wish to thank Directorate of Research and Community Service at the Universitas Indonesia for their funding towards this research through Hibah PITTA. The experimental work was completed at the laboratory of Structure and Material of Universitas Indonesia and the laboratory of Research Centre and Development for Road and Bridge – Ministry of Public Works Republic of Indonesia.

References

[1] Capitalo SD, Picado-Santos LG, Martinho F. 2012. Pavement engineering materials: Review on the use of warm-mix asphalt. *Construction and Building Material*. 36:1016–24.
[2] Sengoz B, Topal A, Gorkem C. 2013. Evaluation of natural zeolite as warm mix asphalt additive and its comparison with other warm mix additives. *Construction and Building Material.* 43:242–52.
[3] Su K, Maekawa R, Hachiya Y. 2009. Laboratory evaluation of WMA mixture for use in airport pavement rehabilitation. *Construction and Building Material.* 23(7):2709–14.
[4] Rizal M, Ali NUR, Bahar T. 2016. Influence of laboratory short term aging on tensile strength of porous asphalt mixture. 3:21–4.
[5] Hadiwardoyo SP, Sinaga ES, Fikri H. 2013. The influence of Buton asphalt additive on skid resistance based on penetration index and temperature. *Construction and Building Material.* 42:5 –10.
[6] Wang H, Dang Z, You Z, Cao D. 2012. Effect of warm mixture asphalt (WMA) additives on high failure temperature properties for crumb rubber modified (CRM) binders. *Construction and Building Material*. 35:281–8.
[7] Behl A, Kumar G, Sharma G. 2013. Performance of Low Energy Crumb Rubber Modified Bituminous Mixes. *Procedia - Soc Behav Sci* [Internet]. 104:49–58.
[8] Rubio MC, Martínez G, Baena L, Moreno F. 2012. Warm Mix Asphalt: An overview. *J Clean Prod* [Internet]. 24:76–84.
[9] Vaiana R, Iuele T, Gallelli V. 2013. Warm Mix Asphalt with Synthetic Zeolite : a Laboratory Study on Mixes Workability. 6(5):562–9.
[10] Read J, Whiteoak D. The Shell bitumen handbook [Internet]. Read J, & Whiteoak D. The Shell bitumen handbook. Thomas Telford. p. 29. 2003.
[11] Huang YH. Pavement Analysis and Design. 2003. 792 p.
[12] Furqon Affandi, Hendri Hadisi. 2011. *Jurnal Jalan dan Jembatan* :264.

Treating domestic effluent wastewater treatment by aerobic biofilter with bioballs medium

R Permatasari, A Rinanti, R Ratnaningsih*

Environmental Engineering Department, Faculty of Landscape Architecture and Environmental Technology, Universitas Trisakti, Jakarta

*Corresponding Author : ratnaningsih@trisakti.ac.id

Abstract. This laboratory scale research aimed to treat wastewater effluent with advanced treatment utilizing aerobic biofilter with bio-balls medium to obtain effluent quality in accordance with DKI Jakarta Governor Regulation No. 122 of 2005. The seeding and acclimatization were conducted in 4 weeks. The effluent were accommodated in a 150 L water barrel supported by a submersible pump. The effluent were treated in two boxes shaped reactors made of glasses with 36 L of each capacity. These reactors were equipped with aquarium aerators, sampling tap is 10 cm from the base of reactors, and bio-balls with 3 cm diameter are made of PVC. Reactors operated continuously with variations of retention time of 4 hours, 8 hours, 12 hours, 18 hours, and 24 hours and also variations of Carbon: Nitrogen: Phosphor = C: N: P ratio were, 100:5:1, 100:8:1, 100:10:1, 100:12:1, 100:15:1. The results showed that the optimum variance of retention time was 24 hours and the ratio of C:N:P was 100:10:1 yielded the largest removal efficiency for 83,33% of COD, 87,33% of BOD, 82,5% of Ammonia, 79,1% of Nitrate, 92% of Nitrite, 84,82% of Oil and Grease. The concentration parameter resulted from outlet biofilter has met the domestic wastewater quality standard of DKI Jakarta.
Keywords: aerobic biofilter, bioball, domestic wastewater, retention time, removal efficiency

1. Introduction

Biofilter is a biological filter with attached biomass on the filter-media. The microorganism attaches on the surface of media and grows up to biofilm will degrade the organic compounds present in the water. For the development of advance biological treatment process, WWTPs need to be more compact, low in operational costs and stable in operation, while at the same time minimize the noise and odor, and space requirements as well as generate high performance. However, with the continuing need to increase wastewater effluent quality, one of the advanced biological treatment systems, the biological aerated filters, proves to be more reliable [1]. This process has several advantages that are easy to operate, less sludge production, and are resistant to fluctuations of flow and concentration of wastewater [2].

In a biological treatment process, the optimum ratio of C: N: P in the aerobic process is 100: 5: 1 (C:N = 20-30 and N:P = 5). Domestic waste water has the C: N: P ratio of 100: 5: 1, which is sufficient for most microorganisms [3]. Aerobic microorganisms work most effective in the pH range of 6.5 - 8.5 [4]. According to study conducted to investigate the effect of using aerobic biological filter system for improving the effluent quality of two-stage anaerobic reactors, average overall removal efficiencies up to 85%, 69%, 88%, 91% and 89% were achieved for COD Total, COD Soluble, BOD_5, TSS and VSS, respectively [5]. The purpose of this research is to treat wastewater effluent with advanced treatment utilizing aerobic biofilter with bio-balls medium to obtain effluent quality in accordance with the effluent quality standard.

2. Research Method

2.1. Reactor Description

This laboratory scale research, two boxes shaped biofilter reactors is made of glasses 8 mm thick. Design dimension is 20 cm x 30 cm x 70 cm with treated capacity of 36 L (Figure 1). Bio-balls medium with 3 cm diameter are made of PVC. There is a buffer from the wire as a media holder at 10 cm up from the based of reactors. The sampling tap is 10 cm from the base of reactors and sludge tap placed under the base of reactors. Other devices that complement the biofilter reactor are waste barrel, submersible pumps, recirculation pumps, and aerators. The effluent were accommodated in a 150 L water barrel provided by submersible pump. Aerators serve to keep the microorganisms in the reactor remains in aerobic conditions.

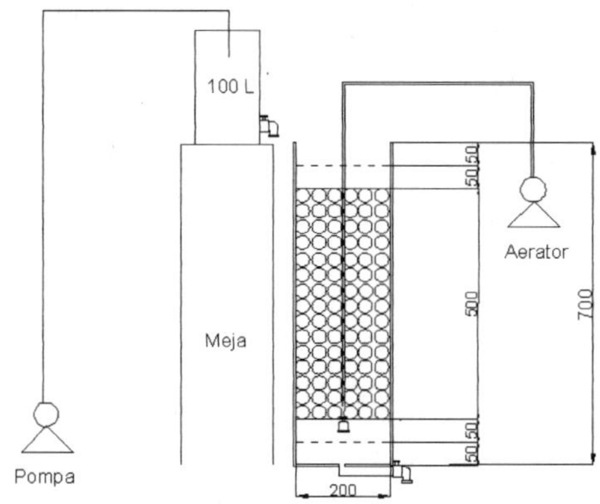

Figure 1. Design of aerobik biofilter reactor.

2.2. Seeding and Acclimatization

Reactors were seeded with microorganisms directly by circulating wastewater. Seeding was used EM_4 diluted with aquadest with a ratio of 1/10 (10%). Use of 10% EM_4 refers to the EM_4 procedure for liquid waste treatment. One liter of EM_4 is mixed with 10 L aquadest and added 5 tbsp of liquid brown sugar, then fermented for 4 days in tightly closed condition with room temperature. This process aims to seed and activate the existing microorganisms in EM_4 so that microorganisms can work optimally when mixed with wastewater in the reactors. EM_4 which has been fermented, mixed into the waste with the amount of 34.2 L of waste water + 1.8 L EM_4 (5%) [6].

Every day nutrients were added to provide food to microorganisms in wastewater based on composition of COD : N: P = 100: 5: 1. Nutrients were added as follows, glucose, NH_4Cl, KH_2PO_4, $MgSO_4.7H_2O$, $CaCl_2.2H_2O$, and $FeCl_2$. During seeding, VSS was analyze to determine the biomass content. The seeding is carried out until the level of VSS ± 2000 mg/L.

Acclimatization started by replacing the wastewater from seeding with the effluent wastewater gradually. The water replacement starts with the ratio of 1/3 to 100% of the original waste water. Acclimatization operated continuously with a residence time of 24 h. Acclimatization was stopped when the concentration of COD removal reached 80%.

2.3. Operation of Reactors

Reactors operated continuously with 5 variations of retention time, started with the longest time as follows, 24 hours, 18 hours, 12 hours, 8 hours, and 4 hours. The flows for each retention time were 1.5 L/h, 2 L/h, 3 L/h, 4.5 L/h, and 9 L/h, respectively.

IOP Conf. Series: Earth and Environmental Science **106** (2017) 012048 doi:10.1088/1755-1315/106/1/012048

After the residence time with the largest removal efficiency was obtained, variation of Carbon : Nitrogen : Phosphor = C:N:P ratios added were 100:5:1, 100:8:1, 100:10:1, 100:12:1, 100:15:1 with the optimum retention time.

2.4. *Sampling Methods and Parameter Analysis*
The representative sampling method was collected with combined sample of time type (composite samples). To get a combined sample of time, each mixed sample should have the same volume. Parameters analyzed were COD, TSS, BOD, Ammonia, Nitrite, Nitrate, Phosphate, and Oil and Grease. COD, Ammonia, Nitrite, Nitrate parameters were analyzed daily for each retention time. As for the parameters of TSS, BOD, Phosphate, and Oil and Grease were analyzed at each transition of retention time.

2.5. *Substrate Degradation Kinetics*
The rate of substrate degradation kinetics aims to determine the extent of substrate reduction in wastewater treatment. The process of removing pollutants is very specific depending on the characteristics of wastewater, biomass colonies, and media characteristics. Microbial growth based on Monod equation [7]:

$$\frac{\partial x}{\partial t} = \mu x \tag{1}$$

$\frac{\partial x}{\partial t}$ = cell growth rate, (mass/(vol.time)), μ = specific growth rate, time^{-1}, x = biomass concentration

The growth coefficient based on Monod Equation developed into *Burk-Lineweaver* Method:

$$\frac{1}{\mu} = \frac{K_s}{\mu_{max}}\frac{1}{S} + \frac{1}{\mu_{max}} \tag{2}$$

μ_{max} = maximum specific growth rate, time^{-1}, Ks = half-velocity constant"—the value of *S* when μ/μ_{max} = 0.5, S = concentration of the limiting substrate for growth

The growth yield (Y) is defined as the amount of biomass increase as a result of the use of substrate amounts. If the microorganisms culture condition is constant, then Y is also constant. The growth yield is calculated by the following formula:

$$x - x_o = Y(S - S_o) \tag{3}$$

x_o, S_o = Initial concentration of biomass and substrate, x, s = Concentration at time of growth

In every process, not all microorganisms cells are in the growth phase, but there are also microorganisms cells in the phase of death and respiration form. These factors effect the cell mass to decrease and proportional to the reduction of microorganism concentration lead to the mortality rate of microorganisms (K_d) [8].

$$\mu = \mu_{max}\frac{8}{K_s+S} - K_d \tag{4}$$

2.6. *Removal Efficiency*
The removal efficiency of wastewater treatment is the ratio between organic content that is set aside through the process of processing with initial concentration. The removal efficiency of wastewater treatment is calculated as follows:

$$\eta = \frac{Co-Ce}{Co} \times 100\% \tag{5}$$

η = Removal efficiency (%), Co = initial concentration (mg/L), *Ce* = final concentration (mg/L)

3. Results and discussion

3.1. Characteristics of Wastewater

The effluent wastewater used in this research was sourced from wastewater treatment in Gedung M, Universitas Trisakti, Jakarta Barat. The characteristics of effluent wastewater were presented in Table 1.

Table 1. Characteristics of effluent wastewater.

Parameters	Unit	Concentration	Governor Reg No. 122/2005	National Government Reg 82/2001 Class IV
BOD	mg/L	112 – 130	50	12
COD	mg/L	320 – 416	80	100
TSS	mg/L	150 – 270	50	400
pH		6,5 – 7,5	6 - 9	5 – 9
Ammonia	mg/L	24.15 – 24.44	10	-
Nitrite	mg/L	0.63		-
Nitrate	mg/L	18 – 19		20
Phosphate	mg/L	2.4		5
Oil and Grease	mg/L	6.8 – 7.2	10	-

3.2. Influence of Retention Time on Removal Efficiency

Aerobic biofilter reactors operated continuously with the variation of retention time used as follows, 24 h, 18 h, 12 h, 8 h, and 4 h.

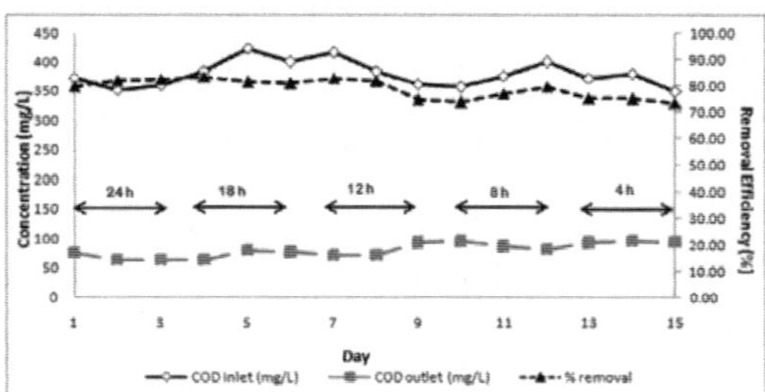

Figure 2. COD Removal at Different Time Retention

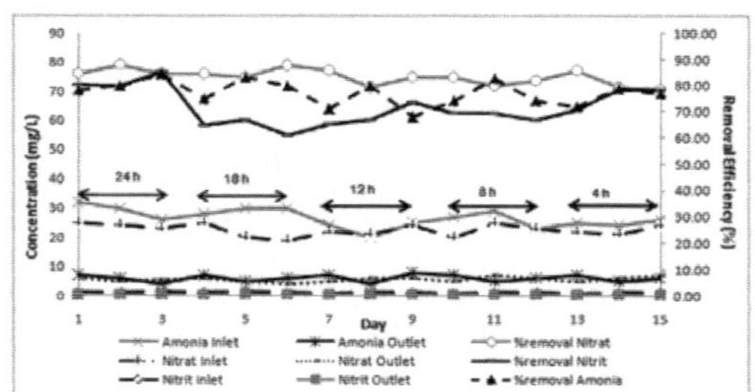

Figure 3. Ammonia, Nitrite, and Nitrate Removals at Different Retention Time

Based on the Figure 2, the largest removal efficiency of COD was obtained at retention time of 24 h with the percentage of 83.33%, then 18 h, 12 h, 8 h, and 4 h with the percentage of removal efficiency were 81.13%, 79.50%, 76.86%, and 73.14%, respectively. The removal efficiency of TSS with the variation of retention time ranged between 79.59% - 83.56% with the largest percentage of 83.56% at the retention time of 24 h. Based on the Figure 3, the largest removal efficiency of Ammonia, Nitrate, and Nitrite at the retention time of 24 h with the percentage 84.62%, 79.17%, and 76.92%, respectively. Concentrations of COD, TSS, and Ammonia had complied with the quality standard of DKI Jakarta in Governor Regulation No. 122 of 2005. Nitrate concentration had fulfilled the quality standard of National Government Regulation No. 82 of 2001.

3.3. Influence of C:N:P Ratios on Removal Efficiency

The variation in the C: N: P ratios used were 100: 5: 1, 100: 8: 1, 100: 10: 1, 100: 12: 1, 100: 15: 1 with optimum retention time 24 h. According to Fig. 4, the largest removal efficiency of COD was 83.33% obtained at C: N: P = 100: 10: 1 ratio, then 79.57%, 81.84%, 80.39 %, and 81.72% were obtained at C: N: P = 100: 15: 1, C: N: P = 100 : 12: 1, C: N: P = 100: 8: 1, and C: N: P = 100: 5: 1 ratios, respectively. The largest TSS removal efficiency was 83.56% obtained at a C: N: P = 100: 10: 1 ratio, then 80%, 82.01%, 82.55%, and 83.14% were obtained at C: N: P = 100: 15: 1, C: N: P = 100: 12: 1 , C: N: P = 100: 8: 1, and C: N: P = 100: 5: 1 ratios, respectively.

IOP Conf. Series: Earth and Environmental Science **106** (2017) 012048 doi:10.1088/1755-1315/106/1/012048

In Figure 5, the removal efficiency of BOD with variation in C: N: P ratios ranges from 75.23% - 82.5% with the largest percentage of 82.5% for C: N: P = 100: 10: 1 ratio. In Figure 6, the largest Ammonia, Nitrite, and Nitrate removal efficiency was obtained in the C: N: P = 100: 10: 1 ratio with the percentage of 82.5%, 79.1%, and 92%, respectively.

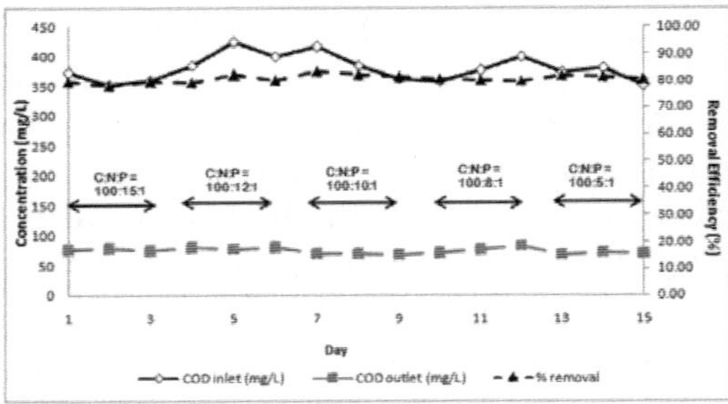

Figure 4. COD removal at different C:N:P ratios.

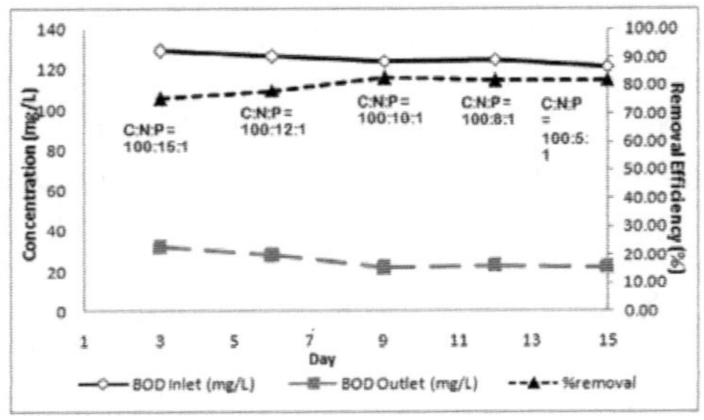

Figure 5. BOD Removal at Different C:N:P ratios.

The removal efficiency of Oil and Grease with variation in C: N: P ratios ranges from 71% - 84.8% with the largest percentage of 84.8% for C: N: P = 100: 10: 1 ratio. In Figure 7, the removal efficiency of Phosphate with variation of C: N: P ratios ranges from 64% to 70.83% with the largest percentage 70.83% for C: N: P = 100: 10: 1 ratio. The concentration of each parameter has fulfilled the wastewater quality standard of DKI Jakarta in Governor Regulation No. 122 of 2005 and National Government Regulation No. 82 of 2001.

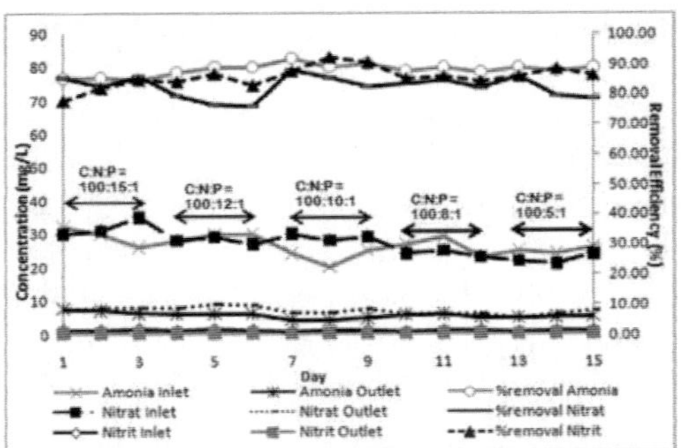

Figure 6. Ammonia, nitrite, and nitrate removals at different C:N:P ratios

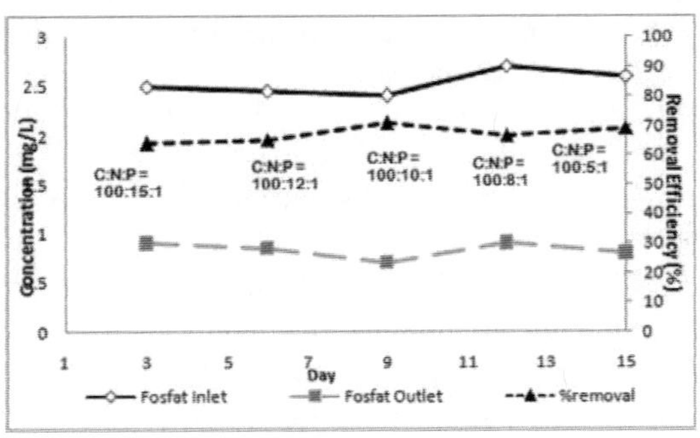

Figure 7. Phosphate removal at different C:N:P ratios

3.4. Reaction Kinetics in Removal of COD

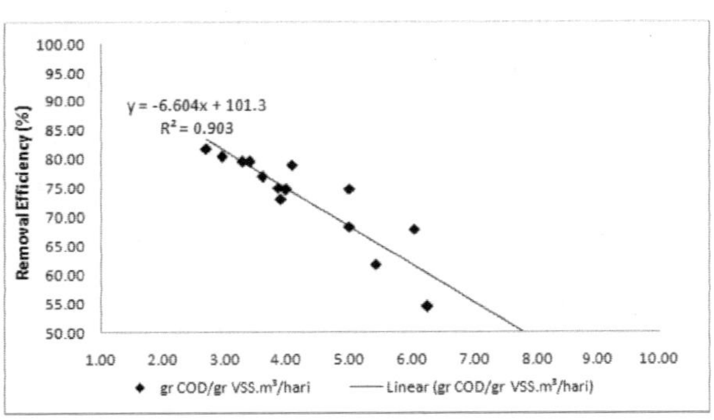

Figure 8. Reaction kinetics in removal of COD.

Figure 8 showed reaction kinetics in degradation of COD obtained linier equation was y = -6.604x + 101.3 with regression value $R^2 = 0.903$, and values of μ_{max}= 0.0767/h, Y=0.19 mg VSS/mg COD, Kd = 0.0006/h, and Ks = 119.14 mg/L.

4. Conclusion

The results showed that the optimum variance of retention time was 24 hours and the ratio of C:N:P was 100:10:1 yielded the largest removal efficiency for 83.33% of COD, 87.33% of BOD, 82.5% of Ammonia, 79.1% of Nitrate, 92% of Nitrite, 84.82% of Oil and Grease, and 70.83% of Phosphate. The concentration parameter resulted from outlet biofilter has met the domestic wastewater quality standard of DKI Jakarta. Reaction kinetics in degradation of COD obtained regression value of 0.903 close to 1 with value of μ_{max}= 0.0767/h, Y=0.19 mg VSS/mg COD, Kd = 0.0006/h, and Ks = 119.14 mg/L.

References

[1] Pramanik B, Fatihah S, Shahrom Z, and Ahmed E. 2012. Biological Aerated Filters (BAFs) for Carbon and Nitrogen Removal: A Review. *Journal of Engineering Science and Technology*. 7(4):428 – 446.

[2] Idaman S. Aplikasi Biofilter untuk Pengolahan Air Limbah Industri Kecil Tekstil (Jakarta: Badan Pengkajian dan Penerapan Teknologi) (in Indonesian version). 2002.

[3] Metcalf & Eddy. Wastewater Engineering Treatment and Reuse (New York: McGraw-Hill). 2003.

[4] Eckenfelder W. Industrial Water Polution Control. 3rd Edition. New York: Mc Graw Hill Series in Water Resources and Environmental Engineering. 2000

[5] El-Gendy A S, Sabry T I, and El-Gohary F A. 2012. The Use of An Aerobic Biological Filter for Improving Effluent Quality of A Two-Stage Anaerobic System. *International Water Technology Journal*. 2(4). Dec.

[6] Pitriani. 2015. The Effectiveness of EM_4 Addition into Biofilter to Reduce of BOD, COD and MPN Coliform of Hospital Wastewater. *IJPRIF*. 8(4):702-708

[7] Reynolds D. Unit Operation and Processes in Environmental Engineering. Boston: B/C Engineering Division. 1982

[8] Tchobanoglous G, Burton F L, and Stensel H D. Wastewater Engineering: Treatment and Reuse. New York: McGraw-Hill. 2003.

[9] Damodhar J, Mapara J V, and Mandar P. 2015. Domestic Waste Water Treatment by Bio-Filtration: A Case Study *International Journal of Science, Environment and Technology*. 4(1):140 – 145

[10] Wei X M, Lina C, Duan N, Penga Y X, Ye Z Y. 2010. Application of aerobic biological filter for treating swine farms wastewater. *Procedia Environmental Sciences*. 2:1569–1584

The essential aspect in sewerage regulation in Indonesia

L Siami*

Environmental Engineering Department, Faculty of Architecture Landscape and Environmental Technology, Universitas Trisakti, Jakarta, Indonesia

*Corresponding Author : lailatus.siami@trisakti.ac.id

Abstract. Several cities in Indonesia have the sewerage local regulation such as Banjarmasin, Bantul, Surakarta and Denpasar. Meanwhile, The National Government have guideline in composing domestic sewerage regulation. Each city have their own characteristic and issues that need to be carried out by the local regulation. By using SWOT analysis, this study tries to figure out several aspect that need to be included in the local regulation. International references from developed and developing countries like Japan, Phillipines, Malaysia and Thailand were also used as benchmark without neglecting the local conditions of cities in Indonesia. Several crucial aspect of local regulation are institutional authority, composition on-site and off-site system, tariff, evaluation and monitoring, as well as punishment and rewards. Both tariff and evaluation aspects need to be narrowed down into specific regulations.
Keywords: domestic sewerage, local regulation, oversead benchmark, SWOT

1. Introduction

Of all sewerage components, regulation is one of most fundamental aspects since all the implementations will rely on it. However, many studies discuss the technical components compared to the conventional methods, activated sludge, heavy metal treatments, as well as biological processes using algae [1]. Other studies consider social, economic, and environmental components in assessing sustainable technology for urban sewage sludge treatment [2]. There are also specific studies about the management of domestic sewage sludge disposal [3]. In Indonesia, the central authority of sewerage management is in the hand of Ministry of Public Works and Housing (MPWH). Under Directorate of Sanitation Development (PPLP), there is a guidance of composing local sewerage regulation [4]. The scheme can be seen in Figure 1.

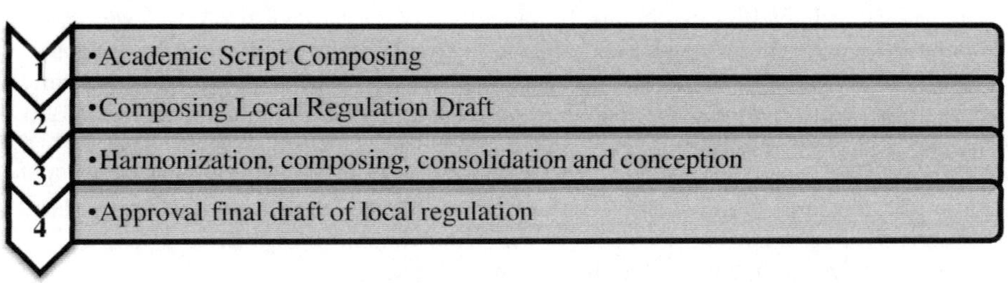

Figure 1. Indonesia sewerage regulation scheme [4].

IOP Conf. Series: Earth and Environmental Science **106** (2017) 012049 doi:10.1088/1755-1315/106/1/012049

Several cities that have a sewerage regulation are Banjarmasin City, Klungkung Regency, Tangerang City, and Bantul [5], [6], [7], [8]. Nevertheless, some of those regulation needs to be updated since the development of the city and regency will bring in different conditions which then alters the sewerage management, both technically and non-technically. In fact, some of these towns make an advancement by having their regulation of particular aspects such as tariff scheme, retribution [6], and regular desludging [7]. In contrary, some cities are still on the progress of composing local regulation concerning sewerage management just like Jakarta.

Devising a local law takes a lot of time indeed since the process of enactment has to go through many steps. Thus, it needs to be simplified. This study is intended to act as a simplification attempt of the necessary aspects of a local sewerage regulation.

2. Research Method

In this study, we try to define the actual condition including the issues and obstacles in each city by using best practices of local regulation in Indonesia. SWOT analysis is a method used to scrutinize the strategic planning. It takes strength, weakness, opportunity, and threat into account of a problem framework so that the best way to implement the strategy can be identified. The internal factors that will offer the potentials to tackle the issue are considered as strengths. The internal factors that will probably become additional problems are weaknesses. The external factors that can serve as subsidiary solutions are opportunities. The external factors that may aggravate the problem and weaknesses are threats. In here, we can find each point of SWOT analysis. The followings are several criteria that need to be considered in using this method:

1) Relevance to the local condition
2) The sewerage system
3) Public awareness

As a reference, Japan seems to be the most suitable developed country. Having relatively similar state with Indonesia, Malaysia, Thailand, and Philippines are set to be the benchmarks. The regulation aspects that are examined by using SWOT analysis and benchmarking are the institutional authority, composition (sharing) of the on-site and off-site system, tariff, evaluation and monitoring, as well as rewards and punishment.

3. Results and Discussion

As seen in Table 1 below, the comparisons of each aspect by considering several criteria mentioned before are listed for its strengths, weaknesses, opportunities, and threats.

Local government systems in many cities in Indonesia are no different except for the special regions such as Aceh and Yogyakarta. Those systems will affect the institutional authority, whereas a particular institution is required in the future for sewerage management, either in the form of public services or local enterprises e.g. the one in Denpasar (BLUPAL) and Jakarta (PD. PAL Jaya). For short term plan, water resources department will be modified, and smaller unit specially intended for wastewater management will be added. Compositions of the on-site (decentralized) and off-site (centralized) systems are often set aside in the local regulation, whereas it will determine the other aspects since many technologies can be used as an alternative treatment for the on-site system [9]. In contrary to business and tidy residential section, a dense area with a complicated geographic condition is more likely to be provided by an on-site system. A city or district with an economically diverse structure will apply a different plan compared to the economically homogenous one. The tariff scheme will need a specific regulation addressing the number or coefficient matter. The evaluation and monitoring aspect implicates the effluent criteria and one's expertise in standardization. This aspect also needs a particular regulation such as a domestic wastewater standard. Thus, the effluent quality will not exceed the requirements. Lastly, reward and punishment as an inseparable aspect of the law enforcement can be directly mentioned in the local regulation.

In general, a sewerage system can be either centralized or decentralized. Decentralized here can consist of either a single system or many [10]. The common type in Indonesia is the communal and ITP (Individual Treatment Plant) units. A sewerage system is usually devised gradually in the fifth year period of mid-term planning RPJMD. In many cities, due to the rapid development led by the population growth, the centralized wastewater systems exceed the limit of their capacities [11]. Hence, it will notably affect the composition of the on-site and off-site system. The remained criteria, the public awareness (willingness to pay), is less affected by the regulation aspects because the government can carry out a socialization or other approaches, even though it needs lots of efforts and takes time. This criterion has an impact on the on-site and off-site composition since a broader public involvement is required in a decentralized system [12].

Table 1. SWOT analysis for sewerage regulation aspect.

Citeria	Aspect	Stengths	Weaknesses	Opportunities	Threats
Relevant with local condition (social, economy and geographic)	Institution Authority	Clearly define the coverage area	Need government capacities improvement	Requirement one particular institution	• Irrelevant with other aspect • Need alteration in government bodies
	On-site and Off-site system Composition	Clearly define type of treatment	Need wide area of Water Treatment Plant (WTP) for Off-site/ decentralized system	Zero in on public necesstity and local characteristic	Require new technology treatment
	Tariff	Subsidize construction, operational and maintenance cost	Require law enforcement	Need other specific regulation	Low willingness to pay
	Evaluation and Monitoring	Improve effluent quality	Need government capacities improvement	Need other specific regulation on standard effluent	Low result and implementation of evaluation and monitoring
	Rewards and punishment	Well regulated	Willingness t o accept by public is low	Increase public awarness	Reluctant to do house connection
Sewerage system	Institution Authority	Define the coverage area	Lack of public participative	Transfer technology	Rely on one specific type of wastewater treatment
	On-site and Off-site system Composition	Zero in on technical condition	Low law enforcement for communal and ITP system	Need other specific regulation	• Dominancy of off-site system • Lack of coverage in on-site system for dense area
	Tariff	Influence to specific tarrif	Need public awarness increasing	Need other specific regulation in tariff scheme	• Tend to refuse by public • Become obstacle for sewerage construction
	Evaluation and Monitoring	Define the system of monitoring	low disciplinary of monitoring in communal system	Need monitoring standardization	Low control in ITP

Citeria	Aspect	Stengths	Weaknesses	Opportunities	Threats
	Rewards and Punishment	Increase public awarness	Low law enforcement	Additional funding for operational cost	Extra monitoring for ITP
Public awarness (Willingness to pay)	Institution Authority	Define the service system	Lowered sense of belonging	Combined with other retribution, e.g. water, electricity, or solid waste charge	Increase the public wariness
	On-site and Off-site system Composition	Increase sense of belonging, if on-site dominated	Lowered sense of belonging, if off-site dominated	People are willing to pay with specified tariff	People reluctant or even refuse to pay
	Tariff	Subsidize sytem from ITP to communal system	Lower revenue from communal system	Get higher tariff for ITP	There is no revenue communal system
	Evaluation and Monitoring	Increase public awarness	Does not make a deterrent effect	Increase the awarness to meet the standard	No effect to public
	Rewards and Punishment	Increase public awarness	Burdening public	Public feel included	No effect to public

Table 2 provides the benchmarking using the best practices in several Asian countries. Japan is one of leading countries in sewerage management in Asia starting from 2,200 years ago using a domestic wastewater treatment under Johkasou Law as the on-site system. In 1990's, Japan provided a technological transfer of field-constructed Johkasou for Indonesia. Manila has implemented an effective and efficient sewerage management by separating sewer and combining sewer-drainage systems. This application is feasible for Manila, considering its decent public capability and willingness to pay. The implemented strategy is by upgrading communal septic tanks (CST) into package-sewage treatment plants (STP). Another way is by using a lift station that delivers wastewater into larger, centralized sewage treatment plant. Meanwhile, the sewer and drainage systems are combined by constructing interceptors (collector pipes) in the drainage toward the STP.

Previously in 1950's, Malaysia emphasized more on primary treatments, i.e. latrine and septic tanks. Nowadays, IWK (Indah Water Consortium) is more focused on addressing diverse types and dimensions of sewerage to improve the operation and maintenance. The characteristics of sewerage system in Malaysia is that the regulator, as well as the operation and maintenance, are provided by one door namely National Water Services Commission (SPAN). The integration of water and wastewater services enables the law enforcement and eases the implementation of the regulation. It can be seen by the large increment of houses connected from 5% to 70% in 2010. Bangkok's sewerage master plan had been drafted using centralized wastewater treatment since 1968. It shifted afterward in 1990's into communal wastewater treatment plants. The government began to improve public awareness through information campaigns in educational programs as well as to expand the human resources capacity.

Table 2. Overseas best practice as benchmarking of sewerage management.

Country	Characte-ristic (coverage area)	Aspect				
		Institution Authority	**On-site and Off-site system Composition**	**Tariff**	**Evaluation and Monitoring**	**Rewards and Punishment**
Japan	65%	There is clear and firm authority distribution between national governemnet and city	Communal system refer to Jokashou law. There are several area that service by on-site system and dominated by off-site system	Firm tariff collection and calculation based on water consumption	• By management in registration private sector (plumber, engineer, factory) • Obligation in operator to do regular effluent monitoring.	Penalty ≤ 50,000 JPY for particular violation
Manila, Filippines	10%	Metropolitan Manila Developmen t Autorithy (MMDA)	Combined sewer for not doable area for seperate system	Progrssive Tariff with environmen-tal and sewerage charge	Follow domestic wastewater effluent standard	Fines of 10,000 – 200,000 Php for every day of violation;
Kuala lumpur, Malaysia	68,8%	On the hand of IWK	Shifting from Individual Septic Tank (IST) to Communal septic tank (CST)	Flat national tariff with regulated and non-regulated business service	Program Certification for facilities and human resources	Penalty called as "Sewerage Control funds"
Bangkok, Thailand	21%	Department of Drainage and Sewerage, Bangkok Metropolitan Administrati on (BMA)	Dominated by sewerage system (cluster WWTP)	Using 'polluter-pay-principle' and 'service-pay-principle'	• On progress program Certification for facilities and human resources • Enforcement of effluent stan dards	Implementation of wastewater discharge fee

4. Conclusion

This paper discusses the local sewerage regulation in general by emphasizing that implementing an appropriate wastewater management is fundamental. There are several crucial aspects of local regulation that need to be considered, particularly concerning the fact that each city in Indonesia has distinct characteristics and issues in implementing the wastewater management. Those aspects are the institutional authority, composition of the on-site and off-site system, tariff, evaluation and monitoring, also punishment and rewards. Both tariff and evaluation aspects need to be narrowed down into specific regulations.

References

[1] P Rajasulochana, V Preethy. 2016. Comparison on efficiency of various techniques in treatment of waste and sewage water – A comprehensive review. *Resouce Efficient Technologies.* 2:175-184.

[2] R Jingzheng, L Hanwei, C Felix T S, Urban sewage sludge, sustainability, and transition for Eco-City: Multi-criteria sustainability assessment of technologies based on best-worst method, p. 11, 2016.

[3] R Claudia Bruna, G Daniele, Sustainability of Domestic Sewage Sludge Disposal. Sustainability. [Internet]. 2014 [Cited 2017 July 17] : 6: 1 [about 11p]. Available from: MDPI.

[4] Panduan Penyusunan Rancangan Peraturan Peraturan Daerah Pengelolaan Air Limbah Domestik 2016. Direktorat PPLP Ditjen Cipta Karya Kementerian Pekerjaan Umum dna Perumahan Rakyat.

[5] Peraturan Daerah Kota Banjarmasin Nomor 5 Tahun 2014 Tentang Pengelolaan Air Limbah Domestik. 2014.

[6] Peraturan Daerah Kabupaten Klungkung Nomor 20 Tahun 2012 Tentang Retribusi Pengolahan Limbah Cair. 2012.

[7] Peraturan Daerah Kota Tangerang Nomor 10 Tahun 2000 Tentang Retribusi Penyedotan dan Penggunaan Fasilitas Pengeolahan Lumpur Tinja. 2000.

[8] Peraturan Daerah Kabupaten Bantul Nomor 15 Tahun 2010 Tentang Pengelolaan Air Limbah. 2010.

[9] J M Galbraith, C E Zipper, R B Reneau Jr. On-Site Sewage Treatment Alternatives. [Internet].[Cited 2017 February 16] : 7 [about 22p]. Available from: ext.vt.edu. 2015.

[10] L Giovanni, G Annamaria Volpi, P Francesco Avezzù G. 2012. To centralise or to decentralise: An overview of the most recent trends in wastewater treatment management. *Journal of Environmental Management*. 94:61-68.

[11] T George. The role of satellite and decentralized strategies in water resources Management. 2009. *Journal of Environmental Management*. 90:144-152.

[12] Decentralized approaches to wastewater treatment and management: Applicability in developing countries. *Journal of Environmental Management*. 90:652-659. 2009.

IOP Conf. Series: Earth and Environmental Science **106** (2017) 012050 doi:10.1088/1755-1315/106/1/012050

The applied technologies to access clean water for remote communities

I B Rabindra

Landscape Architecture Department, Faculty of Landscape Architecture and Environmental Technology, Universitas Trisakti, Jakarta, Indonesia

Corresponding Author : ib.rabindra@trisakti.ac.id

Abstract. A lot of research is done to overcome the remote communities to access clean water, yet very little is utilized and implemented by the community. Various reasons can probably be made for, which is the application of research results is assessed less practical. The aims of this paper is seeking a practical approach, how to establish criteria for the design can be easier applied, at the proper locations, the simple construction, effectively producing a volume and quality of clean water designation. The methods used in this paper is a technological model assessment of treatment/filtering clean water produced a variety of previous research, to establish a model of appropriate technology for remote communities. Various research results collected from the study of literature, while the identification of opportunities and threats to its application is done using a SWOT analysis. This article discussion is looking for alternative models of clean water filtration technology from the previous research results, to be selected as appropriate technology, easily applied and bring of many benefits to the remote communities. The conclusions resulting from the discussion in this paper, expected to be used as the basic criteria of design model of clean water filtration technologies that can be accepted and applied effectively by the remote communities.
Keywords: access, applied technology, clean water, less practical, remote communities

1. Introduction

Overcoming the limitations of public access to get clean water, has been very much a design model research of filtering dirty water into clean water (water filtration) which is offered for the people that need it (see: figure 01)[1].Among them there are some draft model from the previous study already applied and is functioning quite effectively so that the benefit of a particular community. On the other side, not a few are also the design of the model of the research results are applied but less successfully give optimum benefits or at least meet the expectations of its users sake then left not working. The results of the research-based chemicals, such as technology known as Colloidal Applications i.e. wrapping with deposition or blob substances (coagulant) in conventional filtration process, relative simple technology as commonly used by regional drinking water company (PDAM) in Indonesia [2]. In the current study did not mention how the composition of the substance deposition such as alum, activated carbon, chlorine/chlorine, lime and ***tohor*** sand needed for unit volume of dirty water that are processed, and how the resulting clean water volume in units of time[2].

Model research based on the results of the process of gravity through the three stages of Deposition Tank – Filtering Tank – Shelter Tank. On the deposition tank contains material Stone tubs/coral as the material required to be able to deposit the sludge and other heavy material; On the tub filled various filtering materials respectively Palm fiber, smooth sand, Palm fiber, smooth sand, coconut shell

Charcoal, gravel and stone/coral (source: Kemdikbud document/water filter by natural material) [3], [4]. The model is simple and environmentally friendly real relative easy applied, because its material and natural system, widely available and easy to obtain. Resistance on the uncertainty of processed water volume in comparison with the volume of the filter material is required, the speed of the process of filtration and system maintenance [4]. Model Sieve Rock, the results of other studies which have already successfully applied on a community in the village of Kerobokan – Bali, is the water treatment systems that use a single material i.e. the Padas are local Stone called *Jempeng*. With the basic material '*jempeng*' Balinese-sieve rock/stone who has special natural characteristics with the diameter of the pores of the standard filtering clean water, developed two variants of the model form of ' U ' and ' W ' shape of the model [5].Although water filtration applications are rated quite effective, it turns out that this model was developed in other regions is difficult outside the Bali Island [6].

This paper aims at finding the reason why such a very view water filtration research model applied by the people who need it, from the very large number of such alternative and variant of the model water filtration research results are offered? Identification of problems occurred will be relied upon to improve the draft clean water filtration model which will be easily applied by the people who need it. The study will include a search problem of occurrence of resistance of the application model in field, then formulate a solution through a more comprehensive approach was not limited to technical issues alone.

2. Research Method

Previous research on the system and the process of water treatment is generally done in special cases by the method which is also very special, for example, in a case of water pollution at the time of the flood where it needs to be done in emergency water treatment process by using a chemical process. This research was conducted to find instead a model system and a water treatment process can be applied in General in cases of difficulty of remote communities to get access to clean water. The characteristics of the subject of the observation is certainly more common in nature, which means that in many cases the onset of difficulty community get access to clean water. More specifically the research subject is the need of access to clean water for communities in remote areas or isolated, where the characteristic of their social, economic, and cultural into consideration major approach to solving the problem.

Data collection was done through literature review, good secondary data regarding the condition of the social, economic and cultural community, including a variety of background conditions of geographic, topographic, geological, hydrographic as well as its physical environment natural climatologic. Data on social, economic and cultural community desolate analyzed by using SWOT analysis, which identified the various strengths, weaknesses, opportunities and threats facing remote communities to be able to accept, apply, utilize and manage optimally clean water filtration technology model for the fulfilment of his needs would be clean water. Data concerning the various results of previous research and several cases of success and failure of the application model research in the community, the use of secondary data is also done with a review of the literature. Data on the alternative specifications model water filtration technologies tested its opportunity to applied on remote communities, through the analysis of S-O which identify opportunities based on the strengths – social, economic and cultural community. So instead, via the analysis of W-T identified a variety of barriers to acceptance, application, utilization and management model of water filtration technology in remote communities.

3. Results and Discussion

In general the results of previous research to divide the water source being processed into two levels: (first) The base material that is processed is salt water or brackish water, which through the process of with filtration system RO; or through the process of distillation with steam generator system; or through ion exchange process with heat exchange system, to produce standard quality drinking water. (second) The basic ingredients of water, fresh water is treated with acid/quality peat moss or turbid river water through a process of combination of coagulation-filtration produced water with clean water quality standards. The basic ingredients of water, smelling the soil contains iron and manganese, carbon filtration through a process of manganese to be produced water with clean water quality. As well as the basic ingredients of groundwater are ' hard ' is treated through system hard 'softener ' could resulting water quality clean water (see Figure 1).

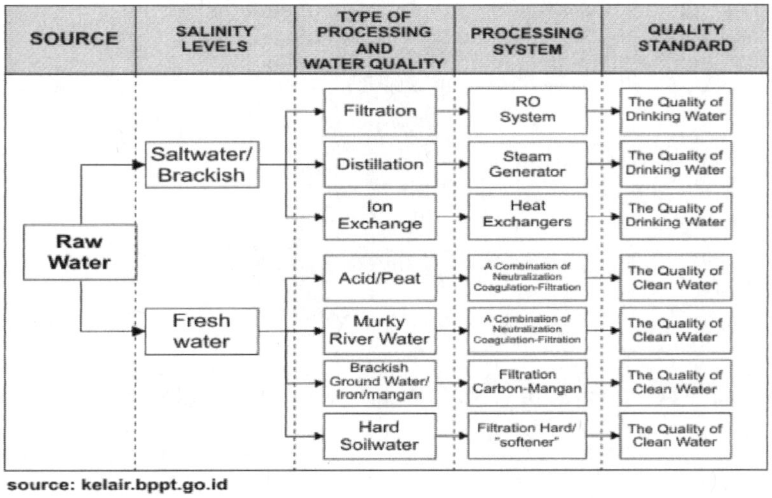

Figure 1. Water quality processing system.

Model the process of filtering dirty water into clean water or drinking water as the development of various research results has been also applied. Some of them can be run effectively as planned, but not a few who failed and subsequently abandoned, because it judged less in accordance with the various application conditions. System of processing through a chemical approach, generally less applicable regions, particularly remote areas with communities in social-economical and cultural importance relative to traditional. One example of a popular relative model was developed by the Ministry of education and culture i.e. gravitational system with multi-level filtering through the various layers of a natural materials filter, such as: Palm fiber - fine sand - Palm fiber - fine sand – coconut shell charcoal – stone – pebbles (Figure 2). The same is also done a system process called Gravity-fed Water Filtering System, which uses basic ingredients of rock and sand as filtering materials (Figure 3).

Jempeng 'U' model is usually placed on the areas of standing water, like a river or pond in conditions always submerged, so can always available fresh water in it. The advantages of Jempeng 'U' is a very simple construction, pipeline system are not required for dirty water or clean water, jempeng 'U' weakness is clean water surface were never higher than the surface of the water place soaked. Jempeng ' W ' models are usually built with mostly embedded in the land site of the House, it takes water pipe network system for dirty water inlet or clean water pipe outlet. In relatively more complex construction, but the volume capacity of clean water produced can be more.

Some models of research results in the form of technology devices in the form of processing units of clean water and drinking water has also been created, including by the environmental technology center (PTL) BPPT to overcome need clean water on a flooded area in Jakarta. Despite the proven processing unit can produce more than 300 cubic meters of clean water and decent drink in six days of operation (source: Humas BPPT), this unit is also not easy to implement a remote region. The cause is this unit has not been mass-produced so that the price is expensive, requires a trained operator, as well as other technical requirements such as the use of membrane ultra-filtration up to 0.0001 microns for Desalinisation plant reverse osmosis (RO).Other models even had enough detail, is already in the form of image work (Detail Engineering Design); Although the maximum capacity of processing units of about 100 cubic meters/day, but not the model dimension calculation formula outlined in relation to the desired capacity, so difficult to measure how large the model units can meet the needs of clean water a community. The model also does not indicate the use of specifically, besides the quality and feasibility of the use the clean water produced from each stage of the process (Figure 4).

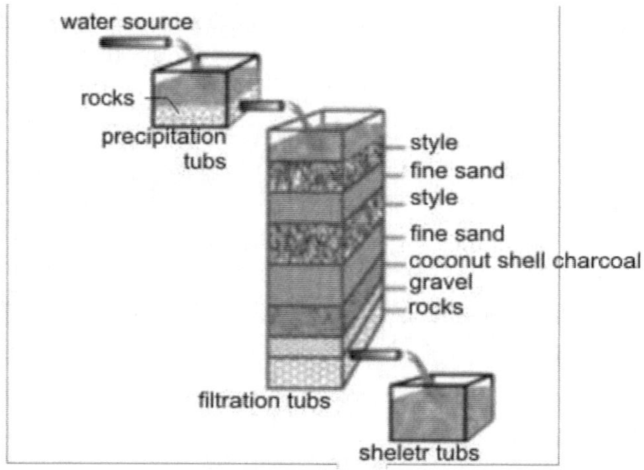

Figure 2. The principle of physics of materials in water treatment process.

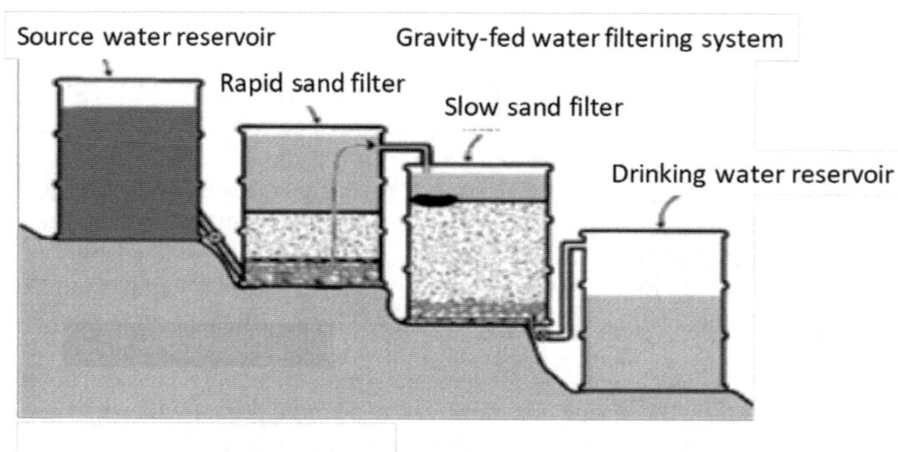

Figure 3. Gravity-fed water filtering system.

Remote communities are groups of people who lived in an area that is difficult to reach by any mode of transportation, even though difficult to achieve enough on foot. Community groups this can be in the number of population which is pretty much that shaped village, but often that is only made up of dozens of individuals who comprise some families only.Location and geographical condition of natural environment is generally at a remote region (far from the city or village) and relative topographic area of undulating, with soil conditions are generally fertile, most close to the flow of the River, but rarely on dry land far from rivers. Living in dry areas or away from the flow of water of course they desperately needed water for life, even living near river flow is sometimes difficult to get a decent water consumed.

Against the background of such an environment, social living conditions they are generally included in the category of pre-prosper. Their education level is generally very low, very little of which had education bench, although elementary school level though. However, they were included in a society that is able to survive in situations of hardship and deprivation, the community also has a very close kinship system also reflects.

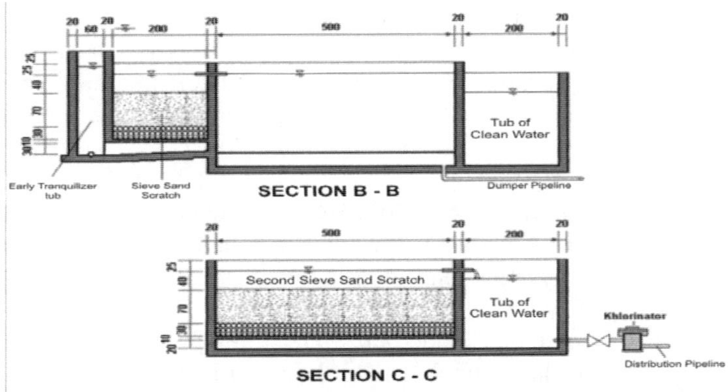

Figure 4. Sedative tank beginning initial sand filter model [3].

In a variety of limited knowledge and understanding of technology, remote communities would be very difficult to accept advanced technology as an alternative so that they gain access to clean water. The difficulty in understanding the system and work process models, of course, will have difficulty in making application, use or maintenance. Economic conditions that fall into the category of pre-prosperous, they will of course deny that require model construction materials that must be drawn away from their communities. Model that uses a material that must be imported from outside the territories of the community they surely will be rejected because it is very expensive and the value of economical becomes very low. With an abundance of natural materials such as stone, sand, Palm fiber, coconut shell charcoal and so on, of course they would prefer clean water filtration technology model that more use of natural materials. Nevertheless the design model construction need to be tailored to the capabilities of the technology that they mastered, as well as little as possible the use of materials that must be imported, such as cement, pipes network and so on. The cement can be replaced with a mixture of clay and limestone, while the pipes can be replaced with bamboo that generally widely available.

The greatest strength of the remote communities of the society is the physical fitness in the form of a strong power and endurance of body height (not easily tired) as well as staunchly moral because it is very usual the face of hardship. On the other hand they also reflects such a sense of community, is another strength where they very humble ready to help each other, as well as the form of cooperatives or other sub communities according to the wishes and expectations of them. Based on the identification of opportunities and barriers to the results of the SWOT analysis against remote communities such as the above characteristics, then the assessment of some alternative model water filtration technologies that are offered are as follows: (1) Model no chemical-based technology-based; (2) no Model-based advanced technology or use fuel/electrical power; (3) Construction does not use ingredients that are hard to come by in to location; (4) the draft construction easy to read, understood and applied; (5) the construction and materials as much as possible the use of materials that are widely available on location; (6) how to use, management and maintenance of the required no cost; (7) the use, management and maintenance of the construction based on nature of community and kinship in the community; (8) the determination of the construction site are submitted on a decision shared by the community.

4. Conclusion

Clean water filtration model acceptable, to apply, it is managed and maintained by the community, in accordance with the various identification of constraint that occur due to distance and travel time to, the degree of knowledge and technical skills in the community, the limitation of individual and collective economic society, is as follows: (1) Model should be built, operated and maintained using only local ingredients, not the product that was imported; (2) the technology used are not advanced technology and have to be appropriate, in order to be easily built, managed and maintained by a community self-help basis. Whereas in accordance with the identification of strengths and

IOP Conf. Series: Earth and Environmental Science **106** (2017) 012050 doi:10.1088/1755-1315/106/1/012050

opportunities owned by the community who have the ability and endurance of physical excellence and are also reflects, are: (3) models and technologies that can only be built, managed and maintained use together; (4) the placement of the location of the application model should be on the middle point of the mileage each Member of the community, in order to awake community colectifity.

References

[1] Departemen Permukiman dan Prasarana Wilayah, Departemen Kesehatan, Departemen Dalam Negeri, & BAPPENAS. Kebijakan Nasional Pembangunan Prasarana Dan Sarana Air Minum Dan Penyehatan Lingkungan Berbasis Lembaga. 2003.
[2] http://kelair.bppt.go.id.[4, 5 and 15 July 2017].
[3] http://www.google.com/ DokumenKemdikbud: water filter by natural material.[5 July 2017].
[4] http://www.sribd.com/. Makalah Pengolahan Air Bersih. [5 July 2017].
[5] http://www.google.com/.[4, and 5 July 2017].
[6] http://aimyaya.com/id, accessed on 4th July 2017.
[7] Kementerian Pekerjaan Umum Kajian Keterpaduan Pengembangan Air Baku, Air Bersih Dan Sanitasi, 2010.
[8] Peraturan Menteri Kesehatan Republik Indonesia Nomor 492/MENKES/PER/IV/2010 tentang Persyaratan Kualitas Air Minum. 2010.
[9] Vol. V, No. 07/I/P3DI/April/2013, Introduction to Water Treatment TL 4001, Rekayasa Lingkungan 2009, Program Studi Teknik Lingkungan ITB, 2009.
[10] Wikantiyoso, Respati, Tutuko Pindo, Local Wisdom: In Urban Planning and Design, to Realize Sustainable Urban Architecture, Universitas Merdeka, Malang, 2009.
[11] Sayuti S A, Towards a Cultural Aware Situation: Between 'the Other' and Local Wisdom, 2005.

Effect of aluminizing on hardenability of steel (S45C)

D Prayitno*, R Sugiarto

Mechanical Engineering, Faculty of Industrial Technology, Universitas Trisakti, Jakarta, Indonesia.

*Corresponding Author : dodyprayitno@trisakti.ac.id

Abstract. The objective of research is to know the effect of aluminizing on hardenability of steel (S45C). The research methodologies were as follows. The Steels (S45C) were machined into the Jominy test samples. Next the samples were preheating at 700 ^{o} C for 30 minutes and then the samples were dipped into the molten of aluminium for 3 minutes as a hot dip aluminizng method. The aluminium molten was 700 ^{o} C. Then the samples were cooled into room temperatures. Finally the samples were into the jominy tested. The results show that the aluminizing (include the preheating process) increases the hardenability of steel (S45C).
Keyword : alumunizing hardenability, jominy test, steel

1. Introduction

High-temperature corrosion of carbon steel in N2/0.1% H2S mixed gas at 600–800 °C for 50–100 h was studied after hot-dipping in the aluminum molten bath. Hot-dipping resulted in the formation of the Al topcoat and the Al-Fe alloy layer firmly adhered on the substrate. The Al-Fe alloy layer consisted primarily of a wide, tongue-like Al_5Fe_2 layer and narrow Al_3Fe layer. When corroded at 800 ^{o}C for 100 h, the Al topcoat partially oxidized to the protective but non-adherent -Al_2O_3 layer, and the interdiffusion converted the Al-Fe alloy layer to an ($Al_{13}Fe_4$, $AlFe_3$)-mixed layer [1].

The Hot Dip Alumunizing result intermetalic coating 0.02-0.10 mm for 1-15 minutes on the steel surface. The best time for immersing steel into the molten of aluminium is 3 menit at 1.023K (750 ^{0}C) [2]. In 1989 Daimler Benz AG, Stuttgart, introduced hot dip aluminized steel sheet for this application in passenger vehicles. By now, after 10 years' experience, most of this manufacturer's current models are equipped with fuel tanks made from aluminized steel sheet. This material incorporates a combination of a specially designed deep drawing If steel with the excellent corrosion protection properties of Al coatings. The two methods for joining the upper and lower halves of the tank body are resistance roller seam welding with an additional wire or laser beam welding. The first method was applied to the fuel tank bodies illustrated in this paper.In laboratory corrosion test series, aluminized steel sheet performed very well with regard to fuel resistance, including resistance to fuels containing methanol and rape-methyl-ester (bio diesel) [3].

The hot dip aluminizing using the molten of aluminium alloy (Al-53% Cu) at 800 ^{o}C while the dipping times are 3 minutes, does not work to prevent die soldering on the steel, H420 J2 [4]. Hot dip aluminizing succed to reduce the corrosin rate of welding area of LPG tube for 3 kg. The rate corrosion reduces from 0.2531 mmy to 0.0734 mmy [5]. The Jominy end quench test measures the effects of microstructure, such as grain size and alloying, on the hardenability of steels. The main alloying elements including Cr, Mn, Mo, Si and Ni; and Boron [6]. The aim of the reseach is to invastige the effect of aluminising on the hardenability of steel (S45C).

2. Research Method

The research flow chart is shown in Figure 1. The steel (S45C) were cut and machined into the jominy test sample. Next the samples were divided into 2 groups i.e non-aluminizing group and aluminizing group where the each group had 5 pieces samples. The non-aluminizing group were not aluminized. The non-aluminizing group were directly to the jominy test. While there were some process for the aluminizing group. The aluminizing group were firstly preheating at 700 °C for 30 minutes. Then the aluminizing group were immersed into the molten of aluminium (700 °C) for 3 minutes as the aluminizing process. Next, the aluminizing group were air cooling and following the Jominy test (heated at 900°C).

Next the samples were air cooled and then the hardness of samples were measured. Finally the data were analized to make conclusions

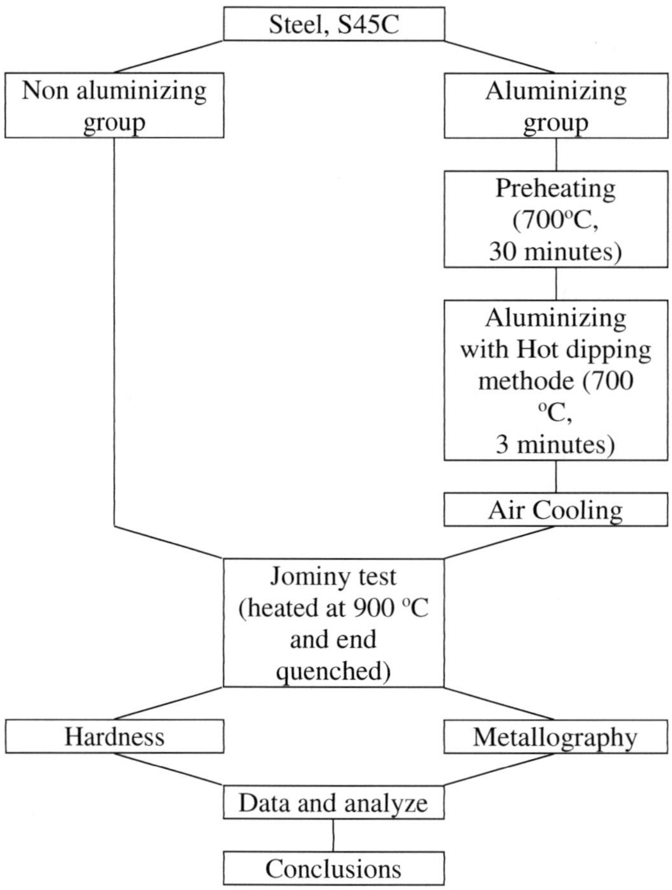

Figure 1. Research flow chart.

3. Results and Discussion

3.1. Sample performance before jominy testing.
The first group is the aluminizing group that the steel S45C was aluminized. The aluminizing group were firstly preheated at 700 °C for 30 minutes and then were hot dip aluminized at 700°C for 3 minutes. Finally the aluminizing group were jominy end quench tested.

The second group is the non aluminizing group or the intial of steel S45C. The non aluminizing group were not preheated and hot dip aluminized. The non aluminizing group of samples were directly jominy end quench tested. Figure 2 shows the samples of aluminizing group and the non aluminizing group before jominy testing. There are aluminium stick on surface of the aluminizing samples group but non on the non-aluminizing samples group. The aluminium stict on the surface afater aluminizing prosecess like a die soldering defect. Because it happen in the die casting where a mollten aluminum stick on the die.

(a) The aluminizing group

(b) The non aluminizing group

Figure 2. The group samples were before Jominy testing: (a) the aluminizing group (b) the non aluminizing group.

Based on the previous research the die soldering mechanism have 5 stage as follows [7] :
a) Erosion of phase boundaries of the steel surface;
b) Pitting of the steel surface;
c) Formation of iron-aluminiium compounds and formation of pyramid-shaped structure of intermetallic phases;
d) Adherence of aluminium on the pyramids of intermetallic phases ; and
e) Straighteniing of erosion pits and intermetallic phases

3.2. Hardness distribution of Jominy test bar
The first group is the aluminizing group and the second is the non aluminizing group. The both of group were jominy tested. Then the hardness of the two groups of samples were measured as a function of the distance from the quenched end to demonstrate the different hardenability of the two groups of sampel. The hardness test results are shown as rockwell hardness (Figure 3).

The Figure 3 shows that he average hardenss of the aluminzing group are higher than the non aluminizing group from J0 to J50. But the hardness of the both groups are relative similary from J50 to J70. The highest average hardness of the aluminizing group reduced sharply is 55 HRC at the end of quench or J0. The hardness reduces gradually from 55 HRC to 38.3 HRC at J50. From J50 to J70, the hardness reduce form 38.3 HRC to 35.8 HRC. So the lowest hardness of the aluminizing group is 35.8 HRC at J70.

The higher average hardness of the non aluminizing is 46 HRC at J0. The hardness of Non Aluminizing group reduce gradually from 46 HRC at J0 to 38.1 HRC at J50. The average hardness of non aluminizing groips reduce from 38.1 HRC (at J50) to 35.9 HRC at J70. Figure 3 shows measured hardness of jominy distance from the end quench (J0) to 70 mm (J70) for the aluminizing group and the non aluminizing group. The average hardness of the aluminizing group are higher than the non aluminizing group from J0 to J50. The hardness results show that the aluminizing process in this research improve the hardenability of steel S45C.

Figure 3 shows the hardenability result from the previous researchers [8], and [9]. The steel samples were heated by [8] at 870 °C for 30 minutes before Jominy testing. While the steel samples were annealed by [9] to get full pearlit. Comparing the hardenability of aluminized steel to to the previous reseach as Figure 2, shows that the hardenabillity of the aluminizing group is higher than the [7] and [8] results.

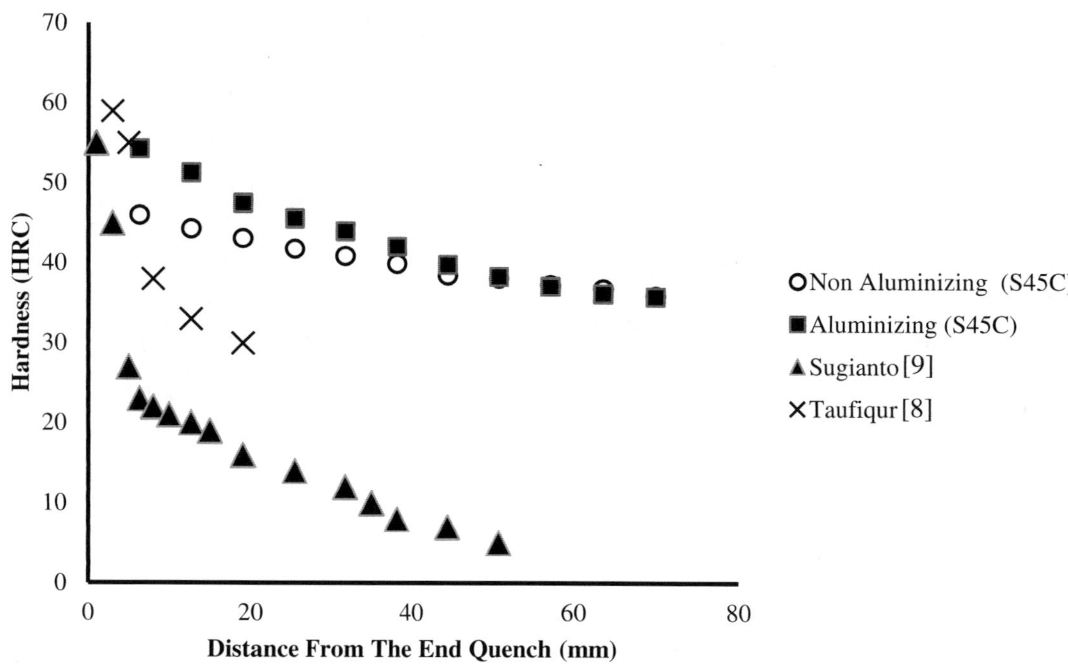

Figure 3. Jominy test result of the aluminized s45C compareing the previous results. [7], [8]

3.3. Microstructure

The microstructure at the end quench were observed using an optical microscope. The microstructures are shown in Figure 4. The observation on microstructure start from the end quench to 1.0 mm. (mm) microstructure of the non aluminizing and the aluminizing group was observed start from the end quench (J0) to 1.0 mm (J10).

(a) Non aluminizing sample (b) Aluminizing sample

Figure 4. Microstructures at J10 for the non aluminizing sample (a) and the aluminizing sample (b).

The grain size of the non aluminizing samples and aluminizing sample increase with increasing the distance from J0 to J10. The grain size of the non aluminizing samples are smaller than the aluminizing samples at the all observation location, i.e Figure 4. For the aluminizing group, the preheating process at 700°C for 30 minutes and the aluminizing process at 700°C for 3 minute gives much time to the grain to grow up. When the aluminizing group were heated (900°C for 30 minutes) before quench, the grain grow up more to become a bigger austenit grain. The nucleation of ferrite and pearlite take places at heterogeneous sites such as the austenite grain boundaries. Therefore, increasing the austenite grain size decreases the available nucleation sites, which pauses the rate of the ferrite/pearlite phase transformation. Thus increasing the austenite grain size increases the hardenability of steels.

4. Conclusion

The hardenability of steel S45C can increase by preheating at 700 °C for 30 minutes and following by the hot dip aluminizing 700 °C for 3 minutes.

References

[1] Muhammad Ali Abro and Dong Bok Lee, Effect of Al Hot-Dipping on High-Temperature Corrosion of Carbon Steel in N2/0.1% H2S Gas, Metals, MDPI. 2016.

[2] R, Rajendran, S, Venkataswamy, U, Jaikrishna , N, Gowrishankar, A, Rajadurai B.S.A., Effect Of Process Parameters In Hot Dip Aluminizing Of Medium Carbon Steel. Crescent Engineering College, Chennai 600048, India, 2IP Rings Ltd, D 11/ 12, Industrial Estate, Maraimalainagar 603 209, India, 3 Anna University, Chennai 600 044, India, Student Papers from FISITA. 2006.

[3] Warnecke W, Froeber J, and Bode R. 1999. Hot Dip Aluminized Steel Sheet - An Excellent Material for Fuel Tank Application. *SAE Technical Paper* 1999-01-0022. doi:10.4271/1999-01-0022.

[4] Fathony Arief Rhamdhany, Dody Prayitno. 2016. Studi Metode Almunizing Untuk Mencegah Die Soldering Pada Baja H420 J2, *Prosiding Seminar Nasional Cendekiawan 2016*, p. 8.1-6. http://www.trijurnal.lemlit.trisakti.ac.id/index.php/semnas/article/view/880. 2016.

[5] David Ray. 2015. Semnas Cendekiawan, Pengaruh Aluminizing (Al-Cu) Terhadap Laju Korosi Sambungan Las Busur Rendam Tabung Gas Elpiji 3 Kg, *Prosiding Seminar Nasional Cendekiawan.* p. 331-341. http://www.trijurnal.lemlit.trisakti.ac.id/index.Php/lemnas/article/view/154.

[6] R.W.K Honeycombe and H.K.D.H. Bhadeshia. Edward Arnold. Steels: Microstructure And Properties. 1995.

[7] A. Sugianto 1, M. Narazaki 2, M. Kogawara 2, and A. Shirayori 2. 2008. Prediction Of Hardness Distribution Of Water-Quenched S45c Steel Parts By Computer Simulation, *Proceedings of the 8th Asia-Pacific Conference on Materials Processing*. June 15-20. Guilin -Guangzhou, China.

[8] Taufiqur Rokhman. 2015. Perancangan Alat Uji Kemampukerasan Jominy Test, Jurnal Ilmiah Teknik Mesin, 3(1) Februari 2015. http://ejournal.unismabekasi.ac.id.

Urban farming model in South Jakarta

E Indrawati

Environmental Engineering, Faculty of Architecture Landscape and Environmental Technology, Universitas Trisakti, Jakarta, Indonesia

Corresponding Author: ettyindra@trisakti.ac.id

Abstract. The development of infrastructure rapidly, large of population and large of urbanization. Meanwhile, agricultural land is decreasing and agricultural production continues to decline. The productive crops is needed for consumption and it is also to improve the environment from oxygen provisioning, antidote to air pollution and to improve soil conditions. The use of yard land for horticultural crops (vegetables, fruits and ornamental plants), spices, medicines, herbs etc. can benefit for the owners of the yard particularly and the general public. The purpose of this research is to identify the model of home yard utilization, mosque yard, office, school, urban park and main road and sub main road, which can improve environmental quality in Pesanggrahan district. The method of analysis used descriptive analysis method by observation. Then analyzed the percentage of the use of yard with productive crops as urban farming. The results showed that the most productive crops were planted in Kelurahan Pesanggrahan 67% which compared with in Kelurahan Ulujami 47%, and in Kelurahan Petukangan Utara 27%. The most types of productive crops were grown as fruit trees and vegetable crops.
Keyword: horticultural, produktive crops, urban farming, yards

1. Introduction

Very rapid infrastructure development, large population and high levels of urbanization. Meanwhile, agricultural land is decreasing and agricultural production continues to decline. The need for productive crops other than for consumption is also to improve the environment from oxygen provisioning, antidote to air pollution, improve soil conditions. The use of yard land for horticultural crops (vegetables, fruits and ornamental plants), spices, medicines, herbs etc. can benefit the owners of the yard in particular, as well as the benefits to the general public.

Reduced agricultural land converted into settlements and industrial land, has become a threat and challenge for Indonesian to become an independent nation in the field of food. The food security situation in our country is still weak. This is indicated by: (a) the number of food insecure population (consumption level <90% of recommendations 2,000 kcal / kap / hari) and very food insecurity (consumption level <70% of recommendations) is still quite large, 36.85 million and 15.48 million people for the year 2002; (B) malnourished children under five are still quite large, ie 5.02 million and 5.12 million people for 2002 and 2003 [1].

BPS population data shows that the percentage of urban population. In 1990 reached 30.9%, in 2010 reached 49.8%, and this year has reached 56%. Indonesian Economic Vision 2025 estimated at 65% of Indonesian population are living in the city. Acceleration of urbanization and city development in Jakarta will certainly cause various problems. Significant increases in urban population without support and offset by the amount of food supply, employment, housing, facilities and infrastructure, law enforcement officials, etc. are issues that must be handled properly.

State Regulation No. 26 of 1997 on spatial planning, the increase of agricultural land conversion due to urban development is very worrying and a threat to agricultural business. Urban development can not be prevented, so it is necessary to think that agricultural activities can be integrated into other sectors within the urban development plan. Integration of agricultural development in urban development can create better urban ecosystems and can play a role in urban poverty reduction.

Urban farming can be one solution because it not only makes empty land useful but also provides cheap and flexible solutions for people who are financially difficult [2].

Sustainable agricultural development seeks to achieve sustainable agricultural production, rural economic sustainability and long-term environmental sustainability. Based on the definition of sustainability, the development of sustainable urban agriculture is the development of urban agriculture that integrates the economic, social and environmental aspects in an integrated manner in order to achieve economic sustainability, social sustainability and environmental sustainability within the framework of urban development [3].

The purpose of research is to identify the model to use of home yards, mosque yards, office yards, school yards, community park, main road yards and side road yards, three sub district in Pesanggrahan district.

2. Research Method

This research was conducted in Pesanggrahan sub district of South Jakarta, with three samples of Pesanggrahan, Ulujami and North Petukangan sub-districts. Primary data was obtained by field survey on home yard, school yard, mosque, office, green space and roadside green open space, to see the type of plants planted in each yard. The method of analysis used is descriptive analysis method by observation. The data were analyzed whether the planted plants were productive crops or non-productive crops. Productive crops planted in the form of fruit trees, vegetable crops, food crops and medicinal plants. Planting of productive crops planted in the yard, indicating urban farming (urban farming). Secondary data obtained from the searching of Pesanggrahan distric.

3. Results and Discussion

The development of the city is closely related to infrastructure development, the existence of toll road construction passing Pesanggrahan district greatly affect the land use, so that land conversion occurs. In general, land conversion is divided into four zones [4]:

a. The downtown zone, the zone in a predominantly urban center of urban buildings and infrastructure, such as government buildings, squares, department stores, trade and services. In this zone, the conversion rate of agricultural land to non-agricultural is very high (> 90%) is characterized by agricultural activities dominated by landscaping and ornamental farming trade for gardening. In this zone can be developed a market center that sells agricultural products.
b. City frame zone, This zone is directly adjacent to the city center. Most of these zones are residential areas where conversion rate of agricultural land to non-agricultural land reaches > 75%, 25% remains can be used agricultural activities. It is a land that is not ready for housing. Some of this land is still used agricultural land with high value agricultural commodities, especially horticultural crops (vegetables, fruits and ornamental plants).
c. The city-village zone. This zone is a mixed area between urban and rural situations. Urban situation is seen in the growth of modern residential buildings mixed with traditional relics of the house with yard and surrounding rice fields. The conversion rate of agricultural land in this zone ranges from 25-75%.
d. Village zone. This zone is an area that describes the countryside with level of land use for farming > 75%. Agricultural activities dominate the business community, with diverse agricultural commodity business. This area is relatively close to the city, so the development of commodities cultivated is strongly influenced by demand of the city.

IOP Conf. Series: Earth and Environmental Science **106** (2017) 012052 doi:10.1088/1755-1315/106/1/012052

Figure 1. Pesanggrahan district with kelurahan Pesanggrahan, Ulujami and Petukangan Utara.

The existence construction of Outer Ring Road infrastructure will reduce agricultural land, green open space, yard, settlement and other open spaces. On the other hand new housing develops newcomers in different customs with indigenous people in managing the yard. The existence of the construction of the Outer Ring Road infrastructure will reduce agricultural land, green open space, yard, settlement and other open spaces. On the other hand new housing develops with newcomers and customs are different from the indigenous people in managing the yards. In the big yards of the population usually plant various types of fruit trees. South Jakarta is one of the largest fruit producers in Jakarta. With the outer ring road, fruit garden and large yards turned into housing. New housing with different characters utilizes pages with productive and non productive plants.

The yard as a plot of land that has a certain boundary, on which there are residential buildings and have a functional relationship both economic, biophysical and sociocultural with its inhabitants [6].

The yard as a land use which is an additional small-scale food production system for and by household members and is a layered canopy ecosystem. The yard is a neighborhood around the house that can be cultivated for agricultural commodities. Planting productive crops such as horticultural crops (fruits, vegetables and ornamental plants), spices, medicines, spices etc. can provide benefits that can meet physical and spiritual satisfaction [5].

Based on observation in Pesanggrahan district, the productive crops planted in Pesanggrahan sub district are 67% at most, compared with 47% in Ulujami sub district and 27% in Petukangan Utara sub district.

Figure 2 shows that Pesanggrahan sub district has the most yard planted with productive crops such as medium house yard, community park and roadside yards (100%). In Ulujami sub district of big yards house and community park is planted with productive plants 100%, while in Petukangan Utara sub district only community park in planted with productive plants (100%). In general, the yard of small houses, yard of mosques, school yard, office yard planted trees and decorative plants only a few plants grow productive plants.

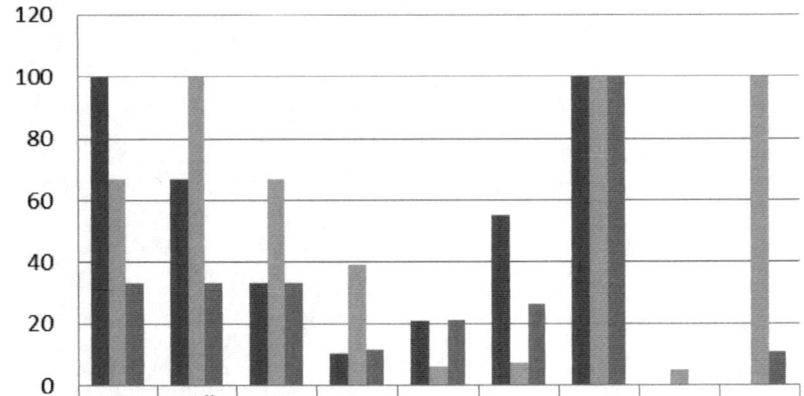

	Big yards house	Medium yards house	Little yards house	Mosque yards	School yards	Office yards	Community park	Main Road Yards	Side Road Yards
■ Ulujami sub district (%)	100	66.7	33.3	10.5	21	55	100	0	0
■ Pesanggrahan sub district(%)	66.7	100	66.7	39	6.2	7.5	100	5	100
■ Petukangan Utara sub district (%)	33.3	33.3	33.3	11.7	21.5	26.5	100	0	10.7

Figure 2. Percentage of Urban Farming in Pesanggrahan district

Community parks in three sub district are planted with productive crops. Roadside yards in Pesanggrahan sub district is also planted with productive crops. Community parks, main road yards and roadsite yards, are common areas, indicating that people use land for primary human needs. The main roads yards in the three sub district are planted with shade tree, such as *Polyanthian longifolia* and *Pterocarpus indicus* (angsana), but there are *Mangifera indica* (mango) and *Artocarpus heterophyllus* (jackfruit) in Pesanggrahan sub district. The yard is quite prospective in supporting people's lives on an ongoing basis. The sustainable benefits of the yard system can be obtained from several aspects [9]:

1) Maintaining and improving crop yields sustainably
2) Supplying energy derived from local resources, especially firewood
3) Produce a wide range of ingredients that can be utilized to meet day-to-day or market-sold needs, including wood, vegetables, toga, fruits and others
4) Protection and simultaneously improve the quality of the environment, especially land, water, flora and fauna
5) Improving socio-economic condition of farmers in accordance

Table 1 shows the types of plants grown in three sub district dominated by fruit plants such as *Musa paradisiaca* (Pisang), *Carica papaya* (Papaya) and *Mangifera indica* (Mangga). Generally fruit trees grown on a large land that is on the big yard of a house and community park. In the small yard and medium yards house planted many vegetables and herbs. Like *Brassica rapa* (mustard), *Solanum lycopersicum* (tomato). *Capsicum annuum* (chili), *Cymbopogon citratus* (lemongrass), *Zingiber officinale* (Ginger) etc. Fruit trees are grown also in accordance with existing land. Variety of business yard and the number of perennials show enough diversity of business which of course also require various treatment from farmer. There is a tendency that the more household-dominated land the more trees and types of timber trees are cultivated [8].

IOP Conf. Series: Earth and Environmental Science **106** (2017) 012052 doi:10.1088/1755-1315/106/1/012052

Table 1. Urban farming model in Pesanggrahan district.

Yard	Ulujami sub district	Pesanggrahan sub distric	Petukangan Utara sub district
Big yards house	*Musa paradisiaca, Carica papaya, Mangifera indica, Manihot utilisima, Brassica juncea, Curcuma longa*	*Musa paradisiaca, Carica papaya, Nephelium lappaceum*, ornamental plants, *Capsicum annum, Casuarina*	*Bambusa sp, Musa paradisiaca, Tectona grandisi,Mangifera indica, Oleana sysygium*
Medium yards house	*Musa paradisiaca, Carica papaya, Nephelium lappaceum*, ornamental plants	*Gnetum gnemon, Syzygium aqueum, Musa paradisiaca, Ipomoea reptana, Brassica juncea, Solanum lycopersicum, Alpinia galanga, Cybopogon citratus, Zingiber officinale*	*Bambusa sp, Nephelium lappaceum, Diffenbachia sp*, Ornamental plans, *Dypsis lutescens, Piper betle*
Little yards house	Ornamental plants, *Momusops elengi, Mangifera indica, Passiflora edulis, Syzygium aqueum*	*Bambusa sp, Piper betle,Manihot utilisima,Mangifera indica, Oleana sysygium, Palmae, Polyanthia longifolia*	*Carica papaya,Manihot utilisima, Psidium guajava, Plumeria, Oleana sysygium, Palmae*
Mosque yards	Ornamental plans, *Cocos nucifera, Palmae*	Ornamental plans, *Tectona grandis, Mangifera indica*	*Oleana sysygium,Palmae, Roystone regia, Plumeria, Piper betle, Mangifera indica*
School yards	Ornamental plans, *Psidium guajava, Mangifera indica, Cocos nucifera, Star Fruit*	*Mangifera indica, Polyanthia longofolia, Bambussa sp, Plumeria, Oleana sysygium, Cordiline fruticosa*	*Mangifera indica, Oleana sysygium, Plumeria, Lavender, Tamarindus undica, Panax, Cymbopogon citratus*
Office yards	Ornamental plans, *Palmae, Mangifera indica, Capsicum annuum, Psidium guajava*	*Mangifera indica, Oleana sysygium, Helianthus annuus, Schefflera actinophylla, Capsicum annuum, Ficus benyamina*	*Oleana sysygium, Ptichosperma macarthurii, Plumeria, Syzygium aqueum, Polyscias scutellaria*
Community park	*Musa paradisiaca, Nephelium lapaceum, Carica papaya, Sandoricum koetjape*	Ornamental plans, *Psidium guajava, Mangifera indica*	*Musa paradisiaca, Areca catechu, Carica papaya, Erythrina variegata, Tectona grandis, Mangifera indica*
Main road yards	*Pterocarpus indicus, bambusa sp, Palmae*	*Mangifera indica, Artocarpus integra,Polyanthia longifolia*	*Polyanthia longifolia, Palmae*
Side road yards	*Samanea saman*	*Mangifera indica, Cocos nucifera, Musa paradisiaca, Averrhoa*	Ornamental plants, *Aloe vera,dracaena, Samanea samani, Prunus avium*

IOP Conf. Series: Earth and Environmental Science **106** (2017) 012052 doi:10.1088/1755-1315/106/1/012052

4. Conclusion

The development of the Toll Outer Ring Road and the community's need for agricultural products can make changes to the planting model in south Jakarta, especially in Pesanggrahan district. Productive crops planted in Pesanggrahan sub district are 67% at most, compared with 47% in Ulujami sub distric and 27% in Petukangan Utara sub district.

The big yard house and the community park are generally planted productive plants in the form of fruit trees. The yard of the small house, the yard of the mosque, the yard of the school, the yard of the office is planted by a protective tree and the ornamental plant produces only a few productive crops. Green open space in Pesanggrahan sub district, has applied the model of urban farming by planting fruit trees and productive plants.

References

[1] Cahaya L Darmawan. 2014. Kajian Peran Pertanian Perkotaan dalam Pembangunan Perkotaan Berkelanjutan (Studi Kasus: Pertanian Tanaman Obat Keluarga di Kelurahan Slipi, Jakarta Barat). *Forum Ilmiah.* 11(3).

[2] Juli M Slabinki. 2013. From Waste land To Oasis: How Pennsylvania can Appropriate Vacant Urban Land Into Functional Space Via Urban Farming. *Widener Law Journal*. 22:253-287,

[3] Li M. 2009. The choice of Sustainable development model of China's. *Agriculture Asian Social Science*. 5:91-93,

[4] Rachmat M, 2011, Potensi Lahan Pertanian Perkotaan dalam Penyediaan Pangan. Dalam Buku MembangunPengelolaan Kemampuan Lahan Pertanian Pangan Keberlanjutan. Badan Penelitian dan Pengembangan Pertanian Kementerian Pertanian. IPB Press

[5] Adiyoga, 2003. Prospek Pengembangan Pertanian Urban (perkotaan). Disampaikan pada Diseminasi Prospek Penembangan Sayuran di Perkotaan, Balai Penelitian Tanaman Sayuran, Lembang, Bandung, 11-13 Agustus 2003.

[6] Hubert De Bon, et al, 2010, Sustainable Urban Agriculture in Developing Countries", A Review, Agronomy for Sustainable Development, 30:21-32

[7] Novitasari E, 2011, Studi Budidaya Tanaman Pangan Di Pekarangan Sebagai Sumber Ketahanan Pangan Keluarga (Studi Kasus di desa Ampel Gading Kecamatan Tirtoyudo Kabupaten Malang). Skripsi. Universitas Brawijaya Malang.

[8] Penny D H, and M Ginting, Pekarangan Petani dan Kemiskinan. Gajah Mada University Press. Yayasan Agro Ekonomika, 1984.

[9] Sutanto R, 2002, Penerapan Pertanian Organik Pemasyarakatan dan Pengembangannya. Kanisius. Yogyakarta, 2002.

[10] FAO, Urban and Peri Urban Agriculture, Report to The FAO Committee on Agriculture (Coag), Meeting from January 25-26, FAO, Rome, 1999.

IOP Conf. Series: Earth and Environmental Science **106** (2017) 012053 doi:10.1088/1755-1315/106/1/012053

Ground water potential assessment of Jatijajar-Ayah-Karangbolong mountain area, South Gombong

U Sumotarto

Geology Engineering Department, Faculty of Earth Technology and Energy, Universitas Trisakti, Jakarta, Indonesia

*Corresponding Author : untung.s@trisakti.ac.id

Abstract. South Gombong is an area of central Java having quite potential natural resources including ground water. A geohydrological survey has been conducted to explore and analyze ground water potential of South Gombong mountain area. Geological knowledge of the area indicates the presence of Karst topography generated from limestone rocks exposed to the earth surface and tropical climatological process. The ground water resource and reserve that formed particularly in the Karst area has been utilized for various purposes, such as for agricultural land, use for drinking water, and others. The survey, as well as data and information analysis, have been performed, outputting figures of how significant the ground water resources in the area are. It is concluded that detailed study and understanding about Karst and its ground water formation are important to develop and to preserve the existence of ground water in the area for living.
Keywords: ground water, karst topography, resource and reserve

1. Introduction

A geological survey has been conducted to collect geohydrological information around Jatijajar, Buayan, Ayah, and Karangbolong areas, located in the hills and mountains in South Gombong. As part of the task, the survey team also conducted geological observations and research in the area.

In general, the communities in South Gombong have diverse livelihoods and development of the area. Among others are agricultural activities (planting rice, vegetables, corn, cassava, etc.), as well as mining igneous, volcanic rocks, and sediment (limestone) as building materials for tourism ventures (Jatijajar Cave, Ayah Beach), etc. These activities require water resources for various purposes, such as for rice cultivation, vegetables, drinking water sources, etc. It becomes a problem when the water source is disrupted due to the specific geological conditions of the soil and rocks around the area.

Geologically, the area of South Gombong is a hill formed by volcanic rocks (breccias, igneous rocks) and sedimentary rocks (limestones, sandstones, napal, etc.) that form geological conditions with distinct groundwater potentials, especially in the areas where stratigraphically limestone sedimentary rocks are located on the surface of the earth. This specific geological condition of the area is interesting to be studied as a mean to exploit and develop the potential of groundwater resources (geohydrology), particularly to acquire, develop efforts, and conserve groundwater sources for various activities requiring water as the basic need of life. Groundwater in this area has been utilized for numerous purposes, such as water sources for rice paddy fields, vegetables, drinking water, small industries, etc.

By understanding the geological condition, especially geohydrological properties, it is expected to be able to maintain, develop, and preserve the groundwater sources in this area for various needs.

Geohydrological research requires to be developed and deepened as a significant consideration in the discovery, mining, and utilization of groundwater sources for the people in this area.

2. Research Method

2.1. Regional Geology

The geological map areas are located in South Gombong (Kebumen District) covering five main areas, i.e.surrounding Ijo Tunnel, Jatijajar Cave, Ayah Beach, Karangbolong, and Buayan District. Figure 1 depicts the region's topographic appearance from an aerial photograph.

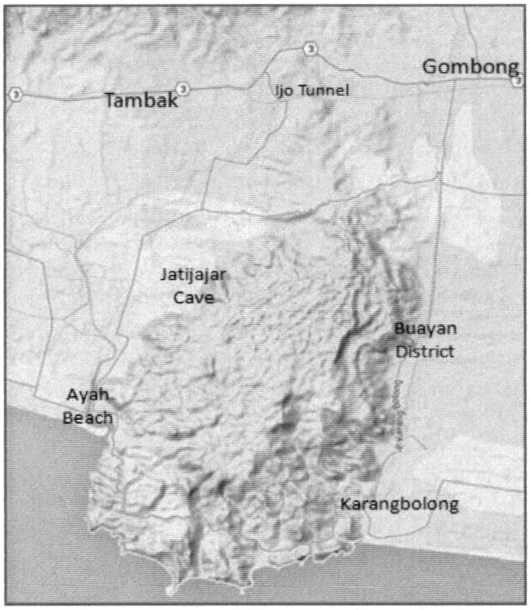

Figure 1. Geological map of South Gombong area, Kebumen District.

Volcanic activities have affected the Gabon Formation, resulting in the andesite rock intrusions. Above the Gabon formation, sedimentary rock deposits were formed around Central Miocene and are grouped into the Kalipucang Formation. Lithologically, the rocks constituting this structure include coral (non-clastic) limestone, local clastic limestone, and formed bitumen shaleat the bottom. When this lithology was exposed above the sea level to turn into the mainland,it developed karst topography.

In the final Miocene, sediments were deposited and formed coalescence of sandstone, claystone, marl, and tuffs with breccia inserts. The deposition process is influenced by turbidity currents and underwater seepage. These rocks are grouped in the Halang Formation, stretching along the northern part of this area. Unconformably, in the Quaternary period until now, alluvial deposits are reserved, composed by loose deposits of clay, silt, sand, gravel, and crust with various compositions of older rock formations.

2.2. Geology and Geohydrology of South Gombong Area

2.2.1. Local Geology

Observations, investigations, and comprehending aspects of geohydrology of the area were also conducted along with the geological surveys. The agricultural activities of the community around South Gombong and the exploitation of drinking water by a Drink Water Company (PAM, Perusahaan Air Minum) are appealing to be developed further. Therefore, an advanced research is highly feasible and recommended to maintain the supply of water resources for agricultural activities, the provision of drinking water, and other activities such as for small industries (livestock, building material production, cave tours, etc.).

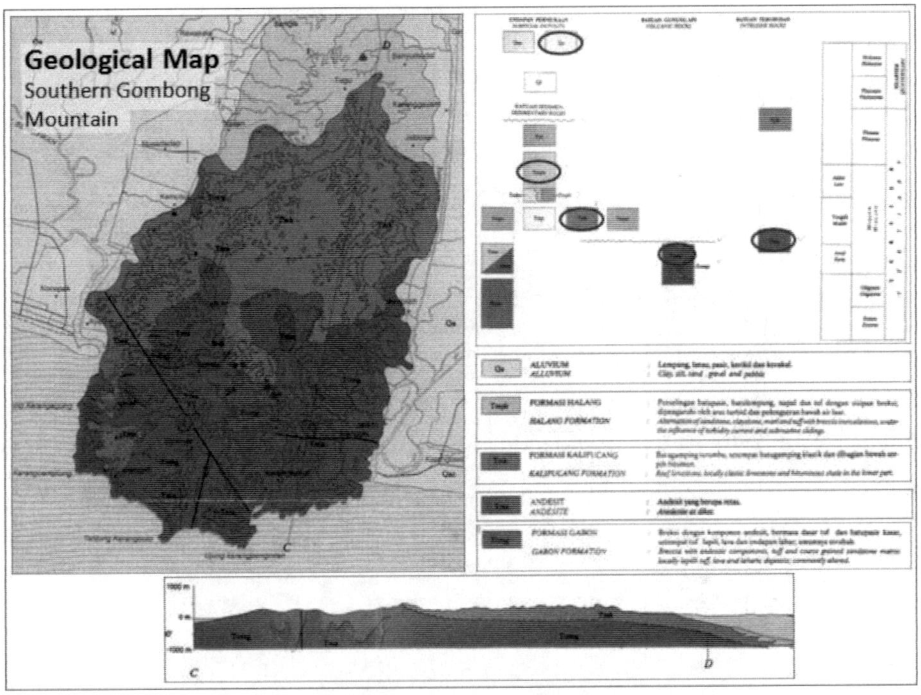

Figure 2. Regional geological maps and stratigraphy as well as geological sections of South Gombong area [1].

The observation of geohydrological aspects of this area is carried out by considering the appearances of morphology, rock composition, and stratigraphic learning to understand the genesis and the formation process of rocks found in the area. Along with these geological mapping activities, observations of issues related to the use of water resources were also covered, such as problems concerning irrigation of rice fields and vegetable plantations, provision of drinking water, livestock, and mining of construction materials (igneous rock blocks, limestone, sand, etc.).

Based on the observations of comprehending the community activities, it is clear that the need for water depends on the presence of the springs, especially in the hills. Rice fields and vegetable crops in the rather high areas use the water source coming from the springs of the surrounding. Figure 2 shows a geological map on the plateau around Buayan-Karangbolong-Ayah region. The agricultural area on the plains is quite high up, but it can still grow because the source of irrigation comes from the springs in the vicinity. Appendix of this paper shows photographs around the plateau with springs as the source of agricultural water, drinking water, etc. throughout the area of Buayan, Karangbolong to Ayah.

2.2.2. South Gombong Geology and Ground Water Potential

Based on the observations and early geological research around South Gombong area, an overview of geological state can be derived to interpret the subsurface conditions. Figure 4 shows a model of the geological map of the "X" area in South Gombong that is depicted based on data and information from the field. Figures 3 and 4 show the initial geological analysis gained from field observations, illustrated by the interpretation results. Real and accurate conditions can be obtained by drilling, for example, to collect data and information related to subsurface conditions, in particular, to know more specific situations about the geohydrology reservoir in this area.

The process of forming other sediments continued until the end of the Miocene, subsequently underwent geological tectonic processes such as lifting, folding, fracturing, and various structural manifestation. The raised rocks then suffered weathering, erosion, transportation, and sedimentation which until now form alluvial, eluvial, and colluvial deposits.

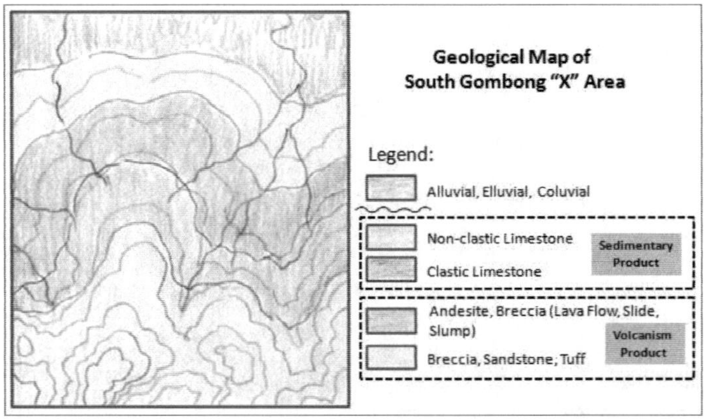

Figure 3. The geological map of South Gombong "X" area as the model of the study area, drawn based on field data and information.

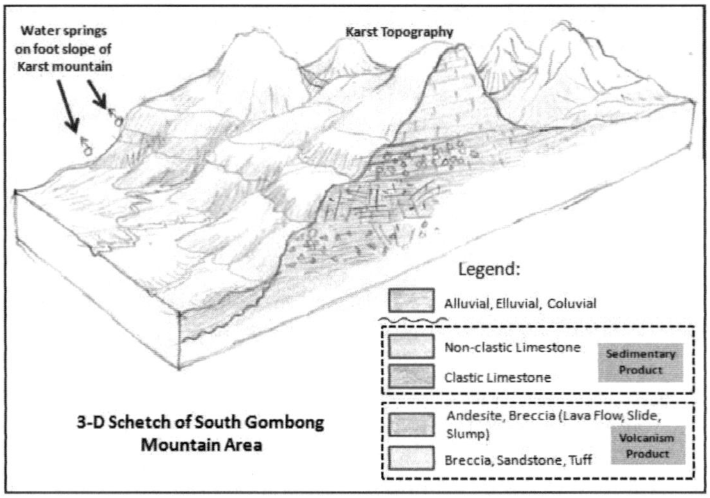

Figure 4. A three-dimensional view showing the cross-section model of the geological map of South Gombong "X" area

3. Results and Discussion

3.1. Geohydrology Assessment of South Gombong Area

Due to the influence of weather (tropical) in the presence of rain, the stretch of clastic limestone, especially in the area of Buayan to the south coast, forms the topography of Karst. Karst is a German word derived from Slovenian (Kras) which means rocky, arid land. This term in its home country is not really related to limestone and the dissolution process, but nowadays 'Kras' has been adopted for the landform of the process of solubility. Researchers define karst as a field with typical hydrological conditions as a result of soluble rocks and has well-developed secondary porosity:

Karst is characterized by:

1) the presence of a closed basin and or a dry valley in various sizes and shapes,
2) the scarcity or absence of drainage/river surface, and
3) the presence of caves from the underground drainage system.

IOP Conf. Series: Earth and Environmental Science **106** (2017) 012053 doi:10.1088/1755-1315/106/1/012053

Characters 1, 2, and 3 are evident on the limestone hillsides of South Gombong mountain area which in geomorphological units have been referred to as Karst hill units. Character 1 in this mapping area forms a valley as a plateau with springs at the base of the hillside slopes. The highlands have been managed by local people into the agricultural land. The results are interpretations as expressed and described in Figures 3, 4, 5, 6

3.2. Karst Hydrology and Utilization of Water Resource

Concerning the utilization of the highlands of this area into agricultural land as well as the acquisition of drinking water sources from groundwater emerging at the bottom of the hill slopes, in-depth research is required for the development and preservation of groundwater sources. The basic hydrological and geological knowledge of the area needs to be studied more carefully for the development of such utilization.

Hydrology is specifically categorized as the study of water phenomena on land/earth, descriptions of the effect of soil properties on water, the physical influence of water on the terrain, as well as the study of the relationship of water to life. Karst is known as a unique area and is characterized by exocarst topography such as karst valley, doline, uvala, polje, karren, cone karst, and the development of a much more dominant underground drainage system than its surface flow system [2].

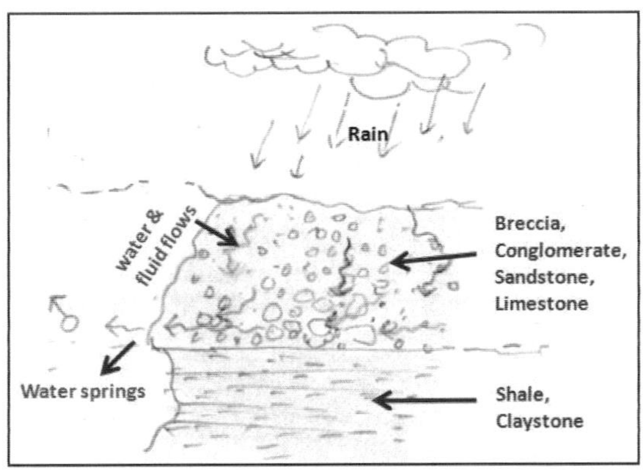

Figure 5. Natural groundwater formation and the emerging water springs.

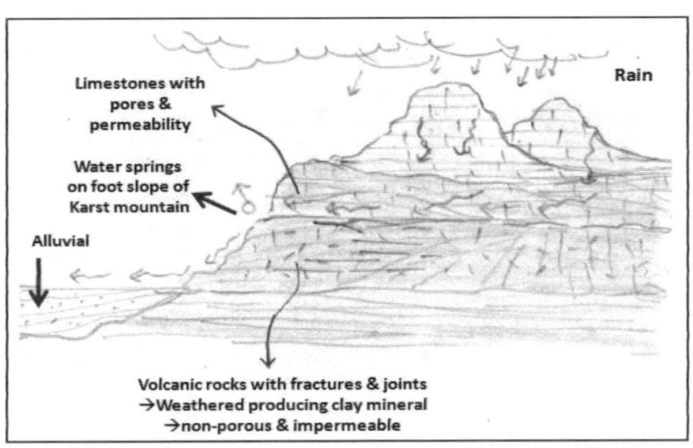

Figure 6. A cross-section of geological model of the South Gombong "X" area, potentially streamed by ground water, emerging as springs and rivers in the highlands.

The hydrologic cycle is the circulation of water on earth from the atmosphere, on the surface of the earth, and below the earth's surface. During the cycle, the water can change its form to solid, liquid, or gas phase, depending on the environmental conditions of the hydrological cycle. The amount of water

IOP Conf. Series: Earth and Environmental Science **106** (2017) 012053 doi:10.1088/1755-1315/106/1/012053

in the hydrological cycle is always fixed and only changes its position from time to time due to weather effects [3].

Researcher [7] describes that there are three main components of the karst hydrological system: aquifers, surface hydrological systems, and subsurface hydrological systems. In karst, the subsurface basin can be identified by looking for links between swallow holes and springs. This subsurface basin can be correlated with the surface flow drainage (watershed) if the fluid alleyways below the surface are primarily sourced to surface streams entering through the ponor. However, in general, the boundaries between the surface and subsurface watersheds are not the same. Subsoil systems, especially those with low groundwater slopes, can have multiple paths and outlets (springs). Furthermore, as the development of the dissolving process continues, the groundwater, aquatic, and underground rivers in karst aquifer can also change over time.

Based on the aforementioned description, it is apparent that the utilization and development of groundwater sources in this area requires in-depth studies, including subsurface surveys (geochemistry, geophysics, and even exploratory drilling) because thorough knowledge is remarkably useful in the utilization of groundwater resources, not only for agricultural land needs but also more importantly, as a source of drinking water.

4. Conclusion

Based on the survey, analysis, and evaluation of data obtained from the hills and mountains in South Gombong area, several things can be concluded: Geologically, this area is composed by a number of rocks from volcanic species to sediments that can be used as a basis for understanding specific things such as potential aspects of water resources.The limestone layer as one of the units showing the clastic and non-clastic species lies above the volcanic rock, covered by other types of sedimentary rocks formed by different compositions and genesis.Due to the removal of layers of rock to become a flat land, the process of weathering, erosion, transportation, and deposition, a specific geomorphological topography and appearance has formed. In particular, the limestone distribution forms the topography of Karst.The observations and field survey results show the existence of utilization of karst highlands for agricultural land as well as the use of springs for drinking water.Concerning the growing need of water resources as a source of irrigation for agricultural land, drinking water, and other purposes such as small industries, it is necessary to develop water supply procurement from existing groundwater sources.This deep understanding of the hydrogeology of the region is expected to facilitate the development of water supply procurement for various purposes and conservation of water resources in this area.

References

[1] Asikin S, Handoyo A, Prastistho B, & Gafoer S, Peta Geologi Lembar Banyumas Jawa, Lembar 1308-3, Skala 1:100.000, Pusat Penelitian dan Pengembangan Geologi, 1992.
[2] Adji TN, Haryono E, Suprojo SW.Kawasan Karst dan Prospek Pengembangannya di Indonesia. *Prosiding Seminar PIT IGI di Universitas Indonesia.* 26-27 Oktober 1999.
[3] AdjiTN, Suyono, Bahan Ajar Hidrologi Dasar, Fakultas Geografi UGM, un-published, 2004.
[4] Ford D and Williams P Karst Geomorphology and Hydrology, Chapman and Hall, London. 1992.
[5] Runkel A C, Tipping R G, Alexander E C Jr, Green J A, Mossler, J H, and Alexander S C. 2003. Hydrogeology of the Paleozoic bedrock in southeastern Minnesota: Minnesota Geological Survey. *Report of Investigations 61.* 105 p. 1 map in pocket.
[6] Todd D K, Groundwater Hydrology, 2nd Ed, John Wiley & Sons, 1998.
[7] Jankowski J, Hydrogeochemistry, Short Course Note, School of Geology, University of New South Wales, Sydney, Australia (un-published). 2001.

Improving the result of forcasting using reservoir and surface network simulation

R S Hendri[1]*, J Winarta[2]

[1]Oil Engineering Department, Faculty of Earth Technology and Energy, Universitas Trisakti, Jakarta, Indonesia
[2]Special Taskforce for Upstream Oil and Gas Business Activities, Jakarta, Indonesia

*Corresponding Author : rsitaresmi@trisakti.ac.id

Abstract. This study was aimed to get more representative results in production forcasting using integrated simulation in pipeline gathering system of X field. There are 5 main scenarios which consist of the production forecast of the existing condition, work over, and infill drilling. Then, it's determined the best development scenario. The methods of this study is Integrated Reservoir Simulator and Pipeline Simulator so-calle as Integrated Reservoir and Surface Network Simulation. After well data result from reservoir simulator was then integrated with pipeline networking simulator's to construct a new schedule, which was input for all simulation procedure. The well design result was done by well modeling simulator then exported into pipeline simulator. Reservoir prediction depends on the minimum value of Tubing Head Pressure (THP) for each well, where the pressure drop on the Gathering Network is not necessary calculated. The same scenario was done also for the single-reservoir simulation. Integration Simulation produces results approaching the actual condition of the reservoir and was confirmed by the THP profile, which difference between those two methods. The difference between integrated simulation compared to single-modeling simulation is 6-9%. The aimed of solving back-pressure problem in pipeline gathering system of X field is achieved.
Keywords: integrated reservoir simulation, surface network simulation, tubing head pressure

1. Introduction

X Field is a gas field in South Sumatera that has been producing from five wells with IGIP estimation 80.06 BCF with reservoir structural configuration at 1450-1850 meters deep illustrated in Figure 1. In this paper, the authors analyze the producing wells in X Field; cumulative production from all 5 wells is 2.27 BSCF with recovery factor estimation approximately 2.8%.

There are several main scenarios which consist of the production forecast of the existing condition (do nothing), additional intervention activities such as Work over and Infill well. By doing a reservoir simulation (simulation model has 215 x 149 x 19 cells, or total grid cells of 608,665) including fluid properties (reservoir pressure 2259 psig, gas viscosity 0.01682 centipoise, gas Formation Volume Factor 0.0078, density of gas 0.2559 gr/cc) and then applying all five scenarios to the model by integration with the pipeline networking or, it is expected that this simulation would give an accurate prediction and also comparison between scenarios so that X Field can be best developed.

IOP Conf. Series: Earth and Environmental Science **106** (2017) 012054 doi:10.1088/1755-1315/106/1/012054

2. Research Method

Feasibility for additional X Field development began with reservoir simulation and integrated simulation where the expected result would represent actual conditions. Simulation process began with structural modeling, modeling reservoir condition by integrating existing data (geology and reservoir) mathematically to get reservoir performance for several well's condition and production profiles. Simulation model has 215 x 149 x 19 cells, or total grid cells of 608,665. In this study, Black Oil Reservoir Simulator and Pipeline Simulator were used. One of the required data for the reservoir simulation input is the Depth Structure Map made by reservoir modeling. Area shown in yellow is the prospect hidrocarbon (Figure 1).

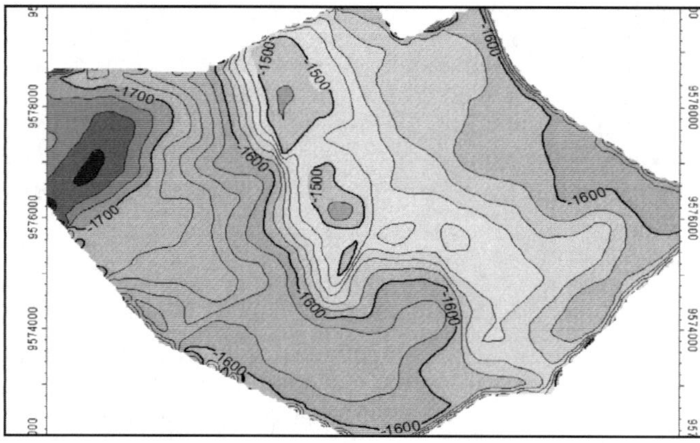

Figure 1. Depth structure map.

Reservoir simulation process began with several steps; data preparation,building model and grids, initialization, history matching and prediction. Initialization process included matching IGIP or total initial hydrocarbons in place with hydrocarbon control volume via volumetric method. History matching was done to ensure the model design represented actual reservoir conditions by changing the relative permeability and transmisibility (Figure 2).

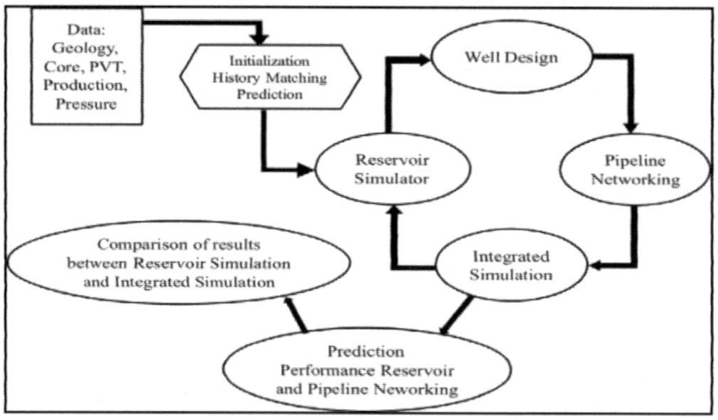

Figure 2. Workflow in integrated simulation.

Advanced processing included integration of reservoir simulation with pipeline networking of in-field well design with well modeling simulation. Combinations of all simulations were accomplished using Paralel Virtual Machine.

The methods of this study is Integrated between Reservoir Simulator and Pipeline Simulator so-called Integrated Reservoir and Surface Network Simulation. Advanced procedure included integration of reservoir simulation with pipeline networking of infill well design with well modeling simulation.

After result of well data from reservoir simulator was then integrated with pipeline networking simulator's to construct a new schedule, which was input for all simulation procedure. The well design result which was done by well modeling simulator was then exported into pipeline simulator. Reservoir prediction depends on the minimum value of Tubing Head Pressure (THP) for each well, where the pressure drop on the Gathering Network is not necessary calculated. The same scenario was done also for the single-reservoir simulation. Integration Simulation produces results approaching the actual condition of the reservoir and was confirmed by the THP profile which differ between those two methods.

3. Results and Discussion

Data for each well was input to well modeling simulator, which includes the process synchronization of well tests data (VLR/IPR Matching), needed to generate accurate output. In the synchronization of VLP of the Bottom Equipment (casing size, tubing depth, and tubing size) specification data has to be changed in order to match the simulation output and the well test results. Matching result between Cumulative Gas Production (Simulation) and Gas Production (Actual) of each well in X Field are displayed in Figures 3. This figure shown in blue is the result plot and red is straight line 45°. It means that the simulation result has already matched with the actual result.

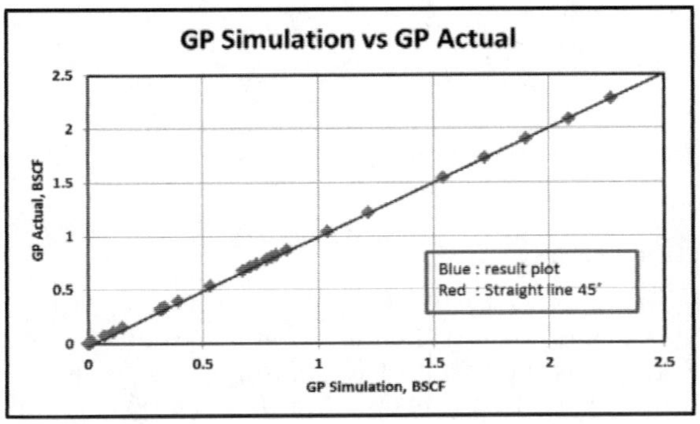

Figure 3. Cumulative gas simulation and actual.

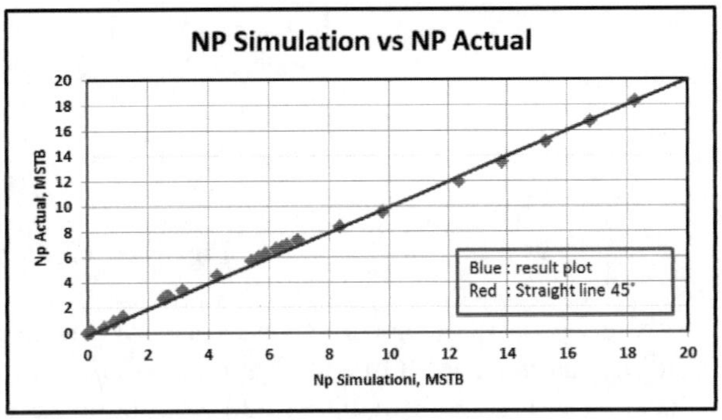

Figure 4. Cumulative oil between simulation and actual.

Matching result between Cumulative Oil Production in Simulation and Oil Production in Actual condition of each well in X Field is displayed in Figures 4. This figure shown in blue is the result plot and Red is Straight line 45°. It means that the simulation result has already matched with the actual result.

The network that was designed depends on field data whereas the output continued into the connecting software for integration with the reservoir *Source.* Integration was done by using the connection software collaboration of pipeline simulator and reservoir simulator. Surface network model is built by using pipeline simulator called Pipeline Network Map (Figure 5).

All the part of surface network have to be connected each other (Figure 6) including reservoir system from reservoir simulator. From those methode, it is known that the back pressure problem from each well to the surface tools such as separator and etc.

Well data from reservoir simulator was then integrated with pipeline simulator's well data to make a new schedule, which was input for full-simulation process. The same scenario was done for the single-simulation (Reservoir simulator only). X Field simulation comprises of five scenarios: Scenario I = Base Case (BC) + Work over, Scenario II = Scenario I + 1 Infill well, Scenario III = Scenario II + 1 Infill well, Scenario IV = Scenario III + 1 Infill well, Scenario V = Scenario IV + 1 Infill well.

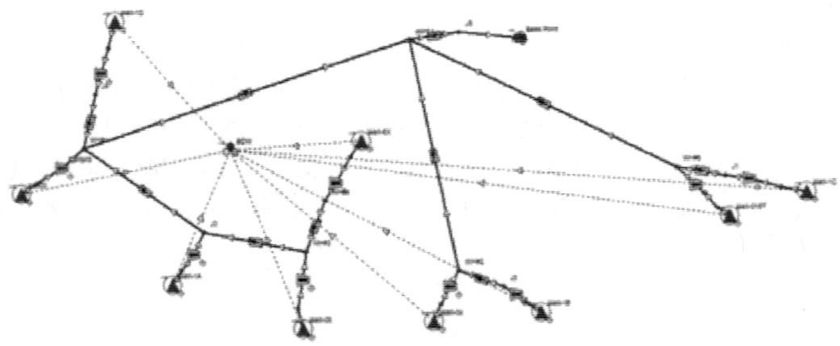

Figure 5. Pipeline network map.

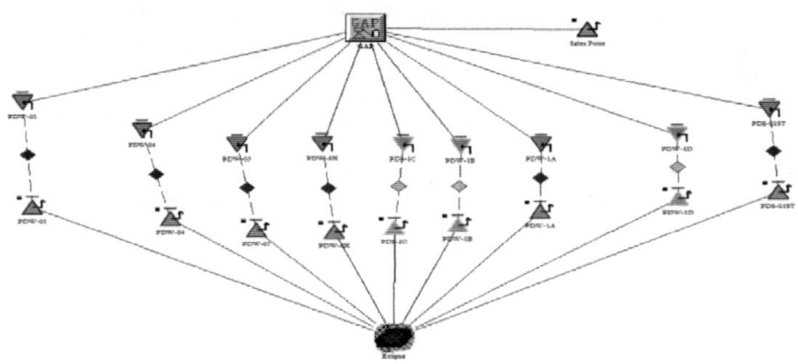

Figure 6. Conjunction network.

The Result of scenario I (Base Case+Work Over) is displayed in Figures 7. Base Case means do nothing that is no infill drilling but only Work over. Red line is the result of Reservoir simulator Rate, Blue line is the result of Integrated Simulator Rate. From 2013 in 12 MMSCFD to 2016 the rate of reservoir simulator declined whereas the integrated simulator rate only started to decline 5 years later. For reservoir simulator cumulative also less than integrated simulator cumulative. The integrated simulator cumulative has 4300 MMSCF.

The Result of scenario II (Scenario I + 1 Infill well) is displayed in Figures 8. Red line is the result of Reservoir simulator Rate, Blue line is the result of Integrated Simulator Rate. From 2013 in 12 MMSCFD to 2018, the rate of reservoir simulator declined whereas the integrated simulator rate only started to decline 6 years later. For reservoir simulator cumulative also less than integrated simulator cumulative. The integrated simulator cumulative has 4500 MMSCF.

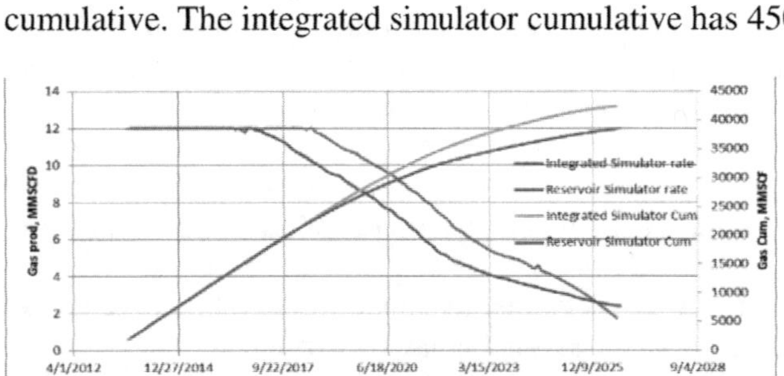

Figure 7. Gas flow rate-cumulative gas production difference (scenario I).

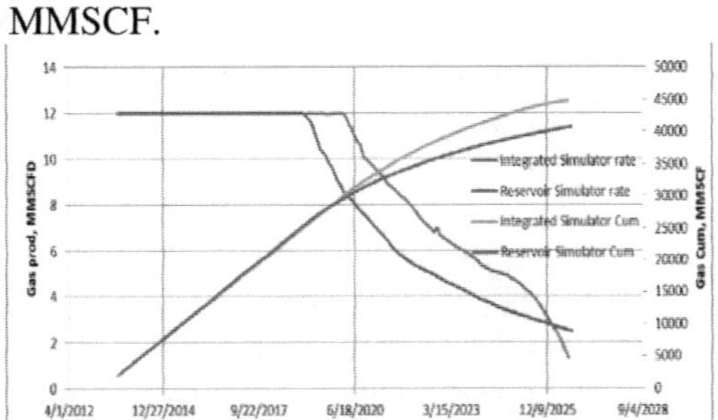

Figure 8. Gas flow rate-cumulative gas production difference (scenario II).

The Result of Scenario III (Scenario II + 1 Infill well) is displayed in Figures 9. Red line is the result of Reservoir simulator Rate, Blue line is the result of Integrated Simulator Rate. From 2013 in

IOP Conf. Series: Earth and Environmental Science **106** (2017) 012054 doi:10.1088/1755-1315/106/1/012054

12 MMSCFD to 2019 the rate of reservoir simulator declined whereas the integrated simulator rate only started to decline almost 8 years later. The integrated simulator cumulative has 4400 MMSCF.

The Result of Scenario IV (Scenario III + 1 Infill well) is displayed in Figures 10. Red line is the result of Reservoir simulator Rate, Blue line is the result of Integrated Simulator Rate. From 2013 in 12 MMSCFD to 2019 years later the rate of reservoir simulator declined whereas the integrated simulator rate only started to decline 8 years later. For reservoir simulator cumulative also less than integrated simulator cumulative. The integrated simulator cumulative has 4500 MMSCF.

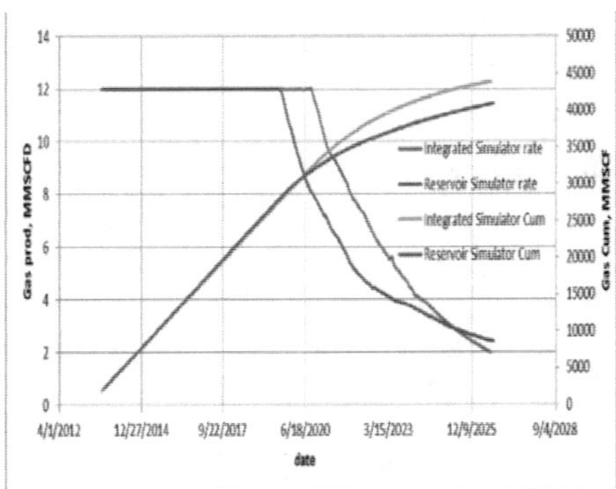

Figure 9. Gas flow rate-cumulative gas production difference (scenario III)

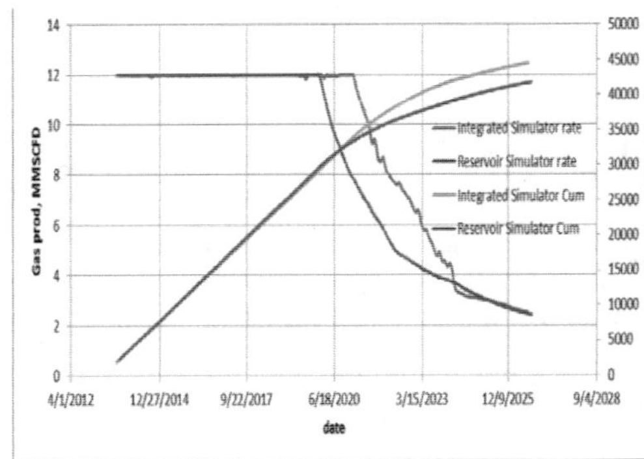

Figure 10. Gas flow rate-cumulative gas production difference (scenario IV)

The Result of Scenario V (Scenario IV + 1 Infill well) is displayed in Figures 11. Red line is the result of Reservoir simulator Rate, Blue line is the result of Integrated Simulator Rate. From 2013 in 12 MMSCFD to 2019, the rate of reservoir simulator declined whereas the integrated simulator rate only started to decline almost 9 years later. The integrated simulator cumulative has 4600 MMSCF.

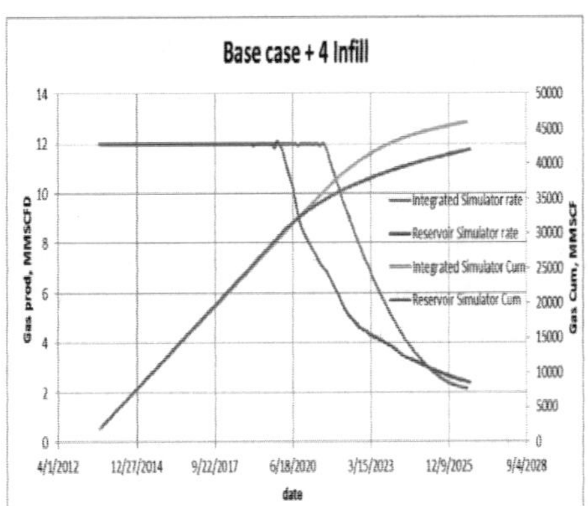

Figure 11. Gas flow rate cumulative gas production difference (scenario V).

From charts shown previously, it can be drawn the conclusion that there is a difference between these two methods. On average the difference ranges from 6% to 9% of cumulative gas production (also with the different plateau of 2 years). Reservoir simulation prediction depends on the minimum value of THP for each well, where pressure drop on the Gathering Network is not calculated. On the other side, the integrated simulation method covers all of the reservoir parameters until the Gathering Network is put into calculation. The difference of Tubing Head Pressure value is illustrated on Figure 12.

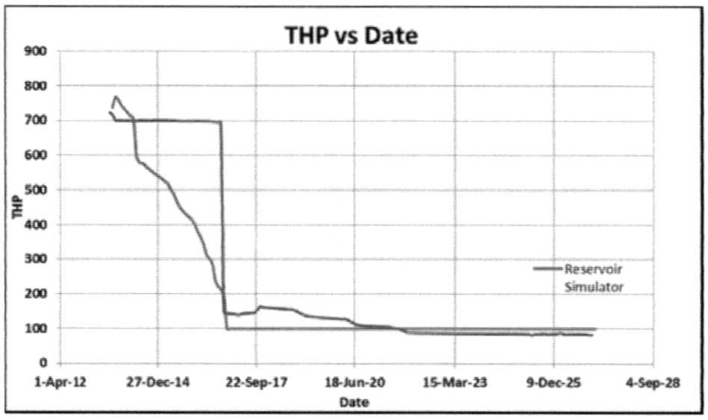

Figure 12. THP difference between single reservoir simulation & integrated simulation.

Integration of the reservoir and pipeline networking is accomplished by reservoir simulator and pipeline simulator. Before initiating integration, first step by making a well design for each existing well then combine with the pipeline networking from previous simulation (Table 1). The results are the same across 5 scenarios that have differences. This confirms single-simulation modeling has a limitation in determining Tubing Head Pressure (THP), but not the integrated modeling. Therefore, the aimed of solving back-pressure problem in pipeline gathering system of X field, is achieved.

Table 1. Comparison of scenarios.

	Comparison		
Scenario	Description	Int Cum(MMSCF)	Difference (%)
First	Base-Case + Work over	4300	9.1
Second	Scenario 1 + 1 infill	4500	9.0
Third	Scenario 1 + 2 Infill	4400	6.7
Fourth	Scenario 1 + 3 infill	4500	6.1
Fifth	Scenario 1 + 4 infill	4600	8.6

4. Conclusion

The difference between integrated reservoir and surface network simulation compared to single-simulation was about 6-9%. Five simulated scenarios show the same result that confirm single-simulation modeling has a limitation in determining Tubing Head Pressure (THP), but not the integrated modeling. Integration reservoir and surface network simulation produces results approaching the actual condition of the reservoir; this was confirmed by the THP profile difference between those two methods.

Acknowledgment

We would like to thank to Special Taskforce for Upstream Oil and Gas Business Activities, Jakarta and Universitas Trisakti for making this paper possible.

References

[1] Barroux C C, Samier P, dkk, 2000, Linking reservoir and surface simulators: how to improve the coupled solutions, SPE 65159, France.
[2] Gonzales F, Bertoldi L, dkk. 2010. A Fully Compositional Integrated Asset Model for a Gas-Condensate Field. *SPE Annual Technical Conference Exhibition*, Italy.
[3] Hayder E, Dahan M, Dossary M. 2008. Production Optimization Through Coupled Facility/Reservoir Simulation. SPE 100027, Saudi Aramco.
[4] Winarta, Jeffier. 2013. Optimisasi Produksi Lapangan Gas X dengan Integrasi Simulasi Reservoir dan Jaringan Pemipaan. Universitas Trisakti, Jakarta.
[5] Ursini F, Rossi R, Pagliari F. 2010. Forecasting Reservoir Management Through Integrated Asset Modelling. *SPE North Africa Conference and Exhibition*, Egypt.

IOP Conf. Series: Earth and Environmental Science **106** (2017) 012055 doi:10.1088/1755-1315/106/1/012055

Factors affecting quality of social interaction park in Jakarta

N I Mangunsong

Landscape Architecture Department, Faculty of Architecture Landscape and Environmental Technology, Universitas Trisakti, Jakarta, Indonesia

Corresponding Author : nurintan@trisakti.ac.id

Abstract. The existence of social interactions park in Jakarta is an oasis in the middle of a concrete jungle. Parks is a response to the need for open space as a place of recreation and community interaction. Often the social interaction parks built by the government does not function as expected, but other functions such as a place to sell, trash, unsafe so be rarely visited by visitors. The purpose of this study was to analyze the factors that affect the quality of social interaction parks in Jakarta by conducting descriptive analysis and correlation analysis of the variables assessment. The results of the analysis can give an idea of social interactions park based on community needs and propose the development of social interactioncity park. The object of study are 25 social interaction parks in 5 municipalities of Jakarta. The method used is descriptive analysis method, correlation analysis using SPSS 19 and using crosstab, chi-square tests. The variables are 5 aspects of Design, Plants composition: Selection type of plant (D); the beauty and harmony (Ind); Maintenance and fertility (P); Cleanliness and Environmental Health (BS); Specificity (Drainage, Multi Function garden, Means, Concern/Mutual cooperation, in dense settlements) (K). The results of analysis show that beauty is the most significant correlation with the value of the park followed by specificity, cleanliness and maintenance. Design was not the most significant variable affecting the quality of the park. The results of this study can be used by the Department of Parks and Cemeteries as input in managing park existing or to be developed and to improve the quality of social interaction park in Jakarta.
Keywords: beauty, cleanliness, maintenance, park, quality, specificity

1. Introduction

The quality of social interaction park in Jakarta is causing for concern, so few visitors and its function is not as expected. On the one hand, people need their park for social interaction but unfortunately the park provided by the government are not used by the residents of the city. This study was conducted to analyze the factors most affecting visitors in assessing a social interaction park quality. The existence of neighborhood parks as green open space that is comfortable, safe, and unexploited as a heterogeneous city where citizens can interact absolutely necessary. Its presence gives a positive value on the lives of citizens and the quality of the urban environment.

A social interaction park or the neighborhood park became a favorite and frequented the public would have factors with certain environmental quality into its appeal. According [1] 8 critical factors that affect the environmental quality of the city of London are (1) Human scale and compactness, (2) Structure, legibility and identity, (3) Cleanliness and safety, (4) Urban management, (5) Visual richness, (6) Activity and mixed use, (7) Public spaces and special places, (8) Moving about and pedestrian friendliness.

IOP Conf. Series: Earth and Environmental Science **106** (2017) 012055 doi:10.1088/1755-1315/106/1/012055

While [2] suggests a community open space is a place to share and exchange ideas with communities that are typically located outdoors, located near the residence. The open space has value if its construction followed the rules and have good drainage can increase the natural growth in the region and greening. Researcher [3] stated that the park is developed as recreational facilities and other functions should show the character feasible physical footprint for both active and passive activities such as a multifaceted vegetation, topography, good drainage with attention to the security and safety of users. Thus a need to consider various aspects of the park from the standpoint of the environment and the quality of the facilities available to those aspects of the security and safety of users.

Results of the study is to provide an overview of social interaction parks as needed by the relevant agencies that can be used in the proposed development of the park in Jakarta.

2. Research Method

The object of study are 25 Social Interaction Parks in 5 municipalities of Jakarta: located in East Jakarta, North Jakarta, West Jakarta, South Jakarta and Central Jakarta. The survey was conducted in June 2016. Variables assessed and the percentage of valuation can be seen in the table below. Score ranged from 70 to 100.

Table 1. Variable and assessment.

No	Aspect	Percentage	Score	Final Score
1.	Design, Plants composition, Selection Type of Plant (D)	25 %		
2.	Beauty and harmony (Ind)	10 %		
3.	Maintenance and fertility (P)	20 %		
4.	Cleanliness and Environmental Health (BS)	20 %		
5.	Specificity (Drainage, Multi Function garden, Means, Concern/Mutual cooperation, in dense settlements, etc.) (K)	25 %		
Total Score		100 %		

The method used is descriptive analysis method, the correlation analysis using SPSS 19 (Statistics Programme Social Science 19) and using crosstab, chi-square tests. Data were collected by survey and field observation and study of literature. Greater number obtained variables indicating the growing influence of the object of study assessment.

3. Results and Discussion

Quality of desire and satisfaction is a factor to consider. If visitors feel less satisfied in the city park, the desire of visitors to come to this park to be reduced. Factors that affect the sustainability of this park include Design, Plants composition, Selection Type of Plants, Beauty and harmony, cleanliness and others.

Figure 1. Cipinang Muara social interaction park, East Jakarta.

Figure 2. Taman Aries social interaction park –West Jakarta.

Table 2 below show the result of linkage analysis how big the influence of these factors on the quality of parks in Jakarta. While Table 3 show the correlations between each factors.

Table 2. Descriptive statistics.

	Mean	Std. Deviation	N
N	79.4520	2.80255	25
Dsn	19.3260	1.70093	25
Ind	8.0320	.26571	25
P	15.8240	.99217	25
BS	15.8000	.96954	25
K	20.0900	.77015	25

Table 3. Correlations.

		N	Dsn	Ind	P	BS	K
Pearson Correlation	N	1.000	.694	.829	.440	.398	.758
	Dsn	.694	1.000	.229	.027	-.006	.110
	Ind	.829	.229	1.000	.579	.537	.907
	P	.440	.027	.579	1.000	.970	.540
	BS	.398	-.006	.537	.970	1.000	.505
	K	.758	.110	.907	.540	.505	1.000
Sig. (1-tailed)	N	.	.000	.000	.014	.024	.000
	Dsn	.000	.	.135	.449	.488	.300
	Ind	.000	.135	.	.001	.003	.000
	P	.014	.449	.001	.	.000	.003
	BS	.024	.488	.003	.000	.	.005
	K	.000	.300	.000	.003	.005	.
N	N	25	25	25	25	25	25
	Dsn	25	25	25	25	25	25
	Ind	25	25	25	25	25	25
	P	25	25	25	25	25	25
	BS	25	25	25	25	25	25
	K	25	25	25	25	25	25

From Table 3, the results of the correlation analysis above shows that there is a significant correlation at the 0:01 level (2-tailed) between Value garden with desian (.694 **). Significant correlation between Value garden with beauty (.829 **). The correlation between the park with Special Values (Drainage, Multi Function garden, Means, Concern/Mutual cooperation, in dense settlements, etc.) (.758 **). There is a significant correlation at 0.05 level (2-tailed) between Value garden maintenance (.440 *). The correlation between the value of a garden with hygiene and environmental health (.398 *). The highest correlation between the value of the park with all of the variables is the beauty of ratings (.829 **). Significant correlation between beauty with specificity values (.907 **). Correlation beauty with hygiene and environmental health (537 **). Correlation with the cleanliness and beauty of Maintenance (579 **).

Table 4 below show the model summary, while Table 5 is ANOVA result and Table 6 is correlations between models.

Table 4. Model summary.

Model	R	R Square	Adjusted R Square	Std. Error of the Estimate
1	.988[a]	.976	.970	.48818

a. Predictors: (Constant), K, Dsn, BS, Ind, P

Table 5. ANOVA b.

Model	Sum of Squares	Df	Mean Square	F	Sig.
Regression	183.974	5	36.795	154.394	.000a
Residual	4.528	19	.238		
Total	188.502	24			

a. Predictors: (Constant), K, Dsn, BS, Ind, P
b. Dependent Variable: N

Table 6. Correlations.

	Model	Unstandardized Coefficients		Standardized Coefficients	t	Sig.
		B	Std. Error	Beta		
1	(Constant)	3.065	3.140		.976	.341
	Dsn	.944	.063	.573	15.023	.000
	Ind	3.950	.974	.374	4.054	.001
	P	-.214	.433	-.076	-.493	.627
	BS	.289	.430	.100	.672	.510
	K	1.256	.316	.345	3.977	.001

4. Conclusion

The research concluded that the factors that influence the value and quality of interaction in a Jakarta park the most significant is the beauty Specialities (Drainage, Multi Function garden, Means, Concern/Mutual cooperation, in dense settlements, etc.) and maintenance. Aspects of the design was not held leadership positions in a quality park. Thus, people prefer the beauty of the city equipped with good drainage, clean and well maintained not design.

References

[1] Carmona, Matthew and Wunderlich, Matos Filipa. Capital Spaces: The Multiple Complex Public Spaces of a Global City. p.34. Routledge Taylor & Francis Group. London & New York. 2012.
[2] Simonds J O. Landscape Architecture: A Manual of Site Planning and Design. 3rd Ed. New York. 1997.
[3] Mertes, James D. and Hall, James R. Park, Recreation, Open Space and Greenway Guidelines. A Project of The National Recreation and Park Association and the American Academy for Park and Recreational Administration. p.102. 1995.
[4] Gold, Seymor M. Recreation Planning and Design. New York. 1980.
[5] Icor, Seely H. Outdoor Recreation and the Urban Environment. Mac-Milan. London. 1973.
[6] Lawson, Fred and Manuel Band-Bovy. Tourism and Recreation Development: A Handbook of Physical Planning. The Architectural Press, Ltd. 1972
[7] Piaw, Yan Chua. Asas Statistik Penyelidikan: Analisis Data Skala Ordinal dan Skala Nominal. p.316-317. McGraw Hill. 2008.
[8] Wurman, Richard S. The Nature of Recreation. MIT-Press. USA. 1972.

IOP Conf. Series: Earth and Environmental Science **106** (2017) 012056 doi:10.1088/1755-1315/106/1/012056

Water quality assessment for self-sufficient water resources for DKI Jakarta

R Yanidar[1*], D M Hartono[2], S S Moersidik[2]

[1]Environmental Engineering Department, Faculty of Landscape Architecture and Environmental Technology, Universitas Trisakti, as a Doctoral Student, Department of Civil Engineering, Faculty of Engineering, Universitas Indonesia, Depok 16242, Indonesia.
[2]Environmental Engineering Departement, Faculty of Engineering, Universitas Indonesia, Depok 16242, Indonesia

*Corresponding Author: ramadhani@trisakti.ac.id

Abstract. DKI Jakarta, Indonesia (106.22'42" S and 106.58'18"E) is passed by the 13 major rivers. However, only Krukut River, in South Jakarta has contributed 5.7% for supply the 60,27% coverage service of the10.178 million populations (2015). This paper is preliminary research to find the potential water resources for self-sufficient water supply. This study used multivariate analyzed of Factor Analyzed (FA) and Cluster Analysis (CA) for the archive river water quality data set covers the period from 2010 to 2015 from the Environment Department of DKI Jakarta. They selected the main pollution parameter and the potential raw water resources which necessary to study severely. The FA identified the two groups of water quality variables, which explain majority of the experimental data, and resulted organic pollutant as main factor that influenced DKI Jakarta river water quality. Therefore, BOD and COD used as main parameter for water quality further assessment. CA grouped 43 sampling sites into four clusters. 5 sampling locations on 4 rivers need further study for additional raw water sources.
Keywords: BOD, COD, cluster analysis, factor analysis, self-sufficient, water resources

1. Introduction

Jakarta is the capital of Indonesia which is a metropolitan city with 10.178 million populations of 2015. A River as a water resource is one of the natural resources which has a multipurpose function for human life and livelihood. Even though, Jakarta has 13 rivers in its administrative area, but there are no water resources meet the standard of raw water quality. However, only Krukut River, in South Jakarta has contributed 5.7% supply for the 60.27% coverage service of the 10.178 million DKI Jakarta's populations (2015). It needs to search a new raw water sources that can meet the standard raw water quality for additional production capacity to reach 100% coverage service water supply area by 2030 as National and SDGs goal. This paper is preliminary research to find the potential rivers that cross through Jakarta can be utilized as water resources for self-sufficient water supply in DKI Jakarta.

The river water has been contaminated because of the waste water discharge containing degradable organics, nutrients, domestic effluent, and agricultural waste [1]. The quality of surface water is threatened by a variety of pollutants from both natural such as precipitation inputs and erosion, and from anthropogenic sources [2], [3], [4].

Factor Analysis (FA) and Cluster Analysis (CA) technique as a multivariate analysis [5]. General Purposes of FA technique is to summarize the information contained in several original variables of a complex data set into new smaller data, with minimum loss information. CA is a reduction statistical method commonly used to classify objects into groups according to their similarity (nearness).

A river's water quality is combination of several interrelated compounds, which are subjected to local and temporal variations and affected by water flow quantity. In the last years, factor analysis (FA) and cluster analysis (CA) as a multivariate techniques, have been widely applied to facilitate in various surface water quality studies, including assessment the analysis of complex datasets to better understand the temporal and spatial characteristic of surface water quality [1], [4], [6], [7], [8], [9], [10]. The aim for this study was selected the main pollution parameter and the potential raw water resources which necessary to study conscientiously to find raw water resources for self-sufficient water supply in DKI Jakarta.

2. Research Method

2.1. Study Site and Collection of Water Quality Data

The DKI Jakarta extends from longitude 106.22'42" E to 106.58'18" E and latitude 5.19'12" S and 6.23'54" S. It is located about 7 m above mean sea level, covering an area of about 7,659.02 km^2. There are water quality monitoring data which consist of 85 sampling location of 19 Rivers published in the Regional Environment Status Book (SLHD) of The Environment Department of DKI Jakarta (*Dinas Lingkungan Hidup DKI Jakarta*) annually. In this research, the selection of physical and chemical of archive water quality data set was conducted. The 43 sampling locations of 11 rivers which allocation as B class for raw water and C class for fishery were chosen, as follows: Ciliwung, Kali Baru, Angke, Krukut, Mampang, Cengkareng, Grogol, Sepak, Pesanggrahan, Mookevart and Tarum Barat river (

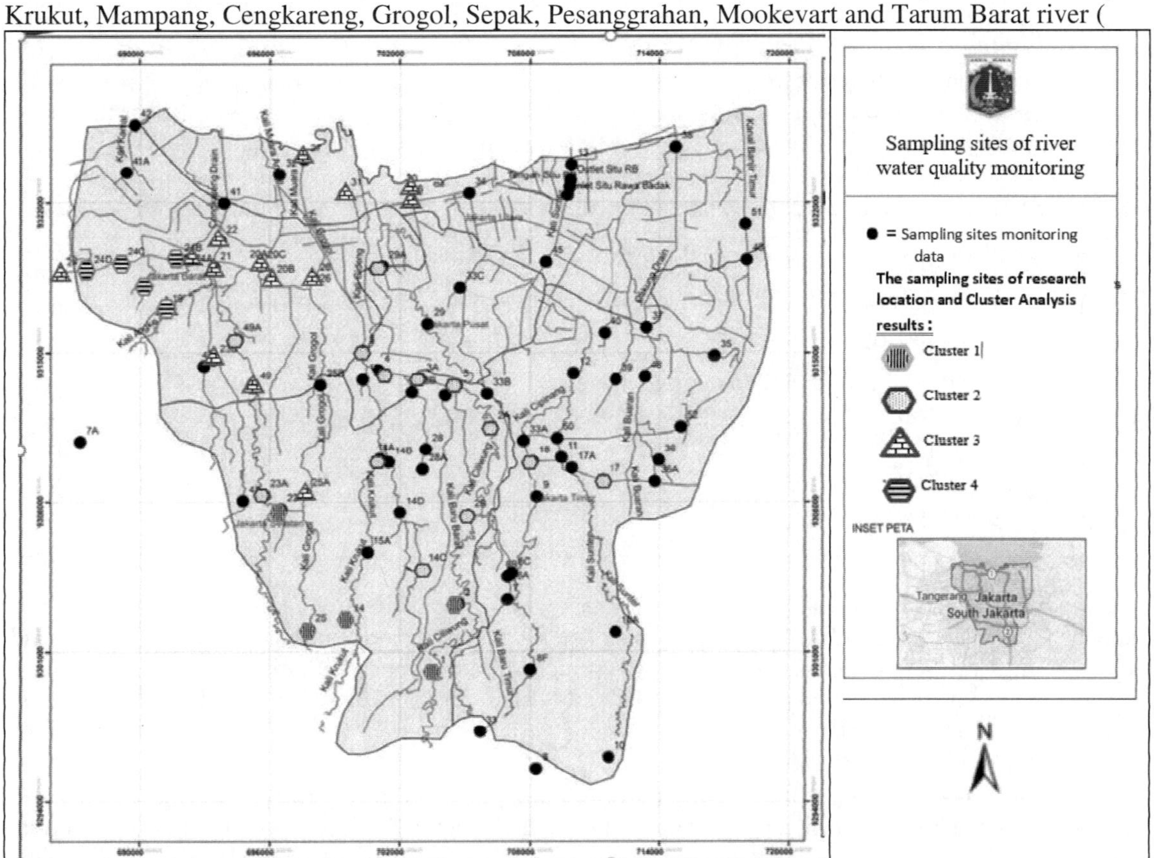

Figure). The archive data set comprise nine (9) water quality variables: Dissolved Oxygen (DO), Biochemical Oxygen Demand (BOD), Chemical Oxygen Demand (COD), Total Dissolved Solid (TDS), Suspended Solids (SS), and pH, Total Phosphor (TP), Ammonia (NH_3) and Nitrite (NO_2) which cover the period from 2011 to 2015 [11].

2.2. Data Treatment and Statistical Analysis

The data processing and multivariate statistical analysis used SPSS (ver. 23.0) with the following steps:

1) The z-scale transformation was applied to standardize the water quality data which expressed in different magnitudes and units.
2) The One-way ANOVA was used to analyze the significant seasonal differences in water quality.
3) The Bartlett's Test of Sphericity and Kaiser-Meyer-Olkin (KMO) was applied for Factor Analysis (FA) recommendation. It was used to explain the significance of the study and thereby illustrate the validity and suitability of the parameters being addressed through the FA. The FA identified the two groups principal component of water quality variables, which explain majority of the experimental data. The first VF, which indicated the most important parameters.
4) The Ward's method of hierarchical agglomerative Cluster Analysis (CA) was carried out to ascertain the multivariate similarity (the parameters were obtained from FA) between different sampling locations. the similarities and dissimilarities are quantified through Euclidean distance measurement and a dendrogram was the output of this CA process.

3. Results and Discussion

Before z-scale transformation, homogeneity of variances test results showed 2 of the variables i.e. TSS and NO_2 parameters were not homogeny and normally distributed, so the research variables which were choosen only BOD, COD, DO, pH, Total Phosphor, TDS and SS. Statistical descriptive for physicochemical characteristics recorded of the selected parameters for water samples in wet and dry seasons is presented in Table 1. Almost all parameters have average concentration exceed the raw water quality standard (class II), consequently, it needs further study to get the river water to meet the standard raw water quality. The mean values of BOD, COD, DO, pH, Total Phosphor, TDS and SS showed that they were not significantly difference between rainy season and dry seasons ($p < 0.05$). Only TP parameters in dry season was significantly higher than rainy season. This could be caused climate change effect that made the rainy and dry seasons cannot be ascertained.

Table 1. Statistical descriptive for physicochemical characteristics recorded in rainy and dry seasons.

Parameter	Units	Dry Season	Wet Season	F	p	Standar Class II	Standar Class III
Biological Oxygen Demand (BOD)	mg/L	22.98 ± 21.04	19.83 ± 13.00	0.698	0.406	3	6
Chemical Oxygen Demand (COD)	mg/L	76.97 ± 52.91	74.82 ± 48.01	0.039	0.844	50	100
Dissolved Oxygen (DO)	mg/L	2.71 ± 0.98	2.61± 0.68	0.401	0.528	4	3
Total Dissolved Solids (TDS)	mg/L	393.69 ± 519.28	428.84 ± 635.09	0.078	0.781	1000	1000
pH		7.19 ± 0.16	7.17 ± 0.19	0.751	0.389	6-9	6-9
Total Phosphorus (TP)	mg/L	1.43 ± 2.56	0.59 ± 0.47	4.423	0.038	0.2	1.0
Ammonia (NH3)	mg/L	8.19 ± 12.94	5.54 ± 6.34	1.450	0.232	-	-

The KMO value is 0.704 and the p-value is 0.000 (<0.05), the value falls in the "*more than enough*" category worthy for the benefit of factor analysis. The FA results corresponding vary factors (VFs) and factor loadings can be seen on Table 2. The two principal components with eigenvalues larger than 1 were retained, which explained almost 75.6% of the total variance in water quality data. The first principal component accounted for 50.34% of the total variance, whereas the other principal components accounted for variances of 25.2%, respectively.

The first VF which the most important parameters has 50.34% strong positive loading on BOD and COD and strong negative loading on DO. Thus, all information about 7 variables pollutant in the original on the 43 monitoring stations collected can be performed in the reduced space by 3 variables. It results organic compound (BOD and COD) as main factor that influenced DKI Jakarta river water quality. The decomposition of the organic pollutant reduced DO levels [12]. The presence of organic compound in water under normal conditions was indicated from BOD_5, COD concentration, then, the organic compound degradation consumed oxygen caused the very low available DO in water (Figure 1).

The second VF has 25.2% strong positive loadings on NH3, TP, pH and moderate TDS. Ammonia and phosphate are natural parameters which accumulates easily in aquatic systems because it is a natural byproduct of organism's metabolism, but in the other hand, an indication that the system is out of balance when the presence of ammonia. The pH correlate positively with TDS, TP and NH3, but correlate negatively with DO in component (FVs) 1. Total Dissolved Solids (TDS) is a measure of the combined content of all inorganic and organic substances present in a liquid. Inorganic materials of ions commonly found in a river water. This may be the reason why they showed the strong positive loading in VF2 but they were not as main factors and support the BOD and COD concentrations. It reflected as natural rivers that was influenced by domestic pollutant activities.

The BOD, COD and DO concentration of each clusters are described on boxplot in Figure 2. It appears that cluster 1 has the lowest BOD and COD values and the highest DO value with small variance. The average BOD of Cluster 1 was 6.42–9.67 mg/L, and COD showed a range of 33.34–39.02 mg/L, and DO 3.63 – 4.23 mg/L. Cluster 2,3 and 4 the BOD and COD concentrations increase, but DO concentrations decrease. The results of this study indicate that the clusters were classified according to the characteristics of pollution sources in the locality or the water quality pollution level of the river [7].

The dendrogram of sampling sites obtained by Ward's method is shown in Figure 3 to analyse the similarities of water quality variation tendencies between the monitoring sites. CA made 43 sampling stations to be grouped into four main clusters. The Cluster 1 indicate 5 locations in 4 river i.e sampling location (SL) 1, and SL 2 in Ciliwung River, SL 14 in Krukut River, SL 23 Pesanggrahan River, and SL 25 in Grogol River which all of them are in upstream of the river. Cluster 2 consists of 16 sampling location in Ciliwung River, 2 SLs in Kalibaru river, 1 SL in Krukut River, Mampang and Sepak River, and each 2 SLs in Pesanggrahan and Tarum Barat River. The cluster 3 consist of 19 sampling locations in 9 rivers, but only 3 sampling locations in Mookevart river in cluster 4. The sampling locations of the Cluster Analysis results can be seen in Figure 4.

Table 2. Loadings of seven (7) water quality variables on two significant vary factors.

Variables	VF1	VF2
BOD	**0,954**	0,107
COD	**0,948**	0,140
DO	**-0,716**	-0,468
TDS	0,242	**0,648**
pH	-0,488	**0,713**
TP	0,451	**0,781**
NH3	0,154	**0,811**
Eigenvalue	3,524	1,767
%Total variance	50,344	25,242
Cummulative % variance	50,344	75,586

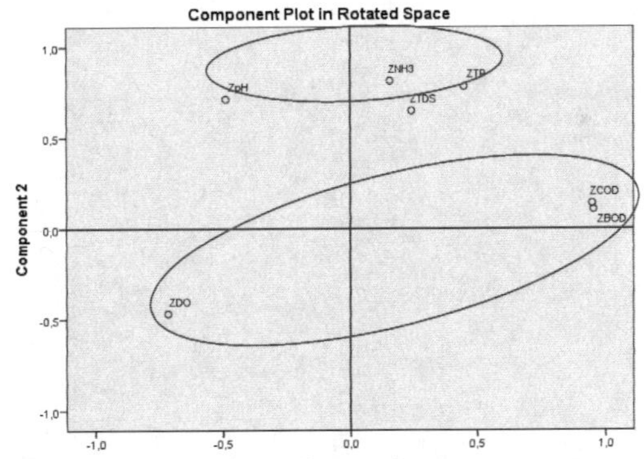

Figure 1. Principal component value of sampling sites data

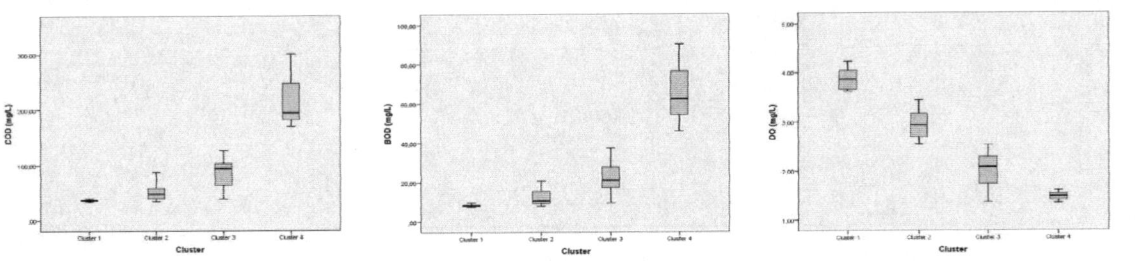

IOP Conf. Series: Earth and Environmental Science **106** (2017) 012056 doi:10.1088/1755-1315/106/1/012056

Figure 2.a) BOD concentration **Figure 2.b**) COD concentration **Figure 2.c**) DO concentration

Figure 2. Boxplot BOD, COD and DO concentration of sampling sites clusters.

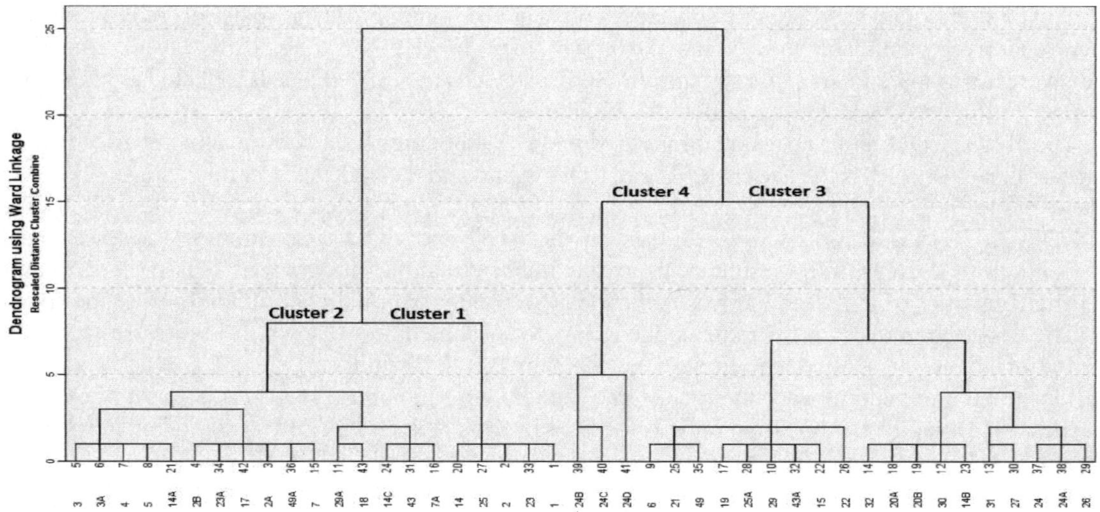

Figure 3. Dendrogram showing hierarchical clustering of sampling sites.

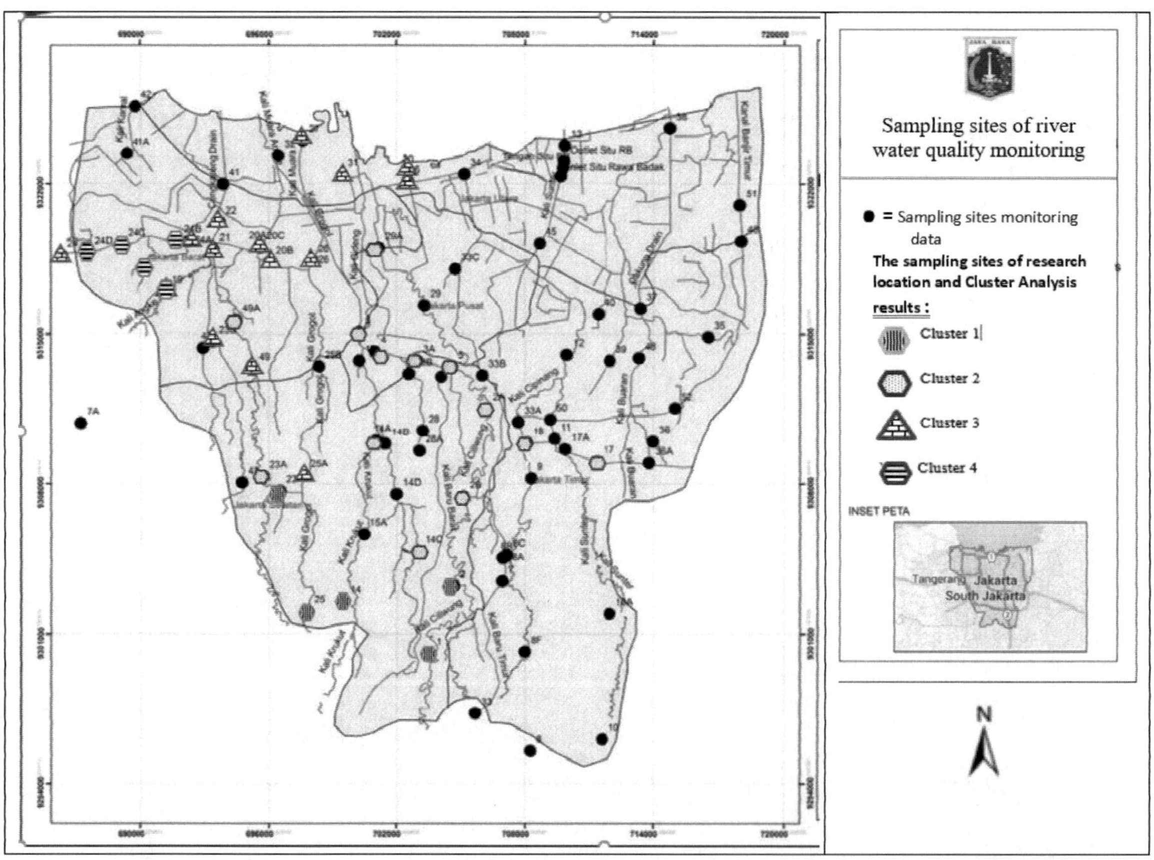

Figure 4. The sampling sites of water quality monitoring [11] and CA research data.

Pollutant sources in DKI Jakarta contribute to the river pollution process because the quality of downstream river water will decrease. Five locations in 4 rivers i.e Ciliwung, Krukut, Grogol and Pesanggrahan River which location in upstream of the river could be proposed as additional raw water sources with further study for quality, quantity, and continuity to meet water supply standard. Based on the above study results, it is necessary to further study the anthropogenic disturbance gradient depending on cluster, which described the water quality pollution level of the river. It is expected that a contribution will be made to data usage for efficient water quality management in terms of future watershed management.

DKI Jakarta will be proposed to adopt a model water supply coverage area by zoning of water services based on the location of the availability of raw water sources location. Depend on the Cluster 1 result, there will be 3 water supply services zones based on the availability of raw water sources location., ie the eastern zone which will be served by Ciliwung River, the middle zone will be served by Krukut and Grogol River, and the southern and western zones to be served by Pesanggrahan River.

4. Conclusion

The FA identified the two groups of water quality variables, which explain majority of the experimental data, and resulted organic pollutant as main factor that influenced DKI Jakarta river water quality. Therefore, BOD and COD used as main parameter for water quality further assessment. CA grouped sampling sites into four clusters. Five sampling locations on 4 rivers (Ciliwung, Krukut, Grogol and Pesanggrahan River) need further study which be proposed as additional raw water sources.

An objective interpretation of surface water physicochemical parameters and sharing as part of the effort identification location of rivers as raw water that necessary to study conscientiously are well provided by multivariate analysis.

DKI Jakarta will be proposed to adopt a model water supply coverage area by zoning of water services based on the availability of raw water sources location. This model will be influenced by the dynamics of urbanization and its impact on self-sufficient water sources which pass through the city. It expects this framework may impact raising public awareness to safe the water resources, in the sense that each resident is responsible for its local water sources.

Acknowledgment

The authors gratefully thanked to Environment Department of DKI Jakarta Province, Indonesia, for their cooperation and for providing river water quality monitoring data, and to University of Indonesia as the sponsor for this research under the project of PITTA 2017 funding scheme.

References

[1] Adamu Mustapha, Ahmad Zaharin Aris, Hafizan Juahir, Mohammad Firuz Ramli, Nura Umar Kura, River Water Quality Assessment Using Environmentric Techniques: Case Study of Jakara River Basin," *Environmental Science and Pollution Reserach 20.8,* pp. 5630-5644, 2013.

[2] Wang, Ruizhao, et al, Effects of land use types on surface water quality across an anthropogenic disturbance gradient in the upper reach of the Hun River, Northeast China, *Environmental monitoring and assessment 185.5 (2013),* pp. 4141-4151, 2013.

[3] Wilbers, Gert-Jan, et al, "Spatial and temporal variability of surface water pollution in the Mekong Delta, Vietnam, *Science of the Total Environment 485,* pp. 653-665, 2014.

[4] Vrebos, Dirk, Olivier Beauchard, and Patrick Meire, The impact of land use and spatial mediated processes on the water quality in a river system," *Science of The Total Environment 601,* pp. 365-373, 2017.

[5] Hair, Joseph F., et al, Multivariate data analysis. Seventh edition, Pearson Education Limited, 2014.

[6] Gatica, Eduardo A., et al, Water quality assessment, by statistical analysis, on rural and urban areas of Chocancharava River (Río Cuarto), Córdoba, Argentina, *Environmental monitoring and assessment 184.12,* pp. 7257-7274, 2012.

[7] Jung, Kang Young, et al, Evaluation of water quality for the Nakdong River watershed using multivariate analysis, *Environmental Technology & Innovation 5,* pp. 67-82, 2016.

[8] Azhar, Shah Christirani, et al, Classification of river water quality using multivariate analysis, *Procedia Environmental Sciences 30,* pp. 79-84, 2015.

[9] Luo, Kun, et al, Using multivariate techniques to assess the effects of urbanization on surface water quality: a case study in the Liangjiang New Area, China, *Environmental monitoring and assessment 189.4,* p. 174, 2017.

[10] Kim, Sung Eun, Il Won Seo, and Soo Yeon Choi, Assessment of water quality variation of a monitoring network using exploratory factor analysis and empirical orthogonal function, *Environmental Modelling & Software 94,* pp. 21-35, 2017.

[11] Anonym, "Status Lingkungan Hidup Daerah DKI Jakarta, Badan Pengelola Lingkungan Hidup Daerah (BPLHD) Provinsi Daerah Khusus Ibukota Jakarta, 2012, 2013, 2014, 2015, dan 2016.

[12] Jouanneau, S., et al, Methods for assessing biochemical oxygen demand (BOD): A review, *Water research 49,* pp. 62-82, 2014.

Potential impact of particulate matter less than 10 micron (PM_{10}) to ambient air quality of Jakarta and Palembang

I Agustine[1*], H Yulinawati[1], D Gunawan[2], E Suswantoro[1]

[1]Environmental Engineering Department, Faculty of Landscape Architecture and Environmental Technology, Universitas Trisakti, Jakarta, Indonesia
[2]Meteorological, Climatological and Geophysics Agency (BMKG), Jakarta, Indonesia

*Corresponding Author: agustineintane@gmail.com

Abstract. Particulate is a main urban air pollutant affects the environment and human wellbeing. The purpose of this study is to analyze the impact of particulate matter less than 10 micron (PM_{10}) to ambient air quality of Jakarta and Palembang. The analysis is done with calendarPlot Function of openair model, which is based on the calculation of Pollutant Standards Index (PSI) or better known as Air Quality Index (AQI). The AQI category of "moderate" dominates Jakarta's calendar from 2015 to 2016, which indicates the impact of PM_{10} is the visibility reduction. There was one day with category "unhealthy" that indicates the impact of dust exposure everywhere in Jakarta during 2015. Similar to Jakarta, the AQI category "moderate" also dominates Palembang's calendar during 2015. However, the AQI category "hazardous" happened for few days in September and October 2015 during forest fires, which indicates the more harmful impacts of PM_{10}, such as reduced visibility, dust exposure everywhere, increased sensitivity in patients with asthma and bronchitis to respiratory illness in all exposed populations. During 2016, AQI category of Jakarta mostly "moderate", while in Palembang was "good". Dominant AQI category from 2015 to 2016 shows higher PM_{10} concentration occurred in Jakarta compared to Palembang.
Keywords: ambient, air quality, particulate, potential impact

1. Introduction

Particulate matter (PM) is a major urban air pollutants that has significant impacts on the environment and human wellbeing. It has high potential to be accumulated in the human respiratory system. The smaller the particulate size, the more likely it is to settle and be inhaled by the human respiratory system. In general, there are two particulate matter types: PM less than 10 micron (PM_{10}) and PM less than 2.5 micron ($PM_{2.5}$). The PM_{10} is inhalable and has the potential to enter the human respiratory system. It can also cause the decrease of visibility, smoke, haze, and smog. However, very limited cities in Indonesia monitor the PM_{10}, and only in recent years have been monitored.

Indonesia is one of the countries in the world with the deadliest air pollution levels [1]. World Health Organization (WHO) has stated that every year there are about 3.2 million cases of death caused by air pollution in the world. A total of 3.3 million people worldwide died in 2010 because of inhaling small dust flying in the air and it is estimated that this number will double by 2050 [2].

Jakarta and Palembang are two metropolitan cities in Indonesia that have various sources of air pollution. Types of pollutants emitted ranging from PM, carbon monoxide (CO), carbon dioxide (CO_2), nitrogen dioxide (NO_2), and others. Currently Jakarta and Palembang are undertaking mass transportation development projects namely Mass Rapid Transit (MRT) and Light Rail Transit (LRT) for Jakarta and LRT for Palembang which indicate the increasing particulate concentration especially PM_{10} in ambient air of both cities.

Monitoring ambient air quality towards PM_{10} can be done by analyzing the status of urban ambient air quality. Status of urban ambient air quality in Indonesia is often represented by the Air Quality Index (AQI). The AQI describes ambient air quality at a particular location and time based on impacts on human health, aesthetic values and other living things. The AQI is presented in the form of color categories that are simply used to inform public about ambient air quality at a particular location and time. The purpose of this study is to analyze the impact of PM_{10} to ambient air quality of Jakarta and Palembang according to AQI from 2015 to 2016 using openair model.

1.1 Literature Review

1.1.1 Openair Model

Openair model is a part of R package in the form of computer programming language developed specially by R-project for data analysis. The advantage of R is that this model comes in the form of free software that can be applied on various platforms from Windows, Mac, and Linux which are available internationally. This model is specifically designed to perform air quality analysis functions along with the atmospheric conditions. Some air quality analysis functions that can be used such as theilSen, windRose, timeVariation, and calendarPlot [3]. Several studies have been conducted using this model such as Munir's research on temporal variation of ozone [4] and Habeebullah's research about characterizing and temporal variability of nitrogen dioxide [5].

1.1.2 Air Pollution

Air as one of the environment component has a special characteristic that is in terms of high level of mobility. Changes in air quality due to high mobility can impact the environmental from local, regional to global scale. Air pollution is the entry or inclusion of substances, energies, and/or other components into the ambient air by human activities, so that ambient air quality drops to some degree which causes ambient air unable to fulfill its function [6]. Air pollution is characterized by the presence of pollutants in the atmosphere in certain concentrations which disturbs the dynamic balance of the atmosphere and has an effect on humans and the environment [7].

1.1.3 Air Quality Index (AQI)

Air Quality Index (AQI) is a tool, introduced by Environmental Protection agency (EPA) in USA to measure the levels of pollution due to major air pollutants [8]. AQI is a non-unit number that describes ambient air quality conditions at a particular location and time based on impacts on human health, aesthetic values, and other living things. In this case, AQI is presented as an air quality report for the public to simply explain how clean or contaminated air quality is and also how it affects human health. Procedure related to the calculation, reporting and information of Indonesia's AQI are regulated by government [9]. For example, a study of ambient air of Vapi City, India by Sarella found the overall AQI category was "moderate" [10].

1.1.4 Particulate Matter 10 (PM_{10})

Particulate Matter 10 (PM_{10}) consists of some elements such as silicate, calcium, and titanium [11]. As PM_{10} includes particles that have aerodynamic diameters less than or equal to 10 micrometers (μm), it is approximately equal to one-seventh the diameter of human hair [12]. The exposure of PM_{10} can be quite far in distance which is varying from less than 1 km to 10 km. The PM_{10} also has a long last age which remains in the air for a long time. A case study by Bigi about long-term trend and variability of PM_{10} concentration in the Po Valley is done in 2014 [13].

The sources of PM_{10} varies from vehicle emissions, dust roads, industry emission, agriculture, construction, building demolition, to airborne dust from burning fossil fuels and potentially cause eye irritation, decreased visibility, respiratory. The concentration of PM_{10} that is inhaled into the respiratory tract is 73.7% and has the potential to cause inflammation and irritation [14].

2. Research Method

2.1 Data Collection
Secondary data used in this research, specifically hourly data of PM_{10} concentration of Jakarta and Palembang from 2015 to 2016. The hourly PM_{10} concentration was collected from Indonesia's BMKG which is exactly from meteorology station Kemayoran for Jakarta and climatology station Kenten for Palembang. The PM_{10} exposure in ambient air is measured by the sampling and analysis method using an instrument called Particulate Monitor. The type of Particulate Monitor used in meteorology station Kemayoran is Beta Attenuation Monitor (BAM) and in climatology station Kenten is Thermo. These instruments run continuously and directly measuring PM_{10} concentration every hour.

2.2 Pre-analysis Data
The hourly PM_{10} concentration is validated using the standard of World Health Organization (WHO). If there is no PM_{10} concentration for at least 12 hours (50%) in a day, the entire concentration data on that day is deleted because it is considered invalid [15]. The validated hourly PM_{10} concentration is averaged using Microsoft Excel into the daily PM_{10} concentration for both Jakarta and Palembang. Then, the daily PM_{10} concentration data is processed and synchronized with AQI number to analyze the color index and the AQI category each day in a calendar. The color and category of AQI in this daily calendar will be used to estimate the impact of PM_{10} exposure.

2.3 Data Analysis
The daily PM_{10} concentration is processed into a color index representing ambient air quality status using calendarPlot function in the openair model. This function is often used to visualize data by showing daily pollutant concentration (diurnal) on calendar format. This function will convert the daily average concentration to a color index or a symbol with a range of AQI number and certain pollution level categories. The final result of this function analysis is an image of calendar with the AQI color to estimate the impact of PM_{10} exposure based on the range of its AQI number and category.

3. Results and Discussion
The average daily PM_{10} concentration is interpolated with the AQI number to identify its daily AQI category on the calendar. This interpolation is intended to obtain information for the AQI category, its index color, its symbol and the potential impact. Recapitulation of the interpolation between PM_{10} concentration and AQI can be seen in Table 1.

Table 1. Interpolation between PM_{10} concentration and AQI.

PM_{10} Concentration ($\mu g/m^3$)	AQI Number	AQI Category	Index Color	Symbol	Potential Impact
0 – 50	0 – 50	*Good*	Green	□	No effects
51 – 150	51 – 100	*Moderate*	Blue	=	Decreased visibility
151 – 350	101 – 199	*Unhealthy*	Yellow	/	Decreased visibility and dust exposure everywhere
251 – 420	200 – 299	*Very Unhealthy*	Red	×	Increased sensitivity in patients with asthma and bronchitis
421 – more	300 – more	*Hazardous*	Black	●	Dangerous levels for all exposed populations

IOP Conf. Series: Earth and Environmental Science **106** (2017) 012057 doi:10.1088/1755-1315/106/1/012057

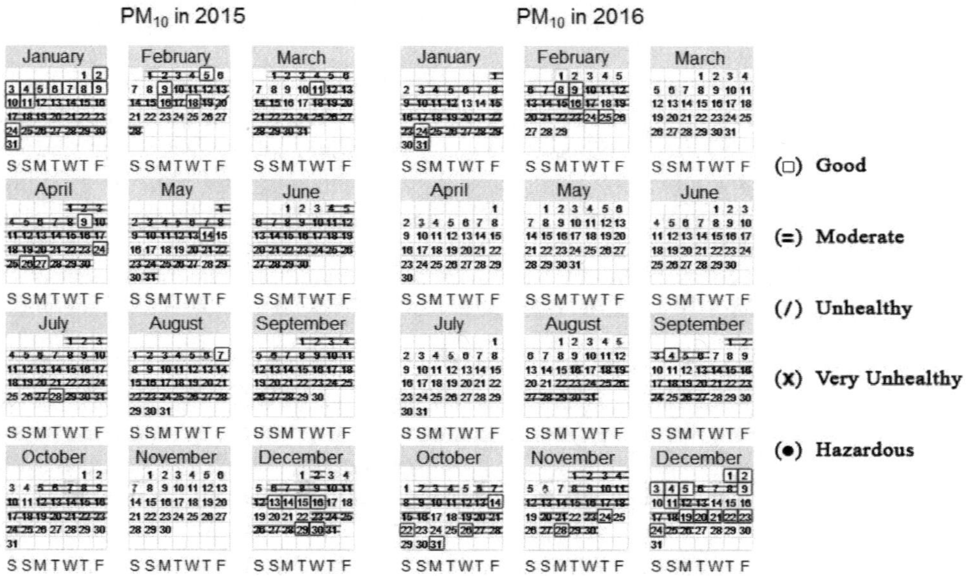

Figure 1. Calendar with AQI category for PM_{10} in Jakarta.

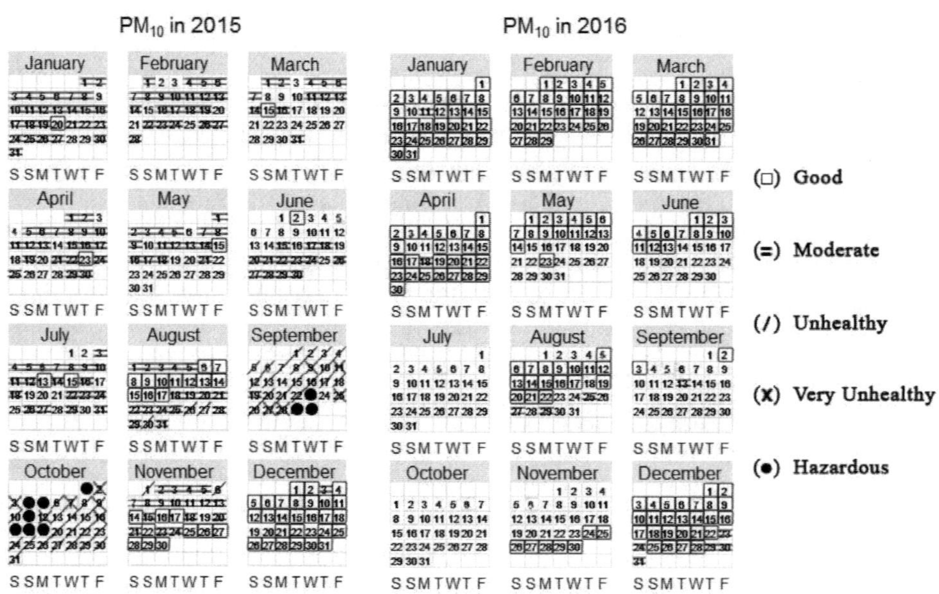

Figure 2. Calendar with AQI category for PM_{10} in Palembang.

3.1 Jakarta

The worst AQI category in Jakarta was "unhealthy" which was only occurred on February 20th, 2015 with PM_{10} concentration 150.6 μg/m^3 (symbol "/" and index color "yellow"). The impact of this PM_{10} concentration includes the decrease of visibility, along with dust exposure everywhere. The "good" days are recorded in 30 days during 2015 and 27 days during 2016 meanwhile the rest are "moderate".

Generally, the status of ambient air quality of Jakarta from 2015 to 2016 was "moderate". It is concluded by its index color which was dominated by "blue" or symbol "=" from 2015 to 2016 (Figure 1). Index color "blue" means AQI category is "moderate" with the impact of PM_{10} exposure is the decrease of visibility. The upper limit PM_{10} concentration of the AQI category "moderate" is 150 μg/m^3 which is the standard value of ambient air quality standard in Indonesia for PM_{10}. Therefore, the index color in AQI is only limited to "green" and "blue" to classify as a good ambient air quality.

3.2 Palembang

Figure 2 shows the calendar with AQI category for PM_{10} in Palembang for 2015 and 2016. There were few days with the worst AQI category from "unhealthy" to "hazardous" in 2015 (42 "unhealthy" days, 11 "very unhealthy" days, 10 "hazardous" days). The highest concentration 606.0μg/m^3 occurred on September 28th, 2015 with the impact such as visibility reduction, dust exposure everywhere, increased sensitivity in patients with asthma and bronchitis to respiratory illness in all exposed populations. Those were expected to be the result of forest fires on Sumatra Island during El-Nino phenomenon.

Based on domination of AQI category and index color, air quality status of Palembang during 2015 was quite good with AQI category "moderate" or symbol "=". The impact was limited to the decrease of visibility. In 2016, the air quality status of Palembang was getting better which was dominated by the index color "green" (symbol "□") and some "blue" (symbol "=").Overall, the "good" days were recorded in 56 days during 2015 and dominated in 192 days during 2016. The status of ambient air quality of Palembang from 2015 to 2016 was getting better from "hazardous" to "good".

3.3 Comparison between Jakarta and Palembang

Table 2. Highest PM_{10} concentration alongside its Index Color and AQI category

Year	Jakarta		Palembang	
	Index Color (AQI Category)	**PM_{10} Concentration (μg/m^3)**	**Index Color (AQI Category)**	**PM_{10} Concentration (μg/m^3)**
2015	*Yellow (Unhealthy)*	150.6	*Black (Hazardous)*	606.0
2016	*Blue (Moderate)*	134.9	*Blue (Moderate)*	76.4

During 2015 when Indonesia experiences El-Nino phenomenon, the highest average daily PM_{10} concentration of Jakarta reached 150.6 μg/m^3 while Palembang reached 606.0μg/m^3 (Table 2). Both of these concentrations have passed the permitted ambient air quality standard that is below150μg/m^3. The gap between these two concentrations have resulted a significant difference of AQI category and its index color. The highest concentration in Jakarta has the symbol "/" or index color "yellow" with AQI category "unhealthy" while in Palembang has the symbol "●" or index color "black" with AQI category "hazardous". This condition indicates that higher PM_{10}exposure occurs in Palembang with more harmful impact than in Jakarta. If the impact in ambient air of Jakarta is limited to the decrease of visibility and dust exposure everywhere, the impact on Palembang City's ambient air is more dangerous such as decreased visibility, dust exposure everywhere, increased sensitivity in patients with asthma and bronchitis to respiratory illness in all exposed populations.

Table 3. Comparison of Dominant AQI index color

Year	Dominant Index Color	
	Jakarta	**Palembang**
2015	Blue (=)	Blue (=)
2016	Blue (=)	Green (o)

During 2016, the highest average daily PM_{10} concentration in both Jakarta 134.9 μg/m^{3}and Palembang76.4μg/m^3 has the same symbol "=" or index color "blue" with AQI category "moderate". The impact of pollution that occurs is similar which the decreased of visibility. However when analyzed based on the highest PM_{10} concentration during 2016, PM_{10} exposure tends to be higher in Jakarta compared to Palembang.

Table 3 shows comparison of index color on AQI calendar which respectively dominates Jakarta and Palembang from 2015 to 2016. During 2015, the dominant index color of Jakarta and Palembang were "blue" or symbol "=" with the AQI category "moderate" which indicates PM_{10} exposure and impacts that happens to be the same. As the result, the impact of pollution of PM_{10} that occurs in both city was only limited to the decrease of visibility. While during 2016, dominant index color was found differently in Jakarta ("blue" or symbol "=") and Palembang ("green" or symbol "□"). The differences indicate that the exposure of $PM_{10}$was higher in the city of Jakarta compared to Palembang during 2016. Overall based on the dominant index color for the entire year 2015 until 2016, higher concentration of PM_{10} occurred in Jakarta than in Palembang.

4. Conclusion

The highest PM_{10} concentration in Jakarta and Palembang has exceeded the ambient air quality standard (150 $\mu g/m^3$). Based on daily analysis of Indonesia's AQI, the worst impact of PM_{10} in Jakarta is visibility reduction and dust exposure everywhere. Whereas in Palembang tends to have higher daily PM_{10} concentration that more harmful impacts such as the decrease of visibility, dust exposure everywhere, increased sensitivity in patients with asthma and bronchitis as well as respiratory illness in all exposed populations. However dominant AQI category from 2015 to 2016 shows higher PM_{10} concentration occurred in Jakarta compared to Palembang. It is needed to continue monitor PM_{10} to reduce the negative impacts with air pollution mitigation program.

References

[1] Randall T. 2015. Breathing is Deadliest in These 15 Countries (New York: Report of Bloomberg Business)
[2] Lelieveld J, Evans JS, Fnais M, Giannadaki D and Pozzer A. 2015. The Contribution of Outdoor Air Pollution Sources to Premature Mortality on a Global Scale. *Journal Nature.* 525:367
[3] Carslaw D. 2015. The Openair Manual Open-Source Tools for Analyzing Air Pollution Data (London: King's College) p 69
[4] Munir S. 2015. An Analysis into the Temporal Variation of Ground Level Ozone in the Arid Climate of Makkah applying k-means Algorithms. *Journal of Environment Asia.* 8(1):53–60
[5] Habeebullah TM. 2015. Characterising NO_2, Its Temporal Variability and Association with Meteorology A Case Study in Makkah, Saudi Arabia. *Journal of Environment Asia.* 8 (2):37–44.
[6] *Government Regulation No. 41/1999 on Air Pollution Control, Central Government* Indonesia
[7] Mukono HJ. 2003. *Pencemaran Udara dan Pengaruhnya terhadap Gangguan Saluran Pernapasan.* (Air Pollution and Its Effects to Respiratory Illness) (Surabaya: Universitas Airlangga Press)
[8] Mamta P and Bassin JK. 2010. Analysis of Ambient Air Quality Using Air Quality Index – A Case Study. *International Journal of Advanced Engineering Technology I.* II:106–114
[9] *Decree of the Head of the Environmental Impact Control No. 107/1997 on Calculation, Reporting and Information of Air Quality Index* Indonesia
[10] Sarella G and Khambete A K. 2015. Ambient Air Quality Analysis using Air Quality Index – A Case Study of Vapi. *International Journal for Innovative Research in Science & Technology.* 1(10).
[11] Tecer L et.al. 2012. Metallic Composition and Source Apportionment of Fine and Coarse Particles Using Positive Matrix Factorization in the Southern Black Sea Atmosphere *Atmospheric Research.* 118:153–169
[12] EPA's Report on the Environment Particulate Matter Emission. 2017. (Washington DC: Environmental Protection Agency)
[13] Bigi, A and Ghermandi G. 2014. Long-Term Trend and Variability of Atmospheric PM_{10} Concentration in the Po Valley *Atmos. Chem. Phys.* 14 p 4895-4907.
[14] Surjanto. 2007. Hubungan antara Pajanan Total Suspended Particulate (TSP) dan Particulate Matter 10 µm (PM_{10}) di Udara Ambien dengan Gangguan Saluran Pernafasan (Correlation between TSP and PM_{10} Exposures in Ambient Air to Respiratory Illness) Thesis (Depok: Universitas Indonesia)
[15] Tiwary A and Colls J. 2010. Air Pollution: Measurement, Modeling, and Mitigation 3rd Edition (New York: Routledge)

IOP Conf. Series: Earth and Environmental Science **106** (2017) 012058 doi:10.1088/1755-1315/106/1/012058

Influence of motorcycles lane to the traffic volume and travel speed in Denpasar, Indonesia

A M Mulyadi

Institute of Road Engineering, Kementerian Pekerjaan Umum dan Perumahan Rakyat, Bandung, Indonesia.

*Corresponding Author: agah.muhammad@pusjatan.pu.go.id

Abstract. The large number of motorcycles population was causing a decrease in road performance. In order to overcome these problems, motorcycles lane was build to make homogeneity movement of the motorcycle on the road. The purpose of this study is to analysis the influence of motorcycles lane to the traffic volume, travel speed and density. In this study, data was collected at three times segmentation times i.e. morning, noon and afternoon in four locations, there were Puputan road, Cok Agung Tresna road, Sudirman road heading south, and Sudirman road heading north in Denpasar. The data were collected are traffic volume and travel speed on motorcycles lane and non motorcycles lane. According to the paired t test of travel speed and traffic volume and density there was significant differences between motorcycle lane and without motorcycle lane.
Keywords: four-wheeled vehicle, motorcycles lane, traffic volume, traffic jammed, transportation mode

1. Introduction

Motorcycles as the most popular of transportation mode in Indonesia. Motorcycles have some advantages such as the relative easyly move by using narrow gap between other vehicles, so that the motorcycle made the main choice to travel, especially when traffic jammed occurs. Ownership of motorcycles increased year by year and dominated the movement of traffic on the road with motorcycle population reaching 98.20 million units or 80.5 percent of the overall traffic composition of vehicles in Indonesia [1]. The motorcycle population is shown in Figure 1.

The high composition motorcycle's ownership and changes riding behavior of motorcycle travel from short distance to long distance caused an impact to the traffic condition. The major impact of the problem is decrease of road capacity and increase the number of accidents. Issue of the Government Regulation no 22 in year 2009 is the regulating to force the movement of motorcycles should be on the left lane of the road [2] in order to reduce traffic conflicts with four-wheeled vehicles that generally have higher speeds.

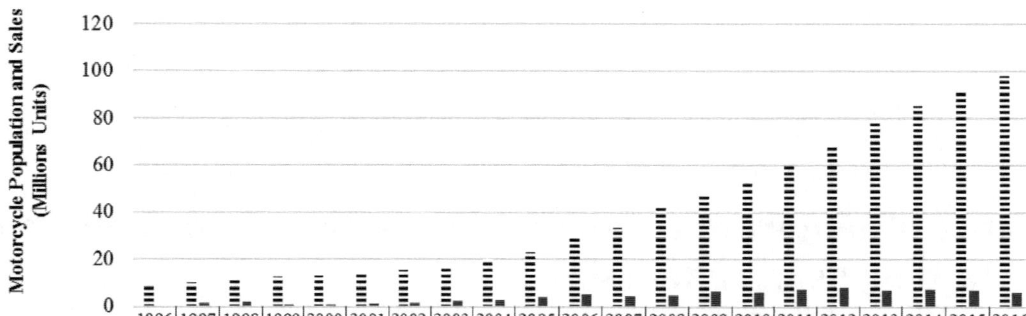

	1996	1997	1998	1999	2000	2001	2002	2003	2004	2005	2006	2007	2008	2009	2010	2011	2012	2013	2014	2015	2016
Population	8.71	10.1	11.7	12.6	13.1	13.6	15.5	17	20	23.1	28.6	33.4	42	47.7	52.4	59.8	67.8	77.7	85.4	92.3	98.2
Sales		1.38	1.8	0.43	0.49	0.86	1.58	2.27	2.81	3.9	5.07	4.43	4.69	6.22	5.88	7.4	8	6.65	7.22	6.88	5.93

Figure 1. Population and sales of motorcycles.

The concept of homogeneity of traffic movement by separate road's lane for motorcycles is one alternative solution to reduce the traffic conflict [3]. Therefore, to force the motorcycle to be on the left lanes, it is necessary to provide motorcycle lane.

Motorcycles lanes are divided into two types of inclusive motorcycles lanes and exclusive motorcycles lanes (Figure 2). Inclusive motorcycles lane is a special lane used for motorcyclists, which serves to separate motorcycles with other vehicles using traffic signs and road markings. Meanwhile, exclusive motorcycles lane is a special lane used for motorcycle, which separate the motorcycle with other vehicles by using a road separator or curb. Both exclusive or inclusive motorcycles lane are expected to reduce the number of accidents involving motorcycles because it is physically separate from other types of vehicles .

Motorcycle lanes are required when the number of motorcycles is greater than 600 motorcycles/ hour/ lane [4]. This number is obtained from determining the headway between motorcycles when driving on urban roads is at least 6 seconds. If converted into hours, then get 600 motorcycles. Providing motorcycles lanes is to reduce accidents. If the amount of accidents is greater than 40 accidents/km/year or the accident rate 100, then it has to have motorcycle lane [5].

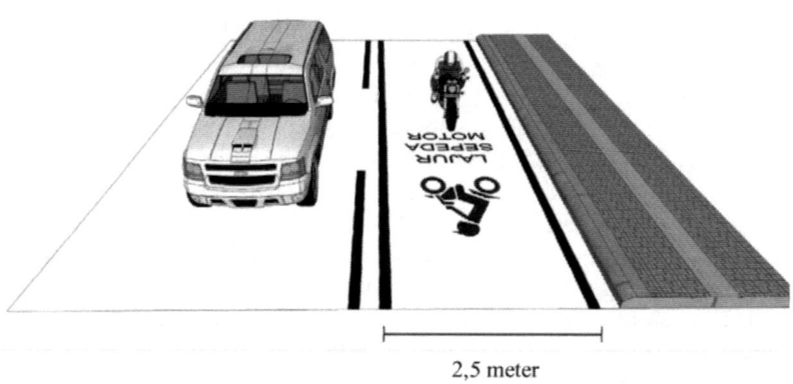

Figure 2. Motorcycles lane.

The placement of motorcycles lanes must meets with same criteria, i.e. motorcycles lanes are priority lane than other vehicles lane on the road, motorcycles lanes in urban arterial roads or suburban arteries which have urbanized travel characteristics. Motorcycles lanes in the urban arterial road are on the left side and should be separated from the lane of four-wheeled vehicles by using road marking or curb separator. This study to analyzes the difference of traffic volume, travel speed and traffic density between a roads with motorcycles lane and without motorcycles lane facilities. It can be seen whether the motorcycles lanes affect the traffic volume and travel speed and traffic density on two different condition with the same criteria.

2. Research Method

2.1. Methods of Collecting Data

Traffic volume and travel speed data collection was conducted in mid July 2016 in four places, there were Puputan Niti Mandala Renon Road (Road with motorcycle lane), Cok Agung Tresna Road (road without motorcycle lane) and Sudirman Road way to south (road with motorcycle lane) and Sudirman Road way to the north (road without motorcycle lane) in Denpasar City. Data were collected during morning, noon and afternoon sessions. Each sessions was taking for two hours of data with intervals of five minutes. The data was taken by the surveyor by using traffic counter eqquipment to collect traffic volume data and speed gun eqquipment to collect travel speed data. Motorcycles lanes on Puputan road and Sudirman road which have width 2.5 meter were divided into 2 lanes in one direction. In addition, there are 2 lanes for other types of vehicles with a width of 3.5 meters. The data collection surveys is shown in Figure 3 and the roads survey condition is shown in Table 1.

Table 1. The roads survey condition.

Location	Criteria	Number of Lanes	Roads Classification [6]	Road Width	Remarks
Puputan Niti Mandala Renon Road	with motorcycle lane	3 lanes	Arterial Secondary	9 meters	Location 1
Cok Agung Tresna Road	without motorcycle lane	3 lanes	Arterial Secondary	9 meters	
Sudirman Road way to the south	with motorcycle lane	2 lanes	Local Primary	8 meters	Location 2
Sudirman Road way to the north	without motorcycle lane	2 lanes	Local Primary	8 meters	

2.2. Method of data analysis

Data analysis in this research was done by using paired sample t test. Paired sample t test is an analytical technique to compare one or more independent variables. This technique is used to test whether a particular value differs significantly or not with an average of two samples. The t test as a technique of testing the descriptive hypothesis has two criteria on each parameter analysis that is volume, speed and density.

Table 2. One-Sample Kolmogorov-Smirnov Test.

		Unstandardized Residual
N		228
Normal Parameters[a,b]	Mean	0E-7
	Std. Deviation	180,18704216
Most Extreme Differences	Absolute	0,273
	Positive	0,273
	Negative	-0,142
Kolmogorov-Smirnov Z		1,337
Asymp. Sig. (2-tailed)		0,556
a. Test distribution is Normal.		
b. Calculated from data.		

IOP Conf. Series: Earth and Environmental Science **106** (2017) 012058 doi:10.1088/1755-1315/106/1/012058

This research uses 5% significance level (0.05) based on research hypothesis that has been determined. Hypothesis using statistical test paired sample t test [7] was conducted to test the pairs of traffic parameters data as follows:

- If the value is significant <0.05 then H0 is accepted (no difference / significant)
- If the value is significant> 0.05 then H0 is rejected (there is a difference / not significant)

Prior to analysis t test of paired sample should be tested normality data, so it can get good value. Normality test aims to test whether the data used in the modelling has been distributed normal or not. To detect data normality, it can be done by One Sample Kolmogorov-Smirnov Test. If the significance value of the Kolmogorov-Smirnov One-Sample Test> 0.05 [10], then the assumption of normality is met as shown in Table 2.

The normality test results obtained significance value of the Kolmogorov-Smirnov One-Sample Test results of 0.556, then the assumption of normality for the data used in this study has been met.

3. Results and Discussion

The traffic volume data were processed by converting traffic volume data to passenger car unit (pcu/ hour) [8]. After that was divided into traffic volume per lane [9].

Table 3. Result of Paired T-test statistic test at Location 1.

Time Session	Parameter Analysis	Criteria	Sig (2 Tailed)	Level of significance	Explanation
Morning	Traffic Volume	Motorcycle Lane	0.492	0.05	Not Significant
		Non- Motorcycle Lane			
	Travel Speed	Motorcycle Lane	0.284	0.05	Not Significant
		Non- Motorcycle Lane			
	Traffic Density	Motorcycle Lane	0.532	0.05	Not Significant
		Non- Motorcycle Lane			
Day	Traffic Volume	Motorcycle Lane	0.097	0.05	Not Significant
		Non- Motorcycle Lane			
	Travel Speed	Motorcycle Lane	0.086	0.05	Not Significant
		Non- Motorcycle Lane			
	Traffic Density	Motorcycle Lane	0.491	0.05	Not Significant
		Non- Motorcycle Lane			
Noon	Traffic Volume	Motorcycle Lane	0.037	0.05	Significant
		Non- Motorcycle Lane			
	Travel Speed	Motorcycle Lane	0.263	0.05	Not Significant
		Non- Motorcycle Lane			
	Traffic Density	Motorcycle Lane	0.411	0.05	Not Significant
		Non- Motorcycle Lane			

From the results of data processing, it was found that the traffic volume on the road with motorcycle lane facilities is greater than in the road without motorcycle lane traffic in location 1 (Table 3). At Jalan Raya Puputan (with motorcycle lane), the traffic volume in the morning and afternoon session sessions is 722 pcu/ hour/lane and 560 pcu/hour/lane which higher than Jalan Cok Agung Tresna (without motorcycle lane) by 536 pcu/hour/lane and 457 pcu/hour/lane respectly. In the afternoon seassions the traffic volume on Jalan Cok Agung Tresna (without motorcycle lane) is 636 pcu/hour/lane was greater than the traffic volume on Jalan Raya Puputan (motorcycle lane) which is 586 pcu/hour/lane.

In the location 2, the traffic volume on the road with motorcycle lane facilities has a higher than on the roads without motorcycle Lane in the morning, afternoon and also afternoon sessions. The traffic volume on the road with motorcycle lane facilities is 25% larger than on the road without the motorcycle lane facilities. The largest traffic volume is on the road with the motorcycle lane facility in the morning session by of 1109 pcu/hour/lane. As for the largest traffic volume on the road without motorcycle lane is 885 pcu/ hour/lane.

Table 4. Result of Paired T-test statistic test at Location 2.

Time Session	Parameter Analysis	Criteria	Sig (2 Tailed)	Level of significance	Explanation
Morning	Traffic Volume	Motorcycle Lane	0.284	0.05	Not Significant
		Non- Motorcycle Lane			
	Travel Speed	Motorcycle Lane	0.073	0.05	Not Significant
		Non- Motorcycle Lane			
	Traffic Density	Motorcycle Lane	0.597	0.05	Not Significant
		Non- Motorcycle Lane			
Day	Traffic Volume	Motorcycle Lane	0.335	0.05	Not Significant
		Non- Motorcycle Lane			
	Travel Speed	Motorcycle Lane	0.181	0.05	Not Significant
		Non- Motorcycle Lane			
	Traffic Density	Motorcycle Lane	0.662	0.05	Not Significant
		Non- Motorcycle Lane			
Noon	Traffic Volume	Motorcycle Lane	0.293	0.05	Significant
		Non- Motorcycle Lane			
	Travel Speed	Motorcycle Lane	0.064	0.05	Not Significant
		Non- Motorcycle Lane			
	Traffic Density	Motorcycle Lane	0.520	0.05	Not Significant
		Non- Motorcycle Lane			

Data processing speed at location 1 has the largest spread distribution of data, espesially the data on Cok Agung Tresna (without motorcycle lane) that have an average travel speed of 62 km/h during the afternoon session. Meanwhile on the Puputan road with the motorcycle lane facilities, the average travel speed is 55 km/h in morning and afternoon session.

Result of data processing of travel speed in location 2 has the largest data spread distribution data that is at the location of Sudirman road (north direction) that have an average speed of 50 km/hour in morning and afternoon seassion. Meanwhile on Sudirman Road (south direction) without motorcycle lane facilities th average speed is 37 km/h in the afternoon seassion.

From the test results it was obtained significance value for the difference of analysis parameters ie traffic volume, traffic speed and traffic density on the road with motorcycle lane and without motorcycle lane > 0.05 then H0 is rejected (there are differences/not significant). Result of data test at location 1 has got difference of data between two criterions of location at all time session and test parameter in the afternoon session there were traffic volume with a significance value of 0.037. This value shows that there is no significant difference in the traffic volume value between road with motorcycle lane and non-motorcycle lane at that time.

The results the analysis data was taken at location 2, it's obtained the significance value for the different parameters of analysis i.e. traffic volume, traffic speed and traffic density on road with motorcycle lane and without motorcycle lane location criteria >0,05 then H0 rejected (there is difference/not significant). The volume of traffic, travel speed and traffic density between motorcycle Lane and without Motorcycle Lane can be concluded differently as shown in Table 4.

4. Conclusion

The result of hypothesis test by Paired t test shows that traffic volume at location 1 shows that there is significant difference between traffic volume in morning and afternoon session time. Meanwhile in the afternoon there were no insignificant differences between with motorcycle lane and without motorcycle lane. For the analysis of travel speed and traffic density, there were significant differences at all time sessions. The result of hypothesis test by Paired t test shows that the traffic volume at location 2 shows that there are significant differences of traffic volume, travel speed and traffic density and at all time session. Based on the results of the research which was conducted, recommendations can be given that the presence of motorcycle lane could improve road performance significantly. Separation of movement between motorcycles and other types vehicles give more road space to other vehicles so that can increase the travel speed.

References

[1] Board of Statistic Centre. Vehicle Development Based on Types of Vehicle. Jakarta. 2016.
[2] Government of the Republic of Indonesia Law Number 22 Year 2009 on Road Traffic and Transportation Article 108 paragraph 3. Indonesia.
[3] Muhammad Idris. Criteria for Motorcycle Lines For Secondary Arterial Road. Bandung: Pustlitbang Jalan dan Jembatan. 2010.
[4] Institute of Road Engineering. Design of Motorcycles Lane in Urban Artery. Institute of Road Engineering. Bandung. 2011.
[5] Institute of Road Engineering. Scientific Script Design of Motorcyclez Lane in Urban Arteries. Institute of Road Engineering. Bandung. 2011.
[6] Denpasar Local Regulation number 27 Year 2011. About Spatial Planning Denpasar City Year 2011 - 2031. Denpasar-Bali.
[7] Ott R L, Longnecker M, An Introduction to Statistical Methods and Data Analysis. Fifth edition, Duxbury 511 Forest Lodge Road Pacific Grove, CA 93950, USA. 2001.
[8] Ministry of Public Works. Indonesia Road Capacity Manual (MKJI). Jakarta. 1997.
[9] A Caroline Sutandi, Wimpy Santosa. Immanuel Krisnaldi Loen. The Influence of the Lane of Motorcycle on The Performance of Dokter Djunjunan Road. Bandung. 2012.
[10] IBM Knowledge Center. Legacy Dialogs (Nonparametric Tests). https://www.ibm.com/support/ knowledgecenter/en/ SSLVMB_23.0.0/spss/ base/idh_ntk1.html. Accessed on august 2016. 2012.

IOP Conf. Series: Earth and Environmental Science **106** (2017) 012059 doi:10.1088/1755-1315/106/1/012059

River water pollution condition in upper part of Brantas River and Bengawan Solo River

D Roosmini[1*], M A Septiono[2], N E Putri[3], H M Shabrina[3], I R S Salami[1], H D Ariesyady[1]

[1]Environmental Technology and Management Research Group, Faculty of Civil and Environmental Engineering, Institut Teknologi Bandung, Bandung, Indonesia
[2]Bali Fokus Foundation, Sesetan, Denpasar, Bali, Indonesia
[3]Undergraduate Program of Environmental Engineering, Faculty of Civil and Environmental Engineering, Institut Teknologi Bandung, Bandung, Indonesia

*Corresponding Author: dwinaroosmini@yahoo.com

Abstract. Wastewater and solid waste from both domestic and industry have been known to give burden on river water quality. Most of river water quality problem in Indonesia has start in the upper part of river due to anthropogenic activities, due to inappropriate land use management including the poor wastewater infrastructure. Base on Upper Citarum River Water pollution problem, it is interesting to study the other main river in Java Island. Bengawan Solo River and Brantas River were chosen as the sample in this study. Parameters assessed in this study are as follows: TSS, TDS, pH, DO, and hexavalent chromium. The status of river water quality are assess using STORET method. Based on (five) parameters, STORET value showed that in Brantas River, Pagerluyung monitoring point had the worst quality relatively compared to other monitoring point in Brantas River with exceeding copper, lead and tin compared to the stream standard in East Java Provincial Regulation No. 2 in 2008. Brantas River was categorized as lightly polluted river based on monitoring period 2011-2015 in 5 monitoring points, namely Pendem, Sengguruh, Kademangan, Meritjan and Kertosono.
Keywords: Brantas River, Bengawan Solo River, hexavalent chromium, polluted river, STORET, solid waste

1. Introduction

River pollution has been an immense problem to most rivers in Java Island. Pollution loads from human activities, both from domestic, industry, wastewater and solid waste. Previous study [4] reported that in several West and Central Java EPA monitoring location measured several exceeding parameters compared to stream standard. Citarum River, specifically in Koyod monitoring point, assessment of water quality monitoring during 2011 – 2014 showed that certain organic substances, heavy metal and *E. coli* exceeding the standards as written in city, provincial and/or government regulation. Bengawan Solo River lays on Central and East Java Province also had decreasing trend of DO value along 2011 to 2014. Registered industries in both rivers discharged wastewater into it, mostly textile industries. In Citarum river basin laid 74% of 561 textile industries while 41.5% of 253 textile industries in Bengawan Solo river basin.

IOP Conf. Series: Earth and Environmental Science **106** (2017) 012059 doi:10.1088/1755-1315/106/1/012059

Industrial wastewater is one of the contributing sector in river pollution. Prior researches [3], [4], [5] showed that wastewater was still toxic and exceeding the effluent standard when discharged into the river. Some industries were discharging wastewater that below the effluent standard with relatively high toxicity level [1].

This research was aimed to assess river water quality in upper part of river for two main rivers in Java Island Brantas and Bengawan. The assessment are using STORET method and water quality standard are using local government regulation and national regulation.

2. Research Method

Water quality data or upper part Brantas River and Bengawan Solo River collected from regular monitoring data in Provincial Environmental Protection Agency and data collected by field sampling in several monitoring point as showed in Table 1.

Table 1. Monitoring stations in Upper Citarum River, Bengawan Solo River and Brantas River.

No	River	Monitoring Points	Remarks	Stream Standards
1	Bengawan Solo River	**BS-1**: Outlet of Gajah Mungkur Dam	Wonogiri Regency	Government Regulation No. 81 in 2001
		BS-2: Taman Jurug Bridge	Surakarta City	
		BS-3: Jalan Arteri	Dusun Ngringgo Palur, Karanganyar Regency	
		BS-4: Kliwonan Bridge	Dusun Pulang, Masaran Disctrict, Sragen Regency	
		BS-5: Dusun Gawan Bridge	Tanon District, Sragen Regency	
2	Brantas River	Pendem Bridge	Malang City	• Government Regulation No.82/2001 • East Java Provincial Regulation No. 2/2008
		Dinoyo Bridge	Malang City	
		Gadang Bridge	Malang City	
		Sengguruh Bridge	Malang City	
		Kali Pare Bridge	Malang Regency	
		Kademangan Bridge	Blitar Regency	
		Ngujang Bridge	Tulungagung Regency	
		Meritjam Bridge	Kediri	
		Kertosono Bridge	Kediri	
		Ploso Bridge	Jombang Regency	
		Padangan Bridge	Mojokerto Regency	

Table 2. Scoring for STORET scoring.

Amount of Sample	Value	Parameter Type		
		Physical	Chemical	Biological
<10	Maximum	-1	-2	-3
	Minimum	-1	-2	-3
	Average	-3	-6	-9
≥ 10	Maximum	-2	-4	-6
	Minimum	-2	-4	-6
	Average	-6	-12	-18

Table 3. Status of river based on STORET scoring.

Status	STORET Score
Lightly Polluted	-1 to -10
Moderately Polluted	-11 to -30
Heavily Polluted	> -30

IOP Conf. Series: Earth and Environmental Science **106** (2017) 012059 doi:10.1088/1755-1315/106/1/012059

3. Results and Discussion

3.1. Upper Bengawan Solo River

Bengawan Solo River located on two different provinces, namely Central Java and East Java, which also involves two different Provincial-EPA in monitoring program. This study is held in upper part of the river, under the authority of Central Java Provincial. Monitoring data collected from Provincial EPA are shown in Figure 1. During 2010-2015, showing that TDS and pH are comply to local government standard. The chromium hexavalent concentration exceeded the standard on May and July 2012 in BS-2 monitoring point. TSS value varies from 21 mg/L (Q1) to 42 mg/L (Q3), however the maximum TSS value is 519 mg/L as the peak in BS-1 on October 2013. The high concentration of TSS was also discovered on the same year in BS-2 (May and July), BS-3 (May and July) and BS-4 (May and July).

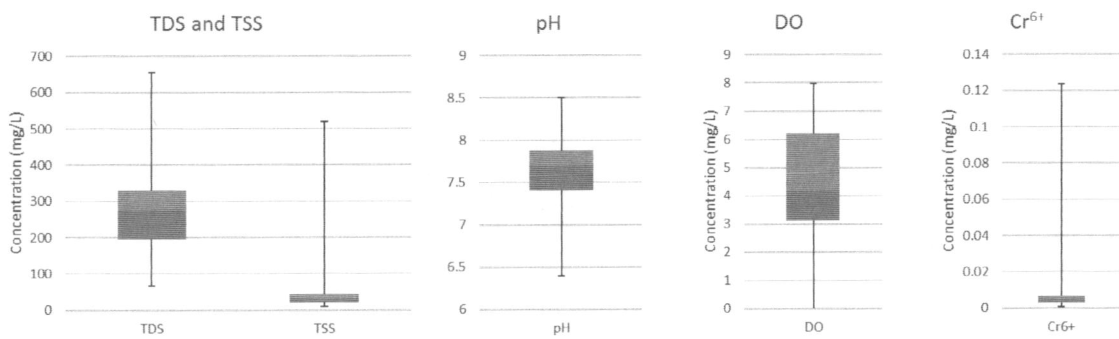

(a) TDS and TSS (b) pH (c) DO (d) Cr^{6+}

Figure 1. Bengawan Solo River monitoring results in BS-1, BS-2, BS-3, BS-4, BS-5
(Developed from Central Java EPA data, 2010 – 2015)

DO was the most frequent river water quality parameter that easily violated stream standard, showed the Figure 2 fluctuated DO value in each monitoring station, the lowest concentration identified at BS-4 reaching 0 mg/L on August 2011, September 2012 and November 2015.

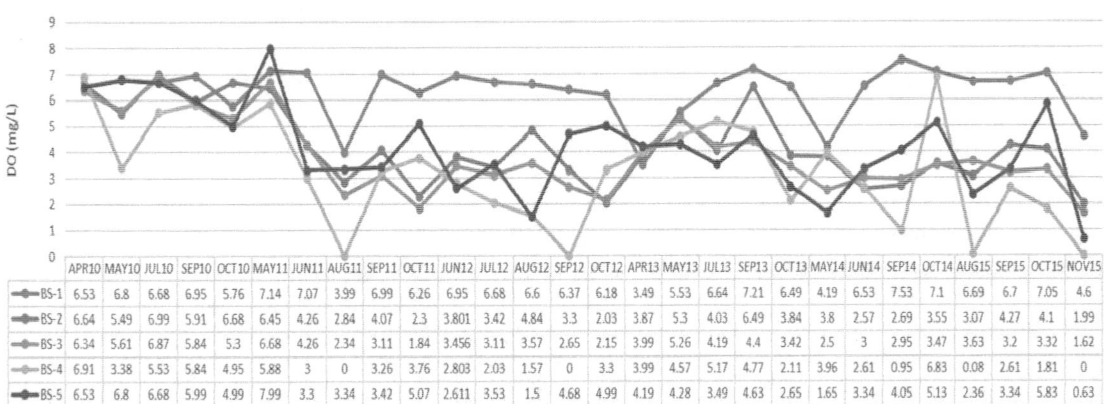

	APR10	MAY10	JUL10	SEP10	OCT10	MAY11	JUN11	AUG11	SEP11	OCT11	JUN12	JUL12	AUG12	SEP12	OCT12	APR13	MAY13	JUL13	SEP13	OCT13	MAY14	JUN14	SEP14	OCT14	AUG15	SEP15	OCT15	NOV15
BS-1	6.53	6.8	6.68	6.95	5.76	7.14	7.07	3.99	6.99	6.26	6.95	6.68	6.6	6.37	6.18	3.49	5.53	6.64	7.21	6.49	4.19	6.53	7.53	7.1	6.69	6.7	7.05	4.6
BS-2	6.64	5.49	6.99	5.91	6.68	6.45	4.26	2.84	4.07	2.3	3.801	3.42	4.84	3.3	2.03	3.87	5.3	4.03	6.49	3.84	3.8	2.57	2.69	3.55	3.07	4.27	4.1	1.99
BS-3	6.34	5.61	6.87	5.84	5.3	6.68	4.26	2.34	3.11	1.84	3.456	3.11	3.57	2.65	2.15	3.99	5.26	4.19	4.4	3.42	2.5	3	2.95	3.47	3.63	3.2	3.32	1.62
BS-4	6.91	3.38	5.53	5.84	4.95	5.88	3	0	3.26	3.76	2.803	2.03	1.57	0	3.3	3.99	4.57	5.17	4.77	2.11	3.96	2.61	0.95	6.83	0.08	2.61	1.81	0
BS-5	6.53	6.8	6.68	5.99	4.99	7.99	3.3	3.34	3.42	5.07	2.611	3.53	1.5	4.68	4.99	4.19	4.28	3.49	4.63	2.65	1.65	3.34	4.05	5.13	2.36	3.34	5.83	0.63

Figure 2. DO concentration in BS-5 monitoring point.
(Developed from Central Java EPA data, 2010 – 2015)

BS-5 and BS-4 were monitoring point which representing high pollution load from seven cities/regencies, namely Wonogiri, Sukoharjo, Klaten, Surakarta City, Karanganyar, Boyolali and Sragen. Compared to the first three monitoring stations (Figure 2). Based on STORET results on each monitoring point shows that BS-3 and BS-4 are the most polluted status, classified as moderately polluted (Table 5). The location of each monitoring point of Central Java Provincial EPA are shown in Figure 3.

Table 4. Exceeding parameters (percentage) during monitoring 2010 – 2015 of Bengawan Solo River (Central Java segment).

Parameters	Exceeding
TDS	0.00%
TSS	17.14%
pH	0.71%
DO	47.14%
Cr6+	1.43%

Table 5. Pollution level of Bengawan Solo River (Central Java segment) monitoring stations using STORET.

Points	Score	Pollution
BS-1	-6	Light
BS-2	-16	Moderate
BS-3	-24	Moderate
BS-4	-24	Moderate
BS-5	-12	Moderate

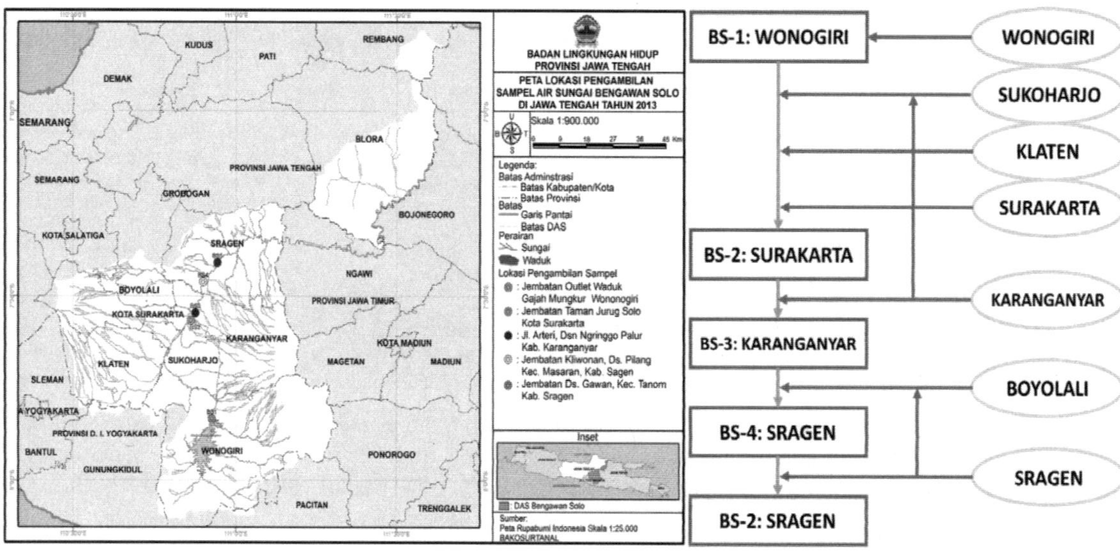

Figure 3. Upper Bengawan Solo River in Central Java Province (left) and simplified flowchart on the pollution load on each monitoring station (right).
(Modified from Central Java EPA, 2015)

3.2. Upper Brantas River

Brantas water basin (Figure 4) is the biggest river in East Java Province, and receiving waste water not only from 293 registered industrial sector, also agriculture and livestock as well. Agriculture and livestock will also contribute the pollution load into the river. From Upper Brantas River collected water sample are from 11 monitoring point along the upper part of the river as shown in Figure 4. Base on East Java EPA monitoring data on data from 2010-2015, parameter of TDS, TSS, pH and Chrome hexavalent are chose and analyzed using descriptive statistic to evaluate the most polluted point in Upper Brantas River. The mean and boxplot of data from 2010-2015 are shown in Figure 5. Most of the data are fluctuated, the worst condition identified in monitoring point of Kertosono up to Padangan as it the downstream of the studied river. Some high concentration of Chrome hexavalent are found in Padangan monitoring point. The pH range are still comply to river water standard regulation along the river from upstream to downstream.

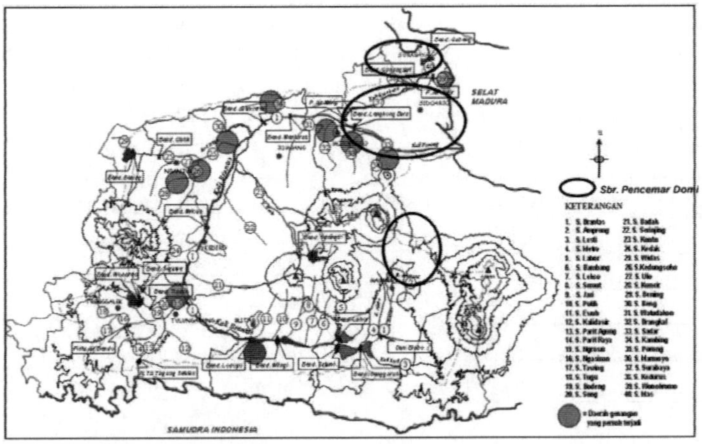

Figure 4. Brantas River water basin map.
(Source: *Balai Besar Wilayah Sungai Brantas*)

TDS and TSS concentration mostly were below the standard, except for once in June 2012 for both parameters. TDS and TSS value are in correlation to each other due to precipitation in mechanism in river water. Hexavalent chromium exceeding the standard at Kertosono (October 2013) and Padangan (April 2013).

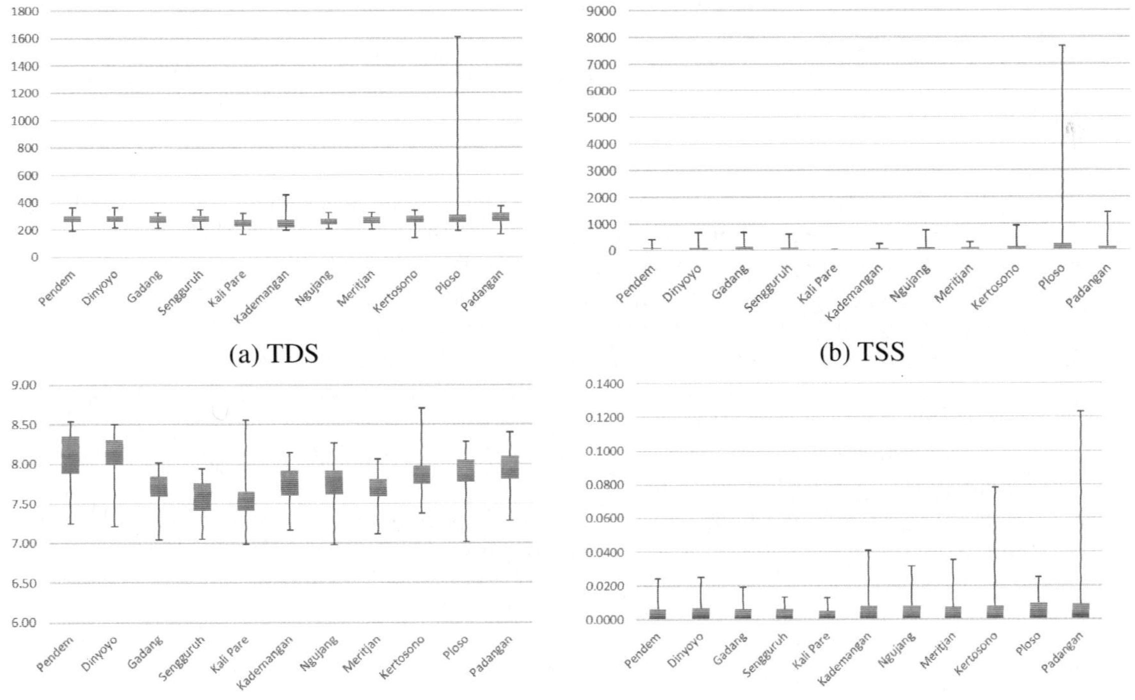

Figure 5. Several parameters on monitoring point during 2010-2015 in Brantas River
(Data source: East Java EPA, 2010-2015

Table 6 showed the river water quality parameters that exceeding/violated the standard of river water quality. STORET results showed that based on stream standard of Regulation No. 82 in 2001 Upper Brantas River has been polluted in some monitoring point as showed in Table 7.

Table 6. Exceeding parameters (percentage) during 2010-2015 of Brantas River.

Parameters	Exceeding
TDS	0.17%
TSS	34.47%
pH	0.00%
DO	4.85%
Cr^{6+}	0.45%

Table 7. STORET result for Brantas River monitoring point.

Points	Score	Pollution
Jemb. Pendem	-14	Moderate
Jemb. Dinoyo	-10	Moderate
Jemb. Gadang	-10	Moderate
Jemb. Sengguruh	-10	Moderate
Jemb. Kali Pare	-16	Moderate
Jemb. Kademangan	-6	Light
Jemb. Ngujang	-10	Moderate
Jemb. Meritjan	-10	Moderate
Jemb. Kertosono	-10	Moderate
Jemb. Ploso	-16	Moderate
Jemb. Padangan	-10	Moderate

4. Conclusion

The Bengawan Solo River and Brantas River has been polluted from the upstream, this condition are related with the industrial and domestic activities in upper watershed. Upper Bengawan Solo River classified as lightly and moderately polluted, with the most exceeding stream standard parameters were TSS and the least was TDS. Brantas River was classified as lightly and moderately polluted, with the exceeding parameters in order from the most amount to the least were TSS, DO, Cr^{6+}, TDS, pH.

Acknowledgment

Ministry of Research Technology and Higher Education for research funding in MP3EI research scheme, 2015-2016.

References

[1] Aprilia T, & Salami I R. 2013. Uji Toksisitas Akut Limbah Cair Industri Pulp dan Kertas Terhadap Daphnia magna (Studi Kasus: PT. Aspex Kumbong).
[2] Perindustrian K. 2015. Rencana Strategis Kementerian Perindustrian 2015-2019.
[3] Roosmini D, Salami I, Suharyanto, Soedomo A, Hadisantosa, F. 2006. Biomarker as an Indicator of River Water Quality Degradation. (pp. 114-122).
[4] Septiono M A, Roosmini D, Salami I R, Ariesyadi H D, & Lufiandi. 2016. Industrial Acitivities and Its Effects to River Water Quality (Cast Study Citarum, Bengawan Solo and Brantas), An Evaluation for Java Island as an Economic Corridor in Master Plan of Acceleration and Expansion of Indonesia Economic Development (MP3EI). *The 12th International Symposium on Southeast Asian Water Environment*. Hanoi.
[5] Wiryanto. 2016. Spatial and temporal description of water pollution status of Gajah Mungkur Reservoir Wonogiri, Central Java, Indonesia. *Biodiversitas.*17:888-893.

IOP Conf. Series: Earth and Environmental Science **106** (2017) 012060 doi:10.1088/1755-1315/106/1/012060

Transformation of traditional houses in the development of sustainable rural tourism, case study of Brayut Tourism Village in Yogyakarta

V R Vitasurya[1,3*], G Hardiman[2], S R Sari[2]

[1]Doctoral Program of Universitas Diponegoro, Semarang, Indonesia
[2]Universitas Diponegoro, Semarang, Indonesia
[3]Architecture Department, Universitas Atma Jaya Yogyakarta, Indonesia

*Corresponding Author : reni792003@yahoo.com

Abstract. This paper aims to reveal local values used by Brayut villagers to maintain the existence of the traditional house as a dwelling. The transformation of traditional houses goes as time passes, influenced by internal aspects related to the needs of residents and external aspects related to the regional development by the government. Traditional Javanese house as a cultural identity of Javanese people, especially in the village, has also experienced the transformation phenomenon. Modernization affects local residents' needs and the Government's Development Program for tourism village influences demands of change. An unfocused transformation can lead to a total change that can eliminate the cultural identity of the rural Java community. The method used is the case study by taking three models of Javanese house in Brayut Village. Brayut Tourism Village is a cultural tourism village that relies on tradition as a tourist attraction. The existence of traditional Javanese house is an important asset for retaining its authenticity as a dwelling. Three models taken as the case studies represent the traditional Javanese house types. The result obtained is that the family bond is a major factor in preserving the traditional Javanese house in Brayut Village, Yogyakarta.
Keywords: family bond, preservation, rural tourism, traditional Javanese house, transformation, tourism village

1. Introduction

The traditional house is a culture which becomes a local identity of a community in a region with a function consistent with its type [1]. Traditional houses are transforming, and the common transformation is of function, following by form as required [2]. Housing transformation not only improves the housing environment but also contributes to cultural development [3].Traditional communities, especially those living in rural areas, cannot escape the phenomenon of transformation that happens due to pushes of internal factors and pulls of external factors [4], [5].

There are five types of traditional Javanese architecture by the roof shape and design [6], [7], i.e., Tajuk with roof narrowing to one point for worship place; Joglo with two layers of roof narrowing upward for *pendopo*, the house of nobilities or community leaders [8]; Limasan with widening roof for *pendopo*, house of merchants or rich people [8]; Kampung with gable roof, house of commoner [8]; and Panggangpe with half-gable roof for non-permanent functions. The transformation of traditional Javanese houses has happened since the 11th century.

Based on the relief on Borobudur Temple, in the beginning, traditional houses had limasan roof and shingle roof tile, then a few types were developed, e.g., Joglo, Kampung, Tajuk, and Panggangpe [9]. Past transformation aimed to accommodate functional needs and self-actualization by adding architectural elements, starting from form, material, ornament, to room proportion [10]. Today, the transformation aims to accommodate changes of functional needs and modernization. Modernization is an external factor triggering the transformation, including the development of a rural area into tourism village. Modern transformation encourages homeowners to make practical decisions based on economic considerations [11], e.g., changing the layout for advanced functions and using modern building materials. Without any clear direction, the transformation of traditional houses can remove the local identity of the traditional community. Unfocused transformation can damage tourism attraction which relies on the natural environment, including architecture as a product of the local culture.

Cultural heritage tourism embraces ecotourism and culture at the same time and emphasizes on the conservation of cultural heritage itself [12]. World Tourism Organization defined sustainable tourism development as an effort to fulfil the present needs of tourists while protecting and promoting opportunities for the future [13].

A study on transformation is required to develop tourism village according to the local wisdom derived from the local ancestors of conserving the environment, including the architecture. A slow transformation in the physical shape of the building and environment can be an indicator of a successful conservation in a region, especially a tourism village. This article aims to explore local wisdom elements in the transformation of Brayut's traditional houses that protect the original environment and architecture.

2. Research Method

This research employed a case study method focusing on original traditional houses in Brayut Tourism Village. This approach was chosen as a case study can answer the questions of "why and how" [14]. Brayut Tourism Village was selected because it has a unique environment as well as exceptional traditional architecture, both are well-maintained. The data collection technique was participator yaction research in the perspective of conservation of traditional architecture in a tourism village.

Participatory action research (PAR) is associated with the community empowerment [15] that is required to maintain the local wisdom. Participatory action research should be supported by other approaches, i.e., observation, in-depth interview, and focus group discussion (FGD) to achieve the study purpose [14].

3. Results and Discussion

Brayut Tourism Village is one of 35 tourism villages in Sleman Regency [16]. The transformation of Brayut from an agricultural region into a tourism village has happened slowly. Based on the joint decision of UNEP and UNWTO [17], sustainable tourism is tourism that takes all of its current and future economic, social, and environmental impacts, addressing the needs of visitors, the industry, the environment, as well as host communities.

The community maintains village assets, sense of belonging, and emotional bonds which direct the development of a tradition-based tourism village. In Brayut, the traditional building is the primary tourist attraction. From five typologies of Javanese buildings, three categories are used for dwellings, i.e., Joglo, Limasan, and Kampung [18]. The ancestors of Brayut were landlords, so the dominant house type is Limasan.

Traditional houses are well-maintained since they are used as dwellings and places for activities. The identification to understand the transformation was performed on 14 traditional houses based on the function and shape as responses of transformers in making changes [19]. The transformation category was based on the owner's reason, the alteredform and function,as well as the ownership status (single or multiple) to transform. The identification is shown in Table 1.

Table 1. Categorization of traditional houses in Brayut.

Identification (Owner)	Existing functions		Transformations				Motivation	Transformer	Category
	House	Others	Forms		Functions				
			Un-changed	Changed	Un-changed	Changed			
Joglo 1 Suyadi		Lodging	X			X	Modernization	Single ownership	1
Joglo 2 Martono	dwelling	Homestay	X			X	Modernization, change function	Single ownership	2
Joglo 3 Sri	dwelling	Art per-formance	X			X	Modernization, change function	Single ownership	2
Limasan Handiyo	dwelling	Art per-formance	X			X	Modernization, change function	Single ownership	2
Limasan Mugiwiyarto	dwelling	Homestay	X			X	Modernization, change function	Single ownership	2
Limasan Soe-giyopranoto		Kinder-garten	X			X	Modernization	Single ownership	1
Sinom Sastro	dwelling	Art per-formance	X			X	Modernization, change function	Single ownership	2
Kampung Darmadi		Office and shop	X			X	Modernization	Single ownership	1
Sinom Tris	dwelling	Homestay	X			X	Modernization, change function	Single ownership	2
Limasan Sutarmin	dwelling	Homestay		X		X	Modernization, inheritance	Multiple ownership	3
Limasan Susmiarto	dwelling	Art per-formance		X		X	Modernization, inheritance	Multiple ownership	3
Limasan Mujiman	dwelling	Homestay		X		X	Modernization, inheritance	Multiple ownership	3
Limasan Sutarman	dwelling	Lodging		X		X	Modernization, inheritance	Multiple ownership	3
Limasan Darmadi	dwelling	Office		X		X	Modernization, inheritance	Multiple ownership	3

Based on the identification, there are three categories of transformation:

- Non-dwelling traditional house. Transformation occurs to accommodate modern functions, such as functional changes from bedroom into bathroom and replacing materials by modern building materials.
- Dwelling and tourist attraction. The maintenance is performed by the single heir, so the transformation is to accommodate modern needs, such as functional changes and modern building materials utilization.
- Communal dwelling house and other functions. The maintenance is performed by several heirs due to the distribution of inheritance. The house can be divided by the number of heirs. The transformation follows the needs of each heir but the main area, i.e., *pendopo*/porch and *ndalem,* are maintained as the dedication to their parents.

The transformations of three traditional house models were selected to represent the three transformation categories in Brayut Village.

3.1. Joglo 1 – Non-dwelling traditional house

Joglo 1 is the oldest building in Brayut, constructed in early 1800 by the first village head, Ki Demang Mertodimedjo. Joglo 1 was built to be a dwelling and office of Brayut government, so the form is a complete Joglo [6], [7]. The transformation in 2005 was changing *sepen* (room for prisoners) into a bathroom. Other parts of Joglo 1 are original and well-maintained, so it received the Decree of

Cultural Heritage Award from the Governor of DIY in 2015 [20]. Joglo 1 then used by tourism village managers as an administrative office and the main attraction as a lodging. The exterior is preserved, and the transformation occurred slowly in the interior. There are not many changes because the heir follows their ancestor's direction to protect the originality of the house. The building no longer serves as a dwelling. The transformation pattern shows that the external factors, the development of tourism village and implementation of cultural heritage, do not dominate the transformation.

3.2. Joglo 2 – Dwelling and tourism traditional house

Joglo 2 is the house of the second village head of Brayut, Mr. Kartopiyogo. When he was the village head, the house was enlarged to be an office. Joglo 2 was built in early 1900. The first stage of transformation happened after the ownership was transferred to his heir, Mr. Martono. Some materials of floor covering were replaced, and a bathroom was added. After the death of Martono, Mrs. Arini (Mr. Martono's wife) and the family of her daughter lived in *Joglo 2*.

In this era when the tourism village has developed, Mrs. Arini participated in it by making her house a homestay. The transformation was adding bedrooms in *Gandhok* (side area of the house). Transformation of function does not necessarily change the building form so that the exterior form can be preserved. Another transformation is replacing building materials.

The single heir assignment becomes the determining factor of the building preservation to follow the parents' direction. In this model, the tourism and conservation aspects work together. The transformation pattern shows that the influence of the external factor (development of tourism village) and the internal factor (the owner's will) are balanced. In terms of form, there is not much transformation. Functionally, changes happened in combining dwelling and tourism purposes.

3.3. Mr. Sutarman's Limasan – Communal dwelling and other functions

Mr. Sutarman's Limasan is a house whose ownership is divided into three parts, i.e., the front area (*pendopo* and *ndalem*) which belongs to Mr. Sutarman (the oldest child), the back area that belongs to Mrs. Kisyantini, and the side area which belongs to Mr. Djoko Santoso. The division was made by the previous owner, Mr. Pawirodarsono. The house was built at the end of 1800 by Mbah Bayan (Mr. Pawirodarsono's parent).

His position as "Bayan", a village officials in 1890 – 1900 era, made the house a part of the upper-middle-class category. *Limasan* form was selected because the owner was not allowed to build a *Joglo* (for village leaders) [8]. There were two stages of transformations. The first happened during the distribution of inheritance in around 1990. Mr. Sutarman was ordered to maintain the main building to be as the original. Mrs. Kisyantini built a modern house at the back and Mr. Djoko Santoso at the side. In the era of tourism village development, Mr. Djoko Santoso built bedrooms for tourists.

The Old Pendopo becomes an attraction for tourists. All heirs live in Mr. Sutarman's *Limasan* house which is integrated with his siblings' houses. The division is to maintain the harmony of the family so that their future generations can still experience the traditional atmosphere. The efforts to follow the ancestor's direction preserve the building.

The transformation combines modern elements in the form and function of dwelling and lodging, with traditional elements of the original main house. The transformation pattern shows that the influence of the external factor (development of tourism village) is smaller than the internal factor (owner's needs). Functionally, transformation occurs in the combination of dwelling and tourism function, so most of the form is still original. The schematic description of the transformations is shown in Table 2.

Table 2. Comparison of transformation patterns of traditional houses in Brayut.

Layout transformation	Change and Unchanged Architectural Elements	Factors	Transformation Pattern
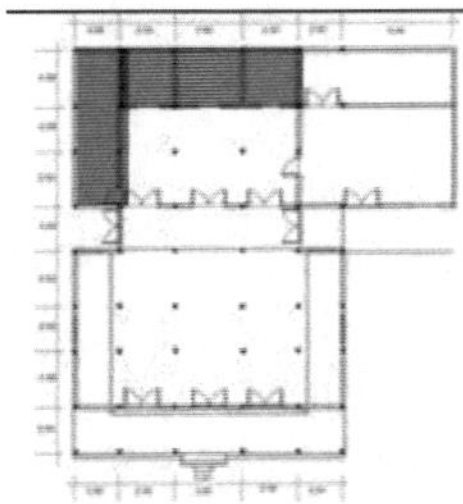	 interior  interior  exterior	**Internal:** requirement to follow the ancestor's direction **External:** as tourism village and cultural heritage	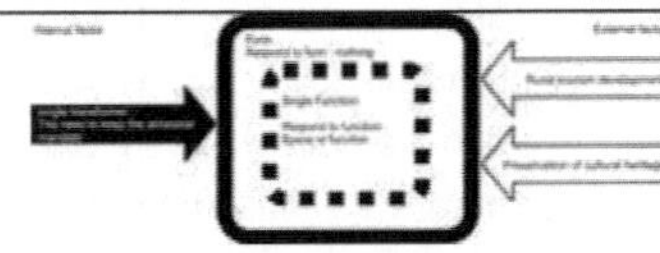 The internal factor has a smaller influence than external factors. Transformation happens to function, permanent form.
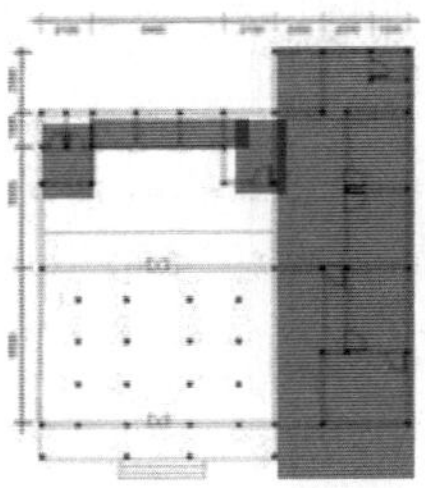	 interior  interior  exterior	**Internal:** the need for a place to live and to follow ancestor's direction **External:** as tourism village	 Internal factors are balanced with the external factor. Transformation happens to function, permanent form.
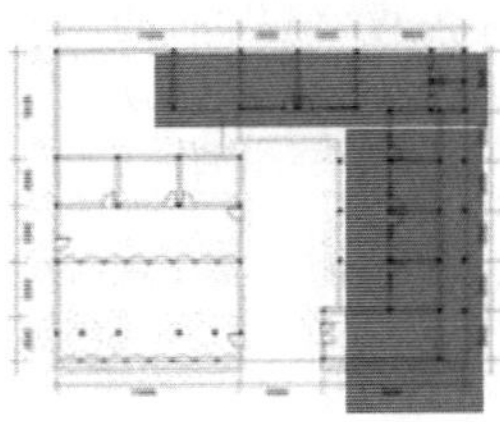	 interior  interior  exterior	**Internal:** the need for a place to live, work, and to follow ancestor's direction **External:** as tourism village	 Internal factors are bigger than the external factor. Transformation happens to function, combine traditional and modern form.

4. Conclusion

The transformations of traditional houses in Brayut is partial, implying that the form is maintained as a whole, but the function changes. The transformations are slow, meaning that the wholeness of form and originality of the main function is preserved as a dwelling. The external factors of the development of tourism village and the establishment of cultural heritage do not influence the transformations.

Meanwhile, the internal factor of the owner's need which related to the number of heirs influences the transformation. Factors affecting transformation are the family agreement and bond. The transformation is bigger when there is a house/plot division policy and becomes smaller when the house is handed to a single heir.

Acknowledgment

The authors would like to thank Brayut Village community, the family of Mr. Suyadi, the family of Mr. Martono, the family of Mr. Prawiro, Bapak Sudarmadi as the coordinator of Brayut Tourism Village, and all members of the lecturer team of PPLK Lab – UAJY during the survey.

References

[1] Rapoport A, House, Form and Culture, Prentice Hall Inc., London, 1969.
[2] Sueca N P. 2004. Home Transformation: Its Prospects for Improving the Home in Indonesia (A Preliminary Study). *Jurnal Natah*. 2(1):10 - 16.
[3] Sueca N P. Housing transformation: improving environment and developing culture in Bali. University of Newcastle upon Tyne – Dissertation, 2003.
[4] Antoniades A, Poetics of Architecture, Van Nostrand Reinhold, New York, 1990.
[5] Najoan S, & Mandey J. 2011. Transformation As A Design Strategy. *Media Matrasain*. 8(2): 117 - 130.
[6] Dakung S, Traditional Architecture of Yogyakarta Special Region, Yogyakarta, 1982.
[7] Ismunandar R, Joglo Javanese Traditional House Architecture *(3 ed.),* Dahara Prize, Semarang, 1990.
[8] Raffles T S, *The History Of Java*, Narasi, Yogyakarta, 2008.
[9] Prie M M, *Pancaran Limasan - The Brilliance of Limasan*, Red & White Publishing - Tembi Rumah Budaya, Yogyakarta, 2016.
[10] Ronal A, Man and House of Java, Gadjah Mada Press, Yogyakarta, 1988.
[11] Vitasurya V R, & Pudianti A, The Influence of Traditional Culinary Industy in the Development of Tourism in Lopati Village, Yogyakarta. *Proceedings Promoting Cultural and Heritage Tourism*, Universitas Udayana, Denpasar, Bali, 2016, p. 280-287.
[12] Pederson A, Managing Tourism at World Heritage Sites:a Practical Manual for World Heritage Site Managers, UNESCO World Heritage Center, Paris, 2002.
[13] WTO, Global Code of Ethic for Tourism, WTO, Santiago, Chile, 1999.
[14] Yin R K, Case Study Research - Design and Method (3th ed.), Sage Publisher, California, 2003.
[15] Mikkelsen, B., Metode Penelitian Partisipatoris dan Upaya - upaya Pemberdayaan (terjemahan) 2nd Ed., Yayasan Obor Indonesia, Jakarta, 2001.
[16] Fauzy A, & Putra A. 2015. Mapping of Potential Tourism Village in Sleman Regency 2015. *Jurnal Inovasi dan Kewirausahaan*. 124-129.
[17] UNEP-UNWTO, Making Tourism More Sustainable - A Guide for Policy Makers, 2005.
[18] Kartono J L. 2005. The Concept of Traditional Javanese Space in Cultural Context. *Dimensi Interior*. 3(2):124 - 136.
[19] Khan T H, Spontaneous Transformation as an Integral Phenomenon of Inhabitation in Living with Transformation, Springer International Publishing, 2005, p. 21-33.
[20] Laboratorium PPLK-UAJY, Architecture of Brayut VillageI, Graha Ilmu, Yogyakarta, 2016.

Outdoor thermal comfort and behaviour in urban area

I Inavonna[1*], G Hardiman[1], A B Purnomo[2]

[1]Architecture Department, Universitas Diponegoro, Semarang, Indonesia
[2]Architecture Department, Universitas Trisakti, Jakarta, Indonesia

*Corresponding Author: inavona@trisakti.ac.id

Abstract. Outdoor comfort is important due to the public spaces functions. Open spaces provide thermal comfort and a pleasant experience to improve the city life quality effectively. The influence of thermal comfort in outdoor activities is a complex problem. This paper presents a literature review and discussion of aspects of physical, psychology, and social behaviour toward outdoor thermal comfort. The valuation is determined not only by the "physical state" but also by the "state of mind". The assessment is static and objective (i.e., physical and physiological characteristics) that it should be measured. Furthermore, an effective model to provide the knowledge of climatic conditions, as well as the dynamic and subjective aspects (i.e., psychological and social characteristics and behaviour), requires a comprehensive interview and observation. The model will be examined to describe the behaviour that is a reflection of perception and behaviour toward the environment. The adaptation process will constantly evolve so that it becomes a continuous cause between human behaviour and the spatial setting of the formation, which is eventually known as places and not just spaces. This evolutionary process is a civic art form.
Keywords: behaviour, perception, outdoor, thermal comfort, thermo-adaptive

1. Introduction

Regarding the thermal comfort in outdoor space, the role of microclimate is very dominant since users will experience an outdoor space directly in contact with elements of the microclimate such as temperature, humidity, solar radiation, and wind speed. Many studies revealed that the microclimate parameters greatly affect the thermal sensation, but approaches or methods used have been only accounted for around 50% of the statistics—a data variation between the objectively and subjectively comfort measurements [1]. There is barely an approach or method to obtain the thermal comfort value by considering all aspects, i.e., the physical, physiological, psychological, and social factors. The physical aspect that is considered as extremely influential is the environmental condition of the outdoor space itself, e.g., the configuration of the 'void', closing materials, building types, as well as other elements such as plants and others. Factors mentioned above affect microclimate conditions. The microclimate characteristics then influence the thermal comfort value, especially physiologically. The human body is quite sensitive to environmental conditions since there is a need to maintain body temperature.

Several studies revealed that physiological approach alone is not sufficient to be able to determine the exact value of outdoor thermal comfort since many other factors are affecting this value. Another researched thermal comfort using the physiological method, and the result turned out to be in contrary to previous theories [2]. Thus, it is required to assess the psychological aspect of individuals toward environmental conditions. In previous study stated that regards to the "thermal history" as well as

IOP Conf. Series: Earth and Environmental Science **106** (2017) 012061 doi:10.1088/1755-1315/106/1/012061

"memory and expectation" are recommended besides paying attention to physiological elements [1]. It was based on the assumption that there are other influential psychological factors such as naturalness, past experience, perceived control, time of exposure, environmental stimulation, and expectations. Researcher developed a network that connects the factors that psychologically affect the thermal adaptation [3]. The contradiction between behavioural aspects of psychological phenomena and the thermal physiology can be bridged by the environmental behaviour learning theory.

Based on results of earlier research, the psychological approach will provide answers to the contrary matters besides the physiological and microclimatic aspects. Various models to predict the thermal comfort are discussed in this article. The outcome can be developed to an accurate model as the criteria in determining the design of outdoor space that is functional, healthy, and comfortable to promote public activities. In the context of urban planning, how the thermal sensation affects human behaviour and the utilization of outdoor space is exciting stuff to be reviewed. Given the variety of literature concerning this subject, framework guidelines on assessing the behaviour of thermal comfort in outdoor space are highly useful for researchers and practitioners of planning and design. Such framework has not yet been the subject of many in-depth discussions [3], so it was further examined in this study. This research aims to find criteria or parameters which can serve as a manual on how to determine the thermal comfort value of outdoor space that is associated with the local microclimate as well as physical, psychological, social, and behavioural aspects.

2. Research Method

This paper discusses how far research has been carried out on obtaining the proper measurement instrument or methods to assess the thermal comfort, both quantitatively and qualitatively, related to the microclimate and subjectively appraised social behaviour. The measurement methods must be able to process the information of environmental conditions in detail based on the customized time and variation, as well as to generalize the analysis results to reveal the interconnectedness among variables. This study employed a descriptive theory of deductive method. According to [4], a logically argumentative research has a rationale that everything can be viewed as a system. The research objective of applying this method is to create a framework system based on logic.

3. Results and Discussion

3.1. Measurement of thermal comfort

Thermal comfort is defined as the condition of mind that expresses satisfaction with the thermal environment [5]. Thermal comfort is a subjective psychological perception based on the mechanism of physiological thermoregulation as well as when the human body is exposed to a combination of environmental factors including air temperature, air humidity, wind speed, and radiation [2]. It can be said that thermal comfort is the satisfaction level of the human condition in responding to the thermal environment, both physically and psychologically.

Physiological Equivalent Temperature (PET) is a way of measuring the thermal comfort developed by Mayer and Hoppe in 1987 from the Munich energy-balance model for individuals [3]. PET is the "temperature dimension index" which is expressed in Celcius unit (°C). It is used in-depth for certain settings to determine the skin and body temperature as well as the sweat rate balance. Index of Thermal Stress (ITS), Fuzzy-PMV, OUT-SET, and the model of COMFA thermal outdoor space are the analysis tools to assess the thermal environmental response of the outdoor space users [3]. Research studied thermal comfort using OUT-SET and PET indexes as the primarily designed instrument for outdoor use [2].

3.2. Physiological aspect of thermal comfort

The basic concept of the physiological thermal comfort is the occurrence of heat balance. The process of metabolism yields by products in the form of body heat. The heat wave is then continuously released to the environment. Body heat production rate should be balanced to prevent heat exhaustion [5]. Heat dissipation is present in the form of radiation, conduction, convection, and evaporation [5]. The human body is quite sensitive to environmental conditions since it has to maintain the body temperature even though the surrounding temperature has changed much. The adaptation process will occur by relying on the mechanism of physiological thermoregulation as described by [2].

Thermal comfort in the psychological point of view is a state of mind that represents satisfaction with the thermal environment. In the thermo-adaptive (psychological) approach, the achievement of a comfortable thermal condition is related to the thermal neutrality—a condition that causes a person to prefer neither a warmer nor cooler circumstances to the present one [6]. The assessment of thermal comfort is demonstrated by the human response to the thermal environment. Thus, this response is a form of a relationship between the human psychologies with the surroundings, implying that the psychological approach is capable of studying the thermal comfort.

As already known, the relationship between the human psychologies with the environment has initiated by the conceiving of perception. It occurs along with the process of sensing the environment. Concerning thermal comfort, the sensing begins with the presence of physical stimuli which are perceived by individuals with their own distinct characteristics. The individual characteristics include the ability or level of adaptation, psychological control, personal factors related to social connections with the community, exposure duration to the stimulus, and so on. It is in line with the statement of previous study [1] that the psychological approach consists of six parameters, i.e., naturalness, expectations, experience (short or long term), time of exposure, perceived control and environmental simulation. Understanding the parameters as well as their interconnectedness and effects on the thermal comfort is crucial in determining the role of design besides promoting its innovation.

3.3. Social and behavioural aspects

Human behaviour is a form of coping with environmental stress, including the open and closed behaviour. Open behaviour is behaviour that can be observed visually, while the closed behaviour is the opposite—cannot be examined visually—such as attitudes, tendencies, and preferences that can be assessed through direct or indirect inquiries. In this context, 'coping' is a human process of dealing with stress [6]. Social and behavioural aspects are crucial in determining the level of thermal comfort. Research established that the psychological approach alone is not sufficient to describe the conditions of thermal comfort in outdoor space [1]. They advised paying attention to the "thermal history" and "memory and expectation", with three levels of adjustment to formulate thermal adaptations: physical, physiological, and psychological extents. According to other study [3], though the concept is quite interesting, the relationships between these parameters have not been considerably affected the design alternatives because of the complexity of the interconnectedness among parameters. Hence, they developed a model that can be used as a generic framework which describes valuations of outdoor thermal comfort based on the behavioural aspect (Figure 1).

3.4. Assessment of thermal perception from behavioural and social perspectives

Based on the above statements, determining the individual level of thermal comfort relies heavily on various parameters whose relations are complicated as well. Factors such as gender are variables that can influence the thermal comfort [2]. Previous research asserted that women have a low tolerance for the conditions of heat and sunlight exposure. On the contrary, the results of this study indicate that physiologically, women have a higher thermal tolerance than men. This finding can be conceded by looking at it from the psychological point of view. In the theory of environmental behaviour, this phenomenon is indeed a contradiction between the psychological behaviour and the physiological aspect of autonomously thermal functions. Women's desire of radiant skin tone is induced by social learning processes, such as observation and education, which are then reflected in their perceptions of psychological factors (fear of heat and sun exposure) and behavioural adjustments (bringing an umbrella or seeking shade). Therefore, the phenomenon of a unique psychological and behaviour cannot be directly explained by the autonomous mechanism of physiological thermoregulation. Outdoor thermal comfort in urban environments is a complex issue involving many affecting aspects. Hence, the assessment to assess public perception of thermal comfort must be carried out on at least four levels of physical, physiological, psychological, and social/behavioural aspects. Social and behavioural aspects in outdoor space greatly influence the built environment, which personally presented on the physical environment.

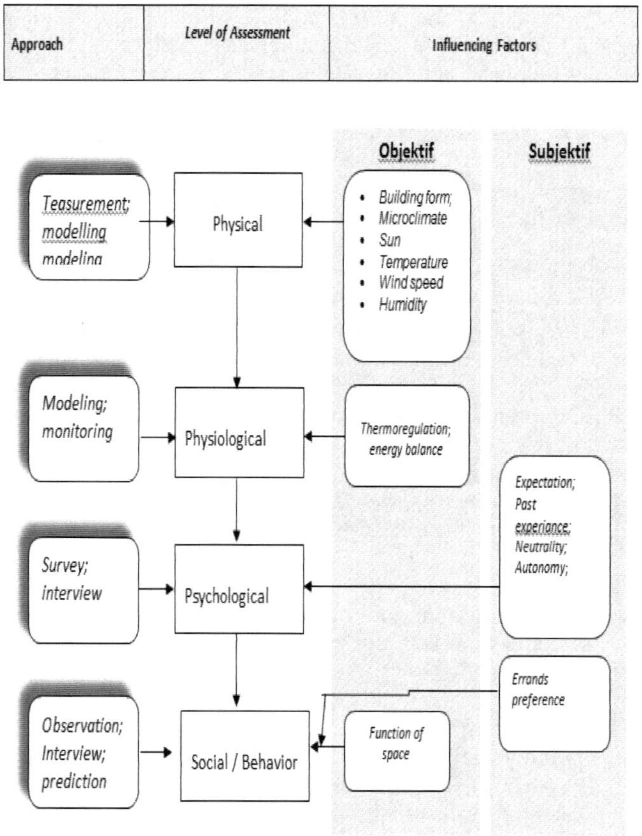

Figure 1. A general framework for outdoor thermal comfort assessment based on the behavioral aspect [3].

Based on the framework guidelines developed by previous study [3], the assessment to determine a person's thermal comfort starts on the physical level that defines the physical condition of the built environment as a consequence of microclimate. This physical condition will then influence the thermal comfort on the physiological level. Someone can respond and adapt to the surrounding thermal condition through balancing the body heat by relying on the thermoregulation mechanism. At this stage, a problem in determining the thermal comfort often emerges since its definition is not limited to the physical state only, but also by the state of mind that expresses one's satisfaction of the thermal environment. Thus, the response toward thermal comfort is a form of human psychological reaction to the environment based on the thermo-adaptive approach—psychology concerning the thermal neutrality. Research proved that there is a difference of thermal comfort perception between men and women, while contrariwise, the previous study employing a physiological approach stated that women *have a lower heat tolerance than men* [2].

Research defined that environmental perception is an interpretation of a setting by individuals. It is based on their cultural backgrounds and experiences [6]. This statement is in line with the research conducted by other study [7] explaining that cultural and climatic differences affect the thermal sensation. It can be said that the perception has a crucial role in the formation of space or physical settings, including the idea of outdoor thermal comfort.

According to [8] in the book of architectural environment and behaviour, conceptually, the approach accentuates that human behaviour is thinking being who has the perception and judgment of its interaction with the environment. Behavioural approaches emphasize the linkages between outdoor space and humans or societies utilizing or inhabiting the space. A setting of a life that is already formed will directly produce values. Those values will be the foundation of the relationship or be a reference to public behaviour [8]. Behaviour is a reflection of human perception and the behaviour toward the environment will experience adaptation and continuously evolve so that it becomes a continuous cause between the behaviour and the formation of its spatial setting which is eventually known as places instead of *just* space. This evolutionary process is a form of civic art. The success achieved by a place through the efforts of civic art is not only physically beautiful to be looking at but also work well for the present and the future [8]. Therefore, the assessment framework of thermal comfort perception developed by other study [3] can be modified into Figure 2.

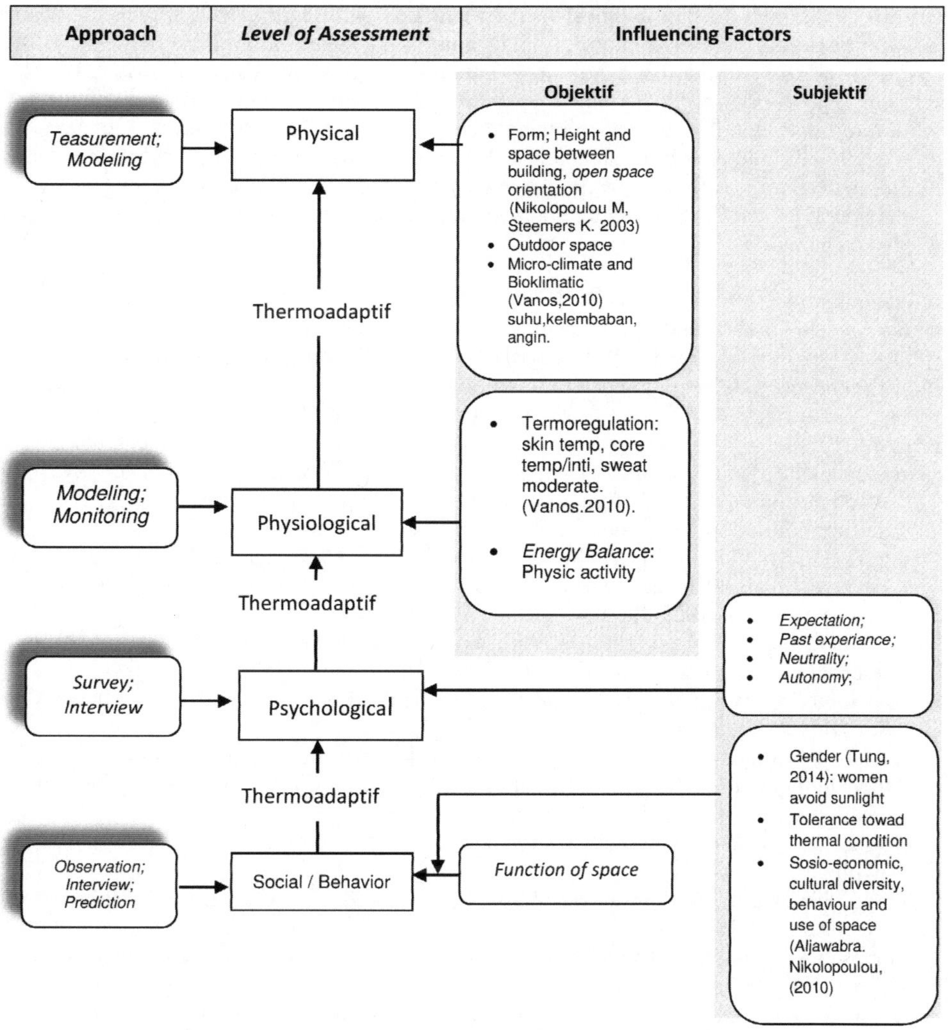

Figure 2. A framework model of outdoor thermal comfort with the approach of physical, physiological, psychological, behavioral, and social aspects [3].

4. Conclusion

Outdoor thermal comfort in urban environments is a complex issue involving many affecting aspects. Environmental stimulus (e.g., the local microclimatic condition) is the most influential factor on the thermal sensation and individual convenience level. Thermal comfort is not only determined by the physical state but also by the state of mind. Hence, the assessment to describe the public perception of thermal comfort must work on at least four levels: physical, physiological, psychological, and social/behavioural extents. The perception begins to form when the environmental sensing process occurs. Concerning thermal comfort, the sensing starts with the presence of physical stimuli which are then perceived by individuals with their own distinct characteristics. Those individual characteristics include the ability or level of adaptation, psychological control, personal factors related to social connections with the community, and prolonged exposure to the stimulus. This adaptability can bridge the four aspects through the physiological thermoregulation with thermal neutrality as the adaptation process. The assessment of thermal comfort is dynamic and subjective: dynamic in the sense that the adjustment to the ambient thermal condition is naturally progressive, also subjective given that the thermal sensation is primarily determined by personal experiences and subjective evaluation that thermal comfort does not always refer to climatic conditions. In other words, the static and objective aspects (i.e., physical and physiological characteristics) should be measured to provide "knowledge on climate" through an effective model, while the dynamic and subjective aspects (i.e., psychological and social/behavioural characteristics) require a comprehensive interview and observation to provide "knowledge on human".

Acknowledgment
The authors would like to send gratitude to Universitas Trisakti in Jakarta for the assignment and opportunity of learning at the doctoral program in Urban Architecture and Engineering Graduate School of Diponegoro University in Semarang, Central Java.

References

[1] Nikolopoulou Marialeha, and Steemers Koen. 2003. Thermal comfort and psychological adaption as a guide for designing urban spaces. *Energy and Buildings*. 35: 95-101. Elsevier Science.
[2] Tung C H, Chen C P, Tsai K T, Kantor N, Hwang R L, Matzarakis A, Lin T P. 2014. Outdoor Thermal Comfort Characteristics in the hot and humid region from a gender perspective. *Int J Biometeorol.* 58:1927-1939.
[3] Chen Liang, Edward Ng. 2012. Outdoor Thermal Comfort and Outdoor Activities: A review of research in the past decade. Cities 29:118-125. Elsevier.
[4] Grout Linda N, Wang David, Architectural Research Methods-Second Edition. John Wiley & Sons, Inc. New Jersey. 2013
[5] Vanos.J K, Warland J S, Gillespie T J, Kenny N A. 2010. Review of the physiology of human thermal comfort while exercising in urban landscapes and implications for bioclimatic design. *Int J Biometeorol*. 54:319-334.
[6] Sugini, Thermal Comfort Room, Concept and Application on Design, Graha Ilmu, Yogyakarta, 2014.
[7] Haryadi Setiawan B, Architectural Environment and Behavior, Gadjah Mada University Press, Yogyakarta, 2010.
[8] Thompson W, Catharine, Travlou, Penny, Open Space People Space, Taylor and Francis Inc., New York, 2007.
[9] Wibawa M, and Dedes Nur Gandarum, Design of Urban Settlements and Socio-Spatial Control, Penerbit Universitas Trisakti, Jakarta, 2014.

IOP Conf. Series: Earth and Environmental Science **106** (2017) 012062 doi:10.1088/1755-1315/106/1/012062

Influence of food kiosk attraction on the road's level of service

I W Agustin

Urban and Regional Planning Department, Faculty of Engineering, Universitas Brawijaya, Malang, Indonesia

Corresponding Author: immasaitama@ub.ac.id

Abstract. The main purpose of the research is to analyze the influence of food kiosk attraction on the road's level service and to find a solution to decrease congestion at Mayjen Sungkono Road in Surabaya City. The development of commercial sector has profoundly contributed to the land use change into business and residential areas. There is an increase in the number of vehicles passing through Mayjen Sungkono Road per year based on the data from Department of Transportation in Surabaya. In 2014, the number of light vehicles (LV) passing through Mayjen Sungkono Road was as many as 131,620 units, heavy vehicles (HV) reached 248 units, while motorcycles (MC) recorded at 187,371 units. The research uses the road's level of service analysis, multiple linear regression analysis, and do-something analysis. The results show that the influential variables comprise of the number of visitors (X_{32}), parking area (X_{30}), and building area (X_{29}). The food kiosk attraction model is $Y_{food\ kiosk} = 51.827 + 0.723(X_{32}) + 5.859(X_{30}) + 0.072(X_{29})$, and the result of attraction movement derived from food kiosk is 7,670.71 pcu/day. Widening the road can raise the average ofthe road's level of service (LOS) at Mayjen Sungkono of formerly F value increased to D.
Keywords: commercial area, food kiosk attraction, multiple linear regression, the road's level of service

1. Introduction

There is a tendency that the development of a city coincides with the aggravating transportation problems so that these issues overshadow the urban growth. One of the problems that often occur in big cities the traffic density or congestion, caused by the inadequacy of infrastructures compared with the greater transportation need. Changes in the activity schemes will affect the network structure through alterations in service levels on the movement system [1]. Being the economic center and capital of East Java Province, Surabaya City gives the biggest contribution to the formation of GRDP of East Java [2]. It is the largest city in East Java with an area of 32,637.08 Ha. In 2016, the population in Surabaya City is 2,977,520 people, and its growth is annually increased by 1.7%. The economic structure of Surabaya City is supported by the tertiary sectors such as trade, hotel and restaurant, transportation and communication, and other services sectors which directly contribute to Gross Domestic Product (PDRB) with an average value of 62.28% per year [3]. Dominated by private vehicles with a percentage of 93.8%, the average number of vehicles in the City of Surabaya in 2010 amounted to 122,087 units.It increased by 4.2% in 2011 to 132,802 units [4].

The growth rate of private transport modes in Surabaya tends to increase annually by 8.3% in line with the population growth, while the increment of public transportation there is only 0.9% per year. As a comparison, the percentages of private vehicles and public transports are 73.41% and 26.58%, respectively [5]. The satellite development unit is directed to carry out the main functions as a residential

IOP Conf. Series: Earth and Environmental Science **106** (2017) 012062 doi:10.1088/1755-1315/106/1/012062

area, trade and services, as well as special areas. Based on current trends, commerce and services dominate the Mayjen Sungkono Road which is the main access in this region [6]. Along with the progression of infrastructures development in the City of Surabaya, the generation, attraction, and movement patterns are shifting toward the east-west. Mayjen Sungkono Road, an addition that connects West Surabaya with Surabaya City Center, is one of the main streets in this corridor that experiences congestion at rush hour [7]. Technically, the function of Mayjen Sungkono Road as one of the areas in the secondary activity structure is serving the internal part of Surabaya as trade and service area [8].

Mayjen Sungkono Road is one of the streets in South Surabaya where there are many office center, high-rise buildings, and many spread trade sectors [9]. An increase in the number of vehicles passing through Mayjen Sungkono Road each year is based on the data from Surabaya's Department of Transportation. In 2014, the number of light vehicles (LV) passing on Mayjen Sungkono Road was as many as 131,630 units, while for heavy vehicles (HV) and motorcycles (MC) were 248 units and 187,371 units, respectively. The rapid development of trade and services sectors has affected the conversion of land use into commercial and settlement areas. This condition becomes a transportation problem that must be resolved because of the effect on the movement that occurred. Therefore, this study was conducted to determine the influence of the foodkioskattraction on theroad'slevel ofservice at Mayjen Sungkono Road in Surabaya City.The presence of this research is expected to boost the movement of vehicles to be more effective and efficient to break the congestion so that the traffic of the city is not inhibited. Thus, it canprovide valuable input for the development of adequate infrastructures at Mayjen Sungkono Road.

2. Research Method

Mayjen Sungkono Road connects West Surabaya and Surabaya Center. This road has administrative boundaries adjacent to Sawahan Sub-district, Dukuh Pakis Sub-district, and Wonokromo District (Figure 1). Mayjen Sungkono Road has a length of ± 2.58 km and six lanes of 3.5 meters width each, also equipped with a median strip. It includes secondary arterial road with two-ways and street status belongs to Surabaya City.

2.1. Data collection

The types of data required for this research are primary data (taken from the field observation) and secondary data (taken from the related institutions, literature, and electronic media). The data collection methods are divided into two as well, i.e., primary survey and secondary survey. The primary surveys cover the land use characteristics (building area, parking area, number of visitors, number of employees), road characteristics, food kiosk attraction characteristics (origin of movement objectives, travel times, and modes used), traffic volume, plate matching, interviews, and documentation techniques. The traffic volume calculation had been beingconducted for one week to determine the peak hour on weekdays and the weekend.The peak days of the movement on Mayjen Sungkono Road are Tuesday for weekdays and Sunday for the weekend. Secondary surveys are obtained from books, dissertations, journals, and other supporting literature.

2.2. Sampling

The sampling was carried out using Isaac and Michael tables [10], where the proportion for attraction is 139 samples. Generation and attraction for other land uses such as housing, education, health, and office do not use the sampling method since there are only a few of them along Mayjen Sungkono Road, occupied by the entire population.

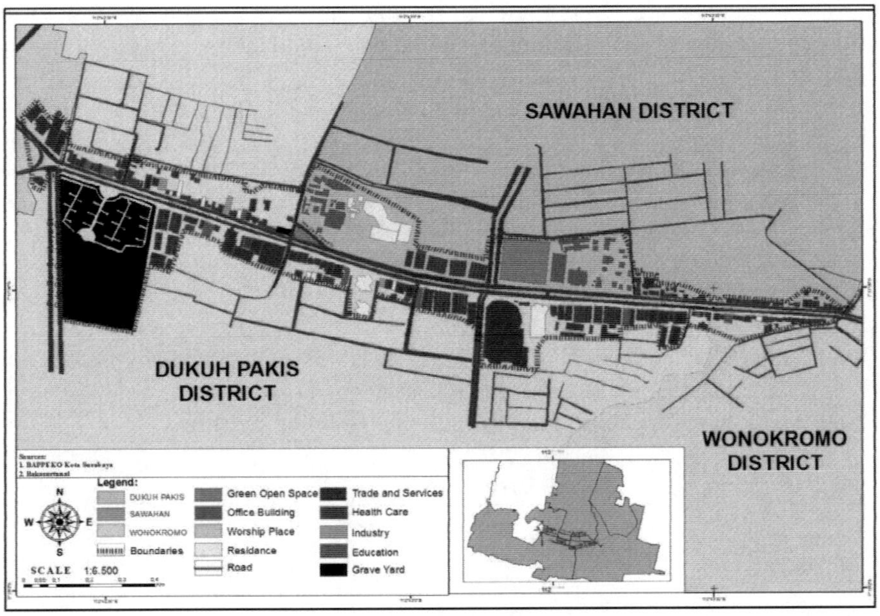

Figure 1. Research location.

2.3. Multiple linear regression analysis

The calculation of the attraction of food kiosk movement uses a multiple linear regression analysis, an analytical technique that links one dependent variable to two or more independent variables that are considered or may affect the observed dependent variable changes [11]. The independent and bounded variables used in this research are:

$$Y = A + B_1X_1 + B_2X_2 + \cdots + B_nX_n \qquad (1)$$

$Y_{Food\ kiosk}$= Total movement, X_{29}= Building area, X_{30}= Parking area, X_{31}= Number of Employees, X_{32}= Number of Visitors, A= Constants, $B_{29} \ldots B_{32}$= Regression coefficient.

2.4. Model of land use interaction and road network

Before calculating the total interaction between land use and road network, it is necessary to know the relation between the model of the attraction of movement in a region with land use variables, between the road network capacity model in one area and the land use variables, as well as the relation between the model of the generation/attraction with road network capacity model designated by the formula [12]:

1. Total volume of vehicle influence from generation/attraction of land use ($V_{internal}$):
   ΣY_I = $Y_1 + Y_2 + Y_3 + Y_4 + \ldots\ldots Y_n$ (volume of vehicle movement per day)
   ΣV_I = $e_1Y_1 + e_2Y_2 + e_3Y_3 + e_4Y_4 + \ldots\ldots e_nY_n$ (volume of vehicle/hour movement in main road corridor)
   E_I = V_I / Y_I = Ratio of volume of vehicle movement out / incoming from land use during certain hours compared to total volume of vehicle movement/day.

2. Total volume of external vehicle movement ($V_{external}$)
   ΣV_{eks} = External vehicle volume / clock on the main road
   ΣV_{eks} = $V_{eks-1} + V_{eks-2} + \ldots\ldots + V_{eks-n} + V_{eks-5} + V_{eks-6}$

Based on the above frameworks to compute the internal and external volumes, the formula derived for finding out the interaction model of land use in this study, especially regarding food kiosks on the road network [12] is:

$$V_{total} = \Sigma V_{external} \quad (2)$$

V_{total} = Total volume of vehicle/hour in a corridor, $\Sigma V_{internal}$ = Total volume of vehicle / hour movement from the generation/attraction of land use, $\Sigma V_{external}$ = The amount of movement of an existing external vehicle/clock on a corridor of the vehicle/hour movement volume of environmental streets or alleys plus the continuous volume of vehicle/clock movement on a corridor.

2.5. The Road's level of service analysis

The degree of saturation (DS) or the road's level of service (LOS) is the ratio of volume (flow) of traffic to the capacity. The degree of saturation is used as the main factor in determining the service level of the road. The degree of saturation can be calculated by the following formula.

$$DS = \frac{Qpcu}{C} \quad (3)$$

Q_{pcu} = Total flow (pcu/h), C = Capacity (pcu/h)

After calculating the degree of saturation, the determination of LOS value was conducted using the boundaries of service level indicator [13]. Scope limitation is useful in classifying the road's level of service according to the stability characteristics of vehicles passing on Mayjen Sungkono.

3. Results and Discussion

3.1. Road Characteristics

Geometrics of Mayjen Sungkono Road can be seen from the ownership of the number of lanes that is 6/2 D or six lanes divided by the condition of cement concrete pavement and the splitting system separating the two directions. The width of the lane at Mayjen Sungkono is 10.5 meters. This road also includes high side barriers due to the use of commercial and service lands along the street (Figure 2).

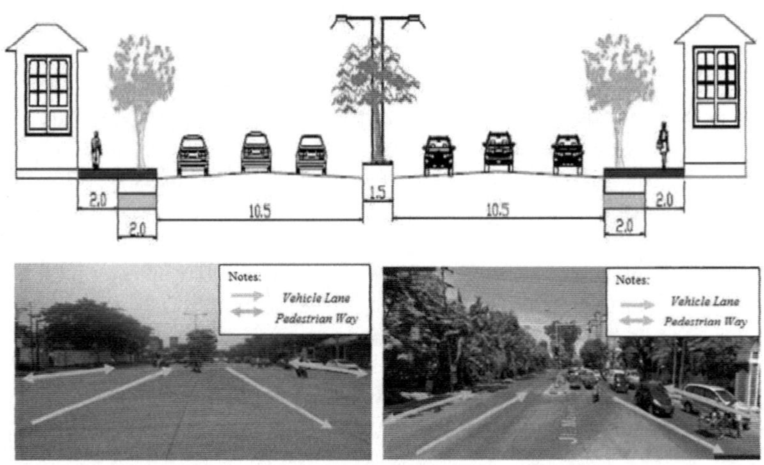

Figure 2. Cross-section of Mayjen Sungkono road.

Mayjen Sungkono is the connecting road between West Surabaya and Surabaya City Center. Its concrete pavement is in good condition, although there are several damages at some points. Mayjen Sungkono is included in the secondary arterial road, categorized as class II. The management of this

IOP Conf. Series: Earth and Environmental Science **106** (2017) 012062 doi:10.1088/1755-1315/106/1/012062

road is under the auspices of Surabaya City Government. In addition, Jalan Mayjen Sungkono is equipped with zebra cross, pelican crossing, and traffic signs.

3.2. Land use characteristics
The existing commercials on Mayjen Sungkono Road are dominated by houses and food stalls, clothing and sports stores, electronics and computer stores, as well as dealers and workshops, whereas for services dominated by Banks and ATMs, also design and photography. As much as 85% of the existing lands on Mayjen Sungkono Road is trade and services, so the in-and-out movement of vehicles from land use is also a factor causing congestion. The continuous movement on Mayjen Sungkono Road is very high because this road has the main function as a liaison between West Surabaya with Central Surabaya, as well as a connecting road to Surabaya-Gempol Toll.

3.3. The Road's level of service
The value of LOS on Mayjen Sungkono Road is obtained from knowing the volume of the vehicles and calculating the capacity of the road. The average LOS on Mayjen Sungkono Road with a value of 9,108 capacity on the working days (weekdays) as well as on holidays is F. The LOS value F meanspoor road service conditions, sluggish roads, and congestion occurs. During peak hours, the volume of vehicles on Mayjen Sungkono Road increases because in the morning (08.00–09.00), the average riders start their travel activities, resulting in the built up of vehicles. On the other hand, during the day (12:00–13:00) is the hours of office break for workers and school hours for students, so the flow of vehicles increases again. Similar to the morning and afternoon, the condition of Mayjen Sungkono Road in the evening (5:00 p.m. to 18:00 pm) also increased because most riders or road users end travel activities of either purpose of work, school, shopping, or other activities.

3.4. Influence of Food Kiosk Attraction

3.4.1. Implementation of Generation/Attraction Model of Mayjen Sungkono Road
Food kiosks are located on Mayjen Sungkono as many as 25 units with an average of visitors as many as 73 people, the mean parking area of 10 m^2, and the average building area of 270 m^2. Based on the above, daily movements are obtained as follows:

$$Y_{\text{Food Kiosk}} = 51.827 + 0.723(X_{32}) + 5.859(X_{30}) + 0.072(X_{29})$$
$$Y_{\text{Food Kiosk}} = 51.827 + 0.723(73) + 5.859(10) + 0.072(270)$$
$$Y_{\text{Food Kiosk}} = 183 \text{ pcu/day}$$

Based on the results of these calculations, the amount of attraction that comes from Food Kiosk, can be obtained from this equation:

$$Y_{\text{Food Kiosk}} = Y_{\text{Food Kiosk}} \text{X Total of Population}$$
$$Y_{\text{Food Kiosk}} = 183 \text{ X } 42$$
$$Y_{\text{Food Kiosk}} = 7686 \text{pcu/day}$$

3.4.2. Influence of food kiosk on the road's level of service
The peak contribution of the highest movement volume affected by food kiosk is at 17.00–18.00 with a value of 6.97% due to the movements of many riders at that hour on Mayjen Sungkono Road, mostly the workers who come home from offices. The contribution of the lowest movement is at 21.00–22.00 with a percentage of 2.73% since there is neither commercial, education, nor office activities at that time anymore.

4. Conclusion
The contribution of food kioskland use to traffic conditions can be seen from the generation and attractionvalue of the resulting movement compared to the value of the road's level of service. The road's level of service at Mayjen Sungkono is on average F value. Based on the research results, it can

be seen that the highest volume of movement influenced by the food kiosk activities is at 8–9 am, 12.00–13.00, and 5–6 pm The highest contribution value of tug is at 5–6 pm and equals to 6.97%, whereas the contribution value of the lowest attraction there is at 9–10 pm and equals to 2.73%. The high contribution is influenced by the operation of land use, commercial, office, health, and education. A recommended effort to handle the congestion is to broaden the road of initially 10.5 m width to 12 m (extending the road body of 0.5 m per lane). Widening the road can raise the average ofthe road's level of service (LOS) at Mayjen Sungkono of formerly F value increased to D.

Acknowledgment
The author wants to acknowledge all members of EIIS Laboratory Department of Urban Regional Planning, Engineering Faculty, Universitas Brawijaya, for their supports, having helped kindly on this research, especially for Syavitri as the LSP team. Thank You very much.

References

[1] Petersen, Rudolf. 2011. Urban Spatial Planning and Urban Transportation. Issue 4a. Translation. Harya Satyaka.
[2] Iswanto, Denny. 2015. Inequality of Inter-District/City Revenue and Economic Growth in East Java Province. 4: 41-66.
[3] Saraswati, Sulistyaningrum, & Cahyono, Hendri. 2014. The Influence of Education and Health Level to PDRB Perkapita of Surabaya City. *Jurnal Ilmiah Universitas Negeri Surabaya.*
[4] Alim, Miftahul, et.al. 2012. Relationship of Physical Environment Condition with Traffic Jam in Surabaya City. *Journal of Universitas Negeri Surabaya.*
[5] Prakoso, Bagus R M, & Sardjito. Conformance of Transit Area of Transtop Surabaya Mass Rapid Transit with Transit Oriented Development Concept. *ITS Engineering Journal.* 5(1).
[6] Ariyaningsih, & Sulistyarso. 2012. Directives Intensity Utilization of Commercial Space Services Based on Telecommuting Opportunities. *Journal of ITS Engineering.* 1: C11-C15.
[7] Isa, Muhammad Hidayat. Tanpa Tahun. TOD alternative solution in Overcoming Congestion Problems in Surabaya City.
[8] Nugroho, Bayu Adjie, & Sardjito. 2013. Optimization of Ciputra World Mall Activities to Level of Road Service at Major General Sungkono. *Journal of Civil Engineering and Planning.*
[9] Budi, Onny Antika. 2014. Study of Shopping Center Locations in South Surabaya Region. *Journal of Universitas Negeri Surabaya.*
[10] Sugiyono, Statistics for Research, Alfabeta, Bandung, 2010.
[11] Miro, Fidel. Transportation Planning. University of Bung Hatta Padang. Padang. 2005.
[12] Waloejo, Budi Sugiarto. Interaction Model of Land Use and Road Network. Dissertation. Univeristas Brawijaya. Indonesia, 2013.
[13] Directorate of Urban Road Development. Indonesia Road Capacity Manual. Directorate General of Highways. Jakarta. 1997.
[14] Department of Transportation, Surabaya Vehicle Volume Data. Department of Transportation. Surabaya. 2014.

Study of locations'characteristics for stabilization of street vendors in Surakarta City

M J Rahayu[1,4]*, I Buchori[2], R Widjajanti[3]

[1,2,3]Doctoral Program of Architecture and Planning Engineering, Universitas Diponegoro, Semarang, Indonesia
[4]Urban and Regional Planning Department, Universitas Sebelas Maret, Surakarta, Indonesia

*Corresponding Author: mjanirahayu@gmail.com

Abstract. Street vendors are usually centered around particular areas where they have made adaptations based on their activities, instead of scattered all over the city. These locations tend to be close to street vendors' primary activities and consumers, occupy the public lands, and are productive and accessible. The Municipality of Surakarta has managed the street vendors in the form of stabilization and relocation. In Surakarta, there are 24 areas used for stabilization of street vendors varying in characteristics and quantities. This study uses a spatial analysis aiming to explain the characteristics of locations used for stabilization of street vendors and to identify their physical condition and arrangement. Furthermore, all of the research results will be used as the criteria in choosing the new locations that are suitable for the street vendors' characteristics and consumers' preference. The results show that the locations are on the roadsides of the neighborhood, local, and secondary collector roads. The traffic conditions in the stabilization areas are ranging from the very busy until relatively busy level since they are close to activities for trade, education, recreation, offices, and settlements.
Keywords: characteristics of locations, management, stabilization

1. Introduction

Surakarta City (often called Solo) is located between longitude 110°45'15"–110°45'35" east and between latitude 7°36'–7°56' south. Surakarta City is one of the major cities in Central Java that directly borders Boyolali district in the North, Karanganyar district in the east, as well as Sukoharjo district in the South and the West. This city has a size of 44.04 km^2 in width, consisting of five subdistricts, i.e., Laweyan, Serengan, Pasar Kliwon, Jebres, and Banjarsari subdistrict. Most of the land in this city serves as settlements (65%) and economic areas (16.5%).

The use of public lands in Surakarta is allocated to the street vendors (PKL) including the use of settlements and economic areas. The management of street vendors in Surakarta City is carried out in two ways namely relocation and stabilization, and the latter is more dominant than the first. According to study [1], street vendors prefer stabilization to relocation. Stabilization has been applied by the government since 2006 by issuing land occupation permits in public areas and uniformly granting trading infrastructures and basic facilities. All of been regulated based on the Local Regulation (*Peraturan Daerah*/*Perda*) of Surakarta City No. 3/2008. Recently, there have been 24 street vendor locations, and according to the interview with the Section Head of Street Vendors Management 2014, they were chosen for stabilization because they were close to the previous locations used by the street vendors to sell their goods. This fact accommodates the statement of researched study [2] that street vendors choose locations near their houses so that the selected areas for the management of street vendors would not be far from the previous ones. This way can both minimize the rejection from the street vendors and control their growth. Study [3] about two organized locations for street vendors in Semarang City, also share the same opinion.

According to [4], stabilization has changed the city's face to be more beautiful, well-arranged and more attractive to visitors. The city has also been more conducive for street vendors to do their businessessince they do not have to worry about being chased by municipal police who used to try to remove their stalls. Economically, not all street vendors experience an increasing income, but they believe that their earnings will increase gradually while the stabilization is being conducted.

The purpose of this research is to identify characteristics of organized locations for street vendors that are used as criteria in choosing the spots that suit the street vendors' activities and their visitors/consumers. Furthermore, the criteria can be used as a reference in selecting new locations for stabilization of street vendors. It is important to fulfill these criteria so that the process of determining locations for stabilization can be more sustainable and the population of street vendors can be easily controlled.

2. Research Method

The research uses a descriptive qualitative approach as the method of analysis, and it is supported by the spatial technique to explain even more comprehensively. The data were collected through observations and interviews with the street vendor community leaders at each location. The triangulation was carried out by interviews with the officials at the Office of Trade, particularly Street Vendors Management Section.

3. Results and Discussion

3.1. Macro Distribution inside the City

From the identification of primary data collected through observations and interviews as well as secondary data taken from Street Vendors Management Office, it was found that the stabilization of street vendors in Surakarta City spreads across 24 locations. An area called Galabo is alternately used by street vendors working on days and at nights. To understand the macro distribution of all sites for stabilization of street vendors, a comparison of distributions among subdistricts and their distance from the downtown were drawn in the following explanation.

3.1.1. The Comparison of Distributions among Subdistricts

The results of identification indicate that the locations for stabilization of street vendors are not equally distributed among subdistricts in Surakarta. Jebres Subdistrict and Laweyan Subdistrict have quite many stabilization areas of 7 sites spread in each. On the other hand, Banjarsari Subdistrict has 6 locations, and Pasar Kliwon Subdistrict has 4 locations. However, there are also subdistricts that do not have any area of stabilization like Serengan Subdistrict. In the comparison of the population density, it can be seen that Pasar Kliwon and Serengan are the subdistricts with the highest density in Surakarta City, yet they have less number of locations of stabilization among other subdistricts. On the contrary, Jebres, Laweyan and Banjarsari that are less dense than those subdistricts mentioned above, actually have more locations of stabilization. This fact proves that population density is not a mere reference for choosing a trading site for street vendors. Based on [5], the location of street vendors orientates the highly concentrated area attended by many consumers. This evidence also conforms to the research conclusion by [4], [6] stating that the principal criteria for location management for street vendors are close to their houses, close to productive activities, full of visitors, and accessible. Having positioned near productive functions or areas with many consumers coming, street vendors can gain many profits from selling complementary products for consumers' activities. As seen from its distribution, the locations are aimed to attract passing customers who do their functional activities from their homes to their destinations [6],even though there are some opinions from the consumers stating that their visits at the locations are merely an act of coming by and it is not their primary purpose.

Researcher [7] explains that the availability of lands, especially the ones belonging to the government, becomes the crucial factor in managing street vendors. The condition of areas at Pasar Kliwon and Serengan Subdistrict, meanwhile, are insufficient since they have been fully utilized by settlements and commercial functions. Therefore, it is difficult to find locations for the arrangement of street vendors because the lands have been dominated by the formal sector activities. The previous explanation indicates a sequence of criteria for the management of street vendors' locations, i.e., the availability of public lands, the proximity to the productive activities, the accessibility, and the proximity to street vendors' houses. The availability of lands, provided by the government for street vendors, has become a key to avoid conflicts and havoc. The proximity to productive activities will drive the users of the activities to visit the street vendors' stabilized locations. The accessible locations will make consumers visit them easily. Meanwhile, one of the criteria, the proximity to the street vendors' houses, is somewhat ignored recently since most of these street vendors have vehicles. Besides, the kinds of goods sold by the street vendors also determine the characteristics/criteria of locations for the management, and related to this fact, researcher [8] explained that street vendors have typology in choosing locations for their goods. In general, the stabilization way is more focused on street vendors selling fast food.

3.1.2. The Comparison of Location Distance with Productive Activities and the Downtown

From the field observation conducted by the researchers, it has been found that the stabilization of street vendors tends to approach productive activities in downtowns. This study, taking place in Surakarta City, also shows several productive activities most favored by the society around Manahan Stadium, Nonongan, Coyudan, Solo Grand Mall, Solo Square, as well as a golden triangle area called Gladag and Kentingan at UNS. The locations of stabilization that are in Banjarsari Subdistrict are grouped in Manahan, Sumber, and Mangkubumen administrative villages ('Kelurahan') that have the furthest distance of 3 km from Manahan Stadium. The locations are also found in Kelurahan Karangasem with its furthest distance of 2 km from Solo Square. The next stabilization areas are in Laweyan Subdistrict, spreading across several villages such as Pajang, Sondakan, Sriwedari, Punggawan, and Penumping. However, the distance of these previous locations is only 0–2 km from the downtown areas like Grand Mall or Solo Square. In Pasar Kliwon Subdistrict, the locations are distributed in several areas namely the northern and the southern regions. The furthest distance of these locations to Gladag is is 0.7–3 km. In Jebres, the locations are relatively near the epicenters of this subdistrict, Administration Office of Kelurahan Mojosongo and Universitas Sebelas Maret, and their distances are 0.7–2 km. Although they are not nearby the city centers in Surakarta, these locations are close to other activities that are very productive, and this situation confirms the existence of street vendors as an "activity support" for those other hustles [9].

3.2. The Quantity Connection of Street Vendors Number

Besides the macro distribution in the city, researchers also identified the number of stabilized street vendors in the city based on the activities approached, the road function, and the traffic level.

The activities approached by the stabilization locations are one of the subjects of this discussion. Based on the field observation, most of the locations are near trading activities and settlements (13 sites). The next activities that also draw the street vendors are office activities (11 locations). Recreation activities fairly attract street vendors as well of 6 surrounding locations. There are also locations of stabilized street vendors that are near education and sports centers as many as 3 locations for each. The other activities like warehouses, health, and industries were chosen as well though the numbers of vendors are not as many as the previously listed sectors. Trading and settlements have the most identified vendors, so it shows that those primary activities drive high and diverse movements and businesses. Hence, they are considered as the most appropriate locations for the street vendors management, conforming to study [4], [6].

Based on the identification of site arrangement seen from road functions, there are 11 out of 24locationsfor stabilization of street vendors in Surakarta City that are established at neighborhood roads, whereas the other 13 locations are at local roads and secondary collector roads (Table 1).

The locations do not have to be on roads with a higher class since they have been situated at major streets. Considering that the settlements have highly concentrated movement even though settled at local and neighborhood roads, the locations of street vendors are quite possibly at the local or neighborhood streets. This condition is relevant to the accessibility criteria and fits the typical rapid movement of the society, so the locations at the collector roads are still well-matched with street vendors' characteristics, intercepting people in running their functional activities [6].

The condition of stabilization locations in Surakarta City complies with the previously mentioned theory. Researchers [5] stated that the number of street vendors in one location highly determines the chosen strategy. The number of traders observed based on the road types in each site can be compared with each other. There are various locations containing a number of merchants. The size can be of 21–40 traders, 41–70 traders, or even more than 70 traders, and these locations are situated at local and neighborhood roads. The sites containing 11–20 merchants tend to occupy the roads that have higher function—in this case are collector roads, while the locations with 10 traders or less occupy most of the neighborhood streets. Therefore, the tendency is that street vendors with a larger number are settled at the local and neighborhood roads while the ones with a smaller number are located at secondary collector roads. This fact is also caused by the availability of lands, and according to [7], it often becomes the substantial reason whether the management of street vendors is necessary or not. Upon that account, the availability of lands for the management of street vendors at both local and neighborhood roads are more than the ones at collector roads. The velocity of vehicles at collector roads that is higher than the one at local and neighborhood roads is also the reason why the street vendors are difficult to survive due to the unwillingness of consumers to stop. Although it is not directly related, the traffic level aspect and the high velocity of vehicles will bring disadvantages for the street vendors. Another side of the traffic parameter is also discussed in the following paragraph.

Based on the observation, the locations of stabilization are established at various roads ranging from the ones with relatively busy level to the ones with highly busy level. As seen in Table 1, there are 5 locations with highly busy level, 10 locations with busy level, and 9 locations with relatively busy level. The traffic level of those locations exists in the roads with varied functions. The locations of stabilization at secondary collector roads are dominated by highly busy road level, the locations of stabilization at roads having local functions are dominated by busy road level, and the locations of stabilization at neighborhood roads are dominated by relatively busy road level.

Table1. Locations of stabilization of street vendors in Surakarta City.

No	Locations	Characteristics
1	Shelter Near Pasar Notoharjo	There are 16 traders, the activities approached are trading activities, located at neighborhood roads, the traffic condition is relatively busy, the distance from Surakarta City Hall is 3km.
2	Shelter at Silir Notoharjo	There are 33traders, the activities approached are trading activities, located at neighborhood roads, the traffic condition is relatively busy, the distance from Surakarta City Hall is 3 km.
3	Shelter at Galabo Malam	There are 25traders, the activities approached are trading activities particularly inservices, located at local roads, the traffic condition is highly busy, the distance from Surakarta City Hall is 1 km.
4	Shelter at Lapangan Mojosongo	There are 20traders, the activities approached are settlements, located at secondary collector roads, the traffic condition is highly busy, the distance from the Administration Office of KelurahanMojosongo.
5	Shelter at Komplang	There are 23traders, the activities approached are trading activities particularly inservices and office activities, located at secondary collector roads, the traffic condition is highlybusy, the distance from Manahan Area is 3 km.
6	Shelter at PDAM	There are 11traders, the activities approached are office activities and settlements, located at neighborhood roads, the traffic condition is relatively busy, the distance from Manahan Area is 4 km.
7	Shelter at Kolang Kaling	There are 26traders, the activities approached are office activities and settlements, located at neighborhood roads, the traffic condition is busy, the distance from Manahan Area is 4 km.
8	Shelter at Sekartaji	There are 7traders, the activities approached are recreational and trading activities, located at neighborhood roads, the traffic condition is relatively busy, the distance from UNS is 2.5 km.
9	Shelter at Pedaringan	There are 7traders, the activities approached are activities in warehouses, located at neighborhood roads, the traffic condition is relatively busy, the distance from UNS is 2.5 km.
10	Shelter at Timur Jurug	There are 6traders, the activities approached are recreational activities and settlements, located at local roads, the traffic condition is busy, the distance from UNS is 2 km.
11	Shelter at Galabo	There are 12traders, the activities approached are trading activities and settlements, located at secondary collector

No	Locations	Characteristics
	Pucang Sawit	roads, the traffic condition is busy, the distance from UNS is 1.5 km.
12	Shelter at Pucang Sawit	There are 6traders, the activities approached are trading activities and settlements, located at secondary collector roads, the traffic condition is busy, the distance from UNS is 1.5 km.
13	Shelter at Pucang Urban Forest	There are 4traders, the activities approached are settlements and industrial activities, located at neighborhood roads, the traffic condition is relatively busy, the distance from UNS is 2 km.
14	Shelter at Solo Square	There are 60traders, the activities approached are trading activatesparticularly in services andsettlements, located at neighborhood roads, the traffic condition is busy, the distance from Solo Grand Mall is 3 km.
15	Fruit Shelter at Purwosari	There are 5traders, the activities approached are trading activatesparticularly in services andsettlements, located at neighborhood roads, the traffic condition is relatively busy, the distance from Solo Grand Mall is 2 km.
16	Shelter at Jl. Supomo	There are 24traders, the activities approached are office activates andsettlements, located at local roads, the traffic condition is busy, the distance from Solo Grand Mall is 1 km.
17	Culinary Shelter at Taman Sriwedari	There are 20traders, the activities approached are recreational activates, located at secondary collector roads, the traffic condition is relatively busy, the distance from Solo Grand Mall is 1 km.
18	Shalter at Timur Telkom	There are 10traders, the activities approached are trading activatesparticularly in services andrecreational activities, located at local roads, the traffic condition is busy, the distance from Solo Grand Mall is 1 km.
19	Shelter at Galabo Siang	There are 54traders, the activities approached are trading activatesparticularly in services, located at local roads, the traffic condition is highly busy, the distance from Surakarta City Hall is 1 km.
20	Shelter at Manahan KS Tubun	There are 37traders, the activities approached are recreational activities, sport activities, and office activities, located at neighborhood roads, the traffic condition is busy, the distance from Manahan Area is 0 km.
21	Shelter at Manahan Menteri Supeno	There are 76traders, the activities approached are recreational activities, sport activities, and office activities, located at local roads, the traffic condition is busy, the distance from Manahan Area is 0 km.
22	Shelter at SD Kristen Manahan	There are 3traders, the activities approached are educational activities and settlements, located at neighborhood roads, the traffic condition is busy, the distance from Manahan Area is 0.5 km.
23	Shelter at Hasanudin	There are 30traders, the activities approached are settlements and trading activatesparticularly in services, located at local roads, the traffic condition is relatively busy, the distance from Manahan Area is 1 km.
24	Shelter/Arcade at Kota Barat	There are 20traders, the activities approached are sport activities, educational activities, trading activities, settlements, and office activities,located at secondary collector roads, the traffic condition is highly busy, the distance from Manahan Area is 1 km.
25	Shelter at Wahidin	There are 13traders, the activities approached are trading activities particularly in services, office activities, and health care activities,located at secondary collector roads, the traffic condition is highly busy, the distance from Solo Grand Mall is 0.5 km.

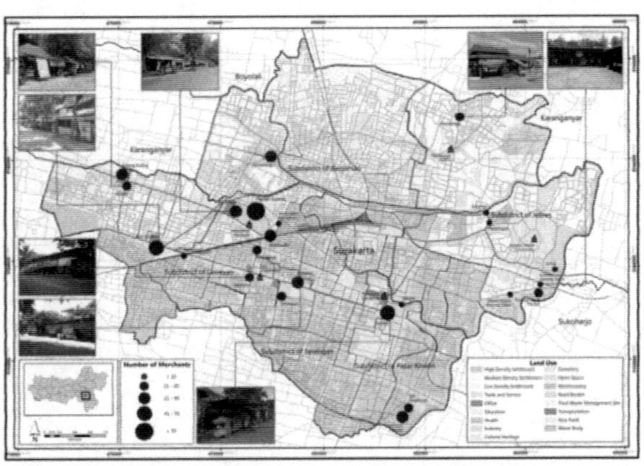

Figure1. The distribution map of locations for street vendor's stabilization in Surakarta City.

4. Conclusion

The management of street vendor locations in Surakarta City considers the availability of public lands, the proximity to the town center, the proximity to the main activities (particularly trading and settlements), the traffic levels, and the accessibility. However, the number of stabilized street vendors and the distance of the stabilization locations to the city centers and street vendors' houses are less considered in this case although they are included in the primary criteria.

Acknowledgment
The researchers would like to express their gratitude to the LPDP for granting the opportunity to write this article, as well as the promoter and co-promoter for their continuous and tireless guidance, street vendors and their communities, and officials in Street Vendors Management of Surakarta City for the significant contributions.

References
[1] Rahayu M J, Putri R A, and Andini I. 2015. Typology of Urban Hawkers Location Preference *Proc. on Procedia—Social and Behavior Science(Surabaya).* 227:239–246.
[2] McGee T G, and Yeung Y M. 1977. Hawkers in Southeast Asian Cities: Planningfor the Bazaar Economy. (Ottawa: International Development Research Center).
[3] Hanifah U, and Mussadun. 2014. Penilaian Tingkat Keberhasilan Relokasi PKL di Kawasan Pasar Warudan Simpang Lima. *SMGM PWK Undip Jurnal Perencanaan Wilayah dan Kota.* 25:228–242.
[4] Werdiningtyas R, Rahayu M J, and Musyawaroh. 2012. Hawkers Behaviour and Characteristic as an Important Factor to Empower Local Economy. *Proc. Int. Conf. on 2nd CONVEEESH & 13th SENVAR, Yogyakarta.* 2:1-10, Yogyakarta: Architecture Department DWCU).
[5] Owusu A B, Abrokwah S, and Frimpong S. 2013. Analysis of the Spatial and Temporal Dynamics of Street Hawking: A Case Study of the Accra Metropolitan Area. *Journal of Geography and Geology.* 5:4.
[6] Widjajanti R. 2015. *Perilaku Berlokasi PKL di Ruang Publik Kota* (Bandung: ITB).
[7] Blackburn S. 2011. *Jakarta Sejarah 400 Tahun* (Jakarta: Yayasan Komunitas Bambu).
[8] Rahayu M J, Musyawaroh, and Werdiningtyas R. 2013. Stabilisasi sebagai Bentuk Penataan PKL Makanan Siap Saji di Kota Surakarta. *Jurnal Tata Loka.* 15(1):41–54.
[9] Shirvani H. 1985. The Urban Design Process (New York: Van Nostrand Reinhold Company).

Environmental performance and financial report integrity: challenges for the mining sector in Indonesia

S Mayangsari

Faculty of Economic and Business, Universitas Trisakti, Jakarta, Indonesia

Corresponding Author : sekar_mayangsari@trisakti.ac.id

Abstract. This study investigates the influence of environmental performance on the financial report integrity. The statistics used were primary data from interviews with senior members of the mining sector regarding environmental issues, as well as secondary data using Financial Report 2016. The samples were listed mining companies with semester data. Questionnaires were used to measure their perceptions of the challenges concerning climate change faced by the mining sector. The results of this research show that regulatory interventions will be critical to environmental issues. This study employed KLD as a proxy for environmental performance, correlated with other variables regarding the integrity of disclosure. The outcome indicates that environmental issues will increase the integrity of financial reports.
Keywords: environmental issues, integrity financial report, KLD, mining firms

1. Introduction

This study investigates the implications of environmental risks on the financial report integrity. The conceptual framework is based on the view that environmental practices influence the quality of disclosure by determining their exposures to legal, reputational, and regulatory risks. Firms that engage in environmental misconducts can incur costly penalties and evoke strong negative reactions from stakeholders, each of which affects their default risks and consequently impairs the integrity of financial reports. For instance, the oil spill of Deep water Horizon in the Gulf of Mexico illustrates the tremendous adverse impacts an environmental incident can have on the quality of financial report using the corporate credit rating as the proxy for financial report integrity.

The differences with prior studies are, firstly, that this study uses an argument that financial report users want to protect themselves against environmental performance. Secondly, concerning the environmental issues, this paper shows that losses require a better understanding of how different corporate environmental activities relate to financial risks, as well as its influence on any decision. This study aims to provide a better insight into the impacts and consequences upon the firms if they neglect the environmental issues.

Based on Figure 1, this study put concerns on the right side; it investigates the benefits for firms pertaining to environmental issues on the market. Moreover, the market in this study correlates with the integrity of financial reports which intends to detect the equity statement by external users.

In examining the hypothesized relation, the analysis of this study took advantage of the expanding coverage and popularity of independent rating agencies that specialize in the impact of the corporate environmental performance [4], [6].

IOP Conf. Series: Earth and Environmental Science **106** (2017) 012064 doi:10.1088/1755-1315/106/1/012064

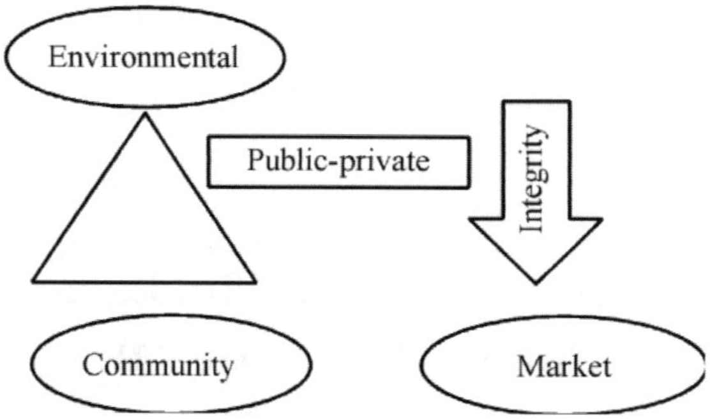

Figure 1. KLD state of the art.

This research begin from researched by [3], they suggested to integrate environmental disclosure and social disclosure in an annual report. Some prior empirical study show that environmental and social disclosure influence on earnings quality [9], [10], but all of them use manufacture companies. This paper only use mining companies to test the benefit of environmental and sosial disclosure for firms. Why? Because in Indonesia, both of the disclosure is mandatory since 2007. This research can increase the generalization of the influence of environmental and social disclosure on financial report quality from external side, while others [9], [10] test from internal side.

2. Research Method

2.1. Research Design

The type of this research design is causality. The variables were environmental performance and financial report integrity. The definition of environmental management performance was based on the instrument from Kinder, Lydenberg, and Domini Research & Analytics (KLD) [3]. KLD is an independent investment research company that specializes in the assessment of corporate environmental management, social performances, governance standards, and product qualities. Table 1 shown indicators of KLD, both KLD environmental strengths and KLD environmental concerns [3].

Table 1. Description of KLD environmental ratings.

KLD environmetal strengths
1. **Beneficial products and services**. The company derives substantial revenues from innovative remediation products, environmental services, or products that promote the efficient use of energy, or it has developed innovative products with environmental benefits.
2. **Pollution prevention**. The company has notably strong pollution prevention programs including both emissions reductions and toxic-use reduction programs.
3. **Recycling**. The company either is a substantial user of recycled materials as raw materials in its manufacturing processes,or a major factor in the recycling industry.
4. **Clean energy** (previously called Alternative fuels). The company has taken significant measures to reduce its impact on climate change and air pollution through use of renewable energy and clean fuels or through energy efficiency. The company has demonstrated a commitment to promoting climate-friendly policies and practices outside its own operations.
5. **Communications**. The company is a signatory to the CERES Principles, publishes a notably substantive environmental report, or has notably effective internal communications systems in place for environmental best practices.
6. **Property, plant, and equipment**. The company maintains its property, plant, and equipment with above-average environmental performance for its industry.
7. **Other strength**. The company has demonstrated a superior commitment to management systems, voluntary programs, or other environmentally proactive activities.

IOP Conf. Series: Earth and Environmental Science **106** (2017) 012064 doi:10.1088/1755-1315/106/1/012064

KLD environmental concerns

1. **Hazardous waste.** The company's liabilities for hazardous waste sites exceed $50 million, or the company has recently paid substantial fines or civil penalties for waste management violations.
2. **Regulatory problems**. The company has recently paid substantial fines or civil penalties for violations of air, water, or other environmental regulations, or it has a pattern of regulatory controversies under the Clean Air Act, Clean Water Act, or other major environmental regulations.
3. **Ozone-depleting chemicals**. The company is among the top manufacturers of ozone-depleting chemicals such as HCFCs, methyl chloroform, methylene chloride, or bromines.
4. **Substantial emissions**. The company's legal emissions of toxic chemicals (as defined by and reported to the EPA) from individual plants into the air and water are among the highest of the companies followed by KLD.
5. **Agricultural chemicals**. The company is a substantial producer of agricultural chemicals, i.e., pesticides or chemical fertilizers.
6. **Climate change**. The company derives substantial revenues from the sale of coal or oil and its derivative fuel products, or the company derives substantial revenues indirectly from the combustion of coal or oil and its derivative fuel products.
7. **Other concern**. The company has been involved in an environmental controversy that is not covered by other KLD ratings.

The closed-type questionnaire was used in this study. Respondents were required to answer a list of questions using the Likert scale of 1–5. The other independent variable is corporate environmental responsibility with indicators as shown in Table 2.

Table 2. Corporate Environmental Responsibility (CER) indicators.

No.	CER	Indicators
1	Climate Change	a. Clean energies b. Climate change
2	Product Services	a. Beneficial product services b. Ozone-depleting chemicals c. Mining chemicals
3	Operation Management	a. Pollution prevention b. Recycling c. Management system d. Hazardous waste e. Regulatory problems f. Substantial emissions
4	Others	a. Audit reports b. Environmental internal control

Source: KLD

2.2. *Mathematical Equation*

The corporate rating was measured using the following equation:

$$Rating = \alpha_{it} + + \beta_1 CERit + \beta_2 SIZE_{it} + \beta_3 CI_{it} + \beta_4 INT_{it} + \beta_5 ROA_{it} + \beta_6 LOSS_{it.} + \varepsilon_{it} \quad (1a)$$

$$Rating = \alpha_{it} + \beta_1 ENVStrengths_{it} + \beta_2 ENVConcerns_{it} + \beta_3 CERit + \beta_4 SIZE_{it} + \beta_5 CI_{it} + \beta_6 INT_{it} + \beta_7 ROA_{it} + \beta_8 LOSS + \varepsilon_{it.} \quad (1b)$$

where:

ENVStrenghts	=	Environmental strengths
ENVConcerns	=	Environmental concern
CER	=	Corporate environmental responsibility
Size	=	Log Total Assets
CI	=	Capital Intensity Ratio is fixed assets divided by number of workers [2]
INT	=	Interest Coverage is earnings before interest and tax divided by interest Expense [5]
ROA	=	Return on Asset is measured by net income divided by total assets
LOSS	=	Nominal scale; 1 if firms profit and 0 otherwise

For integrity financial reporting, this paper use credit rating as measurement in line with [7] (see Table 3).

Table 3. Rating measurements.

Symbol	Description
idAAA	The obligors' capacity to meet its long-term obligation is excellent.
idAA	An obligor has a decent capacity to meet its long-term financial commitments, relative to that of other Indonesian obligors.
idA	The obligor has a strong capacity to fulfill the commitments but is more susceptible to economic conditions.
idBBB	The obligor has an adequate capacity to meet its long-term obligation.
idBB	The obligor has a weak capacity to meet its long-term financial commitments, relative to that of other Indonesian obligors.
idB	The obligor has a weak capacity to fulfill the commitments and is more susceptible to economic conditions.
idCCC	The obligor is currently vulnerable, depending on favorable conditions.
idD	The obligor **has failed** to pay one or more of its financial obligations.

2.3. Data Collection

Primary data were collected from the manager of each listed mining company to construct aggregate measures of the corporate environmental strengths and concerns. Subsequently, the regression of the cost of debt and credit ratings was performed as the measurement of integrity of financial statements. The number of samples was 82 corporate semester dataset from 41 firms of the observation year of 2016. Data were also gathered from [7], the Indonesia credit rating agency.

2.4. Data Analysis

Tobit regression model was used in the data analysis because the dependent variable (RATING) is measured by limited numbers as 1=idAAA, 2=idAA, 3=idA, 4=idBBB, 5=idBB, 6=idB, 7=idCCC, and 8=idD. Regression is conducted twice, first without environmental variables including strengths and concerns, and second, with both variables.

3. Results and Discussion

Table 4 below shows the detailed results, summarizing the effect of environmental management performance on credit ratings. The results presented herein are consistent with the prediction that the corporate environmental management has implications on the integrity of financial reports. The higher environmental concern and strengths firms, the higher credit rating which is indicate stronger credit profile and lower risk of debt. The results were economically meaningful and statistically significant in depicting the relations between environmental performance measures and both representatives of financial report integrity, and corporate credit ratings. The corporate activities underlying the documented relations are mainly attributable to regulatory and climate change issues. They serve as efforts to reduce environmental risk exposure and enhance cash flows by supplying innovative products and services with environmental benefits. The findings of this study extend earlier research by showing that the credit market does not only respond to the potential misreporting of existing environmental liabilities but also to costs of corporate environmental management in anticipating associated losses.

Table 4. Results of credit rating regressions.

Variables	**Coefficients (1a equation)**	**Coefficients (1b equation)**
ENVStrengths		0.1622* (1.93)
ENVConcerns		0.1925** (2.18)
CER	-0.0169 *** (-3.93)	-0.0165*** (-3.90)
Size	0.5708*** (4.68)	0.6164*** (5.99)
Capital Intensity	0.0067*** (3.21)	0.0069*** (3.19)
Interest Coverage	0.0321** (2.29)	0.0323** (2.31)
ROA	0.0739*** (3.88)	0.0746*** (4.23)
Loss	-0.9782*** (-4.14)	-0.9814*** (-4.02)
Pseudo R^2	0.33	0.34
Observations	82	82

This regression models to test the influence of environmental variables and corporate environmental responsibility with integrity financial report. T-statistics (in parentheses) are compared with t-tables to get the significance results. *, **, and ***denote statistical significance at the 10%. 5%, and 1% levels, respectively.

This study presents evidence that a proactive corporate environmental engagement is associated with a lower credit risk. However, since this relation is primarily reflected in bond prices, it may indicate that credit rating agencies have not yet fully accounted for this effect. The results are consistent with the view that regulatory implications of climate change have sensitized lenders to the downside risk of poor environmental practices. Furthermore, the results show that an assessment of past and ongoing corporate environmental activities enhances the accuracy in the pricing of corporate bond issues, by enabling bond investors to evaluate the risk of environmental-performance-related losses. These findings contribute to a better understanding of how corporate environmental activities affect the integrity of financial reporting, which is in this research is surrogated by credit rating. Nowadays, environmental performance has been considerably increasing over time. Firms' credibility and going concern can be reached when current needs are met with future generation's needs. The awareness of green notion leads to the development of environmental accounting [1].

Accordingly, this paper show that environmental management accounting (EMA) give many benefits to business not only for global warming issues but also for increasing firm's credibility. These benefits will improve firm's reputation from environmentally-friendly products into the market and performing firm's activities, including exploration and exploitation, with less harmful effects on surrounded environment. Some Indonesia mining companies revealed that they have employed EMA to improve firms' value and also shareholder value. Unfortunately, some companies still perceive that implementation of EMA is cost burden. This paper also confirms the benefit of EMA implementation

IOP Conf. Series: Earth and Environmental Science **106** (2017) 012064 doi:10.1088/1755-1315/106/1/012064

around the world. For instance, [2] show the benefits of EMA implementation as to eco-efficiency in some companies in Japan, and the impact on firm's value. Some companies around the world not only mining industries but also including many kind of industries, such as pharmaceutical, chemistry and automotive, implement EMA voluntary and then find their value increase. In Lithuania, the result of evaluation on EMA program revealed that EMA assisted companies in reduce their operating costs, lower product's price, higher revenue, higher profit and higher market value.

4. Conclusion

This paper reports extensive evidence that the environmental management of public corporations has credit risk implications on bond investors. The underlying fundamental hypothesis is that environmental practices affect the solvency of borrowing firms, by determining their exposure to potentially costly legal, reputational, and regulatory risks. Aggregate measures were constructed for the environmental strengths and concerns of firms. Their relations with the yield spread of newly issued bonds, bond ratings, and long-term issuer ratings were then examined. The analysis shows that companies with environmental concerns have higher credit ratings assigned to them. The corporate activities that underlie this relation are mainly related to regulatory and climate change issues. Particularly, higher bond rating are associated with the supply of innovative products and services with environmental benefits, as well as the firm's efforts to reduce their impacts on climate change and air pollution through the use of clean energy, energy efficiency, or their commitments to climate-friendly policies and practices. However, our results indicate that environmental management practices have become increasingly relevant to bond investors over the recent decade. This outcome is robust to control many credit risk determinants, various model specifications, and industry membership.

References

[1] Burrit R L, Hahn T, & Schaltegger S. 2002. Towards a comprehensive framework for environmental management accounting- Links between business actors and environmental management accounting tools. *Australian Accounting Review* 12 (2): 39-50

[2] Burrit R L, & Saka C. 2006. Environmental management accounting application and eco-efficiency: Case studies from Japan. *Journal of Cleaner Production* 14: 1262-1275

[3] Chatterji A K, David I L, & Michael W R. 2008. How well do sicial ratings actually measure CSR? *Journal of Economics & Management Strategy*

[4] Kempf A, and Osthoff P. 2007. The effect of socially responsible investing on portfolio performance. *European Financial Management.* 13:908–922.

[5] Lim David. 1976. On The Measurement of Capital Intensity. *Review of World Economics.* 760-766

[6] Statman M, and Glushkov D. 2009. The wages of social responsibility. *Financial Analysts Journal*. 65:33–46.

[7] PEFINDO. 2015. Pemeringkat Efek Indonesia. www.pefindo.com.

[8] Subramanyam, Wild. 2014. *Financial Statement Analysis*. Mc-Graw Hill Edition

[9] Staden,Chris Van, Erica Yio & Steven Cahan. 2011. Corporate Social Responsibility Reporting And Earnings Management: The Role of Political Costs Australasian Accounting, *Business and Finance Journal.* 3: 17-34

[10] Walter, A, Denis C & Michael Magnan. 2008. Corporate environmental disclosure, financial markets and the media: An international perspective. Ecological Economics 64: 643-659

The prospects of the domestic water equity indicators in Indonesia: a review

A Nastiti[1]*, A Komarulzaman[2], A Sudradjat[1]

[1]Environmental Management Technology Research Group, Faculty of Civil and Environmental Engineering, Institut Teknologi Bandung, Bandung, Indonesia
[2]Economics Department, Faculty of Economics and Business, Universitas Padjajaran, Bandung, Indonesia

*Corresponding Author: anindrya@tl.itb.ac.id

Abstract. Despite the major progress achieved by the domestic water supply sector since the commencement of the Millennium Development Goals (MDGs), there is still a concern that access towards water does not distribute evenly among citizens in different geographical areas or diverse economic groups. The Sustainable Development Goals (SDGs) strive for a universal water target that highlights the sustainable access to safe and affordable water supply for all. Hence, the ensuing challenge is how to comprehensively report the progress of achieving water equity in relation to the SDGs target. This paper reviews the current research and policy papers on equity metrics in the water supply sector. This study has identified that water inequity may manifest in the variations of the level of access, the dimensions of access, and the impacts of poor water supply—spatially, socially, economically, or the combination thereof. This paper also presents challenges related to the application of equity measurements in the context of Indonesia. The results will be useful in designing appropriate tool to inform decision making in water sector policy.
Keywords: domestic water supply, equity, indicator, metrics, SDGs

1. Introduction

Despite the major progress achieved by the domestic water supply sector since the commencement of the Millennium Development Goals (MDGs), the WHO/UNICEF's Joint Monitoring Programme (JMP) has recognized that the emphasis on macro-level progress masks the problems of inequity [1]. The JMP further reported that access to improved drinking water sources is lower in rural areas compared to that in urban areas in almost all countries, and is doubled among the richest of the population than among the poorest quintile group [1, 2]. This statement is particularly true in Asian regions that still suffer from the incomplete water service delivery. For example, urban Southeast Asian have 50% coverage of piped water; much higher than the coverage of piped water service in the rural Southeast Asian of only 13% of the total rural population [3]. Sustainable Development Goals (SDGs), the replacement development framework for MDGs, set an ambitious goal: "By 2030, achieve universal and equitable access to safe and affordable drinking water for all". 'Universal' means access to water in all settings, including households, schools, healthcare facilities, and workplaces. This paper will only discuss the water supply in households setting. 'Equitable' indicates a progressive reduction and elimination of inequity between population subgroups, not only in term of physical access to water but also service levels. The normative interpretation of 'for all' in Target 6.1 of SDGs is that the water provided should be suitable for use by all citizens without discrimination—men, women, girls and boys of all ages, including people with disabilities. It is clear that equity is a significant issue in the SDGs water target. Moreover, Goal 17 (Target 17.18) of the SDGs also emphasizes to increase, by 2020, the high-quality, timely, and reliable data disaggregated by various geographic and socio-economic characteristics. This point can serve as a potential lead-in to start integrating the concept of equity in reporting the development progress. Hence, the ensuing questions are: Does equity refer to merely an even distribution of access to water? How can we be sure whether water equity is achieved? This paper aims to provide an overview of the current research and practices regarding domestic water equity. Research findings and policy papers have been reviewed to gain a better understanding of water equity. Afterwards, the Indonesian survey instruments were scrutinized to seek for potential indicators for equity and equality. Lastly, the discussion contains the prospective application of such indicators in representing the concept of water equity. Like inequality, inequity can be measured as well [4], but there should be a sense of fairness issue in the measurements. For example, Gini coefficient is a measure of equality, not equity [4] since it measures the degree of variation represented in a set of values without an elaboration of how these values vary among different groups of income, gender, or ethnicity in an instance. Similarly, Researchers [5] use the term 'inequality' in their paper that talks about the geographic heterogeneity of the use of improved drinking water sources within countries. In the domestic water supply sector, however, the issue of fairness is inherent, primarily because water is a prerequisite for not only life but also domestic and economic activities. Equity is related to Amartya Sen's freedom and capability metrics [6]. Different access levels and circumstances of water can elevate people out of poverty or condemn them to it [7]. It often works perpetually; in his book, Development as Freedom, Sen points out that the economic poverty robs people of the freedom to satisfy their needs of clean water [8]. Poverty influences the technology of household water supply, which in turn affects the quantity of water used by households for consumption, basic hygiene, other domestic purposes, and other potential income earning activities [9]. The lack of access to water can hinder a various range of individual abilities, not only to survive but also to thrive of their full potentials.

It is imperative also to understand that the definition of equity depends on the history or social values and the conditions of a country or region as well, and such definition often stems from, as Karan and Wilkinson mention, "generally accepted human values of fair play and justice at that time" [10]. In an instance, a recent study in a peri-urban area of Bandung City, Indonesia, focusing on how local water entrepreneurs, together with citizens, work towards cultivating equity in providing access to spring water network [11]. Although not everyone can get a similar type of access to the spring water network, conflicts related to the unfairness are mostly avoided through a shared consciousness that 'not all can get water from the communal tank' and 'the water will not be enough' [11]. Therefore, assessing the real water equity is not always straightforward.

2. Water Equity Metrics

There are many ways to measure the state of water inequity in the domestic setting. In scholarly works, the metrics of inequity mainly talks about the uneven distribution of access to improved water sources across social groups or geographical regions. For example, Dungumaro measured the inequity of access to water supply based on income, dwelling unit, household size, and economic status index [12]. Dungumaro assessed what in this paper is called **social inequity**. Social inequity represents an uneven distribution of access to control over and use the water among different social groups and their members (e.g., income level, ethnicity, gender) [13]. Figure 1 shows an example of the type of social inequity where access to improved water sources is more prominent in Indonesian rich households than in poor ones, both in urban and rural areas.

A study by [5] assessed the distribution of the proportion of population reporting use of improved water sources protected from outside contamination, based on administrative areas, countries, and urban/rural types [5]. It reveals another category of water inequity: spatial inequity, that refers to an unequal distribution of access to control over and use the water among people living in different regions. Figure 2 illustrates an example of the spatial inequity of access to water in the Southeast Asian region. Some spatial inequity may also have some socioeconomic, even political backgrounds.

IOP Conf. Series: Earth and Environmental Science **106** (2017) 012065 doi:10.1088/1755-1315/106/1/012065

For example, the problem of underserved peri-urban communities in many postcolonial cities in the developing world, such as Bandung and Jakarta, roots from the colonial segregation in the past. Piped water connections were built to serve those resided in the city centres: local elites and expatriates who were considered as first-class citizens [14]. In an instance, access to piped water in Bandung City mostly higher in the central urban districts, although the overall coverage is still considered low [17]. However, there are cases, like in Jakarta, in which several population subgroups currently residing in the slum pockets of the city centres, who supposedly have good connections due to the colonial legacy, still do not have access to piped water network [15].

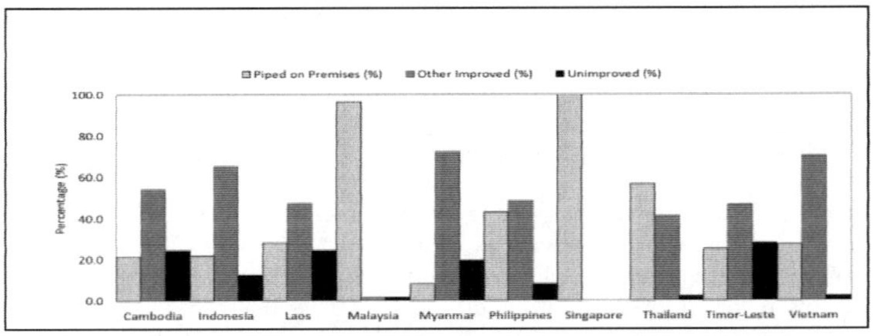

Figure 1. Trends in drinking water coverage by Indonesian rural and urban wealth quintile in 2012. The coverage of piped water and improved water sources increases along with the increase of wealth. Graph modified from data of the JMP [2].

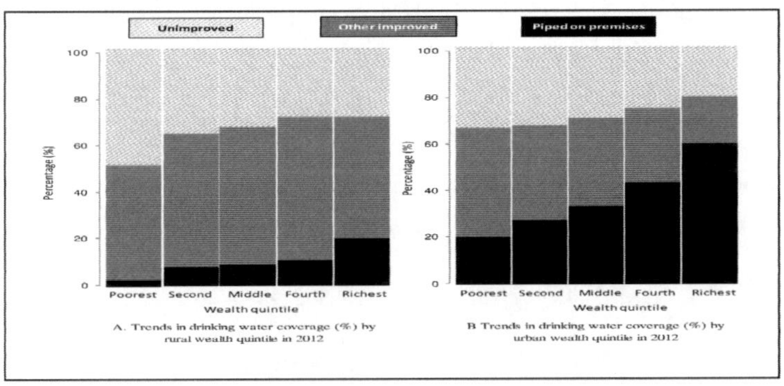

Figure 2. Drinking water coverage (%) in Southeast Asian countries. Access to piped water on premises in Singapore and Malaysia is much higher than that of other countries (data extracted from the JMP [16]).

Uneven distributions of water often manifest, not only in the access level but also in service level or the variations in the dimensions of access, such as collection time, water quality, water quantity or consumption, water reliability, and water affordability or expenditure (for instance, see [12], [18], [19], [20]). The examination towards the concept of equity is getting complex with a question that Goff and Crow posed: "Is equity connected to improved water sources access, or to the quantity of potable water that households can collect?" [10]. Later, the authors suggested that the two ideas of equity not be sufficient to measure equity since, as the authors conclude, "...the most vulnerable people in the world have not benefited from progress." Access to water is only an intermediate target [9]. Results should be focused on expanding actual water use in order to achieve the intended development outcome. For example, the increasing quantities of potable water used by households for hygienic purposes may help reduce the rates of under-five mortality related to diarrheal diseases. Therefore, there is also a possibility that water inequity is measured in the context of uneven distributions of various impacts due to poor water supply. In an instance, opportunity costs for collecting water and diarrheal diseases are associated with poor drinking water quality. Even so, reliable and continuous data related to such matters are difficult to be found in the nation-wide survey. Thus, firm causal-effect relationships may be challenging to accomplish. Based on the previous discussion, an illustration of different types of water inequity and potential water equity indicators was then devised (Figure 3). Meanwhile, Table 1 shows water inequity studies and what types of water inequity these studies have focused on.

Table 1. Studies on water inequity.

Study	Operational definition on equity	Focus	
Dungumaro, 2007, South Africa [12]	Distribution of household source of water supply (by income; by dwelling unit; by household size; by economic status index)	SOC	A
	Distribution of time spent to water source (by income)	SOC	DoA
Pullan et al., 2014, Sub-Saharan Africa [5]	Distribution of the proportion of population reporting use of improved water source, that is protected from outside contamination (by administrative areas; by countries; by urban/rural type)	SPAT	A
	Distribution of the proportion of population reporting use of improved water source, that is protected from outside contamination (by quintile within countries)	SOC	A
	Distribution of the proportion of population reporting use of accessible, improved water source, that is can be accessed within 1 km or 15 minutes of the household (by administrative areas; by countries)	SPAT	DoA
	Distribution of the proportion of population reporting use of accessible, improved water source, that is can be accessed within 1 km or 15 minutes of the household (by quintile within countries)	SOC	DoA
Sebri, 2015, Tunisia [18]	Distribution of basic/minimum water consumption per household AND average residential water consumption per household (by governorate; by income).	SOC	DoA
	Affordability Index (AI) for BOTH basic and average water consumption. AI is calculated as basic OR average household water charges or bill divided by the average household income.	SOC	DoA
Sun et al., 2015, China [19]	Distribution of water expenditure (by income; by city scale (small-, medium-, and large-cities))	SPAT	DoA
Yang et al., 2012, Bangladesh, Peru, Ethiopia, Jordan, Nicaragua, Nigeria, Tajikistan [20]	Distribution of household using improved water sources (by socio-economic status)	SOC	DoA
	Distribution of tubewell with <50 ppb of arsenic (by socio-economic status)	SOC	DoA
	Distribution of piped water with >0.1 mg/l free residual chlorine (by socio-economic status)	SOC	DoA
	Distribution of compliant improved water source, in term of water quality (by socio-economic status)	SOC	DoA

Note: SOC = social inequity, SPAT=spatial inequity, A= inequity in access, DoA=inequity of the dimensions of access.

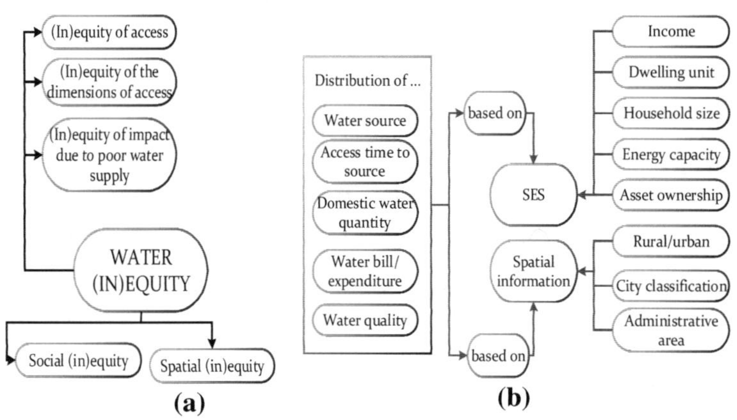

Figure 3. **(a)** Different types of water inequity. Inequity can be assessed by looking at the variations of the level of access, dimensions of access, and impacts due to poor water supply. These variations can also be based on the socio-economic characteristics and geographic location; **(b)** Potential indicators for water equity metrics (SES = socioeconomic status).

3. Measuring equity in Indonesia: application and challenge

After scrutinizing various indicators that were mentioned in the previous studies, the survey instrument of SUSENAS, the Indonesian annual nationwide socioeconomic survey, was then observed to see whether potential indicators for water equity may already be available in the surveys. SUSENAS has been used as the basis for MDGs' monitoring and will continue to become the main data supplier in tracking progress towards SDGs. Table 2 shows the SUSENAS questions related to the potential water equity metrics. Since there is no nationwide water quality analysis, the type of water source available in a household and the distance to such a source to the nearest wastewater containment facility serve as a proxy to safe water. These two questions, the type of water source and the distance to wastewater containment facility, are presently used to classify whether households have access to a certain 'improved' or 'unimproved' water source. Note that SUSENAS records what kind of water source that households have for three different uses: for drinking, cooking, and bathing purpose.

Table 2. Relevant questions in SUSENAS.

Issue	Question in SUSENAS
Water source	• What is the main source of water used for drinking, cooking, and bathing? (branded bottled water, refill water, metered piped water, resale piped water, borehole, protected well, unprotected well, protected spring, unprotected spring, surface water, rainwater, others)
Water bill	• How much money does each household spend for water?
Water quality	• If the main source of water used for drinking is borehole/well/spring, how far is the source of water to the nearest waste water containment facility? (<10 m, ≥10 m) • If the main source of water used for cooking is borehole/well/spring, how far is the source of water to the nearest waste water containment facility? (<10 m, ≥10 m) • If the main source of water used for bathing/washing is borehole/well/spring, how far is the source of water to the nearest waste water containment facility? (<10 m, ≥10 m)
Water quantity	Is the water used for bathing/washing originate from the networked piped water or a public water terminal? (networked piped water, public terminal)

The behaviour of using multiple water sources for various domestic purposes are commonly found in Indonesian households. The implications of such multiple water sources use towards the water equity needs to be scrutinized. One thing to consider, the use of multiple water sources by households indicates that the primary water source they have is 'lacking' [21]. Thus, the rate of combined use of 'improved' and 'unimproved' water sources needs to be taken into account. Moreover, there is no direct question related to the domestic water quantity consumed by households.

However, data on water bill are limited to households who are subscribed to piped water companies or buy water to vendors. Water expenditure and water quantity consumed by private groundwater users will be difficult to estimate. A question in SUSENAS also focuses on whether water for bathing and washing is obtained from in-house water connection of public water terminal. It can serve as a potential proxy for water quantity as there is a limit to how much water can be collected by household members when the source of water is not located on premises. As Larson et al. have recorded, households with private water connections use water in a quantity that is six times higher than households that need to collect water from somewhere else [9].

4. Conclusion

The widely used approach of aggregating data has masked serious inequity problems that policy makers and water supply managers need to address. The main challenge in overcoming such inequity problems is to understand how unequal our water supply service among different population groups or across administrative boundaries. This study has identified that water inequity may manifest spatially, socially, economically, or the combination thereof. This paper also highlights that not only the level of access, the dimensions of access and the impact of poor water supply are also crucial in revealing the real persisting gap in water service. There are currently missing elements in the SUSENAS to properly measure the water inequity, such as water expenditure for piped and non-piped water users, water quantity consumed/used by households, and the duration of daily water availability. Consequently, the focus of future research should not stop in designing appropriate water equity metrics for Indonesian context. The perception of users and policy-makers concerning the water inequity and the production of water inequity should also be on the agenda of water equity research. Finally, the discussions in this paper serve as a base for future studies in understanding the pattern of water inequity in Indonesia and how to overcome it.

Acknowledgment

This study is funded by the Program Penelitian, Pengabdian kepada Masyarakat, dan Inovasi (P3MI) Kelompok Keahlian Institut Teknologi Bandung 2017.

References

[1] WHO & UNICEF, Drinking Water Equity, Safety and Sustainability, Thematic Report on Drinking Water 2011, 2011.
[2] JMP. 2015. https://www.wssinfo.org/fileadmin/user_upload/resources/JMP-Wealth-Quintiles-Indonesia.xlsx.
[3] Cronin A A, Badloe C, Torlesse H, & Nandy R K. 2015. Water, Sanitation and Hygiene: Moving The Policy Agenda Forward in The Post-2015 Asia. *Asia & the Pacific Policy Studies*, 2(2): 227-233.
[4] Cullis J, & Van Koppen B, Applying The Gini Coefficient to Measure Inequality of Water Use in The Olifants River Water Management Area, Vol. 113, IWMI, South Africa, 2007.
[5] Pullan R L, Freeman M C, Gething P W, & Brooker S J. 2014. Geographical Inequalities in Use of Improved Drinking Water Supply and Sanitation Across Sub-Saharan Africa: Mapping and Spatial Analysis of Cross-Sectional Survey Data. PLoS medicine. 11(4): e1001626.
[6] Lu F, Ocampo-Raeder C, & Crow B. 2014. Equitable Water Governance: Future Directions in The Understanding and Analysis of Water Inequities in The Global South. Water International. 39(2):129-142.
[7] Goff M, & Crow B. 2014. What is Water Equity? The Unfortunate Consequences of A Global Focus on 'Drinking Water'. *Water International.* 39(2):159-171.
[8] Sen, A. 2001. Development as Freedom. Oxford Paperbacks.
[9] Larson B, Minten B, & Razafindralambo R. 2006. Unravelling The Linkages Between The Millennium Development Goals for Poverty, Education, Access to Water and Household Water Use in Developing Countries: Evidence from Madagascar. *The Journal of Development Studies*. 42(1): 22-40.
[10] Karar E, & Wilkinson M, Water Equity Dimensions in a Developmental State, In Jägerskog, A., Clausen, T. J., Holmgren, T., and Lexén, K. (eds.), Water for Development – Charting a Water Wise Path, Report No 35. SIWI, Stockholm, 2015.
[11] Nastiti A, Meijerink S V, Oelmann M, Smits A J M, Muntalif B S, Sudradjat A, & Roosmini D. 2017. Cultivating innovation and equity in co-production of commercialized spring water in peri-urban Bandung, Indonesia. Water Alternatives. 10(1):160.
[12] Dungumaro E W. 2007. Socioeconomic differentials and availability of domestic water in South Africa. *Physics and Chemistry of the Earth*, Parts A/B/C. 32(15): 1141-1147.
[13] Phansalkar S J. 2007. Water, equity and development. *International Journal of Rural Management*. 3(1):1-25.
[14] Prasetiawan T, Nastiti A, & Muntalif B S. 2017. 'Bad' piped water and other perceptual drivers of bottled water consumption in Indonesia. Wiley Interdisciplinary Reviews: Water.
[15] Bakker K, Kooy M, Shofiani N E, and Martijn E J. 2008. Governance Failure: Rethinking the Institutional Dimensions of Urban Water Supply to Poor Households. World Development. 36(10):1891–1915. www.dx.doi.org/10.1016/j.worlddev. [15 September 2007].
[16] JMP. 2015. https://www.wssinfo.org/data-estimates/tables/.
[17] Bandung Municipal Water Company Tirtawening. 2013. www.pambdg.co.id. [11 October 2013].
[18] Sebri M. 2015. Water affordability and social equity in Tunisian governorates: a distributive approach. Water Policy. 17(1): 26-45.
[19] Sun C, Zhang Y, Peng S, & Zhang W. 2015. The inequalities of public utility products in China: From the perspective of the Atkinson index. Renewable and Sustainable Energy Reviews. 51:751-760.
[20] Yang H, Bain R, Bartram J, Gundry S, Pedley S, & Wright J. 2013. Water safety and inequality in access to drinking-water between rich and poor households. *Environmental science & technology*. 47(3):1222-1230.
[21] Nastiti A, Muntalif B S, Roosmini D, Sudradjat A, Meijerink S V, & Smits A J M. 2017. Coping with poor water supply in peri-urban Bandung, Indonesia: Towards a framework for understanding risks and aversion behaviours. *Environment and Urbanization*. 29(1):69-88.

IOP Conf. Series: Earth and Environmental Science **106** (2017) 012066 doi:10.1088/1755-1315/106/1/012066

The spatial distribution of suspended sediment analysis along Krueng Cut River, Banda Aceh

M Irham[1*], Y Fadhla[1], I Setiawan[1,2]

[1]Faculty of Marine and Fisheries, Universitas Syiah Kuala, Banda Aceh, Indonesia
[2]Center for Marine and Fisheries Study of Universitas Syiah Kuala, Banda Aceh, Indonesia

*Corresponding Author : irham@unsyiah.ac.id

Abstract. The spatial distribution of suspended sediment research at an estuarial area was carried out in Krueng Cut River. The study employed a purposive sampling method at seven stations, and the data were analyzed by using the spatial distribution analysis of suspended sediment. The samples were taken at three places in each station during high tide and low tide conditions. The purpose of the research is to understand how the concentration of suspended sediment is distributed throughout Krueng Cut estuarial area during high tide and low tide conditions. The results informed that during high tide condition, the concentration of suspended sediment was mostly circulated downstream near the mouth of the river. In contrast, during low tide condition, the suspended sediment concentration was distributed predominantly in the upstream of the study area.
Keywords: estuarial area, Krueng Cut, suspended sediments, spatial distribution

1. Introduction

The Krueng Cut estuarial area is located at Alue Naga sub-district, Banda Aceh. This estuarial area has multifunctional activities that are frequently used by the fisherman community for their livelihood. Such activities have changed the existence of this estuarial area spatially and temporally. Beside the communal activities, the hydro-oceanographic parameters such as current, tide, and wave in this region have also influenced the environmental conditions [1], [2] so that the surrounding area can be positively or negatively impacted [5], [15]. Some of those impacts are the addition of land area as well as the occurrence of soil erosion along the river bank because of sedimentation and erosion [3], [11].

The occurrence of sedimentation in this area is due to the accumulation of upstream eroded, weathered rock materials carried away by the water and settled in the location where the materials cannot be transported anymore by the flow of the water [4], which is called sediment transport. The factors affecting sediment transport on this occasion are the properties of sedimentary materials such as grain size, cohesiveness, and grain distribution [8], [9], [12], [13], as well as hydrodynamics such as wave and current [7], [10], [14]. The combination of these properties resulted in water turbidity that affects the distribution of suspended sediment load spatially even though the dynamics of current and wave are relatively small [3], [6].

Krueng Cut River has a high level of turbidity because of the number of sedimentary particles hovering the surface area of this canal. At some points, the turbidity of this river is higher than usual. However, to assess it requires a specific study, both qualitatively and quantitatively. Therefore, the

IOP Conf. Series: Earth and Environmental Science **106** (2017) 012066 doi:10.1088/1755-1315/106/1/012066

amount of suspended sediment and their distribution pattern can be determined by analyzing the profile and the spatial distribution of suspended sediments.

2. Research Method

The research was conducted in Krueng Cut estuarial area located at 5°35'16.42"N and 95°21'01.79"W. The data were collected from February to April 2017 at seven stations by using the purposive random sampling (Figure 1). For the suspended sediment analysis, the field samples were examined in Marine Chemistry Laboratory of Marine and Fisheries Faculty of Syiah Kuala University.

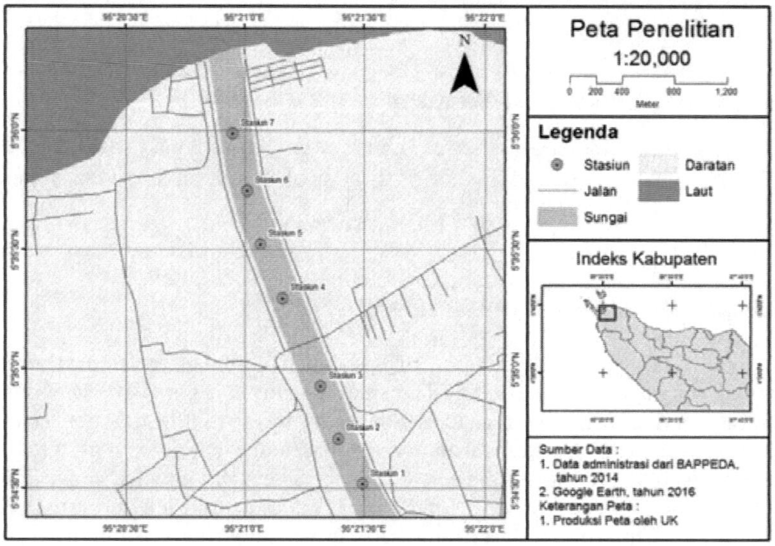

Figure 1. Map of sampling site of Krueng Cut River (the estuarial area).

Each of the seven stations was divided into three points of sampling located in the west, central, and east part of it. Samples were taken during low tide and high tide conditions, subsequently brought to the laboratory for a screening at their perspective volumes. The sediment analysis was performed in the laboratory using a filter paper sized 125 mm Ø. The profile analysis was applied by employing the sediment weight based on the suspended sediment examination result, while the spatial analysis was conducted by obtaining the spatial space interpolation.The suspended sediment concentration was calculated using the following formula:

$$\text{Concentration (g/ml)} = \frac{\text{Sediment weight (g)}}{\text{Sample volume (ml)}} \tag{1}$$

3. Results and Discussion

The results of suspended sediment concentration analysis for high tide and low tide conditions are listed in Table 1and Table 2, respectively, showing that some samples could not be obtained due to the prevailing circumstances—the water was either too shallow or dangerous for the sampling. The sediment profile result can be seen in Figure 2, indicating a significant difference between the suspended sediment concentrations during the high tide and those during low tide condition.

Table 1. Suspended sediment concentrations during high tide condition.

Station	Point	Sample (ml)	Sediment + filter (g)	Sediment weight (g)	Sediment concentration (g/ml)	Average (g/mL)	Sediment concentration (g/L)
	1	200	0.16	0.05	0.00025		0.25
1	2	200	0.16	0.05	0.00025	0.0002	0.25
	3	200	0.16	0.05	0.00025		0.25
	1	200	0.17	0.06	0.0003		0.3
2	2	200	0.16	0.05	0.00025	0.0002	0.25
	3	200	0.16	0.05	0.00025		0.25
	1	N/A	N/A	N/A	N/A	N/A	N/A
3	2	N/A	N/A	N/A	N/A	N/A	N/A
	3	N/A	N/A	N/A	N/A	N/A	N/A
	1	100	0.12	0.01	0.0001		0.1
4	2	100	0.17	0.06	0.0006	0.0003	0.6
	3	100	0.13	0.02	0.0002		0.2
	1	100	0.12	0.01	0.0001		0.1
5	2	100	0.13	0.02	0.0002	0.0001	0.2
	3	100	0.13	0.02	0.0002		0.2
	1	100	0.15	0.04	0.0004		0.4
6	2	100	0.15	0.04	0.0004	0.0003	0.4
	3	100	0.13	0.02	0.0002		0.2
	1	100	0.16	0.05	0.0005		0.5
7	2	100	0.15	0.04	0.0004	0.0005	0.4
	3	100	0.18	0.07	0.0007		0.7

Table 2.Suspended sediment concentrations during low tide condition.

Station	Point	Sample (ml)	Sediment + filter (g)	Sediment weight (g)	Sediment concentration (g/ml)	Average (g/mL)	Sediment Concentration (g/L)
	1	100	0.15	0.04	0.0004		0.4
1	2	100	0.13	0.02	0.0002	0.0002	0.2
	3	100	0.13	0.02	0.0002		0.2
	1	100	0.15	0.04	0.0004		0.4
2	2	100	0.14	0.03	0.0003	0.0003	0.3
	3	100	0.15	0.04	0.0004		0.4
	1	100	0.13	0.02	0.0002		0.2
3	2	100	0.13	0.02	0.0002	0.0002	0.2
	3	100	0.13	0.02	0.0002		0.2
	1	N/A	N/A	N/A	N/A	N/A	N/A
4	2	N/A	N/A	N/A	N/A	N/A	N/A
	3	N/A	N/A	N/A	N/A	N/A	N/A
	1	N/A	N/A	N/A	N/A	N/A	N/A
5	2	N/A	N/A	N/A	N/A	N/A	N/A
	3	N/A	N/A	N/A	N/A	N/A	N/A
	1	100	0.13	0.02	0.0002		0.2
6	2	100	0.13	0.02	0.0002	0.0002	0.2
	3	100	0.13	0.02	0.0002		0.2
	1	100	0.15	0.04	0.0004		0.4
7	2	100	0.14	0.03	0.0003	0.0003	0.3
	3	100	0,15	0,04	0,0004		0,4

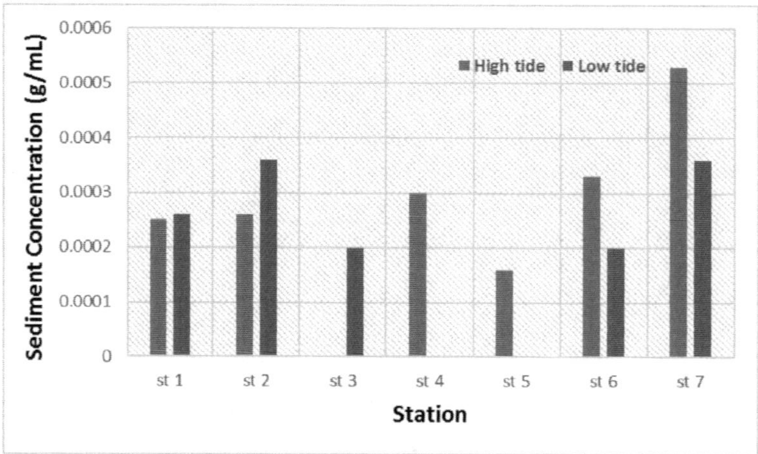

Figure 2. Sediment concentrations during high tide (*pasang*) and low tide (*surut*) conditions.

The results of suspended sediment distribution during high tide and low tide conditions are illustrated in Figure 3A and 3B. The distribution depicts the dominant influence of water dynamics in the estuarial area. During the high tide condition, suspended sediment distribution was strongly dominated by the dynamics of the sea. On the contrary, during low tide condition, the flow of the river outnumbered the sea dynamics so that the river influence on the suspended sediment was dominating.

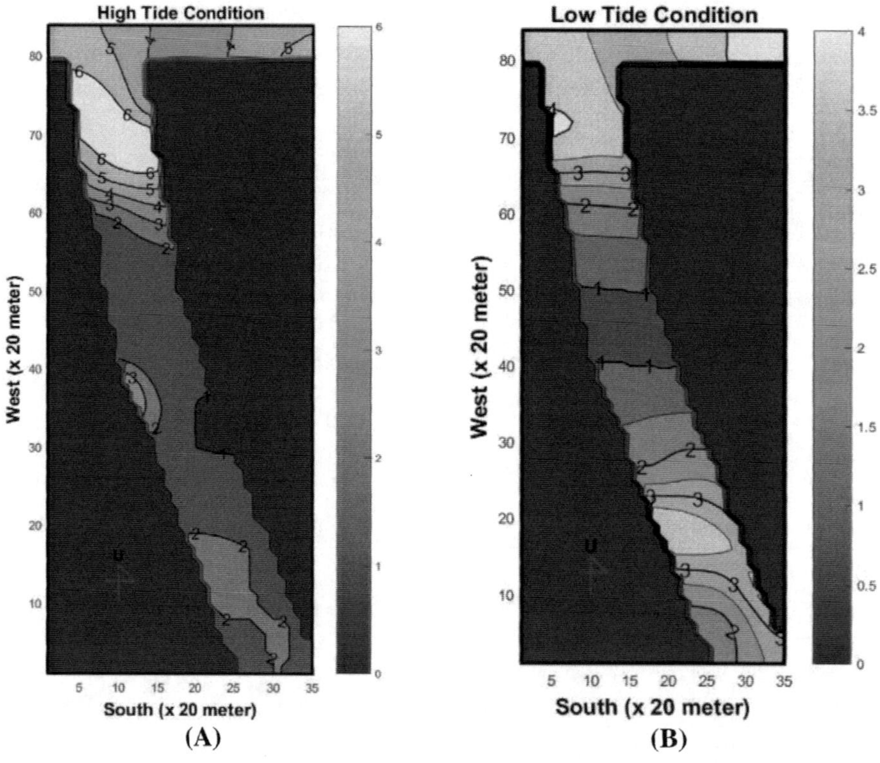

Figure 3. Spatial distributions of suspended sediment concentrations in mg/L during high tide condition (A) and low tide condition (B).

The results of data analysis showed that the spatial distribution of suspended sediment concentrations was ranging from 1 to 6 mg/L during high tide season and 0.5 to 4 mg/L for the low tide condition. These differences occur due to the increasing and decreasing water volume in the estuarial area [15]. The concentration of suspended sediment in the estuary area increases during high tide condition because of the dominant influence of hydro-oceanographic dynamics at the river mouth compared to the river dynamics (Figure 3A) [16]. The profile analysis also showed that the concentration of suspended sediment during low tide condition increases in the upstream area due to the predominate river flow during this season. The results indicated that the circulation region (in the middle of the study area) tended to settle the sediments.

In a shallow area where suspended sediments flow from the upstream toward the sea, the influence of sea currents and waves is more dominant than that of the river flow and river discharge [8], [10]. The mixing between current dynamics and sediments in shallow water increases the suspended sediment concentration so that it is higher during high tide condition compared to the low tide season. The sediment distribution stated that high water discharge led to the higher suspended sediment concentration near the coastline area. It is because the coast region is shallow so that the interaction between the currents and the sediments can transport the sediment upwards, augmenting the water turbidity [6].

The results also showed that the significant suspended sediment concentrations were encountered in the river mouth (Figure 3A). It is on account of the higher interaction between ocean current sand bed sediments compared to the upstream area where the river flow starts to weaken, and the current interaction with bed sediments decreases as the depth of water increases [7]. On the other hand, during low tide condition, the dominant suspended sediment concentrations occurred in both upstream and estuary areas due to the interaction of currents and bed sediments, especially in the upstream regionwhere bed sediments were facilitated to transport upward.

4. Conclusion

The spatial distribution analysis of suspended sediments in Krueng Cut estuarial area has two major differences between the high and low tide conditions. During the high tide season, the concentration of suspended sediments was heavily distributed downstream near the river mouth due to the mixing between riverbed sediments and hydro-oceanographic dynamics which then resurfaced the sediments. On the other hand, during low tide condition, the concentration of suspended sediments was circulated mostly in the upstream of the study area.

References

[1] Bayram A, Kankal M, Tayfur G, and Onsoy H. 2014. Prediction of suspended sediment concentration from water quality variables. *Journal of Natural Computing and Aplications*. 24(5):1079-1087.

[2] Barnard P L, D H Schoellhamer, B E Jaffe, and L J McKee. 2013. Sediment transport in the San Francisco Bay Coastal System: An overview. *Mar. Geol.* 345:3-17.

[3] Dai Z J, Chu A, Li W H, Li J F, and Wu H L. 2013. Has suspended sediment concentration near the mouth bar of the Yangtze (Changjiang) estuary been declining in recent years? *Journal of Coastal Research*. 29(4):809-818.

[4] Warrick J A, Madej M A, Goni M A, and Wheatcroft R A. 2013. Trends in the suspended-sediment yields of coastal rivers of northern California, 1955-2010. *Journal of hydrology*. 489:108-123.

[5] Irham M. 2016. The spatial distribution of bed sediment on fluvial system: A mini review of the Aceh meandering river. *Aceh International Journal of Science and Technology*. 5(2):82-87.

[6] Bian C, Jiang W, Quan Q, Wang T, Greatbatch R J, and Li W. 2013. Distribution of suspended sediment concentration in the Yellow Sea and the East China Sea based on field surveys during the four seasen of 2011. *Journal of Marine System*. 121-122:24-35.

[7] Cáceres R A, Zyserman J A, Perillo G ME. 2014. Analysis of Sedimentation Problems at the Entrance to Mar del Plata Harbor. *Journal of Coastal Research*. 32(2):301-314.

[8] Giglou A N, Sahaberi A J, Shakibaeinia A, and Borghei S M. 2016. An experimental study of sediment transport in channel confluences. *International journal of Sediment Research*. 31(1):87-96.

[9] Higgins A, Restropo J C, Ortiz J C, Pierini J, and Otero L. 2016. Suspended sediment transport in the Magdalena River (Colombia, South America): Hydrologic regime, rating parameters and effective discharge variability. *International Journal of Sediment Research*. 31(1):25-35.
[10] Liu J H, Yang S L, Zhu Q, and Zhang J. 2014. Controls on suspended sediment concentration profiles in the shallow and turbid Yangtze Estuary. *Continental Shelf Research*. 90:96-108.
[11] Irham M, Giardino J r, and Guneralp I. 2013. Two dimensional morphological dynamic modeling of a complex fluvial system: The spatial distribution of the bed sediment of the Brazos River, TX. *GSA Annual Meeting*, Minnepolis.
[12] Irham M, and Setiawan I. 2017. The study of flow resulting from wave on Lhoknga Beach, Aceh Besar. *Omni-Akuatika*. 13(1):5-12.
[13] Ma G, Chou Y, and Shi F. 2014. A wave-resolving model for near shore suspended sediment transport. *Journal of Ocean Modelling*. 77:33-49.
[14] Delpey M T, F Ardhuin P Otheguy, and A Jouon. 2013. Effects of waves on coastal water dispersion in a small estuarine bay. *J. Geophys. Res*. 119:70-86.
[15] Davidson S K, Hartley A J, Weissmann G S, Nichols G J, and Scuderi L A. 2013. Geomorphic elements onmodern distributive fluvial systems. *Geomorphology*. 180-181: 82-95.
[16] Irham M, F Abrar and V Kurnianda. 2017. Analisis BOD dan COD di perairan estuaria Krueng Cut, Banda Aceh. *Depik*. 6(3): 199-204.

The numerical model of the sediment distribution pattern at Lampulo National fisheries port

M Irham[1]*, I Setiawan[1,2]

[1]Faculty of Marine and Fisheries, Universitas Syiah Kuala, Banda Aceh, Indonesia
[2]Center for Marine and Fisheries Study of Universitas Syiah Kuala, Banda Aceh, Indonesia

*Corresponding Author : irham@unsyiah.ac.id

Abstract. The spatial distribution of sediment pattern was studied at Lampulo Fisheries Port, Krueng Aceh estuarial area, Banda Aceh. The research was conducted using the numerical model of wave-induced currents at shallow water area. The study aims to understand how waves and currents react to the pattern of sediment distribution around the beach structure in that region. The study demonstrated that the port pool area had no sedimentation and erosion occurred because the port was protected by the jetty as the breakwater to defend the incoming waves toward the pool. The protected pool created a weak current circulation to distribute the sediments. On the other hand, the sediments were heavily distributed along the beach due to the existence of longshore currents near the shoreline (outside the port pool area). Meanwhile, at the estuarial area, the incoming fresh water flow responded to the coastal shallow water currents, generating Eddy-like flow at the mouth of the river.
Keywords: Lampulo Fisheries Port, longshore current, sediment distribution, wave-induced current

1. Introduction

Lampulo National Fishery Port located at 5°35'11.11"N and 95°19'8.44"E (Figure 1) is an area where all fishery activities are conducted, providing the multiservice fishery business as well as absorbing a large workforce of both industrial and economic sectors. The function of this fishery port is to support the management as well as to utilize fishery resources and environment, ranging from pre-production, production, processing, to marketing, implemented in a fishery business system. However, such utilization of the port and the establishment of a pool protected area have created a distribution pattern of sedimentation and erosion outside the pool protected region, bringing impacts on changes in sediment distribution patterns [1], [2].

During the recent years, activities conducted in this area have been increasing in quantity as well as in variety, leading to some changes that may have unwanted side effects such as environmental damages [3]. Plans that can limit and control these adverse effects have been becoming more important than ever, including thoughts about interactions of waves propagating into coastal areas which lead to the sediment deposition and erosion that interact with the coastal structures [4], [5].

The knowledge of wave behavior near coastal areas is then becoming quite significant. Run-up waves on coastal structures, for example, determine the design of breakwater. The wave speed and energy influence the strength of the coastal structures or floating bodies [6]. A study of such wave behavior can be directly applied in engineering practices [7], [8].

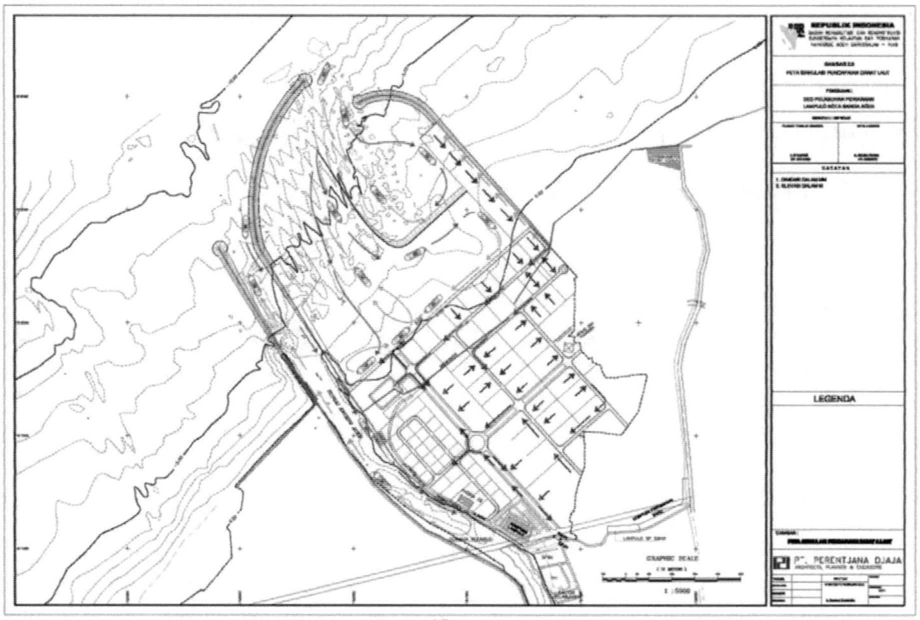

Figure 1. Map of Lampulo national fisheries port of Banda Aceh.

This paper presents mathematical modeling and analysis for wave-induced currents on coastal structures of Lampulo fisheries port that lead to sediment transport distribution. Based on this model, a numerical code has been designed using 2D fully non-linear equations of the wave. This article also presents a procedure for calculating the sediment transport distribution in this area.

2. Research Method

2.1 Hydro-dynamic Model and Sediment Transport

The wave-induced current model uses 2D depth-averaged momentum conservation and mass conservation which can be written as:

$$\frac{\partial \bar{u}}{\partial t}+\bar{u}\frac{\partial \bar{u}}{\partial x}+\bar{v}\frac{\partial \bar{u}}{\partial y}=-g\frac{\partial \zeta}{\partial x}-\frac{C_f\bar{u}\sqrt{\bar{u}^2+\bar{v}^2}}{(h+\zeta)}+R_x+M_x \tag{1}$$

$$\frac{\partial \bar{v}}{\partial t}+\bar{u}\frac{\partial \bar{v}}{\partial x}+\bar{v}\frac{\partial \bar{v}}{\partial y}=-g\frac{\partial \zeta}{\partial y}-\frac{C_f\bar{v}\sqrt{\bar{u}^2+\bar{v}^2}}{(h+\zeta)}+R_y+M_y \tag{2}$$

$$\frac{\partial \zeta}{\partial t}+\frac{\partial(\bar{u}(h+\zeta))}{\partial x}+\frac{\partial(\bar{v}(h+\zeta))}{\partial y}=0 \tag{3}$$

where t is time, (x,y) is the Cartesian coordinate in the horizontal plane, (u,v) is the component of velocity, C_f is basic friction coefficient, (R_x, R_y) is radiation stress in x- and y-direction, g is gravitational acceleration, h is water depth, and ζ is the water elevation.

The momentum changes caused by the turbulent currents that tend to spread due to the influence of the wave force exceeding the area of the breaking wave, then the lateral mixing can be written as follows [9]:

$$M_x=\frac{\partial}{\partial x}\left(\varepsilon\frac{\partial \bar{u}}{\partial x}\right)+\frac{\partial}{\partial y}\left(\varepsilon\frac{\partial \bar{u}}{\partial y}\right) \tag{4}$$

$$M_y = \frac{\partial}{\partial x}\left(\varepsilon \frac{\partial \overline{v}}{\partial x}\right) + \frac{\partial}{\partial y}\left(\varepsilon \frac{\partial \overline{v}}{\partial y}\right) \tag{5}$$

with:

ε $= N l \sqrt{g(h+\zeta)}$ [8]

N = the constant value less than 0.016

l = the distance toward offshore $=(h+\zeta)/\tan\beta$

$\tan\beta$ = the bathymetric slope

As for the irregular slope, the lateral mixing can be written as:

$$M_x = A_H\left(\frac{\partial^2 \overline{u}}{\partial x^2} + \frac{\partial^2 \overline{u}}{\partial y^2}\right) \tag{6}$$

$$M_y = A_H\left(\frac{\partial^2 \overline{v}}{\partial x^2} + \frac{\partial^2 \overline{v}}{\partial y^2}\right) \tag{7}$$

where A_H is the horizontal viscosity coefficient.

The component of radiation stress is defined as the momentum flux due to the existence of wave motion and has the same dimension as the momentum flux formed by two factors, namely the velocity of water particles caused by waves and the pressure. If the waves come near the coast by forming an angle to the shoreline, then the radiation force is as follows [10]:

$$S_{xx} = \frac{\overline{E}}{2}(2n-1) + \overline{E}n\cos^2\theta \tag{8}$$

$$S_{yy} = \frac{\overline{E}}{2}(2n-1) + \overline{E}n\sin^2\theta \tag{9}$$

$$S_{xy} = \frac{\overline{E}}{2}n\sin 2\theta \tag{10}$$

with $\overline{E}$ is $\frac{1}{8}\rho g H^2$. Therefore, the force causing the occurrence of longshore currents is proportional to the radiation of stress gradient [8] which can be written as:

$$R_x = \frac{1}{\rho(d+\overline{\eta})}\left(\frac{\partial S_{xx}}{\partial x} + \frac{\partial S_{xy}}{\partial y}\right) \tag{11}$$

$$R_y = \frac{1}{\rho(d+\overline{\eta})}\left(\frac{\partial S_{xy}}{\partial x} + \frac{\partial S_{yy}}{\partial y}\right) \tag{12}$$

The total sediment transport, in this case, is defined as the sum of sediment transport, both bed and suspended sediments with a simple approach as follows:

$$q_t = 0{,}05U \frac{\tau_{cw}{}^2 C}{\rho^2 g^{5/2}\left[\frac{(\rho_s-\rho_w)}{\rho_w}\right]^2 D_{50}} \tag{13}$$

where:

ρ_w = water density

ρ_s = sediment density

D_{50} = 50% of the sediment size

IOP Conf. Series: Earth and Environmental Science **106** (2017) 012067 doi:10.1088/1755-1315/106/1/012067

$\tau_c = \rho g U^2 / C^2 + \frac{1}{2}\left(\xi \frac{\hat{u}_b}{U}\right)^2$ = the friction between waves and currents

$\hat{u}_b = (\pi H / T)(\sinh kh)^{-1}$ = bottom wave velocity amplitude

$\xi = \frac{c}{\sqrt{(2g)}}\sqrt{(f_w)}$ = bottom roughness

(f_w) = wave friction coefficient

Hence, the mathematical relationship for the sedimentary equilibrium of the sediment movement is written as:

$$\frac{\partial \zeta_b}{\partial t} + \frac{\partial q_{tx}}{\partial x} + \frac{\partial q_{ty}}{\partial y} = 0 \quad (14)$$

with q_t represents the total sediment transport [12].

3. Results and Discussion

The wave-induced current model was applied based on the bathymetry shown in **Figure 2**, covering 1,600 m along the coastal line and 1,350 m toward the offshore. The simulated model employed bottom stress $C_f = 0.01$, horizontal viscosity coefficient $A_H = 10$ m^2/s, $\Delta x = 10$ m, $\Delta y = 10$ m, and the interval time dt = 0.45 second. The applications of sediment transport model and bottom morphological changes were similarly designed to the current model with a sediment diameter D_{50} = 0.1 mm. The input of the sediment transport rate was based on the results of the wave-generated current model, completed simultaneously after the simulating the current model. Likewise, the bottom morphological changes were calculated from the sediment transport rate results.

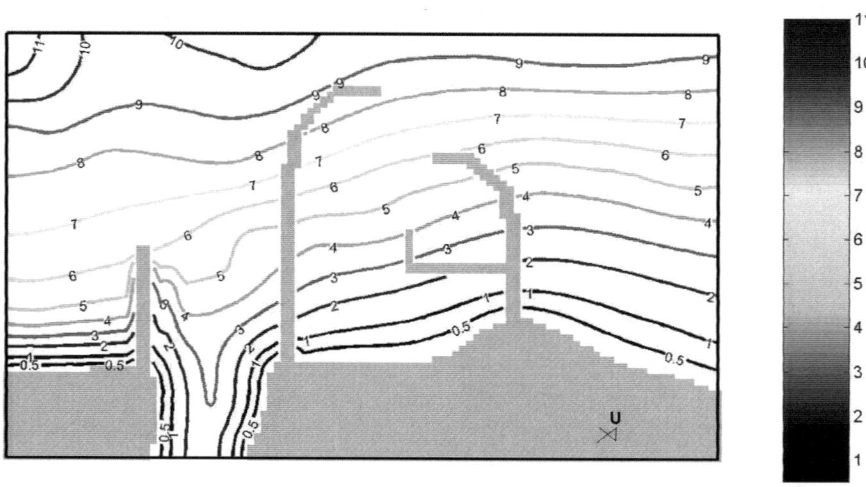

Figure 2. The bathymetry of Lampulo national fisheries port, Banda Aceh.

The results of the model showed that the pattern of current distribution was affected by the diffraction and reflection brought about by the coastal structures. Figure 3 depicts that the western area of coastal structures or around the mouth of the river (estuarial area) generated current vortexes with the largest magnitude of 6.9 cm/s. This Eddy-like circulation pattern is due to the elevation difference between the inner and outer sides of the coastal structures with the smaller flow velocity compared with the protected area [6], [13]. This whirl is also influenced by the river flow from the upstream of the estuarial area toward offshore, triggering the water particles to move faster and provide a boost to the slow-moving water particles [7]. On the other hand, the current pattern showed that currents moved eastward at the left and right of the coastal structures, while currents in the pool protected area merely had a small movement, implying that almost no flow occurred in this zone.

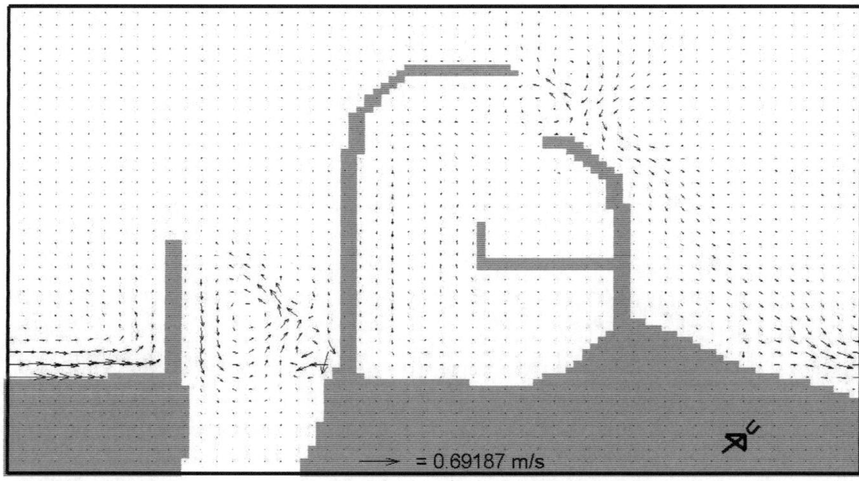

Figure 3. The average current distribution in Lampulo National Fisheries Port.

The application results of the sediment transport model in this area can be seen in Figure 4 below. This sediment transport model, as well as the current model (Figure 3), was carried out for a month. The results showed that on the left side of coastal structures near the river mouth, the erosion was more dominant than the sedimentation. The largest sedimentation occurred in the western part where the jetty that protects the estuary has a large contribution to settling the sediments and deposits in this area [14]. Figure 4 illustrates that the erosion predominantly took place on the eastern coast, while neither sedimentation nor erosion occurred in the pool protected port since the weak current pattern formed in this area could not transport the bottom sediments upward to create a sedimentation and erosion.

Figure 4. The profile and distribution pattern of erosion and sedimentation in Lampulo National Fisheries Port

4. Conclusion

The model employed in this study has been able to simulate the wave-induced currents representatively, depicting the longshore current pattern. Qualitatively, the erosion and sedimentation model in this coastal area has shown a correspondence with the field measurements that both

IOP Conf. Series: Earth and Environmental Science **106** (2017) 012067 doi:10.1088/1755-1315/106/1/012067

phenomena predominantly occurred outside the protected port area by the coastal structures, while neither sedimentation nor erosion took place in the pool area because of the weak current pattern. On the other hand, the sediment distribution was displayed along the beach due to the existence of longshore currents near the shoreline. Meanwhile, in the estuarial area, the outcoming river flow toward the offshore responded to the coastal shallow water currents that generated Eddy-like flow at the mouth of the river.

References

[1] Setiawan I, Supriyo T, and Mihardja D K. 2006. Modeling sediment transport due to current generated by waves in the coastal waters of Bai island, Bengkulu. *Jurnal Geofisk.* 1(2):79-90.

[2] Setiawan I, Yuni, S M, Mariana, and Ilhamsyah Y. 2017. Modeling the distribution of waste concentration P. T. Pupuk Iskandar Muda. *Proceeding Seminar Pasca Sarjana Unsyiah.* A70:A74.

[3] Cáceres R A, Zyserman J A. Perillo G M E 2014. Analysis of sedimentation problems at the entrance to Mar del Plata Harbor. *Journal of Coastal Research.* 32(2):301-314.

[4] Barnard P L, D H Schoellhamer, B E Jaffe, and L J McKee. 2013. Sediment transport in the San Francisco Bay coastal system: An overview. *Mar. Geol.* 345:3-17.

[5] Setiawan I, Ilhamsyah Y, Miswar E, and Haddrevi A. 2008. A prediction study of wave propagation before and after tsunami in UleeLheue coastal waters, Aceh, Indonesia. *Proceeding for International Symposium Land Use after the Tsunami*, supporting education, research and development in Aceh Region, Aceh – Indonesia, 249:254.

[6] Gomes E R, Mulligan R P, Brodie K L, and Mc Ninch J E. 2016. Bathymetric control on the spatial distribution of wave breaking in the surf zone of a natural beach. *Coastal Engineering*. 116: 180-194.

[7] Irham M, and Setiawan I. 2017. The study of flow resulting from wave on Lhoknga Beach, Aceh Besar. *Omni-Akuatika*. 13(1):5-12.

[8] Giglou A N, Sahaberi A J, Shakibaeinia A, and Borghei S M. 2016. An experimental study of sediment transport in channel confluences. *International journal of Sediment Research.* 31(1):87-96.

[9] Horikawa K, Nearshore Dynamics and Coastal Processes, University of Tokyo Press, Tokyo, 1988.

[10] vanRijn L C, Principles of fluid flow and surface waves in rivers, estuaries, seas, and oceans, The University of Utrecht, Department of Physical Geography, 1990.

[11] Longuet-Higgins M S. 1970. On the longshore currents generated by obliquely incident sea wave. *Journal of Geophysics*, 75(2).

[12] Koutitas C G, Mathematical models in coastal Engineering, Pentech Press Limited, London, 1988.

[13] Medina G G, Haller H T O, and Rugiero P. 2014. Wave resource assessment in Oregon and southwest Washington, USA. *Reneable Energy*, 64:203-214.

[14] Soomere T, and Vika M. 2014. Simulated wave-driven sediment transport along the eastern coast of the Baltic Sea. *Journal of Marine Systems*, 129:96-105.

The potential of household solid waste reduction in Sukomanunggal District, Surabaya

I D A A Warmadewanthi*, S Kurniawati

Environmental Engineering Department, Faculty of Civil and Environmental Engineering, Institut Teknologi Sepuluh Nopember, Surabaya, Indonesia

*Corresponding Author : warma@its.ac.id

Abstract. The rapid population growth affects the amount of waste generated. Sukomanunggal Subdistrict is the densest area in West Surabaya which has a population of 100,602 inhabitants with a total area of 11.2 km^2. The population growth significantly affects the problem of limited land for landfill facilities (final processing sites). According to the prevailing regulations, solid waste management solutions include the solid waste reduction and management. This study aims to determine the potential reduction of household solid waste at the sources. Household solid waste samplings were performed for eight consecutive days. The samples were then analyzed to obtain the generation rate, density, and composition so that the household solid waste reduction potential for the next 20 years could be devised. Results of the analysis showed that the value of waste is 0.27 kg/person/day, while the total household solid waste generation amounted to 27,162.58 kg/day or 187.70 m^3/day. Concerning the technical aspects, the current solid waste reduction in Sukomanunggal Subdistrict has reached 2.1% through the application of waste bank, composting, and scavenging activities at the dumping sites by the garbage collectors. In the year of 2036, the potential reduction of household solid waste in Sukomanunggal Subdistrict has been estimated to reach 28.0%.
Keywords: community participation, household solid waste management, reduction

1. Introduction

Surabaya City has a strategic position and becomes the development center of the eastern part of Indonesia. The rapid population growth considerably affects the amount of solid waste generated each day. Sukomanunggal district is the densest region in West Surabaya where the landfill (the final processing site) for solid waste resides. Subdistrict Sukomanunggal including in the area of West Surabaya, with a population of at most that was recorded in year 2013 as many as 98.401 inhabitants [1]. The development plan Surabaya contained regulation [2] states that regional Development Unit (UP) VIII Dukuh Pakis used for settlements, trade and services, industry, national defense and security, with the area of the development center located at Segi Delapan , Segi Delapan of the areas included in the District of Sukomanunggal.

The rapid population growth a major effect on the amount of waste generated each day. The increasing amount of garbage produced each year, it will be a big problem on landfill capacity and the difficulty to expand the landfill area. Waste management has a complex and lengthy process, not only regarded as a technical problem, but also as a management paradigm [3].

Based on [4], solid waste management solutions include the solid waste reduction and management. The solid waste reduction can be accomplished through the restriction of waste, waste recycling, and waste reuse. This study aims to determine the existing condition of household solid waste management and reduction potential in Sukomanunggal Subdistrict. This research can provide benefits in the form of information related to the potential reduction of household waste and the improvement efforts that can be developed.

Waste Bank is one of the community-based waste management systems which enable citizens to participate actively in managing the environment. According to regulation [5] explained that the bank is a sorting trash and waste collection that can be recycled and or reused with economic value. Community-based waste management in the city of Surabaya conducted by processing organic waste into compost, inorganic waste sold to collectors or used into amterial recycling, as well as the development and improvement of home composting [6] . Sukomanunggal sub-district currently has 10 units of waste bankk with an active status located in 3 villages. Moreover, in the District of composting reactor Sukomanunggal have had as many as 11 reactors. Based on the results of analysis of waste in the number of waste bank customers 2016, there were 492 customers, waste bank customers in the district Sukomanunggal has increased from previous years. As with the composting activities, community participation in the activities of composting is still low, it is evidenced by the few number Sukomanunggal composter in the district that is not functioning properly. One of them, in the Village Sukomanunggal that there are several units composter help from an agency that is not working anymore. The solution to this problem it is necessary to further study the causes of decrease in community participation.

2. Research Method

This study began with the data collection which aims to gather all the information related to the research. Data were collected in the form of primary and secondary statistics. Primary data were obtained by direct observations in the field through distributing questionnaires, conducting interviews, measuring the household solid waste generation, and assessing the reduction of household solid waste. The measurement of waste generation has been carried out for eight days. Sampling was conducted on 110 families with the assumption that one family produces 1 kg of solid waste per day. Wet volume (m^3/ day) and wet weight (kg/day) were used as the units of the solid waste.

The second step was the calculation of household solid waste composition to determine the characteristics of the solid waste, performed after the measurement of solid waste generation by taking as many as 100 kg samples of waste. The sampling method referred to [7] about the Method of Measurement Sample Collection and Urban Waste Generation and Composition. The solid waste can be characterized by sorting the type of materials including organic waste, paper, plastic (e.g., HDPE plastic bottles, aluminum, LDPE, PET, styrofoam, and PP), glass, metal, fabric, wood, and others.

The reduction can be assessed in advance after collecting data on the number and type of reduction facilities. A survey was then conducted to determine the amount of solid waste that can be reduced. The subjects of domestic solid waste reduction in Sukomanunggal Subdistrict are waste bank and scavengers in TPS.

The data of community participation in reducing solid waste were obtained from questionnaires related to the will of society. The number of respondents was 110 people, which had been adapted to the calculation of the required number of solid waste samples. The sampling was conducted by the random sampling system. Public participation in reducing solid waste can be judged from the interest in sorting and reusing household solid waste. Moreover, the community contribution to solid waste management activities was assessed as well. As for the secondary data, they were gathered from relevant agencies. Finally, when all the information had been collected, the analysis and discussion were performed to reach a conclusion. The purpose of this study is to determine the existing condition of the reduction of household solid waste in Sukomanunggal Subdistrict, obtained from the data of technical aspects. Subsequently, it also aims to determine the potential reduction of household solid waste based on the community participation in terms of behavior and attitude.

IOP Conf. Series: Earth and Environmental Science **106** (2017) 012068 doi:10.1088/1755-1315/106/1/012068

3. Results and Discussion

The questionnaires were analyzed to assess the score weight based on the Likert scale. It was examined based on the technical aspects as well using the method of dynamic models with Stella 9.1.3 software. The discussion below presents each aspect analyzed by the dynamic system optimization of Stella 9.1.3 software.

3.1. Analysis of household waste generation

The analysis of solid waste generation was conducted by collecting the household waste for eight consecutive days (Table 1). The sampling was carried out from the same sources of residential solid waste. Furthermore, the analysis of the amount of solid waste was calculated referring to [2] about the Method of Measurement Sample Collection and Urban Solid Waste Generation and Composition. The results showed that the average of solid waste generation per person per day in Sukomanunggal Subdistrict is 0.27 kg/person/day. The calculation of household solid waste generation can be seen in Table 1.

Table 1. Household waste generation.

Day	Total Weight of Solid Waste (kg)	Total Sample	Solid Waste Generation (kg/person/day)
1	108.49	387	0.28
2	98.79	387	0.26
3	110.70	387	0.29
4	108.20	387	0.28
5	98.90	387	0.26
6	99.76	387	0.26
7	109.40	387	0.28
8	99.80	387	0.26
Average			0.27

The solid waste generation per person per day was used as the basis for determining the household solid waste generated in Sukomanunggal Subdistrict. The total population in this area in the year of 2017 is 100,602 inhabitants. This number was multiplied by the number of solid waste generation of 0.27 kg/person/day to calculate the amount of household waste generated in Sukomanunggal Subdistrict of 27,162.58 kg/day.

3.2. Analysis of the composition and density of household solid waste

The reference used in the calculation of the solid waste composition was [2] about the Method of Measurement Sample Collection and Urban Solid Waste Generation and Composition. This analysis had been performed for eight consecutive days as well. Results of the field measurements showed that the composition of household solid waste in Sukomanunggal Subdistrict was predominantly food scraps and garden waste with the total percentage of 71.68% (67.99% food scraps and 3.69% garden waste) (Figure 1). The garden waste was yielded partly from individual street cleaning activities by residents and janitors.

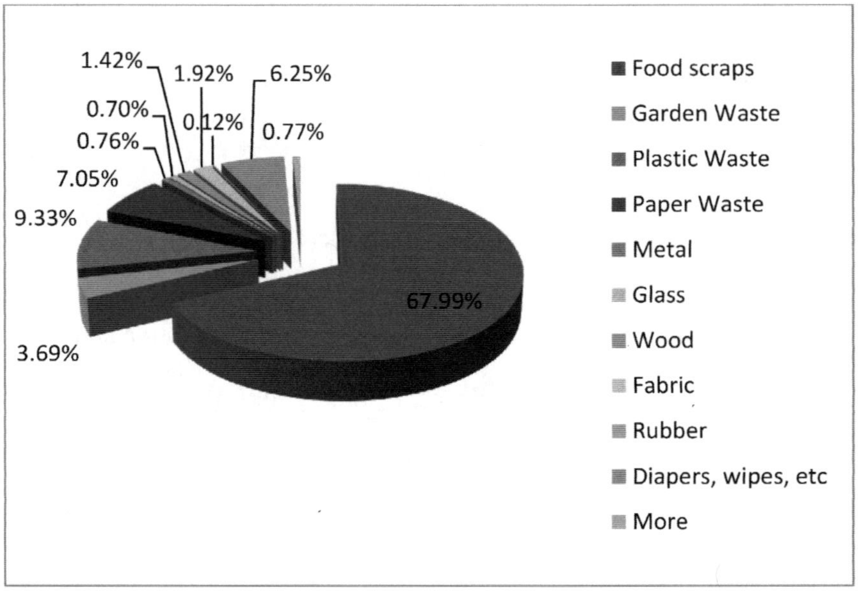

Figure 1. Percentages of household solid waste composition in Sukomanunggal Subdistrict.

Density measurement of household waste in the District Sukomanunggal conducted over three days. The purpose of measuring the density of household waste is to determine the volume of waste generated at the source of waste and the volume of waste that goes to the polling stations. Based on field sampling in mind that not all rubbish weighing 100 kg, can be entered into the box density. This is due to the type of household waste is currently very diverse and has a large volume, such as bottles, plastic cups, and the rest of other packaging. The density of household waste in the District Sukomanunggal of the analytical results ranged between 134.61 kg/m^3 to 147.98 kg/m^3 with an average density of household waste amounted to 142.03 kg/m^3.

3.3. Reduction of household waste

The composition data can be used to determine the reduction value of household solid waste in Sukomanunggal Subdistrict. The analysis of household solid waste reduction was carried out by observing and calculating the amount of waste sorted by the community before being transported to TPS and TPA.

Potential reduction of waste at source can be determined from the data waste bank. Subdistrict Sukomanunggal own trash banks since 2012. The majority of the waste bank active status. The number of banks in the district trash Sukomanunggal has increased every year. Currently there are 10 waste bank with an active status, but not every village has a waste bank.

Data reduction of waste from the waste bank obtained from interviews with bank officials related trash. Waste bank in the District Sukomanunggal most have good administration system, data on the number of customers and the amount of waste sorting has been recorded periodically. On average recapitulation conducted every month. The amount of waste is reduced by waste banks can be seen in Table 2.

Table 2. Total Waste Reduction by Waste Bank.

No	Name Waste Bank	Address	The number of customers (kk)	Weight reduction of waste (kg/month)	Weight reduction of waste per household (kg/kk/month)
1	Tanjung Mandiri	Tanjungsari	37	196.40	5.31
2	Karya Mulia	Simomulyo Baru	25	78.20	3.13
3	Tanjung Jaya	Tanjungsari	25	157.90	6.32
4	Wartun Sejahtera	Simomulyo Baru	245	368.70	1.50
5	Karya Usaha Bersama	Simomulyo Baru	35	147.80	4.22
6	Cahaya Asa	Simomulyo	25	216.73	8.67
7	Mandiri	Simomulyo	25	69.60	2.78
8	Prima	Simomulyo	25	91.40	3.66
9	Pelangi	Simomulyo	25	73.20	2.93
10	Amanah	Simomulyo	25	120.50	4.82
Amount			492	1520.43	43.34
Average			49,2	152,04	4.33

Potential reduction of household waste is known from recycling activities/waste bank and composting. This reduction was planned within a period of 20 years. Reduction in this scenario is to describe the magnitude of the reduction of the value of household waste when it is done with recycling and composting activities. Reduction value of each type of waste is influenced by the rate of increase in *PF* (*Participation Factor*) and *RF* (*Recovery Factor*).

Table 3. Potential reduction by recycling and composting.

Year	Waste bank reduction (ton/year)	Composting reduction (ton/year)	Total reduction (ton/year)	Total Generation (ton/year)	Residues in TPA (ton/year)	% reduksi
2016	115.55	0.56	116.11	9730.50	9614.39	1.2%
2017	150.19	62.79	212.98	9914.34	9701.36	2.1%
2018	187.87	130.92	318.79	10098.18	9779.39	3.2%
2019	228.62	204.93	433.55	10282.02	9848.47	4.2%
2020	272.41	284.85	557.26	10465.87	9908.61	5.3%
2021	319.27	370.65	689.92	10649.71	9959.79	6.5%
2022	369.18	462.34	831.52	10833.55	10002.03	7.7%
2023	422.14	559.93	982.07	11017.39	10035.32	8.9%
2024	478.16	663.41	1141.57	11201.23	10059.66	10.2%
2025	537.23	772.79	1310.02	11385.07	10075.05	11.5%
2026	599.36	888.05	1487.41	11568.92	10081.51	12.9%

Year	Waste bank reduction (ton/year)	Composting reduction (ton/year)	Total reduction (ton/year)	Total Generation (ton/year)	Residues in TPA (ton/year)	% reduksi
2027	664.54	1009.21	1673.75	11752.76	10079.01	14.2%
2028	732.78	1136.26	1869.04	11936.60	10067.56	15.7%
2029	804.07	1269.21	2073.28	12120.44	10047.16	17.1%
2030	878.42	1408.04	2286.46	12304.28	10017.82	18.6%
2031	955.83	1552.77	2508.60	12488.12	9979.52	20.1%
2032	1036.29	1703.39	2739.68	12671.97	9932.29	21.6%
2033	1119.80	1859.91	2979.71	12855.81	9876.10	23.2%
2034	1206.37	2022.31	3228.68	13039.65	9810.97	24.8%
2035	1295.99	2190.61	3486.60	13223.49	9736.89	26.4%
2036	1388.67	2364.80	3753.47	13407.33	9653.86	28.0%

The results show that in 2016 the reduction of household waste recycling and composting activities reached 1.2% and increased in 2036 to reach 28%. This increase was influenced by the value of people's behavior in the conduct of the recycling / waste banks and composting. Recycling has been done but the community composting activities are still few who do. Based on the results of the questionnaire known why people who do composting at least one of them is they do not understand about how to compost and equipment that they have not had. However, some places are already doing composting activities also could not run because of lack of assistance and motivation given to the public.

The respondent data of the questionnaires can be analyzed to determine the potential reduction of solid waste in Sukomanunggal Subdistrict. The analysis results stated that the current reduction of household solid waste in Sukomanunggal Subdistrict reached 2.1%. Based on the analysis of public participation, it is known that the increased willingness of Sukomanunggal residents to reduce the household solid waste will potentially lead to the value of waste reduction of 28.0% in 2036.

4. Conclusion

Based on the results and discussion above, it can be concluded that the total household solid waste generation in Sukomanunggal Subdistrict in 2017 amounted to 27,162.58 kg/day. The total reduction of household solid waste in Sukomanunggal Subdistrict today reached 2.1% and the total reduction of household waste in Sukomanunggal Subdistrict in 2036 was estimated to reach 28.0%.

References

[1] Badan Pusat Statistik Kota Surabaya. 2015. *Kecamatan Sukomanunggal dalam Angka*
[2] Surabaya City Spatial Plan, 2014
[3] Wijayanti, D. R., dan Suryani, S. 2015. Waste Bank as Community-based Environmental Governance: A Lesson Learned from Surabaya. *Procedia-Social and Behavior Science*. 184: 171-179.
[4] Undang-Undang No.18 on Waste Management. 2008.
[5] Peraturan Menteri Lingkungan Hidup No.13 Tahun 2012 tentang Reduce, Reuse, and Recycle Implementation Guidelines
[6] Surabaya Environmental Status Report, 2012
[7] SNI 19-3964-1994. Methods of Measuring and Sample Measurement of Waste Generation and Urban Composition. 1994

Cleaner production options for reducing industrial waste: the case of batik industry in Malang, East Java-Indonesia

M Sirait

Industrial Engineering Department, Faculty of Engineering, Universitas Brawijaya, Malang, Indonesia

Corresponding Author: marudut@ub.ac.id

Abstract. The aim of this research is to conduct cleaner production options for improving the environmental performance during the production of batik industry, the case of UKM batik, Malang, East Java. Batik industry is one of small and medium textile industry which has contribution to economic growth in Malang. However, during production the batik, it generates wastewater that has potential to decrease the environmental performance. Wastewater from Celaket batik industry has BOD, COD, TSS, and pH level is far larger than the threshold of water quality standard as a result of use chemical substance during the dyes processing. In order to prevent generating wastewater, this study utilized cleaner production options, such as substitution of input material.Substitution of input material for dyes process was implemented by replacement chemical dyes (e.g.indigosol, nafthol, rapid) with natural dyes (e.g. Indigofero Tintoria). Modifying of technology/equipment was conducted by developing wastewater treatment equipment to reduce waste of batik productionThe implementation of this strategy was carried out by changing input material from chemical dyes with natural dyes. The CP uptake could reduce significantly the environmental impact in term of reduction of COD, BOD, and TSS.
Keywords: batik industry, cleaner production, nature dyes, wastewater

1. Introduction

Batik Industry is one of the small and medium enterprises (SMEs), which have a significant contribution to economic growth in Indonesia. In 2002-2006 SMEs have contributed to gross domestic product (GDP) of Indonesia IDR 104,638 trillion and have employed 5.4 million per year. In the year 2006, the creative industry has exported IDR 82 billion or 9.13 % from total national export. On the other hand, the production of batik generates waste during the life cycle of batik product. Batik and textile industries produce of wastewater generating from the dye process. Also, batik wastewater contains synthetic ingredients, which are difficult to be degraded. Wastewater produced from batik or textile industries is generally a non-biodegradable organic compound, which can cause the environmental pollution, especially to the aquatic environment. Therefore, the main environmental problems for the batik industry or textile are the wastewater with high pollutant level [1]. Textile wastewaters generally contain surfactants, dyes, pigments, resins, chelating agents, dispersing agents, inorganic salts, heavy metals, biocides, etc. and therefore they are heavily loaded with chemical oxygen demand (COD), color and salt [2], [4]. Textile waste can cause significant environmental problems if they do not manage properly [5]. The most dyes commonly used in textile industry are red and golden yellow and remazol black [9]. In fact, during coloring process, only 5 % can be used these compounds while 95% will be released as a liquid waste to the environment. Moreover, the chemical

IOP Conf. Series: Earth and Environmental Science **106** (2017) 012069 doi:10.1088/1755-1315/106/1/012069

dyes in nature are very difficult to be degraded because they are stable enough and for high concentrations these compounds are dangerous to the environment because they can increase the COD and BOD levels in water [1], [9].

The current method conducted in order to reduce industrial waste is end of pipe method. However, this method has some weakness, for example, this method has reactive response to pollutant and waste after they are produced, pollutant are controlled by waste treatment tools and methods. Therefore, Industries require an innovative method to prevent in generating waste. Currently, cleaner production is a preventive method for generating waste and this method has used widely in the world in order to minimize industrial waste. Cleaner production is a continuous application of an integrated preventive environmental strategy applied to processes, products, and services to increase overall efficiency and reduce risks to humans and the environment [1]. Furthermore, CP is a preventive method to generate waste from industrial activities. In the last decades, this method has been of main concern in the control environmental impact of industrial activities [3]. This pollution control method requires high investment and operational costs and causing to decline market competitiveness. Based on the cleaner production approach, Waste/pollution is a result of project, inefficiency, failure, use of resource and ineffectiveness in production processes [4]. Cleaner production method is associated with green production/industry, sustainable production and eco-efficiency [2]. United Nations Environment Programme (UNEP) defines cleaner production as "The continuous application of an integrated environmental strategy to processes, products and services to increase efficiency and reduce risks to humans and the environment" [5]. Therefore, the implementation of cleaner production can increase the economical, Technical, and environmental performance of industrial companies.

Many researches show that CP is effective tool to improve the environmental performance in some SMEs such as milk processing [6], poultry slaughterhouse [3], and fish processing [7]. They implement CP strategies such as modified technology, change equipment, chemical change, automated water management, off-site reuse, etc. However, none of these studies focus on small industry, which are lack of education, expertise, initiation and financial limitations. One of the opportunities used in the application of CP is input substitution without investment or very low investment. Therefore, the object of this paper is to apply cleaner production option in order to improve the environment performance in terms of wastewater reduction. Therefore, this paper proposes cleaner production options to reduce waste from industrial activities, particularly for small and medium industries.

2. Research Method

2.1. The cleaner production concept

CP is relatively new approach to improve the environmental performance of process industries [5]. It describes a preventive approach to environmental management, which can be defined as *"the continuous application of an integrated preventive environmental strategy to processes, products and services to increase overall efficiency, and reduce risks to humans and the environment"* [5]. Moreover, according to UNEP (2001) CP can be implemented to: Production processes: Reducing energy, water and raw materials; eliminating dangerous raw materials and toxic; and minimizing waste during the manufacturing process and the quantity of toxicity of all emissions. Products: minimizing the environmental risk, health, and safety impact of products along the life cycles of product, from raw materials extraction, manufacturing and use, to the disposal of the product. Services: incorporating environmental concerns into designing and delivering services.

Furthermore, the aim of CP is to carry out more efficient for use of natural resources (raw materials, energy, and water) and to minimizing the production of wastes and emissions at the source [5]. The most common way to do such thing is through five prevention practices al. 1991 USEPA 1992 [8] those are: Substitution of input materials: using environmentally friendly (e.g. renewable) materials in products and processes. Substitute input materials by less toxic or by renewable materials or by adjunct materials, which have a longer service lifetime in production. Modification of the product: modifying the product characteristics in order to minimize the environmental impacts of the product during or after its use (disposal) and also to minimize the environmental impacts of its production. Modification of technology: changing the equipment's and technology to make the process more efficient and to reduce waste and emissions. Good housekeeping: adopting standardized /improved operation and maintenance procedures to avoid waste and emission, such as leaks, spills, etc. This may include improved worker training. On-site recycling: recovery and reuse of waste materials, water and heat at the company where they are generated.

2.2. CP procedure
CP options were conducted on small-scale Batik industry located in Malang. The research was based on the CP methodology prescribed by UNEP [5]. The method was implemented in five phases as follows:

Phase 1 : *Planning and organization.* At this phase, the owner and the worker have confirmed their participation and commitment to implement CP.

Phase 2 : *Pre-assessment (qualitative review).* This phase is carried out to know basic information about the enterprise. This phase is to acquire and identify all information for all stages of life cycle batik production. Review was conducted by interviewing the owners and workers. Also, inspection during the life cycle of batik industry has done.

Phase 3 : *Assessment (quantitative review)* includes measuring resource usage and waste generation throughout the process, identifying causes and solutions, and generating of CP Option.

Phase 4 : *Feasibility analysis.* Each CP option is evaluated to determine the feasibility of the economic, technical and environmental aspect. Economic feasibility study was carried out by assessing the amount of the investment and Net Present Value (NPV) for each CP option.

Phase 5 : *Implementation and continuation.* The selected Cleaner production options are further implemented in Celaket batik industry to reduce the concentration and the amount of wastewater generated. The result of CP option implementation was measured using the environmental performance parameter based on ministerial decree of the State Minister for Environment, the Republic of Indonesia No. 3 on 15 January, 1998 about the quality standard of wastewater for industrial area

3. Results and Discussion

3.1. Wastewater of Celaket Batik Industry
Celaket Batik Industry located in Malang, East Java is one of the small and medium enterprises producing traditional batik. The production process of batik consists of five steps: The first step, the wax is applied over the penciled-in outline of the pattern. The original cloth is white or beige. The second step, the cloth is dyed in the first dye bath. The area of the cloth where the wax was applied in step one will remain white. The third step, second application of wax is applied. In this case it is a dark brown color. The darker color helps to differentiate it from the first wax applied. The third step, the cloth is dyed in the second bath. In this case it is a navy blue. Any areas that are not covered by wax will become dark blue. The next step, all the wax that has been applied thus far is removed. Finally, the fabric is submerged in the final dye bath. The finished cloth after all of the wax has been removing. During the production of batik, it generates wastewater from all steps particularly from dye process. Batik Industry generate wastewater which 95% coming from the dyes process. Wastewater is generated by 200 liters per day for producing 4 m batik and it is discharged directly to environment without any treatment. In order to determine the quality of water, assessment of wastewater was conducted in laboratory. The result shows in Table 1 that wastewater has the BOD and COD and TSS level far larger than the threshold of quality standard determined by government of Malang, East Java (see Table 1).

Table 1. Characteristic of Celaket batik industry wastewater.

Parameter	Characteristic of batik waste water	Threshold (mg/L)
BOD	5226 mg/L	75
COD	20900 mg/L	200
TSS	2036 mg/L	100
pH	6-9	11.8

The high concentration of COD, BOD and TSS in the wastewater during the production batik is caused by containing high chemical substance and organic matters. The characteristics of such wastewater could potentially occur in batik industrial as raw materials in the form of organic matter. The organic matter is decomposed by microorganisms that cause levels of COD and BOD in the wastewater become high. In addition to the raw materials, chemical substance during the dye process can increase the level of COD and BOD of water. Traditional batik industry, such as Celaket Batik Industry uses chemical dyes for coloring process. As a result, this process generates wastewater with high concentration of BOD, COD and TSS (see Table 1).

3.2. Cleaner production alternative

In order to reduce environmental impact in term of wastewater in producing batik, this paper utilized cleaner production. Cleaner production options, such as substitution of input material and modifying technology have reviewed (technical, economic and environmental benefit). Substitution of input material for dyes process was implemented by replacement chemical dyes (e.g.indigosol, nafthol, rapid) with natural dyes (e.g. Indigofero Tintoria). Modifying of technology/equipment was conducted by developing wastewater treatment equipment to reduce waste of batik production.

The feasibility of the implementation of this option was reviewed from the economic, technical, and environmental performances. Economic feasibility analysis showed that both alternatives feasible to be implemented. Feasible alternatives that have a high priority would be effective for improvement of environmental performance. This study was applied option 1 due to economic feasibility rate higher compared with option 2. Another reason is the option 1 easies to do due to limited time and lack of knowledge of the owner and employees of batik industry.

Table 2. Cleaner production option for reducing wastewater.

No	Cleaner Production alternatives	Strategies	Investment	Benefit	Economic feasibility (NPV)
1	Substitution of input material	Replecement chemical dyes (Remazol and Vinyl Sulphone) with natural dyes (e.g.IndigoferoTintoria)	IDR 2.610.000	Rp 3.378.000	Positive
2	Modification of technology/equipment	Development wastewater treatment on site	IDR 2.278.791	Rp 2.808.000	Positive

3.3. Implementation of subtitution of material input

The impact of applying substitution of input material strategy can be seen in table 3. The parameters of BOD, COD, TSS and pH decrease significantly after treatment with improvement around 85 %, 89%, and 98 % respectively (see Table 3). A decrease in the level of BOD, COD, and TSS show an increase in the water quality. Also, the use of natural dyes is not harmful to the environmental. In addition, waste of batik industrial generating from natural dye does not decrease the viability of soil bacteria

[10]. However, the improvement of environmental performance of batik industry requires more effort to meet the level of threshold for wastewater before release to environment.

Table 3. Characteristic of batik wastewater before and after CP treatment.

Parameter	Characteristic of batik wastewater before implementing CP	Characteristic of batik wastewater after implementing CP	Threshold (mg/L)
BOD	5226 mg/L	738.3 mg/L	75
COD	20900 mg/L	2930 mg/L	200
TSS	2036 mg/L	38.6 mg/L	100
pH	9	10.8	11.8

4. Conclusion

The implementation cleaner production has a significant contribution to improve the environmental performance during the production of batik. Batik industry generates wastewater with high concentration of BOD, COD that has potential to contaminate environment. In order to improve the environmental performance during the industrial activities, cleaner production option was implemented. Substitution of material input strategy has applied by replacement chemical dyes with natural dyes during the coloring process. The result indicates that the CP option (substitution of input material) significantly improves the environmental performance in terms of BOD, COD, and TSS reduction by 85%, 89%, and 98% respectively.

References

[1] Dasgupta J, Sikder J, Chakraborty S, Curcio S., Drioli E. 2015. Remediation oftextile effluents by membrane based treatment techniques: a state of the artreview. J. *Environ. Manag*. 147: 55-72.

[2] Moore S B, Ausley LW. 2004. Systems thinking and green chemistry in the textile industry: concepts, technologies and benefits. *J. Clean. Prod*. 12: 585-601

[3] Kist L.T, Moutaqi S E, Leandro E M. 2009. Cleaner Production in The Management of Water Use at A Poultry Slaughterhouse of Vale do.

[4] Hillary R. 2004. Environmental Management Systems and The Smaller Enterprise. *Journal Cleaner Production.*12(6): 561-569.

[5] United Nations Environment Programme (UNEP). 1996. Cleaner Production: a Training Resource Package, Industry and Environment. United Nations Environment Programme, Nairobi, Kenya. United Nations Industrial Development Organization (UNIDO) 2010 Global Industrial Energy Efficiency Benchmarking-an Energy Policy Tool Working Paper.

[6] Ozbay A, Demirer GN. 2007. Cleaner Production Opportunity Assessment for A Milk Processing Facility. *Journal of Environmental Management*. 84: 484–493.

[7] Thrane M, Holm E N, Christensen P. 2009. Cleaner Production in Danish Fish Processing – Experiences, Status and Possible Future Strategies. *Journal of Cleaner Production*. 17: 380–390.

[8] United States Environmental Protection Agency (USEPA). 1998. Preliminary Industry Characterization: Fabric Printing, Coating, and Dyeing. http:// www.p2pays.org/ref/01/00166.pdf. [12 May 2017]

[9] Reddy N, Chen L, Zhang Y, Yang Y. 2014. Reducing environmental pollution of the textile industry using keratin as alternative sizing agent to polyvinyl alcohol. *J. Clean. Prod.* 65:561-567.

[10] Yanisworo, Wijaya Ratih, Purwono Budi Santosa, and EniMuryani. Pengaruh limbah industry batik menggunakan pewarna alami dari desa wukir sari terhadap viabilitas bakteri tanah.

Granulometric analysis at Lampulo Fishing Port (LFP) substrate, Banda Aceh, Indonesia

S Purnawan[1*], I Setiawan[1], H A Haridhi[1,2], M Irham[1]

[1]Marine Science Department, Faculty of Marine and Fishery Sciences, Universitas Syiah Kuala, Banda Aceh, Indonesia.
[2]Earth System Science Program, Taiwan International Graduate Program (TIGP), Academia Sinica and National Central University, Taiwan.

*Corresponding Author: syahrulpurnawan@unsyiah.ac.id

Abstract. The study of sediment granulometry was completed at Lampulo fishing port (LFP). The LFP is a main fishing port in Aceh Province, Indonesia, located at 5°34'35" N; 95°19'23" E. The purpose of the research is to study and construct the environment condition of the bottom substrate. The data was taken by incorporating coring method at 10 stations using purposive random sampling. The wet sieve method was used to analyze the grain size for geostatistical analysis. The geostatistical parameters analysis in this study is classified as mean, sorting, skewness and kurtosis. The result informs that the types of sediments are sand, sandy clay and clayey sand for all stations. Station 1, however, is found as the coarsest compares to the other stations. All of the sediment collected at each station displays moderately sorted to poor sorted, while kurtosis values may be categorized as very leptokurtic. The results of the sediment parameters indicate that the environment of harbor pool was in a stable state, related to a sheltered condition.
Keywords: grain-size, granulometry, Lampulo fishing port, sediment

1. Introduction

Grain size is the most fundamental property of sediment clastic particles. The knowledge of particle size distribution contains numerous clues to their origin, hydrodynamic process, and the environment in which they were deposited. Material transport and deposition can be affected by tidal flow, wave activity, currents and anthropogenic activities [1], [2], [3], [4], [5], [6], [7], [8]. Therefore grain size analysis provides important information to the sediment provenance, transport history and depositional conditions, which can be done with a quantitative approach of the clast types present [9], [10]. Furthermore, sediment budgets reflect the origin, weathering processes, erosion or abrasion and the transportation and material deposition process [11], [12], [13], [14], [15].

The Lampulo Fishing Port (LFP) became main loading and unloading site for many fishing boats from around Aceh Province, Indonesia, causing vessel traffic level is high. On this occasion, the information about sediment aspects is very limited, since it was just operated on January 7th 2014, yet it is crucial for the activity within the port. Determinations of the sediment distributions are necessary for predicting and controlling sedimentation in harbors [16], [17], [18], [19], [20], [21], [22]. Sediment transport in harbors can be important for maintaining navigation channels, dredging harbors, and preventing coastal erosion [23], [24]. Sediment analysis in LFP also described by [25] in mini thesis, but performed by different methods as ours. That study carried out sediment samples using the

IOP Conf. Series: Earth and Environmental Science **106** (2017) 012070 doi:10.1088/1755-1315/106/1/012070

sediment-grab sampler and sieved by dry technique, compared to coring method and wet sieves technique as ours, which ensures more appropriate sampling process.

The objective of the present study was to obtain insight into the grain-size parameters of LFP sediments and their statistical relationships. Detailed investigation concerning spatial variability is required in order to have a better understanding of the sedimentation processes in LFP. Granulometric analysis performed to assess the statistical parameters of the sediment grain size as well as mean, standard deviation (sorting), skewness, and kurtosis.

2. Research and Method

This study located at LFP, Aceh, Indonesia (5°34'35" N; 95°19'23" E). LFP area of approximately 111 hectares and has seawall by its surroundings. The data collection was done during October 2013, at ten stations that indicated by number in Figure 1. The determination of each station location is based on a representative condition at LFP. Station 1 represents natural beach facies, station 2 located in berthing pool, while station 9 and 10 represent open-coast facies.

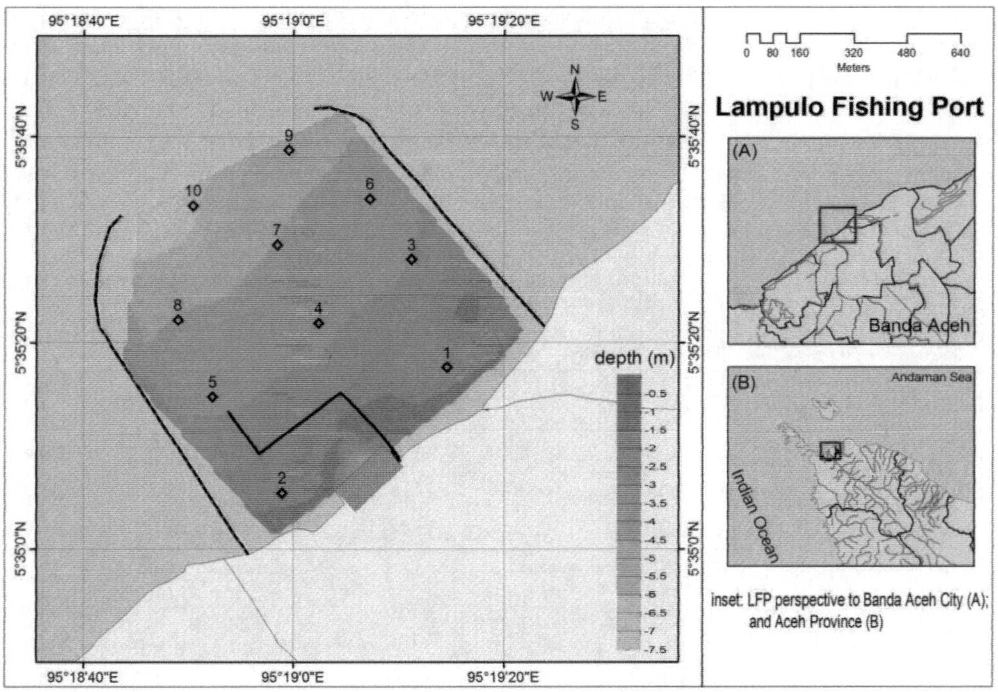

Figure 1. Sampling station at Lampulo Fishing Port (LFP). The map contain information of bathymetry in LFP.

Coring method was used to collect the sediment sample, and it is done by a diver at all stations. The core sample has a diameter of 2.5 inches, and the sediment thickness was set to 15 cm to be collected in each coring.

Six sets of sieves were used to analyze the particle size distribution of sediment sample, i.e. 1 mm, 0.5 mm, 0.25 mm, 0.125 mm, 0.075 mm and 0.038 mm, referred to Udden-Wentworth scales. Wet sieving technique was applied to all samples based on American Society for Testing Materials (ASTM) procedure, D4823-95 [26]. The sediment types classification of samples is derived from folk triangle. Sedimentological characteristics were analyzed by statistical method by using of moment method. The statistical analyses were used in measuring the grain sizes are mean (m_z), sorting (σ_a), skewness (Sk_a), and curtosis (K_a), with formulas as follows:

$$m_z = \frac{\sum f m_m}{100} \quad (1)$$

$$\sigma_a = \sqrt{\frac{\sum f(m_m - m_z)^2}{100}} \quad (2)$$

$$Sk_a = \frac{\sum f(m_m - m_z)^3}{100\sigma_a^3} \quad (3)$$

$$K_a = \frac{\sum f(m_m - m_z)^4}{100\sigma_a^4} \quad (4)$$

Where, f is weight frequency (%) in each class, m_m is median of each class in *phi* value ($\varphi = -\log_2 d$).

3. Results and Discussion

3.1. Sedimentological characteristic

Table 1 shows the sieve result of the grain size distribution at each station. In general, we found that the sediment samples were divided into sand to silt. The weight percentage of each class determines the type, which is follows the Folk triangle. Through all the stations, we found that sand was found as predominant in LFP bottom, where as some portion of silt, compared to [25] which was found a considerable portion of silt. We suggest that different sampling method led to the different result, where sediment-grab only carried sediment from thin upper layer in the bottom. Most of our sediment samples were categorized as silty sand which is found at six stations ,i.e. station 2, 3, 4, 7, 8 and 9 respectively, sand in stations 1 and 10, where as sandy silt in stations 5 and 6.

Table 1. Granulometric analysis.

St.	Percentage of sediment weight							Types
	>1.00	0.500-1.00	0.250-0.500	0.125-0.250	0.075-0.125	0.038-0.075	<0.038	
1	41.90	18.10	19.05	9.52	2.86	6.67	1.90	Sand
2	3.00	1.00	1.00	50.00	27.00	9.00	9.00	Silty sand
3	0.99	0.99	1.98	30.69	33.66	26.73	4.95	Silty sand
4	2.44	4.88	4.88	35.37	13.41	18.29	20.73	Silty sand
5	0.00	1.32	6.58	26.32	9.21	10.53	46.05	Sandy silt
6	1.02	6.12	19.39	10.20	7.14	10.20	45.92	Sandy silt
7	1.16	2.33	1.16	11.63	39.53	25.58	18.60	Silty sand
8	1.69	0.85	0.85	60.17	8.47	22.03	5.93	Silty sand
9	1.09	3.26	4.35	42.39	13.04	21.74	14.13	Silty sand
10	12.16	16.89	28.38	35.81	4.05	2.03	0.68	Sand

References: classification of grain size follows the Udden-Wentworth scale. Very coarse sand (>1.00 mm); coarse sand (0.500-1.00 mm); medium sand (0.250-0.500 mm); fine sand (0.125-0.250 mm); very fine sand (0.075-0.125 mm); coarse silt (0.038-0.075 mm); and silt (<0.038 mm). Sediment types were determined based on the Folk diagram.

3.2. Statistical analysis of sediment grain size

Mean grain size is the arithmetic average of grain size in represented sediment samples. The variation of mean grain size, defines the differential energy condition, which is effected to their deposition. Mean grain size value at station 1 indicates the coarsest grain of the entire station (Table 2). Based on the field observations, it is known that station 1 is located near to the coastline (SE of the LFP), where the wave height is tend to slightly increase and breaks as it came to shallower water (see bathymetry in Figure 1) [27]. High energy level may transport a smaller grain size to move from its original locations

[28]. While at the other stations generally can be described as having low energy level conditions, as the LFP has seawalls in almost surrounding area.

Table 2. Sediment statistical parameters.

Station	Mean (φ)	Sorting	Skewness	Kurtosis
1	0.8022	1.5711	1.0370	3.2262
2	2.9573	1.1305	-0.1799	5.0436
3	3.2576	0.9667	-0.5244	4.8704
4	3.2364	1.4828	-0.2850	2.6965
5	3.9273	1.4539	-0.4209	1.7354
6	3.6443	1.8067	-0.5074	1.7446
7	3.6641	1.1457	-0.7900	4.8678
8	2.9694	1.0576	0.0829	4.3264
9	3.1790	1.2808	-0.0517	2.9057
10	1.5200	1.1850	-0.0784	2.8767

The examining of sorting parameter shows that entire LFP was categorized as poorly sorted, while station 3 was the only categorized as moderately sorted. The lower sorting (σ_a) values correlated to a better sorting process of the sediment sample. Those sorting values indicate that there are fluctuating energy in the LFP. Since the LFP condition is a semi-closed water, with the breakwater at its surroundings, it is supposed that the entrance in the north that is still quite wide opened and provide energy input from the offshore. This sorting condition indicates mixing between coarse grains and fine grains on the bottom of the water [29], allegedly caused by changes in water conditions that were previously in the form of coastal waters which then changed into a harbour. Coarse grains are generally found from coastal waters while the accumulation of fine granules as a result of the small grain deposition process [30]. Deposition of fine grains indicates that the energy retained in LFP waters is relatively quiet, in line with semi-enclosed LFP since the breakwater is created around the LFP. The existence of this breakwater might have been eliminated the energy transferred from the open sea.

Descriptions of skewness and kurtosis result analysis also support the previous statement. Most locations have a skewness value that was negative (coarse skewed), where several have near symmetrical, and only station 1 was positively skewed. The numbers of locations that have negatively skewed conditions argue that there is a fine grain settling process in LFP. Although there are still positively skewed conditions in a number of locations, the semi-closed port conditions allow for the accumulation of fine grain deposits which will then redirect the curve of sediment distribution to a negative direction. This trend is also confirmed by the leptokurtic sediment character, which indicates that the distribution of sediments is highly concentrated around the mean value, which results in a very peaked distribution curve.

In general, it can be said that sediment characteristics in LFP indicate a stable condition, although at station 1 indicated there is still a slightly larger energy. The addition of the breakwater to the still-open area can add stability levels of water and sediments within the harbour pool.

4. Conclusion

Sediment grain size analysis that conducted at LFP observed the sediment is classified into three types, i.e. sand, silty sand, and sandy silt. The statistical analysis of sediment grain size shows the coarsest grain size is observed within station 1 with a mean value of 0.8022. Then it is stated that most of the stations in LFP were in a poor sorted, negative skewed and very leptokurtic, in which conclude as LGP environments support the occurrence of fine sediment deposition as a result of low-energy aquatic environment.

Acknowledgment
We would like to thank the UPTD LFP and Marine and Fisheries Department of Aceh that allow us to collect several samples at the LFP sites. Thanks also to Irwansyah, who help a lot along the data collection and analysis.

References

[1] Abdulkarim R, Akinnigbagbe E A, Imo D O, Titocan M I, Ibitola M P, Faleye B R, Shonde O O, Jimoh O R, and Abe O B 2011 Sedimentological variation in beach sediments along the barrier lagoon coastal system, Lagos, South West Nigeria *Nature and Science* 9 19-26
[2] Barnard P L, Schoellhamer D H, Jaffe B E, and McKee L J 2013 Sediment transport in the San Francisco Bay Coastal System: An overview. *Marine Geology* 345 3-17
[3] Amrouni O, Hermassi T, Jaouad S A and Messaoudi S 2014 Contribution of Grain-size Trend to Sediment of a Microtidal Beach. Case of the Gulf of Tunis Bay (Cape Ferina-Cape Gammarth, Tunisia) *Research Journal of Environmental Sciences* 8 161-77
[4] Purnawan S, Haridhi H A, Setiawan I, and Marwantim 2015 Parameter statistik ukuran butiran pada sedimen berpasir di muara Kuala Gigieng, Kabupaten Aceh Besar *Jurnal Ilmu dan Teknologi Kelautan Tropis* 7 15-21
[5] Li Y, Wolanski E and Zhang H 2015 What processes control the net currents through shallow straits? A review with application to the Bohai Strait, China *Estuarine Coastal Shelf Science* 158 1-11
[6] Kim H J, Suh S W, Seok J S, and Park W K 2017 Sedimentation for a Flood-dominant Estuarine Harbor Induced by Anthropogenic Activities *Journal of Coastal Research* SI79 339-343
[7] Clayton J A, and Pitlick J 2008 Persistence of the surface texture of a gravel-bed river during a large flood *Earth Surface Processes Landforms* 33 661–673 doi:10.1002/esp.1567
[8] Van Maren D S, and Winterwerp J C 2013 The role of flow asymmetry and mud properties on tidal flat sedimentation *Continental Shelf Research* 60S S71-S84
[9] Wachecka-Kotkowska L, and Kotkowski P 2011 Grain-size distribution analysis of quaternary sediments from the southern part of the Lodz region in Poland: a computational-methods approach *Geologos* 17 205–19
[10] Mao L, Cooper J R, and Frostick L E 2011 Grain size and topographical differences between static and mobile armor layers *Earth Surface Processes Landforms* 36 1321–34 doi:10.1002/esp.2156
[11] Jafarzadeh M, and Hosseini-Barzi M 2008 Petrography and geochemistry of Ahwaz Sandstone Member of Asmari Formation, Zagros, Iran: implications on provenance and tectonic setting *Revista Mexicana de Ciencias Geológicas* 25 247-60
[12] Saniah, Purnawan S, and Karina S 2014 The characteristics and mineral content of coastal sand from Lhok Mee, Beureunut and Leungah, Aceh Besar District *Depik* 3 263-70
[13] Armstrong-Altrin J S, Ramasamy N, Yong I L, Juan J K and Leslie P C 2014 Geochemistry of sands along the San Nicolás and San Carlos beaches, Gulf of California, Mexico: implications for provenance and tectonic setting *Turkish Journal of Earth Science* 23 533-558
[14] Purnawan S, Setiawan I, and Muchlisin Z A 2015 Sediment grain-size distribution in the Lake Laut Tawar, Aceh Province, Indonesia *AACL Bioflux* 8 404-10
[15] Chikin A L, and Chikina A L 2011 Modeling the silt transport and deposition in approach channels in the Taganrog Bay *Vestnik Yuzhnogo Nauchnogo Tsentra RAN* 7 45–48
[16] Kian R, Velioglu D, Yalciner A C and Zaytsev A 2016 Effects of Harbor Shape on the Induced Sedimentation; L-Type Basin *Journal of Marine Science and Engineering* 4:55
[17] Caceres R A, Zyserman J A, and Perillo G M E 2014 Analysis of Sedimentation Problems at the Entrance to Mar del Plata Harbor *Journal of Coastal Research* 32 301-14
[18] Gorlova A A, and Makarov K N 2017 Sediment Deposition in Artificial Harbors and Approach Channels *Power Technology and Engineering* 50 473–9

[19] Kuklev S B, Divinskii B V, and Kozachinskii Y S 2012 Predicting sediment deposition in maritime approach channels by mathematical simulation. *Gidrotekhnika* 3 55 –57

[20] Delile H, Mazzini I, Blichert-Toft J, Goiran J P, Arnaud-Godet F, Salomon F, and Albarède F 2014 Geochemical investigation of a sediment core from the Trajan basin at Portus, the harbor of ancient Rome *Quaternary Science Reviews* 87 34-45

[21] Lepland A, Boe R, Lepland A, and Totland O 2009 Monitoring the volume and lateral spread of disposed sediments by acoustic methods, Oslo Harbor, Norway *Journal of Environmental Management* 90 3589–98

[22] Coch N K 2016 Sediment Dynamics in the Upper and Lower Bays of New York Harbor *Journal of Coastal Research* 32 756-67

[23] Supiyati Suwarsono, and Setiawan I. 2011. Sediment transport cause of Port Superficiality of Bengkulu Baai Island with diskritisation model of oceanography dynamics *Dinamika Teknik Sipil* 11 172-80

[24] Van Maren D S, Van Kessel T, Cronin K, and Sittoni L 2015 The impact of channel deepening and dredging on estuarine sediment concentration *Continental Shelf Research* 95 1-14

[25] Rinanda S, Study of Bottom Sediment Characteristics for Sediment Transport Potency in Fisheries Port of Lampulo *Thesis Universitas Syiah Kuala* p 37 2015

[26] ASTM. D4823–95. *Standard Guide for Core Sampling Submerged, Unconsolidated Sediments* (West Conshohocken: ASTM International) p 14. 2008

[27] Measures R, and Tait S 2008 Quantifying the role of bed surface topography in controlling sediment stability in water-worked gravel deposits *Water Resources Research* 44 W04413 doi:10.1029/2006WR005794

[28] Delpey M T, Ardhuin F, Otheguy P, and Jouon A 2013 Effects of waves on coastal water dispersion in a small estuarine bay *Journal of Geophysical Research* 119 70-86

[29] Venkatramanan S, Ramkumar T, Anithamary I, and Ramesh G 2011 Variations in texture of beach sediments in the vicinity of the Tirumalairajanar river mouth of India *International Journal of Sediment Research* 26 460–70

[30] Gibson S, Abraham D, Heath R, and Schoellhame R D 2009 Vertical gradational variability of fines deposited in a gravel framework *Sedimentology* 56 661–76 doi:10.1111/j.1365-3091.2008.00991.x

Potential reduction of non-residential solid waste in Sukomanunggal district West Surabaya

I D A A Warmadewanthi*, S A Reswari

Environmental Engineering Department, Faculty of Civil Environmental and Geo Engineering, Institut Teknologi Sepuluh Nopember, Surabaya, Indonesia

*Corresponding Author: warma@its.ac.id

Abstract. Sukomanunggal district a development unit 8 with the designation as a regional trade and services, industrial, education, healthcare, offices, and shopping center. The development of this region will make an increasing solid waste generation, especially waste from non-residential facilities. The aims of this research to know the potential reduction of waste source. The method used is the Likert scale questionnaire to determine the knowledge, attitude, and behavior of non-residential facilities manager. Results from this research are the existing reduction of non-residential solid waste is 5.34%, potential reduction of the waste source is optimization of plastic and paper waste with the reduction rate up to 19,52%. The level of public participation existing amounted to 46.79% with a willingness to increase recycling efforts amounted to 72.87%. Efforts that can be developed to increase public awareness of 3R are providing three types of bins, modification of solid waste collection schedule according to a type of waste that has been sorted, the provision of the communal bin.
Keywords: likert, potential reduction, solid waste

1. Introduction

Sukomanunggal district was the region with the highest population density in West Surabaya, located in 7.2643° S, 112.6973° E. District Sukomanunggal is development unit 8 which attended for trade and services, office, education, healthcare, industrial and shopping center. There is currently no related data contribution of non-residential solid waste in contributing to the amount of waste in the District Sukomanunggal. In the handling of non-residential solid waste should each source of waste both residential areas, commercial areas, industrial areas, special areas, public facilities, social facilities and other facilities required to carry out the sorting of waste and in collecting required to provide transfer station, 3R facilities and or transporter to transfer station or transported to the landfill if waste generated more than 30 m^3/month [1], [2] but in the implementation are still many sources of non-residential facilities that burdening collection facilities and public transfer station. The aims of this research to know the potential reduction of waste source.

2. Research Method

There are three subdistricts were selected to be the location of the study, including the subdistrict of Simomulyo, Simomulyo Baru, and Sukomanunggal. Public facilities used as the sample is education, worship house, healthcare, offices, industrial, sweep streets and commercial areas consisting of a hotel, restaurant, minimarket and traditional market. To determine the potential reduction which may occur in any source of waste is carried out with a Likert scale questionnaire. The number of samples to be determined by the formula slovin in equation 1 with the result 163 units samples.

$$n = N / (1 + N e 2) \quad (1)$$

Information:
n = number of samples
N = number of population
e = Limit fault tolerance

Using a four-point Likert scale or more questions to re-measure the behavior of individuals to respond to the 5 point choice on any of the questions, strongly agree, agree, not deciding, disagree, and strongly disagree combined to form a value that represents the nature individual, knowledge, attitudes, and behaviors [3], [4].

3. Results and Discussion

3.1. Existing of non-residential solid waste reduction in Sukomanunggal District

The analysis conducted in this study is an analysis of existing conditions and simulation reduction efforts in accordance with the will of society. An existing condition that is needed in this research includes generation conditions, the composition and the reduction of non-residential solid waste.

3.1.1. Total Generation and composition of Non-Residential Solid Waste. Waste generation is the amount of waste generated from waste sources expressed in units of mass or volume or weight per capita or per unit time or per unit area [5]. Total waste generation type of household waste is 21109.707 kg/day, with details such as Table 1. The composition of waste that most of the organic waste 57.31%, the second largest is plastic waste 19.98% and the third largest is paper waste 14.20%. The reduction of the waste composition will be significant if done on organic waste by composting or recycling of plastic and paper. Diagram composition of non-residential solid waste presented in Figure 1.

Tabel 1. Total waste generation of non-residential solid waste.

Public Facilities	Solid waste generation (Kg/day)	Density (Kg/m^3)
Worship House	79,11	92,15
Education	877,16	140,26
Office	7719,99	107,14
Hotel	1828,58	107,22
Restaurant	1302,50	140,95
Minimarket	132,29	81,82
Traditional market	6830,95	210,84
healthcare	230,41	105,06
Industry	1076,99	185,64
Street sweeping	1031,73	214,14
Total	21109,71	138,52

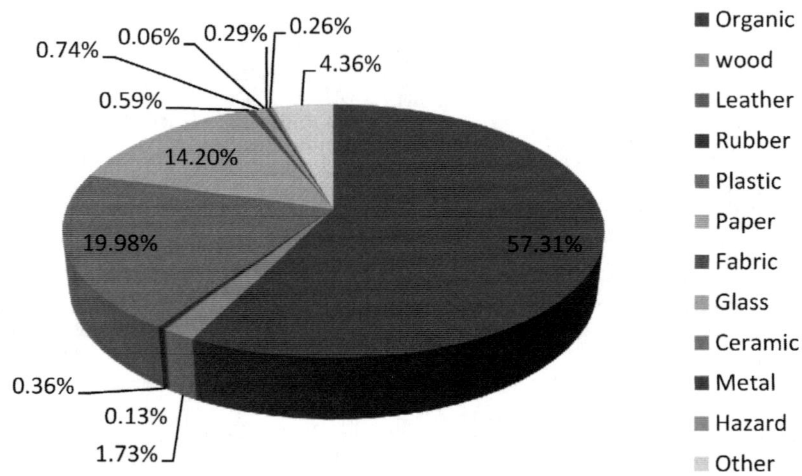

Figure 1. The composition of non-residential solid waste.

3.1.2. Existing Reduction of Non-Residential Solid Waste.

Waste processing that can be done by waste sources such as composting and recycling material [6], [7]. Recycling can reduce the amount of waste generated by the community, recycling efforts are quite prominent and generally involve the informal sector [8]. From the research, it can be seen that waste reduction activity is done in the sources and in transfer stations. The reduction done in the waste source is composting and recycling plastic and paper. Reduction done in transfer station is the sorting of plastic waste and paper by informal sector to be sold to the collectors. The existing level of reduction in Sukomanunggal District is 5.34% where the reduction in resources 1.57% and 3.77% reduction in the transfer stations. Reduction of non-residential solid waste is summarized in Table 2.

Table 2. Existing reduction of non-residential solid waste.

Public Facilities	Solid waste generation (Kg/day)	Total reduction (Kg/day)	
		source	transfer station
Worship house	79.11	1.59	3.73
Education	877.16	34.31	48.11
Office	7719.99	178.18	50.77
Hotel	1828.58	25.08	31,29
Restaurant	1302.50	38.79	120.21
Minimarket	132.29	12.85	9.91
Marker	6830.95	-	399.57
Healthcare	230.41	-	4.71
Industry	1077.00	41.13	56.59
Street sweeping	1031.73	-	71.85
Total	21109.71	331.93	796.73

Recovery Factor

Recovery factor is the percentage of segregated garbage that can be utilized on the total weight of waste overall. Public facilities with potential for composting are educational facilities, offices, markets, healthcare and street sweeping. Garbage can be reduced by composting activities amounting to 37.35% of a total non-residential solid waste in Sukomanunggal.

Plastic waste can be reduced include PET (1) such as beverage bottles, HDPE (2) such as *shampoo* bottles, PP (5) such as glass bottled mineral water, Other (7) such as *shampoo* packaging plastic *sachets*. Based on the interview almost all public facilities are willing to do the sorting and utilization of plastic waste garbage can be reduced by recycling plastic for 7.74% of a total non-residential solid waste in Sukomanunggal.

Types of paper waste that can be reduced include cardboard, paper, cardboard, newspapers. As well as plastic waste, almost all public facilities willing to conduct utilization of waste paper bins except sweep the street waste that can be reduced with the utilization of paper waste amounted to 5.32% of the total waste generated by the public facilities in the district Sukomanunggal.

Simulation recovery of these factors produces the optimum reduction effort in which all public facilities would perform reduction in the form of composting and recycling plastic and paper waste. The reduction that can be achieved next 20 years is 50.20%.

Potential Reductions by the will of society.

The existing reduction is too small to achieve the reduction in accordance with the recovery factor, for that, we need a scenario with a reduction in composting or recycling plastic and paper waste. From the result of Likert scale questionnaire known that the average attitude towards the source of waste reduction efforts with a composting plan ranges from 52.98% or less agree with the composting plan. Attitude toward the sources of waste plastic recycling effort and paper reached 72.87% or agree with the reduction of recycling plastic and paper.

Judging from existing conditions, a source of garbage composting at 24.52% while the source of garbage to recycle as much as 46.79%. Comparison of public attitudes toward waste reduction efforts plans presented in Figure 2.

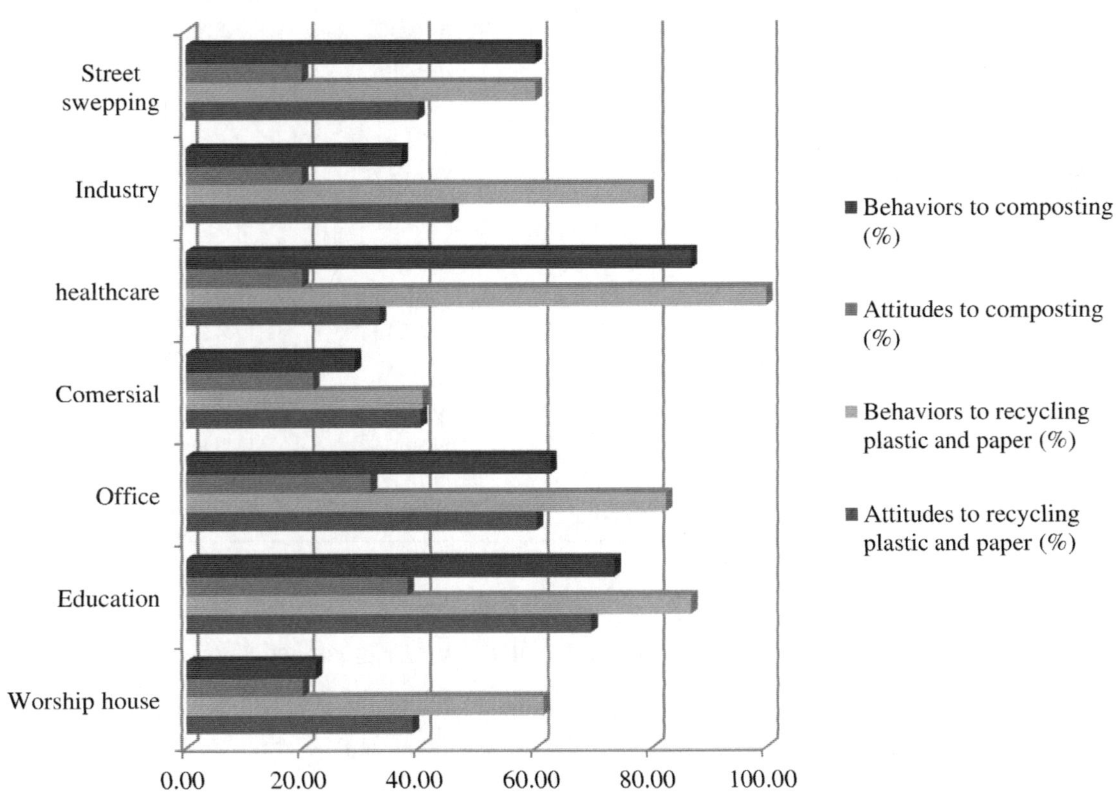

Figure 2. Comparison of public attitudes and behaviors to reduce solid waste.

Based on the willingness of the community to do the approximate reduction of over 20 years using composting and recycling plastic and paper. The calculations show that with the development of the reduction effort using waste composting that can be reduced is 28.13%, while the reduction with plastic and paper recycling is 19.52%. Composting reduction level is greater with plastic and paper recycling, this is caused by the composition of wet garbage that can be composted is much larger than the composition of the plastic and paper waste that can be recycled. However, based on the will of society with a reduction potential of recycled plastic and paper is greater when compared to composting. Given the reduction potential that you want to upgrade is the potential at the source of waste then the best alternative that can be done to achieve a sustainable reduction is by recycling plastic and paper waste. Table 3 shows the potential reduction every year with plastic and paper reduction efforts.

Table 3.Potential reduction by recycling plastic and paper.

Year	Total Generation (Ton/Year)	Residues in TPA (Ton/Year)	Reduction (Ton/Year)	Reduction (%)
2017	7705.04	7293.44	411.59	5.34
2018	8262.90	7557.88	618.25	7.56
2019	8889.61	7987.23	715.11	8.22
2020	9594.10	8463.34	827.08	8.90
2021	10386.49	8991.84	956.12	9.61
2022	11278.23	9579.75	1103.78	10.33
2023	12282.34	10232.8	1273.95	11.07
2024	13413.59	10959.26	146.,17	11.82
2025	14688.73	11767.87	1692.92	12.58
2026	16126.79	12668.33	1949.2	13.33
2027	17749.39	13671.49	224.61	14.09
2028	19581.09	14789.46	2578.44	14.85
2029	21649.83	16039.1	2959.41	15.58
2030	23987.35	17452.49	3375.52	16.21
2031	26629.76	19030.12	3851.12	16.83
2032	29618.17	20796.69	4389.49	17.43
2033	32999.35	22746.04	5028.37	18.10
2034	36826.56	24976.13	5705.31	18.60
2035	41160.47	27517.35	6430.02	18.94
2036	46070.18	30418.64	7198.69	19.14
2037	51634.47	33594.42	8147.81	19.52

Strategy to Increase Recycled Plastics and Paper

Basically, there are five technical waste management criteria such as waste reduction, waste treatment location, human resource availability, ease of application of technology and economic value [9]. The efficiency of waste treatment site and application of technology so as to bring economic value can increase public participation in waste reduction. The strategy that can be done is by modifying the operational pattern of integrated waste management which consists of warehousing, collection, transportation, final disposal [10], [11]. Warehousing waste is a temporary waste collection activity in a waste source comprised of individual and communal bins [10]. Modification bins can be done by requiring each source of waste to provide three types of waste bins for waste paper, plastic, and others.

IOP Conf. Series: Earth and Environmental Science **106** (2017) 012071 doi:10.1088/1755-1315/106/1/012071

Communal bins can be provided at public facilities such as markets and parks, these communal bins are also separated for paper, plastic, and others. The recommended strategy in the collection is the scheduling of waste collecting according to the type of waste that has been sorted. This Collecting has been carried out since the removal of garbage from the bins to the transfer station while still guaranteeing the separation of waste according to the type of waste [12]. To ensure the sustainability of the program, the government should conduct public awareness programs on the importance of waste reduction for an environmental and public health [13].

4. Conclusion

Based on result data and discussion, the conclusion of this research is the existing non-residential solid waste reduction rate in Sukomanunggal district is only about 5.4% and potentials that can be developed to increase the reduction of non-residential solid waste in Sukomanunggal district are recycled paper and plastic waste with the reduction potential of 19.52%.

References

[1] The government of Surabaya City. Surabaya City Regulation No.5 of 2014 about Waste Management and Hygiene in Surabaya City.
[2] Public Working Ministry. Minister Regulation No. 03 of 2013 about Implementation of Infrastructure and Solid Waste Facility, in handling residential and non-residential solid waste.
[3] Likert RA. Technique for the measurement of attitudes. Archives of Psychology. 1932; 140:1-55.
[4] Budiaji W. Skala Pengukuran dan Jumlah Responden Skala Likert. Jurnal Ilmu Pertanian dan Perikanan. 2013;02:125-31.
[5] Ruslinda Yenni, Indah Shinta, Laylani Widya. Studi Timbulan, Komposisi dan Karakteristik Sampah Domestik Kota Bukit Tinggi. Jurnal Teknik Lingkungan UNAND. 2012;09(1):01-12
[6] Fadhilah A., Sugianto H., Hadi K., Firmandhani S.W., Murtini T.W., Pandelaki. E.E. Kajian Pengelolaan Sampah Kampus Jurusan Arsitektur Fakultas Teknnik Universitas Diponegoro. Jurnal Modul. 2011;11:02.
[7] Susilo F. Pengelolaan Sampah Terpadu Sebagai Peluang Bisnis Rumah Tangga di Kota Medan. Jurnal Agrobio. 2011;03:01.
[8] Damanhuri and Padmi. Diktat Kuliah Pengelolaan Sampah. 1st Engl. Ed. Bandung. Environmental Engineering Department FTSL ITB. 2010. 37 p.
[9] Widawati Enny, Tanudjaja H., Iskandar I., Budiono C. Kajian Potensi Pengolahan Sampah. Jurnal Metris. 2014;15:119-26.
[10] Tamod Z. Partisipasi Masyarakat dan Teknik Pengelolaan Sampah di Permukiman. Jurnal Formas. 2008;01(04):277-83
[11] Dwiyanto Bambang M. Model Peningkatan Partsipasi Masyarakat dan Penguatan Sinergi dalam Pengelolaan Sampah Perkotaan. Jurnal Ekonomi Pembangunan. 2011;12(02):239-56
[12] Gallardo A., Bovea M.D., Colomer F.J., Prades M. Analysis of Collection System for Sorted Household Waste in Spain. Waste Management. 2012;32:1623-33.
[13] Pai Rajesh R., Rodrigues Lewellyn L.R., Mathew Asish O., Herbbrar S. Impact of Urbanization on Municipal Solid Waste Management: a System Dynamics Approach. International Jurnal of Renewable Energy and Environmental Engineering. 2014;02(01):31-7

IOP Conf. Series: Earth and Environmental Science **106** (2017) 012072 doi:10.1088/1755-1315/106/1/012072

Network integration modelling of feeder and BRT(bus rapid transit) to reduce the usage of private vehicles in Palembang's suburban area

D Nur'afalia*, F Afifa, L Rubianto, K D M E Handayeni

Urban and Regional Planning Department, Institut Teknologi Sepuluh Nopember, Surabaya, Indonesia

*Corresponding Author: diannur.afalia@yahoo.co.id

Abstract. The aim of this research is to determine the optimal feeder network route that integrates with BRT (Bus Rapid Transit). Palembang, a high growing population city with unresolved transportation demand sector. BRT as main public transportation could not fulfill people's demand in transportation, especially in Alang-Alang Lebar sub-district. As an impact, the usage of private vehicles increases along the movement toward the city center. The concept of Network Integration that integrates feeder network with BRT is expected to be a solution to suppress the rate of private vehicles' usage and to improve public transportation service, so that the use of BRT will be increased in the suburban area of Palembang. The method used to identifying the optimal route using Route Analysis method is route analysis using Tranetsim 0.4. The best route is obtained based on 156 movement samples. The result is 58,7% from 199 mobility's potency of private vehicle usage's can be reduced if there is a feeder network's route in Alang-Alang Lebar's sub-district. From the result, the existance of integration between feeder network and BRT is potential enough to reduce the usage of private vehicles and supports the sustainability of transportation mobility in Palembang City.
Keyword : BRT, private vehicles, tranetsim

1. Introduction

The main problem in a developing city is transportation [1]. According to that, public transportation has a big role in a developing city to solve transportation problems. The development of public transportation in Palembang, South Sumatera, has started in 2002. However, congestion and private vehicle user are increasing. To solve it, in the beginning of 2010, a public transportation of Bus Rapid Transit (BRT) called Bus Trans Musi has developed. Despite that, the aim of BRT as the main transportation in Palembang City is not yet optimal. Based on the previous research, 33% of the transportation user used motorcycle as the main vehicle and 31% used *Angkutan Kota (Angkot)*, because it is easier for mobilization and time efficiency of using private vehicle [2]. Around 41.040 of Alang-Alang Lebar sub-district's resident have the potency of mobilization to the city center and the the congestion from suburban area to the city center [3]. The change of transportation mode from private vehicle to BRT as mobilization potency in Alang-Alang Lebar sub-district will be able to reduce the city problem.

To reduce the number of private vehicle usage in suburban area of Palembang, needs a handling concept. In connection with the development of the Palembang's density and the high demand of transportation, according to researcher [1], the integration of public transportation is the most rational solution to solve the urban transportation problem. This also supported by strategy and policy of the transportation development in RPJMD Palembang 2013-2018 to integrate the public transportation in Palembang, as also mentioned by the head of Palembang's Transportation Department [4] about the instruction of feeder usage in 2016 and public transportation usage in the suburban area of Palembang in 2016.

The ideal integrated public transportation concept that should be applied in Alang-Alang Lebar sub-district is Network Integration. The concept of Network Integration is able to increase the usage of public transportation in London by 40% from 1982 to 2001 with a developed road network [5]. However, an analysis of the best network to suppress the usage of private vehicle in Alang-AlangLebar sub-district is needed. According to the problems that have been described, a study of private vehicles' reduction in suburban area of Palembang by applying public transportation feeder to BRT, which can be the main public transportation in Palembang to suppress the transport density in Palembang, will be conducted.

The concept of integrated public transportation modes has been applied in big cities such as Curitiba, Brazil. With the concept of integrated public transportation, transportation's flow is getting well although the population in the city is increasing every year. The integrated transportation system in Curitiba has increased the resident's quality of life and also public transportation user [6]. Integrated public transportation modes can defined as a combination of two or more types of transportation modes in serving theresident that mobilizes in one way of origin-destination [1], [7].

In a few big cities in Indonesia, the stagnation of network integration indicated by overlapped path of public transportation and it's inability to reach every area in the city. Therefore, private vehicle becomes the main option in mobilization. As mentioned by Gwilliam (2008) in [7], it will be better if public transportation network applies the service hierarchy based on the main mode, branch mode, and feeder mode; so that the distribution of service area becomes clearer and the integration of public transportation's service network can be implemented easily, so the determination of transportation mode's transfer point and the stop can be more optimal. However, to determine the compatible network for public transportation optimization, optimal network analysis is required to increase the usage of public transportation in a city.

2. Research Method

This research is descriptive and experimental which based on positivist paradigm. The population of this research is the head of household in Alang-Alang Lebar sub-district, with the number of 17,527 head of households. Sample determination is differed by the goal of this research. The first and fourth goals are the identifying resident's mobility characteristic and calculating the potency of private vehicle usage's reduction, using purposive sampling technique with sampling criteria:

a) Private vehicle user in Alang-Alang Lebar sub-district
b) The mobility's destination is to the Palembang City center
c) Resident's mobility destination have been accommodated by BRT or other feeder.

Then, in determining the sample for the first and fourth goal, random sampling technique is used to limit the sample quantity in Alang-Alang Lebar sub-district with calculation using Slovin's equation, as follows:

$$n = 17527/(1 + 17527.(0.08)^2) \quad (1)$$
$$n = 156$$

From 156 samples, sample determination process based on each *kelurahan* in Alang-Alang Lebar sub-district. In determining sample quantity for the second goal, stakeholder analysis are used to identify related stakeholders in determining feeder network's routein Alang-Alang Lebar sub-district. The chosen stakeholders are Palembang's Department of Transportation, Palembang's Department of Binamarga, PT SP2J, Palembang City and *stakeholder* in Alang-Alang Lebar sub-district. The method used for identifying the optimal route is route analysis using Tranetsim 0.4.

IOP Conf. Series: Earth and Environmental Science **106** (2017) 012072 doi:10.1088/1755-1315/106/1/012072

3. Results and Discussion

3.1. Identifying Resident's Mobility Characteristic in Alang-Alang Lebar sub-district, Palembang

The population's mobility characteristics are obtained from samples' questionnaires. From 156 head of households' samples, it is known that there are 239 mobilities. Mobility is being dominated in *Kelurahan* Karya Baru, with the total of 100 mobilities. From 239 mobility samples, 233 mobilities are using private vehicle, which most of them are motorcycle. Most of the resident goes to the city center to work with the samples of 72.5%. Meanwhile 199 of 233 samples that use private vehicle are willing to use BRT if there is a feeder network, that is become next step samples. There are 19 of 63 city road becomes road segment based on the sample's route to other place.

3.2. Weighting the Indicator, Variable and Sub-Variable From the Determination of Integrated Feeder Network with BRT

The next step is to weighting the indicator, variable and sub-variable. Whereas indicator, variable and sub variable are obtained from the result of referecnces synthesis. The process of determining the weight is interview with stakeholders, then, the result is being processed using Expertchoice 11. The results of indicator's weighting, variable and sub-variable are listed in Table 1.

Table 1. Weight of indicator, variable and sub-variable.

Indicator	Variable	Sub-Variable
Area's Characteristics (0.568)	Land Use (0.457)	Residential (0.343)
		Commercial (0.202)
		Office and Institutional (0.213)
		Industrial (0.182)
		Community Facilities and Public Lands (0.06)
	Road Network (0.543)	Local Road (0.707)
		Collector Road (0.209)
		Arterial Road (0.085)
Coverage Area (0.432)	Public Transport Demand Pattern (0.333)	-
	Population (0.359)	-
	Service Area (0.317)	-

Based on the table above, it could be known that indicator with highest weight is area's characteristics (0.568%). However, sub-variable with highest weight is local road (0.707%). The approach in this step is the higher the weight then the road segment's potency to become the most optimal feeder network is also higher, based on indicator, variable and sub-variable.

3.3. Determining Feeder Network Route that Integrates with BRT

In this process, researcher determines the feeder network route based on indicator, variable, and sub-variable using Tranetsim 0.4. The early stage of the determination is identifying road in Alang-Alang Lebar sub-district. It is identified that 39 road segment's point and 110 road segment will be the input data to Tranetsim 0.4. Road segment is the intersection that samples go through. Road segment that becomes the origin is road segment's point with the distribution of samples that's covering it's area. On the other hand, road segment that becomes the feeder network's destination is BRT stop in Alang-Alang Lebar sub-district.

After submitting road segment's point as the input of Tranetsim 0.4, then valuate the road segment. The valuation process of road segment is done by calculating feeder network's variable and sub-variable. After the buffering process in 500 m radius, sub-variable for each road segment is being calculated. Here are the examples of sub-variable's calculation in each road segment (Table 2).

Table 2. Example of land use's calculation in road segment 10-11.

Variable	Buffer500 m's Area (ha)
Residential	19,8
Commercial	11,4
Office and Institutional	0
Industrial	0
Community Facilities and Public Lands	0

After calculating each road segment based on indicator, variable dan sub variable, then standardize the values with the equation as follows:

$$NX_1 = (X_1/Xa_{max}) \times 100 \quad (2)$$

N_{X1} = Variable Value of Road Segment 1
X_{1a} = Variable a's Value of Road Segment 1
Xa_{max} = Maximum Value of Variable a

The aim of the standardization is to calculate the value of each variables and sub-variables as a whole. Afterwards, the next step is calculating each road segment. The example of road segment 10-11's calculation are listed in Table 3.

Table 3. Example of road segment 10-11's calculation.

Variable and Sub-Variable	Value	Standardization Value	Sub-Variable's Weight	Variable's Weight	Indicator's Weight	Sub-Variable's Total
A	B	C	D	E	F	G
• Residential	19.8	15.1	0.343	0.457	0.568	1.3
• Commercial	11.4	38.9	0.202	0.457	0.568	2.0
• Office and Institutional	0.0	0.0	0.213	0.457	0.568	0.0
• Industrial	0.0	0.0	0.182	0.457	0.568	0.0
• Community Facilities and Public Lands	0.0	0.0	0.060	0.457	0.568	0.0
• Arterial Road			0.085	0.543	0.568	0.0
• Collector Road			0.209	0.543	0.568	0.0
• Local Road	1.0		0.707	0.543	0.568	0.2
• Public Transport Demand Pattern	2.0	1.0		0.333	0.432	0.3
• Service Area	31.2	21.2		0.317	0.432	2.9
• Population	602.2	15.1		0.359	0.432	2.3
Total						**9.1**

After the value of each segment is acquired, another standardization process needs to be done so the value will be the input data to Tranetsim 0.4. The difference between each value's approach in segment based on indicator, variable, and sub-variable, will be equalized by the standardization process. Standardization's equation for the input of Tranetsim 0.4 is:

IOP Conf. Series: Earth and Environmental Science **106** (2017) 012072 doi:10.1088/1755-1315/106/1/012072

$$NY_1 = (Y_{1min}/Y_1) \times 100 \quad (3)$$

NY_1 = Tranetsim's input data of road segment 1
Y_{min} = Minimum value of variable and sub-variable
Y_1 = Total value of variable and sub-variable in road segment 1

The input data for Tranetsim 0.4 are the origin and destination point. Then, Tranetsim will determine the route with the lowest value. The lowest value is the value for each road segment that has been inputted inTranetsim system. Tranetsim will determine the route with the lowest value and the most optimal BRT stop based on the lowest value. The lowest value will become the best route for integratedfeeder network with BRT. As seen in Tranetsim 0.4, mobility from segment point 1 to segment point 18 is the destination, which is BRT stop. In this process, feeder network route is acquired from sample's origin to feeder network's destination. 13 (thirteen) road segment's points are the origin of the mobility. The destination or chosen stop is road segment's point, which is a BRT stop. As the result of Tranetsim's process, there are 13 best route which will be the base calculation of the potency of private vehicle usage's reduction in Alang-AlangLebar's sub-district.

3.4. Calculating the Potency of Private Vehicle Usage's Reduction Based on Integrated Feeder Network with BRT

In this stage, calculation process of the potency of private vehicle usage's reduction will be executed. The research's approach is listed in Table 4.

Table 4. Research's approach to identify the potency of private vehicle usage's reduction.

Potential	Not Potential
Maximum distance based on community's preference≥distancetothe nearest BRT (hate) using feeder's route	Maximum distance based on community's preference ≤distance to the nearest BRT (hate) using feeder's route

Maximum distance based on community's preference is acquired from the questionnaire in Step 1. On the contrary, the distance to the nearest BRT stop is calculated from sample's origin point by feeder network that has been determined in the previous step. Here is the example of the calculation of potency of private vehicle usage's reduction (Table 5).

Table 5. Example of calculation of potency of private vehicle usage's reduction.

No Sample	Shortest Distance Using Feeder Network (km)	Community's Preferred Distance (km)	The Potency of Private Vehicle Usage's Reduction
1	1.08	2	Potential
	1.08	2	Potential
	1.08	2	Potential
3	1.27	1	Not Potential
	1.27	1	Not Potential

As the result of the previous calculation, 58.7% is potential to private vehicle usage's reduction, while the other 41.2% is not potential because of the long distance between the origin and BRT stop. Therefore, the concept of integrated feeder network with BRT can be developed to reduce the usage of private vehicle in Alang-Alang Lebar sub-district.

IOP Conf. Series: Earth and Environmental Science **106** (2017) 012072 doi:10.1088/1755-1315/106/1/012072

4. Conclusion

From the research above, the are 4 conclusions. First, the majority of mobilities in Alang-Alang Lebar sub-district is using private vehicle. The high number of private vehicle usage is caused by the public transportation that have not been able to reach mobility's origin in Alang-Alang Lebar sub-district. Second, Based on AHP analysis using Expertchoice 11, sub-variable with highest weight in indicator: area's characteristic is local road network (0.707%). On the other hand, for indicator: coverage area, sub-variable with highest weight is population (0.359%). That weight determines best road as a feeder network route in Alang-Alang Lebar sub-district. Third, the outcomes of road's valuation are 13 road segments. Those road segments are Jalan Lorong Masjid, Kol. H. Burlian, Mitra Hj, Jenderal Murod, HBR Motik, Sulaiman Amin, Perumdam 1, Srijaya, Mandi Api, Bypass Alang-AlangLebar, Jalan Sepakat, Jalan Mahoni Raya and Jalan Kelapa Gading 5. Fourth, after determining the best route, estimating the potency of private vehicle usage's reduction with comparing distances from samples' origin using integrated feeder network with BRT to the preferred distance by mobility's sample. The result is 58.7% from 199 mobility's potency is the potency of private vehicle usage's reduction if there is a feeder network's route in Alang-AlangLebar's sub-district.

References

[1] Adisasmita S. 2011. Transportasi Dan Pengembangan Wilayah. Yogyakarta : Graha Ilmu
[2] Arliansyah J. 2013. Analisis Feeder Menuju Halte Transmusi II Koridor VI Kota Palembang .
[3] Yudhaprawira R. 2014. Studi Pengadaan Park And Ride Di Terminal Alang-Alang Lebar Kota Palembang.
[4] Utomo U T 2013. Mewujudkan Transportasi Kota Terintegrasi Untuk Atasi Macet. Palembang: antara news
[5] Nielsen G. 2005. Public Transport Planning The Networks.
[6] Octora. 2013. Transportation Case Study Curitiba Brazil
[7] Miro. 2012. In Pengantar Sistem Transportasi (p. 93). Jakarta : Erlangga.

IOP Conf. Series: Earth and Environmental Science **106** (2017) 012073 doi:10.1088/1755-1315/106/1/012073

Proposing water balance method for water availability estimation in Indonesian regional spatial planning

A T Juniati[1]*, D Sutjiningsih[1], H Soeryantono[1], E Kusratmoko[2]

[1]Civil Engineering Department, Faculty of Engineering, Universitas Indonesia, Depok, Indonesia

[2]Geography Department, Faculty of Mathematics & Natural Sciences, Universitas Indonesia, Depok, Indonesia

*Corresponding Author : atietrij@gmail.com

Abstract. The water availability (WA) of a region is one of important consideration in both the formulation of spatial plans and the evaluation of the effectiveness of actual land use in providing sustainable water resources. Information on land-water needs vis-a-vis their availability in a region determines the state of the surplus or deficit to inform effective land use utilization. How to calculate water availability have been described in the Guideline in Determining the Carrying Capacity of the Environment in Regional Spatial Planning. However, the method of determining the supply and demand of water on these guidelines is debatable since the determination of WA in this guideline used a rational method. The rational method is developed the basis for storm drain design practice and it is essentially a peak discharge method peak discharge calculation method. This paper review the literature in methods of water availability estimation which is described descriptively, and present arguments to claim that water balance method is a more fundamental and appropriate tool in water availability estimation. A better water availability estimation method would serve to improve the practice in preparing formulations of Regional Spatial Plan (RSP) as well as evaluating land use capacity in providing sustainable water resources.
Keywords: hydrological model, regional spatial plan, spatial and temporal variability, water availability, water balance

1. Introduction

Nature research field has been lately in focusing on the full integration of water resources aspects into spatial planning. Establishing strong connection between regional spatial plans and water resource management has been found to be able to prevent undesirable consequences of poor water resource management [1], [2], [3], [4], [5]. Spatial planning has a major impact in directing river basin management and is crucial in solving to water related problems [6].

Water related environmental problems; e.g. flood, landslides, and droughts; are often the effects of changes in land use as a consequence of regional spatial planning that lacks consideration to environmental carrying capacity [7]. Integrating water resource management pattern and regional spatial planning is expected to reduce such problems that accommodate competing land use interests. The relationship between water management patterns and spatial plan is dynamic and one should be able to adjust according to the other. The pattern used to manage water resources should be a key consideration in developing or revising spatial plans, likewise, water policies should be founded on existing spatial plans [4].

To address this problem, as well as to fulfill the legal mandate of the Explanandum of Article 25 of Law No 26/2007 on Spatial Planning, there is a need to set a Guideline to Determine Environmental Carrying Capacity in Regional Spatial Planning. The main purpose of this Guideline is an attempt in estimating regional carrying capacity. However, it will also be useful in the review and evaluation of actual spatial use, to be aligned with environmental carrying capacity.

The Guideline to Determine Environmental Carrying Capacity in Regional Spatial Planning (addendum to the Ministrial Regulation No 17/2009) states that environmental carrying capacity is considered to contain two components: environment's capacity to supply resources (supportive capacity) and; its capacity to assimilate refuse and waste (assimilative capacity). Carrying capacity is essentially the region's capacity to fulfill needs and therefore directly related to the availability and needs of land and water in an area/region. Since natural resource capacity depends on ability, availability, and needs upon land and water; the assessment of environmental support capacity in The Guideline is performed using three approaches, i.e. i). Regional land capacity to accommodate space utilization, ii). Ratio of land availability vis-à-vis land needs, iii). Ratio of water availability vis-à-vis water needs.

The integration of water management pattern with regional spatial plans is rely in the third approach. Water support in a regional system is considered in surplus when the water available in the system is sufficient to balance the water needs of the region, and considered in deficit otherwise [7].

Water availability is determined using a runoff coefficient method based on land use and annual rainfall information. Meanwhile, water need is calculated using a conversion result against decent living necessities. The calculation using runoff coefficient method as a modification from rational method uses this following equation:

$$C = \Sigma\ (ci \times Ai) / \Sigma Ai; \ R = \Sigma\ Ri / m; \ \ SA = 10 \times C \times R \times A \tag{1}$$

Where, SA = water availability (m^3/year), C = weight runoff coefficient, Ci = land use runoff coefficient I, Ai = area of land use i (ha), R = average of the regions annual rainfall (mm/year), Ri = yearly rainfall on observation station I, m = the number of rainfall observation station, A = area of the region (ha), 10 = converting factor from mm.ha to mm.m^3

However, the method of determining the supply and demand of water on these guidelines, need to be improved, since the determination of water availability (SA) in this guideline uses a rational method.

The rational method has served as the basis for United States storm drain design practice since the turn of the twentieth century, where water condition is characterized by small drainage areas and a short time of concentration [8, 10]. It is essentially as a peak discharge method [11, 12].

Based on these problems, literature study will be conducted to study the alternative methods used by previous researchers to calculate water availability.

This paper argues the use of rational method as a means of determining water supply and demand is not suitable due to several reasons and that Guideline can be improved through the use of different methodologies.

2. Research Method

To write this paper, a review of literature of national and international journals will be performed, specifically on water availability research, is then described descriptively. As the first step of the research activities, the library study will be conducted from various scientific publications, focused on the aspects of water availability definition to the method of estimating WA.

3. Results and Discussion

Recent concept of water availability distinguishes two kinds of water sources: blue and green water. Blue water is defined as the water parts in the river, lakes, and ground water reserves available and directly consumable by people, either for agriculture, industry, or household consumptions. While green water is defined as water stored in the soil as ground moisture, available to be used by plants and vegetations before getting into the air through evapotranspiration processes [13] , [14], [15], [16].

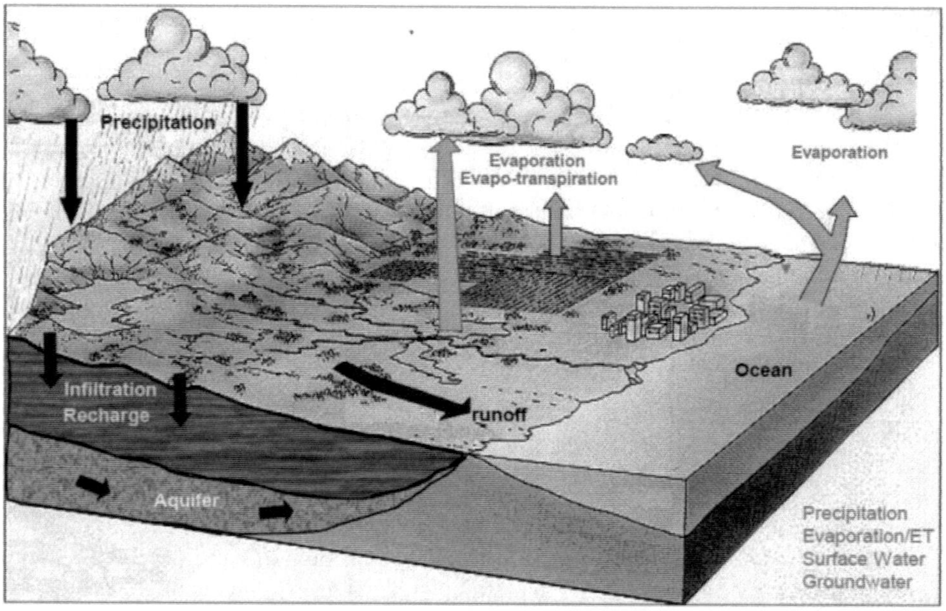

Figure 1. Blue water and green water [13], [16].

Water availability in river basin is affected by several factors, such as climate, soil topography and geology, and variety of vegetation lies above ground. Water availability in river basin is the minimum flow velocity during rain and dry seasons, measured in the river basins output/outlet [17].

Water measured in the outlet point is called surface runoff, whereas it consists of surface water and ground water that will be considered potential water source to be used as water supply for various human needs [18]. Further, water availability is defined by [19] as the total volume of river flow from water catchment area. Accordingly, water balance equation is often used for basic analysis of water availability in a region, e.g. to estimate the amount of water available for drinking or irrigation, Q or runoff is used [18].

To ensure availability of adequate water for human use, water resource managers need to be able to estimate the amounts of water that enter, pass through, and leave individual watersheds. This is a challenge because the relative magnitudes of the transfers of individual components in the hydrologic cycle can vary greatly [20]. In the hydrologic cycle, water availability can be estimated from the amount of water that can be utilized/storage, Figure 2 [21].

In the global hydrologic cycle a given molecule of water is always in one of the storage components of Figure 2. The term "storage" often connotes a static situation, water is always moving through any control volume. Water in the hydrologic cycle is always in motion and always in storage, and any hydrologic control volume represents storage. In many hydrologic reservoirs, such as lakes, segments of rivers, ground-water bodies, and watersheds, the outflow rate increases as the amount of storage increases. This research is focused on water needs for spatial planning purpose, the water availability to be studied is only surface water and groundwater storage.

3.1. Surface Water Availability Estimation

To date, there is a limited but growing number of model approaches dealing with water availability analysis. Most of these models are based on hydrologic water balance approach with time steps ranging between daily, monthly to annually. Among the most well known models is SWAT.

The Oregon Water Resources Department (Department or OWRD) has created and maintains a database of the amount of surface water available for appropriation for most waters in the state. This database is used to evaluate applications for new uses of water [22]. Water availability (WA) is obtained from natural stream flow (QNSF) by subtracting existing storage (ST), out-of-stream consumptive uses (CU) and in-stream demands (IS). The water availability methodology defines three types of expected demands: (1) Storage, (2) Consumptive uses, and (3) In-stream demands. Then water availability can be defined in this equation: WA = QNSF – ST – CU – IS. Where, WA= the water available, QNSF= the 80-percent exceedance natural stream flow at a specified point on the stream, ST= Storage in or from the stream and its tributaries upstream from the specified point, CU= Consumptive uses from the stream and its tributaries upstream from the specified point, IS = In-stream flow demands for a stream reach that includes the specified point.

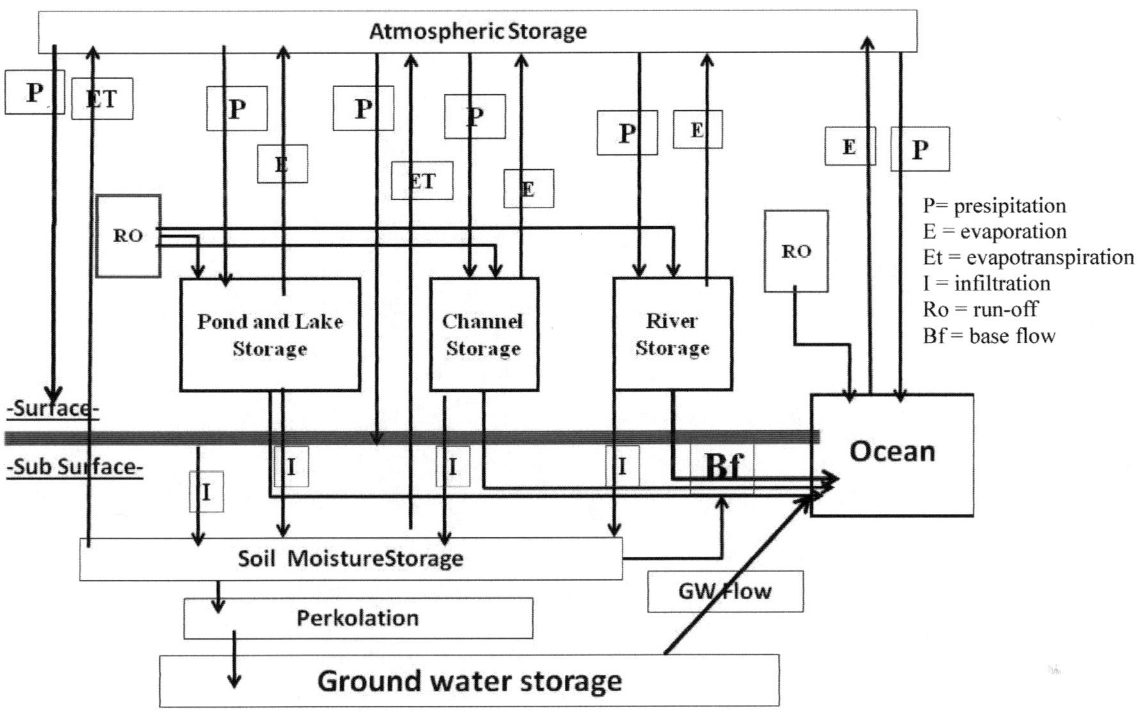

Figure 2. The principal storages (boxes) and pathways (arrows) of water in the global hydrologic cycle.

A case study in Tasmania carried out by [19] used for conducting large-scale assessments of current and future water availability and use. This study describes a robust methodology for determining current surface and groundwater availability and use, as well as future changes due to climate and land use changes. The methodology assesses current water availability through the application of rainfall–runoff and river models, and recharge and groundwater models. These were calibrated to stream flow records and groundwater levels, and parameterised using estimates of current surface and groundwater extractions and use.

3.2. Water Availability Estimation

Researchers in other countries such as [19, 23, 24] also in Indonesia [25, 31] defined water availability as river discharge, while [14] defined water availability as surface and ground water, [15] defines it as blue water. Most researchers calculate the water availability using the principle of water balance (WB) by hydrologic water balance. The WB principles are then used in the model, such as the distributed hydrological model " Soil and Water Assessment Tool " (SWAT). Some researchers were used SWAT to calculate water availability, such [14, 23, 24, 25, 28] while others used hydrological models such as

Mock, Nreca, Rainrun [26, 27, 29, 32]. In addition, some other researchers [30, 33, 34] were used long-term discharge data to predict water availability.

4. Conclusion

Water availability in a river basin, according to hydrology cycle, is storage. Thus, water availability for spatial planning is surface water and ground water. Surface water availability according to several researchers is river discharge and ground water. The most fundamental tool for calculating water availability is the water budget. A common method used by Indonesian researchers to estimate water availability is water balance method such as SWAT, Mock, EnReca, and Rainrun. Further research would be better aimed towards a search for a feasible and suitable water balance method to be used in Indonesia.

References

[1] G. Bouma, and A Slob. 2014. Risk-Informed Management of European River Basins, Handb. Environ. Chem. 29, DOI 10.1007/978-3-642-38598-8_12, ©Springer-Verlag Berlin Heidelb. 29:321–345.
[2] G Jeremy. 2007. Spatial planning, water and the Water Framework Directive : insights from theory and practice. *Geogr. Journal; ProQuest.* 173(4):330–342.
[3] S K, and D Shaw. 2007. Integrated Water Resource Management and Instituional Integration: Realising tht Potential of Spatial Planning in England. *Geogr. Journal.* 173(4):312-329.
[4] P R Sianipar. Mitigasi Resiko Bencana Sumber Daya Air. p. 1–10. 2011.
[5] J Woltjer, and N Al. 2007. Integrating Water Management and Spatial Planning: Strategies Based on the Dutch Experience. *Am. Plan. Assoc. J. Am. Plan. Assoc.* 2:211.
[6] J Alahuhta, V Hokka, H Saarikoski, and S Hellsten. 2010. Practical integration of river basin and land use planning: lessons learned from two Finnish case studies. *Geogr. J.* 176(4):319–333.
[7] Anonim, Lampiran Peraturan Menteri Negara Lingkungan Hidup Nomor :17 Tahun 2009, tentang Pedoman Penentuan Daya Dukung Lingkungan Hidup Dalam Penataan Ruang Wilayah. Indonesia: Indonesia, Kementrian Lingkungan Hidup, p. 1–31, 2009.
[8] V Te Chow, D Maidment R, and L Mays W, Applied Hydrology, McGraw-Hill, Inc., p. 548. 1988.
[9] S A Thompson, Hydrology for Water Management. USA: A.A. Balkema Publishers, Old Post Road, Brookfield, VT 05036-9704, USA, p. 377. 1999.
[10] R H McCuen, Hydrologic Analysis and Design, Second Ed., Prentice Hall, Inc, Upper Saddle River, New Jersey 07458, p. 833, 1998.
[11] ASCE, Hydrology Handbook (Manual No. 28), Second. ASCE 345 East 47th Street. New York, NY 10017-2398, p. 825, 1996.
[12] H M Ragunath, Hyrdology, Principle, Analisys, Design., Second ed., New Delhi: New Age International (P) Ltd., Publishers, p. 477, 2006.
[13] M Falkenmark. 2006. The New Blue and Green Water Paradigm : Breaking New Ground for Water Resources Planning and Management. *J. Water Resour. Plan. Manag. ASCE.* 6:129–132.
[14] J Schuol, K C Abbaspour, R Srinivasan, and H Yang. 2008. Estimation of freshwater availability in the West African sub-continent using the SWAT hydrologic model, *J. Hydrol.* 352(1–2):30–49.
[15] L Menzel, and A Matovelle. 2010. Current state and future development of blue water availability and blue water demand : A view at seven case studies. *J. Hydrol.* 384(3–4):245–263.
[16] C P Kumar, Water Balance Analysis, Natl. Inst. Hydrol. Roorkee – 247667, vol. 247667.
[17] F Irsyad, Analysis of Cidanau River Discharge using SWAT Application (Analisis Debit Sungai Cidanau dengan Aplikasi SWAT). Institut Pertanian Bogor. Bogor, Indonesia, 2011.
[18] C S Ojha, R Berndtsson, and P. Bhunya, Engineering Hydrology, Oxford University Press, p. 440. 2008.

[19] D A Post, F H S Chiew, J Teng, N R Viney, F L N Ling, G Harrington, R S Crosbie, B Graham, S Marvanek, and R McLoughlin. 2012. A robust methodology for conducting large-scale assessments of current and future water availability and use: A case study in Tasmania, Australia, *J. Hydrol.,* 412–413:233–245.

[20] S K Gupta, Modern hydrology and sustainable water development, First. UK and USA: John Wiley & Sons, p. 464. 2011.

[21] S L Dingman, Physical Hidrologi, Third Ed. United States of America: Waveland Press, Inc., p. 657. 2015.

[22] R M Cooper, Determining Surface Water Availability in Oregon, State Oregon Water Resour. Dep. 2002.

[23] T B Le, River Basin Scale Hydrologic Modeling For Prediction of Water Availability, Dissertation. Presented to the Graduate Faculty of The University of Texas at San Antonio, in Partial Fulfillment of the Requirements for, the Degree of Doctor of Philosophy in Environmental Science and Engineering. 2014.

Pathok Negoro mosque as the form of territorial defense region of Mataram Kingdom of Islam Java in Jogjakarta

E Setyowati [1,2*], G Hardiman[2], T W Murtini[2]

[1]Universitas Teknologi Yogyakarta, Yogyakarta, Indonesia
[2]Doctoral Program in Architecture and Urban Design, Universitas Diponegoro, Semarang, Indonesia

*Corresponding Author : endang.setyowati@uty.ac.id

Abstract. Territory is the space boundary of ownership. Defense concepts of an area also include territory. The defense concept of a territory of power or ownership is usually in the form of fortress, ditch, river, road, alley or gate. Many of these are seen in the forms of old city of Dutch's inheritances. Javanese Islamic Kingdom in Yogyakarta has territory in the form of mosque. The mosque is called the Pathok Negoro as consists of 4 mosques. This territorial not same with another Islamic kingdoms. The research problem is why mosque used as a form of defense? What is the uniqueness?The purpose of the research is to reveal the form of territory of Javanese Islamic Kingdom. The benefit of this research is to get the characteristic form of territory of Javanese Islamic Kingdom in Yogyakarta. Their located in Ploso Kuning, Mlangi, babadan, Dongkelan, and Wonokromo. The research method that was used was qualitative method with grounded research, by using history and map search.In the analysis was carried out identification and comparison between mosques.The results show that there are characteristics on the location , the building orientation , the landscape, the area pattern and the shape of the building.
Keywords: Javanese Islam, kingdom, mosque, territory

1. Introduction

Pathok Negoro Mosque is a building used as a form of territory region Yogyakarta State Sultanate. The purpose is as a fortress. This pattern was designed by Sultan Hamengkubuwana I, based by advice of KyaiWelit. KyaiWelit is a religious figure in Ploso Kuning village, Yogyakarta. Because of the political situation in facing Dutch colonialism, then was established state defense system with *Mancapat*pattern (*Kiblatpapat limo pancer*/Java) [4].

The defense system of territory includes the defense system of state, by using many fortress buildings [2], [8] canals and apprentices [16], [17], garrison city [4], alley way and public open space [13], [18]. Yogyakarta city used the mosque building for territorial concept. So the following questions in this research are Why is the concept of territory using mosques? and How is the existence Pathok Negoro mosque as a territorial object?

The research Purposes are Knowing the role of Pathok Negoro mosque role as a territorial component and knowing the characteristics of Pathok Negoro Mosque existence as a territorial object.

The research target is revealing the characteristics of Pathok Negoro Mosque existence as a territory. And the benefits of research is provide a knowledge to the architectural and urban areas about the concept of territory by using the building of worship.

IOP Conf. Series: Earth and Environmental Science **106** (2017) 012074 doi:10.1088/1755-1315/106/1/012074

2. Research Method

This research uses naturalistic paradigm whith qualitatif approach [5], [6], [10], [11]. Strategic research is inductive research and the methode is grounded research, because with this method can produce Grounded Theory as Local Theory or Pattern Theory which is open-ended and can be expanded indefinitely [20]. Focus observation on five Pathok Negoro mosque buildings located in Ploso Kuning, Mlangi, Dongkelan, Babadan, Wonokromo and Mosque of Gedhe Kraton mosque Yogyakarta. This method is used to build the theory of phenomena that exist in the field. Technique of data collecting done on object of research which is certain, from six mosque building. Detailed observation on territory-specific elements is used as an analytical material to achieve research objectives. The typologic generic approach is used to obtain a description of the structural configuration, the basic characteristics and elements of the mosque's identity so as to obtain a certain type of grouping [19].

2.1. Overview of Territory

Territory is the boundary that is controlled or become the right of a person or a certain group as their own area [3]. The tendency to form social space embodied in physical space is ultimately not only through the process of habitat selection, but also through the creation process of culture. This is one of the terms of territory in the abstract space [7], [14], [15].

2.1.1. Mosque in territory concept

Mosque is building of Muslim worship place. The mosque was built based of the needs of places for worship and other religious activities. Mosque always has congregation. So in mosque environment will form the basic power of religion by congregation.

2.1.2. Pathok Negoro Mosque in the territory concept

The placement concept of Pathok Negoro mosque in the urban spatial layout of Javanese Islamic Kingdom in Yogyakarta is as manifestation of the transcendent concept [1].

The concept of territory is cosmologically used by Majapahit kingdom. Majapahit kingdom uses the concentric boundaries of Java Nagaragung [12]. In the spatial concept of the Sultanate of Yogyakarta, the basic concept of Mancapat (which means orientation four five centers/Javanese) is used. The mosques are placed as a place in the four corners of the wind and a mosque in the center of the royal city (Figure 1).

By the existence of the building of the mosque, then in the defense is formed naturally with the basis of religion. This form of defense is made by students at traditional Muslim school, who live in the area around the mosque. The base of this power is physically invisible. But transcendently is formed a large defense force in every corner of the kingdom of Yogyakarta Sultanate.

This strategy is used against Dutch colonialism. The power of transcendence cannot be detected by the Dutch, but Sultan Hamengkubuwana I quietly has great power.

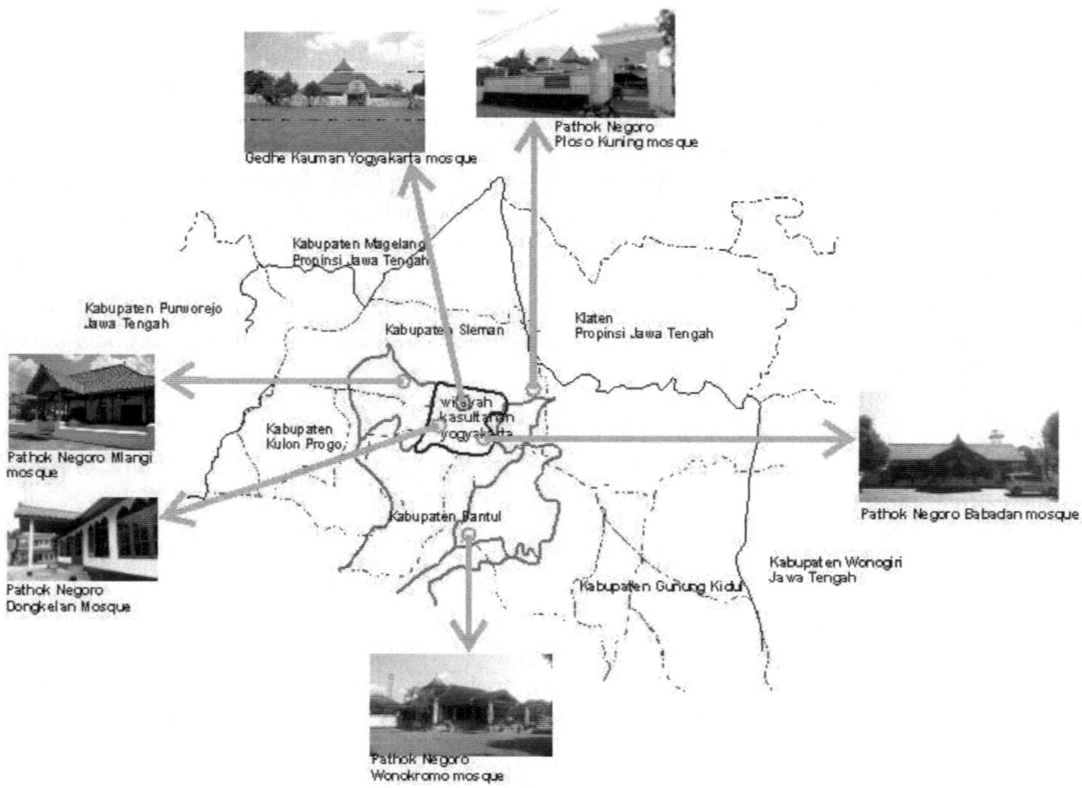

Figure 1. The position of Pathok Negoro mosque in Yogyakarta map.
Source : Google map, redrawing 2017.

3. Results and Discussion

3.1. Typology Of Mosque Location

The location of the mosque has type of location boundary by always using river on one or two sides of its location. Two mosques are inside the inside circle (ring road), there are Dongkelan Mosque and Babadan Mosque. Two mosques position are outside the ring road circle, Mlangi Mosque and Ploso Kuning Mosque (Figure 1).

In relation with its existence as a territorial boundary area, the mosques outside the ring road serve as the outer boundary. These mosques are located in the northern part of the royal kingdom. While the other two mosques are in the ring road are used as inner territory. This happens because in the south is bordered by the southern sea which is the outer boundary of the southern region.

3.2. Typology Of Area Patterns

The pattern of the mosque area is always equipped with the establishment of Islamic school studies for children and young people. There are more than three Islamic school. Most of all is Islamic school studies for children and young people is in Mlangi mosque (17 Islamic school for children and young people). Besides Islamic school for children and young people, in every mosque is always equipped with a sanctuary.

By the existence of Islamic school for children and young people, then transcendently formed the power with base on religious in each mosque area.

The sanctuary becomes the form of respect to the ancestors, especially the founder of the mosque. Sanctuary forms a transcendent force of the ancestral spirits in the souls of the congregation. The history of Kyai or the founder of the mosque gives its own meaning to the congregation.

3.3. Tipology Landscape

Pathok Negoro Mosque has the following landscape types:

a) The courtyard of the mosque

All of Pathok Negoro mosques have a large front yard. The courtyard surrounds the mosque building.

b) Gate

Pathok Negoro Mosque always has more than one gate, i.e. main door, side and back gate. The main gate serves as the entrance and orientation of the building. The side gate serves to connect with the surrounding environment and the back gate is the gateway to the tomb complex that is always present in every mosque. There are several types of gates (Table 1)

Table 1. Type of location gate.

Type	Location	Mosque
Type 1	Main gate on the east side, side gate on the north and south side	Gedhe Kraton mosque
Type 2	Main gate on the east side, side gate on the north side	Mlangi and Babadan mosque
Type 3	Main gate on the east side, side gate on the south side	Ploso Kuning and Dongkelan mosque

Besides of its location, there are types of gate shape (Table 2).

Table 2. Type of Gate shape.

type	Gate shape	mosque
Type 1	the shape of the gate with the architecture of Java	Gedhe Kraton mosque
Type 2	the shape of gates such as temple buildings, intercropping overlap and in the top section there are decorative crown-shaped flora	Ploso Kuning, Mlangi, Babadan mosque

This typology shows the influence of Java, Islam and Pre-Islam on the mosque buildings.

a) *Pool or water element*

The water element in the form of a pool that surrounds the mosque building is always present in all Pathok Negoro mosques. Only the width and the depth are different.

The use of a pool or often called trade, is also a form of territory. The concept of urban development by Dutch is Canal Concept, especially for coastal cities. Channels are used as city boundary as well as water transport. Other outskirts are fortress. The canal is around the citadel. As in Semarang city, Semarang is planned by Karsten (1800s) with fortress and canal completeness. The fortress is named Vijheek. On the fortress is made entrance gates. These gates include the Berok Bridge (named De Zuider Por), which is still visible [2].

b) *Typology Of Building Orientations*

There are two types of building orientation (Table 3).

Table 3. Type of building orientation.

Type	Orientation	Mosque
Type 1	Orientation of buildings to the west - east	Gedhe Kraton, Babadan, wonokromo mosque
Type 2	The orientation of the building towards Qiblah	Ploso Kuning, Mlangi and Dongkelan mosque

c) *Typology Of Building form*
Building type is taken as formas the roof, there are two types (Table 4).

Table 4. Type of building form.

type	orientation	mosque
Type 1	Building with Tajug roof is composed of three	Gedhe Kratonmo mosque
Type 2	Building with Tajug roof is composed of two	Ploso Kuning, Mlangi, Babadan and Dongkelan mosque

4. Conclusion
The results show that there are characteristics on the location of the mosque, the orientation of the building, the landscape, the pattern of the area and the building form.

Acknowledgment
Acknowledgments are conveyed to the students of Architecture Program of Yogyakarta University of Technology who have helped the field survey, data documentation and depiction, and also to Mr. Like Suryadi, *abdi dalem* Yogyakarta Palace, as a source of information, Mr. Agus, *takmir* Ploso Kuning mosque, Mr. Bardi, Mr. Topo, who have provided enough information about Pathok Negoro mosque.

Reference
[1] A Mulyadi, Makrifat Jawa,Galang Press, Yogyakarta, 2017.
[2] Adji, Murtomo B. 2008. Concept of City Development by the Netherlands with Channel System. *Jurnal Ilmiah Perancangan Kota & Permukiman*. 7(2):69-79. ISSN.1412-7768.
[3] Barliana Rr. Putri, Pangarsa, Galih Widjil, dan Ernawati, Jenny. 2012. DIMENSI. *Journal of Architecture and Built Environment*. 39(2): 65- 76 December 2012. ISSN 0126 - 219X.
[4] Handinoto. 2004. Political and Economic Policy of the Dutch Colonial Government Influential on City Morphology in Java. *Jurnal Dimensi.* Juli 2004:19-27.
[5] Hasbinsyah. 2008. Phenomenology Approach: Introduction to Research Practice in Social Sciences and Communicatics, *Jurnal MediaTor*. 9(1). e-journal.unisba.ac.id.
[6] John D. Cresswell, Qualitative Inquiry and Research Design Choosing Among Five Approach, Sage Publication, California, USA, 2007.
[7] Kurniadi Fery, Pramitasari, Diananta, Wijono, Djoko, Konsep Perilaku Teritorialitas di Kawasan Pasar Sudirman Pontianak, URI: http://repository.polnep.ac.id/xmlui/handle/123456789/99, Date: 2013-02-18), 2013.
[8] Purwoto L M F. 2005. Old Colonial City of Semarang, Overview of City Development History. *Jurnal Dimensi*. 33(1).
[9] Lestariningsih, Dwi Jati. 2010.Tinjauan Kota Abad Pertengahan, *Jurnal Teodolita*, 12(1): 12-31. Juni 2010.
[10] Lexy J. Moleong, Metode Penelitian Kualitatif, Remaja Rosda Karya, Bandung, 2000.
[11] Nindito, Stefanus. 2005. Phenomenology Alfred Shutz, Study on the Construction of Meaning and Reality in Social Sciences. *Jurnal Ilmu Komunikasi.* 2(1). eISSN. 2548-8643. DOI. http://dx.doi.org/10.24002/jik.v2i1.245.
[12] Nugroho W, Uji M A. Coastal Importance in Realizing Indonesia as A maritime Axis of the World, Lawatan sejarah Regional BPNB DIY, 2016. Tema : Menelusuri jejak maritim di Pantai Utara Jawa, BPNB, 2016, editor : Darto Handoko, 2016.
[13] Ramelan, Rubiyanto, Sri Handayani. 2008. Gang Kampung Kota As Multi-Function Public Room. *Jurnal TERAS*, 8:1. Juli 2008.
[14] Retna H, Parmono A. 1999. Pesantren Pabelan Study on Delinization of Pesantren Territory. *Teknosains*, 12 (2). Mei 1999.
[15] Retno Hardiyanti. 2002. Sedulur Concept as a Barrier Factor for the Establishment of Exclusive Space in Samin Settlements. *Journal of Architecture and Built Environment.* 30(2).

[16] Sarinastiti Ajeng, Siti Rukayah, Titin Woro M. 2015, *ejurnal UNDIP*, http://ejournal.undip.ac.id. 36(2).
[17] Setyowati E. Regional Space Characteristics In Beteng Baluwarti Kraton Yogyakarta. *Teknosain.* UGM. 1:September 2010.
[18] Supritanto A dalam Johan Silas, 1998. Mataram Dalam Angka, Kerjasama Badan Pusat Statistik Kodya Mataram dengan Bapeda Kodya Dati II Mataram, Wacana H L, Lalu Mulyadi. 2000. Rumah Produktif, *Jurnal Dimensi Tradisional.* ITS Surabaya.
[19] Depari C D A, The Influence of Islam Against Yogyakarta City Plan. In Konservasi Arsitektur Kota Yogyakarta, pp. 23–39, Yogyakarta: Laboratory of Environmental & Regional Planning & Design, Prodi Arsitektur Fakultas Teknik Universitas Atma Jaya Yogyakarta, 2013.
[20] Noeng Muhajir, Metodologi Penelitian Kualitatif, Edisi IV, Jogjakarta, Penerbit Rake Sarasin, 2000.

Activities of four bus terminals of Semarang City gateway and the related GHG emission

H S Huboyo*, I W Wardhana, E Sutrisno, L S Wangi, R A Lina

Environmental Engineering Department, Universitas Diponegoro, Semarang, Indonesia

*Corresponding Author : huboyo@undip.ac.id

Abstract. The activities of the bus terminal, including loading-unloading passengers, bus idling, and bus movements at the terminal, will emit GHG's emission. This research analyzes GHG emission from four terminals, i.e., Mangkang, Terboyo, Penggaron, and Sukun in Semarang City. The emission was estimated by observing detail activities of public transport means, especially for moving and idling time. The emission was calculated by Tier 2 method based on the vehicle type as well as fuel consumption. The highest CO_2e during vehicle movements at Sukun area was contributed by large bus about 2.08 tons/year, while at Terboyo terminal was contributed by medium bus about 347.97 tons/year. At Mangkang terminals, the highest emission for vehicle movements was attributed by medium bus as well of about 53.18 tons/year. At last, Penggaron terminal's highest GHG emission was attributed by BRT about 26.47 tons/year. During idling time, the highest contributor to CO_2e was the large bus at the three terminals, i.e., Sukun of 43.53 tons/year, Terboyo of 196.56 tons/year, and Mangkang of 84.26 tons/year, while at Penggaron, BRT dominated with CO_2e of 26.47 tons/year. The management of public transport in terminals is crucial to mitigate the emission related to bus terminals activities.
Keywords: air pollution, bus, CO_2e, Semarang, terminals, vehicle

1. Introduction

The demand for travel in cities worldwide is increasing. This travel is closely related to economic and social needs of the people. In Indonesia, the growth rate of motorization is about 7–12% per year which far beyond the rate of road development. Other developing nations share the same condition where road capacities fail to compensate the traffic growth [1]. Transport activities consist of road and non-road activities. In the case of public transport, road activities mean serving passengers to their specific destination, while non-road transports are usually related to the activities of vehicles at terminals where they will substantially emit air pollutants due to small movements, idling, and engine starting (hot-cold start). In the case of running mode, the fuel consumption commonly will increase as the vehicle engine speed escalates [2]. During idle time, the fuel consumption is higher than in running operation [3]. The engine cold start mode means initiating the vehicle after 6–12 hours of being shut down [4]. In general, cold start mode will emit much higher air pollutants compared with the hot start mode [5]. In fact, there are many other factors which significantly affect the vehicle's emission while it is in running mode, e.g., vehicle size, engine displacement, and driving conditions [6].

The emission of air pollutants will affect the workers as well as the passengers in bus terminals. The cyclist with low movements, long idling time, and cold start engine will simultaneously add up to the air pollutants near the vehicles. Cheng [7] studied that the ultrafine particles (UFP) concentration inside the bus terminal is more than ten times of ambient urban background UFP. Thus, knowing the potential emission of air pollutants in the bus terminal is a fundamental part to mitigate the air pollution comprehensively. This study aims to identify the potential emission of GHG due to bus terminal activities, particularly from vehicle movements and idling time.

2. Research Method

2.1. Location of study
The study took place at four bus terminals in Semarang City as listed in Table 1.

Table 1. Bus terminals location.

Bus Terminals	*Lat Long Coordinates*	*Corridor Connection*	Remarks
Mangkang	6°58'06.0"S 110°17'22.8"E	West	Gateway fromWest
Penggaron	7°01'03.2"S 110°29'36.8"E	East	Gateway from East (Purwodadi)
Sukun	7°03'43.9"S 110°24'48.3"E	South	Gateway from South not intended as an official bus terminal
Terboyo	6°57'04.8"S 110°27'45.5"E	East	Gateway from East (Demak)

2.2. Data collection
The primary data were collected by observation as well as inspection of vehicles while they were running or idling. This idle time means the engine was on, but the vehicle did not move, waiting for passengers to get on it. The vehicles' time and route length to travel within the terminal were also recorded. We define the vehicle route as the course where the vehicle was beginning to enter the bus terminal, passing through the retribution fee, collecting passengers, until leaving the terminal. The data of vehicles' engine capacity were gathered as well to estimate the idling emission. The emission calculation related to idle time [8]:

$$\text{Idling fuel use (L/year)} = (\text{idling fuel flow}) \times (\text{idling time per day}) \times (\text{total vehicle in a year}) \quad (1)$$

$$\text{Idling emission} = (\text{idling fuel use}) \times (\text{GHG emission factor}) \quad (2)$$

The number of vehicles entering the terminal for the whole year was gathered from Local Transportation Agency to be compared with those we counted manually on the spot. In this case, the vehicles were classified by the year of manufacturing, i.e., ≤2010, 2011, 2012, 2013, and 2014.

2.3. Emission calculation
The GHG emission was estimated using an equation derived from IPCC [9] as the basis of calculation. The formula is as follows:

$$\text{Emission} = \text{Activity data x Emission Factor} \quad (3)$$

$$\text{Emission} = \Sigma_{a,b,c}[\text{Fuel}_{a,b,c} \text{ x } \text{EF}_{a,b,c}] \quad (4)$$

$\text{Fuel}_{a,b}$: fuel consumptions of fuel a and b EF : emission factor (g/L)

The relation between fuel consumption and the vehicle speed was derived as seen in Table 2 using the method researched by JICA through SITRAMP [10]. This approach is useful when the vehicle speed varies. However, it should be noted that these formulas have a limited application, particularly for a very slow speed of the vehicle.

IOP Conf. Series: Earth and Environmental Science **106** (2017) 012075 doi:10.1088/1755-1315/106/1/012075

Table 2. The relation between vehicle speed and fuel consumption.

Vehicle Types	Formulas
PC (private car)	y = 7E-05x2 - 0.0077x + 0.2579
MC (motorcycle)	y = 1E-05x2 - 0.0009x + 0.0601
SB (small bus)	y = 3E-05x2 - 0.0029x + 0.1285
MB (medium bus)	y = 5E-05x2 - 0.0056x + 0.2961
Patas-AC, LB (large bus)	y = 3E-05x2 - 0.0029x + 0.1533
S/MT (small/medium truck)	y = 5E-05x2 - 0.0053x + 0.2771
LT (large truck)	y = 5E-05x2 - 0.0060x + 0.3147
x: vehicle speed variable (km/h)	y: fuel consumption (L/km)

The emission factors acquired from IPCC and other studies were then combined. The emission factors used in this study are listed in Table 3.

Table 3. Emission factors used in this study.

Vehicle Types		Emission Factors (g/L)					
		Gasoline			Diesel		
		CO_2[a]	CH_4[a]	N_2O[b]	CO_2[a]	CH_4[a]	N_2O[b]
Light vehicle	Paratransit	2780.5	0.3243	0.041			
	Mini bus				4,586.2	0.1157	0.022
Heavy vehicle	Bus				1,593.7	0.0804	0.051

Notes : a[9], b[11]

The observation schedule was arranged sequentially and simultaneously to capture the habitual situation in the field. The observations were conducted twice on weekdays and the weekend for each terminal. Sampling time was set at daytime where the vehicles usually have daily activities (Table 4).

Table 4. Observation schedule at the terminals.

Terminals	Categories	Dates	Sampling Time
Mangkang	Weekdays	16 May 2016	05.30 – 19.00
		27 May 2016	05.30 – 19.00
	Weekend	21 May 2016	05.30 – 19.00
		29 May 2016	05.30 – 19.00
Penggaron	Weekdays	20 May 2016	05.30 – 18.00
		23 May 2016	05.30 – 18.00
	Weekend	22 May 2016	05.30 – 18.00
		28 May 2016	05.30 – 18.00
Sukun	Weekdays	23 May 2016	06.00 – 17.30
		27 May 2016	06.00 – 17.30
	Weekend	14 May 2016	06.00 – 17.30
		22 May 2016	06.00 – 17.30
Terboyo	Weekdays	16 May 2016	06.00 – 17.30
		3 June 2016	06.00 – 17.30
	Weekend	15 May 2016	06.00 – 17.30
		21 May 2016	06.00 – 17.30

3. Results and Discussion

3.1. The condition of the terminals their activities

Mangkang, as an A class terminal, serves transfer passengers from outer cities (Westward) to Semarang. This terminal was initially operated in 2002, and it is more than 21.000 ha wide. Based on the observation, the vehicle speeds within this terminal were around 7–30 km/h. For large bus, the primary idling time was 15 min, while for the small bus was around 5 min. On the other hand, BRT bus had idling time around 20 min, while small bus had 5 min and paratransit had 4 min of idling time.

Penggaron terminal, as a B class terminal, connects passengers from East cities such as Purwodadi, Blora, and Cepu, by smaller vehicles compared to Mangkang. Penggaron terminal, which has been operating since 1997, has an area of 5.7 ha. Due to the lower occupancy of vehicles, the vehicle speeds were a bit higher than those for Mangkang, reaching 12–30 km/h. In this terminal, the BRT, large bus, small bus, and paratransit had the idling time of 17 min, 7 min, 3 min, 3 min, and 3 min, respectively.

Sukun terminal acts as pseudo terminal since it was not intended as an official vehicle terminal. However, due to its strategic location for passenger transfers, many drivers, passengers, and vehicle enterprises make use of this location as a terminal. Beyond idling time, the vehicle speed passing through this terminal was around 10–45 km/h. The idling time for many vehicles was around 2–10 min due to the lacking capacity of the parking area. The large bus had an idling time of 10 min, the medium bus had 4 min, the small bus had 3 min, and paratransit had the smallest duration of 40–50 seconds.

Terboyo terminal, as an A class terminal, serves passengers from or to Semarang from the East such as Demak Regency, Kudus Regency, and Jepara Regency. The vehicle speeds within Terboyo terminal were recorded a bit higher of 20–50 km/h. This terminal has a problem of the diurnal sea water rise. Thus, every time the sea floods occurred, many passengers were reluctant to await the vehicles inside the terminal. The idling time of the large bus, medium bus, and small bus was around 20 min, 15 min, and 5 min, respectively.

From the data provided by Local Transportation Authority, the numbers of vehicles entering these terminals were recorded as follows (Table 5):

Table 5. Numbers of vehicles entering the terminals in 2015.

Terminals	Large Bus	Medium Bus	Small Bus	Paratransit	BRT
Mangkang	63,264	31,671	35,539	64,962	17,160
Penggaron	6,648	5,989	3,021	19,087	17,160
Sukun (*pseudo* terminal)	42,684	15,276	14,556	10,152	9,732
Terboyo	101,242	58,507	33,864	-	-

3.2. Fuel consumption and GHG emission

After inspecting the travel time for each vehicle inside the terminal, the speeds could be predicted, then the fuel consumption for each vehicle could be estimated. The classification refers to the grouping used by Local Transportation Agency (see Table 3). The estimates of fuel consumption are shown in Table 6.

Table 6. Estimated average fuel consumption (L/day) based on vehicle types.

Vehicle Types	Mangkang	Penggaron	Sukun	Terboyo
Large Bus	0.157	0.234	0.030	0.657
Medium Bus	0.417	0.155	0.011	1.423
Small Bus	0.117	0.066	0.004	0.414
BRT	0.100	0.111	0.006	-
Paratransit	0.161	0.159	-	-
Shuttle travel	-	-	0.016	-

It is clear that when the terminal is quite busy, then the fuel consumption also rises. Terboyo terminal is the busiest terminal connecting major cities at the East. The highest emission of moving vehicles was recorded at Terboyo terminal, followed by Mangkang and Penggaron. Interestingly, during idling time, the emission of vehicles at Sukun terminal was higher than that of Penggaron. Based on Figure 1, it is concluded that the moving mode of vehicles does not always yield higher GHG emission compared to idling mode. In general, the large bus has higher emission related to idling mode rather than moving mode. It is due to the long distance routes that these buses had to take so that they took longer idling time awaiting passengers. The emission ratio of moving to idling is even small in Sukun terminal since no regulation prohibits the duration of vehicles to stay at the terminal.

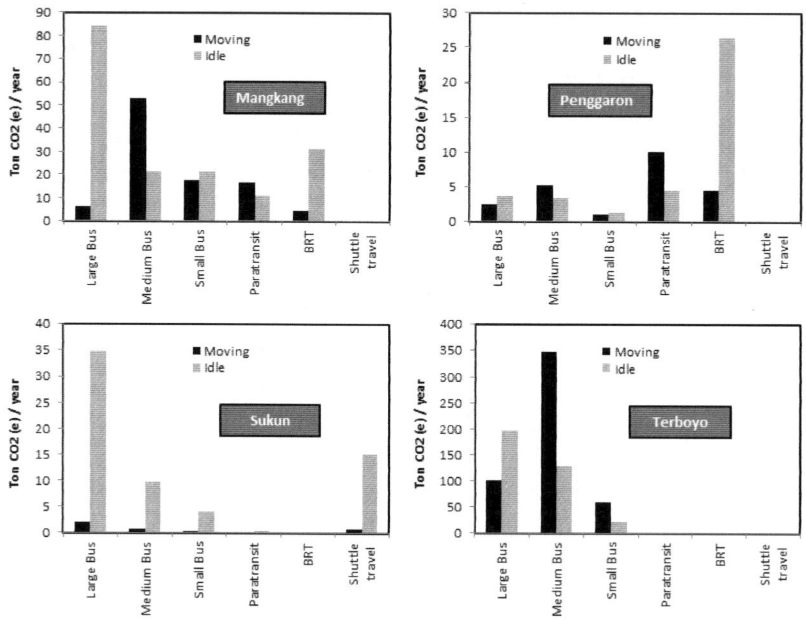

Figure 1. Comparison of moving emission and idling emission.

Based on the field observation and the account that vehicle emissions are closely related to the distance travelled, vehicle speed, the number of vehicles entering the terminal, engine displacement, idling time, and driver behavior, several recommendations are made as follows:

1) Reduce the idling time and shut off the engine if the vehicle is going to stop for a long time.
2) Conduct regular check of engine inspection and rejuvenate the vehicle.
3) Improve the road infrastructure to optimize the vehicle speed within the terminal.
4) Improve the parking management in terminal areas, particularly to minimize on-street parking which may disturb the flow of vehicles entering the terminal.
5) Apply smart driving to the drivers.

4. Conclusion

Activities in the bus terminal of major cities are very complex which may pose a high threatofGHG emission. Therefore, the emission inventory for transportation in the city should consider these activities for depictingthe complete inventory. Valuable information of terminal emission was obtained by calculating the emission regarding the moving mode and idling mode of the vehicles entering the terminals (Mangkang, Penggaron, Sukun, and Terboyo) using Tier 2 method. The activities of each vehicle were summarized based on the vehicle type. The highest CO_2 during vehicle movement at Sukunarea is contributed by large bus about 2.08 tons/year, while at Terboyo terminals is contributed by medium bus about 347,97 tons/year. At Mangkang terminals, the highest emission for vehicle movement is contributed by medium bus about 53.18 tons/year. Lastly, at Penggaron terminal, the highest GHG emission is attributed to BRT about 26.47 tons/year. During idling time, the highest contributor to CO_2e emission is the large bus at the three terminals, i.e., Sukun of 43.53 tons/year, Terboyo of 196.56 tons/year, and Mangkang of 84.26 tons/year, while at Penggaron, BRT is dominating with CO_2e of 26.47 tons/year. The management of public transport in terminals is crucial to mitigate the emission related to bus terminals activities.

Acknowledgment
This study was supported by Cluster Research Group on Air Pollution Management at Environmental Engineering Department, Diponegoro University. Semarang City Transportation Agency is acknowledged for providing permission to take primary data at the terminals.

References

[1] Cervero R, Integration of Urban Transport and Urban Planning, Chapter in Book: The Challenge of Urban Government: Policies and Practices, Publisher: The World Bank Institute, Editors: M.~Freire and R.~Stren, 2001, p. 407-427.
[2] Greenwood B R, User and Environmental Effects In HDM-4. Birmingham: ISOHDM, 2001, p. 108-111.
[3] Zahra E, and Driejana. Perbandingan Estimasi Beban Emisi CO dan CO2 Dengan Pendeketatan Konsumsi Bahan Bakar dan Kecepatan Kendaraan, Studi Kasus: Bunderan Cibiru Lembang (in Indonesian). Undergraduate thesis. Bandung Institute of Technology. Bandung, Indoneisa. 2009.
[4] U.S. Environmental Protection Agency (EPA). 1993. Extreme Low-Temperature Cold Starts. Washington DC, U.S.A
[5] Yung- Chen You. 2009. Comparison of Exhaust Emissions Resulting from Cold and Hot-Start. *Journal of the Air & Waste Management Association.*
[6] Wu X, Zhang S, Wu Y, Li Z, Zhou Y, Fu L, Hao J. 2015. Real-world emissions and fuel consumption of diesel buses and trucks in Macao: From on-road measurement to policy implications. *Atmos. Environt.* 120:393-403.
[7] Cheng Y H, Chang H P, Hsieh C J. 2011. Short-term exposure to PM10, PM2.5, ultrafine particles and CO2 for passengers at an intercity bus terminal. *Atmos. Environt.* 45:2034 – 2042.
[8] Taylor, G W R. 2003. Review of the incidence, Energy Use and Costs of vehicle idling. Office of Energy Efficiency, Canada.
[9] IPCC. 2006. Guidelines for National Greenhouse Gas Inventories. Volume 2 : Energy
[10] JICA. Study of Integrated Transportation Master Plan for Jabodetabek (SITRAMPhase II). 2014
[11] Environmental and Climate Change Canada. 2014

IOP Conf. Series: Earth and Environmental Science **106** (2017) 012076 doi:10.1088/1755-1315/106/1/012076

Model of sustainability of vernacular kampongs within *Ngadha* culture, Flores

M B Susetyarto

Architecture Department, Faculty of Civil Engineering and Planning, Universitas Trisakti, Jakarta, Indonesia

Corresponding Author: bambang.s@trisakti.ac.id

Abstract. In the indigenous people of *Ngadha*, Flores (8°52'40.45"South, 120°59'8.18"East), the phenomenon of sustainability could be seen in its very interesting architectural traces in the setting of local factors. The sustainability phenomenon had a high value in their life and it was clearly indicated in daily activities as well as farmers, weavers, or carpenters. The phenomenon was unique and has been successfully created as a model. The research has been done by qualitative method in inductive paradigm. The data collection and comprehensive analysis have done in the field by occasional discussions with some sources of *Ngadha* traditional experts, vernacular architecture researchers, sociologists, anthropologists, and others. The result was a model of sustainability of vernacular kampongs within *Ngadha* culture, namely Tuku nunga lo'a ghera adha *Ngadha*. The concept of sustainability was a cultural event that synergizes the five factors supporting continuously sustainability until the optimum momentum of sustainability occurred in those synergistic conditions. The five factors were natural environment (one nua), indigenous community (mesu mora), vernacular architecture (sa'o bhaga ngadhu ture), economy (ngo ngani), and *Ngadha* culture (adha *Ngadha*). The significance and impact of the research were to provide input for the completeness of sustainability knowledge, especially the vernacular kampongs sustainability model.
Keywords: indigenous, *Ngadha*, vernacular architecture

1. Introduction

The sustainability of vernacular kampongs founded by indigenous peoples of *Ngadha* was the basic concept to maintain the aesthetic of vernacular architecture in *Ngadha* cultural environment. Vernacular architecture was defined as the architecture which has been created by people or community with utilizing traditional technologies [1]. The sustainability of *Ngadha*'s vernacular architecture by the *Ngadha* indigenous people was understood as the totality of cultural events. Cultural events were starting from the clan or tribal community in each kampong. They discussed their idea of doing traditional activities to build vernacular houses (*utu bhou dia sa'o*), and/or vernacular kampongs. The process of development began when people cut the wood (*oja/fai*), took the stones (*watu/tere*), or got building materials derived from their fields (*uma*), or from the mountains (wolo). Then, brought the building materials came into the kampong, carved it, constructed it for the building structure, and gave the name of vernacular houses or vernacular buildings, until to the beginning event for using the houses and/or the buildings. All series of cultural events were acknowledged to be very beneficial for the sustainability of the *Ngadha* culture and traditional life of the community. The sustainability of *Ngadha* culture was not only performed on every customary ritual occasion, but also it was actualized in everyday life, both when they were working in the fields and when they were weaving cloths in vernacular houses. The sustainability was manifested creatively in traditional artwork. For example, in the contents of the poetry of the teachings to build a vernacular house, called *'soka'*, in the form of andce *'ja'i,'* in the music of gong gendang *'laba go'*, on the *'jara'* woven motif, on the 'weti' of the stair construction, and on the front wall of the building 'one.' The event involved the indigenous community since the older generations (*ebu nusi*) up to the young generations (ana woe), both female (*ana fa'i*) and male (*ata saki*), while maintaining the cohesion among tribes or clans in each kampong (woe). Therefore, the sustainability of vernacular kampongs and the *Ngadha* culture were understood by indigenous people of *Ngadha* as a cultural event in the construction works, and/or of daily work activities as well as cultural activities, which were traditionally conducted by indigenous communities. They used local material resources in order to meet the needs of *Ngadha* indigenous communities in the present and the future.

2. Research Method

The research was done by qualitative architectural research methods [2] with the inductive paradigm [3]. The method was formulated by seven stages, i.e., First, researcher emphasized on understanding the *Ngadha* culture and the natural environment of the *Ngadha* region as natural settings. Second, researcher focused on vernacular architectural settings in each kampongs of the *Ngadha* region. In this session, it was included on the architectural design, the graphical design in carving and/or woven motif products. Third, researcher did deep-interviews, discussed the meaning of architecture design, and others which connected to the vernacular kampongs. Fourth, researcher did interpretation and made sense of the provisional findings. Fifth, researcher did comprehensive analysis in the fields by occasional discussions with some sources of *Ngadha* traditional experts, vernacular architecture researchers, sociologists, anthropologists, and others. Sixth, research formulated the findings, a model of the sustainability of vernacular kampongs within *Ngadha* culture. Seventh, researcher verified and predicted the findings, and took into account to be a (new) theory of sustainability in the global knowledge. The important consideration was how using theory and making theory [4].

3. Results and Discussion

The vernacular architecture of the kampongs in the *Ngadha* cultural area was originally a closed world of cultural blindness. The vernacular architecture aimed to meet their own needs for security, mutual protection and complementarities. All of those were implied in the traditional teachings of indigenous peoples in the cultural region of *Ngadha*, which had high authenticity. One of the traditional teachings was 'soka,' which was a chant poem to guide people in the process of building vernacular houses and vernacular buildings, to manifest aesthetic and meaningful architectural artifacts, to make architecture form by traditional dimension, to create a simple space [5], to express hierarchy, symbols and ornaments, according to the traditional values of *Ngadha* culture. However, after vernacular kampongs opened their worlds of life for tourist visits and researchers from outside of *Ngadha* region, the tradition to build vernacular kampongs in the cultural region of *Ngadha* experienced the transmission of knowledge and the cultural transition. This condition brought an impact on the physical embodiment of vernacular architecture and on the contemporary values in the traditional life of the vernacular kampongs in the cultural region of *Ngadha*.

The basic concept of sustainability of vernacular kampongs within *Ngadha* culture actually was found to be pluralistic, opened to correction, not reductionist, and did not pretend to reach an identity or label for a certain period. The basic concept was placed as a model of *Ngadha* cultural production system that was closely linked to the formation of national identity and culture, the paradigm of creative thinking, the ontological and epistemological of the vernacular architecture concepts, and the spirit of preservation and prediction of cultural sustainability in the future, in accordance with the situation and socio-political conditions of society in its period. Therefore, verification and prediction of the findings of the basic concepts of sustainability of vernacular kampongs throughout the *Ngadha* culture region were conducted. The concept of sustainability was a cultural event that synergized with the five factors supporting continuously sustainability until the optimum momentum of sustainability occurred in those synergistic conditions. The five factors mentioned were firstly, natural environment (*one nua*) was in the surrounding vernacular kampongs. It could create a good micro climate for the healthy life. Second, the indigenous community (*mesu mora*), which was the inhabitants of vernacular kampong. Third, vernacular architecture (sa'o bhaga ngadhu ture), which were the living facilities. Fourth, economy (*ngo ngani*), which was an important factor to support their productivity and survival. Fifth, the *Ngadha* culture (*adha Ngadha*), which became the fundament of their spirit and their human behavior. As a result, the basic concepts of sustainability of vernacular kampongs within *Ngadha* culture can be applied to eight vernacular kampongs in the *Ngadha* region. Eight vernacular kampongs had the five factors of sustainability with might be different influences. In these context, it could be imagined how the configuration of five circles with the different sizes and different colors. Therefore, the size of sustainability for each vernacular kampong might be different and might be a lot of variant. However, the entire vernacular kampongs in the *Ngadha* region surely had the five factors which might be created as a certain configuration. The configuration might be occurred an intersection of the five circles, and that was the sustainability factor. Eight vernacular kampongs mentioned were the kampong of Tololela, the kampong of Bela, the kampong of Wogo, the kampong of Langa Gedha, the kampong of Bolozi, the kampong of Gurusina, the kampong of Nage, and the kampong of Bena. Eight vernacular kampongs located in the surrounding valley of the Inerie Mountain. The findings hopefully were an embryo of a (new) theory of architecture sustainability, which had the context of *Ngadha* cultural regionalism [6], [7], especially in terms of the basic concept of sustainability, which could be illustrated as shown below.

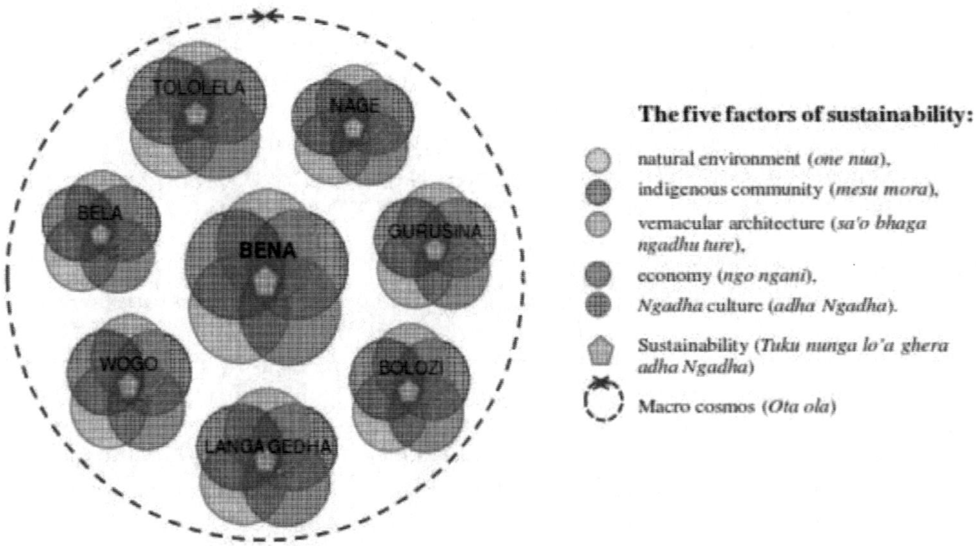

Figure 1. The model of sustainability of vernacular kampongs.

4. Conclusion

The findings of the research were a model of sustainability of vernacular kampongs within *Ngadha* culture, namely Tuku Nunga Lo'a Ghera Adha *Ngadha*. Actually the model was there in each of the vernacular kampongs, but had different sizes and variants. The sustainability concepts was a cultural event that synergized the five factors supporting continuously sustainability until the optimum momentum of sustainability occurred in those synergistic conditions. The five factors were natural environment (*one nua*), indigenous community (*mesu mora*), vernacular architecture (*sa'o bhaga ngadhu ture*), economy (*ngo ngani*), and *Ngadha* culture (*adha Ngadha*). The significance and impact of the research to the architecture science were to provide input for the completeness of sustainability knowledge, especially the sustainability model of vernacular kampongs. The (new) sustainability knowledge should be donated to the global academic society.

Acknowledgments

Author expressed the gratitude, especially to my beloved wife and my mother, and to my colleagues at the Department of Architecture, surveyors and librarians, and to the financial support of Beasiswa Program Pasca Sarjana (BPPS) organization and Trisakti University, and to other friends may not be mentioned his/her name.

References

[1] Oliver, Paul, Built to Meet Needs, Cultural Issues in Vernacular Architecture, Elsevier Ltd, Oxford, 2006, p. 3-105 and 267-286..
[2] Groat, Linda and Wang, David, Architectural Research methods, John Wiley & Sons. Inc, New York, 2002.
[3] Creswell, John W., Researh Design, Pendekatan Kualitatif, Kuantitatif, dan Mixed, (translated by Fawaid, Achmad), Pustaka Pelajar, Yogyakarta, 2010.
[4] King, Peter, Using Theory or making Theory: Can there be Theories of Housing?, Routledge, Leicester, p. 41-52. 2009
[5] Edensor, Tim; Leslie, Deborah; Millington, Steve; and Rantisi, Norma m. (Eds.), Space of Vernacular Creativity, Rethinking the cultural and economy, Routledge, London and New York, p. 1-16 and 183-199, 2010.
[6] Heath, Kingston W. M., Vernacular Architecture and Regional Design: Cultural Process and Environmental Response, Elsevier Ltd, Oxford, p. 3-21, 2009.
[7] Hu, Richard Weixing, Building Asia Pasific Regional Architecture: The Challenge of Hybrid Regionalism, The Brookings Institution, Washington D.C., 2009.
[8] Arndt, Paul, Masyarakat Ngadha, Keluarga, Tatanan Sosial, Pekerjaan, dan Hukum Adat" (translated by Nama, Paul Sabon, editor: Lege, Lukas and Embu, Eman J.), Nusa Indah, Ende, Puslit Candraditya, Maumere, 2009.
[9] Fischer, Michael M.J., 2007. Culture and Cultural Analysis as Experimental Systems, Cultural Antropology, Vol.22, Issue 1, p. 1-65, h p://www.anthrosource.net/.
[10] Nik, Ali Sarrafi. 2011. Sustainability in Vernacular Architecture-Outlooks on Iranian Tribes, *Proceeding of 5 th Symposium on Advances in Science and Technology (SASTech)*, Mashhad, Iran.
[11] Nugraheni, Rini and Suprayitno, Iwan. 2009. Kampung Wogo dan Potensi Pengembangannya, Prosiding *Seminar Regional Jelajah Arsitektur Negeri Seri-6 Nusa Tenggara Timur*, Bajawa: Puslitbang Permukiman-Baand Litbang-Departemen Pekerjaan Umum,.
[12] Pangarsa, Galih W. 2008. Pelestarian Arsitektur Rakyat, Paradigma Partisipasi Sosial, *Prosiding Seminar Regional Jelajah Arsitektur Negeri Seri-6 Nusa Tenggara Timur*, Bajawa: Rapoport,
[13] Amos. 2009. Some Further Thoughts on Culture and Environment, Archnet-IJAR, Vol.2, issue 1, pp. 16-39, h p://www.Archnet-IJAR/ [March 2008].

IOP Conf. Series: Earth and Environmental Science **106** (2017) 012076 doi:10.1088/1755-1315/106/1/012076

[14] Susetyarto M.B., Budihardjo E., Pangarsa G.W., and Hardiman G. 2011. Living in Harmony with the Natural Environment in the Boat Culture; Critical Analysis of Vernacular Architecture in Flores, *Proceedings of The 12th International Conference on Sustainable Environment and Architecture (SENVAR)*: Nusantara"(Local) Wisdom for The Better of Sustainable Architecture, p. A3-23 - A3-29, Malang, Indonesia.
[15] Susetyarto M.B., Budihardjo E., Pangarsa G.W., and Hardiman G., 2011. Sustainability of Vernacular Kampongs of Flores, A Valuable Lesson From The Ancestors for Future Architects. *Sustainable Future for Human security" (SustaiN'2011),* p. 64-65; C-037, Kyoto, Japan.

IOP Conf. Series: Earth and Environmental Science **106** (2017) 012077 doi:10.1088/1755-1315/106/1/012077

The effect of environmental performance and accounting characteristics to earnings informativeness

V Herawaty

Faculty of Economic and Business, Universitas Trisakti, Jakarta, Indonesia

Corresponding Author: vinola.herawati@trisakti.ac.id

Abstract. The objective of this empirical study is to analyze the influence of environmental performance and company's accounting characteristics to earnings informativeness proxied by earnings response coefficient (ERC) on manufacturing companies listed on Indonesia Stock Exchange and consistently follow the PROPER assessment in 2010–2014.One of the company's considerations is to create the green environment reflecting its environmental measures, drawing investors to respond to the company's environmental performance. The data were obtained from Indonesian Capital Market Directory (ICMD), the Indonesia Stock Exchange homepage, the company's annual reports, the decree of the Minister of Environment. The samples used in this research are 27 go public manufacturing companies listed on Indonesia Stock Exchange that consistently follow the PROPER in 2010–2014. The sampling technique used was the purposive method. This research uses multiple regression analysis. The results show that the environmental performance and profitability have a positive influence to earnings informativeness, while leverage has a negative influence to earnings informativeness. Growth opportunities as a control variable has a positive effect on earnings informativeness. This research has proved that the environmental performance is crucial through observing the investors' reaction in the capital market.

Keywords: environmental performance, earnings response coefficient (ERC), leverage, PROPER

1. Introduction

The Indonesian Ministry of Environment explained that the environmental performance is a result of an integrated natural resource and environmental management policy to support the achievement of sustainable development as well as emphasize the green economy (**www.menhl.go.id).** One of the environmental performance measurement tools in Indonesia is the Corporate Performance Rating Program in Environmental Management, abbreviated as PROPER which is one of the leading programs conducted by the Ministry of Environment (KLH), seeking to supervise by public disclosure mechanism that gives incentives or disincentives to the person in charge of the business or activities stipulated in **Regulation of the Minister of Environment No. 5/2011.**

Environmental performance is the company's performance in creating a green environment [1]. The company's environmental performance is measured by colour ranging from the best ones, i.e., gold, green, blue, red, to the worst of black (**www.menhl.go.id**). Through this community, it will be easier to know the arrangement level of management in the company [2]. A higher environmental performance will affect the reaction of investors [3]. In making investment decisions, investors use PROPER as one of the information to be taken into account in investment.

A gauge that can be used to measure the investor reaction toward accounting earnings information is the earnings response coefficient (ERC), which is the correlation between unexpected earnings with the abnormal return of the stock. ERC is defined as the measure of the abnormal return rate of the stock in response to the unexpected earnings component [4]. ERC reflects the market participants' confidence in the quality of earnings announced by the company as well as represents the perspective of a measure of earnings quality based on the market performance. A strong market reactiontoearnings information will be reflected by a high ERC showing are markable quality of corporate income, and vice versa [5].Companies that have a high leverage means they own a larger debt than the capital. If there is an increase in profits, then the beneficiary is the creditor. Thus, the better the condition of the company's earnings, the more negative the response of shareholders since they assume that the profit only benefits creditors [6]. The capital structure significantly affects ERC [6], [7]. ERC is the effect of unexpected earnings on the cumulative abnormal return (CAR), which is shown through the slope coefficient in the regression of the abnormal return of stock with unexpected earnings [8].

Environmental performance will also be achieved at a high level if the company proactively conducts various environmental management actions in a controlled manner. Furthermore, the company's management is encouraged to disclose these environmental management measures in the annual report [9].The objective of this empirical study is to examine and to analyze the influence of environmental performance and company's accounting characteristics to earnings informativeness represented by ERC on manufacturing companies listed on Indonesian Stock Exchange and consistently follow the PROPER assessment at the year 2010–2014. Therefore, the hypotheses of this researchare (1) H_1: Environmental performance has a positive effect on earnings informativeness, (2) H_2: Profitability positively affects the informativeness of earnings, and (3) H_3: Leverage negatively affects earnings informativeness.

2. Research Method

2.1. Data and Sample

This study was conducted on manufacturing companies listed on Indonesia Stock Exchange and consistently followed PROPER. The samples were chosen by the purposive sampling method. Secondary data were obtained from Indonesia Stock Exchange in the form of audited financial statements. This study was conducted by an empirical approach using data recorded during 2010–2014. The data analysis was carried out by the multiple linear regression. The statistical test is necessary to decide whether to accept or reject the proposed hypotheses.

2.2. Variable Measurement

Earnings Informativeness

The earnings response coefficient (ERC) is formulated as follows [10]:

$$\mathbf{CAR_{i,t\,(-5,+5)} = \beta_0 + \beta_{1\,UE_{i,t}} + \varepsilon_{i,t}} \tag{1}$$

$CAR_{i,t\,(-5,\,+5)}$is the cumulative abnormal return of company i in year t ± 5 days from publication of financial report, β_0 is the constant, β_1 is the coefficient of ERC, ($UE_{i,t}$) is the Unexpected Earnings (UE) of company i in year t, εi, and t is the error component.

The first stage is the computation of Cumulative Abnormal Return (CAR). The calculation of cumulative abnormal return at the time of accounting earnings being published, referring to [11], is using the window (time interval) of 11 days, i.e., 5 days before (-5) financial reporting, 1 day (0) and 5 days after (+5) the company's financial report. It is based on the account that the window period of 11 days is considered as neither too short nor too long. The cumulative abnormal return formula is as follows:

$$\boldsymbol{CAR_{i,t\,(-5,+5)}} = \sum_{t=+5}^{t=-5} \boldsymbol{AR_{i,t}} \tag{2}$$

$$ARi,t = Ri,t - Rm,t \quad (3)$$

$$R_{i,t} = \frac{P_{i,t} - P_{i,t-1}}{P_{i,t-1}} \quad (4)$$

$$R_{m,t} = \frac{IHSG_t - IHSG_{t-1}}{IHSG_{t-1}} \quad (5)$$

$CAR_{i,t\ (-5,\ +5)}$ is the cumulative abnormal return of firm i in year t, ($AR_{i,t}$)is the abnormal return of company i in year t, ($R_{i,t}$)is the actual return of company stock i in year t, ($R_{m,t}$) is the return market in year t, ($P_{i,t}$) is the closing price of stock i on day t, ($P_{i,t-1}$) is the closing price of stock i on day t–1, $IHSG_t$ is the composite stock price index on day t, and ($IHSG_{t-1}$) is the composite stock price index on the day t-1.

The second stage is the calculation of unexpected earnings (UE) or earnings surprises as a change from earnings per share of the company in the current year minus the earnings per share of the company the previous year, and it is scaled to the price per share of the previous year [5]. The formula for unexpected earnings is as follows:

$$UE_{\ i,t} = \frac{EPS_{i,t} - EPS_{i,t-1}}{P_{i,t-1}} \quad (6)$$

$UE_{i,t}$ is unexpected earnings of company i in year t, $EPSi_{,t}$ is earnings per share of company i in year t, $EPS_{i,t-1}$- is earnings per share of company i in year $_{t-1}$, and $P_{i,t-1}$ is the stock price $_{t-1}$.

The third stage is the regression of the cumulative abnormal return (CAR) and unexpected earnings (UE) to find the value of earnings response coefficient (ERC).

Environmental Performance

Based on the research [12], the measurement of environmental performance is conducted by giving a score on the PROPER rating obtained from the company. PROPER's performance rating system includes the company's ranking in five colors, i.e., Gold is Very Very Good with Score = 5, Green is Very Good with Score =4, Blue is Good with Score =3, Red is Bad with Score =2, and Black is Very Bad with Score =1.

- Profitability is proxied by Return on assets that can be calculated by dividing a company's annual earnings by its total assets.
- Leverage is represented by Debt to equity ratio that can be calculated by dividing a company's total liabilities by its stockholders' equity.
- Growth Opportunities areproxied by Price to book value that is the ratio of the company's share market price over its book value of equity.

Population and Sample of the Research

The population used in this research is the manufacturing sector companies listed on the Indonesia Stock Exchange and follow the PROPER program. The sample selection criteria are manufacturing companies listed on Indonesia Stock Exchange (BEI) in 2010–2014 and follow the Corporate Performance Rating Program in the 2010–2014 Environmental Management, as well as a manufacturing company that is consistent with the Company's Performance Rating Program in Environmental Management for five consecutive years (2010–2014) and publishes financial reports during 2010–2014.

ERC is obtained from the α_1 slope on the CAR relationship with the UE. The multiple regression model is as follows:

$$CAR = \alpha 0 + \alpha 1\, UE + \alpha 2\, Proper + \alpha 3\, ROA + \alpha 4\, DER + \alpha 5\, PBV + \alpha 6\, UE * Proper + \alpha 7\, UE * ROA + \alpha 9\, UE * DER + \alpha 11\, UE * PBV + e \quad (7)$$

IOP Conf. Series: Earth and Environmental Science **106** (2017) 012077 doi:10.1088/1755-1315/106/1/012077

CAR is the cumulative abnormal return of company i during the observation period of ±5 days from the publication of financial reports, UE is unexpected earnings, ROA is return on assets or profitability ratios, DER is debt to equity ratio or ratio of capital structure (leverage), Proper is corporate performance rating program in environmental management, and PBV is price to book value ratio which is a proxy for the company's growth.

3. Results and Discussion

This study uses samples of manufacturing companies listed on the Indonesia Stock Exchange and consistently follow PROPER during the observation period of 2010–2014. From 135 samples, there are 12 outlier data resulted in the study, suspected to potentially cause the whole samples could not meet the classical assumption test that they would have an extreme value. Therefore, during that period, only 123 companies meet the sample criteria for five years (Table 1).

Table 1. Descriptive statistics.

Variables	N	*Minimum*	*Maximum*	*Mean*	*Std. Dev*
CAR	123	-0.160	0.391	0.011	0.068
UE	123	-1.247	1.660	0.022	0.282
ROA	123	-0.619	0.341	0.057	0.106
DER	123	0.040	14.890	1.394	2.424
PROPER	123	1.0	5.0	3.187	0.716
PBV	123	-1,280	8,740	2,112	1,886

The outcome of classical assumption test of the multiple regression model has fulfilled the assumption of normality, heteroscedasticity, and autocorrelation, yet there is a problem with the multicollinearity. In general, moderating regression analysis model will cause a high multicollinearity between variables [13].The existence of this multicollinearity issue is difficult to avoid since the moderation variables are the interaction of independent variables. The assumption of multicollinearity may be ignored if there is at least one independent variable that has a significant effect on the dependent variable at the initial regression result [14].

The hypothesis testing with multiple regression analysis is based on the results of the research model processing. The multiple regression model is as follows:

$$CAR = -0{,}075 + 0.697\,(UE) + 0{,}110\,(ROA) + 0{,}001\,(DER) + 0{,}014\,(PROPER) + 0{,}003\,(PBV) + 0.798\,(UE * ROA) - 0.028\,(UE * DER) - 0.218\,(UE * PROPER) + 0{,}14\,(UE * PBV) \quad (8)$$

Based on the analysis of F-test, the results show that the independent variables have a significant influence on the informative earnings variables. It can be seen from the F-test significance value of 0.000 which is much smaller than the alpha of 5% (0.000 <0.05). Thus, it can be concluded that the research model has been appropriate, then the test could be continued. The value of adjusted $R^2$of 0.316 indicates that the capability of independent variables to explain the influence of corporate social responsibility, profitability, and leverage to earnings informativeness equals to 0.316 or 31.6% (with environmental performance as the moderation variable and growth opportunities as the control variable), while the rest of 68.5% can be explained by other variables not included in the study.

Partial test (t-test) is used to examine the effect of each independent variable on the dependent variable (Table 2).

Table 2. Results of the t-test (partial testing).

Hyphoteses	Pred.	Variables	Beta	t	Sig.	Decisions
H1	+	UE x PROPER	0.218	2.948	0.009	H1 is accepted
H2	+	UE x ROA	0.798	2.584	0.011	H2 is rejected
H3	+	UE x DER	-0.028	-3.039	0.003	H3 is accepted

Based on Table 2, it is known that UE x PROPER has a significance value of 0.009 (<0.05) with a beta value of 2.18. Thus, $H_1$is accepted, implying that environmental performance has a positive effect on earnings informativeness. The companies havingnifty environmental performance will tend to reveal their measures since it provides good news for market participants. Therefore, these companies publish more environmental information both quantitatively and qualitatively, compared with firms having worse environmental performance. The better the environmental performance, the companies' environment will be much more improved and the value of the companies increases, attracting investors to respond to them. Companies with good environmental performance include not only their concern for the environment, but also the product quality, product safety, corporate social responsibility to the surrounding community, even the safety issue as well as the welfare of their workforces [2]. It owes to the fact that investors are more interested in investing their capital in an environmentally friendly company.

Based on the results of partial regression testing shown in Table 2, it is known that the EE x ROA has a significance value of 0.011 (<0.05) with a beta value of 0.798. Thus, H_2 is accepted, stating that profitability has a positive effect on earnings informativeness. In other words, the higher the profitability, the higher the informative earnings. It means that the company's ability to earn profits from available assets will positively affect the earnings response coefficient. The outcome of this study is consistent with the research conducted by [11] and [15] which state that profitability has a positive and significant effect on ERC.

The significance of EU x DER is 0.003 (<0,05) with a beta value equals to –0,028. Thus, H3 is accepted, implying that leverage negatively affects earnings informativeness. Therefore, the higher the level of corporate leverage, the lower the ERC. Aprogressively high level of debt structure or leverage means that the company has a larger debt than its capital. The result of this study is consistent with the studies conducted by [16], [17], and [18] saying that the leverage has a negative and significant effect on earnings response coefficient. Thus, if there is an increase in the net income, the beneficiary is the creditor since the debtor has the belief that the company will be able to make payments on the debt.

The opportunity to grow (PBV) as the control variable has a significance value of 0.001 (<0.05) with a beta value of 0.141. It can be concluded that growth opportunities have a positive effect on earnings informativeness. This study is consistent with the research conducted by [16] and [18] which state that growth opportunities have a positive effect on ERC, meaning firms with greater growth opportunities will have a higher ERC. This condition indicates that the greater the company's growth opportunities, the higher the chances of the company getting or adding earnings in the future and the higher the ERC as well.

4. Conclusion

This study concludes that environmental performance and profitability has a positive effect on earnings informativeness, while the leverage has a negative effect on earnings informativeness. It can assist investors in analyzing the factors affecting earnings informativeness. This research can be used as a reference by investors in considering the decision to invest, sell their shares, and maintain their investment so that the return on investment can be profitable for investors. The results of this study serve as additional information for the company to determine whether there are factors affecting the earnings informativeness. Hence, the revenue generated by the corporation can provide relevant information about its performance. This study can benefit or contribute to the development of economics, especially accounting, on the earnings informativeness, profitability, and leverage. It can provide additional literature for further research development as well. For the community, the results of this research can be supplementary information to know whether the companies have caused environmental pollution and whether they are obedient to environmental regulations. The further study should add other independent variables that can affect the informative earnings such as the company size and the corporate governance to obtain a greater adjusted R-squared value.

IOP Conf. Series: Earth and Environmental Science **106** (2017) 012077 doi:10.1088/1755-1315/106/1/012077

References

[1] Suratno, Ignatius Bondan, Darsono, and Mutmainah S. 2006. Effect of Environmental Performance on Environmental Disclosure and Economic Performance. National Symposium on Accounting 9.

[2] Rakhiemah A N, and Agustia D. 2009. The Effect of Environmental Performance on Corporate Social Responsibility (CSR) Disclosure And Financial Performance of Manufacturing Companies Listed on Indonesia Stock Exchange. *Simposium Nasional Akuntansi XII*, (4-6 November) 2009, Palembang.

[3] Prabandari, K Rina, and Ketut Suryanawa. 2014. The Effect of Environmental Performance on Investor Reaction in High Profile Company Indonesia Stock Exchange. *E-Jurnal Akuntansi* Universitas Udayana.7.2: 298-312.

[4] Scott William R. 2015. Financial Accounting Theory. 7th Ed. Canada: Prentice Hall Canada Inc. 2015.

[5] Sayekti, Yosefa, and Wondabio, Ludovicus S. 2007. The Influence of CSR Disclosure Against Earning Response Coefficient: An Empirical Study on Companies Listed on the Jakarta Stock Exchange. National Symposium on Accounting X.

[6] Rofika. 2015. Factors Affecting Earnings Response Coefficient (ERC) Manufacturing Companies Listed on Indonesia Stock Exchange 2012. *Jurnal Akuntansi,* 3(2)April: 174-183.

[7] Mulyani, Sri, Nur Fadjrih Asyik, Andayani. 2007. Faktor-Faktor yang Mempengaruhi Earnings Response Coefficient pada Perusahaan yang Terdaftar di Bursa Efek Jakarta. *JAAI.* 11(1) Juni: 35-45.

[8] M A Berry, D A Rondine. May 1, 1998. Proactive corporate environmental management: A new industrial revolution. The Academy of Management Executive. 12(2):38-50.

[9] Chaney P K, and D C Jeter. 1991.The Effect of Size on the Magnitude of Long Window Earnings Response Coefficients.*Contemporary Accounting Research* 8. No.2 : 540-560.

[10] Melati, Rosa Aprilia, and Kurnia. 2013. The Influence of CSR Disclosure and Profitability to Earning Response Coefficient (ERC). *Jurnal Ilmu & Riset Akutansi*. 2(12).

[11] Pujiasih. 2013. The Effect of Environmental Performance on Financial Performance with Corporate Social Responsibility (CSR) as Intervening Variable. Skripsi S1. Fakultas Ekonomi Universitas Negeri Semarang.

[12] Wijayanti, N Risma, and Supatmi. 2008. The Effect of Dividend Payment Ratio and Capital Expenditure on Earnings Response Coefficient (ERC) with Cash Flow as Moderating Variables. *Jurnal Bisnis dan Ekonomi* Hal. 1-14, Maret. Universitas Kristen SatyaWacana.

[13] Gujarati, and Porter. 2009. Fundamentals of Econometrics. Jakarta: Salemba Empat.

[14] Setiawati Erma, Nursiam,, and Fitri Apriliana. 2014. Analysis of Effect of Company Size, Growth and Profitability on Profit Response Coefficient (Empirical Study on Manufacturing Company Listed on Indonesia Stock Exchange (BEI) Year 2009-2011). *Seminar Nasional dan Call for Paper* (Sancall).

[15] Subagyo, and Cicilia N. Olivia. 2012. Factors Affecting Earnings Response Coefficient (ERC).*Jurnal Akuntansi.* 12(1)April: 539-558.

[16] Murwaningsari Etty. 2008. Pengujian Simultan: Beberapa Faktor Yang Mempengaruhi Earning Response Coefficient (ERC). Simposium Nasional Akuntansi (SNA) XI, 23-24 Juli. Pontianak.

[17] Mulyani Sri, Nur Fadjrih Asyik, and Andayani. 2007. Factors Affecting Earnings Response Coefficient on Companies Listed on the Jakarta Stock Exchange. *JAAI.* 11(1)Juni: 35-45.

Nitrogen removal on recycling water process of wastewater treatment plant effluent using subsurface horizontal wetland with continuous feed

T Tazkiaturrizki[1]*, P Soewondo[2], M Handajani[2]

[1]Environmental Engineering Department, Faculty of Landscape Architecture and Environmental Technology, Universitas Trisakti, Jakarta, Indonesia
[2]Research Group of Water and Wastewater Engineering, Faculty of Civil and Environmental Engineering, Institut Teknologi Bandung, Bandung, Indonesia

*Corresponding Author: tazkiaturrizki13@trisakti.ac.id

Abstract. Recycling water is a generic term for water reclamation and reuse to solve the scarcity of water. Constructed wetlands have been recognized as providing many benefits for wastewater treatment including water supply and control by recycling water. This research aims to find the best condition to significantly remove nitrogen using constructed wetland for recycling water of Bojongsoang Waste Water Treatment Plan (WWTP) effluent. Using media of soil, sand, gravel, and vegetation (*Typha latifolia* and *Scirpus grossus*) with an aeration system, BOD and COD parameters have been remarkably reduced. On the contrary, the removal efficiency for nitrogen is only between 50–60%. Modifications were then conducted by three step of treatment, i.e., Step I is to remove BOD/COD using *Typha latifolia* with an aeration system, Step II is todecrease nitrogen using *Scirpus grossus* with/without aeration, and Step III isto complete the nitrogen removal with denitrification process by *Glycine max* without aeration. Results of the research show that the nitrogen removal has been successfully increased to a high efficiency between 80–99%. The combination of aeration system and vegetation greatly affects the nitrogen removal. The vegetation acts as the organic nitrogen consumer (plant uptake) for amino acids, nitrate, and ammonium as nutrition, as well as theoxygen supplier to the roots so that aerobic microsites are formed for ammonification microorganisms.
Keywords: constructed wetland, nitrogen removal, recycling water, WWTP

1. Introduction

The high human activity from both domestic and commercial sectors such as industries has resulted in the scarcity of clean water due to water pollution. Wastewater treatment is equipped with wastewater recycling technologies based on the account that the treated water can be recovered for non-potable use with a quality of up to grade 3 [6]. One of the WWTP effluents that can be utilized for recycling technology is the treated water coming from Bojongsoang WWTP which collects the domestic wastewater of eastern and central southern Bandung with the capacity of 596,556 people for the total Bandung's citizen of 2,390,120 people, according to PDAM Tirta Wening.

An alternative to treat the wastewater effluent efficiently and economically is needed to attain a better quality of water that can be used sustainably, and wetland can be a promising option. The wetland will be applied to solve the problems of domestic wastewater because it is a low-cost installation that requires neither sophisticated technology nor specifically skilled operator to manage it [1].

The use of plant and microbial presence certainly has a crucial role in the degradation of N and P compounds [14] explained that compared with the arable crops, wetland plants could obtain a decent removal efficiency Huett et al. because of the presence of nitrifying bacteria on the root surface [14]. P. Kemp and George also describedin their research that the N removal process would be better if it is carried out in two stages of treatment [5]. There is a good correlation between N and P removal in constructed wetland reactors [10]. Combining the legume plant *Glycine max* (soybean) reportedly affects the number of bacteria forming a mutualistic symbiosis with soybean roots to enhance nitrogen fixation [8].

From the previous study, the obtained nitrogen removal efficiency was still low at 50–60% [9]. Therefore, this research will be conducted on wetland modification by using the optimum conditions from prior studies regarding the volume ratio and optimum surface area, the addition of aeration, as well as by combining a legume vegetation *(Glycine max)* for nitrogen removal optimization.

2. Research Method

2.1. Research preparation

Experiments were performed using wastewater effluent of Bojongsoang WWTP from maturation ponds. Reactors were made from resin and poly-ethylene structure which then divided into three zones, i.e., inlet, treatment, and outlet area. The inlet and outlet zone contain gravel with a similar diameter of 2 cm, while the treatment zone containssoil, sand, gravel, and vegetation. The reactor's slope is 0.1%.The wetland aeration system uses aluminum pipes to keep the oxygen supply. There were three kinds of vegetation used in this research, i.e. *Typha latifolia*, *Scirpus grossus,* and *Glycine max*. The reactor volume may vary for each stage: $V_1 = (0.7 \times 0.5 \times 0.4)\ m^3$; $V_2 = (0.4 \times 0.3 \times 0.3)\ m^3$; and $V_3 = (0.35 \times 0.32 \times 0.30)\ m^3$.

2.2. Research variations

The research was carried out using several variations concerning vegetation types and the aeration system. The nitrogen removal was measured with and without aeration using different kinds of vegetation in each sequence. The first sequence, the water from Reactor 1 (*T. latifolia*) was delivered to Reactor 2 that contains*S.grossus,*followed by the aeration system, then conveyed to the last reactor that has *S.grossus*. The second sequence, the water from Reactor 1 was delivered to Reactor 3 that contains*S.grossus,* followed by the aeration system, then brought to the last reactor that has *G.max*. The third sequence, the effluent from Reactor 1 was delivered to Reactor 4 that contains*S.Grossus,* this time without aeration process, then carried to the last reactor that has *G.max*.

All the sequences have different treatment so that the optimum condition for nitrogen and phosphate removal can be deduced from the comparison. Besides, there were also three variations of N/P ratio as well(N/P = 10, 20, 30) to gain more accurateresults and the correlation between the removal of N and P can be identified. The concentration of nitrogen was raised by adding the chemical of NH_4Cl, whereas the phosphate concentration was increased by augmenting the chemical of KH_2PO_4.

From observing the 1st and 2nd Sequence, hasthe *G.max* increased the nitrogen removal efficiency? While from the 2ndand 3rd Sequence, has the aeration contributed to increasing the nitrogen removal efficiency?

2.3. Sampling and analytical methods

The key parameters to measure the water quality consist of physical and chemical parameters. Samplings were conducted daily to analyze the pH and temperature using pH-meter, as well as COD by micro COD. In steady state (when the degradation of carbon had reached a fairly constant point), the analyses of BOD, total solids, ammonium, nitrite, nitrate, and total ammonium (Total Kjeldahl Nitrogen) were conducted using Winkler titration, gravimetric method, spectrophotometer, and distillation.

3. Results and Discussion

The effluent of Bojongsoang WWTP has met the quality standard for the 4th grade stream standards of the water quality classification used in Indonesia.For the 2nd grade standard, there are several parameters that still exceed the threshold for recycled water, i.e., BOD, COD, and nitrite.The typical treatment process of a constructed wetland is as follows: (1) Wastewater is added at the inlet, (2) It flows through the subsurface, (3) It then makes contact with the root system of the plants as the aerobic and anaerobic zones have developed in the growing media, and (4) As the wastewater flows toward the outlet, biochemical reactions remove or transform some of the chemicalcontents [2].

Nitrogen is a compound that has a natural variation in the degree of oxidation. This oxidation rate ranges from +5 to -3, composed of organic and inorganic nitrogen compounds. Nitrogen removal rates correlate with initial concentrations of nitrogen. Inorganic nitrogen and phosphate compounds are necessary for plant growth and are potential to maximize the amount of nutrient removed from wastewater effluents using macrophytes [3]. There are five main principals of the nitrogen transformation: (1) ammonification/mineralization, (2) nitrification, (3) denitrification, (4) nitrogen fixation, and (5) nitrogen assimilation.

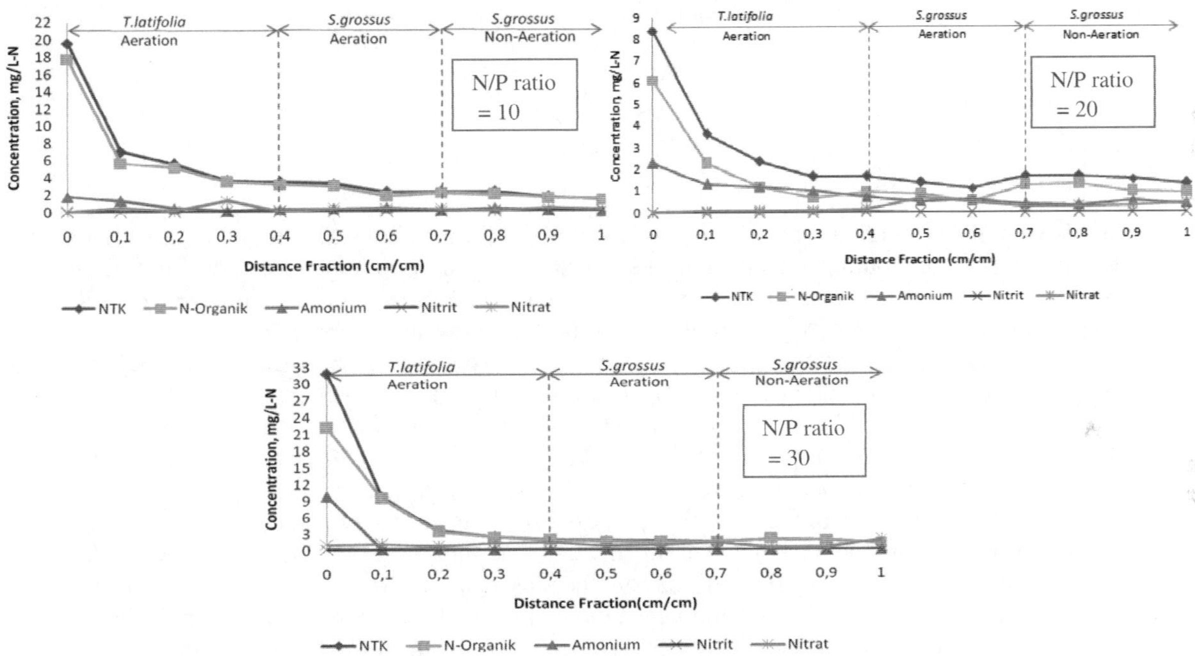

Figure 1. Nitrogen removal of the 1st Sequence.

The inlet concentrations of NTK parameter is 8.4–31.92 mg/L-N, whereas of ammonium is 1.85–9.76mg/L NH_4^+-N. Thus, the concentration of N-organic can be calculated of about 6.09–22.16 mg/L-N (see Figure 1). Most of the nitrogen compounds in water are organic. Those organic nitrogens will transform into ammonia biologically by ammonification process, which is the first step of nitrogen mineralization and called as ammonia volatilization.

The transformation of ammonia into ammonium is affected by pH and temperature condition as well as the availability of dissolved oxygen in the water. The reactor without the second aeration delivered a better removal of organic nitrogen compared with other sequences at 0.6–1.1 fraction of distance (see Figure 1-3). Microorganisms only take the nitrogen if they need it for their growth, they will otherwise take carbon as the main nutrition [11] so that the nitrogen removal occurs slowly.

IOP Conf. Series: Earth and Environmental Science **106** (2017) 012078 doi:10.1088/1755-1315/106/1/012078

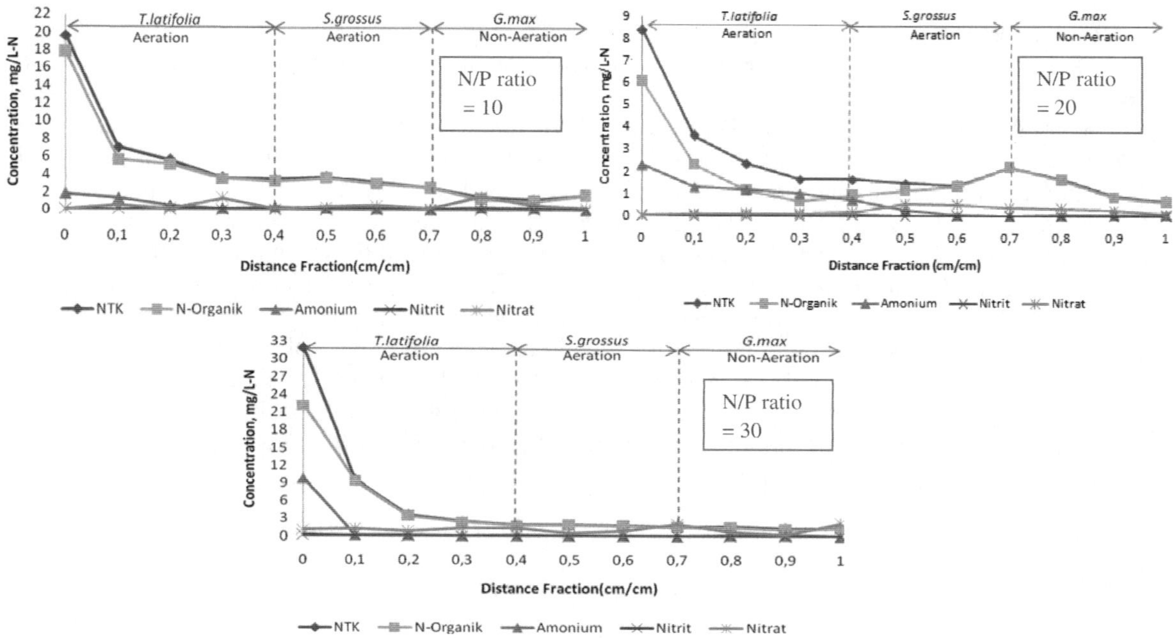

Figure 2. Nitrogen removal of the 2nd sequence.

In this step, the vegetation plays the role as the organic nitrogen consumer (plant uptake) for amino acids, nitrate, and ammonium as nutrition, also as the oxygen supplier to the roots so that aerobic microsites are formed for ammonification microorganisms. The best organic removal condition is found in the sequence usingthe variation of *T. latifolia*, *S.grossus*, and *G.max*.

At the pH and average temperature of 7 and 25°C, respectively, ammonia will be ionized into ammonium of about 99.4% [4], while at pH of 8, about 95% of the total ammonia nitrogen will be ionized into ammonium [12]. Figure 1–3 show that the ammonium removal was better with the efficiency of 99% when the variation of *T. latifolia*, *S.grossus*, and *G.max* was used. On the other hand, the ammonia removal occurred well at the 2nd and 3rd Sequences as the effect of aeration. The decreasing ammonium is due to the oxidation process involving autotroph microbes to yield nitrite and nitrate afterward, and this process is called nitrification. The nitrification process occurred well in the aerobic condition, involving nitrification microorganisms namely *Nitrosomonas* and *Nitrobacter*. At each degradation process, the nitrification needs oxygen as the terminal electron acceptor. Figure 1–3 depict the nitrification process as a function of distance as well as the effect of vegetation variation and aeration.

The aeration system highly affects the nitrification process because oxygen is the terminal acceptor electrons involving autotroph microorganisms. However, autotroph groups cannot keep up with the heterotrophs in using oxygen so that autotrophs will be active when the carbon concentration in water is very low. Therefore, an extended aeration is needed to allow a thorough nitrification process to occur.The final step is denitrification, which is by increasing the electron toward nitrite and nitrate, gas phase products will be formed due to the reaction, e.g., nitrogen (N_2), nitrous oxide (N_2O),or nitric oxide (NO). Denitrification will occur when there is no free oxygen or known as an anoxic condition like in Reactor 5 (*S.grossus*), Reactor 6 (*G.max*), and Reactor 7 (*G.max*). The microorganisms involved are *Bacillus, Enterobacter, Micrococcus, Pseudomonas,* and *Spirilium* [4].

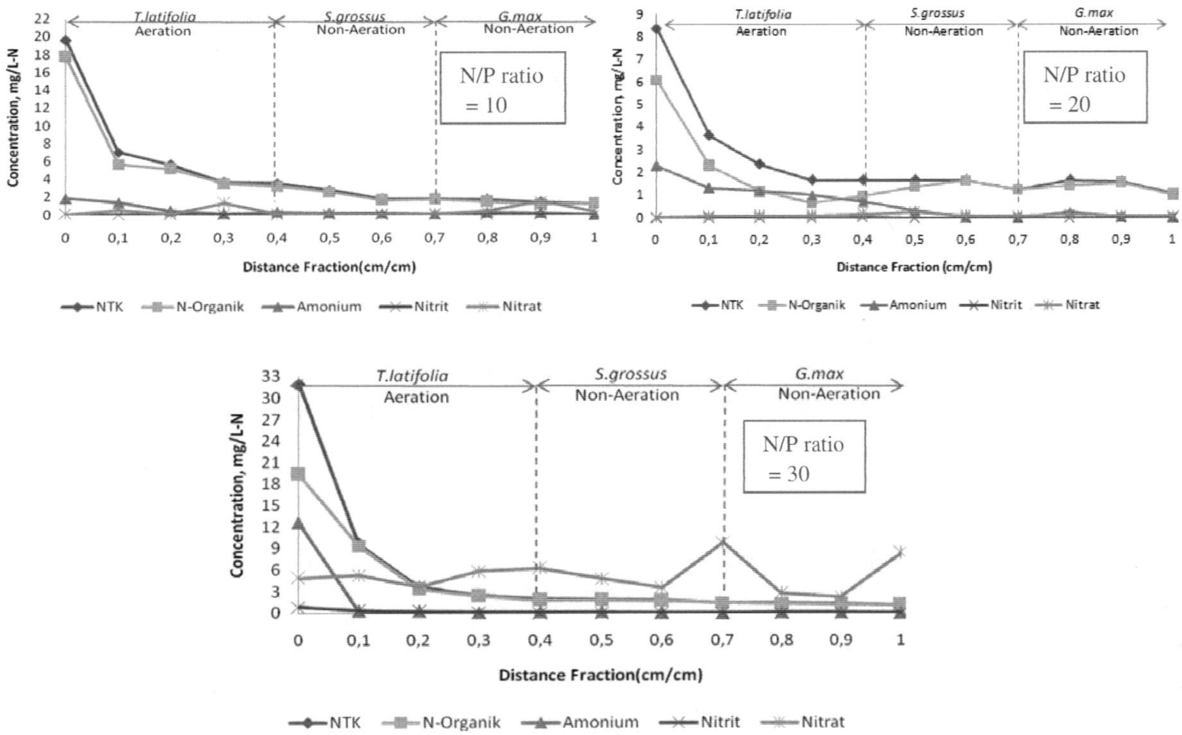

Figure 3. Nitrogen removal of the 3rdSequence.

The total nitrogen removal was measured by Total Kjeldahl Nitrogen (NTK) which is the sum of total ammonium and organic nitrogen. The inlet concentration of NTK is 8.4–31.92 mg/L. Figure 1–3 shows that the decreasing of the total nitrogen concentration occurred in the constructed wetland is an effect of the variation of vegetation and aeration system. Reactor 6 (*Glycine max*) gave the best results than others. There is nevertheless no significant effect of the aeration system on the total nitrogen level because nitrogens are present in many forms and have different responses to the oxygen consumption in water.

4. Conclusion

Concerning the removal efficiencyand energy consumption, the third sequence (*T.latifolia* + aeration, *S.grossus*, *G.max*) delivered the best performance compared with others. The contaminant concentration in the effluent after the treatment comprised ammonium ranging between 0.002–0.4 mg/LNH_4^+-N, nitrite of 0.0001–0.0009 mg/L NO_2-N, nitrate of 0.08–1.89 mg/L NO_3-N; NTK (Total Nitrogen Kjeldahl) ranging between1.06–1.26 mg/L-N; N-Organic of 1.05–1.25 mg/L-N; Total Phosphate in the range of 0.01–0.08 mg/L-P; COD of 6–17 mg/L; and BOD of 1–4 mg/L. According to PP 82/2001, those parameters have met the quality standard for the 2nd grade that can be used as recycled water.

References

[1] Al-Omari A, and Fayyad M. 2003. Treatment of Domestic Wastewater by Subsurface Flow Constructed Wetlands in Jordan. *Journal of Elsevier. Desalination*, 155:27–39.

[2] He Q, and Mankin, K. Dec. 2002. Performance Variations of COD and Nitrogen Removal by Vegetated Submerged Bed Wetlands. *Journal of the American Water Resources Association*. ProQuest. 38(5):1679.

[3] Iasur-Kruh L, Hadar Y, Milstein D, Gasith A, and Minz D. 2010. Microbial Population and Activity in Wetland Microcosms Contructed for Improving Treated Municipal Wastewater. *Journal of Springer Science. Microb. Ecol.* 59:700–709.

[4] Kadlec R H, and Knight R L. 1996. Treatment Wetlands. Boca Raton: CRC, Lewis, p.893.

[5] Kemp P, Michael C, George, Dennis B. 1997. Subsurface Flow Constructed Wetlands Treating Municipal Wastewater for Nitrogen Transformation and Removal. *Proquest Journal International*. Water Environmental Research, p.1254.

[6] Indonesia Goverment, 2001. Peraturan Pemerintah Nomor 82/2001 about Water Quality Management and Water Pollution Control.

[7] West Java Governor. 2005. Peraturan Gubernur Jawa Barat Nomor 69/2005 about Pedoman Penentuan Status Mutu Air.

[8] Puspita M. 2011. Effect of Leguminosa Plant in Constructed Wetland with Vertical Sub-surface Flow to remove organic compound of Domestic Wastewater. Thesis: Master Program of Environmental Engineering ITB.

[9] Rinarti A. 2011. Effect of Aeration and Width of Media Reactor in Organic Removal of WWTP Effluent Using Wetland in Laboratorium Sclae (Case Study: Bojongsoang WWTP). Thesis: Master Program of Environmental Engineering ITB.

[10] Stottmeister, Wieβner, Kuschk, Kappelmeyer. 2003. Effects of Plants and Microorganisms in Constructed Wetlands for Wastewater Treatment. *Elsevier Journal*. Biotechnology Advances 22: 93–117.

[11] Vymazal J. 2008. Fractionation, Bidegradability and Particle-Size Distribution of Organic Matter in Horizontal Subsurface-Flow Constructed Wetlands. Journal of Springer Science.

[12] Vymazal J, and Kropfelova L. 2008.Wastewater Treatment in Constructed Wetlands with Horizontal Sub-Surface Flow. Springer: Czech Republic.

[13] Wisjnuprapto D. 2004. Bologica Process and Engineering. Department of Environmental Engineering ITB, Bandung.

[14] Zemanova K, Picek T, Dusek J, Edwards K, and Stanruckova H. 2010. Carbon, Nitrogen, and Phosphorus Transformations are related to Age of Constructed Wetland. *Journal of Springer Science*. Water Air Soil Pollut. 207:39–48.

The determination of ground water quality based on the presence of *Escherichia coli* on populated area (a case study: Pasar Minggu, South Jakarta)

Y Rohmah, A Rinanti, D I Hendrawan*

Environmental Engineering Department, Faculty of Landscape Architecture and Enviromental Technology, Universitas Trisakti, Jakarta, Indonesia

*Corresponding Author: dianahendrawan@trisakti.ac.id

Abstract. This study aims to determine the quality of groundwater in densely populated areas in the Pasar Minggu district, which focuses on examination of *Escherichia coli* bacteria as biological parameters. *E.coli* analysis was conducted in 3 sub-districtin the Pasar Minggu district, South Jakarta (106° 45'0"E - 6° 15'40"S), which included Pejaten Barat, Jati Padang and Kebagusan sub-district based on topography. The method used is Most Probable Number (MPN), referring to SNI 01-2332.1-2006. The results showed that the presence of *E. coli* in the highest groundwater ranged from 1100 MPN/100 ml to > 1100 MPN/100 ml, present in Pejaten Barat sub-district (06° 16'39 "S-106° 49'48" E) whereas the lowest number ranged from 11.5 MPN/100 ml to 23 MPN/100 ml occurred in Jati Padang sub-district (06° 17'6 "S - 106° 49'48" E).The correlation test results show that the amount of *E.coli* in ground water is affected by the distance of the septic tank to the groundwater well. The amount of *E.coli* contained in groundwater is also increasing, 1100 MPN/100 ml in groundwater samples at <10 m and < 20 MPN/100 ml in groundwater samples at> 15 m depth. This study provides information that topography and water flow affect the amount of *E. coli*.
Keywords: *Escherichia coli*, most probable number, pasar minggu district, septic tank, surfer

1. Introduction

Water is an abundant chemical in the human body and plays a central role in the regulation of nutrient transport, toxic waste disposal, regulation of body temperature and digestion, organ function and metabolic activity. However, if drinking water is contaminated with feces, it will spread the disease to large numbers of consumers [1].

A Pipeline distribution network that has not been maximized in Jakarta, causing many residents to use ground water as a source of drinking water. The election of Pasar Minggu district as the object of the research is based on data from Agency of Environmental Management Jakarta 2015, indicating the ground water in Pasar Minggu Disrict (Located in South Jakarta) has been populated by *E. coli* bacteria [2].

Various bacteria, parasites, and viruses, known to be pathogenic, have the potential to cause health problems if they enter the human body. The Environmental Protection Agency (EPA) considers Total Coliform useful as an indicator of other pathogens in drinking water. Total coliform is used to determine adequate water treatment (eligibility) and integrity in the distribution system [3]. Therefore, the maximum limit of contaminant of pathogenic bacteria such as Total coliform and Escherichia coli is 0 for drinking water.

Coliform is naturally present in the environment, but Fecal coliform and E. coli are derived from human and animal waste [4]. Thus, the presence of Fecal coliform and E. coli on a material or water indicates that the water is contaminated by human or animal waste. E. coli causes disease (pathogens) in this waste can cause diarrhea, cramps, nausea, headaches. This pathogen may pose a special health risk for infants, young children, and people with impaired immune systems [5].

Referring to the description, this research is conducted to know the quality of ground water of densely populated area in the Pasar Minggu district, emphasized on examination of Escherichia coli bacteria as biological parameter. The relationship between the number of E. coli bacteria in groundwater with well distances to septic tanks was studied to determine whether ground water is suitable for use as a source of drinking water or not.

2. Research Method

Sampling was conducted in 3 sub-district in Pasar Minggu district, Pejaten Barat, Jati Padang and Kebagusan sub-district (Table 1) and detail location figure out at map (Figure 1). This location has a decreasing contour from north to south where Kebagusan sub-district is with the lowest contour. Sampling is done from April - June, which represents the data of rainy and dry months. In addition to biological parameters such as the presence of E. coli bacteria, physic and chemical parameters including pH, temperature and DO are also measured during sampling.
Bacterial contamination does not alter the physical appearance, odor or taste of the water. Testing by the laboratory is the only way to know if water contains *Coliform bacteria.* [6] The method used in analyzing the number of *Escherichia coli* bacteria in the sample water using the Most Probable Number (MPN) matrix refers to SNI 01-2332.1-2006 on the determination of *coliform* and *Escherichia coli* in fishery products. The groundwater sampling method refers to SNI 6989.58: 2008 on Water and wastewater - Section 58: Groundwater Sampling Method. Analysis of physical, chemical and biological parameters was performed in the Microbiology Laboratory of the Faculty of Landscape Architecture and Environmental Technology.

Table 1. Sampling location coordinate.

Sub-district	No	Location	Coordinate
Pejaten Barat	1	Siaga Raya Street RT 12/ RW 4 no 8 Pejaten Barat	06°16’12”S ; 106°50’15”E
	2	WarungJati Barat Street no 98 Pejaten Barat	06°16’39”S ; 106°49’48”E
Jati Padang	3	Jatiwangi Street no 7 Jati Padang	06°17’6”S ; 106°49’48”E
	4	Raya Ragunan Street no 4 Jati Padang	06°17’6”S ; 106°50’15”E
	5	Damai Street no 62 RT 2/RW 1 Ragunan	06°17’33”S ;106°49’21”E
Kebagusan	6	Kebagusan Street 1 No 9 Kebagusan	06°18’0”S ;106°49’48”E
	7	Kebagusan Street 4 Kebagusan	06°18’54”S ;106°49’48”E

The MPN is a procedure to estimate the population density of viable microorganisms in the test sample. This is based on the application of probability theory to the number of positive growth responses observed in a series of standardized sample inoculum dilutions placed on a number of culture media tubes. Positive growth response after incubation can be demonstrated by observations such as gas production in the fermentation tube or turbidity seen on the broth tube, depending on the type of media used [7].

The verification indicating the presence of total coliform bacteria is indicated by the production of gas in Lactose Broth (LTB) media followed by positive LTB inoculation into the Brilliant Green Lactose Broth (BGLB). Furthermore, for total coliform testing, positive and negative bacteria should be retested on LTB / BGLB media. For testing the presence of positive and negative *E.coli* bacteria bacteria were tested again using LTB / EC Broth (EC-MUG) [8].

IOP Conf. Series: Earth and Environmental Science **106** (2017) 012079 doi:10.1088/1755-1315/106/1/012079

The results of the analysis will be compared with drinking water quality standard that is the Regulation of the Minister of Health in 2010 where the maximum presence of *E. coli* bacteria in drinking water is 0 (amount per 100 ml sample). From the data will be analyzed the relationship of well distance to septic tank with the number of *E. coli* bacteria in ground water, due to the possibility of leakage septic tank.

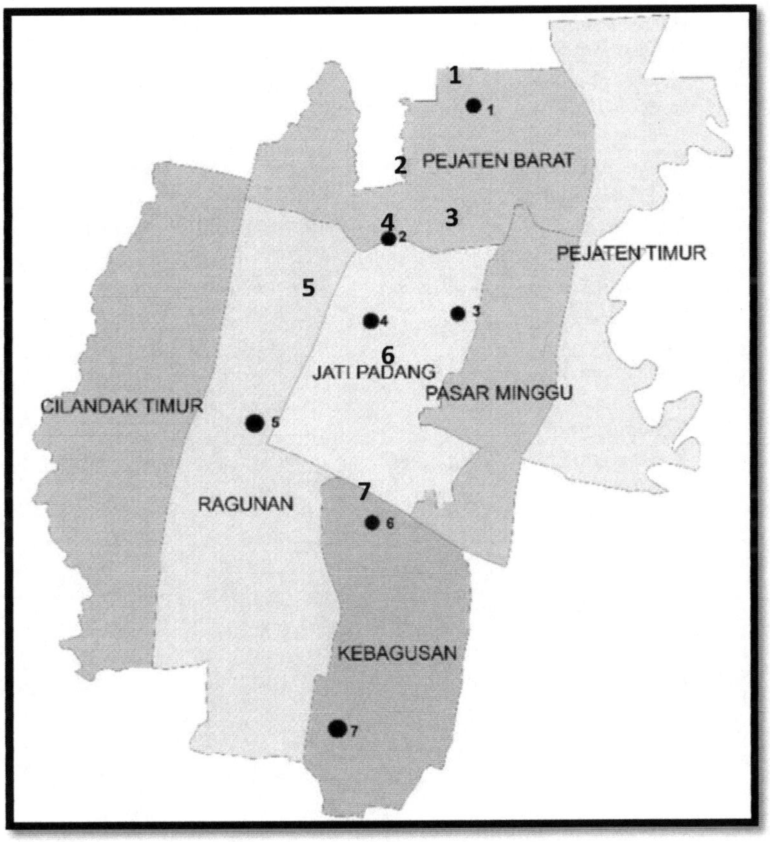

Figure 1. Sampling location map.

3. Results and Discussion

Amount of presence of E.coli in MPN/100 ml, where the results were obtained from sampling in May (sampling 1), June (sampling 2), and July 2017 (sampling 3) see Table 2. Based on the average number *E. coli* was found that the highest value of 738 MPN/100 ml located in Pejaten West village is Warung Jati Barat Street. The lowest value is in Jati Padang village that is Jatiwangi street 13 MPN/100 ml. With the average data of *E. coli* at third sampling times, the standard deviation score is 453. With a large standard deviation value indicates that the data obtained show high diversity (heterogeneous).

Table 2. Amount of *E.coli* bacteria at each sampling location.

Sub-district	No Sampel	Location	Sampling 1	Sampling 2	Sampling 3	Average	Quality Standard
			(MPN/100 ml)				
Pejaten Barat	1	Siaga Raya Street RT 12/ RW 4 no 8 Pejaten Barat	138 ± 1,442	<3 ± 0	1100 ± 0	414	0
	2	Condet Pejaten Street RT 1/RW 7 no 7	>1100 ± 0	1100 ± 0	< 3 ± 0,141	738	0
Jati Padang	3	Jatiwangi Street no 7 Jati Padang	12 ± 0,636	23 ± 0	<3 ± 0,967	13	0
	4	BP 7 Belakang Balai Street	>1100 ± 0	40,5 ± 0,353	< 3± 0,263	385	0
Ragunan	5	Damai Street no 62 RT 2/RW 1 Ragunan	1100 ± 0	780 ± 4,525	122 ± 4,030	664	0
Kebagusan	6	Kebagusan 1 Street no 9 Kebagusan	14 ± 0	195 ± 6,363	225 ± 2,121	159	0
	7	Kebagusan 4 Street Kebagusan	29 ± 1,485	195 ± 0,636	195 ± 0,636	150	0

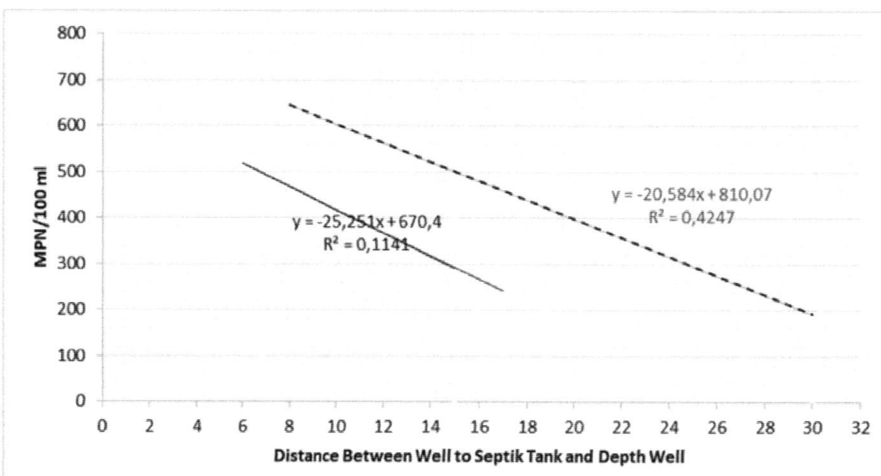

Figure 2. Correlation between amount of *e.coli* on the distance of wells to septic tank and depth well.

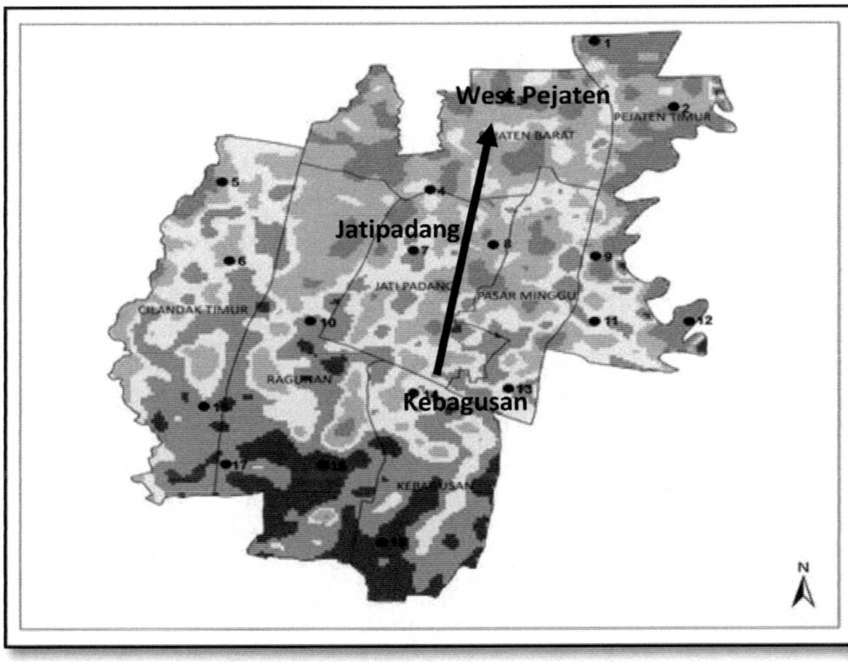

Figure 3. Map contour of Pasar Minggu district.

The relationship between the number of *E.coli* to the well distance to the septic tank (Figure 1) gives the regression coefficient 0.4247, which means there is a weak linear relationship between the number of *E.coli* with the well distances in the septic tank. And for the depth of the well gives a regression coefficient 0,11417 means a weak relationship between the amount of *E. coli* with the depth of well. So it can be concluded that the presence of *E. coli* in groundwater in Pasar Minggu subdistrict is influenced by the distance of the well to septic tank and the depth of the well but the influence is not significant or weak.

Based on the contour of Kebagusan subdistrict area is in the south of Pasar Minggu district and has the highest topography, then West Pejaten Village is in the northern part of Pasar Minggu and has the lowest topography (Figure 3). Based on the value of MPN analysis results in 3 the urban subdistrict, Kebagusan Village has the lowest MPN value, then the village of Jati Padang and West Pejaten village has the highest MPN value. It can be concluded that topography influences the amount of *E. coli*. The topography of the area is related to groundwater flow, ground water flows from high to low. This is supported by previous research that Winata (2013) proves that the pattern of *E. coli* distribution is increasing in the Bajiro village to Gembiraloka Residential following the flow direction of Gajah Wong from north to south, so that the *E. coli* bacteria distribution follows the direction of water flow [9].

4. Conclusion

With regression value obtained shows that there is a weak relationship between the number of bacteria *E.coli* (MPN/100 ml) with the well distance to septic tank. The presence of *E. coli* in ground water is influenced by groundwater flow based on the topography of the region. Based on the data obtained found that the amount of *E. coli* on the groundwater is all above the maximum value of the quality standard. Thus, this research is not recommended for the use of well water as the main source of drinking water in the area.

References

[1] Nakade D B. 2013. Assessment of Bacteriological Quality of water in Kolhapur City of Maharashtra, India; *International Research Journal of Environment Series*. 2(2):63-65 (http://www.isca.in/IJENS/Archive/v2/i2/10.ISCA-IRJEvS-2012-002.pdf)

[2] Badan Pengelolaan Lingkungan Hidup Daerah DKI Jakarta. Data Kualitas Air Sumur Periode 1 (September-Oktober) SD.16. 2015.

[3] United States Enviromental Protection Agency. 2013. Revised Total Coliform Rule and Total Coliform Rule (https://www.epa.gov/dwreginfo/revised-total-coliform-rule-and-total-coliform-rule)

[4] United States Enviromental Protection Agency. 2009. National Primary Drinking Water Regulations. (https://www.epa.gov/ground-water-and-drinking-water/national-primary-drinking-water-regulations#Microorganisms)

[5] United States Enviromental Protection Agency. 2009. National Primary Drinking Water Regulations(https://www.epa.gov/sites/production/files/201606/documents/npwdr_complete_table.pdf)

[6] Saha U, Sonon L, Kissel D. 2011. Total Coliform and Escherichia coli Bacteria in Georgia Private Wells : Relationships of Age and Depth of the Wells. *Proceeding of the 2011 Land Grant and Sea Grant Natonal water Conference*. January 31-February 1, 2011. Washington DC (http://extension.uga.edu/publications/detail.cfm?number=C858-7)

[7] United States Department of Agriculture Food Safety and Inspection Service, Office of Public Health Science. 2014. Laboratory Guide Book Notice of Change (Most Probable Number Procedure and Tables) (https://www.fsis.usda.gov/wps/wcm/connect/8872ec11-d6a3-4fcf-86df-4d87e57780f5/MLG-Appendix-2.pdf?MOD=AJPERES)

[8] United States Environmental Protection Agency. 2010. EPA Microbiological Alternate Test procedure (ATP) Protocol for Drinking Water, Ambient Water, Wastewater, and Sewage Sludge Monitoring Methods (https://nepis.epa.gov/Exe/ZyPDF.cgi?Dockey=P100IK92.txt)

[9] Winata, Ekrar and Hartantyo, Eddy. 2010. *Quality of Ground Water along Gajah Wong River reviewed from the Pattern of Escherichia coli (Study Case : Umbulharjo District)* (http://pdm-mipa.ugm.ac.id/ojs/index.php/jfi/article/view/814)

Distribution patterns study of *Escherichia coli* as an Indicator for ground water quality at Matraman District, East Jakarta

U Anisah, B Iswanto, A Rinanti*

Environmental Engineering Department, Faculty of Landscape Architecture and Environmental Technology, Universitas Trisakti, Jakarta

*Corresponding Author: astririnanti@trisakti.ac.id

Abstract. The purpose of this study was to determine the distribution pattern *E.coli* as indicators of ground water contamination by *fecal coliforms* at Matraman District, East Jakarta, Indonesia ($106^0$49'35"E $06^0$10'37"LS) consisting of six sub-districts (Utan Kayu Selatan, Utan Kayu Utara, Kayu Manis, Pal Meriam, Kebon Manggis and Pisangan Baru). The presence of *E.coli* examined by the method of Most Probable Number (MPN) according to SNI 01-2332.1-2006 about the determination of *coliforms* and *E.coli* in fishery products which consists of three stages of tests in a row, the estimation, determination or confirmation and complementary or morphology test. Matraman District is a densely populated area and the distance between septic tank and well is inadequate, both with their own's septic tank or with neighbors's septic tank. The result of sample analysis shows that *E.coli* counts exceeds water quality standards in accordance with regulation Health Ministry No.492/2010 concerning the drinking water quality requirements. The lowest MPN number is 3 MPN/100 ml and the highest number of *E. coli* is more than1100 MPN/100 ml. The temperature, pH and DO_0 on average are 27.52^{0}C, 5.59 and 2.24 mg/L. This study is expected to be a reference in the use of ground water for daily activities in the district Matraman.
Keywords: e.coli, fecal coliform, ground water

1. Introduction

Clean water is very much needed by living things but not so easy to obtained. Based on [1] and [2], Indonesia is one of the 10 countries that nearly two-thirds of the population have no access to obtain a source of clean drinking water and that diarrhea is still the main cause of death of children aged under five years old, while 946 million people still defecating outside, like in urban drainage or open channel [1], [2]. The use of drinking water refers to the criteria of quality standards according to [3] on the requirements of the drinking water quality. Drinking water that contaminated by fecal bacteria coli, namely *Escherichia coli* (*E. coli*) could inflict disorder against human health.

E. coli is one of indicator of ground water pollution and in general may found living in the ground or soil. The existence of fecal coliform in well water or ground water indicates the contamination of ground water caused by waste water from toilet [4]. There are some of the total and fecal coliform in ground water covering agricultural runoff, out of septic tank or drainage, and absorption of human waste and animal which also containing fecal [5].

E. coli is usually found in waste water that when directly or indirectly consumed would cause diarrhea, a disease characterized by increasing the frequency of excreting more than 3 times each day, accompanied by changes of form and the consistency of fecal matter. Because one of the factors greatest in water pollution of the land was the construction septic tank itself so the construction should fulfill the SNI (Indonesian National Standard) 03-2398-2002 [6] regarding the procedures of the

IOP Conf. Series: Earth and Environmental Science **106** (2017) 012080 doi:10.1088/1755-1315/106/1/012080

designing the septic tank with a system of absorption. In the regulation, it is explained that distance septic tank absorption to the plane to buildings must be 1.5 m, to clean water wells must be located up to 10 m and 5 m with absorption wells rain water. Technical requirements it gets into a building the septic tank covering material buildings must be strong, resistant to an acid and is impermeable to water, building material can be selected for the base of the building. The lid and channeling pipe waste water are the times, red bricks, adobe, reinforced concrete, concrete without a bone, pvc, ceramic, plate metals and plastics

Referring to this problem, a research has done to know the quality of ground water in Matraman district, East Jakarta (106⁰49'35" E 06⁰10'37" LS) who would be emphasized to the examination of the presence of *Escherichia coli*. For supporting this analysis, examination parameter physical chemistry such as temperature, pH, Dissolved Oxygen (DO). Pattern to scatter bacteria *Escherichia coli* needed to provide recommendations whether the ground water at Matraman district, East Jakarta worthy used as the source of drinking water or not.

2. Research Method

Measurement of the pH using pH indicators, with the Winkler methods modified by Alsterberg for remove the effect of ions nitrites in the water by means titration, while water temperature measured by thermometer. The amount of bacteria Escherichia coli calculated by a method of Most Probable Number (MPN) based on SNI 01-2332.1-2006 [7] about how to test microbiology section 1 namely the determination of coliform and Escherichia coli on fisheries products. A method of MPN it was split into three (3) stages examination or the determination of the presumptive test, confirmation test and complete test. The results obtained will be compared with the Decree of Health Ministry No.492 year 2010 about the requirements the quality of drinking water that bacteria *E. coli* is must be 0 MPN/100 ml.

The sampling point of ground water determined by stratified random sampling methods on the six sub-district at Matraman district, based on the wide of the area and the possibility of a great height of *Escherichia coli* like near the market, close to river and slums area (Table 1).

Table 1. Sampling point and coordinate.

Sampling Point	Location (Subdistrict) and Coordinate	Sampling Point	Location (Subdistrict) and Coordinate
1	Sub-District Utan Kayu Utara 106⁰51'47" E ; 6⁰11'36" S	10	Sub-District Kebon Manggis 106⁰51'21" E ; 6⁰12'15" S
2	Sub-District Utan Kayu Utara 106⁰52'13" E ; 6⁰11'36" S	11	Sub-District Kebon Manggis 106⁰51'8" E ; 6⁰12'28" S
3	Sub-District Utan Kayu Utara 106⁰52'26" E ; 6⁰11'49" S	12	Sub-District Palmeriem 106⁰51'34" E ; 6⁰12'28" S
4	Sub-District Kayu Manis 106⁰51'34" E ; 6⁰11'49" S	13	Sub-District Kayu Manis 106⁰51'47" E ; 6⁰12'28" S
5	Sub-District Palmeriem 106⁰51'21" E ; 6⁰12'2" S	14	Sub-District Utan Kayu Selatan 106⁰52'13" E ; 6⁰12'28" S
6	Sub-District Palmeriem 106⁰51'34" E ; 6⁰12'2" S	15	Sub-District Pisangan Baru 106⁰52'0" E ; 6⁰12'41" S
7	Sub-District Utan Kayu Selatan 106⁰52'0" E ; 6⁰12'2" S	16	Sub-District Palmeriem 106⁰51'47" E ; 6⁰12'41" S
8	Sub-District Utan Kayu Selatan 106⁰52'13" E ; 6⁰12'2" S	17	Sub-District Kebon Manggis 106⁰51'21" E ; 6⁰12'41" S
9	Sub-District Utan Kayu Selatan 106⁰52'0" E ; 6⁰12'15" S	18	Sub-District Pisangan Baru 106⁰52'26" E ; 6⁰12'54" S

IOP Conf. Series: Earth and Environmental Science **106** (2017) 012080 doi:10.1088/1755-1315/106/1/012080

3. Results and Discussion

3.1. The Measurement Result

The measurement result of groundwater in all points of sampling indicate the average DO values is 2.24 mg/l, the average temperature is 27.52^0 C and the average of pH is 5.59. The existence of DO in ground water could assure bacteria to survive, while for pH is below the requirement and for temperature must be ± 3^0C air temperature. Escherichia coli is growing on temperature between 10-45^0C with the optimum temperature is 37^0C and optimum pH of 7-7,5 but able to survive at pH 4 (minimum) and 9 (maximum) [8] to the steady state for E.coli to grow well and be expand. So it can be concluded that E.coli can still growing based on the results of research finding and the quality of water did not fulfill the criteria for drinking water.

3.2. The Measurement Result of E. Coli at Matraman District

The amount of E.coli in all sampling points show in Table 2. In Table 2 appears that the presence of bacteria E. coli on groundwater in Matraman District within range <3 MPN/100 ml up to >1100/100 ml. The distribution patterns of E. coli is shown in Figure 1a for 1st period, Figure 1b for 2nd period and Figure 1c for 3rd period. The number of E. coli in this research it would probably influenced by several factors such as distance between septic tank and well, the depth of well, climate, human activities, and sanitation.

The distance between septic tank and well must be more than 10 m and the depth of the well should be 7 m based on the Regulation of Ministry for Public Works and Public Housing No.27 year 2016. Climate, heavy rainfall and flood will cause the impact, especially severe for sanitation system that still bad in slum areas. *E.coli* cause diarrhea that can be contagious or through the intermediary water. Based on Jakarta Environmental Management Board Report Year 2004-2005 that 75 ground water well in Jakarta have high pollutants organic and inorganic. It means that groundwater in Jakarta has not fulfill water quality standard, especially for drinking water. In East Jakarta, 45% ground water have been contaminated by fecal coliform and 80% by *E. Coli* [9]. The other research in Ciracas District, East Jakarta report that coliform and *E.coli* in ground water within range of 18-26000 MPN/100 mL samples and 1.8-6790 MPN/100 mL, respectively [10].

Table 2. Sampling point and number of *E. coli.*

Sample point	Depth of well (m)	Distance between well and septic tank (m)	1st Period	2nd Period	3rd Period	Regulation Standard [3]
			MPN/100 ml	MPN/100 ml	MPN/100 ml	
1	25	5	1100 ± 0	220 ± 9.8	>1100 ± 0	
2	12	3	7.4 ± 0	195 ± 6.3	>1100 ± 0	
3	25	2	43 ± 0	43 ± 0	1100 ± 0	
4	10	13	<3 ± <3	43 ± 0	695 ± 5.70	
5	15	3	>1100 ± 0	>1100 ± 0	>1100 ± 0	
6	15	10	21.5 ± 2.12	1100 ± 0	>1100 ± 0	
7	14	1.5	93 ± 0	210 ± 0	>1100 ± 0	
8	15	10	10.1 ± 1.27	23 ± 0	350 ± 5.50	
9	22	9	1100 ± 0	240 ± 0	780 ± 15.00	0
10	14	20	3 ± 0	<3 ± <3	<3 ± <3	
11	20	5	<3 ± <3	23 ± 0	<3 ± <3	
12	14	8	>1100 ± 0	>1100 ± 0	>1100 ± 0	
13	12	1.2	>1100 ± 0	1100 ± 0	<3 ± <3	
14	14	2	23 ± 0	6.4 ± 3.96	1100 ± 0	
15	15	10	3.6 ± 0	58 ± 4.9	>1100 ± 0	
16	15	15	18 ± 4.24	<3 ± 1.4	251.5 ± 4.90	
17	10	5	39 ± 5.66	3.3 ± 0.42	<3 ± 0.39	
18	30	11.4	460 ± 0	126.5 ± 4.70	>1100 ± 0	

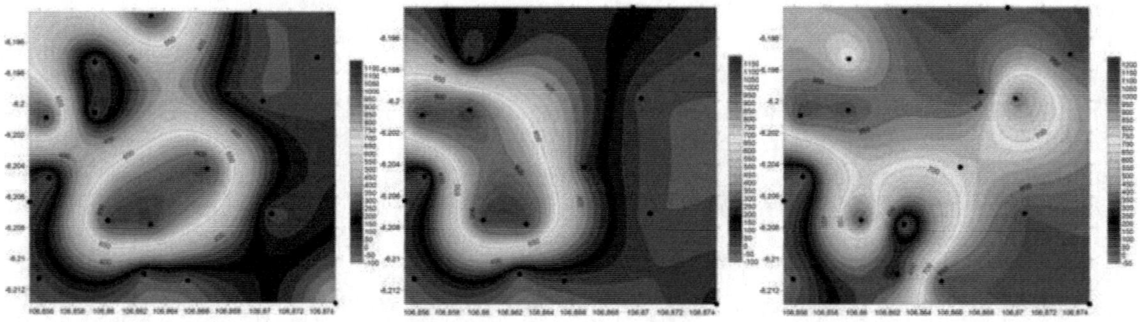

Figure 1. The distribution patterns of *E. coli* a) 1st Period, b) 2nd Period, c) 3rd Period.

In this research, the depth of well and the distance of the septic tank to the well also be a concern. The correlation between the number of bacterial cells of *E.coli* and the depth of well presented in Figure 2. Based on the Figure 2 it can be concluded that there is a relation between number of *E. coli* against the depth of well as much as 6.5% for 1st period, 3.2% for 2nd period and 9.9 for 3rd period. The correlation between the number of bacterial cells of *E.coli* and the distance between septic tank and well is presented in Figure 2. Based on the Figure 2 it can be concluded that there is a relation between number of *E. coli* against the distance between septic tank and well as much as 6.3% for period I, 6.3% for period II and 8.7% for period III. The graph shows that although the distance septic tank and well far enough (>10m) and the depth of well is 7 m, the number of *E.coli* was high.

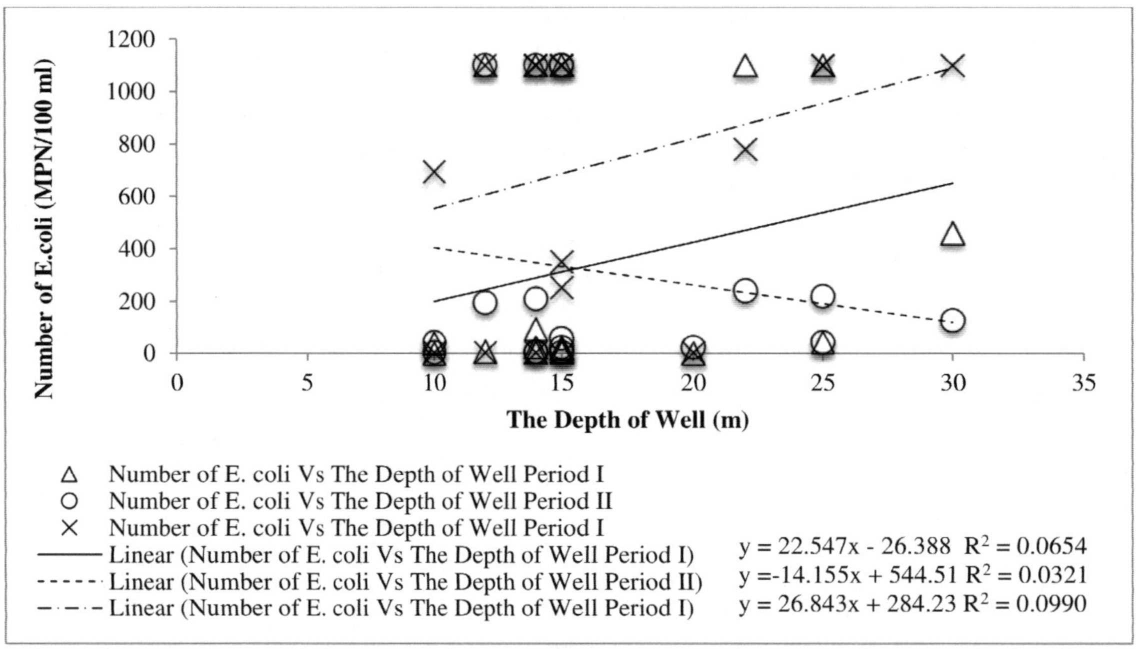

Figure 2. Number of *E.coli* (MPN/100ml) vs the depth of well.

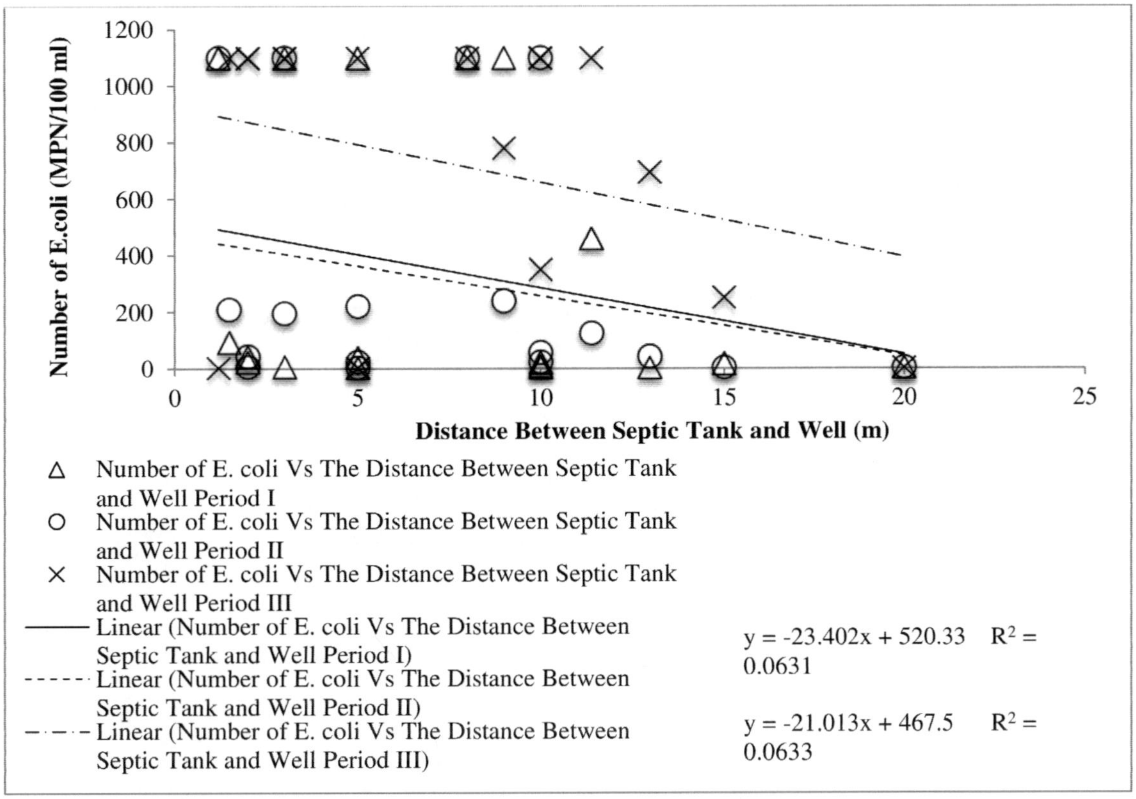

Figure 3. Relationship between number of *E.coli* and the distance between septic tank and well

Allegedly building construction of septic tank that not met the standard that causes the leak of septic tank. Considering Matraman District is very dense residential population so the distance between the septic tank and the well is near. Although wells and septic tank which is private property have met standard but the condition of wells and septic tank belonging to neighbors could be not qualified then there is still found high numbers of *E.coli*. This is in accordance with the research of [11] states that there is relationship between the depth of the wells and septic tank distance. On the contrary, although the distance septic tank close enough (<10m) but found very little *E.coli*. Allegedly building septic tank around the sample has met standard construction so there is no leakage in the septic tank to well.

4. Conclusion

The depth of well and distance septic tank to well affecting the amount of *E.coli* in ground water. Deeper well contained a few bacteria *E.coli*. The same case with the distance septic tank and well, the far the distance the amount of bacteria *E.coli* will descrease. With the result that have obtained, ground water at Matraman District not meet the quality standard of drinking water that had been determined.

References

[1] Unicef Indonesia. 2012. Ringkasan Kajian Air Bersih, Sanitasi dan Kebersihan.
[2] World Health Organization. 2016. Sanitation.
[3] Health ministry regulation of the Republic of Indonesia No.492, year 2010. The requirements the quality of drinking water.
[4] Emmanuel Aboh. 2015. Microbiological assessment of well waters in Samaru, Zaria, Kaduna,State, Nigeria. *Annals of African Medicine*. 14(1). Nigeria.
[5] The British Columbia Ground Water Association. 2007. Total, Fecal and E.coli Bacteria in Groundwater. Columbia

IOP Conf. Series: Earth and Environmental Science **106** (2017) 012080 doi:10.1088/1755-1315/106/1/012080

[6] SNI (Standar Nasional Indonesia) 03-2398-2002.
[7] SNI 01-2332.1-2006
[8] Faridz Raden. 2007. Analisis Jumlah Bakteri dan Keberadaan Escherichia coli pada Pengolahan Ikan Teri Nasi di PT.Kelola Mina Laut Unit Sumenep. *Embryo.* 4(2).
[9] National Environmental Management Institution. Ground Water Quality : Status of groundwater quality at Jakarta year 2004-2005. Jakarta.
[10] Putri, lily Surayya Eka Putri. 2013. Effect of Land Use on Ground Water Quality (A Case Study from Ciracas Sub District, East Jakarta, Indonesia). *International Journal of Bioscience, Biochemistry and Bioinformatics*. 3(1).
[11] Fachrul, Melati Ferianita. 2009. Contamination and Distribution Pattern of Escherichia coli in Ground Water at Kembangan Distric, West Jakarta, Indonesia.

IOP Conf. Series: Earth and Environmental Science **106** (2017) 012081 doi:10.1088/1755-1315/106/1/012081

Escherichia coli as bioindicator of the groundwater quality in Palmerah District, West Jakarta, Indonesia

M P Dayanti, M F Fachrul*, A Wijayanti

Environmental Engineering Department, Faculty of Landscape Architecture and Environmental Technology, Universitas Trisakti, Jakarta, Indonesia

*Corresponding Author: melati@trisakti.ac.id

Abstract. The aim of thie research is to determine the quality of groundwater in Palmerah District, West Jakarta (6°11'24.32"S 106°47'49.88"E) by correlation between the depth of the well and the distance of septic tank with the distribution of *Escherichia coli*. The presence of *E. coli* is measured by using the Most Probable Number method. The distribution pattern of the *E. coli* was processed by Surfer Program. Research was conducted in April upto July 2017. The highest amount of *E. coli* always found in the Jati Pulo and Palmerah sub-district which is >1100 MPN/100 ml; while the lowest amount of *E. coli* found in the Kemanggisan sub-district and Slipi sub-district which is <3.0 MPN/100 ml; with every samples is obtained on the condition of pH 5 – 7, DO 0.81 – 7.65, and water temperature of 26 – 34°C. Refering to the Ministry of Health Regulation No. 492 of 2010 on the requirements of drinking water quality; it is shown that the groundwater in Palmerah District is not feasible to be directly consumed. This research provides the initial information to local sanitation to the distribution pattern of *E. coli* within the dense residential area.
Keywords: *E. coli*, lactose broth, presumtive test, most probable number

1. Introduction

Groundwater is one source of drinking water of Indonesia people. The usage of groundwater as a source of drinking water should be free from parameters that could jeopardize to human health and in accordance with of quality standard drinking water, based on Ministry of Health Regulation No. 492 of 2010 the amount of bacteria *E. coli* in drinking water is zero. Coliform bacteria were used to assess the general hygienic quality of water. The thermotorelant faecal coliforms, especially the presence of *Escherichia coli* in water, indicate recent faecal pollution [1]. Many waterborne enteric pathogens are transmitted by the faecal-oral route. Faecal contamination can occur from a variety of sources, including: poor sanitation infrastructure, faulty wastewater treatment facilities, leaking septic tanks, domestic and wild animal feces [2].

Consumption of groundwater and well drilling without suitable conditions increase the vulnerability to contamination, related either to anthropogenic or to natural activities. Degradation and contamination of groundwater are related to the depth and the type of well [3], [4]. The lack of groundwater quality monitoring and of regulations governing the drilling of wells result in the population consuming water that lacks proper treatment. Consumption of water without adequate quality control is a public health concern, because it is often a vehicle for spreading disease [5]. Preventive methods such as proper well site selection and construction are the best way to safeguard water supplies against contamination by fecal material [6]. In aquifers, water can move quickly, allowing transport of microorganisms, with insignificant interaction between these microorganisms and the host rock. In porous aquifers, such as gravel or coarse sand aquifers, permeability results from pore space between grains. In this case, although the microorganisms can be easily transported, they also undergo interaction with the sand-gravel matrix, which can reduce pollution loads [7].

Based on the Central Agency on Statistic of West Jakarta City, shows that the density of Palmerah sub-district is the second highest of West Jakarta which is 28,678.03 (person/km^2). The use of well as a source of drinking water in Palmerah sub-district is one of the highest in the area of municipal West Jakarta namely 36,509 head families. Based on the Regional Environmental Agency Province of Jakarta (2015), the groundwater of the Palmerah sub-district have not fulfilled of quality standard.

2. Research Method

The study areas is Palmerah sub district, West Jakarta consisting of 6 sub-district. The sampling was conducted in three period in April up to July 2017), those are Period 1 (April-May), Period II (June) and Period III (July) in 17 sampling points, categorized based on the economy conditions that were high income, middle income, and low income done by questioners. The analysis of research was conducted in the Biology and Microbiology Laboratory of Environmental Engineering, Universitas Trisakti.

Table 1. Location sample points.

Sample Points	Sub-district	Location	Coordinate
1	Jati Pulo	RT. 03/RW.01	6°10'47.04"S 106°47'67.54"E
2		RT.11/RW.05	6°10'45.00"S 106°48'30.37"E
3	Kota Bambu Utara	RT.04/RW.02	6°11'01.69"S 106°48'30.20"E
4		RT.06/RW.06	6°11'01.00"S 106°48'30.15"E
5	Kota Bambu Selatan	RT.10/RW.04	6°11'15.80"S 106°48'29.00"E
6		RT.08/RW.09	6°11'17.00"S 106°47'60.00"E
7	Slipi	RT.03/RW.02	6°11'30.03"S 106°48'04.00"E
8		RT.10/RW.06	6°11'44.72"S 106°48'15.15"E
9		RT.02/RW.04	6°12'00"S 106°48'03.00"E
10	Kemanggisan	RT.12/RW.06	6°11'15.38"S 106°47'28.64"E
11		RT.03/RW.09	6°11'31.00"S 106°46'59.69"E
12		RT.01/RW.06	6°11'30.59"S 106°47'30.00"E
13		RT.07/RW.02	6°11'30.29"S 106°47'45.03"E
14	Palmerah	RT.08/RW.13	6°11'44.71"S 106°47'00"E
15		RT.02/RW.12	6°12'03.01"S 106°47'15.00"E
16		RT.01/RW.09	6°12'15.09"S 106°47'29.28"E
17		RT.03/RW.04	6°11'58.00"S 106°47'44.90"E

The parameters that were analyzed in physical parameters is temperature, the chemical parameters is pH and DO (dissolved oxygen). The *Escherichia coli* that wre analyzed byh the Most Probable Number (MPN) methods using 3 series tube, referring to SNI 01-2332.1-2006, they were the presumptive test, confirmed test and completed test, with three times repetition (triplo).

3. Results and Discussion

The result show, that people in Palmerah sub-district use well pump to take the groundwater as the source of clean water. The condition of housing from questioner show that 17.64% depth of the well is <10 m, the wells with 10 – 20 m depth were 35.3% and the wells with >20 m depth were 47.06%. The distance from wells to septic tank are 88.24% <10 m, is not comply with SNI 03-2398-2002.

IOP Conf. Series: Earth and Environmental Science **106** (2017) 012081 doi:10.1088/1755-1315/106/1/012081

Table 2. Measurement results of well depth and well distance towards septic tank on total *E.coli.*

Sample Location	Sub-district	Depth of wells (m)	Distance towards Septic Tank (m)	MPN / 100 ml			Standard*)
				I	II	III	
1	Jati Pulo	8	6.5	>1100 ± 0	>1100 ± 0	>1100 ± 0	0
2		8	8.0	180 ± 2,1	166.5 ± 2.4	121.5 ± 1.5	0
3	Kota Bambu Utara	4	3.25	>1100 ± 0	305 ± 4.6	33 ± 0.4	0
4		20	2.0	276.5 ± 4.6	225 ± 2,4	166,5 ± 2.4	0
5	Kota Bambu Selatan	25	5.0	86.5 ± 1.5	>1100 ± 0	68 ± 0.9	0
6		10	10	33 ± 0.4	>1100 ± 0	195 ± 2.4	0
7	Slipi	40	3.0	5.4 ± 0.4	69 ± 7,1	<3.0 ± 0	0
8		30	5.0	6.4 ± 0.4	225 ± 2,4	3.6 ± 0	0
9		20	10.0	93 ± 4	65,5 ± 9,3	>1100 ± 0	0
10	Kemanggisan	30	5.0	16,1 ± 0	5,5 ± 2,7	33 ± 0	0
11		30	5.0	93 ± 9.8	9,2 ± 2,5	>1100 ± 0	0
12		30	4.0	>1100 ± 0	240 ± 0	460 ± 0	0
13		30	4.0	<3.0 ± 0	43 ± 0	43 ± 0	0
14	Palmerah	20	5.2	>1100 ± 0	>1100 ± 0	>1100 ± 0	0
15		8	6.5	180 ± 2.1	43 ± 0	121.5 ± 1.5	0
16		8	8.0	93 ± 0	150 ± 0	195 ± 2.4	0
17		4	3.25	93 ± 0	35 ± 0,4	150 ± 0	0

*) the Ministry of Health Regulation No. 492 of 2010 on the requirements of drinking water quality

The highest amount of *E. coli* is >1100 MPN/100 ml which on period I is found in sample point 1, 3, 12, and 14, meanwhile the lowest amount of *E. coli* is found in point 13 with <3,0 MPN/100 ml. On period II, the highest amount of *E. coli* is also found in sample point 1, 14, also 5 and 6, meanwhile the lowest amount of *E. coli* is found in point 10 which is 5.5 MPN/100 ml. On period III the highest amount of *E. coli* is found in sample point 1, 14, also 9, with the lowest amount of *E. coli* is in point 7 which is <3 MPN/100 ml.

The range of temperature show in Table 3 is 27 – 31°C in period I, 27 – 34°C in period II, and 26 – 29°C in period III with average is 28.76 °C. The pH range is 4 – 7 in period I, 5–7 in period II and III which the average is 5.67. Dissolved oxygen range is 1.01 – 3.02 mg/l in period I, 0.81–7.65 mg/l in period II, and 0.95 – 3.66 mg/l in period III with average is 2.42 mg/l.

Escherichia coli grows in temperatures between 10 - 45 °C, with an optimum temperature is 37°C, and has an optimum pH for its growth is 7 -7.5, a minimum pH is 4 and a maximum pH is 9 [8], so that E. coli can still grow in water wells in Palmerah Sub-district. Referring to the quality standard by Ministry of Health Regulation No. 492 of 2010, indicating that ground water in Palmerah Sub-district is not feasible to be consumed directly as drinking water.

Table 3. Measurement results on physical and chemical parameters.

Sample Location	Sub-district	Temperature (°C)			pH (unit)			Dissolved Oxygen (mg/l)		
		I	II	III	I	II	III	I	II	III
1	Jati Pulo	27	27	27	5	7	6	1.41	2.01	1.81
2		28	28	27	5	6	6	1.01	0.81	0.95
3	Kota Bambu	31	29	28	5	6	6	2.42	3.22	2.62
4	Utara	28	27	28	4	6	6	1.41	3.62	2.44
5	Kota Bambu	30	29	29	5	7	7	2.01	7.65	3.66
6	Selatan	29	28	27	6	7	6	2.42	4.63	3.51
7	Slipi	28	29	29	5	6	7	2.01	1.21	1.61
8		27	30	28	5	6	6	2.62	4.63	3.42
9		27	28	28	7	6	6	1.81	3.22	3.03
10	Kemanggisan	28	32	28	4	6	6	3.02	2.42	2.72
11		27	29	28	5	6	6	2.01	2.01	2.05
12		29	30	26	6	6	6	1.01	2.62	1.23
13		27	29	27	5	5	5	2.21	4.23	2.43
14	Palmerah	31	29	27	5	6	6	1.81	1.21	1.51
15		31	34	27	6	6	6	2.01	1.81	1.91
16		28	28	27	5	7	7	1.01	1.81	1.51
17		28	28	26	5	6	6	1.41	3.42	2.44

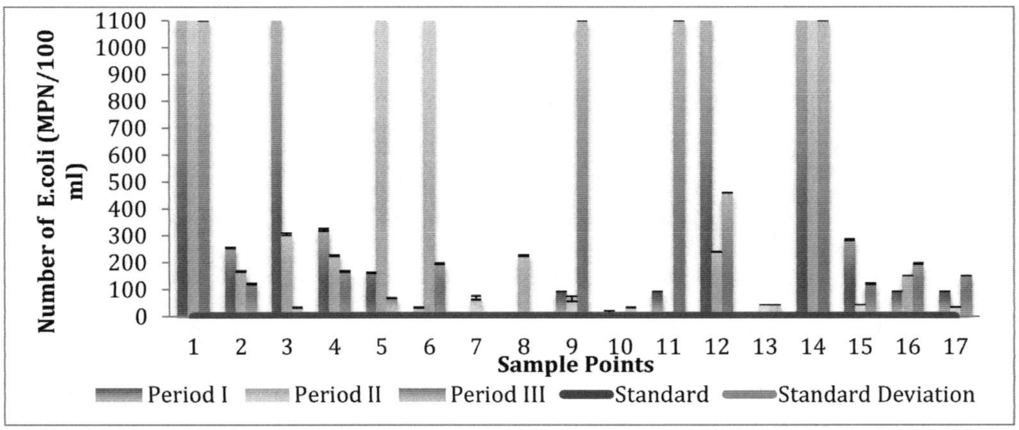

Figure 1. Numbers of *E.coli*.

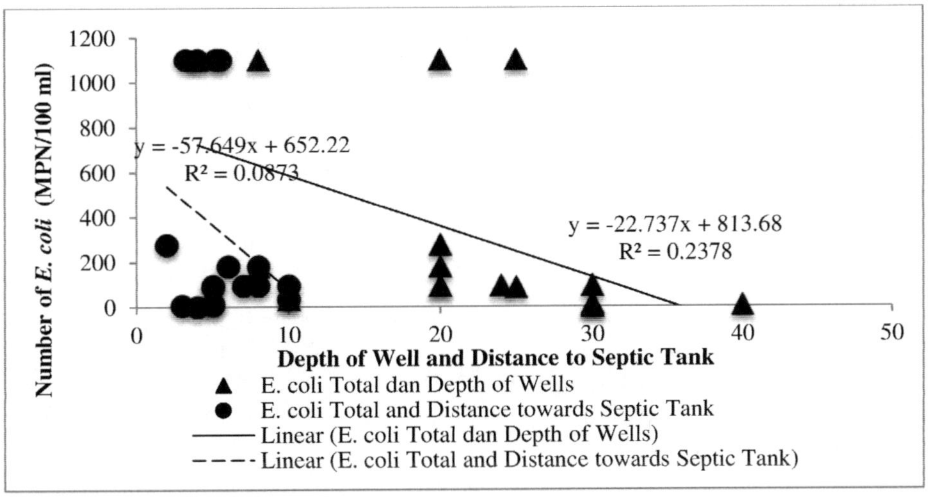

Figure 2. Relationship among number of *E.coli* and depth of well and distance to septic tank period I.

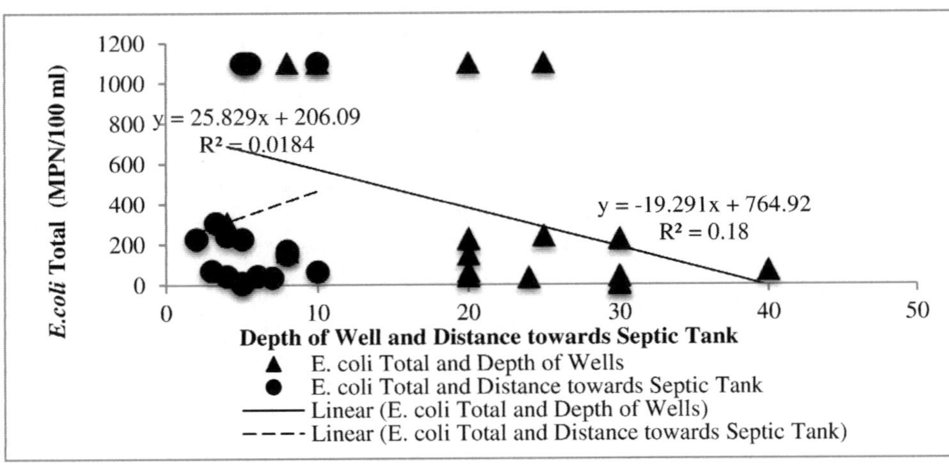

Figure 3. Relationship among number of *E.coli* and depth of well and distance to septic tank period II.

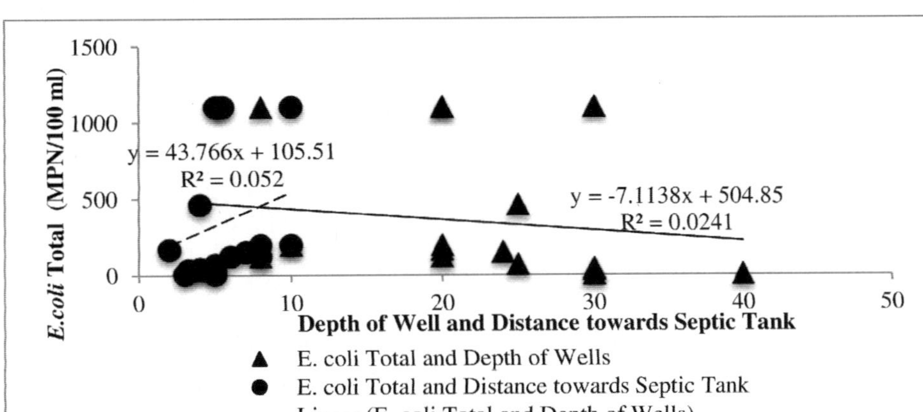

Figure 4. Relationship among number of *e. coli* and depth of well and distance to septic tank period III

Coefficients correlation in graphics show that the strength and direction of a relationship between two continuous variables [9]. The Figure 2 show R^2 0.087 and 0.237, Figure 3 R^2 is 0.018 and 0.18,

IOP Conf. Series: Earth and Environmental Science **106** (2017) 012081 doi:10.1088/1755-1315/106/1/012081

and Figure 4 R^2 is 0.052 and 0.024 shows that there is a weak correlation between the number of *E.coli* with the depth of the well and the distance to the septic tank. As the research conducted by [10] on the number of *E. coli* in Kembangan sub-district, West Jakarta, which had the highest *E. coli* number is 1100 MPN/100 ml and the lowest is 11 MPN/100 ml, with R^2 was 0.629 in the amount *E. coli* to the depth of the well and 0.643 in the number of *E. coli* to septic tank distances. This indicates that the deeper the well and the longer the distance to the septic tank, the higher the amount of *E. coli* contained in groundwater [10].

4. Conclusion

Result of this research shows that there is a weak correlation between the number of *E. coli* bacteria with well depth and well distance to septic tank in Palmerah sub-district. The relationship shown in is the closer the well distance to the septic tank and the lower the depth Well the higher the amount of *E. coli*. Based on the result of the research, all samples analysed have *E. coli* amount above the standard so it is not feasible to be consumed directly.

References

[1] Momba M N B, Makala N, W R Kelly, and T Mazomba. 2008. Possible Threat of the Microbiological and Chemical Contamination of the Mpofu. Training Centre Borehole to the Health and Surrounding Community (South Africa: WISA), p 139.

[2] Neil L, 2008, Tracking Faecal Pollution in Environmental and Drinking Water Resources: Assessing the Potential of Emerging Culture and Library-Independent Source-Tracking Technologies, South Africa: WISA, p 188.

[3] Hynds P, Misstear B D, Gill L W, Murphy H M. 2014. Groundwater source contamination mechanisms: Physicochemical profile clustering, risk factor analysis and multivariate modelling. *J. Contam. Hydrol.* 159:47–56. doi: 10.1016/j.jconhyd.2014.02.001.

[4] Azizullah A, Khattak M N K, Richter P, Häder D P. 2011. Water pollution in Pakistan and its impact on public health—A review. *Environ. Int.* 37:479–497. doi: 10.1016/j.envint.2010.10.007.

[5] Maran N H et al., 2016. Depth and Well Type Related to Groundwater Microbiological Contamination *.United States: Int J Environ Res Public Health.*

[6] The British Columbia Ground Water Association, 2007, Total, Fecal and E.coli Bacteria in Groundwater (Columbia).

[7] Berger P, 2008. Viruses in Ground Water. In Dangerous Pollutants (Xenobiotics) in Urban Water Cycle. Part of the series NATO Science for Peace and Security Series. Hlavinek P et al. (eds), Springer, 131–149.

[8] Faridz R, Hafiluddin. and Anshari M. 2007 Analisis Jumlah Bakteri dan Keberadaan Escherichia coli pada Pengolahan Ikan Teri Nasi di PT. Kelola Mina Laut Unit Sumenep, Indonesia, vol 4 p 94-106.

[9] Schneider, Astrid., et al. Linear Regression Analysis. https://www.ncbi.nlm.nih.gov/pmc/articles/PMC2992018/. Diunduh pada 1 July 2017

[10] Fachrul M F, 2009, Contamination and Distribution Pattern of Escherichia Coli in Ground Water at Kembangan District, West Jakarta, Indonesia, pp 16-21.

Development of salt production technology using prism greenhouse method

G Guntur, A A Jaziri, A A Prihanto, D M Arisandi, A Kurniawan*

Faculty of Fisheries and Marine Science, Universitas Brawijaya, Malang, Indonesia

*Corresponding Author: andi_k@ub.ac.id

Abstract. The main problem of salt production in Indonesia is low productivity and quality because the technology used commonly by Indonesian salt farmers is traditional method. This research aims to increase production of salt by using the prism greenhouse method. The prism greenhouse method is a salt production system with a combination of several salt production technologies, including geomembrane, threaded filter, and prism greenhouse technology. This research method used descriptive method. The results of this study were the productivity increased threefold, and the quality of salt produced also increased in terms of the content of NaCl from 85% to 95%. In addition, salt production with the prism greenhouse method has several advantages, such as faster harvest time, weather resistance, easy to use, and higher profit than traditional methods.
Keywords: geomembrane, prism greenhouse, salt, threaded filter

1. Introduction

Salt is an important food ingredient in life. Indonesia's government had the priority program in the field of food sovereignty targeting salt production around 3.6 million tons in 2016. The General Director of Marine, Coastal and Small Islands at the Ministry of Marine and Fisheries Affairs stated that the need for salt of household and industrial consumption is about 3.6 million tons. It consists of 1.48 million tons for household consumption and 2.12 million tons for industrial consumption. However, in real, salt production only reached 144,009 tons or 4.6 percent of national needs [1]. Through the Ministry of Trade, the government imported salt from other countries to meet the need of the salt. Ironically, Indonesia is a leading salt producer country with the second longest coastline in the world of about 99.093 km^2. In addition, Indonesia has potential salt field of 29,000 ha, which should be able to meet the national salt needs [2].

On the other hand, The Ministerial Regulation no. 125 of 2015 showed that the provisions on the import of salt is contradictory to the government's priority program, for instances 1) the absence of importer obligations to buy salt from farmer at least 50% of total production; 2) there is no provision on the price determination of salt produced by farmer; and 3) prioritizing the import of salt, either household or industrial consumption. Implementation of salt importing policy is considered not in favor of farmers since salt produced by farmers is commonly low quality.

The salt production is mostly carried out by independent farmers, and only a small number of salt productions is conducted by private companies and PT. Garam, a state-owned salt producing firm. The salt production technology used by Indonesian salt farmers, in general, is still traditional. Traditional method has been developing in Indonesia since the 19th century. The salt production process in traditional method consist of evaporation, precipitation, and crystallization. The evaporation process spends 7-10 days. Afterthat, the evaporated seawater is transferred into the salt table for precipitation process. This process take 10-15 days, then the precipited seawater flowed into the salt table for crystallization process. The crystallization process will take 4-10 days relying on the weather [3]. The farmers normally produce and harvest the salt three times a year.

The criteria of salt production using traditional method have relatively low in terms of productivity and quality. The productivity of land used to salt production is 60 ton/ha/year. The quality of salt production contains sodium chloride (NaCl) and other impurities, i.e. magnesium, calcium, etc. The NaCl is around 80-85%, and it is under Indonesia National Standarization (SNI). The color of salt product is russet and the taste is slightly bitter [4]. Because of low quality, the price of the salt product is low. Hence, the development of salt production technology is needed to increase productivity and quality of salt product. Many researchers reported that geomembrane and filtering-threaded technology could increase productivity and quality of salt product.

Geomembrane technology is the method used to produce salt by coating the base of pond with geomembrane plastic. The main characteristics of geomembrane as a liner in salt ponds are economic value, durability, and resistance to degradation by biological, chemical and ultra violet (UV). The benefits of salt production through geomembrane technology are the productivity of salt increased, the evaporation process is shorter, the turnaround time is improved, the harvest and the quality of salt is improved and the long lifetime of geomembrane will reduce the preparation works [5].

Filtering-threaded technology (TUF) is the modified technology of salt production integrating geomembrane technology through the serial plots. The main principle of TUF is the evaporation process of seawater flowed through the serial plots in the salt tables. In addition, natural materials are added as filters to purify seawater [6]. The advantages of implementing TUF are the productivity could attain 80 ton/ha/cycle [7], the color of salt product is white and pure, the purity of salt (NaCl) and price are higher than traditional process products [6].

Although the geomembrane and the filtering-threaded technology are better than the traditional technology, its development still requires more innovative breakthrough and more effective technologies to deal with unstable weather in Indonesia. In 2016, the productivity of salt production throughout area in Indonesia decreased drastically because Indonesia was suffered tropical wet drought, so salt production process needs longer time. Therefore, one of alternative technologies is prism grenhouse. Prism greenhouse is a method to produce salt by using greenhouse model formed a prim structure. The principle of this method is integrating geomembrane and filtering-threaded technology. In this study, we focus on productivity and quality of salt produced through the prism greenhouse method. The study was performed in Sidayulawas village, Brondong District, Lamongan Regency, East Java.

Lamongan is one of the regency in Indonesia that has the potential of salt producing area. The potential area of salt production reaches 350 ha with total salt production reaches ~30,000 tons per year. However, the salt production in 2016 only reached 80 tons although it was higher than the previous year. The productivity of salt in Lamongan should be able to be improved every year. The central of salt productions in Lamongan located in the district Brondong, including Lohgung, Labuhan, Sedayulawas, Brengkok and Sidomukti villages.

2. Research Method

The research was conducted on January 3rd to February 28th, 2017 in Arifin's salt pond located in Sedayulawas, Paciran, Lamongan, East Java, as described on Figure 1.

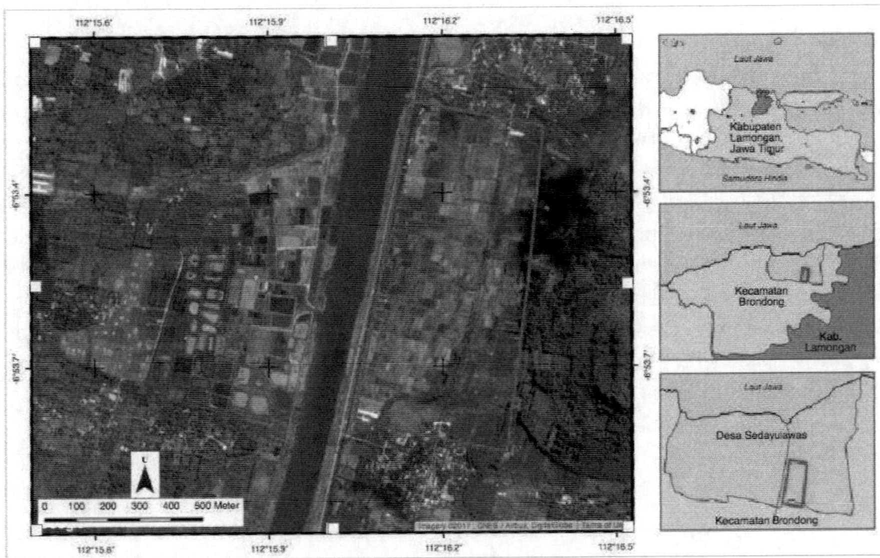

Figure1. Map of research on salt producing technology.

This study used the descriptive research with observational and survey methods [8]. The survey performed include measurements of water or brine temperature, physical observation and interview. The physical observation was carried out to know the salt producing process using prism greenhouse method as showed on Figure 2.

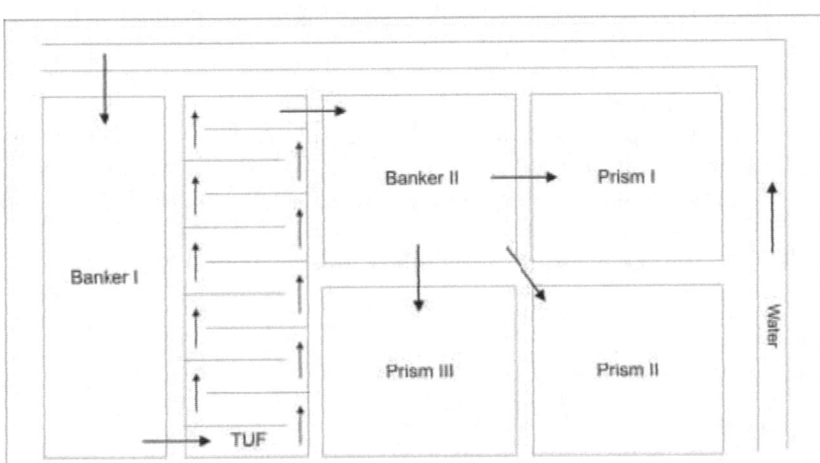

Figure 2. The schematic of salt producing technology in the prism greenhouse method.

The interview conducted to tabulate the problems facedby farmer during producing salt in the prism method, especially to increase productivity of salt. The productivity can be determined by comparing the area with yield of production. The quality of salt can be measured by titration using Volhard's method [9].

3. Results and Discussion

The salt producing technology in Indonesia was much less developed since 19th century. Nowadays, with increasing concerns of salt production, the development of salt producing technology is urgently required. One of the alternative technologies is prism greenhouse method. The prism greenhouse method is an integration of pre-existing salt producing technologies included geomembrane [6] dan filtering-threaded technology [7]. The process of salt production using the prism greenhouse method is (1) banker I, as a reservoir pond, to collect seawater with concentration of 7-8 °Bé (°Baumé); (2) filtering-threaded pond, as a evaporation pond, to evaporate seawater until reach concentration of 20-22 °Bé; (3) banker II, as a precipitation pond, to precipitate seawater with concentration 25-27 °Bé; and (4) prism greenhouse, as a crystallization pond, to crystallize seawater. The process of salt producing technology in the prism greenhouse method can be saw on Figure 3.

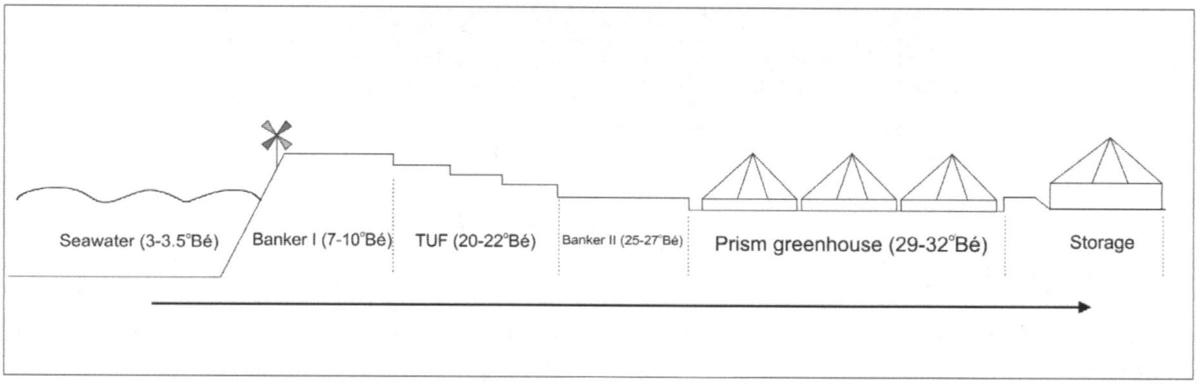

Figure 3. The salt producing technology in the prism greenhouse method.

3.1. The productivity of salt production

The productivity is very important to determine the sustainability of salt production. The salt productivity describe the ability to produce a lot of salt by using a narrow field. In this study of productivity, we compared traditional to prism greenhouse method, as described on Table 1.

Table 1. The comparison of traditional and prism greenhouse method.

Parameters	Traditional Method	Prism Grenhouse Method
Productivity (kg/ha)	30	100
Sodium chloride (NaCl) (%)	85	95
Color	Slight brown	White
Product price (IDR/kg)	250-300	500

The Table 1showed that the salt productivity with prism greenhouse method generated salt around 100 kg/ha. It wasthreefold higher than using traditional method of 30 kg/ha. The prism greenhouse method indicated this technology more effective in producing salt than others.Santoso [9] reported that the salt productivity using filtering-threaded technology could increase twofold than conventional method. The increase of productivity in prism greenhouse method occuredbecause of integration betweengeomembrane and filtering-threaded technology, so the process of making salt through evaporation, precipitation, and crystallization pond could produce salt in the shorter period.

3.2. The quality of salt production

The quality of salt production is determined to know the characteristic of salt. The characteristic can be analysis by comparing color and sodium chloride (NaCl) concentration. The color of the salt produced by using prism greenhouse method is significantly whiter than the salt produced by traditional method, as shown on Figure 4 and 5.

Figure 4. The salt produced by prism greenhouse.

Figure 5. The salt produced by traditional.

Furthermore, the purity of salt (NaCl) produced by using prism greenhouse method is 95%, which is higher than traditional method of 85%. The purity was effected by the filters located on the filtering-threaded pond, so that the NaCl produced by prism greenhouse method is significantly higher purity than traditional method. Because of higher purity, the salt produced by prism greenhouse is more expensive as described at Table 1. Hence, the salt producing technology using prism greenhouse technology has significant impact both on product quality and quantity. In addition, salt production with the prism greenhouse method has several advantages, such as faster harvest time, weather resistance, easy to use, and higher profit than traditional methods.

4. Conclusion

The salt producing technology using prism greenhouse method could beconcluded that the productivity and quality was significantly higher than traditional method. As the results, the price of salt poduced by prism greenhouse method also more expensive than traditional method. The prism greenhouse technology is suitable for producing salt in Indonesia to attain national target of salt product.

Acknowledgment

The authors are grateful to the Faculty of Fisheries and Marine Science and Coastal and Marine Research Center, Brawijaya University, Malang, Indonesia for financial support through research grant.

References

[1] Central Bureon of Statistic. Salt Production Report. 2016.
[2] Ministry of Fisheries and Marine Affairs.Annual report of Fisheries and marine Science Affair. 2016.
[3] Mustofa Turjono E. Optimum Analysis on Farmer Salt Activities Through Ultimate Approach in Mining. Probolinggo J. Wiga 5.
[4] Sudarto. 2011. Process Technology Saltern in Indonesia. *J. Triton* 7:13-15.
[5] Suhendra A. 2016. Increasing The Productivity of Salt Trought HDPE Geomembrane Indonesia Case History in Salt Evaporation Pond. *EJGE* 11:4272-4280.
[6] Susanto H, Rokhati N, Gunawan W Santoso. 2015. Development of Traditional Salt Production Prosess for Improving Product Quality dan Quality in Jepara District, Central Java,

Indonesia. *Proceeding Environmental Science.* 23:175-178.
[7] Bramawanto R, Sagala S L, Suheimi I R, and Prihatno. 2015. Structure and Composition od Salt Pond Using Filtering Threaded Technology to Increasing Salt Production. *J Segara* 11:1-11. 2015.
[8] Arikunto Suharsimi. Research Procedure. Rineka Cipta Jakarta. 2010.
[9] Iacoban C, Macoveinu M. 2005. A Comparison in Precipitation Samples Romanian Society of *Biological Sciences.* 10:2105-2109.

IOP Conf. Series: Earth and Environmental Science **106** (2017) 012083 doi:10.1088/1755-1315/106/1/012083

Constructing counterproductive behavior for supporting evironmental management system research

Tiarapuspa[1*], D L Indyastuti[2], W R Sari[1]

[1]Faculty of Economic and Business, Universitas Trisakti, Jakarta, Indonesia.
[2]Faculty of Economic and Business, Universitas Jendral Sudirman, Purwokerto, Indonesia

*Corresponding Author: tiara.puspa@trisakti.ac.id

Abstract. This study aims to explore the definition of counterproductive behavior based on supervisors' and sub ordinaries' perceptions. Recently, environmental management system is a strategic tool to gain a competitive advantage. Human resource is the vital factor for successful environmental management system. Counterproductive behavior will destroy environmental management system. Unfortunately, the construct of counterproductive behavior is still debatable. Different culture show different dimensions and indicators of counterproductive behavior. The unclear construct results ambiguous empirical evidence. This study results that many items are included of counterproductive behavior, such as come late, impolite communication, playing gadget in working time, and the other negative behaviors.
Keywords: counterproductive work behavior, environmental management system, human resource, supervisor and subordinates perceptions.

1. Introduction

Today, we are in a global environment where money, intelligence, is and products flow easily from one country to another country. In globalization era, environmental management system becomes a great issue in scientific discussion [1], [2]. For reducing global crisis, firms need implement the effective environment management system [3]. The productivity of environmental management system increase firm performance [4].

Despite of the constructing factors of effective environmental management system, there are some destructing factors of effective environmental management system. One of the vital destructing factors of effective environmental management system is counterproductive work behavior. "Green" human resource management (HRM) positively affects environmental management system. Green HRM is managing human resource that fulfills firm's current need and social at large as well as confirming the fulfillment of future need. It implies that counterproductive behavior will destroy environmental management system [5]. Although many scholars posit that counterproductive work behavior negatively effect on effective environmental management system, but the measurement and construct of counterproductive work behavior is still debatable. Different countries have different construct of counterproductive work behavior.

Counterproductive behavior is the behavior of the organization and its members. Various kinds of counterproductive work behaviors are delays, theft, sabotage, and evil in both verbal and physical forms [6]. The key characteristic of counterproductive action, the action must have a purpose and not an accident. That is, a person must have malicious intent (e.g., deliberately damaging company's equipment), or a malicious behavior without intent (e.g. employees refuse equipment or security procedures, do the accident itself is not expected) [7]. Employees are failure to complete the job because it does not have the ability or accidents that occur even though it is safe, excluding counterproductive behavior. [8] Behavioral counterproductive work is termed differently by the researcher. Different terms are given by researchers such as counterproductive behavior, workplace deviance, and aggression at work, antisocial behavior, retaliation and violence [9], [10].

The construct of counterproductive behavior is much discussed in Western literature, such as America. Some researchers in Indonesia [11], [12], [13] also conduct counterproductive behavioral studies. These studies [14], [15], [16] show that counterproductive work behavior is an important aspect of employment in Indonesia. Nevertheless, these studies generally use counterproductive behavioral items from Western countries; do not examine the items of behavior of counterproductive work across different cultures.

2. Research Method

To get the items that are the factors of the counterproductive behavior dimension, two steps are taken. The first step of a group of Indonesian managers and supervisors was asked to mention critical events. From these steps generated various events, but has not established the dimensions of behavior. Furthermore, a group of other Indonesian managers are invited to sort the incidents of counterproductive behavior that have been done in the first step. The order is grouped into several categories. Grouping of categories based on the dimensions selection of this dimension because the dimensions of Robinson and Bennet fit the typology of Buss's typology is the most common form of classification for counterproductive behavior [10].

The process is detailed as follows, step 1, raises the behavioral counterproductive behavioral items in Indonesia. Participants for step 1 are 170 respondents who work in service and manufacture industry in Jakarta. Procedure for Step 1, participants are invited to participate in the study on a voluntary basis. Participants presented a broad definition of counterproductive behavior followed by two examples of behavior. The definition of counterproductive behavior is voluntary behavior that violates organizational norms and harms organizational well-being [10]. This information is presented to the participants in Bahasa Indonesia. Participants were then asked to describe five incidents in the workplace that behave counterproductive. Respondents write it in Indonesian. Of the 170 respondents will produce approximately 850 events of counterproductive work behavior [10].

The next three-step process is used to analyze the 850 events. First, researchers review all events. This is to ensure that the event has a clear meaning in the Indonesian language and is behavioral. Events that are ambiguous or difficult to interpret are deleted and are not analyzed further. Second, researchers review each event and are grouped according to similar behavior. Examples of organizational statistics and organization fraud scams are grouped under organizational fraud. After cleaning 850 events will be obtained behavior to be in the next analysis. Third, summarize the entire event in a single table for further analysis (Table 1 for subordinate counter-behavioral behaviors assessed by the leadership and Table 2 for counterproductive behaviors assessed by subordinates) [10].

Furthermore, the work forms of counterproductive work that have been studied will be included in the dimensions of contraproductie behavior of work. These dimensions consist of property deviance, production deviance (denial of formal prohibitions and norms adopted by companies, by doing quality and quantity of work lower than they otherwise are), political deviance (involvement in social interaction that puts individuals in Misfortune personnel or political), and personal aggression (hostility with other individuals) [10].

Step 2: aims to determine the dimensions of counterproductive behavioral items in Indonesia. Participants for step 2 are a different group of 3 managers in Indonesia and asked to do sorting tasks. Procedure for Step 2, participants are invited to participate in this step on a voluntary basis [10]. Participants presented a set of instructions in which they were asked to sort a number of incidents in different categories based on behavioral similarities. Participants were labeled behavioral categories according to the dimensions, and were asked to carefully read each behavior and place it in a category that they believed had behavioral similarities. After each participant completes the sorting task, a research assistant records the number of categories that each participant uses to sort the events of counterproductive behavior. The information in this second step is used as the basis for the analysis. Data analysis was analyzed using qualitative descriptive analysis [10].

Data have been derived from distributing questionnaires to 170 respondents consisting of 81 respondent's questionnaires for employees who assess the leaders and 89 respondent's questionnaires for leaders who assess the employees.

Table 1. Descriptive statistics of the respondent.

Description	Age (years)	Working period in the organization (years)	Work period in this position (years)
Respondent's questionnaires for employees who assess the leaders			
N	81	81	81
Mean	39	12	6
Std. Deviation	13	10	5
Respondent's questionnaires for leaders who assess the employees			
N	89	89	89
Mean	30	6	4
Std. Deviation	11	7	3

Table 1 showed that the number of respondent's questionnaires for employees who asses the leader is 81 with the age average are 39 years and standard deviation 13, for the working period in the organization average are 12 years and standard deviation 10, and also for the work period in this position average are 6 years and standard deviation 5. The number of respondent's questionnaires for leaders who assess the employees is 89 with the average are 30 years and standard deviation 11, for the working period in the organization average are 6 years and standard deviation 7, and also for the work period in this position average are 4 years and standard deviation 3.

3. Results and Discussion

3.1. The list of counterproductive work behaviors described for employees by leaders

Based on interviewing respondents, we get four factors that form counterproductive work behavior of employee (Table 2). First factor is Production deviance. Production deviance have 37 item. Second factor is property deviance. Property deviance have 10 items. Third factor is political deviance. Political deviance 16 items. The forth factor is personal aggression. Personal aggression have five items.

Some of the actions that often arise when dealing with Indonesian workers are: first, behaviors such as thinking or making decisions by using more feelings, with the approach of the heart, and not including logic. Second, the attitude of restraint, prominent in interpersonal relations among the Javanese, and such values are widely embraced throughout Asia. While Americans have an emotional tendency and are not afraid of conflict. Yet Indonesia has a tendency to avoid conflict. Indonesian people do not like a direct confrontation because they do not like to embarrass others. Third, get respect from others. A manager who behaves and does justice to employees is a manifestation of respect done to employees. When leaders expect employees to perform better, then it's only natural that their business is rewarded. Respect is not only done to social relations, but also to regulations in state institutions and policies. Fourth, attitude wants to be known and viewed about who he is. This is very influential when a foreign country manager negotiates. The process will run smoothly if the negotiator has previously known who he is dealing with. Fifth, the attitude of the Indonesian people who always avoid conflict, thus affecting the communication style. Indonesians are reluctant to say "no" and many say 'yes' when they mean no, because they say what others think they want to hear, and do not want to give bad news.

Table 2. Four factors that form counterproductive work behavior of employee.

Counterproductive Work Behaviour	
Production Deviance:	**Property Deviance:**
1. Come late for many reasons	1. Behavior of sabotage
2. Rest longer than the time specified	2. Office expenses without approval
3. Homework earlier than the time specified	3. Using office facilities for personal use
4. Not coming to work for no reason	4. Data manipulation
5. Postpone the job	5. Stealling
6. Work not in accordance with the SOP	6. Corruption
7.Busy with gadgets in working hours	7. Do additional work if given extra money
8.Out of office during work hours for personal matters	8. Leaking information to unauthorized parties
9. Ask for a big salary but not performance	9. Falsify receipts for personal use
10. Leave the office for longer than the specified time	10. Always negative thinking about the company
11. Playing game in working hours	
12.Joking in working hours	**Political Deviance:**
13. Not responsible for the work	1. Impolite communication
14. Not attending the meeting without permission	2. Not entering the meeting without permission
15, Did not clean up the laboratory	3. Ignore command of leader
16.Sleeping in working hours	4. Not attending the meeting without permission
17. Delegate tasks to other divisions	5. Telling lie
18. Do not use work uniform	6. Delegate tasks to other divisions
19. Error creating report	7. Not following regular training
20. Not following regular training	8. Did not provide a full report when representing the meeting
21. Do not close the door of air-conditioned room	9. Not much contribution in the group
22.Do not use safety clothes	10. Bribery
23,Eat in working hours	11. Illegal charges
24.Streaming YouTube in working hours	12. Hiding the problem
25.Gossip in working hours	13. Lack of communication
26.Field employees did not back to the office after the visit	14. Gratuities
27. Did not update technology	15. Not sharing information
28. Do additional work if given extra money	16. Choose your boss's favor to your subordinates
29. Document archiving is not systematic	
30. Not keeping the office clean	**Personal Aggression:**
31. Less careful in carrying out the work	1. Arguing with colleagues
32. Slow down the work process so that the target is not	2. Commit acts of violence to other employees
33. Lack of creativity in work	3. Affects other employees to act counterproductive as well
34.Working with no systematic	4. Behavior attacks others
35. Did not spend all his ability in duty	5. Disturbing colleagues
36. Slow in handling problem solving	
37. Doing the task is not wholehearted	

Table 2 showed that the leaders usually evaluate their employee doing counterproductive behavior which is more subtle like (didn't update techonology, lack creativity). They choose these behavior because the consequence more lenience than property deviance. These fenomana also happened to leader, they choose to do behavior which addressed to action with vague consequences, such as production deviance because by doing these, they could express their dissatisfaction with the organization but without fear of being explicitly known.

3.2. The list of counterproductive work behaviors described for leaders by employee

Based on interviewing respondents, we get four factors that form counterproductive work behavior of leaders (Table 3). First factor is production deviance. Production deviance have 12 item. Second factor is property deviance. Property deviance have 5 items. Third factor is political deviance. Political deviance 20 items. The forth factor is personal aggression. Personal aggression have 2 items.

Table 3. Four factors that form counterproductive work behavior of leaders.

Counterproductive Work Behaviour	
Production Deviance:	**Political Deviance:**
1. Coming late	1. Making policies regardless of the opinions of others
2. Homework earlier than the time specified	2. Ignoring the Deputy Director's instructions for not having too
3. Busy with gadgets in working hours	3. Less of coordination between leaders with subordinates
4. Out of office during work hours for personal matters	4. Do not appreciate the work of employees
5. Falsify receipts	5. Choose your boss's favor to your subordinates
6. Gratuities	6. Not fair in giving assignments
7. Bribery	7. Giving too many tasks without thinking about the ability to work
8. Making policies that do not comply with the rules	8. Less communication to employees
9. Conducting meetings beyond the working hours	9. Not convey information to subordinates
10. The work instructions provided are inconsistent and confusing	10. Not coordinate subordinate well
11. Violate work rules	11. Not appreciate the work and contribution of employees
12. Violate decisions already made	12. Payroll is not timely
	13. Do not know employees personally
Property Deviance:	14. Unfair in treating its employees
1. Using office facilities for personal use	15. Create an uncomfortable environment for employees who are not needed by the company
2. Data manipulation	16. Never pay attention to employee performance
3. Corruption	17. Employee salaries are rarely reviewed so that the increase is long-time
4. Often giving work out of working hours	18. Cut employee salaries
5. Provide company confidential information	19. Not guiding subordinates
	20. Forged a signature of payment to get a big turn
Personal Aggression:	
1. Angry for no reason and looking for the faults of his subordinates	
2. Often argue with employees	

Indonesia is known as the most difficult and frustrating task place. Such a situation occurs because between a foreign country, such as America has a different value with the state of Indonesia. Controlling undesirable behavior is a common characteristic of government in certain societies, including Indonesia. From the perspective of convergence theory, some counterproductive behaviors identified in Western workplaces, such as not stealing from the organization, will also occur in various cultural environments such as in Indonesia. However, from the perspective of divergence theory, people in different social and cultural contexts may have different conceptions of actions that are counterproductive behavior [11]. In other words, social and cultural contexts can form different specific information about counterproductive behavior.

Indonesia is one of the countries that uphold the law. Although Indonesia is known as a state of law, it does not mean that Indonesia is a safe, orderly, and far-away country from behaviors that deviate from the norms of the rule of law. One of the most commonly perverted behaviors is corrupt behavior. Corruption is not only money, but also time and energy. This obviously counterproductive deviant behavior is detrimental to the organization and even the state. The experience of foreign managers working in Indonesia as mentioned above provides for the development of counterproductive behavior.

Table 3 showed that the leaders who are judged by their employees are more likely to engage in counterproductive behavior, namely political deviance. This can be understood because leaders are more often faced with others.However, all forms of counterproductive behavior conducted by leaders or subordinates tend to be selected behavior of a subtle nature. For employee these process which aimed to organization referred as production deviance. Meanwhile for leader these process aimed to personal referred as political deviance.

4. Conclusion

Behavioral counterproductive work influences the planned environmental management system susceptibility. Behavioral counterproductive behavior perceived by subordinates to the leadership and perceived by the leadership of subordinates. This study still requires further research, some of which include the process of inserting the description into dimensions. The process is still using a qualitative approach, for further research using Multidimension Scalling Technique can be an alternative to sort out the description in a dimension.

References

[1] Marcus W. 2009. Business Strategy and the Environment. *National Culture, Regulation and Country Interaction Effects on the Association of Environmental Management Systems with Environmentally Beneficial Innovation.*18:122–136

[2] Scoot S, Katherine A D, and Jane E D. 2014. It's Not Easy Being Green: The Role of Self-Evaluations in Explaining Support of Environmental Issues. *Academy of Management Journal*. 57(1):7–37

[3] Alin E A. 2011. Reducing The Effects of The Global Crisis on SME's by Implementing an Environmental Management System. *Economic Studies.* 4:137-140

[4] Gustavo L, Javier G B, and Oscar G B. 2015. Business Strategy and the Environment. Input, Output, and Environmental Management Productivity: Effects on Firm Performance. 24:145-158

[5] Marcus W. 2013. Do They Matter as Determinants of Environmental Management System Implementation? *Journal of Business Ethics 'Green' Human Resource Benefits*. 114(3):443-456

[6] Everton W J et al, 2007. Be nice or else: understanding reasons for employee's deviant behaviors. *The Journal of Management Development*. 26(2):117.

[7] Jennifer R, Suzanne V, and Ben E. 2011 Bullying in early adolescence and its association with anti-social behavior, criminality and violence 6 and 10 years later. *Criminal Behavior and Mental Health*. 21(2):117-127

[8] Gorod B J. 2007. Rejecting 'Reasonableness': A New Look at Title VII's Anti-Retaliation Provision American University Law Review. 56(6).

[9] Aytolan Y, and Dilek E. 2007. Mobbing in the workplace by peers and managers: Mobbing experienced by nurses working in healthcare facilities in... *Journal of Clinical Nursing*. 16: 1444–1453

[10] Maria R, and Jia L X. 2008. Understanding the domain of counterproductive work behavior in China. *The International Journal of Human Resource Management*. 19(5):856-877

[11] Christopher M B, Nichelle C C, and Clare L B. 2012. A Do other-reports of counterproductive work behaviour provide an incremental contribution over self-reports? Meta-analytic comparison. *Journal of Applied Psychology*. 97:613-636

[12] Castiglione A. 2010. Counterproductive work behaviors: The role of employee support policies, envy, and narcissism. (Doctoral dissertation) Available from ProQuest Dissertations and Theses database (UMI No. 1493097)

[13] Charbel J C J, and Fernando C A S. 2008. Relationships between human resource dimensions and environmental management in companies: proposal of a model. *Journal of Cleaner Production*. 16:51-58

[14] Tiarapuspa. 2015. Pengingkaran Kontrak Psikologis dan Pengaruhnya Terhadap Perilaku Kerja Kontraproduktif. Unpublished Dissertation, Gajah Mada Univeristy, Yogyakarta, Indonesia.

[15] Ariani D A 2013. The Relationship between Employee Engagement, Organizational Citizenship Behavior, and Counterproductive Work Behavior. *International Journal of Business Administration*. 4(2):46-56

[16] Agustin N, and Seger H. 2013. Hubungan Antara Keadilan Distributif dan Perilaku Kerja Kontraproduktif dengan Mengontrol Leader Member Exchange (LMX). *Jurnal Psikologi Industri dan Organisasi*. 2(3):183-190

The linkage between household water consumption and rainfall in the semi-arid region of East Nusa Tenggara, Indonesia

J J Messakh*, D L Moy, D Mojo, Y Maliti

Study Program of Civil Engineering Education, Universitas Nusa Cendana, Kupang, East Nusa Tenggara, Indonesia

*Corresponding Author: yapmessakh@gmail.com

Abstract. Several studies have shown that the amount of water consumption by communities will depend on the factors of water consumption patterns that are influenced by social, cultural, economic and local climate conditions. Research on the linkage between rainfall and household water consumption in semi-arid areas of Indonesia has never been done. This study has been conducted on 17 regions in NTT, and case study has taken samples in one town and one village. The research used survey and documentation method. The results show that the average amount of household water consumption in semi-arid region of East Nusa Tenggara is 107 liters / person / day. Statistical test results using Pearson correlation found r = -0.194 and sig = 0.448. This means that there is a negative correlation between rainfall and household water consumption. The greater the rainfall the smaller the consumption of water, or the smaller the rainfall the greater the consumption of water, but the linkage is not significant. Research shows that the amount of household water consumption will be influenced by many interrelated factors and none of the most dominant factors, including the size of the rainfall that occurs in a region.
Keywords: East Nusa Tenggara, household water consumption, rainfall, semi-arid

1. Introduction

UNICEF and WHO estimate that 1.1 billion people lack access to improved water supplies [1]. Until now, the problem of clean water is still a complicated problem in a number of regions in Indonesia. The reach of water services nationwide reaches about 76% by the end of 2015 through piped and non-piped networks. Most of the Regional Water Company (PDAM) as drinking water service operators are still having problems with their performance. Report of [2] recorded about 20% of 383 PDAMs in Indonesia are categorized as 'sick', 30% 'less healthy' and 50% 'healthy'. Whereas the government target at the end of 2019 access to drinking water services must reach 100%, while the rate of water demand is higher due to population growth and development activities. Clean water supply is affected by many factors including consumption patterns are influenced by the social, cultural and economic [3]. Research conducted [4] indicates that income, housing type, members per household, the presence of outdoor uses (garden and swimming pool), the kind of species planted in the garden and consumer behaviour towards conservation practices play a significant role in explaining variations in water consumption. Research [5] states that the most significant determinants for increases in water demand are population growth, climate change, and the type of urban development that occurs. The strategy to meet drinking water needs in areas with limited water resources capacity can be done by minimizing the wasteful use of domestic water (households) with public facilities that can be utilized together [6]. The study conducted at Regional Kupang by researcher [7] states that the potential of existing water sources, is difficult to meet the needs of community drinking water in the long term. Therefore, the water resource management paradigm for multisector needs to be adjusted according to the semi-arid region specificity, with the main priority for drinking water needs. Potential rainfall that occurs should be optimized utilization as a source of raw water, through the construction of reservoirs, other rain water storage facilities and accompanied by conservation efforts. Previous study [8] said that extreme changes in the availability of surface water would be a serious problem in sustainable water management in semi-arid climate areas. The NTT area which is the study area is a semi-arid climate area [9], [10], [11], which continues to be problematic to date in the provision of clean water for its citizens. This shows that there is a relationship between climate and water availability in a region. But the extent to which rainfall factors affect the level of clean water consumption for household needs is not yet known and is an interesting problem that needs to be studied, especially in semi-arid NTT. Research on rainfall conducted by study [12] at 20 rain stations in Iran shows the variability of rainfall in arid and semi-arid regions. Research studied [13] rainfall in China revealed that each rain station has different characterizations. The results of this study demonstrate the recognition of the characteristics of rainfall from recorded time series data which is essential for sustainability of water availability and future management of water resources. A study of the rainfall characteristics in Kupang by study [10] found that rainfall occurs in high intensity in a short time duration. In general there is an increasing trend of rainfall, number of rainy days, and maximum daily rainfall. Rainfall in the wettest month gets wetter and dries up the driest moon, resulting in increasing run-off and smaller infiltration of water into the soil, resulting in flooding during the rainy season and drought during long droughts. The aim of this research is to know the amount of household water consumption in semi-arid area of East Nusa Tenggara, which is reviewed in 17 districts through 15 PDAMs as drinking water operators in the region. Therefore, the amount of water used by the community from other sources outside the PDAM pipeline network is not a part of this study. The two areas that have been studied for in-depth research are Ende City representing rural areas, and Hane Village represents poor areas. The study also examined the correlation between rainfall and household water consumption in each of the studied areas.

2. Research Method

The study area is East Nusa Tenggara Province (NTT) which is generally a semi-arid region with rainfall ranging from three to four months, even in some areas, rainfall only lasts for two months of the year [7], [11]. Average rainfall that recorder in meteorology/climatology stations was between 600 - 2700 mm [9], [14]. Areas in NTT have different temperature. From 10 meteorology/climatology stations in NTT, the highest temperature in 2015 was 37.4^0C and the lowest was 8.8^0C. Typically, NTT is a hot area with average temperature of 26 - 28^0C all year in 2015 with some exception areas that have average temperature of 19.9^0C. NTT is an archipelago province with 1.192 numbers of island, 432 island with names, and 44 inhabitants island. Big islands in NTT usually called Flobarmorata (Flores, Sumba, Timor, Alor, and Lembata). The area of NTT lands is 47,931.54 km^2. NTT Province has 21 regency and one city, 306 districts and 3268 villages. Astronomically, East Nusa Tenggara Province is located between 8° - 12° South Latitude, and between 118° - 125° East Longitude [14].

The data were collected by survey method to assess the household water consumption in Hane Village, TTS Regency and Ende City, Ende Regency. The sample size is 22 - 30 households in each region selected by cluster random sampling. Reserched study [15] argue that household surveys are a useful method for assessing water needs and demand. In urban areas, conducted by grouping the population according to the economic level of the family. In this study the economic level is based on the type of house construction, the lower economy is indicated by the simple building, the middle economy is indicated by a semi-permanent and permanent building, and a capable economy indicated by the type of permanent house building with storey building. The selected sample is household using only clean water from PDAM and recorded through water meter. In the village area, grouping according to the level of community education. Group 1 is the education of the head of household

IOP Conf. Series: Earth and Environmental Science **106** (2017) 012084 doi:10.1088/1755-1315/106/1/012084

(KK) graduated from elementary school or not. Group 2 finished junior high school, while group 3 graduated undergraduate education. Next, each group was randomly selected for each of the 7-10 samples. Data retrieval time is done for one month, in 2017. Study [5] found that water consumption dependent on household size, seasonal use, education level and the size of the property's outdoor space. While study [15] discusses how household consumption of water is affected by income. Documentation method was used to obtain data on household water consumption from 15 PDAMs in 17 regions in NTT through study [2]. Rainfall data obtained from the Agency of Meteorology and Geophysics of NTT (BMKG NTT) with five-year data length from 2012 - 2016.

The analysis is done by statistic of mean test to know the average number of household water consumption in liters/person/day. To know the linkages between rainfall and household water consumption used Pearson-corellation statistical analysis using SPSS ver.21.

3. Results and Discussion

3.1. Rainfall

Through the results of data analysis of average rainfall for five years obtained the fact that, the average of rainfall in East Nusa Tenggara is 1445 mm/year. However, there are several wet climates with high rainfall of 2893 mm/year in Manggarai Regency, and East Sumba with relatively dry climate with average rainfall of 863 mm/year. The amount of rainfall in the NTT region by region, as shown in Table 1.

Table 1. Average rainfall in NTT, by regency/city (2012 – 2016) [18].

No	Regency/City	Annual rainfall (mm)	No	Regency/City	Annual rainfall (mm)
1	Kupang City	1486	9	Manggarai	2893
2	Kupang Regency	1884	10	Manggarai Barat	1196
3	Timor Tengah Selatan	1610	11	Ngada	1321
4	Timor Tengah Utara	1651	12	Ende	1261
5	Belu	1480	13	Sikka	976
6	Rote-Ndao	1721	14	Flores Timur	1478
7	Sumba Timur	830	15	Lembata	856
8	Alor	1027		**Average (mm)**	**1445**

Researched studied [10] in Kupang found that rainfall occurs in high intensity in a short time duration. The average number of rainy days in Kupang is 80 days a year. The average amount of rainfall in Indonesia ranges from 1000 - 5000 mm per year. Study [11] said that Indonesian rainfall patterns are influenced by several factors, such as, monsoon, Inter-tropical Convergene Zone (ITCZ), El Nino - Southern Oscillation (ENSO), and other regional circulations in Pacific and Indian oceans. Indonesia rainfall patterns are divided into three main areas, they are: (i) monsoon region (type A) is the dominant pattern in Indonesia, because it covers almost the entire territory of Indonesia. The area has a peak in November to March (NDJFM), which is influenced by wet northwest monsoon, and a trough in May to September (MJJAS), that is affected by the dry southeast monsoon, so it can be distinguished clearly between dry and rainy seasons, (ii) equatorial region (type B) has two peaks in October to November (ON) and March to May (MAM). This pattern is influenced by a shift to the north and south from ITCZ or equinox point (culmination) of the sun and (iii) local climate region (type C) has a peak in June to July (JJ) and a trough in November to February (NDJF). This pattern is the opposite of pattern A.

3.2. Household Water Consumption

The household water consumption study was conducted in 17 regency/city in NTT served by 15 PDAMs operating in the area. The results showed that the total household water consumption was 107 liters/person/day, lower than the minimum water requirement according to Indonesian Government standards of 122 liters/person/day. Details of household water consumption by regency/city in NTT as shown in Table 2.

IOP Conf. Series: Earth and Environmental Science **106** (2017) 012084 doi:10.1088/1755-1315/106/1/012084

Table 2. Household water consumption by regency/city in NTT.

NO	Name of PDAMs/Region	Population served (persons)	Water consumption (liters/day)	Water consumption (liters/person/day)
1	Kupang City (parts of Kupang city)	23,256	1,505,434	65
2	Kupang Region (Kupang City, Kupang regency and Sabu regencey)	200,552	20,178,202	101
3	Timor Tengah Selatan	41,719	3,930,250	94
4	Timor Tengah Utara	35,240	4,144,694	118
5	Belu (include Malaka regency)	200,552	18,413,568	92
6	Rote-Ndao	10,530	1,224,029	116
7	East Sumba	45,312	8,756,986	193
8	Alor	23,310	3,684,010	158
9	Manggarai	81,310	8,981,453	110
10	West Manggarai	19,078	2,351,722	123
11	Ngada	39,576	4,703,530	119
12	Ende	53,262	6,440,083	121
13	Sikka	70,492	5,128,531	73
14	East Flores	82,453	3,587,846	44
15	Lembata	12,304	996,624	81
	Average (liters/ person / day)			**107**

From Table 2, it is known that the highest household water consumption in East Sumba regency is 193 liters/person/day and lowest in East Flores is 44 liters/person/day. Interesting data is indicated by the amount of household water consumption in Kupang City, the capital city of NTT province, which is categorized by big city with population > 500 thousand people, that is water consumption only reaches 65 liters/person/day served by *PDAM Kota Kupang* or only Reaches 83 liters/person/day if included which is served by *PDAM Kabupaten Kupang*. Whereas the standard of clean water consumption is 170 liters/person/day. Study [7] reported that the fulfillment of drinking water needs in Kupang City and surrounding areas, mostly obtained through dug wells whose water quality has been contaminated. Study [17] states that over-exploitation of ground water not only causes a reduction in the availability of water but also result in the degradation of water quality. These data indicate that the Government through PDAM has not been able to meet the community's need for clean water. The result of research data analysis proves that the fulfillment of household clean water needs in the semi-arid region of NTT is still experiencing a complicated problem, because most of the region has not been able to meet the needs of clean water for the community. Only three areas (East Sumba Regency, Alor Regency and West Manggarai Regency) are able to fulfill household water supply. Researchers [1] states that major challenges to providing safe water on a global basis: (1) contamination of water in distribution systems, (2) growing water scarcity and the potential for water reuse and conservation, (3) providing sustainable water supplies for megacities, and (4) reducing global and regional disparities in access to water and developing financially sustainable water services.

The 2015 MDGs target for drinking water is the percentage of households with sustainable access to a rural drinking water source of 65.81% and for urban areas of 75.29% or an average of 68.87%. Targets that have not been achieved in many areas in NTT. Challenges to achieving targeted drinking water services for communities are quantity, quality, continuity, funding aspects including funding sources, affordability, services for the poor, management and institutions.

3.3. Case study of household water consumption in urban and rural areas of NTT

Ende City located on the southern coast of Flores Island. The drinking water service in Ende City is done by PDAM Tirta Kelimutu Ende. Water sources come from: springs, boreholes and streams, with a production capacity of 101 liters/sec. The population in the Ende Regency is 263,201 person. The coverage of drinking water services is only 20% of the total population or the population is served by 53,262 person [2]. The result of data analysis from 22 household samples shows that household water consumption in Ende City is 96 liters/person/day. With a population of 106,584 people, the minimum standard of household water consumption is 150 liters/person/day. This shows that the fulfillment of household water needs has not been fulfilled in Ende City. Although the average amount of water consumption is 96 L/person/day, the data also shows that there is a high deviation in household water consumption in Ende City. Lowest is 40 L/person/day, while the highest is 210 liters/person/day. Study [4] indicates that income, housing type, members per household, the presence of outdoor uses (garden and swimming pool), the kind of species planted in the garden and consumer behaviour towards conservation practices play a significant role in explaining variations in water consumption. The number of households using clean water according to minimum eligibility standards is 6 households or 27% of the total sample, the remaining 73% is below the standard. The results of this study are in line with what is said by UNICEF and WHO estimate that 1.1 billion people lack access to improved water supplies [1].

Hane village is located in TTS Regency. The results of the analysis on the household samples studied show the total consumption of household water consumption in Hane Village is 48 liters/ person/day. With a population of <3000 people the minimum standard of water consumption is 60 liters/person/day. This indicates that the fulfillment of the need for household water in quantity has not been fulfilled in Hane Village. Although the average amount of water consumption is 48 L/person/ day, the data also show that there is a high deviation in household water use in rural areas of Hane. The lowest is 21 L/person/day while the highest is 80 L/person/day. The number of households using clean water based on minimum standards of eligibility by the Indonesia Government is only 7 households or 23% of the total sample, the remaining 77% below the standard. Source of clean water utilized for the fulfillment of household needs comes from: springs, rivers and rural piping networks. The supply of drinking water through the rural pipeline network has not run continuously and there is still rotation, in the dry season the debit decreases drastically and there is no drainage. Similarly, water sources from springs and rivers experience a very large reduction of discharge and even drought so that household water needs become disrupted.

3.4. The linkage between household water consumption and rainfall

Statistical test result using Pearson correlation between rainfall and household water consumption rate is r = -0.194 and sig = 0.448 (Table 3). Mean rainfall is negatively linkage to household water consumption, and the correlation is not significant. The linkage caused is the greater the rainfall of an area the less the level of water use, or vice versa. This study shows that, the amount of rainfall will help minimize water consumption in a region, although not significant.

Table 3. The correlation value between rainfall and the amount of household water consumption.

Descriptions		Rainfall	Water consumption
Rainfall	Pearson Correlation	1	-.194
	Sig. (2-tailed)		.488
	N	15	15
Water consumption	Pearson Correlation	-.194	1
	Sig. (2-tailed)	.488	
	N	15	15

4. Conclusion

The average amount of household water consumption in semi-arid NTT is 107 L/person/day. In urban areas, only 27% of households use water according to minimum standards, as well as in rural areas only 23% of households. Research shows that the fulfillment of household drinking water needs in most areas of NTT has not been met in terms of quantity. There is a negative correlation between the variable of household water consumption and rainfall occurring in the semi-arid region of NTT. The greater the rainfall will be to minimize the consumption of household water consumption, and the smaller the rainfall will result in greater household water consumption, although the statistical test shows that the correlation is not significant. Research shows that the amount of household water consumption will be influenced by many interrelated factors and none of the most dominant factors, including the size of the rainfall that occurs in a region. This research needs to be developed again to obtain more detailed results. But at least the results of this study further reinforce that the problem of water supply is still a joint problem, which needs to be taken prompt and appropriate strategic steps by the Government with the support of all components of society.

References

[1] Moe C L, and Rheingans R D. 2006. Global challenges in water, sanitation and health. *Journal of water and health*, 4(1):41-57
[2] BPPSPAM, 2013, Kinerja PDAM Wilayah 2 tahun 2013. Jakarta
[3] Kim S H, Choi S H, Koo J Y, Choi S I, and Hyun I H. 2007. Trend analysis of domestik water consumption depending upon sosial, cultural, economic parameters. *Water Science and Technology Water Supply*. 7(5-6):61-8.
[4] Domene E, and Sauri D. 2006. Urbanisation and water consumption: Influencing Factors in the Metropolitan Region of Barcelona. *Sage Journals*. 43(9).
[5] Peters H, Lily, Bethany P, and Heejun C. 2010. Effects of urban spatial structure, sociodemographics, and climate on residential water consumption in Hillsboro, Oregon. *Journal of the American Water Resources Association* (JAWRA). 46(3):461-72.
[6] Halper E B, Dall'erba S, Bark R H, Scott, C A, and Yool S R. 2015. Effects of irrigated parks on outdoor residential water use in a semi-arid city. *Landscape Urban Planing Journal*. 134(1): 210-220.
[7] Messakh J J, Arwin, and Hadihardaja I K. 2015. Management Strategy of Water Resources Base on Rainfall Characteristics in The Semi-Arid Region in Indonesia. *International Journal of Scientific & Engineering Research*. 6(8):331-7.
[8] Masih I, Ahmad M, Uhlenbrook S, Turral H, and Karimi P. 2009. Analysing streamflow variability and water allocation for sustainable management of water resources in the semi-arid Karkheh River Basin, Iran, *Physics and Chemistry of the Earth*. 34:329–40.
[9] Messakh J J, Arwin, Hadihardaja I K 2015. Climate change and strategy of reservoir operation in semi-arid area, West Timor. *International Proceeding: The 3rd International Seminar on Sustainable Urban Development*. 112-9.
[10] Messakh J J, Arwin, Hadihardaja I K., and Chalik A A. 2015. A Study on Fulfillment of Drinking Water Need of People in Semi-Arid Areas in Indonesia. *Journal of People and Environment*. 22(3):271-80.
[11] Aldrian E, and Susanto R D. 2003. Identification of three dominant rainfall region within Indonesia and their relationship to sea surface temperature. *International Journal of Climatology*. 23(12):1435-52.
[12] Modarres R, and de Paulo. 2007. Rainfall trends in arid and semi-arid regions of Iran. *Journal of Arid Environments*. 70:344-55
[13] Zhai P X, Zhang H W, and Pan X. 2005. Trends in total precipitation and frequency of daily precipitation extremes over China. *Journal of Climate*. 18(7):1096-1108.
[14] BPPS Nusa Tenggara Timur. 2016. *Nusa Tenggara Timur Province in figures 2016*. Kupang.
[15] Harlan S L, Yabiku S T, Larsen L, and Brazel A J. 2007. Household water consumption in an arid city: affluence, affordance, and attitudes. *Society & natural resources*. 22(8):691-709.
[16] Sigel K, Altantuul K, and Basandorj D. 2012. Household needs and demand for improved water supply and sanitation in peri-urban ger areas: the case of Darkhan, Mongolia. *Environmental Earth Sciences*. 65(5):1561–66.
[17] Malik N, and Simi N S. 2014. Effect of urbanization on groundwater quality in the Gafsa Town South Western Tunisia. *Desalination and Water Treatment*, 52(10-12):2022-31.
[18] Analysis of research data from BMKG 2012-2016.

Towards green loyalty: the influences of green perceived risk, green image, green trust and green satisfaction

K Chrisjatmiko

Faculty of Economic and Business, Universitas Trisakti, Jakarta, Indonesia

Corresponding Author: niachrisjatmiko1@gmail.com

Abstract. The paper aims to present a comprehensive framework for the influences of green perceived risk, green image, green trust and green satisfaction to green loyalty. The paper also seeks to account explicitly for the differences in green perceived risk, green image, green trust, green satisfaction and green loyalty found among green products customers. Data were obtained from 155 green products customers. Structural equation modeling was used in order to test the proposed hypotheses. The findings show that green image, green trust and green satisfaction has positive effects to green loyalty. But green perceived risk has negative effects to green image, green trust and green satisfaction. However, green perceived risk, green image, green trust and green satisfaction also seems to be a good device to gain green products customers from competitors. The contributions of the paper are, firstly, a more complete framework of the influences of green perceived risk, green image, green trust and green satisfaction to green loyalty analyses simultaneously. Secondly, the study allows a direct comparison of the difference in green perceived risk, green image, green trust, green satisfaction and green loyalty between green products customers.
Keywords : green perceived risk, green image, green trust, green satisfaction, green loyalty, green products customers.

1. Introduction

This study uses a holistic approach to green marketing so it is necessary to develop conceptual frameworks of research to understand the effect of green marketing on green customer loyalty. [1], [14], [2]. This study is in line with previous research that is to test the function of green marketing in increasing brand image and its effect on brand loyalty and test the moderate effect of green perceived risk [3], [4], [5]. Hereinafter in this research use both approaches to explore consumer reaction to green product or service that is green marketing and green loyalty [1], [14],[6].

Nevertheless, this study extends previous studies on green marketing and green customer loyalty by adding green perceived risk [1], [7], [8]. There is little previous research that investigates the effect of green perceived risk on green customer loyalty either directly or indirectly [1], [17], [3]. This study shows that green perceived risk negatively affects green image, green trust and green satisfaction [1], [3]. In addition, this study provides empirical evidence that green image, green trust, and green satisfaction mediate the relationship between green perceived risk and green loyalty, which is consistent with previous research [1], [3], [9].

Increased consumer perceptions of green image, green trust, and green satisfaction can help increase customer loyalty to use green products [1], [15], [10]. These results reveal the relevance of environmental issues in the current business context and company needs in their management strategies and policies [1], [16], [11]. The findings suggest that the green issue is a key strategic tool, given its crucial role in building customer loyalty including image, trust and customer satisfaction on green products [1], [17], [3], [6].

Green marketing refers to organizational efforts to design, promote, price and distribute products that do not have an adverse impact on the environment. Green marketing is a concept that includes all marketing activities developed to stimulate and sustain eco-friendly behavior and consumer behavior. To respond to the green needs of consumers, the idea of green marketing has been widely accepted in both practical and academic fields. In green marketing campaigns, the target customers are given information about the environmental impact of the product, which can ultimately affect their buying behavior. So green marketing is a holistic management process to address the needs of the customer and community environment in a profitable and sustainable way [1], [15], [17], [3], [10].

Companies can adopt green marketing activities to perform differentiation strategies and to satisfy customers' needs or wants. As consumers become more concerned about the environment, there are more customers who are more willing to buy green products that have less harmful effects on the environment. Companies implementing defensive green marketing to perform minimum activities in compliance with local government environmental regulations in order to avoid the punishment of fines. When assertive approaches are used, companies are more aggressive in their marketing campaigns. Companies respond not only to market regulations and trends but also to the requirements of their stakeholders [1], [15], [17], [3], [12].

Previous research has analyzed the effect of green marketing on consumers. Furthermore, previous research also confirms that companies should use green marketing strategies to identify the green needs of customers, to promote green products, to group green markets into different niches, to target one or more specialized markets, to formulate a green positioning strategy and apply Green marketing. As a result, this study bridges the gap by investigating how to develop green loyalty using green concept perceived risk, green image, green trust and green satisfaction [1], [15], [17], [3], [13].

2. Research Method

The conceptual framework of this research (Figure 1) has 6 hypothesis that is 1 H1: there is a negative effect of green perceived risk to green image; 2 H2: there is a negative effect of green perceived risk to green trust; 3 H3: there are negative effects of green perceived risk to green satisfaction; 4 H4: there is a positive effect of green image to green loyalty; 5 H5: there is a positive effect of green trust to green loyalty; 6 H6: there is a positive effect of green satisfaction to green loyalty [1], [3], [14], [17].

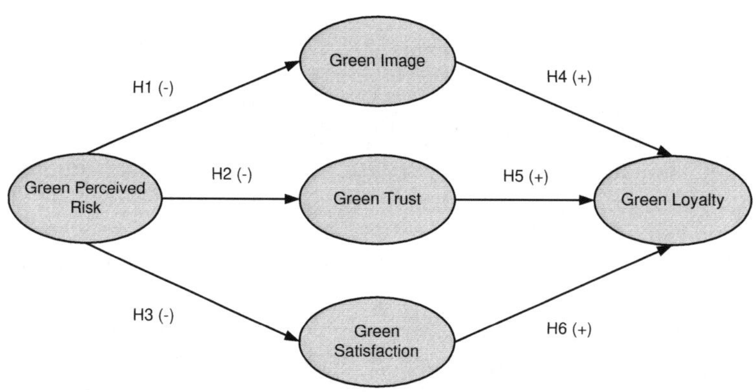

Figure 1. Conceptual framework.

Table 1 is the respondent's profile in this study consisting of mostly female respondents as many as 82 people with age mostly between 26 years to 35 years as many as 72 people. Furthermore, most of the undergraduate education as many as 109 people and most work as employees of 139 people. The frequency of respondents in buying green product between 2 times to 4 times as many as 73 people with the type of green product that many purchased is an organic egg as much as 54 people.

2.1. Sample and data collection

Table 1. Profile of respondents.

Variable	Frequency	Percentage
Gender		
Male	73	47.1
Female	82	52.9
Age years old		
18-25	38	24.5
26-35	72	46.5
36-45	28	18.1
Above 46	17	11.0
Education Level		
High School	20	12.9
Diploma	15	9.7
Undergraduate	109	70.3
Graduate	11	7.1
Occupation		
Employee	139	89.7
Entrepreneur	10	6.5
Part time	6	3.9
Frequency of green purchased		
2-4 times	73	47.1
5-7 times	38	24.5
8-10 times	20	12.9
Above 10 times	24	15.5
Types of green products purchased		
Organic egg	54	34.8
Organic vegetable	34	21.9
Organic rice	24	15.5
LED Television	43	27.7

3. Results and Discussion

All statement items in this study are valid and reliable that shown from Table 2. The average value of all statements show good results with the standard deviation value away from the number 0 which means that the answer of the questionnaires of the respondents have varying values.

IOP Conf. Series: Earth and Environmental Science **106** (2017) 012085 doi:10.1088/1755-1315/106/1/012085

Table 2. Description of the items used to measure the constructs.

Items	Standardized loading	Cronbach's α	Mean	SD
Green perceived risk		0.848		
There is a chance that there will be something wrong with environmental performance of the product	0.806		2.5871	0.71883
There is a chance that the product will not work properly with respect to its environmental design	0.803		2.2645	0.65542
There is a chance that using the product will negatively affect the environment	0.756		2.3355	0.67687
Green Image		0.873		
This product is regarded as the point of reference of environmental commitments	0.859		3.4387	0.69398
This product has a strong environmental reputation	0.830		3.4839	0.68712
This product is successful about its environmental protection	0.723		3.7032	0.68542
This product is well-established about its environmental concerns	0.772		3.5226	0.68694
This product is trustworthy about its environmental promises	0.614		3.7935	0.71801
Green Trust		0.902		
The environmental commitments of this product are generally reliable	0.802		3.8645	0.66525
The environmental performance of this product is generally dependable	0.782		3.7226	0.67934
The environmental argument of this product is generally trustworthy	0.771		3.7677	0.73696
The environmental concerns of this product meet my expectations	0.850		3.7226	0.68883
This product is sincere and honest about its environmental protection	0.828		3.6387	0.72867
Green Satisfaction		0.872		
The choice of this product due to its environmental commitment makes me happy	0.899		3.6194	0.72319
I consider it is correct to buy this product because of its environmental commitment	0.828		3.5097	0.68725
I am satisfied with this product because of its environmental performance	0.775		3.7419	0.68223
Green Loyalty		0.874	3.4839	0.68712
I generally choose this product as my choice option	0.807		3.6968	0.70595
I would buy this product although other competitors had the same environmental attributes	0.676		3.5806	0.69199
I will choose this product as my choice option in the future	0.773		3.6452	0.70011
I would make positive comments about this product to family and friends	0.731		3.4839	0.68712

Table 3 presents the results of hypothesis testing from this study. All hypothesis testing results are supported data. Hypothesis 1 has a coefficient value of -0.335 and t-value of -3.542 with p-value 0.000 means there is a negative effect of green perceived risk to green image. Hypothesis 2 has a coefficient value of -0.543 and t-value of -6.258 with p-value 0.000 means there is a negative effect of green perceived risk to green trust. Hypothesis 3 has a coefficient value of -0.446 and t-value of -4.333 with p-value 0.000 means there is a negative effect of green perceived risk to green satisfaction. Hypothesis

4 has a coefficient value of 0.660 and t-value of 10,262 with p-value 0.000 means there is a positive effect of green image to green loyalty. Hypothesis 5 has a coefficient value of 0.137 and t-value of 2588 with p-value 0.000 means there is a positive effect of green trust to green loyalty. Hypothesis 6 has a coefficient value of 0.230 and t-value of 5.096 with p-value 0.000 means there is a positive effect of green satisfaction to green loyalty. The conclusion that green perceived risk has the greatest negative effect on green trust. While green image has the greatest positive influence on green loyalty.

Table 3. Structural equation model result.

Hypotheses	Structural relationships	Path coefficients	t-value	Results
H1	Green perceived risk → Green image	-0.335	-3.542*	*H1* supported
H2	Green perceived risk → Green trust	-0.543	-6.258*	*H2* supported
H3	Green perceived risk → Green satisfaction	-0.446	-4.333*	*H3* supported
H4	Green image → Green loyalty	0.660	10.262*	*H4* supported
H5	Green trust → Green loyalty	0.137	2.588*	*H5* supported
H6	Green satisfaction → Green loyalty	0.230	5.096*	*H6* supported

NFI=0.773; IFI=0.830; TLI=0.801; CFI=0.828; χ^2=542.729; df=164 $p = 0.000$
Note: *Statistically significant of $p < 0.05$

4. Conclusion

The results showed that green perceived risk had a negative and significant effect on green image, green trust, and green satisfaction. While green loyalty is influenced positively and significantly by green image, green trust, and green satisfaction. Companies should always pay attention to all risks that occur due to business processes so as not to damage the environment. In addition, the company must design a green environment by producing green products for its customers. Companies should eliminate products that would risk damaging the environment. So with the company always pay attention to the green environment then the green marketing activities will run well. Creating a green environment is not an easy thing, but it requires a great effort from the company to create the green.

Companies need to increase their commitment to keeping the green environment. In addition, the company is expected to improve the reputation of its business by running a green marketing. The image formed by creating a green environment will significantly affect customer loyalty in consuming the company's green products. Furthermore, it is necessary to increase the customer's confidence to continue to use the company's green products. Thus customer loyalty will continue to grow with the existence of these beliefs. Increased trust is followed by increased customer satisfaction. With customers satisfied with green products, it will significantly increase customer loyalty in consuming green products. Finally, the company that produces green products is a company that loves its environment.

References

[1] Martínez P. 2015. Customer loyalty: exploring its antecedents from a green marketing perspective. *International Journal of Contemporary Hospitality Management*. 27:896-917.

[2] Chahal H, R Dangwal, S Raina. 2014. Antecedents and consequences of strategic green marketing orientation. *Journal of Global Responsibility*. 5(2):338-362.

[3] Lin J, A Lobo, C Leckie. 2017. Green brand benefits and their influence on brand loyalty. Marketing Intelligence & Planning. 35(3):425-440.

[4] Wu S I, Y C Wu. 2014. The influence of enterprisers' green management awareness on green management strategy and organizational performance. *International Journal of Quality & Reliability Management*. 31(4):455-476.

[5] Narula S A, A Desore. 2016. Framing green consumer behaviour research: opportunities and challenges'. *Social Responsibility Journal*. 12(1):1-22.

[6] Akehurst G, C Afonso, H M Gonçalves. 2012. Re-examining green purchase behaviour and the green consumer profile: new evidences. *Management Decision*. 50(5):972-988.

[7] Moser A K. 2015. Thinking green, buying green? Drivers of pro-environmental purchasing

behavior. *Journal of Consumer Marketing*. 32(3):167-175.
[8] Chaudary S, Z Zahid, S Shahid, S N Khan, S Azar. 2016. Customer perception of CSR initiatives: its antecedents and consequences. *Social Responsibility Journal*. 12(2):263-279.
[9] Zhang H, F Yang. 2016. On the drivers and performance outcomes of green practices adoption: An empirical study in China. *Industrial Management & Data Systems*. 116(9):2011-2034.
[10] Khare A, S Pandey. 2017. Role of green self-identity and peer influence in fostering trust towards organic food retailers. *International Journal of Retail & Distribution Management*.
[11] Chew M Y L, Conejos S. 2016. Developing a green maintainability framework for green walls in Singapore. *Structural Survey*. 34(4/5):379-406.
[12] Wong S K S. 2012. The influence of green product competitiveness on the success of green product innovation: Empirical evidence from the Chinese electrical and electronics industry. *European Journal of Innovation Management*. 15(4):468-490.
[13] Borin N, J L Mullikin, R Krishnan. 2013. An analysis of consumer reactions to green strategies. *Journal of Product & Brand Management*. 22(2):118-128.
[14] Rosenbaum M S, I A Wong. 2015. Green marketing programs as strategic initiatives in hospitality. *Journal of Service Marketing*. 29(2):81-92.
[15] Chan E S W. 2013. Gap analysis of green hotel marketing. *International Journal of Contemporary Hospitality Management*. 25(7):1017-1048.
[16] Suki N M. 2016. Green product purchase intention: impact of green brands, attitude, and knowledge. *British Food Journal*. 118(12):2893-2910.
[17] Chen Y S, C H Chang. 2013. Towards green trust: The influences of green perceived quality, green perceived risk, and green satisfaction. *Management Decision*. 51(1):63-82.

IOP Conf. Series: Earth and Environmental Science **106** (2017) 012086 doi:10.1088/1755-1315/106/1/012086

Leachate characterization and performance evaluation of leachate treatment plant in Cipayung landfill, Indonesia

E Noerfitriyani*, D M Hartono, S S Moersidik, I Gusniani

Environmental Engineering Study Program, Civil of Engineering Department, Universitas Indonesia, Depok, West Java, 16424, Indonesia

*Corresponding Author : eki.noerfitriyani@ui.ac.id

Abstract. The operation of landfill can cause environmental problems due to waste decomposition in the form of leachate production. Cipayung Landfill has a leachate treatment plant using stabilization ponds. The objective of this research is to evaluate the performance of stabilization ponds at Cipayung Landfill. The data were analyzed based on leachate samples from treatment unit's influent and effluent under rainy season condition from April to May 2017. The results show the average leachate quality based on parameters of temperature by 34.81°C, Total Suspended Solid (TSS) of 72.33 mg/L, pH of 7.83, Biochemical Oxygen Demand (BOD) of 3,959.63 mg/L, Chemical Oxygen Demand (COD) of 6,860 mg/L, Total Nitrogen of 373.33 mg/L, and heavy metal Mercury of 0.0016 mg/L. The treatment plant's effluent quality exceeds the leachate standard limit based on Indonesia's Ministry of Environment and Forestry Law No. 59 of 2016. The results of design evaluation show that the anaerobic pond, facultative pond, and maturation pond system do not meet the design criteria. Therefore, a design improvement is needed to increase the performance of the leachate treatment plant and to ensure that the leachate discharged to water bodies does not exceed the standard limit to prevent contamination of the environment.
Keywords: leachate characteristics, leachate treatment plant, stabilization pond

1. Introduction

The population growth in urban areas leads to the high volume of solid waste generated. Inevitably, this development has a significant impact on the increasing amount of solid waste disposed at Cipayung Landfill, Depok City. Cipayung Landfill has been operated since 1984 using a controlled landfill system [1].The increasing amount of waste generation can cause problems in its operation. Moreover, an improper waste management is potentially harmful to human health and can lead to environmental degradation. One potential environmental issue driven by the decomposition of waste in landfills is leachate production. Leachate can be derived from rain water seepage, runoff of waste decomposition, and moisture content of the waste itself [2]. The characteristics of leachate produced depend on several factors, including the degree of stabilization of the dumped waste, solid waste collection system, the type and composition of the discarded waste [3], the solubility of solid waste content [4], site and hydrological factors, solid waste compaction, as well as landfill design and operations [15]. Leachate is a liquid manifestation of solid waste and has been considered a serious polluter that affects natural resources such as water bodies and groundwater, as well as human health. Leachate is a contaminated liquid derived from the bottom of solid waste disposal facilities such as landfills containing dissolved organic compounds and inorganic compounds and suspended solids. The leachate composition depends on the nature of solid waste, chemical, and biochemical processes for the decomposition of waste

IOP Conf. Series: Earth and Environmental Science **106** (2017) 012086 doi:10.1088/1755-1315/106/1/012086

materials, as well as the total moisture content of solid waste [6]. Cipayung Landfill has a leachate treatment plant for treating leachate from landfill using a stabilizing pond treatment system. Stabilization ponds are large basins in which wastewater is treated by natural processes involving microorganisms, and are considered as the most appropriate method for wastewater treatment in developing countries where the climate is favorable for its operation [7]. Stabilization pond system can be classified based on the types of biological activity, i.e. anaerobic, facultative, and maturation ponds [8]. Treated leachate is then discharged to the Pesanggrahan River. In Indonesia, the quality standards for liquid waste disposal are stated in [15] to ensure that the disposal of waste to recipient water bodies will not cause damage to the aquatic ecosystem and its potential economic resources, as well as public health affairs. Of the several contaminants carried in leachate, organic matter, nitrogen, and heavy metals are the primary substances of concern. In general, the concentration of the substances contained in the leachate depends on the biochemical decomposition of solid waste in landfills [9]. The main objectives of this study are to analyze leachate characterization and to evaluate the performance and design of leachate treatment plant in Cipayung Landfill, Depok City, Indonesia. The results of leachate characterization of Cipayung Landfill treatment plant is expected to be useful for landfill management, especially to improve the existing leachate treatment system and prevent contamination of discharged leachate to recipient water bodies.

2. Research Method

2.1. Description of Study Area

This study was conducted in Cipayung Landfill, located in Depok, Indonesia. The climate in Depok City is influenced by monsoon season and is relatively similar, characterized by fairly small rain differences. The conditions of rainfall in all Depok areas are practically the same, with the average rainfall of 3,332 mm/year or 278 mm/month and the mean temperature of 26°C [10]. The leachate treatment system applied in Cipayung Landfill is a stabilization pond system. Figure 1 shows that the Cipayung Landfill stabilization pond consists of an anaerobic pond, facultative pond, and maturation pond. The leachate treatment plant has a pretreatment unit equalization tank that is followed by the stabilization pond system.The treated leachate of the treatment plant is discharged to Pesanggrahan River as recipient water body.

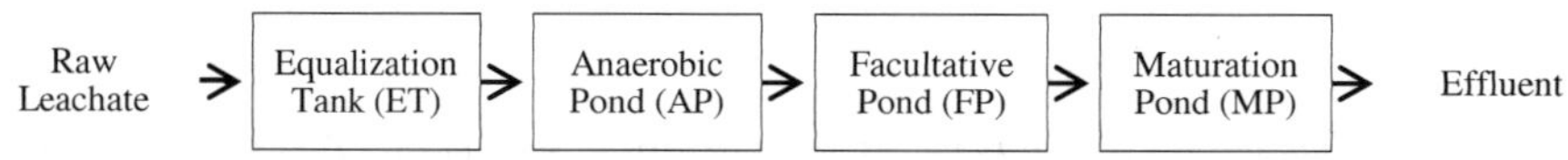

Figure 1. Schematic flow diagram of leachate treatment plant Cipayung Landfill.

2.2. Data Collection and Analysis

In this study, leachate samples were taken every two weeks under rainy season condition from April to Mei 2017. Table 1 describes the sampling locations and shows that samples were taken from the equalization tank, anaerobic, facultative, and maturation ponds effluents. The sampling procedure follows SNI (Indonesian National Standard) number 6989.59:2008 which explains the wastewater sampling method.

The leachate samples were collected using pre-sterilized bottles for physicochemical parameter analysis. The parameters analyzed for the performance evaluation of the leachate treatment plant are water temperature (T), Total Suspended Solids (TSS), pH value, Biochemical Oxygen Demand (BOD), Chemical Oxygen Demand (COD), Total Nitrogen (TN), and Mercury (Hg). The temperature is measured using a digital thermometer (SNI 6989.23:2005), TSS using gravimetric analysis (SNI 6989.3:2004), pH value using a digital pH meter (SNI 06-6989.11-2004), BOD_5 using Winkler with 5 days incubation time (SNI 6989.72.2009), COD using the close reflux titrimetric method (SNI 06-6989.2-2009), TN using spectrophotometer with Nessler (HACH 399), and Mercury using Atomic Absorption Spectrophotometry method (SNI 6989.78:2011).

Table 1. Leachate sampling locations.

Sampling point	Sampling location	Latitude; Longitude
ET	Anaerobic pond influent	6°25'11.60"S; 106°47'14.42"E
AP	Anaerobic pond effluent; Facultative pond influent	6°25'11.78"S; 106°47'13.70"E
FP	Facultative pond effluent; Maturation pond influent	6°25'12.09"S; 106°47'14.15"E
MP	Maturation pond effluent	6°25'11.94"S; 106°47'13.55"E

The methodology developed to study the performance and assess the design of the leachate treatment plant is divided into several steps, comprising of identification and characterization of leachate influent, performance evaluation based on removal efficiency, and design evaluation by comparing the existing plan and physical characteristics to the design criteria.

3. Results and Discussion

3.1. Leachate Characterization

Cipayung Landfill has been operating since 1984 and is classified as a mature landfill. However, the measured average of raw leachate temperature of around 34.81°C, TSS of 72.33 mg/L, pH of 7.83, BOD of 3,959.63 mg/L, COD of 6,860 mg/L, TN of 373.33 mg/L, Hg of 0.0016 mg/L, and BOD/COD ratio of 0.58 indicate that the leachate characteristics resemble an intermediate landfill. This deviation is due to the mixing of young leachate and old leachate since the solid waste dumping system in Cipayung Landfill does not use cell system to produce leachate with the character between young and old leachate (intermediate). The comparison between the leachate characteristics in Cipayung Landfill and the typical leachate is shown in Table 2.

Table 2. Comparison of Cipayung Landfill leachate characteristics and typical leachate.

Parameter	Cipayung Landfill (33 years)	Young landfill (<2 years)	Mature landfill (>10 years)
Temperature (°C)	34.81	-	-
TSS (mg/L)	72.33	200–2.000[11]	100-400[11]
pH	7.83	4.5–7.5[11]	6.6–7.5[11]
BOD (mg/L)	3,959.63	2,000–30,000 [11]	100–200 [11]
COD (mg/L)	6,860	3,00–60,000[11]	100–500[11]
TN (mg/L)	373.33	500–1,500 [12]	50–200 [12]
Hg (mg/L)	0.0016	-	-

The leachate of Cipayung Landfill tends to be alkali which is the typical characteristic of leachate in Indonesia [13]. However, the pH value is still within the range of values 6–9 that is appropriate for biological life. The leachate temperature is in the range of 33.8°C–36.4°C and still conforms to the optimum temperature for microorganisms of 25°C–35°C [11]. The BOD/COD ratio of Cipayung Landfill leachate is 0.58, meaning that it can be treated by biological processing [14].

3.2. Performance and Design Evaluation

The average of raw leachate flow rates entering the stabilization pond system was 91,600 m^3/d. The results obtained for each treatment unit and the overall leachate treatment plant are presented in Table 3. The quality of Cipayung Landfill leachate in each treatment unit shows that the removal efficiency does not comply with the design criteria, implying that the effluent quality of Cipayung Landfill leachate does not meet the quality standard based on [15] for BOD, COD, and TN parameters. The examination results of the anaerobic pond leachate samples state that the TSS concentration of 72.33 mg/L at the influent has been increased to 151 mg/L at the effluent since the mud at the base of the pond was too high. The TSS concentration of the facultative pond effluent was 32.33 mg/L, indicating that the removal of suspended solids at facultative pond processing unit reached 79%. Although the TSS concentration rose back to 60.33 mg/L at the maturation pond effluent, the overall removal efficiency of TSS concentration could reach 17%, and the quality of leachate discharged has fulfilled the requirement. The pH values of leachate in all treatment units are still within the standard quality range. The pH value of the influent leachate is 7.8–7.9 which tends to be alkali. Therefore, the leachate treatment does not require a neutralization process as a pre-treatment because pH 6 is the lowest limit for anaerobic processing [16]. The pH value of the anaerobic pond effluent has increased, indicating that the anaerobic processing has entered the final stage, raising the alkalinity that allows the methane microorganism to live. Volatile acids will be converted to methane and carbon dioxide (CO_2), while organic matter will decrease since its solubility will drop due to the boosted pH value. Supposedly, the pH value will dwindle back in the maturation pond because of the organic matter hydrolysis in the aerobic process, which resulted in the decrease of pH value and mineral dissolution. The rise of pH value to 8.07 in the effluent can be brought by the lack of oxygen concentration in the aerobic treatment within the maturation pond.

Table 3. Average and removal percentage of the parameters in the leachate treatment plant.

Pond type		TSS (mg/L)	pH	BOD (mg/L)	COD (mg/L)	TN (mg/L)	Hg (mg/L)
AP	Influent	72.33 ± 2.517	7.83 ± 0.058	3959.63 ± 993.711	6860 ± 2182.212	373.33 ± 93.350	0.0016 ± 0.00135
	Effluent	151 ± 6.00	8.0 ± 0.173	3894.77 ± 1064.242	9956.8 ± 530.288	793.33 ± 84.120	0.00076 ± 0.00033
	% Removal	**-110%**	**-2%**	**2%**	**-45%**	**-113%**	**53%**
FP	Influent	151 ± 6.00	8.0 ± 0.173	3894.77 ± 1064.242	9956.8 ± 530.288	793.33 ± 84.120	0.00076 ± 0.00033
	Effluent	32.33 ± 2.517	7.97 ± 0.058	3714.63 ± 258.802	6155.1 ± 1605.116	186.70 ± 70.00	0.0031 ± 0.00286
	% Removal	**79%**	**0%**	**5%**	**38%**	**76%**	**-307%**
MP	Influent	32.33 ± 2.517	7.97 ± 0.058	3714.63 ± 258.802	6155.1 ± 160.116	186.70 ± 70.00	0.0031 ± 0.00286
	Effluent	60.33 ± 2.517	8.07 ± 0.289	3399.13 ± 1418.785	9525.6 ± 2040.284	334.43 ± 94.318	0.00072 ± 0.00028
	% Removal	**-87%**	**-1%**	**8%**	**-55%**	**-79%**	**77%**
Total	**% Removal**	**17%**	**-3%**	**14%**	**-39%**	**10%**	**55%**
Leachate standard[15]		100	6 - 9	150	300	60	0,005

The BOD concentration decreased toward the end of treatment with a removal efficiency of 14%. However, the leachate effluent has not met the leachate effluent standard. The high BOD concentration is caused by the organic residues contained in the anaerobic treatment that cannot be degraded in abundant quantity [17]. COD effluent concentration of 9,525.6 mg/L has not met the quality standard. The concentration of COD effluent is increased due to the absence of aeration and the presence of organic material residues. The concentration of TN decreased by 6% removal until the end of treatment. However, the effluent concentration of TN of 334.4 mg/L has not met the leachate standard. The concentration of ammonia nitrogen within the range of 25–30 mg/L can be toxic to anaerobic microorganisms because it inhibits their growth, affecting the removal of BOD and COD [18]. The high concentration of TN in the Cipayung Landfill leachate lessen the efficiency of BOD and COD removal.The concentration of Hg effluent was 0.0007 mg/L, and the removal efficiency was 56%. High concentration of heavy metals can be toxic to *Chlorella* species that are the main species in the stabilization ponds system. The pH value of >8 can cause the metal ions to precipitate, and the treatment process of stabilization ponds can run as usual [19]. Table 4 shows the existing design and physical characteristics of the leachate stabilization ponds system. Anaerobic ponds can be used for waste with a

COD concentration of ≥4,000 mg/L [20]. Loaded with a COD concentration of 6,860 mg/L, the Cipayung Landfill leachate treatment unit has appropriately used an anaerobic pond as the first treatment. BOD removal in an anaerobic pond can reach 90% [21], with the details of 2.5-5 m depth, detention time of 5–50 days [22], and organic loading rate of ≤0.3 kg/m^3.day [23]. Based on the measurement, the anaerobic pond's BOD removal was only 2% and the OLR value of 0.37 kg/m^3.day exceeds the design criteria. OLR values above the design criteria can cause odor around the treatment unit. The 3.5 m depth and 10.7 days detention time in the anaerobic pond treatment have met the design criteria.

Table 4. Physical characteristics of leachate treatment plant Cipayung Landfill.

Parameter	Anaerobic Pond	Facultative Pond	Maturation Pond
Detention time	10.7 day	3.1 day	3.04 day
Useful depth	3.5 m	3.5 m	3.5 m
Area	278.85 m^2	80.29 m^2	79.21 m^2
Organic loading rate (OLR)	0.37 kg/m^3.d	44,273.96 kg/ha.d	42.804,31 kg/ha.d

Facultative pond serves to remove some of the BOD concentration that is still contained in the anaerobic pond effluent with a removal efficiency of up to 95% [24]. However, the efficiency of BOD removal in the Cipayung Landfill facultative pond only reached 5%. The facultative pond's depth of 3.5 m exceeds the design criteria of 1–2 m [21]. The facultative pond's detention time of 3,1 days is too short, compared to the design criteria of 7–50 days [24]. The OLR value of the facultative pond of 44,273.96 kg/ha.day exceeds the design criteria of 15–20 kg/ha.day [21]. OLR values exceeding the design criteria can lead to the incapability of the facultative pond to treat incoming BOD concentration. Maturation pond serves to reduce the concentration of BOD that is still contained in the effluent of the facultative pond, as well as to eliminate the pathogen bacteria contained in the leachate by utilizing sunlight penetration into the pond. The efficiency of BOD removal in the maturation pond is normally 80% [21]. However, the removal efficiency of Cipayung Landfill maturation pond only reached 8%. Its depth of 3.5 m exceeds the design criteria of 0.3–1.0 m [21]. This excess may hinder the entirely aerobic condition throughout the pond since the sunlight cannot penetrate deeply inside the maturation pond, leading to the unavailability of oxygen from algae photosynthesis. The OLR value of 42,804.31 kg/ha.day exceeds the design criteria of 40–120 kg/ha.day [21]. The OLR value exceeding the design criteria causes no removal of BOD concentration in the maturation pond. The detention time of facultative pond of only 3.04 days is too narrow compared to the design criteria of 5–20 days [21].

4. Conclusion

Based on this research, the effluent of Cipayung Landfill leachate treatment plant exceeds the leachate standard limit based on [15] for BOD, COD, and TN parameters. The evaluation results of leachate treatment plant unit in Cipayung Landfill show that the designs of the anaerobic pond, facultative pond, and maturation pond do not meet the design criteria. To improve the quality of leachate effluent, it is necessary to make physical improvements by redesigning the treatment unit according to the design criteria. Establishing a constructed wetland is recommended as well since the climatic condition of the study area implies that this method is suitable and has a high potential to improve the leachate effluent quality by reducingthe nutrients, suspended solids, algae, and BOD concentrations.

Acknowledgment

The authors gratefully acknowledge Universitas Indonesia as the sponsor for this research under the project of PITTA 2017 funding scheme.

References

[1] Buku Putih Sanitasi Kota Depok, Dinas Komunikasi dan Informatika Kota Depok, Depok, 2011.
[2] Renou S, Givaudan J, Poulain S, Dirassouyan F, Mouli. Landfill Leachate Treatment Review and Opportunity. pp. 468-493,2008.

[3] Fitzke B, Blume T, Wienands H, Cambiella A. Hybrid Processes for the Treatment of Leachate from Landfills, Economic Sustainability and Environmental Protection in Mediterranean Countries through Clean Manufacturing Methods. Springer, pp. 107-126, 2013.
[4] Salem Z, Hamouri K, Djemaa R, Allia K. Evaluation of Landfill Leachate Pollution and Treatment. Desalination 220, pp. 108-114, 2008.
[5] Baig S, Coulomb I, Courant P, Liechti P. Treatment of landfill leachates: lapeyrouse and Satrod case studies. Ozone: Science & Engineering, pp. 1-22, 1999.
[6] Fatta D, Papadopoulos A, Loizidou M A study on the landfill leachate and its impact on the groundwater quality of the greater area. Environmental, Geochemical, Health, pp. 175-190, 1999.
[7] Mara D D. Domestic Wastewater Treatment in Developing Countries. London: Earth scan, 2004.
[8] C B Gawasiri. Modern Design of Waste Stabilization Ponds in Warm Climates: Comparison with Traditional Design Methods. UK: University of Leeds, 2003.
[9] Jokela J P Y, Rintala, J A. 2003. Anaerobic solubilisation of nitrogen from municipal solid waste (MSW). *Reviews in Environmental Science and Biotechnology 2*. pp. 67-77.
[10] Balai Data dan Informasi Sumber Daya Air Kota Depok, 2009.
[11] Tchobanoglous G, Hilary Theisen, Samuel Vigil. Integrated Solid Waste Management. Singapore: McGraw Hill, 1993.
[12] J F Crawford, P G Smith. Landfill Technology. UK: British Publishing, 1985.
[13] Damanhuri, Enri, Padmi, Tri. Diktat Kuliah Pengelolaan Sampah Pengurugan (Landfilling) Sampah. Bandung, Program Studi Teknik Lingkungan FTSL ITB, pp. 1-11. 2008.
[14] Grady C P, Leslie Lim, Henry C. Biological Wastewater Treatment Theory and Applications. USA: Marcel Dekker, 1980.
[15] Peraturan Menteri Lingkungan Hidup dan Kehutanan Republik Indonesia Nomor 59 Tahun 2016 tentang Baku Mutu bagi Usaha dan/atau Kegiatan Tempat Pemrosesan Akhir Sampah.
[16] Pescod, Mara. 1998. [Online]. Available: http://stabilizationponds.sdsu.edu/.
[17] Malina, Joseph F, Pohland, Frederick G. Design of Anaerobic Processes for The Treatment of Industrial and Municipal Wastes. USA: Technomic Publishing Company, Inc, 1992.
[18] Sergrist. 1997. [Online]. Available: http://stabilizationponds.sdsu.edu/ 1997.
[19] Murniwati, Camelia Indah. Studi Evaluasi dan Perbaikan Instalasi Pengolahan Lindi TPA (Studi Kasus : TPA Suwung Kota Denpasar. Tugas Akhir S1 Teknik Lingkungan Institut Teknologi Bandung, 2012.
[20] Grady C P, Leslie Lim, Henry C. Biological Wastewater Treatment Theory and Applications. USA: Marcel Dekker, 1980.
[21] Qasim Syed R. Wastewater Treatment Plant, Planning, Design, and Operational. New York: College Publishing, 1985.
[22] Metcalf Eddy, Wastewater Engineering: Treatment and Reuse Fourth Edition. Singapore: McGraw-Hill, Inc, 2003.
[23] WHO. 1987. Wastewater Stabilization Ponds: Principles of Planning and Practice. *WHO EMRO Technical Publication No. 10.*
[24] Benefield, Larry D., Randall, Clifford W. Biological Process Design for Wastewater Treatment. USA: Prentice-Hall, Inc, 1980.

Harvesting of freshwater microalgae biomass by *Scenedesmus* sp. as bioflocculant

A Rinanti[1*], **R Purwadi**[2]

[1]Environmental Engineering Department, Faculty of Lanscape Architecture and Environmental Engineering, Universitas Trisakti, Jakarta, Indonesia
[2]Chemical Engineering Department, Faculty of Industrial Technology, Institut Teknologi Bandung, Bandung, Indonesia

*Corresponding Author: astririnanti@trisakti.ac.id

Abstract. This study is particularly expected to provide information on the diversity of microalgae as the flocculant agent that gives the highest biomass yield. Bioflocculation was done by using one of the flocculating microalgae i.e. *Scenedesmus obliquus* to concentrate on non-flocculating microalgae *Chlorella vulgaris*. The freshwater microalgae *S. obliquus* tested it ability to harvest other non-flocculating microalgae, increased sedimentation rate in the flocculation process and increased biomass yield. The flocculation of biomass microalgae with chemical flocculant as comparison was done by adding alum ($K_2SO_4.Al_2$ $(SO_4)_3.24H_2O$). The addition of alum ($K_2SO_4.Al_2(SO_4)_3.24H_2O$) as flocculant at pH 11 and *S. obliquus* sp. as bioflocculant caused significant alteration of nutrition of microalgae. Overall, the essential content produced by flocculation method with addition of alum or with bioflocculation (%, mg/100 mg dry weight) are lipid 31,64; 38,69, protein 30,79; 38.50%, and chlorophyll 0.6253; 0.8420). Harvesting with bioflocculation methods conducted at the end of the cultivation period increase the amount of biomass significantly and can accelerate the settling time of biomass. Harvesting microalgae cells by bioflocculation method becomes an economically competitive harvesting method compared to alum as a chemical flocculant because of the cheaper cost of flocculant, not toxic so it does not require further water treatment after harvesting due to the use of alum as chemical flocculants.
Keywords: biomass, bioflocculant, freshwater, flocculating microalgae

1. Introduction

The cultivation of tropical microalgae as a source of renewable energy and dietary supplements has constrained the harvesting of the microalgae itself. Small size and density make microalgae difficult to harvest. To improve the economic viability of microalgae in order to obtain a useful final product, there are three important key parameters: high biomass microalgae productivity, high metabolite compound content, and cheap and environmentally friendly harvesting methods. Harvesting of microalgae is an important part in microalgae cultivation system to produce higher biomass. The purpose of harvesting is to obtain microalgae slurry, at least 2-7% of total solids [1]. However, the harvesting of microalgae is often still a constraint. Harvesting microalgae can be done by techniques such as centrifugation, filtration, sedimentation, flocculation, flotation, ultrasonic vibration, and screening [2], [3], [4], [5].

IOP Conf. Series: Earth and Environmental Science **106** (2017) 012087 doi:10.1088/1755-1315/106/1/012087

The main disadvantage of these harvesting techniques is that it requires expensive operational costs due to high energy dependence and less environmentally friendly. In commercial industry, the best biomass harvest can reach between 0.3-0.5 g dry cell / L or 5 g dry cell / L. This makes the harvest of microalgae very difficult and expensive [6]. To minimize energy consumption in the process of harvesting microalgae, an integrated approach is needed.

Flocculation can be done in various ways. Chemical flocculation performed using Zn^{2+}, Al^{3+}, Fe^{3+} or other chemical flocculants has been studied extensively [1].

1) Several chemical flocculants have been applied on an industrial scale, especially in wastewater treatment plants [7]. Although the use of chemical flocculants can be done easily but this method is expensive and not environmentally friendly because the excess caution in the flocculants needs to be removed first before the residual water (medium) of production is discharged into water bodies, thus requiring additional operational costs again [7].
2) Flocculation can also be induced by altering the culture conditions by applying extreme pH, nutrient reductions, temperature changes and changes in dissolved O_2 levels. The flocculation method is not appropriate for pre-harvesting of microalgae on a large scale. Most of these methods are not applicable because flocculation is difficult to control and can lead to undesirable cell composition [8]. All of these growth mediums require pre-treatment to remove chemical flocculant cations when they are reused [7].

Bioflocculation is a technique of harvesting microalgae whose principle is similar to flocculation, which is the differentiator is the flocculant used. Bioflocculation uses living things such as bacteria, fungi, or microalgae as flocculants. Flocculation using bacteria [8] and fungi [1] are alternatives to replace chemical flocculants, so the effects of chemical pollution on culture media can be reduced. However, the use of fungi and bacteria as flocculants requires additional special media as a source of energy for its growth. In addition bacteria and fungi can contaminate microalgae [9].

This research is an advanced research from previous research as an effort to obtain the technique of cheap harvesting and environmentally friendly, so that economic feasibility in microalgae biomass production can be improved. By studying the speed gradient in floc formation, this study is particularly expected to provide information on the diversity of microalgae as the flocculant agent that gives the highest biomass yield. In general, it is expected that this research can provide scientific contribution to the development of microalgae biomass production technology that is more economical, efficient and environmentally friendly.

2. Research Method

2.1. Cultivation of flocculant and non flocculant microalgae

Microalgae were cultivated using an artificial growth medium of *Provasoli Haematococcus* Media (PHM). Microalgae that grow in light and climate are controlled by incubators with a rotational speed of 100 rpm, 25 °C, fed with 2% pure CO_2 with a flow rate of 5 $L \cdot min^{-1}$, illuminated by fluorescent lamps (50 $\mu mol \cdot photons\ m^{-2}\ s^{-1}$) with 16/8 hour light/dark cycle [11], [12]. The *S. obliquus* as flocculating microalgae was tested on its ability to improve the recovery efficiency and the rate of harvesting of the non-flocculating microalgae (*Chlorella vulgaris* and *Ankistrodesmus* sp.).

2.2. Measurement of microalgae growth

Cell concentration was measured as the optical density at 750 nm (OD750) with an Ultraspec 2,000 spectrophotometer (Pharmacia Biotech Ltd. UK) equipped with a temperature controlled carousel cell holder with six positions. Demineralized water served as reference. The microalgal samples were diluted in a 10 × 10 × 45 mm polystyrene cuvette (Sarstedt, DE) using filter-sterilized tap water for the freshwater microalgae.

2.3. The study of flocculation (%) of microalgae by flocculants and bioflocculants

The flocculation of biomass microalgae with chemical flocculant as comparison was done by adding alum ($K_2SO_4.Al_2(SO_4)_3.24H_2O$) with concentration of 120 ppm, pH 10. After the settling is done the filtration or filtration process. The precipitation of microalgae with bioflocculant is obtained from the calculation of OD750 (Optical Density) data using equation 1 [13].

$$recovery\ (\%) = \frac{OD_{750}(t_0) - OD_{750}(t)}{OD_{750}(t_0)} \,.100 \quad (1)$$

$OD_{750}(t_0)$ is the turbidity of sample taken at time zero and OD_{a750} (t) is the turbidity of the sample taken at time t.

2.4. Determination of Recovery Efficiency
To determine the ability of microalgae strain as microalgae flocculant or non-flocculant microalgae it is necessary to calculate recovery efficiency, i.e. recovery by non-flocculant microalgae to flocculating microalgae divided by non-flocculant microalgae recovery without presence of flocculant microalgae (adapted from [1], [10]).

$$recovery\ efficiency\ (\%) = \left[1 - \frac{\frac{OD_{a750}(t)}{OD_{a750}(t_0)}}{\frac{OD_{b750}(t)}{OD_{b750}(t_0)}}\right] .100 \quad (2)$$

$OD_{a750}(t_0)$ and $OD_{a750}(t)$ are the turbidities of samples of non-flocculating microalgae with flocculating microalgae taken at time zero and at time t, respectively. $OD_{b750}(t_0)$ is the turbidity of sample of non-flocculating microalgae taken at time zero and OD_{b750} (t) is the turbidity of the same sample taken at time t.

$OD_{a750}(t_0)$ and $OD_{a750}(t)$ are the turbidities of samples of non-flocculating microalgae with flocculating microalgae taken at time zero and at time t, respectively. $OD_{b750}(t_0)$ is the turbidity of sample of non-flocculating microalgae taken at time zero and OD_{b750} (t) is the turbidity of the same sample taken at time t.

3. Results and Discussion

Increased recovery of non-flocculant microalgae added with flocculant microalgae have been evaluated by calculating percentage of recovery efficiency. For calculating the percentage of recovery efficiency (Equation 2), the measurements are seen from the average turbidity. Sedimentation rates and recovery percentages for all samples tested showed a standard deviation of less than 3.5%.

The flocculant microalgae exhibit higher recovery efficiency when added to higher concentrations, however, doubling the doubled flocculent microalgae concentration does not necessarily result in a twice higher recovery efficiency of the non-flocculant microalgae.

Sedimentation of microalgae suspension was monitored for 8 hours and recovery percentage of microalgae was determined from time to time (Figure 1).The microalgae sedimentation rate in the suspension was calculated by linear regression of the data in the curve of the recovery percentage in time and the use of the linear regression curve.

The initial sedimentation rate of flocculant microalgae measured during the first 2 hours was higher than sedimentation by non-flocculant microalgae (Table 1 and Table 2). Mixing of flocculant microalgae increases the rate of early sedimentation of non-flocculant microalgae. Large clumps formed by flocculant microalgae seem to cause the non-flocculant microalgae trapped. In addition, the sedimentation rate becomes faster than individual non-flocculant microalgae cells without mixing. Furthermore, increasing the ratio of bio-flocculent microalgae to non-flocculant microalgae causes higher sedimentation rate to be higher.

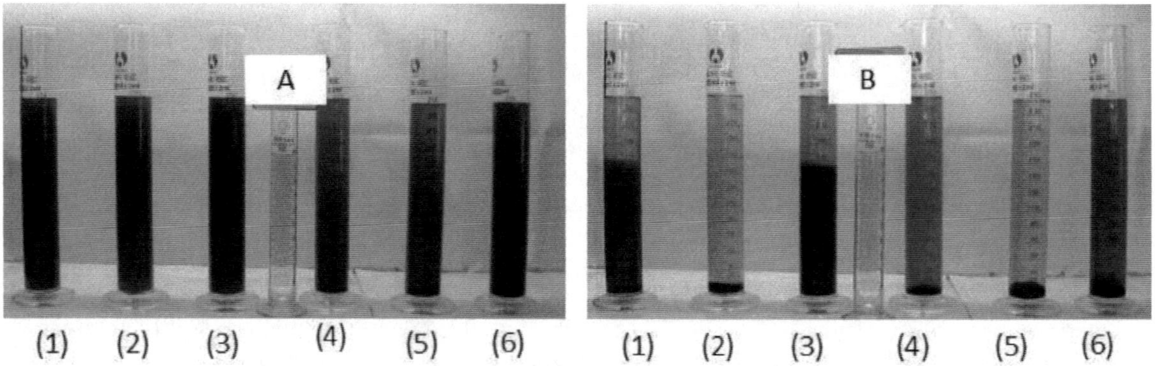

A = t0, B = after 8 hours,
= mix 1 added by $K_2SO_4.Al_2(SO_4)_3.24H_2O$; (2) = *S. obliquus;* (3) *C. vulgaris* (4) C. *vulgaris* added by *S. obliquus;* (5) *Ankistrodesmus* sp added by *S. obliquus;* (6) *Ankistrodesmus* sp.

Figure 1. Sedimentation of microalgae suspension.

Table 1. Optical densities (OD750 (t0)).

	Optical densities (OD750 (t0))			
	A	**B**	**C**	**D**
A				
Mix-1	0.8	0.9	0.8	0.8
Mix-2	0.7	0.5	0.6	0.8
Mix-3	0.5	0.6	0.6	0.6

A = $K_2SO_4.Al_2(SO_4)_3.24H_2O$, B = S. obliquus
C = The non-flocculating microalgae (*V. vulgaris*) with low concentration of added *S. obliquus*
D = The non-flocculating microalgae (*Ankistrodesmus* sp.) with high concentration of added *S. obliquus*

Table 2. Initial sedimentation rate.

	Initial sedimentation rate (% recovery.h^{-1})			
	A	**B**	**C**	**D**
($K_2SO_4.Al_2(SO_4)_3.24H_2O$)				
Mix-1	39.4	16.7	13.3	9.8
Mix-2	34.9	23.7	20.4	14.6
Mix-3	43.8	40.0	35.7	16.9

A = $K_2SO_4.Al_2(SO_4)_3.24H_2O$
B = *S. obliquus*
C = The non-flocculating microalgae (C. *vulgaris*) with low concentration of added *S. obliquus*
D = The non-flocculating microalgae (*Ankistrodesmus* sp.) with high concentration of added *S. obliquus*

Flocculating is a method of harvesting where microalgae linked each other and converge to form a larger mass mass, called floc. Flocculation in this study is a process of aggregating small particles into larger particles resulting from the existence of physical or chemical instability. In some literature, it is mentioned that flocculation is the same as coagulation, because the mechanism of operation is almost the same. However, for some researchers, the two methods are different, where flocculation forms aggregates due to polymers, whereas coagulation results from electrolytes. The aggregates formed are caused by changes in pH or addition of flocculating microalgae as bioflocculants or chemical coagulants.

In the intercellular connection there are two forces involved. The first force is the repulsive force of electrostatic rejection of the cell surface because it has the same charge, which is negatively charged. This is due to the ionization of functional ionogenic groups [9]. This force occurs at a considerable distance relative. At a very close relative distance, there is an interesting attraction due to Van der Waals intermolecular forces. The intermolecular force is larger than the electrostatic force, but only occurs in very close proximity [13].

S. obliquus microalgae is proven to function as a bioflocculant to overcome some of the constraints that occur in the harvesting process. Thus, this study proves that total recovery and sedimentation rates of various non-flocculant microalgae will increase in each addition of different flocculant microalgae. The results showed that the addition of autoflocculant microalgae was able to induce sedimentation rate faster than sedimentation rate of non-flocculant microalgae so as to increase the efficiency of harvesting. Similar positive effects on sedimentation rates and harvesting efficiency were also observed in bioflocculation by bacteria as non-flocculant microorganisms [14].

4. Conclusion

Harvesting of non-flocculating microalgae biomass by *Ankistrodesmus* sp as bio-autoflocculating has been evaluated as an effective and promising alternative method for future harvesting of microalgae. With bioflocculation method, harvesting biomass of microalgae is possible without addition of chemical flocculant. Growth media during the cultivation process does not contain harmless chemicals so that it does not require processing before it is discharged into the environment. Bioflocculation is an alternative to energy-efficient harvesting methods.

Acknowledgment

The authors would like to thank Directorate for Research and Community Services as well as Directorate General for Strengthening Research and Development at the Ministry of Research, Technology, and Higher Education Indonesia for funding this study through Hibah Penelitian Pasca Doktor 2017

References

[1] Papazi A, Makridis P, Divanach P. 2010. Harvesting Chlorella minutissima Using Cell Coagulants. *J ApplPhy*col. 22:349–355.

[2] Wijffels RH, Barbosa M. 2010. An Outlook on Microalgal Biofuels. *J Science*.doi:10.1126/science. 1189003

[3] Zittelli G C, Rodolfi L, Biondi N, Tredici M R. Productivity and photosynthetic efficiency of outdoor cultures of *Tetraselmissuecica* in annular columns. Aquaculture. 2006. 261:932–943.

[4] Del Campo J. A., Garcia-Gonzalez M, and Guerrero M G. 2007. Outdoor Cultivation of Microalgae for Carotenoid Production. Current State and Perspectives, *Appli. Microbiol. Biotechnol.* 74:1163-1174.

[5] Schenk P M, Thomas-Hall S R, Stephens E, Marx U, Mussgnug J H, Posten C, Kruse O, Hankamer B. Second generation biofuels: high-efficiency microalgae for biodiesel production. Bioenergy Res. 2008. 1:20–43. doi: 10.1007/s12155-008-9008-8. SimGoh, and Becker .1988.

[6] Wang B, Li Y, Wu N, and Lan C Q. 2008. CO_2 biomitigation using microalgae. *Appl Microbiol Biotechnol* 79:707–718.

[7] Griffiths MJ, Harrison STL. 2008. Lipid Productivity as A Key Characteristic for Choosing Algal Species for Biodiesel Production. *J Appl Phycol.* 21:493–507. doi: 10.1007/s10811-008-9392-7.

[8] Hulteberg C, Karlsson, H T, Børresen, B T, and Eklund, H. *Final Report on Biodiesel Production from Microalgae.* Presented to Statoil Hydro ASA Oslo, Norway May 16, 2008.

[9] Salehizaden H, Vossoughi M, Alemzadeh I. 2000. Some Investigations on Bioflocculant Producing Bacteria. *J Biochemical Engineering*. 5:39-44.

[10] Lee A K, Lewis D M, Ashman P J. 2009. Microbial Flocculation, A Potentially Low-Cost Harvesting Technique For Marine Microalgae for The Production of Biodiesel. *J Apply Phycol.* 21:559–567.doi 10.1007/s10811-008-9391-8

[11] Rinanti A, Kardena E, Astuti D I, Dewi K. 2013. Improvement of Carbon Dioxide Removal Through Artificial Light Intensity and Temperature by Constructed Green Microalgae Consortium in A Vertical Bubble Column Photobioreactor. *Malaysian Journal of Microbiology*. 10(1):29-37

[12] Rinanti A, Kardena E, Astuti D I, Dewi K. 2014. Biotechnology Carbon Capture and Storage (CCS) by Mix-culture Green Microalge to Enhancing CO_2 Uptake Rate and CO_2 Removal Efficiency with Different Aeration Rates in Closed System Photobioreactor. *Jurnal Teknologi-UTM*. 16(2): 253-258.

[13] Salim, S, Bosma, R, Vermue, M H, Wijffels, R H. 2010. Harvesting of Microalgae by Bio-flocculation. *Journal of Applied Phycology*. 23(5):849-855

IOP Conf. Series: Earth and Environmental Science **106** (2017) 012088 doi:10.1088/1755-1315/106/1/012088

Molecular imprinted hydrogel polymer (MIHP) as microbial immobilization media in artificial produced water treatment

E Kardena[1,3], S L Ridhati[2], Q Helmy[1,3*]

[1]Water and Wastewater Engineering Research Group, Environmental Engineering Department, Institut Teknologi Bandung, Bandung, Indonesia
[2]Environmental Engineering Department, Faculty of Civil and Enviromental Engineering, Institut Teknologi Bandung, Bandung, Indonesia
[3]Bioscience and Biotechnology Research Center, Institut Teknologi Bandung, Bandung, Indonesia

*Corresponding Author : helmy@tl.itb.ac.id

Abstract. Produced water generated during oil and gas exploration and drilling, consists of many chemicals which used in drilling process. The production of produced water is over three fold of the oil production. The water-cut has increased over time and continues to do so because the fraction of oil in the reservoir decreases and it is more difficult to get the oil out from an old oil-field. It therefore requires more sea water to be injected in order to force the oil out; hence more produced water is generated. Produced water can pollute the environment if it is not treated properly. In this research, produced water will be treated biologically using bacterial consortium which is isolated from petroleum processing facility with Molecular Imprinted Hydrogel Polymer (MIHP) for microbial immobilization media. Microbial growth rate is determined by measuring the MLVSS and hydrogel mass, also by SEM-EDS analysis. SEM-EDS analysis is an analysis to evidence the presence of microbe trapped in hydrogel, and also to determine the types and weight of the molecules of hydrogel. From this research, suspended microbial growth rate was found at 0.1532/days and attached microbial growth rate was 0.3322/days. Furthermore, based on SEM analysis, microbe is entrapped inside the hydrogel. Effectiveness of microbial degradation activity was determined by measuring organic materials as COD. Based on COD measurement, degradation rate of organic materials in wastewater is 0.3089/days, with maximum COD removal efficiency of 76.67%.
Keywords: COD, hydrogel, MIHP, produced water, SEM-EDS

1. Introduction

Saline wastes are generated in significant quantities in petrochemical industries such as oil and gas production. These generated wastes contain organic compounds and high concentrations of salt (>1.5%). Generated wastewater is considered as salty water trapped in the reservoir rock and brought up along with oil or gas during production. It subsists under high pressures and temperatures, and usually contains hydrocarbons and metals. Therefore, it must be treated before being discharged to the environment. The discharge of such wastewater containing at the same time of high salinity and high organic content without prior treatment is known to adversely affect the aquatic life, water potability and agriculture [1]. In spite of the detrimental effect of salty produced water on microbial activity, moderate acclimation of activated sludge to high salinity is possible. Acclimation implies the exposure of non-salt-adapted micro-organisms to increasing salt concentrations in order to permit the obtention of satisfactory effluent treatment performance at a given salt concentration. The success of such adaptation depends on several factors, such as the type and growth phase of micro-organisms, as well as the rapid or gradual increase of salt concentration during acclimation [1], [2].

Many studies reported that operation of activated sludge process at salt contents higher than 20 g/L is characterized by poor flocculation, high effluent solids, and a severe decrease in substrate utilization rate [2]. Researcher [3] reported that the nature of pollutants and the high salinity (about 29 g/L) of oil-field generated water has an unfavorable effect on the activated sludge process. High hydraulic loadings (above 2.5 m^3 /m^3.day) increased the wash-out of the activated sludge from the reactor. The addition of Powdered Activated Carbon improved the sludge volume index and increased the rate of biodegradation. This is due to the ability of biofilm formation on the activated carbon surface. The low settleability of sludge is caused by the salt content in produced water that reduces the population of protozoa, resulting in low settle-ability. Salt content in produced water increases the buoyancy forces, causing low sedimentation efficiencies.

Cell immobilization is an efficient strategy to form an artificial biofilm on a substrate. Biofilm formation in a media/substrate will overcome the buoyancy forces of the biomass in a salty environment and avoid the wash-out of the activated sludge from the treatment plant. The advantages of immobilized technique include easy separation of the cells from the carrier material, greater productivity due to high cell concentrations achieved and the protection of cells against harsh environmental conditions [4], [5], [6]. Studies have shown that polymer types play a dominant role in determining the properties of the microspheres. Examples of synthetic gels include poly(carbamoyl) sulphonate, polyhydroxyethylmethacrylate (polyHEMA) and polyacrylamideand polyvinyl alcohol (PVA). Microspheres are almost exclusively produced using water-soluble polymers which provide a high degree of permeability for nutrient sand metabolites with a low molecular weight, thus providing optimal conditions for the functioning of immobilized microbial cells [6], [7].

This study investigated the applicability of an immobilisation technique based on cell entrapment in a Molecular Imprinted Hydrogel Polymer (MIHP), a strategy to increase the ecological competence of degrading microorganisms in treating salty produced water.

2. Research Method

2.1. Bacterial and growth condition

Mixed culture of degrading bacteria wasobtained from culture collection of Environmental Laboratory, Bioscience and Biotechnology Research Center, Institut Teknologi Bandung, Indonesia. All culture were previously isolated from petroleum-contaminated soils by an enrichment culture technique using crude oil as the sole source of carbon and energy. The culture was maintained at 4° C on crude oil basal medium.

The syntheticoilfield-produced water was used throughout the study to determinethe capability of strains of bacteria to degrade organic compounds.The synthetic oilfield-produced water was prepared by mixing crudeoil (1.5 mL/L) and salt water (sea water and tap water), as described in the literature [8]. The C/N/P ratio of the medium was adjusted to approximately 100/5/1 by adding appropriate concentrations of $(NH_4)_2SO_4$ and KH_2PO_4 to the synthetic oilfield-produced water. This synthetic oilfield-produced water was used throughout our study to determine the capability of immobilized cell ofbacteria to degrade organic compounds. The characteristics of the realand synthetic oilfield-produced water samples are listed in Table 1.

2.2. Molecular Imprinted Hydrogel Polymer (MIHP) Synthesis

Reagent used in MIHP synthesis are bis-acrylamide, N-N'methylenbisacrylamide, NaOH 4N, potassium persulfate, tetramethyletilendiamine (TEMED), and buffer pH 6. All chemical are obtained from Merck, Sigma-Aldrich and JT Baker. MIHP produced in the form of dry gel Bis-acrylamide and N-N'methylenbisacrylamide used to synthesize acrylamide-bis 40%, which functions as crosslinker.MIHP synthesized by adding a buffer solution of pH 6 into acrylamide-bis 40%. After homogenized, add potassium persulfate (as initiator) and TEMED (as catalyst) to speed up the polymerization process. The solution was left overnight until a hydrogel is formed. Hydrogel formed are washed by demineralized water until it reaches neutral pH, then dried in the oven with temperature of 110°C. The hydrogel are used as immobilized media and attached media for microbes growth throughout the experiment in the synthetic oilfield-produced water treatment. SEM-EDS analysis was used to determine the hydrogel structure and microbial attachment both in outer and inner of the hydrogel.

2.3. Produced water biodegradation by immobilized cells

In order to assess the biodegradation potential of oilfield-produced water by the immobilized degrader, the bacteria were acclimated first to the intended environment. Acclimation implies the exposure of non-salt-adapted microorganisms to increasing salt concentrations substrate until it reached the desired performance. The degradation potential was based on the growth rate (μ) and chemical oxygen demand (COD) removal efficiency. The microbial growth was obtained by measured the MLVSS over the observation time (t). The initial and final CODs were measured using standard methods (DR-2800, Hach Company). Reactor-1 arrangement (PW+Cell) is produced water + immobilized cell, while reactor-2 (PW-Cell) is a control reactor contains produced water without immobilized cell addition.

3. Results and Discussion

3.1. Cell immobilization

This research focuses on the synthesis of molecular imprinted hydrogel polymer as microbial immobilization medium. The polymer used is non-template polymer, the cavity formed in hydrogel is a cavity of water molecules that evaporate due to heating process. Polyacrylamide is synthesized from bis-acrylamide 40%, which contains a monomer of acrylamide and bis-acrylamide as crosslinker. To initiate polymerization, an initiator is required. The initiator is a material that can immediately turn into free radicals when heated or irradiated by electromagnetic radiation. Initiator and catalyst used in MIHP polymerization is 10% potassium persulfate and tetraethylmethylendiamine (TEMED). Excessive or too low concentrations of ammonium persulfate will not produce MIHP, whereas TEMED has an effect on polymerization time. The ratio between bis-acrylamide 40%, buffer pH 6, initiator, and catalyst must be precise, otherwise polymerization will be difficult and hydrogel is hard to form. The hydrogel structure can be seen in Figure 1.

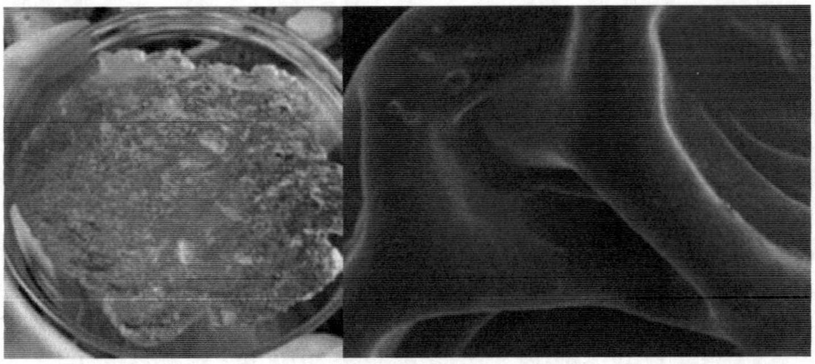

Figure 1. Wet hydrogel structure (left), and SEM image with 1000x magnification (right).

Immobilized cells have advantages of higher activity, higher cell density and longer stability than free cells. Hence, the immobilized cells have the potential to degrade toxic chemicals at higher concentration as compared to the suspended cells. Requirement for an effective and safecarrier for resting a microorganisms should be nontoxic and non-polluting, consistent in quality, have a long shelf life, allow sufficient cell activity and cell density [6], [9]. MIHP hydrogel was introduced to the microbial culture suspension in order to imbedded the cell into the hydrogel structure by using a shrink/swell procedure. The cell suspension will enter and entrapped into the hydrogel pore, also attached on the surface of the hydrogel structure (Figure 2).

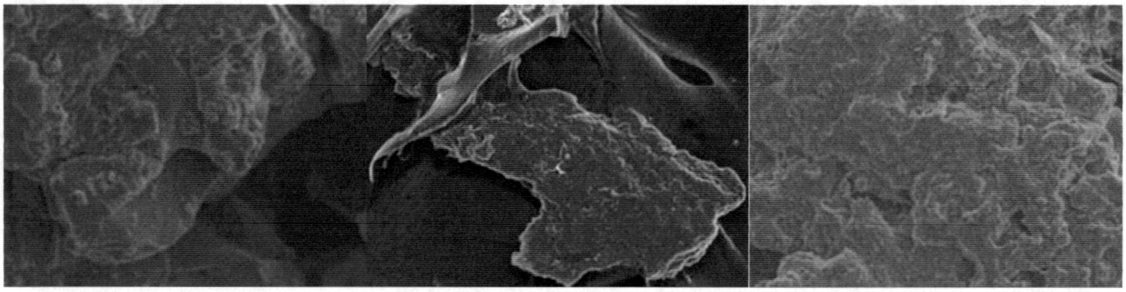

Figure 2. Apparent structure of hydrogel: (left) internal structure of hydrogel indicating an immobilized cells; (middle) cells attached on the surface of hydrogel after 3 days incubation; (right) cells attached on the surface after 5 days incubation.

3.2. Biodegradation of oilfield-produced water by immobilized cells

Performance of immobilized cell in the biodegradation of synthetic oilfield-produced water was studied in the 2L reactor volume that contain initial concentration of 868 mg/L MLVSS degrader microbe. The growth and performance of the immobilized cell in the treatment of oilfield-produced water can be seen in Figure 3. Characteristics of the synthetic oilfield-produced water used in this study and its comparison with typical produced water is described in Table 1.

Table 1. Characteristics of produced water (PW).

No.	Parameter	Synthetic PW	Typical PW*
1	COD, mg/L	1,200	100-4000
2	Oil&Grease, mg/L	-	2-500
3	TSS, mg/L	-	10-200
4	pH	6.89	4-10
5	Conductivity, μS/cm	13,080	4,000-58,000
6	Salinity, %	0.73	0-2.5
7	Phenols, mg/L	0.472	0-23

*adapted from [8], [10], [11]

The growth of degrader microbe in the reactor (PW+Cell) tends to grow and reach the exponential phase without showing any lag phase. This indicates that the strains of bacteria can consume organic compounds in the oilfield-produced water as a carbon source as soon as they come into contact with the wastewater. The specific growth rate (μ) and COD removal of the strains of bacteria tended to increase over time showing the efficacy of the degrader microbe. As observed in Figure 3, the COD removal rate was increased significantly after 2 days of incubation time. On the other side, the control reactor (PW-Cell) where there is no augmented degrader bacteria that added in the system shows a considerable increase in COD removal for up to 33.3% efficiency after 5 days of incubation time. Even without addition of degrader microbe, degradation process will occurred because of the oxidation from aeration process in the reactor.

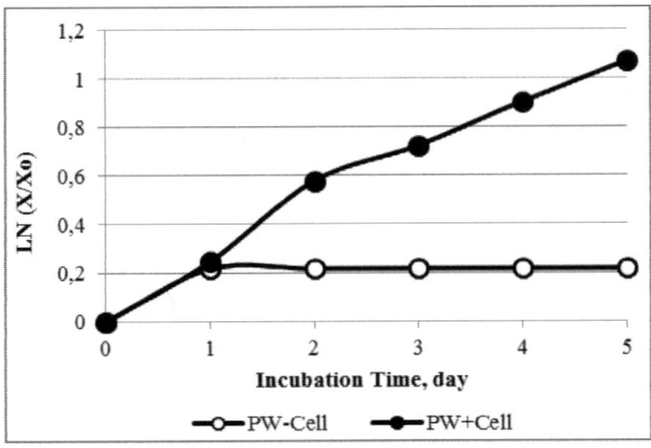

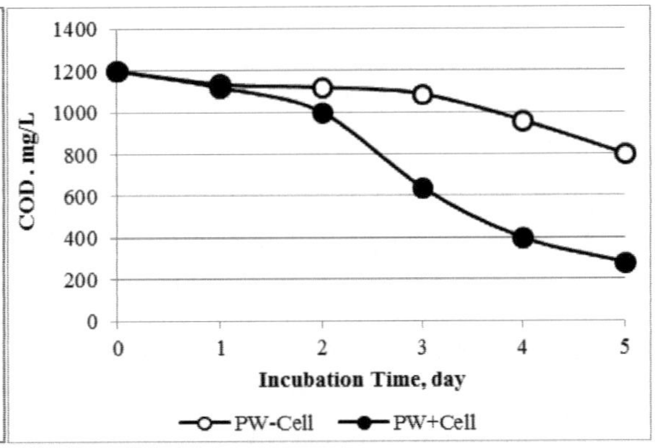

Figure 3. Profile of microbial growth (MLVSS) and COD removal of the syntetic oilfield-produced water in the reactor with addition of immobilized cell (–•–); and control reactor without addition of immobilized cell (–○–).

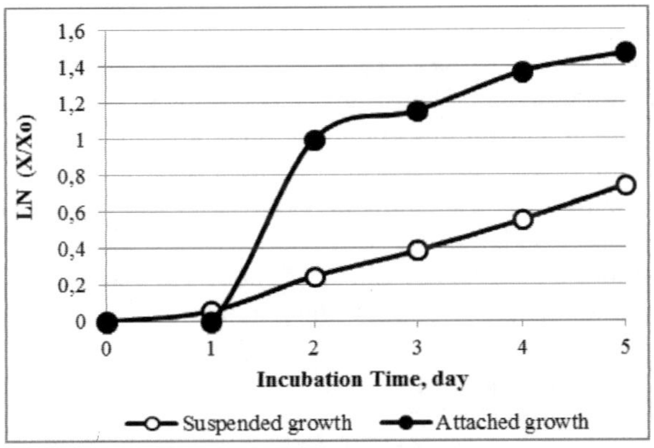

Figure 4. Growth rate of degrader microbe in the PW+Cell reactor, (–○–) suspended growth; and attached growth in the hydrogel (–•–).

The performance of the activated sludge in the treatment period can be seen in Figure 3. It was observed that, from the beginning of lag phase, the concentration of MLVSS increased. However, reactor without addition of microbial starter (PW-Cell) reached a steady-state condition after day 1, while reactor with addition of microbial starter (PW+Cell) continue to increase and became more adapted to the new environment as a function of time. In the fifth days of incubation period, the highest MLVSS concentration was observed in PW+Cell reactor (2.532 mg/L), while the lowest concentration was found in PW-Cell reactor (385 mg/L). COD removal efficiency in the PW+Cell compared to PW-Cell reactor within 5 days incubation time was 76.6% to 33.3%, respectively. As can be seen in Figure 4, attached growth rate is higher than in suspended state in the PW+Cell reactor, indicating that microbe able to attached in the hydrogel and utilized it as a growth support media.

4. Conclusion

The produced water discharged from both onshore and offshore platforms is typically treated by use of chemical and physical systems due to its simple operations. These conventional technologies, however, does not remove the organics such as suspended oil particles and other dissolved compounds. Some chemical treatment technologies also produce hazardous sludge and the cost of those operational processes can be significant. Biological treatment of produced water can be a cost-effective and environmental friendly method, but could be a problem on offshore installation due to the space limitations on the installations. To overcome the space limitation, a compact biological process is needed such as microbial entrapment or immobilization methods can be used. High concentration of microbe/MLVSS can be achieved in a compact reactor compare to conventional suspended activated sludge units that require larger space to achieve the same cell density. Molecular Imprinted Hydrogel Polymer (MIHP) from Polyacrylamide can be used as a support medium for microbial immobilization. This indicated by the SEM analysis, which shows that on the fifth days of incubation, the internal surface of the hydrogel is covered by microbes. The rate of microbial growth in wastewater suspension is 0.1532/day and the rate of microbial growth in hydrogel is 0.3322/day. The rate of microbial growth in hydrogels is higher than in suspended state. This shows that microbes are able to utilized hydrogel as an attached growth media. The advantage of this method is the amount of biomass required in the synthetic oilfield-produced water treatment process fulfilled rapidly, because the microbe grows faster with hydrogel supporting media.

Acknowledgement

The study was performed within the Indonesian Ministry of Research, Technology and Higher Education (Ristek-Dikti) Research Grant fiscal year 2017, also facilitated by Environmental Laboratory of Bioscience and Biotechnology Research Center, Institut Teknologi Bandung.

References

[1] Lefebvre O, and Moletta R. 2006. Treatment of organic pollution in industrial saline wastewater: A literature review. *Water Research*. 40:3671– 82
[2] Burnett W E, 1974, The effect of salinity variations on the activated sludge process. Water & Sewage Works, 12137-55.
[3] Dalmacija B, Karlovic V, Tamas Z and Miskovic D. 1996. Purification of high salinity wastewater by activated sludge process. *Wat. Res.* 30:295-8.
[4] Friel M T, and Mcloughlin A J. 1999. Immobilization as a strategy to increase the ecological competence of liquid cultures of Agaricusbisporus in pasteurized compost. *FEMS Microbiol Ecol.* 3039-46.
[5] Rathoreet S, Desai M P, Liew V C, Lai W C, and Paul W S H. 2013. Microencapsulation of microbial cells. *J Food Eng.* 116369-81.
[6] Wang HQ, Hua F and Zhao YC 2014 Immobilization of Pseudomonas sp. DG17 onto sodium alginate-attapulgite-calcium carbonate Biotechnology & Biotechnological Equipment. 28: 834-42
[7] Horan N J, and Huo C X. 2004. Production and characterization of immobilized-biomass carriers, using polyacrylamide. *Environ Technol.* 25:667-72.
[8] Kardena E, Hidayat S, Nora S, and Helmy Q. 2017. Biological Treatment of Synthetic Oilfield-Produced Water in Activated Sludge Using a Consortium of Endogenous Bacteria Isolated from A Tropical Area. *J Pet Environ Biotechnol.* 8:2-6
[9] Cassidy M B, Lee H, and Trevors J T. 1996. Environmental applications of immobilized microbial cells: a review. *J. Ind. Microbiol Biotechnol.* 1679-101
[10] Lu M, Zhang Z Z, Yua W Y, and Zhu W. 2009. Biological treatment of oilfield-produced water: A field pilot study International Biodeterioration & Biodegradation. 63316–21
[11] Igunnu E T, Chen G Z. 2014. Produced water treatment technologies. *Int. J. Low-Carbon Tech.* 9:157-77

Decolourization of remazol black-5 textile dyes using moving bed bio-film reactor

R Pratiwi[1], S Notodarmojo[2], Q Helmy[2,3*]

[1]Environmental Engineering Department, Faculty of Civil and Environmental Engineering, Institut Teknologi Bandung, Bandung, Indonesia
[2]Water and Wastewater Engineering Research Group, Environmental Engineering Department, Institut Teknologi Bandung, Bandung, Indonesia
[3]Bioscience and Biotechnology Research Center, Institut Teknologi Bandung, Bandung, Indonesia

*Corresponding Author : helmy@tl.itb.ac.id

Abstract. The desizing and dyeing processes in the textile industries produces wastewaster containing high concentration of organic matter and colour, so it needs treatment before released to environment. In this research, removal of azo dye (Remazol Black 5/RB 5) and organic as COD was performed using Moving Bed Biofilm Reactor (MBBR). MBBR is biological treatment process with attached growth media system that can increase removal of organic matter in textile wastewater. The effectiveness of ozonation as pre-treatment process to increase the removal efficiency in MBBR was studied. The results showed that in MBBR batch system with detention time of 1 hour, pre-treatment with ozonation prior to MBBR process able to increase the colour removal efficiency of up to 86.74%. While on the reactor without ozone pre-treatment, the colour removal efficiency of up to 68.6% was achieved. From the continuous reactor experiments found that both colour and COD removal efficiency depends on time detention of RB-5 dyes in the system. The higher of detention time, the higher of colour and COD removal efficiency. It was found that optimum removal of colour and COD was achieved in 24 hour detention time with its efficiency of 96.9% and 89.13%, respectively.
Keywords: decolourization, textile, dye, MBBR, ozonation, RB-5,

1. Introduction

The production processes in textile industry not only utilizes large quantity of water and energy but also generatesa large amount of waste products. Desizing and dyeing processes can generate high organic and dyes pollutants in the textile industry. The effluent generated by the textile industry is an important issue to be treated. Unfortunately, Indonesian government standard for textile wastewater for colour parameter is not regulated so far (The Indonesian Ministry of Environment and Forestry Decree No. 5 Year 2014) [1], [2]. Generally, the synthetic dyes are widely used for dyeing process of textile products in Indonesia. Among of those, azo dyes are the most used ones [1] and estimated that they represent about 60 to 70 % of all dyes commonly used [2]. Azo dyes are characterized by the presence of –N=N- bonds, linked to aromatic groups [3]. About 10 to 15% of the dyes used in dyeing processes is not incorporated into the fabric or product, and therefore, is released in the effluent. Likewise in the desizing process, it can generate high organic pollutants and be a significant environmental pollution source. The concentration of BOD from textile waste is about 80-6,000 mg/L, while the concentration of COD is about 150-12,000 mg/L and the ratio of COD and BOD is about 1.5: 1 to 3:1 [4].

In recent years, studies are focusing on hydraulic systems combining the advantages of suspended growth and the biofilm system. The Moving Bed Biofilm Reactor (MBBR) has been developed as one of the most attractive hydraulic systems [5]. Moving bed biofilm reactor (MBBR) is a type of biological treatment using attached growth processes in media. MBBR process has the advantages of both activated sludge and bio-film systems. MBBR can be operated similar to activated sludge process with the addition of freely moving carrier media. Carrier elements provide a surface for bacterial growth. Carrier elements have lighter density than water so they continuously move along with water stream inside the reactor [6]. This technology has the removal efficiency almost similar to activated sludge system for treating organic matter, but the advantages of this MBBR system is just requires a smaller reactor tank [7]. Other advantages for MBBR like smaller size, no sludge bulking, high concentration of biomass, tolerance to loading impact, high COD loading, no need of periodic backwashing [8]. Despite the widespread use of cost-effective biological processes for the treatment of several types of municipal and industrial wastewaters, studies have shown that aerobic biological processes are not capable of effective decomposition of azo dyes [9]. It is because the textile waste has low BOD/COD ratio (less biodegradable), hence it is necessary to give a pre-treatment before biological process [6]. Studies have shown that the processes of chemical oxidation appeared as suitable alternative for the pretreatment of textile waste containing organic dyes due to their capability of high colour removal. Among the processes of oxidation, the ozonation process is preferably for certain applications, such as dye colour removal. Ozone can rapidly break the azo bond, and remove colour in short time intervals [10]. Meanwhile, ozonation can increase the biodegradability of dyes, enable the use of a subsequent biological process.

In this study, RB-5 dyes will be treated using ozonation as pre-treatment and then using biological treatment by moving bed biofilm reactor (MBBR). The purpose of oxidation using ozone is to increase the biodegradability of RB-5 prior to biological degradation.

2. Research Method

2.1. Materials

Azo Reactive Black-5 (RB-5) dye and starch as carbon were obtained from environmental laboratory, Indonesian center for textile, Bandung. The ozone generator type RSO-25 was used with O_3 output of 400 mg/hour. An artificial wastewater was prepared in the laboratory by mixing ozonated or non-ozonated dye solution with the following compounds: Azo Reactive Black-5 (RB-5) as the main colour source of 100 mg/L and starch as the main carbon source to reach 1,000 mg/L of COD. The concentration of colour is about 100 mg/L and the organic is about 1000 mg/L represented the characteristic of dye and organic in the real condition based on study literature. The media used for MBBR is Kaldness K1 with volume to reactor ratio ($V_{media}/V_{reactor}$) of 20%. The cylindrical-shape MBBR carrier was made of polyethylene with a density of 0.123 g/mL and a diameter of 10 mm with cross inside. The carrier materials are shown in Figure 1.

Table 1. Specification of Kaldnes K1 media.

Specification	Value
Material	Plastic (PE)
Diameter	10 mm
Thickness	0,5-0,8 mm
Shape	cylinder
Density	0.123 gr/mL
Surface area	500 m^2/m^3

IOP Conf. Series: Earth and Environmental Science **106** (2017) 012089 doi:10.1088/1755-1315/106/1/012089

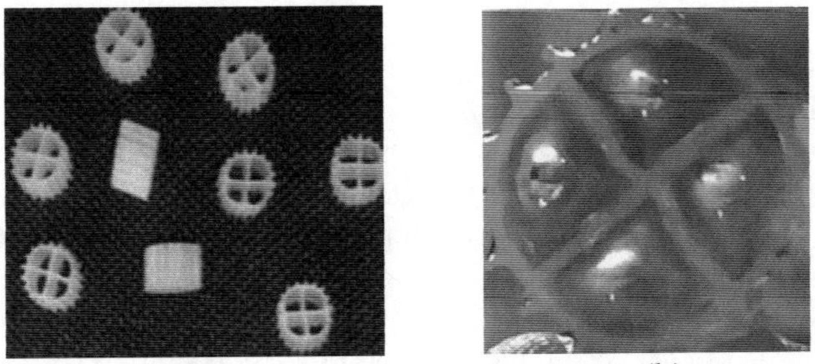

(a) (b)

Figure 1. (a) Kaldness 1 media (K1); (b) Biofilm attached on K1 media.

A laboratory scale acrylic reactor with a total volume of 5 L was used for this study (diameter x height of 14 x 32 cm, respectively). Aeration was provided by a porous air diffuser connected to a compressed air line, located at the bottom of the reactors. The function of aerator is to mobilize the media contained in the reactor and provide dissolved oxygen to keep the system in aerobic conditions (with a minimum dissolved oxygen concentration of 4 mg/L). The microorganism used is a mixed culture obtained from the Textile Wastewater Treatment Installation in Bandung City.

2.2. Experimental procedure and analysis

RB-5 dyes are difficult to treat by biological process only, hence it is necessary to give a pre-treatment before biological process. The purpose of oxidation using ozone is to increase waste biodegradability, make it easier to enter biological process. The ozonation process aims to increase the ratio of BOD/COD to 0.5. While In other set of experiments, the other reactor directly processed using MBBR without ozone oxidation pretreatment.

2.2.1. Ozone pre-treatment. Ozonation was performed in a 2 L glass column. Ozone was generated from a commercial ozone generator (Resun RSO-25) and continuously introduced through a porous diffuser to the system. The applied ozone dose was set at 0.4 g O_3/hour, with airflow of 1.4 L/min. The applied ozone time detention was set for 120 minutes. The ozone concentration and detention time used were obtained based on previous research using the same dye, ozone generator and reactor. No pH adjustment was carried out during the ozonation tests.

2.2.2. Analytical method. The main parameters measured are colour and COD. The COD measurement method is based on 5220D (APHA, 2005) standard method. While the colour absorbance of RB-5 was measured at maximum wavelength (590 nm) using a spectrophotometer JENWAY 6305 type.

3. Results and Discussion

Azo dyes are difficult to treat by biological treatment only. Then, the ozonation process is introduced as pre-treatment before MBBR. The results show that ozonation can remove the organic content from 993 mg/L COD decreased to 397.20 mg/L. Ozone is able to oxidize the organic compounds contained in the carbon source i.e. starch. In addition, it is known that oxidation using ozone can enhance biodegradability of RB5. The process of ozonation can reduce the concentration of COD, so it enhances the ratio of BOD/COD. The obtained ratio of BOD/COD is 0.23 enhanced to 0.42. The ozonation process remove RB-5 dyes with an initial concentration of 121.8 mg/L decreased to 59.1 mg/L. The oxidation using ozone causes the primary degradation process with the changes in the structural of RB-5 compound. Therefore, the biodegradability of RB-5 dye is enhanced [11].

Each MBBR was filled with Kaldness k1 carrier by 20% (v/v). Before receiving dye wastewater, microorganisms were attached to the Kaldness K1 carrier using activated sludge collected from a Textile Wastewater Treatment Installation in Bandung City. COD and colour removal from RB-5 dye were tested on a lab-scale reactor to investigate the decolourisation on both variation used.

3.1. COD removal
In this research, the initial concentration of COD is 928 mg/L. Ozonation process has removed the concentration of COD about 76.90%. It decreases from 928 mg/L to 214.4 mg/L. The COD removal in both reactors is shown in Figure 2. The result shows that MBBR system can decrease COD in the both reactors. Biological treatment is able to treat *organic* compounds sourced from starch (less degradable with the ratio of BOD/COD 0.23). While the other reactor with ozone pre-treatment, the ratio of BOD/COD is increased to 0.42. The result shows that the COD removal efficiency in both reactors is different. R1 is more efficient to remove organic compound than R2. The removal efficiency of R1 is 59.31%, and R2 is 72. 07%. It is because the ozonation process is able to oxidize organic compounds, and increase the COD removal. The result also shows that the efficiency of COD removal and time have a linear relation. The maximum of removal efficiency in both reactors occur at 48 hours. The removal efficiency of COD during contact time for 48 hours on R1 and R2 are 87.59% and 85.86%, respectively.

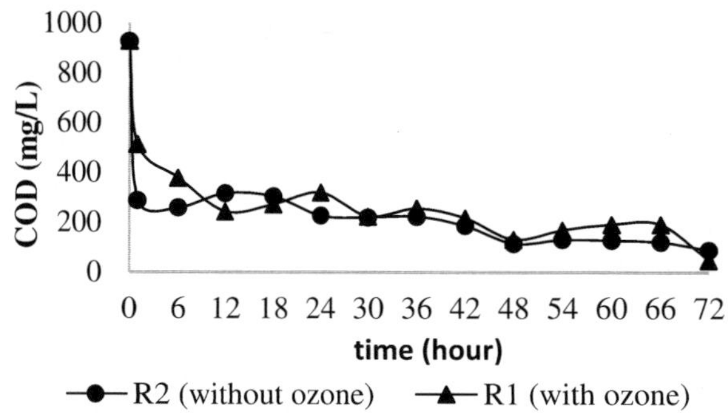

Figure 2. Removal of COD in both MBBR reactors.

3.2. Colour removal
The initial concentration of colour in the batch system is 131.07 mg/L. Ozonation process removes the colour concentration of RB-5 down to 100.59 mg/L. Ozone acts as the powerful oxidising agent because it can easily decompose azo bond on the dyes that causes the colour to disappear. Oxidation in specific compounds is characterized by the occurrence of primary degradation, which there is a structural change in the main compound [11]. The results show that the significant colour removal in both reactors occurs at the first hour. The efficiency of colour removal in R1 and R2 is 86.74% and 68.6%, respectively. However, after 6 hours of incubation time, the concentration of colour was increasing (Figure 3).

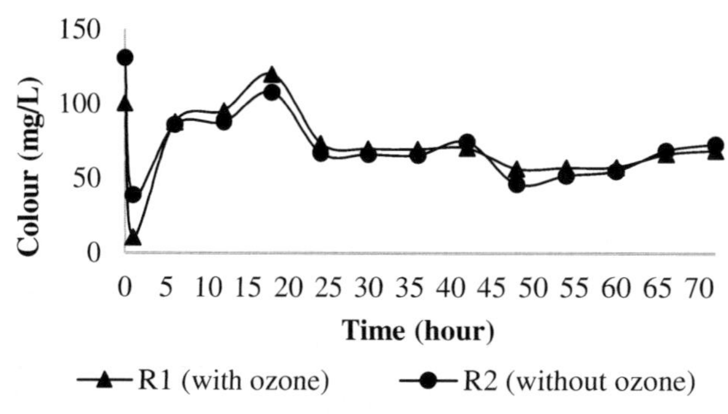

Figure 3. Decrease of colour on RB-5 dyes in both reactors.

This result suggests that biosorption process took places in the first hours of colour removal whereas the RB-5 dye was adsorbed by biomass in the MBBR reactor. The removal of colour is constant after 24 hours indicated by no significant colour change. The visual of colour removal is shown in Figure 4.

(a) with ozonation process (R1) until 72 hours.

(b) without ozonation process (R2) until 72 hours.

Figure 4. Visual of colour gradation in a) MBBR with ozone pre-treatment and b) MBBR without ozone pre-treatment.

3.3. Biological process on continuous system

The application of MBBR on continuous system aims to conduct similar condition with the existing condition. The hydraulic retention time (HRT) was varied to investigate the shock loading resistance of both reactors. The determination of HRT is obtained from the batch system result. Variation of HRTs used is 24 hours, 18 hours, 12 hours and 6 hours.

3.3.1. COD removal. The process of COD removal on the continuous system depends on HRT. The greater the HRT, the smaller the organic loading rate (OLR). The result shows that the COD removal changes significantly in both reactors as the HRT decreases from 24 to 18 h. By further decrease of HRT, the COD removal declined apparently. Reactor with ozone pre-treatment is more efficient compared to reactor without ozone pre-treatment. The average COD removal on both reactors is shown Table 2. The result indicates that the MBBR had a strong resistance to shock loading. The longer duration was beneficial for microorganism to degrade pollutants.

Tabel 2. Performance of the MBBR reactors in COD and colour removal.

HRT (hour)	R1 (+ Ozone + MBBR)		R2 (– Ozone + MBBR)	
	COD	Colour	COD	Colour
24	89.13%	96.95%	83.63%	81.21%
18	67.98%	85.68%	61.06%	76.89%
12	62.56%	82.89%	49.97%	76.37%
6	51.08%	78.83%	37.55%	75.74%

3.3.2. Colour removal. Decolourization of colour in the continuous system utilizes the activity of microorganisms. The result shows that R1 (reactor with ozone pre-treatment) has higher efficiency than R2 (reactor without ozone pre-treatment). The colour removal changes significantly in R1 reactor when HRT was also reduced. The average of colour removal in R1 reactor was 96.95%, 85.68%, 82.89% and 78.83% at 24, 18, 12, and 6 h HRT, respectively. While in the R2, the colour removal is still constant and not affected by HRT. The average of colour removal on the both reactor is shown in Table 2. In the continuous systems, the colour removal process is caused by aerobic bacteria. The microorganism uses carbon from dyes solution as their carbon source to support their growth.

4. Conclusion

A lab scale Moving Bed Biofilm Reactor (MBBR) system was operated to treat RB-5 dye. Biological treatment process was conducted in batch system to treat RB-5 dyes by ozone pre-treatment and without ozone pre-treatment. The pre-treatment process aims to examine the effectiveness of ozone in the removal of COD and colour in MBBR system. The results show that in the batch system, MBBR can decrease COD and colour in both reactors. Significant removal of colour in both reactors occurs at the first hour, but R1 (with ozone) has higher removal efficiency of colour than R2 (without ozone). The removal efficiency of colour in the both reactor was 86.74% and 68.6% respectively. While the result of COD removal shows that R1 is more efficient to remove organic compound than R2. The removal efficiency of R1 and R2 is 59.31%, and 72.07%, respectively. While in the continuous system, pretreated reactor with ozone pre-treatment is more efficient than without ozone pre-treatment. The average removal of colour and COD at 24 h in R1 reactor was 96.95% and 89.13%, while in R2 reactor was 81.21% and 83.63%, respectively. It is known from continuous system that the removal of COD and colour are affected by HRT.

Acknowledgment

The study was performed within the ITB-P3MI research scheme; a research programs on community service, and innovation of ITB research group and co-funded by the Indonesian Ministry of Research, Technology and Higher Education (Ristek-Dikti) Research Grant Fiscal Year 2017.

References

[1] The Indonesian Ministry of Environment and Forestry, Wastewater effluent standard for industries, Decree No. 5, 2014.
[2] Ghaly A E, Ananthashankar R, Alhattab M, and Ramakrishnan V V. 2014. Production, characterization and treatment of textile effluents: a critical review. *J. Chem. Eng. Process Technol.* 51:18.
[3] Wang C, Yediler A, Lienert D, Wang Z, and Kettrup A. 2003. Ozonation of an azo dye C.I. Remazol black 5 and toxicological assessment of its oxidation products. *Chemosphere.* 52:1225–32
[4] Azbar N, Yonar T, and Kestioglu K. 2004. Comparison of various advanced oxidation processes and chemical treatment methods for COD and colour removal from polyester and acetate fiber dying effluent. *Chemosphere*. 55:81-6.
[5] Shin D H, Shin W S, Kim Y H, Han M H, and Choi S J. 2006. Application of a combined process of moving-bed biofilm reactor (MBBR) and chemical coagulation for dyeing wastewater treatment. *Water Sci. Technol.* 54:181-9.
[6] Francis A, and Sasomony K J. 2015. Treatment of pre-treated textile wastewater using Moving bed bio-film reactor. *Procedia Technology*. 24:248 - 55.
[7] Esmaeilirad N, Borghei S M, and Vosoughi M. 2015. Kinetic of ethylene glycol biodegradation in a sequencing moving bed biofilm reactor. *Journal of Civil Engineering and Environment Science*. 1:2-7.
[8] Borkar R, Gulhane M, and Kotangale A. 2013. Moving Bed Biofilm Reactor - A New perspective in Wastewater Treatment. *IOSR Journal of Environmental Science, Toxicology and Food Technology*. 6:15-21.
[9] Forgacs E, Cserháti T, and Oros G. 2004. Removal of synthetic dyes from wastewaters: a review. *Environ Int*. 30:953–71.
[10] Castro F D, Bassin J P, and dezotti M. 2011. Treatment of a simulated textile wastewater the Reactive Orange 16 azo dye by a combination of ozonitation and moving-bed biofilm reactor:evaluating the performance, toxicity, and oxidation by-product. *Environ Sci Pollut Res Int*. 24:6307-16.
[11] Tchobanouglous G, Burton F L, and Stensel H D. 2003. Wastewaster engineering: treatment and reuse, New York: McGraw-Hill.

IOP Conf. Series: Earth and Environmental Science **106** (2017) 012090 doi:10.1088/1755-1315/106/1/012090

Biosorption of heavy metal copper (Cu^{2+}) by *Saccharomyces cerevisiae*

S A J Ririhena, A D Astuti, M F Fachrul, M D S Silalahi, R Hadisoebroto, A Rinanti*

Environmental Engineering Department, Faculty of Landscape Architecture and Environmental Technology, Universitas Trisakti, Jakarta

*Corresponding Author : astririnanti@trisakti.ac.id

Abstract. This research aims to study the optimum effect of contact time and pH adsorption of copper (Cu^{2+}) from electroplating industry waste by dried beer waste *S.cerevisiae*. This research conducted using batch culture with pH variation 2,3,4,5, and 6, contact time variation 60, 90, 120, 150, 180 minutes, 150 rpm at room temperature (± 28°C), initial Cu^{2+} concentration 33,746 mg/l, and biosorbent mass 200 mg & 500 mg. The adsorption of heavy metal ions Cu^{2+} occurs in all variations of pH and contact time at optimum pH. The optimum adsorption occurs at pH 4 with contact time 120 minutes for both 200 mg (41.60%) and 500 mg (61.04%) beer waste biosorbent. Cell morphology seen with Scanning Electron Microscope (SEM) analysis shows the change of cell wall that gets damaged from Cu^{2+} adsorption. It also proved by the decreased concentration of initial high concentration carboxyl groups. The adsorption process of this research complies to Freundlich Isotherm with R^2 value closest to 1 and followed first order kinetic.
Keywords: contact time, copper, pH adsorption, S.cerevisiae

1. Introduction

Various research regarding metal ion reduction treatment has been done, such as oxidation-reduction, precipitation, adsorption, compaction, electrolytic recovery, and ion exchange. These processes are expensive in which some technological problems exist especially when applied to diluted metal solutions. Therefore, the search for clean and competitive technologies is strongly recommended. Biological treatment is usually considered as an effective method and can significantly reduce the quantity of heavy metals in aqueous solutions [1]. Study of biosorption as an alternative treatment considered effective in cost and easy to apply. Biosorption can adsorb more heavy metals than chemical precipitation on bigger pH range [2]. Biosorption capacity is mainly influenced by three kinds of influential factors: metal ionic characteristics (e.g., atomic weight, ion radius, valence) the nature of the biosorbents (e.g., cell age), and biosorption conditions (e.g., pH, temperature, contact time) [3].

This research aims to study the effect of contact time and pH to adsorption of Cu^{2+} ion by biosorbent *S. cerevisiae,* determine the adsorption efficiency of Cu^{2+} ion contain in electroplating wastewater by biosorbent *S. cerevisiae,* and determine functional groups involved in Cu^{2+} adsorption process.

IOP Conf. Series: Earth and Environmental Science **106** (2017) 012090 doi:10.1088/1755-1315/106/1/012090

2. Research Method

2.1. Biosorbent Preparation
S. cerevisiae that found in beer waste can be used as biosorbent. Beer waste centrifuged at 3600 rpm for 60 minutes. Biomass that formed in the bottom of centrifuged tube collected into a petri dish and then put it in the oven with temperature 80°C for 28 hours to let it dry. Dried biomass then grind with mortar and passed through screener (30 mesh) until it becomes powder.

2.2. Industrial Electroplating Wastewater Characterization.
Characterization utilized with *Atomic Absorption Spectrophotometry* (AAS).

2.3. Cu^{2+} adsorption on pH variation
Dried biosorbent of 200 mg and 500 mg were put into 70 mL electroplating wastewater contained Cu^{2+} then solution pH was adjusted to variation 2, 3, 4, 5, and 6. After that the solution was put on shaker 150 rpm for 90 minutes. Then the solution was centrifuged with 4000 rpm for 30 minutes to separate biomass with the solution. The concentration of Cu^{2+} remainder in solution determined by *Atomic Absorption Spectroscopy* (AAS) analysis.

2.4. Cu^{2+} adsorption on contact time variation
Dried biosorbent of 200 mg and 500 mg were put into 70 mL electroplating wastewater contained Cu^{2+} in optimum pH. After that, the solution was put on a shaker with 150 rpm for a specific variation of contact time 60, 90, 120, 150, and 180 minutes. The solution then was entrifuged with 4000 rpm for 30 minutes to separate biomass with the solution. The concentration of Cu^{2+} remainder in solution determined by *Atomic Absorption Spectroscopy* (AAS) analysis.

3. Results and Discussion
The result of electroplating wastewater characterization indicates that it contained 33,764 mg/L Cu^{2+} without any other metals detected.

3.1. Determining Optimum pH
The potential of hydrogen (pH) is one of the main factors that affects biosorption process. Metal ions adsorptions depends on solution pH, which influences electrostatic binding of ions to corresponding functional groups [4].

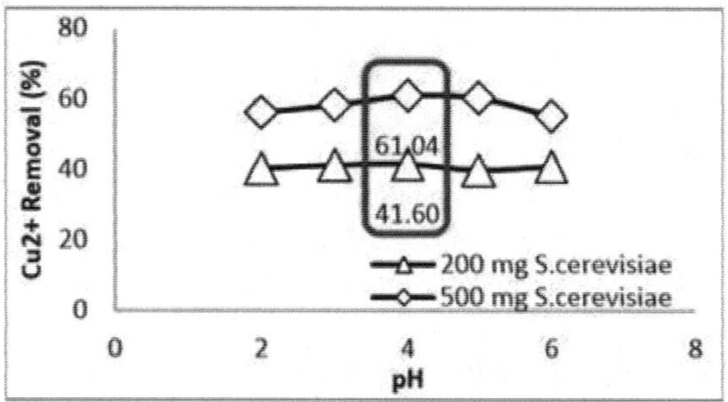

Figure 1. Effects of pH variation towards Cu^{2+} removal (%) at contact time 90 minutes, room temperature (± 28°C), and Cu^2 initial$^+$ concentration 33,746 mg/L.

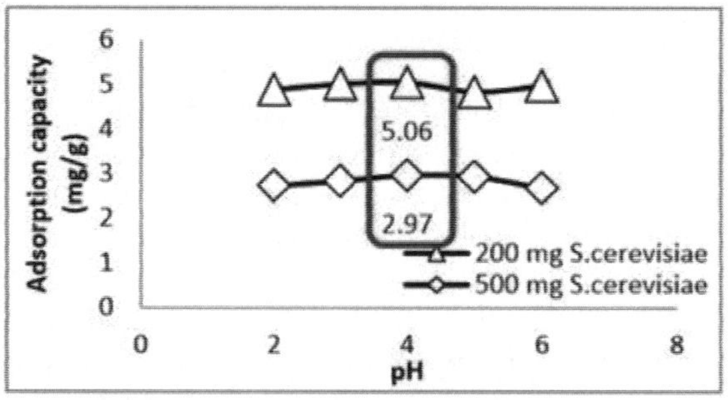

Figure 2. Effects of pH variation towards adsorption capacity (mg/g) at contact time 90 minutes, room temperature (± 28°C), and Cu^{2+}, initial concentration 33,746 mg/L.

Based on Figure 1 Cu^{2+} adsorption efficiency increased from pH 2 until pH 4 which has the highest adsorption efficiency for all biosorbent samples whereas at pH 5 until pH 6 Cu^{2+} removal efficiency decreased and has the lowest removal efficiency. It correspond with previous study that shows the optimum adsorption for Cu^{2+} ion happens at pH 4 until 5 [5]. The adsorption of heavy metal Cu^{2+} occurs because of ions exchange mechanisms between heavy metal ions and carboxyl groups, phosphate, and amino which are the main components of *S.cerevisiae* cell wall, capable of acting as ion exchangers and complex formers with metal ions. At the low pH, concentration of protons is high and the ion exchange sites become solidly protonated, therefore metal adsorption will occur better at low pH [6].

As seen in Figure 2, the highest adsorption capacity also happen at pH 4, which compatible with previous study said the removal efficiency of Cu^{2+} occurs at a maximum of 3 g/L or 3 mg/mL biosorbent concentration but the removal efficiency will decrease if there is additional biosorbent concentration used [7]. This may occur due to the presence of biosorbent bundles that cause a decrease in surface area for adsorption, resulting in decreasing adsorption capacity. Previous study report at pH 4.5 200 mg *S. cerevisiae* adsorb 2,042 mg/g Cu^2 [8]. Based on these things it can be concluded that pH 4 is the optimum pH for biosorbent *S. cerevisiae* to adsorb Cu^{2+}.

3.2. Determining Optimum Contact Time

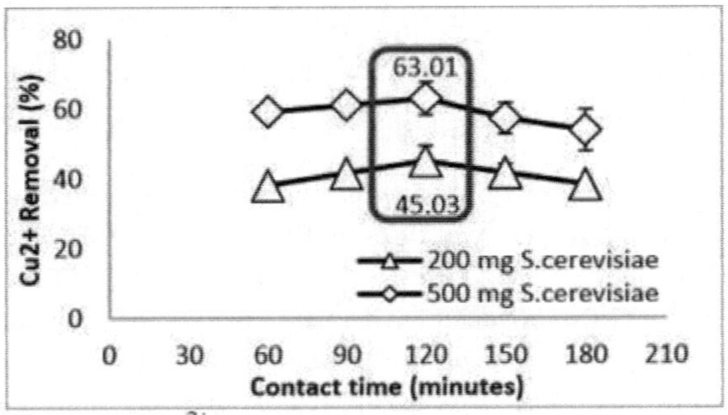

Figure 3. Cu^{2+} removal (%) towards contact time variation (minutes) at pH 4 (optimum), room temperature (± 28°C), and initial Cu^{2+} concentration 33,746 mg/L.

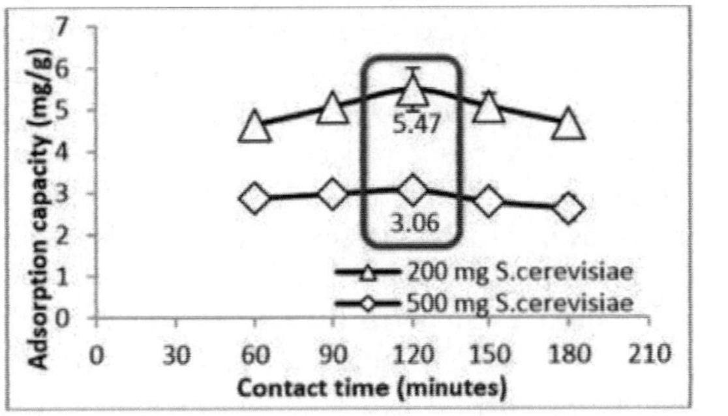

Figure 4. Adsorption capacity (mg/g) towards contact time variation (minutes) at pH 4 (optimum), room temperature (± 28°C), and initial Cu^{2+} concentration 33,746 mg/L.

Based on Figure 3 Cu^{2+} removal efficiency keeps increasing until optimum at 120 minutes and decreasing at contact time 150 minutes. Previous study shows that optimum contact time for Cu^{2+} removal happened at 90 minutes and stay stable until 120 minutes [8]. 500 mg biosorbent *S. cerevisiae* has larger surface area than 200 mg *S. cerevisiae*, therefore 500 mg *S. cerevisiae* has higher Cu^{2+} removal efficiency than 200 mg *S. cerevisiae*.

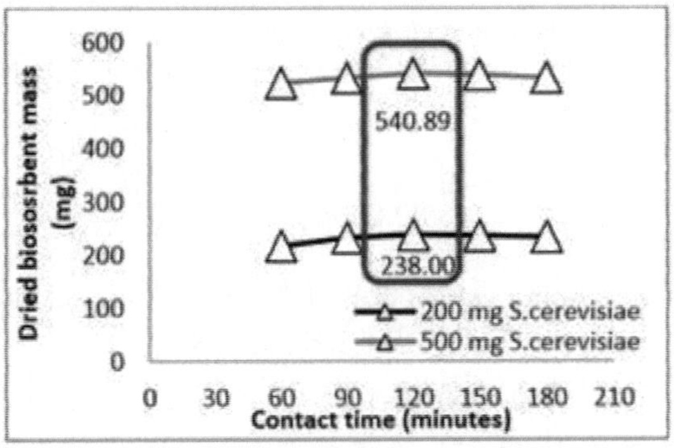

Figure 5. Dried biosorbent mass (mg) towards contact time at pH 4 (optimum), room temperature (± 28°C), and initial Cu^{2+} concentration 33,746 mg/L.

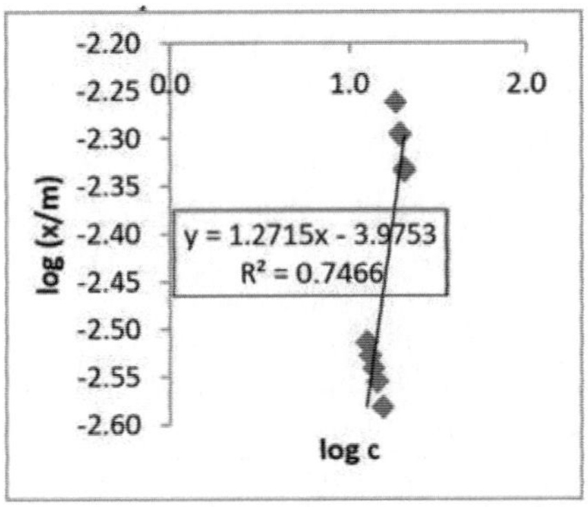

Figure 6. Isotherm Freundlich.

The same tendency can be seen in Figure 4, adsorption capacity for 200 mg biosorbent *S. cerevisiae* is in the range 5-6 mg/g with highest capacity at 5.47 mg/g at and for 500 mg biosorben *S. cerevisiae* is in the range 3 mg/g with the highest capacity at 3.06 mg/g at optimum pH and contact time.

Figure 5 shows a correlation between dried bisorbent mass with contact time. Initial 200 mg dried biosorbent mass increased up until 238.00 mg while initial 500 mg increased up to 540.89 mg. The increasing dried biosorbent weight occurs due to heavy metal ion Cu^{2+} presence from solution that adsorb into biosorbent *S. cerevisiae* in accordance with the quantity Cu^{2+} adsorb from solution into biosorbent *S. cerevisiae* cell wall, other than that there's a chance of impurities in solution that goes through the filter and weighted together with dried *S. cerevisiae* biosorbent. As contact time increase, the weight of dried biosorbent will increase until the saturation point which is 120 minutes and begin to decrease at 150 minutes and 180 minutes.

3.3. Determining Optimum Contact Time

Adsorption isotherm used to determine the occurrence of Cu^{2+} adsorption characteristic, and a curve will be generated to shows the maximum adsorption capacity of *S.cerevisiae* biosorbent based on the characteristic. Calculation of adsorption isotherm value of this study using Langmuir, Freundlich, and BET isotherm. The use of adsorption isotherms is to determine the most representative description of this study biosorption process.

Langmuir isotherm assumes that adsorbent surface has a certain number of adsorption points which is proportional to the surface area of the adsorbent and adsorption only occurs on the monolayer thus the adsorption process can only occur in the adsorbent that has not been filled by any adsorbate. Whereas for Freundlich Isotherm used to explain non-ideal adsorption process on the heterogeneous surface because of functional groups difference in adsorbent (Figure 6). BET isotherm assumed that adsorbent has homogenous surface and adsorbat molecule can form more than one layer on the surface [9].

From calculation results shown in Table 1, adsorption process of this study follows Freundlich Isotherm with the highest R^2 value between Langmuir and BET Isotherm. R^2 value of Langmuir Isotherm is not so different than Freundlich value, according to Freundlich the adsorbent surface is heterogeneous, composed of different classes of the adsorption region, but each class adsorption follows Langmuir isotherm, therefore the adsorption tends to be similar.

3.4. Adsorption Kinetics

Adsorption kinetics is a study to determined reaction order and speed rate of adsorption process that happened. The reaction of the biosorption rate provides important information to create a batch scale biosorption system. The approach used in calculating the adsorption kinetics in this study is first order and second order.

Table 1. R^2 Value calculation of adsorption isotherm and kinetics.

R^2			R^2	
Isotherm Langmuir	Isotherm Freundlich	Isotherm BET	First Order	Second Order
0.71	**0.7466**	0.4731	**0.0448**	0.0351

Based on R^2 value seen in Table 1, the value of first order kinetic adsorption is closer to 1 than second order thus it can be said that the adsorption kinetics that occurs tend to follow the first order. The first-order reaction is a reaction whose velocity depends only on one of the reactive substances which or is proportional to one of the reactant forces [10].

3.5. Analisis Fourier Transfer Infra-Red Spectrophotometer (FTIR)

Biosorbent *S.cerevisiae* has three main functional groups which are carboxyl groups, amino groups, and hydroxyl groups. *S. cerevisiae* cell wall composed of a-mannan and b-glucan, protein, lipid, and inorganic substances particularly phosphate, chitin, and chitosan. Proteins that construct *S.cerevisiae* cell wall are rich in glutamic acid residues, aspartate, serine, threonine, and asparagine. Based on FTIR analysis it is known that biosorbent *S. cerevisiae* has high carboxylic acid, carboxyl group, and amine. In this FTIR results chitin included in acid carboxyl functional groups which in the 3277/cm area, which indicates the peak of O-H.

The carboxyl group in the 1625/cm region indicates the peak presence of C = O, and amine at 1234/cm indicates the peak presence of N-H. The content of carboxylic acid decreased from 92.27% to 92.04%, carboxyl group decreased from 98.04% to 89.77%, Amine from 94.17% to 92.33%, and alkane from 95.93% to 94.37%. This indicates the presence of functional groups involved in the adsorption of Cu^{2+} ions which cause changes in functional groups in the biosorbent of *S. cerevisiae* either in the presence or the content of functional group present in the biosorbent.

3.6. Scanning Electron Microscope (SEM) Analysis

Scanning Electron Microscope (SEM) is a type of electron microscope that uses a focused beam of high-energy electrons to produce images of a sample by creating a variety of signals at the surface of the sample [11]. SEM produces high-resolution images of a sample surface, capturing a complete size of approximately 1 - 5 nm and can produce 3-dimensional shape characteristics so as to provide easier data about the analysed samples [12].

Figure 7. Control *S. cerevisiae* seen with SEM (500x)

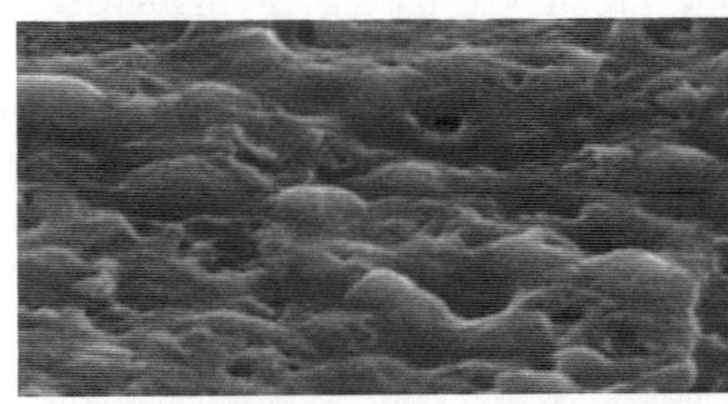

Figure 8. *S. cerevisiae* after contact with Cu^{2+} in optimum condition seen with SEM (500x)

Based on Figure 7 and Figure 8 we can see the difference in cell morphology before and after contact with Cu^{2+}. As seen on Figure 8 *S.cerevisiae* cell wall looks shrivelled and have more holes than before contact (Figure 7) which look more hydrated and plump. This morphological change occurs due to Cu^{2+} ion adsorption process that happened in the cell wall during contact with electroplating wastewater. The change that happened is not too significant because the biosorbent is used in a state of death & dried so no metabolism progress occurs and disturbed by the absorption of Cu^{2+} ions.

4. Conclusion

From this research, we can conclude that contact time and pH affect the adsorption of Cu^{2+} by biosorbent *S.cerevisiae,* with optimum contact time 120 minutes and optimum pH 4. Optimum Cu^{2+} adsorption efficiency happens at pH 4, contact time 120 minutes, 500 mg *S.cerevisiae*, initial Cu^{2+} concentration 33.764 mg/l, at room temperature (± 28°C), shaker speed 150 rpm is 63,01%. This research adsorption process follows Freundlich Isotherm with the highest R^2 value between Langmuir and BET isotherm which is 0.7466. Reaction kinetic of this research follows first order kinetics with R^2 value higher than second order value. Several functional groups involved in adsorption of heavy metal Cu^{2+} are carboxyl acid, carboxyl groups, and amine which proved by the decreased concentration of these functional groups.

Acknowledgment
The authors would like to thank Directorate for Research and Community Services as well as Directorate General for Strengthening Research and Development at the Ministry of Research, Technology, and Higher Education Indonesia for funding this study through Universities Prime Research Grant Program (Penelitian Unggulan Perguruan Tinggi-PUPT) 2016.

References

[1] Peng Qingqing, Yunguo Liua, Guangming Zenga, Weihua Xua, Chunping Yanga, Jingjin Zhanga. 2010. Biosorption of copper (II) by immobilizing Saccharomyces cerevisiae on the surface of chitosan-coated magnetic nanoparticles from aqueous solution. *Journal of Hazardous Materials.* 177:676–682.

[2] Elangovan R, Philip, Ligy, and Chandaraj K. 2007. Biosorption of heksavalent and trivalent chromium by palm flower (Borassus aethiopum). *Chemical Engineering Journal*, 141:99-111.

[3] Cheng Can, and Jianlong Wang. 2007. Influence of metal ionic characteristics on their biosorption capacity by Saccharomyces cerevisiae. *Applied Microbiol Biotechnol*. 74:911–917

[4] Mukhopadhyay M, S B Noronha, and G K Suraishkumar. 2008. Kinetic modelling for the biosorption of copper by pre-treated Aspergillus niger biomass. *Bio resource Technology.* 98:1781 - 1787

[5] Thipeswamy B, Shivakumar C K, Khrisnappa M. 2014. Study on heavy metals biosorption ability of Saccharomyces cerevisiae. *International Journal of Biological Research.* 2(2):106-115.

[6] Volesky, Bohumil. 2007. Biosorption and me. *Water Research.* 41:4017-4029

[7] Akcelik, Oznur, Ayse Tosun, Mubeccel Ergun. 2012. Biosorption of Copper (II) from Aqueous Solution by Saccharomyces cerevisiae. *Gazi University Journal of Science*. 25(3):783.

[8] Wang Yong. 2012. Optimization of Cadmium, Zinc, and Copper biosorption in an aqueous solution by Saccharomyces cerevisiae. *International journal of chemistry*. China.

[9] Anita, Shera Dwi and Edwan Kardena. 2013. Biosorpsi kromium heksavalen oleh mikroalga amobil pada limbah industri pelapisan logam. ITB. Bandung.

[10] Bulut Emrah, Ozacar Mahmut, Sengil Ayhan. 2008. Adsorption of Malachite Green onto Bentonite: Equilibrium and Kinetics Studies and Process Design, *Microporous and Mesoporous Materials, Elsevier Journal*. 115:234-256

[11] Nada Majid Hameed. 2015. Scanning Electron Microscopy. BAO *J Microbio*. 1:1. USA.

[12] Sun, Xiang-Yu Yu Zhao, Ling-ling Liu, Bo Jia, Fang Zhao, Wei-dong Huang, and Ji-cheng Zhan. 2015. Copper tolerance and biosorption of Saccharomyces cerevisiae during alcoholic fermentation. PloS ONE 10(6):e0128611.doi:10.1371/ journal.pone.0128611.Leicester.

Waste utilization of red snapper (*Lutjanus* sp.) fish bone to improve phosphorus contents in compost

S Ramadhani, B Iswanto, P Purwaningrum*

Environmental Engineering Department, Faculty of Landscape Architecture and Environmental Technology, Universitas Trisakti, Jakarta, Indonesia

*Corresponding Author : pramiati@trisakti.ac.id

Abstract. The purpose of this research is to get the idea that bone waste will be the P content enhancer in compost so that the compost produced meets the standard P levels specified in SNI 19-7030-2004 which regulating compost quality standard. Nutrient levels were obtained in fish bone meal (FBM) are C (3.35%), N (0.48%), P (30.90%) and K (0.02%). Effects of fish bone meal to the rising levels of P in the compost has been known. P levels of compost B, C, D, and E increased at 428.57; 542.85; 657.14 and 914.28% against the compost A (blank). FBM ideal addition indicated in compost B, as much as 15 gr, with a P content of 0.37% and has been passed according standards (0.10% for P). C/N ratio decreased over the 21 days period of composting, with the greatest decline was compost E with a ratio of 16:1. Highest nitrogen (N) levels recorded respectively in compost B and C with value of 1.09% and the lowest of recorded N content was compost A, D and E (1.08%). N content in all samples of compost were eligible minimum N of 0.40%. Carbon (C) is the highest recorded in compost B; 20.20% and the lowest in the compost E; 17.34%. Highest and lowest C levels on the compost has met the minimum C of 9.80%. Composting is done in a bucket as an aerobic composter (with air holes), compost pile turnover for each sample is controlled as much as once/2 days. Mesophilic period (23-45^{0}C) occurs during the 21-day period of composting. Compost B has P content of 0.37%, so it has fulfilled the provisions of SNI 19-7030-2004 about the recommended compost standard.
Keywords: compost, fish bone meal, phosphorus

1. Introduction

In the processing of red snapper fillet (*Lutjanus sp.*) the fish bone with the proportion of 13.70% of the total body of fish has not been widely used [1].The company's red snapper fish fillet produces a lot of fish bone waste, which until now has not been well utilized [2]. Marine fisheries production for the snapper species increased in volume from 91339 tons in 2004 to 130301 tons in 2014. In 2014 demersal fish captured reached 1557261 tons/year and red snapper volume contributed 130301 tons/year, the first place after jack trevallies fish (109441 tons/year) and giant catfish (Ariidae family)(102111 tons/year), [3] then it is estimated that the potential of red snapper fish bone waste generated can reach 17851.23 tons/year. Fish bone waste includes solid waste types generated from small-scale production to large fish farming as well as from marine catches. Bone is rich in phosphorus, while phosphorus is commonly known as one of the components of NPK (Nitrogen-Phosphorus-Potassium) fertilizer. As a fertilizer, the ratio of NPK in bone meal is generally 4-12-0 and in steamed bone meal of 1-13-0 [4]. The content of calcium (Ca) in fish bone is 20-35 mg, phosphorus (P) 190 mg, iron (Fe) 1.0 mg and potassium (K) 250-350 mg [5]. An increase in phosphorus content reached about 278.4; 251.3; 278.4 and 819% in tempe factory wastewater which added 10% of phosphorus isolate from red snapper fish bone waste [6].

Biological treatment can be described as biological decomposition of organic waste under controlled conditions, for composting the condition to be in aerobic state and rise of temperature achieved by an exothermic process catalyzed by microbial enzymes. The C/N ratio is an important factor affecting the performance of microorganisms. The carbon element (C) is used as an energy source in the metabolic process of cell multiplication by bacteria. While the element of nitrogen (N) is used for protein synthesis or the formation of protoplasm, optimum pH range for bacterial growth is 6.0-7.5 while fungi prefer an environment in the range of pH 5.5-8.0 [7]. The purpose of this research is to get the idea that bone waste will be the P content enhancer in compost so that the compost produced meets the standard P levels specified in SNI 19-7030-2004 which regulating compost quality standard.

2. Research Method

2.1. Mass Balance

The mass balance is intended to find out how much incoming, recyclable, compostable and residual waste [8]. To know the percentage of garbage composition can be calculated by using formula :

$$\% \text{ Component} = (\text{weight of component/total weight of waste}) \text{ kg x } 100\% \quad (1)$$

The data obtained from this calculation is used to compile a mass balance equilibrium and analyze the potential utilization [9].

2.2. Decomposition Rate

The rate of substrate decomposition can be facilitated by simplifying it using the rate of first order reactions as is often the case in nature [10]. First order reaction rate is widely used in describing biological oxidation processes. The equation of the first order reaction rate can be written as follows :

$$\frac{d(C)}{dt} = -k_d(C)$$

$$\int_{BVS_0}^{BVS} \frac{d[C]}{[C]} = -k \int_0^t dt \quad (2)$$

$$\ln \frac{[C]}{[C_0]} = -kt$$

$$\ln[C] - \ln[C_0] = -kt$$

$$\ln[C] = -kt + \ln[C_0]$$

C = amount of carbon (unit weight), T = time (day), Kd = decomposition rate (per day)

If the equation is represented by a graph, then ln (C) will be on the y-axis, while time is on the x-axis, with the slope being the rate of decomposition. The value of negative decomposition rate indicates that the substrate will decrease over time [10].

2.3. Research Design

Prepare 10 pieces of fermentation equipment, each filled with mixture of 200 ml groundwater, 2 ml EM4 (bioactivator), 2 g sugar and fish bone meal (Table 1). Preparing 10 composter, each filled with leaf waste, sawdust and mixed samples of fermented A to E' and analyzed levels of C, N, P, K, C/N ratio and water content on day 10, 15, and 21. Here are the composting preparations shown in the experimental design form in Table 2.

Table 1. Research design 1.

Treatment	Groundwater (mL)	Sugar (gr)	EM4 (mL)	Fish bone meal (gr)
A & A' (duplo)	200	2	2	0 (blank)
B & B' (duplo)	200	2	2	15
C & C' (duplo)	200	2	2	20
D & D' (duplo)	200	2	2	25
E & E' (duplo)	200	2	2	30

Table 2. Water content addition.

Treatment	Groundwater (mL)	Sugar (gr)	EM4 (mL)
A & A' (duplo)	400	4	4
B & B' (duplo)	400	4	4
C & C' (duplo)	400	4	4
D & D' (duplo)	400	4	4
E & E' (duplo)	400	4	4

EM4 bioactivators and sugar were used to accelerate the maturation of compost with a concentration of (1:1:100) EM4 : sugar : water, while the water content used was the optimum range of 56-57% for the starter of composting process. Effective moisture content for composting is 55-65% [11]. After the mixture in Table 1 and Table 2 is stirred 15 minutes and left for 24 hours, this mixture is fed into the waste of dried leaves and sawdust in Table 3.

Table 3. Research design 2.

Treatment	Leafwaste (gr)	Sawdust (gr)	Mixtures	C/N
A & A' (duplo)	900	200	Table 2.1 & 2.2	30:1
B & B' (duplo)	880	200	Ditto	30:1
C & C' (duplo)	870	200	Ditto	30:1
D & D' (duplo)	860	200	Ditto	30:1
E & E' (duplo)	850	200	Ditto	30:1

The mixture in Table 3 was composted and measured for daily parameters such as temperature, pH, humidity and weight reduction, and measurements of C, N, P, K, C/N and water content were performed on days 10, 15 and 21. The dried leaves used are trembesi tree leaves from Univ. Trisakti garden waste, while sawdust is a mixture of sengon and coconut trees obtained from the janitor of Univ. Trisakti. The use of one type of leaf litter aims to make the compost mix more homogeneous. Leaf waste is first chopped up with a blender and compost chopper machine because the smaller the particles, the larger the amount of surfaces that the microorganisms digest [12], then the sawdust is used for compost bulking. Waste weight adjustment is performed to adjust the optimum C/N ratio in composting, ie 30:1. This composition is consistent so that composting can be more effective at a C/N ratio of 30% to 40% [11].

2.4. Measurement of Water content, N-Total Content, Organic Carbon Content, Phosphor and Kalium

Measurement of water content was done by Gravimetry method, ie the water in the compost sample was evaporated by drying at 105^0C for 16 hours.Measurement of total N-content was done by using Kjeldahl method. The N-organic and N-NH_4 contained in the examples are decomposed with sulfuric acid and selenium mixture to form ammonium sulphate, distilled by addition of the excess base and finally the distillate is titrated, nitrogen in the form of nitrate is extracted with water, reduced by alloy devers, distilled and ultimately titrated.Measurements of organic carbon content were carried outusing Walkley & Black method. The organic carbon in the sample is oxidized by dichromate in an acidic atmosphere. The chromium III formed is equivalent to the oxidized C-organic and measured by spectrophotometry. The measurement of P and K content was done by using Spectrophotometry and Flamephotometry

(AAS) method. Samples were oxidized wet with HNO_3 and $HClO_4$. The extracts obtained were used to measure P elements Spectrophotometrically and K with AAS [13][14].

3. Results and Discussion

3.1. Results of Processing Snapper Fish Bone into Bone Meal
60 kg of processed wastes have been processed to produce three types of derivative waste, namely fish bone waste of ± 5 kg (8.33% of total waste weight), meat and fin waste of 45-46 kg (75-76.67% of Total weight of the waste) and the residue about 9-10 kg (15-16.67% of total waste weight) in the form of water content in the form of fish blood, ice and bone remnants and fish fins removed during the process, as shown in Figure 1.

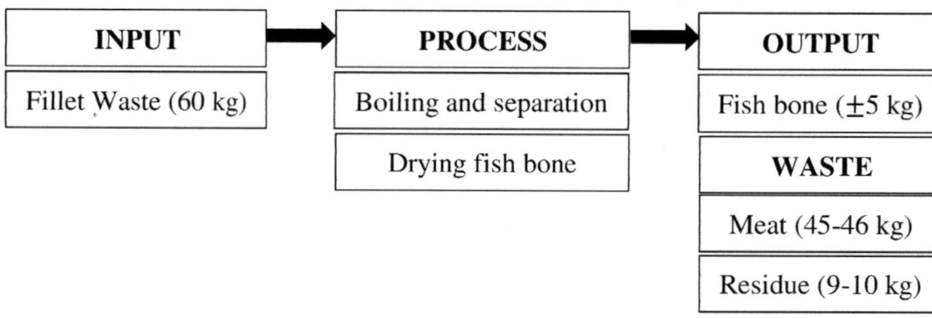

Figure 1. Mass balance of snapper fish fillet.

The fish bone in Figure 1 which has been separated from the meat and fins is taken as many as 10 samples for flouring process. The weight of fish bones in the flouring process decreases due to residue filtering, protein and fat dissolution, and decreased moisture due to the drying process in the oven. Depreciation of snapper fish bone weight can be seen in Table 4.

Table 4. Depreciation of snapper fish bone weight

Sample	Initial weight of bone (gr)	Bone meal (gr)	Weight reduction (%)
1	110*1	75,51	31,35
2	137*1	90,80	33,72
3	132*1	89,54	32,16
4	145*1	104,00	28,27
5	143*1	91,54	35,98
6	127*1	81,68	35,68
7	142*1	95,38	32,83
8	152*1	108,42	28,67
9	122,60	96,00	21,69
10	130,03	99,48	23,49

*1Bone weight weighed with Ohauss manual mechanical balance

In Table 4 the mean values were taken from the samples, so that the average value of fish bone weight reduction was 40.82 gr and for the mean weight reduction percentage was 30.38%. From fish bone meal (FBM) produced is taken as many as 391.73 gr FBM for sieves with sieve shaker. The weight of the largest sieve resulted was a sieve of 170 mesh, weighing 85.26 grams or 21.76% of the total FBM sieved for 60 minutes with 90 amps of vibration. The temperature reached at the end of composting (day 21) is; compost A (32.75^{0}C), compost B (33.25^{0}C), compost C (33^{0}C), compost D (32.75^{0}C) and compost E (32.75^{0}C). This shows that the addition of FBM in the composting process does not affect the composting temperature. The humidity of the 21st day compost sample on a consecutive basis is as follows; Compost A (51.50%), compost B (52.75%) compost C (54.50%), compost D (49.25%) and compost E (51.50%).

In the results obtained, the addition of FBM did not affect the water content of compost.The weight reduction of the compost sample is as follows; compost A (17 g/day), compost B (16 g/day), compost C (14 g/day), compost D (14 gr/day) and compost E (12 gr/day). So it can be concluded that the addition of FBM has no effect in compost weight reduction. The value of compost pH on the 21st day is as follows; Compost A (6.81), compost B (6.66), compost C (6.60), compost D (6.73) and compost E (6.64). In SNI standard 19-7030-2004, minimum pH value is set at 6.80 and a maximum of 7.49. This shows that only compost A samples have met the standard compost pH, whereas compost B, C, D, and E are still slightly below the recommended pH value because of the remaining formation of organic acids by microbes. In composting conducted for 21 days calculated from the 10th day to 21st day there is a decrease in the level of C in compost with the following values; Compost A (16.98%), compost B (14.53%), compost C (23.53%), compost D (20.67%) and compost E (19.95%).The rate of carbon degradation is as follows: compost A (0.14/day), compost B (0.12 /day), compost C (0.16/day), compost D (0.17/day) and compost E (0.16/day).

In composting conducted for 21 days calculated from the 10th day to 21st day increment of nitrogen levels in the compost among others ; Compost A (11.63%), compost B (5.53%), compost C (-1.38%), compost D (12.96%) and compost E (12.50%). Total increase in N is due to the loss of C-organic dry mass as CO^2 during the composting process. In addition, N values may also increase due to nitrogen-wearing bacteria that are generally active at the end of composting [15]. N levels in the compost sample from day 10 to day 21. On the 21st day of composting, the percentage of compost B, C, D and E is bigger than compost A (blank), that is 428.57% (compost B), 542.85% (compost C), 657.14% (compost D) and 914.28% (compost E) to compost A (blank). P levels on all compost samples from day 10 to day 21.

The addition of fish bone meal (FBM) gives significant value to the increase of P content in the compost so that it can fulfill the requirements of SNI 19-7030-2004 which determine the minimum compost P level of 0.10%. The decrease in P content in the composting process is due to the dissolution of P in dissolved organic solution. It is also supported by the addition of EM4 containing phosphate solubilizing bacteria or Phospate Solubilizing Bacteria (PSB). In solubility P bound to Ca, microorganisms through secretion of various types of organic acids such as carboxylic acid and rhizospheric have a mechanism for lowering pHthat separates phosphate bond forms such as $Ca_3(PO_4)_2$ [16]. However, medium buffer capacity reduces the effectiveness of PSB in releasing P from $Ca_3(PO_4)_2$ [16]. The decrease percentage in C/N ratio is as follows; Compost A (27.08%), compost B (19.57%), compost C (21.43%), compost D (29.79%) and compost E (28.89%). Decreasing of C content is because carbon is a basic element of living things that is needed in large quantities by microbes rather than other elements [11].

Based on SNI 19-7030-2004 standard, the minimum value of C/N ratio in compost is 10:1 and maximum 20:1, so it is concluded that the ideal composting time is 21 days because the average C/N ratio of 5 compost samples has value of 17:1. The addition of water made to all compost samples is the same, ie with 300 ml of ground water/compost using sprayer on the 11th day. The upper limit of moisture content is between 60 and 80%, depending on the composite material [11]. Water is added for moisture content reaches 60% (wet base) in the compost mixture[15], then the percentage increase in moisture content in all compost samples from the 10th day to the 21st day is as follows; Compost A (12.53%), compost B (9.70%), compost C (18.07%), compost D (12.16%) and compost E (11.22%). Based on the standard specified in SNI 19-7030-2004 the maximum water content allowed for compost is 50% so it can be concluded that all compost samples that have not met this requirement because the value of water content is still above 50%, this can be overcome by drying the compost pile in hot weather for 1-2 days so that the water content decreases.

4. Conclusion

Compost B (FBM 15 gr) has P content of 0.37%, so it has fulfilled the provisions of SNI 19-7030-2004 about the recommended compost standard of 0.10%, the increase of P content on compost B reached 428.57% compared to the blank (compost A) with P levels of 0.07%. The rate of organic carbon degradation is: compost A (0.14/day), compost B (0.12/day), compost C (0.16/day), compost D

IOP Conf. Series: Earth and Environmental Science **106** (2017) 012091 doi:10.1088/1755-1315/106/1/012091

(0.17/day) and compost E (0.16/day). The average calculation of residual weight of bone fish processing into bone meal form is 30.38% so that the percentage of FBM mass balance is 5.22-9.53% from total weight of whole snapper fish.

References

[1] Wawasto A, R Kusumawati, and Tazwir. 2008. Effect of Immersion in Chloride Acid on Quality of Red Chapped Bone Gelatin (Lutjanus sp.*), Jurnal Pascapanen dan Bioteknologi Kelautan dan Perikanan*, Jurusan Perikanan dan Jurusan Ilmu Kelautan, Universitas Diponegoro, Semarang.
[2] Peranginangi R, S Hadi, and Suryanti. 2006. Extraction of Gelatin from Red Snapper (Lutjanus sp.) , *Jurnal Pascapanen dan Bioteknologi Kelautan dan Perikanan*, IPB, Bogor.
[3] Jenderal Perikanan Tangkap, Sub-Direktorat. 2015. Statistik Perikanan Tangkap Indonesia, 2014, Direktorat Jenderal Perikanan Tangkap, Jakarta.
[4] Card A, D Whiting, C Wilson, and J Reeder. Organic Fertilizers, Colorado State University. 2002.
[5] Peranginangin R, Murniyati, and F.R. Dewi. 2014. Processing Technique of Calcium Flour from Tilapia Fish Bone, Penebar Swadaya, Jakarta.
[6] Prasetya A T, and E B Susatyo. 2013. Utilization of Bone Fish Snap to Increase Phosphorus Level of Liquid Waste Tempe Fertilizer, *Indonesian Journal of Chemical Science*, Jurusan Kimia FMIPA, UNNES, Semarang.
[7] Artola, A., Barrena, R., Font, X., Gabriel, D., Gea, T., Mudhoo, A., Sánchez, A. Composting from a Sustainable Point of View: Respirometric Indices as Key Parameter, Global Science Books, Japan. 2009.
[8] Firmansyah M, and R Noor. Integrated Waste Management Plan for Housing of Citra Graha City, South Kalimantan Province, Program Studi Teknik Lingkungan. Fakultas Teknik Universitas Lambung Mangkurat. Banjarbaru. Kalimantan Selatan. 2016.
[9] Rahayu D E, and Y Sukmono. Study of Potential of Market Organic Rubbish Utilization based on Its Characteristic (Case Study of Segiri Market of Samarinda City), Fakultas Teknik Universitas Mulawarman. 2013.
[10] Amanah F. Effect of Stirring and Composition of Compost Material on Quality of Compost Mixed Sludge, Program Studi Teknik Lingkungan. Fakultas Teknik Universitas Indonesia. Depok. 2012.
[11] Shrivastav A, A M Jana, and S S Gill. Aerobic Bacterial Degradation of Kitchen Waste : A Review, JMBFS, Department of Microbiology and Biotechnology, College of Life and Sciences, Madhya Pradesh, India. 2014.
[12] Salim T, and Sriharti. 2010. Utilization of Garden Trash (Grass-Rumputan) for Making Compost, *Prosiding Seminar Nasional Teknik Kimia "Kejuangan"*, Balai Besar Pengembangan Teknologi Tepat Guna LIPI, Subang.
[13] Horwitz W. 2002. Official Methods of Analysis of AOAC International, Gaithersburg, Maryland.
[14] BSN. 2004. Compost Specification from Domestic Organic Waste, Jakarta Pusat, Indonesia.
[15] Che Jusoh M L, L A Manaf, and P A Latiff. 2013. Composting of Rice Straw with Effective Microorganisms (EM) and its Influence on Compost Quality, *Iranian Journal of Environmental Health Science & Engineering*, Department of Environmental Science, Faculty of Environmental Studies, University Putra Malaysia, Selangor, Malaysia.
[16] Khan A A, G Jilani, M S Akhtar, S M S Naqvi, and M Rasheed. 2009. Phosphorus Solubilizing Bacteria : Occurence, Mechanisms and their Role in Crop Production, *Journal Agriculture Biology Science*, Department of Soil Science, PMAS Arid Agriculture University, Rawalpindi, Pakistan.

The prospect of hazardous sludge reduction through gasification process

R Hakiki*, T Wikaningrum, T Kurniawan

Faculty of Engineering President University, Jl. Ki Hajar Dewantara, Bekasi 17550, Indonesia.

*Corresponding Author: rijalhakiki@president.ac.id

Abstract. Biological sludge generated from centralized industrial WWTP is classified as toxic and hazardous waste based on the Indonesian's Government Regulation No. 101/2014. The amount of mass and volume of sludge produced have an impact in the cost to manage or to dispose. The main objective of this study is to identify the opportunity of gasification technology which can be applied to reduce hazardous sludge quantity before sending to the final disposal. This preliminary study covers the technical and economic assessment of the application of gasification process, which was a combination of lab-scale experimental results and assumptions based on prior research. The results showed that the process was quite effective in reducing the amount and volume of hazardous sludge which results in reducing the disposal costs without causing negative impact on the environment. The reduced mass are moisture and volatile carbon which are decomposed, while residues are fix carbon and other minerals which are not decomposed by thermal process. The economical simulation showed that the project will achieve payback period in 2.5 years, IRR value of 53 % and BC Ratio of 2.3. The further study in the pilot scale to obtain the more accurate design and calculations is recommended.

Keywords: disposal cost, gasification, hazardous sludge, industrial estate, wastewater treatment

1. Introduction

Wastewater treatments involve physical, chemical, or biological process, depend on the wastewater characteristics. According to the Indonesian's Government Regulation No. 101/2014, the biological treatment will be more efficient for wastewater with high organic content. Biological treatment of industrial wastewater utilizes microorganism activities that release an amount of hazardous sludge as a product. On other hand, disposing hazardous substance requires great effort thereby it accounts for additional cost for waste water treatment process [1]. In Indonesia, typical method for hazardous sludge treatment is either by land deposition (landfill) or by reusing it as additional fuel in the industry. The amount of hazardous sludge produced is directly proportional to the cost. Moreover, the increasing number of industries will increase the amount of hazardous sludge produced and it will have an impact on the availability of landfilling area. An additional attempt is required to reduce the amount of hazardous sludge to be managed by landfill method. Gasification technology was chosen for reducing the volume of sludge while as much as possible recover some of its energy content. Gasification is a partial combustion process of solid substance that was optimized for maximum production of gaseous fuel, known as synthesis gas (syngas). Gasification is a faster method for volume reduction compared to biochemical options and it offers advantages over direct combustion method [2], [3]. Gasification was initially developed for converting coal into syngas [3]. With the increasing concern of global warming, gasification research has been focused on the gasification of biomass to produce renewable energy [4],

[5], [6], [7], [8]. There are but few who have discussed the treatment of hazardous sludge by using gasification technology. Researcher [9] conducted the analysis of bio-sludge as an energy source, while [10] performed a study on biofuels from pyrolysis of sewage sludge, and [11] researched on steam gasification of sewage sludge to produce hydrogen and syngas. However, the sludge produced from different treatment plants will certainly have different characteristics. It will have an impact on the effectiveness of gasification unit performance. Therefore, the development of gasification unit based on the characteristics of hazardous sludge is necessary. This preliminary study encompasses initiatory characterization, lab-scale trial, and economic simulation to calculate the feasibility of gasification process.

2. Research Method

As mentioned earlier, this preliminary study aims to estimate the prospect of hazardous sludge (which released from Jababeka's centralized WWTP) reduction through the gasification process. Jababeka has being managed 3000 ha of industrial land in Bekasi Regency that have already occupied by more than 1700 industries. A lab-scale trial has been conducted at Environmental Engineering Research Laboratory, President University, while sludge characterization has been carried out at Jababeka's Environmental Laboratory.

2.1. Hazardous sludge characterization

Hazardous sludge characterization is the early stage of this study to analyze the chemical composition of sludge, especially to measure the organic content. Volatile organic compounds as raw materials for gasification process will be converted into syngas through the thermochemical process. Moreover, other impurities should be recognized to determine the syngas purification unit.

2.2. Lab-scale pyrolysis

There are several simultaneous processes occurred, i.e., redox reaction, pyrolysis, and drying treatment. Pyrolysis process yields syngas that is formed by thermochemical conversion of the organic complex to be less complex compounds. The lab-scale trial focuses on the pyrolysis process as an estimation to find out how the conversion works. A flare test has been conducted to determine the syngas formation. The reactor configuration used in this study consists of a closed-system stainless steel vessel, a bunsen burner that generates heat from butane combustion, flare test equipment, and other required tools.

2.3. Particle size analysis

The reduction of hazardous sludge amount is determined by particle size measurements at the initial and final thermochemical treatments. Particle size measurements were conducted using Fiji-ImageJ software ver. 20151222-1. The sizes of random-shape sludge particles are expressed in Feret diameter, obtained from 2D digital objects that have been adjusted so that they can represent the actual 3D forms.

2.4. Simulation of economic feasibility

The economic aspect is an important parameter to be simulated. Economic feasibility relates to investment cost, profit, and duration of the payback period. Simulation of the economic aspect is based on the analysis results, laboratory experiments, and some required assumptions to calculate the value of PP, IRR, BC ratio and NPV which are commonly used as economic feasibility indicators. The NPV (net present value) leads to better investment decisions than other criteria. In the opportunity-cost concept should identify financial assets with risks equivalent to the project under consideration. In this term, estimation of IRR is needed to calculate the opportunity cost [18]. The value of IRR was very interesting compared to the financial banking interest of 6%. The financial analysis as analyzing estimates possible returns is needed by the investors to determines how certain the expected rate of return on an investment [19].

3. Results and Discussion

The study of pyrolysis as an estimation for gasification process will be the basis of designing the gasification system, including the reactor, separator unit, as well as the mass and energy balance calculation in ideal condition to determine the typical process parameters. The initial characterization of hazardous sludge is required as a preliminary approximation in lab-scale trial design. The reduction of sludge particles is related to the subtracted amount of hazardous sludge. Moreover, an economic simulation should be conducted as a preliminary calculation for pilot-scale trial.

3.1. Hazardous sludge characteristics

Initial and final chemical compositions were analyzed to determine the configuration changes as the results of the thermochemical treatment. Theoretically, several parameters that are supposed to alter following the thermochemical process are organic content as well as water and sulfur content. Organic materials consist of fixed carbon and volatile carbon that will be converted into syngas and other volatile products. Sulfur content that will be transformed into the gas phase and contained in the syngas is an impurity that should be removed to cleanse the syngas. Moreover, sulfur content will release sulphur oxide (SO_x) to the emission. Analytical results show that the organic materials equal to 66% of dry-sludge basis, the water content equals to 43% of the total sludge mass, while sulfur concentration and other minerals content is about 4% of dry-sludge basis.

Table 1. Initial composition of hazardous sludge.

Components	Units	Amount
TSS	mg/L	7,168.0
MLVSS	mg/L	4,732.0
SO_4	mg/g	594.0
Ammonia	mg/g	14.2
Organic content	%	66.0
H_2O	%	43.0

3.2. Hazardous sludge pyrolysis

Pyrolysis process can reduce the amount of hazardous sludge up to 40% of the total sludge mass basis (in a condition that 10% of the total sludge mass is water content). Several stages occurred the inside fixed-bed pyrolysis reactor are as follows: (1) heating process inside the vessel (the measured temperature is up to 550°C); (2) water content evaporation; (3) when the moisture content decreases, the thermochemical conversion begins. Moreover, it has been known that not all of the top product will be stable as a gas when it reaches room temperature. There are some fractions of the top product that will be condensed because of the temperature difference between the inside and outside part of the reactor. A gas-liquid separator unit is required to prevent the condensate carried along with the syngas. This modified separator is equipped with a carbon filter to prevent other impurities contained in the syngas. A sudden drop in gas-velocity leads to the formation of aerosols, then the phase separation occurs. A gas having a lower density tends to move to the top of the separation tank (top product), while the condensate moves to the bottom of the tank (bottom product). The simplified mass balance can be seen in Figure 1.

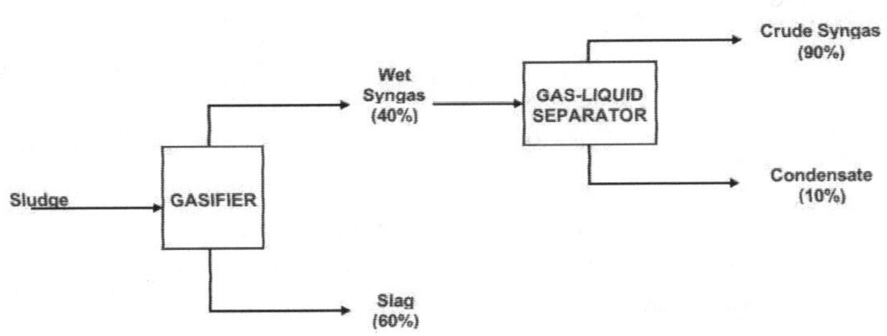

Figure 1. Mass balance of gasification process (lab-scale).

3.3. Particle size reduction

Figure 2 shows the processed image for the particle size determination. The lab-scale gasification trial

IOP Conf. Series: Earth and Environmental Science **106** (2017) 012092 doi:10.1088/1755-1315/106/1/012092

can reduce sludge particle size of up to 60%. It is in line with the results of sludge weighing where there is mass difference (mass reduction occurred is up to 50%) between the initial and final sludge mass after the process. Based on both data, it can be said that the sludge amount reduction occurs due to the thermal degradation, resulting in volatilization of volatile carbon, water content, and other thermal sensitive matters.

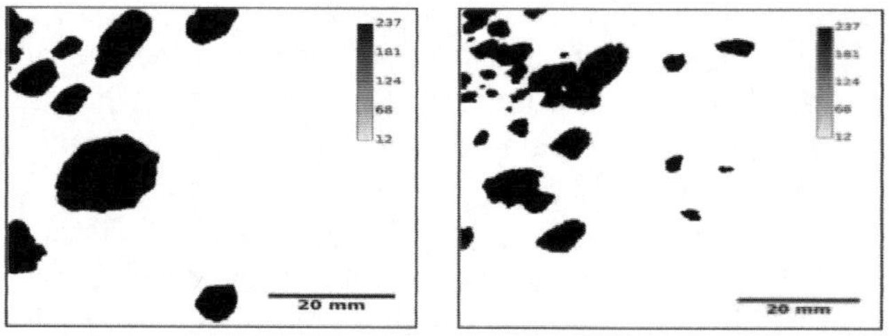

(a) (b)

Figure 2. Feret diameter of particles: (a) sludge before the process (11 mm) and (b) slag after the process (4 mm).

3.4. Economic feasibility

The economic feasibility simulation is based on analysis data, lab-scale trial results, and assumptions required in the calculation. There will be a new proposed business model to be compared with the existing scheme. Figure 3 and Figure 4 show the comparison of both models.

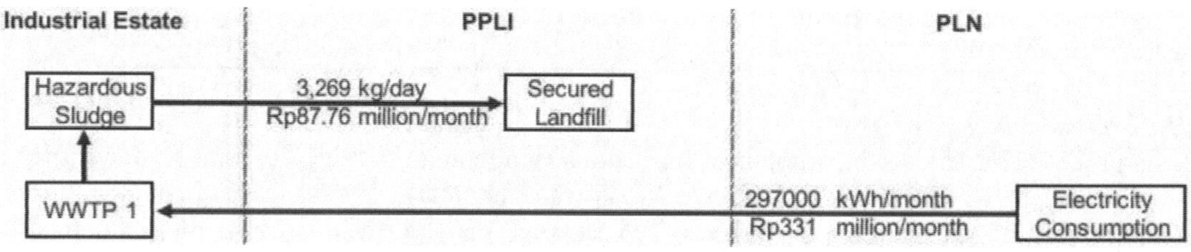

Figure 3. The existing business model.

In the existing business model, the industrial estate spends about 87 million Rupiah/month for sludge disposal and pays out about 331 million Rupiah/month for the electricity consumption. As seen in the proposed business model depicted in Figure 5, the expenditures are only about 43 million Rupiah/month for sludge disposal and 319 million Rupiah/month for electricity consumption. The comparison between both business models shows a cost reduction of up to 40%.

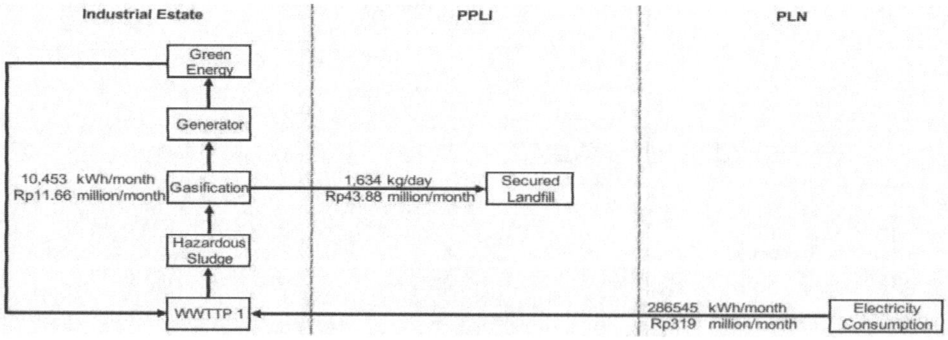

Figure 4. The proposed business model.

Several assumptions taken into account to simulate the economic feasibility and to develop the proposed business model are as follows: (1) The increment of sludge disposal cost is about 5% per year;

IOP Conf. Series: Earth and Environmental Science **106** (2017) 012092 doi:10.1088/1755-1315/106/1/012092

(2) The increment of electricity cost is about 3% per year; (3) The increment Increasement of employee salary is about 10% per year; (4) The increment Accretion of fuel cost is about 10% per year; (5) The increment Increasement of maintenance cost is about 10% per year; (6) Accretion of miscellaneous expenses is about 10% per year; and (7) Inflation rate of about 6% per year. (8) All assumption rate and cost of energy, man power, sludge disposal, mechanical & electrical maintenance, inflation rate were based on data in March year 2017.

Table 2. The financial profit and loss analysis of the proposed business model.

No	Items	Rp/month
1	Revenue	
	Sludge disposal cost saving	35,104,457
	Green energy produced (expressed in Rp1115/kwh)	11,657,317
	Employment cost saving (2 people)	6,400,000
	Total in revenue	**53,161,774**
2	COGS/Expenses	
	Employment cost (4 people)	12,800,000
	Electricity consumption	481,680
	Fuel for start up	1,000,000
	Genset maintenance	5,000,000
	Reactor maintenance	5,000,000
	Miscellaneous expenses	2,000,000
	Depreciation on investment	10,371,528
	Total Cost	**36,653,208**
3	**Profit**	**16,508,566**
4	**% Profit to total cost**	**45%**

Based on several assumptions above, the financial profit and loss analysis is presented in Table 2. Table 3 shows the summary of economic simulation results for the proposed business model.

Table 3. Summary of economic simulation based on the proposed business model.

Items	Units	Value
Capacity	tons/day	2.5
Investment	Rp	775,000,000
Lifetime of gasification unit	years	5
Profit percentage	%	45%
Payback Period	years	2.5
IRR	%	56%
NPV	Rp	574,569,163
BC ratio	-	2.3

4. Conclusion

The results showed that the gasification process is quite effective in reducing the volume of hazardous sludge which then lowers the disposal cost without causing an adverse impact on the environment. The reduced mass contains moisture and volatile carbon which are decomposed by the thermal process, while the residues are fixed carbon and other minerals which are not degraded by the thermal process. The economic simulation shows that the project will achieve payback period in 2.5 years, IRR value of 53%, and BC Ratio of 2.3. The value of IRR was very interesting compared to the financial banking interest of 6%. The further study in the pilot-scale is highly recommended to obtain a more accurate design and calculation.

Acknowledgment
This project is supported by PT Jababeka Infrastructure as a facilitator of the required data and laboratory test for several parameters. We extend our gratitude to Mr. Kukuh Sulaksono, Mr. Susilo, Ms. Istingani, and other parties for the full support and assistance.

References

[1] Fytili D, and Zabaniotou A. 2008. Utilization of sewage sludge in EU application of old and new methods—A review, *Renewable and Sustainable Energy Reviews*. 12:116 – 140.
[2] Umberto Arena. 2012. Process and technological aspects of municipal solid waste gasification. A review. *Waste Management*. 32:625 – 639.
[3] Young, Gary C. Municipal solid waste to energy conversion process: economic, technical, and and renewable comparisons. John Wiley & Sons, 2010.
[4] Akhlas J, Baesso S, Bertucco A, and Ruggeri F. 2015. Coal gasification by indirect heating in a single moving bed reactor: Process development & simulation. *AIMS Energy*. 3(4):635-665.
[5] Kirkels A F, Verbong G P J. 2011. Biomass gasification: Still promising? A 30-year global overview, *Renewable and Sustainable Energy Reviews*.15:471-481.
[6] Kirubakarana V, Sivaramakrishnanb V, Nalinic R, Sekard T, Premalathae M, and Subramaniane P. 2009. A review on gasification of biomass, *Renewable and Sustainable Energy Reviews*. 13:179 – 186.
[7] Sharma S, Sheth P N. 2016. Air–steam biomass gasification: Experiments, modeling and simulation, *Energy Conversion and Management*. 110: 307–318.
[8] Sune Bengtsson. 2011. The CHRISGAS project, *BIOMASS AND BIOENERGY*. 35: 52 – 57.
[9] Wen-Tien Tsai. 2012. An analysis of the use of bio-sludge as an energy source and its environmental benefits in Taiwan. *Energies*. 5:3064–3073.
[10] M E Sanchez, E Lindao, D Margaleff, O Martinez, and A Moran. 2009. Bio-fuel and bio-char production for pyrolysis of sewage sludge. *J. Residual Science & Technology*. 6(1).
[11] N Nipattummakul, Islam I Ahmed, S Kerdsuwan, and A K Gupta. 2010. Hydrogen and syngas production from sewage sludge via steam gasification. *International Journal of Hydrogen and Energy* (Elsevier). 35:1173–1175
[12] Nichholas P G, Lumley, Dotti F Ramey, Ana L Prieto, Robert J Braun, Tzahi Y Catch, and Jason M Porter. 2014. Techno-economic analysis of wastewater sludge gasification: a decentralized urban perspective. *Bioresource Tecnology*. 161:385–394.
[13] T Karayildirim, J Yanik, M Uksel, and H Bockhorn. 2006. Characterisation of products from pyrolysis of waste sludges. *Fuel (Elsevier)*. 85: 1498–1508.
[14] Particle size analysis SOP 2011. Dune sciences, Inc. Particle size analysis-standard operating protocol v1.1.
[15] J W Twidell, and A D Weir, Renewable energy resources, 2nd Edition. New York: Taylor & Francis, Chapter 11, p. 355–374, 2006.
[16] R A Brealey and S C Myers, Principles of Corporate Finance. 4th Edition, McGraw-Hill, Chapter 5. P 73-79,1991
[17] K C Brown, and F K Reilly. Analysis of Investments and Management of portofolios, 9th Edition, South-Western Cengage Learning, Chapter 1, p. 3-25. 2009.

Antecedents of willingness to pay for green products

S I Pratiwi, L A Pratomo*

Faculty of Economic and Business, Universitas Trisakti, Jakarta, Indonesia

*Corresponding Author: luki.adiati@trisakti.ac.id

Abstract. The main purpose of this paper is to examine whether there is a positive influence of pro-environmental behavior and environmental locus of control toward the willingness to pay for green products. The data obtained by distributing online and offline questionnaires, reaching 419 respondents of 18 to ≥55 years old that have the knowledge and already bought a green product. The purposive sampling was used as the sampling technique, and the data were tested by Statistical Equation Modeling (SEM).The results show that environmental locus of control does not positively affect pro-environmental behavior. However, the environmental locus of control and pro-environmental behavior do have a positive influence on the willingness to pay. Based on the findings, it is essential for green product companies to improve customers' pro-environmental behavior and environmental locus of control. To do so, the marketer of green products should increase consumers' concern, awareness, and behavior of conserving nature through activities such as campaigns and demonstrations.

Keywords: environmental locus, green products, pro-environmental behavior, willingness to pay

1. Introduction

Natural resources increasingly scarce and future generations may not be able to relish it [1]. Environmental destruction induces many disadvantages to lots of organisms such as animals (especially the endangered ones) and humans getting tortured, directly or indirectly [2]. Due to those reasons, nowadays people make efforts to live a healthy lifestyle and conserve the environment.

A lot of organizations or individuals who support the healthy lifestyle and green products consumption has begun to produce organic foods as well as recycled (eco-friendly) products including car and other vehicles [3]. Recycled, environmentally friendly products and organic foods are also renowned in Asia due to the environmental destruction and heavy global warming especially in Indonesia, Middle East, India, and China which ruins the temperature, climate, and seasonal changes. Consequently, people appreciate natural resources more and want to move toward a better life [4].

Despite the increasing awareness of environmental issues, green products are more expensive compared to normal products [5]. Indonesia is one of the countries where environmental awareness is increasing but not mainstream yet [6]. Therefore the products manufactured in Indonesia mostly need imported machinery or tools for the production process [7], triggering the higher cost of green products compared to normal ones, and customers will consequently refuse to buy green products unless they have the willingness to pay more for it.

It is crucial for companies to understand why customers want to pay more for green products. The willingness to pay means the maximum price that consumers agree to pay for specific goods or services that satisfy them [8]. Regarding green products, the willingness to pay is affected by pro-environmental behavior (PEB) and environmental locus of control (ELOC) [9].

Pro-environmental behavior (PEB) is the way people do things to minimize the negative impact of their actions on the environment, such as using recycled products, diminishing energy consumption, and reducing waste production [10], [11]. A study showed that pro-environmental consumers would easily practice pro-environmental behavior compared with consumers without pro-environmental commitment [11]. In their study [11], they found that consumer with high degree of cognitive attitude, affective attitude, social norm and personal norm of environmental concerns showed the strong intention of recycling in their daily lives.

Locus of control is divided into two types, i.e., the internal and external categories [12]. The internal environmental locus of control explains the extent to which people believe that they can affect [pro-environmental] outcome through their actions [13].The external environmental locus of control expresses people's belief that environmental destruction or things that happened to the environment is the outcome of fate, or caused by other human beings and not because of themselves [12]. However, the environmental locus of control (ELOC) represents people's beliefs whether the nature-related situation they are facing is caused by themselves (internally) or others (external) [13]. In their study, environmental locus of control divided on four dimensions, namely biospheric altruism, corporate skepticism, economic motivation and individual recycling efforts [13].

Customers' willingness to pay (WTP) for green products is essential for the companies. Previous studies have already tried to examine factors that make customers agree to pay more for their products [8], but only a few have investigated the influence of environmental locus of control as an affecting factor toward customers' willingness to pay [9]. Thus, this study aims to analyze the relationships between ELOC and WTP, ELOC and PEB, as well PEB and WTP for environmentally friendly products.

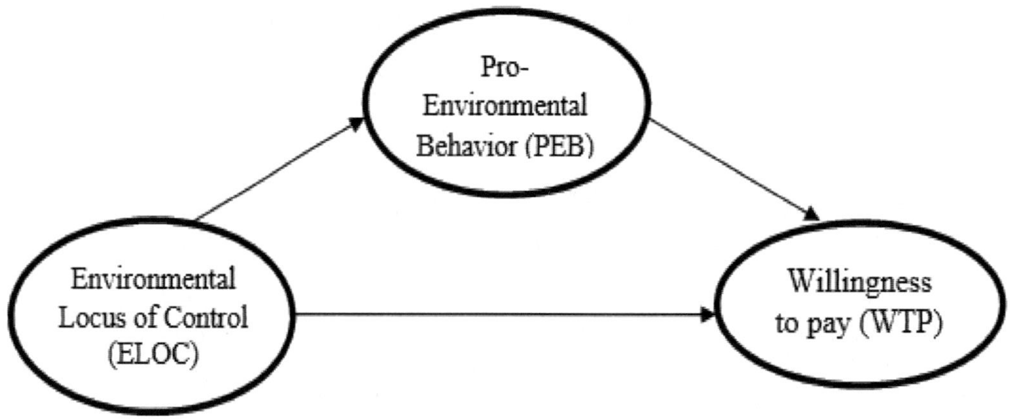

Figure 1. Conceptual framework.

Based on the conceptual framework above, three hypotheses were developed in this research:

Environmental locus of control (ELOC) makes individuals aware that the environment destruction is caused either by them or others. Therefore, they will try to conserve environment more than before in which makes them as the green products consumer and adapting healthy lifestyle [9] that will cause their Willingness to Pay (WTP) to increase. It can hypothesized that :

H1: Environmental Locus of Control (ELOC) is positively related to Willingness to Pay (WTP) for environmentally friendly products.

Pro-environmental behaviour is affected by internal locus of control because individual's motivation to intervene or do something related to environmentally responsible is on behalf of the environment [17]. With the increasing of consumers' environment locus of control, their pro-environmental concern which lead to pro-environmental behavior will also increase in order to reduce

IOP Conf. Series: Earth and Environmental Science **106** (2017) 012093 doi:10.1088/1755-1315/106/1/012093

their guilt for the environmental destruction that caused by either themselves or others by conserving environment and try to make the environment to its former state [9]. It is hypothesized as:

H2: Environmental Locus of Control (ELOC) is positively related to Pro-Environmental Behavior (PEB) for environmentally friendly products.

Environmental concerns in which affect pro-environmental behaviour can also affecting willingness to pay for green products. [13] It is likely, that consumers who exhibit high levels of environmental consciousness make more green purchasing decisions than those exhibiting low levels [9]. Therefore as the level of PEB increases, the willingness to pay for green products will also increase. It is hypothesized as:

H3: Pro-Environmental Behavior (PEB) is positively related to Willingness to Pay (WTP) for environmentally friendly products.

2. Research Method

All of the instruments and variables are adapted from previous research [9]. This study employs hypotheses testing and cross-sectional (one-shot) time horizon where data are gathered just once [16]. The unit used in the analysis is individuals. The sampling method is non-probability with the purposive sampling technique (PST) [15] which in this case, the respondents (samples) were taken with criteria of both knowing and have the experience of buying green products. Both online- and offline-distributed survey questionnaires were translated into Bahasa Indonesia to prevent any misunderstanding arises from language barriers.

The respondent characteristics of this study show that the male respondents exceed females only by 7 responses, aged between 18–25 years (75.2%), of which are mostly students with a high school diploma having a monthly income ranging from IDR 1,500,001–5,000,000. As many as 419 out of 450 respondents (93.11%) have known and purchased green products, ergo can be used for further analysis.

CFA was conducted to check the validity of the measurement used in this study. One item of pro-environmental behavior and five items of environmental locus of control were removed from the final scale due to their large standardized residuals. The reliability of the measurement was analyzed by Cronbach Alpha, and it resulted in a good validity with each construct reliability value of above 0.7 [17]. According to RMR (0.075) and RMSEA (0.754) on the goodness-of-fit statistics, it indicates a good model fit [16].

Central tendency (mean and standard deviation) was used as the descriptive statistics to analyze each variable in this study. It has been found that (1) pro-environmental behavior has an average value of 3.933, implying that the respondents have practiced pro-environmental means of conserving nature; (2) the willingness to pay has an average value of 3.765, expressing that the respondents are willing to purchase green products; and (3) the environmental locus of control has an average value of 3.912, meaning that the respondents are concerned about environmental destruction.

3. Results and Discussion

The hypotheses testing results can be seen in Table 1.

Table 1. Hypothesis testing results.

Hypotheses		Estimates	ρ-values	Decisions
H1	ELOC→ WTP	0.160	0.007	Supported
H2	ELOC → PEB	-0.104	0.113	Not Supported
H3	PEB → WTP	0.677	0.000	Supported

H1: Environmental Locus of Control positively affects Willingness to Pay

The result indicates that the higher consumers' environmental locus of control gets, the higher their willingness to pay. It means that along with the increment of consumers' concerns about environmental destruction such as air, water, and land pollution, their willingness to purchase green products increases as well. This study also shows that although Indonesia is renowned for its forest and other abundant natural resources, the respondents think that recycling means are still needed nonetheless.

Fortunately, even if the price of recycled products is usually higher, the result points out that consumers are willing to pay 10% more for products that are manufactured, processed, and packaged in an environmentally friendly way. Based on the research [13], they found that economic motivation as one of ELOC dimension has stong influence on consumer willingness to pay for environmental products. They[13] found that consumer agree to accept bigger financial responsibility in order to prevent the ecological impact of their consumption. It is supported by previous research [14], [9] who found that people with a high environmental locus of control are expected to buy green products on adaily basis.

H2: Environmental Locus of Control has no positive influence on Pro-Environmental Behavior

In this study, the second hypothesis is not supported, which means that people's environmental locus of control cannot determine their behavior toward nature. According to a study in 2005 [13], PEB is a situational-specific variable which has a complex reaction, denoting that consumers' decision regarding pro-environmental behavior might change in any occasion despite their concerns for the environment.

Based on they research [13], they found that corporate skepticism as one dimension of ELOC, has the least influenced on pro-environmental behavior. Consumers tend to think that most of the companies actually are not environmentally concerns and responsible. Additionally, the locus of control is originated from a few types of characteristics namely self-esteem, belonging, and personal control [17]. Out of those three, only personal control directly has a positive relationship with pro-environmental behavior without any intercession [17].

H3: Pro-Environmental Behavior positively affects Willingness to Pay

The third hypothesis is supported by this study, implying that the increasing number of behavioral things people have done and about to do will also raise their willingness to purchase green products because they have the behavior of conserving nature by consuming green products. Consumers in Indonesia tend to do necessary procedures to minimize the damage to the environment caused by air and waste pollution. For example, most of the respondents prefer walking to using vehicles to go to nearby stores, buy products with recycled packages, and keep their vehicles well-tuned to reduce the emission.

This outcome is supported by previous studies, such as the result of one reseach on 2013 in Turkey [14] that consumer who have high level of environmental concerns showed stronger intention to become e-invoice subsriber that more environmentally responsible. Another study in India [9] found consumer with high degree of pro-environmental behavior such as using public transport, turn off electronic equipment when they did not use it and others pro-environmental behaviors have the willingness to buy more expensive bulb that more energy saving [9]. Based on those studies, the higher the customers' pro-environmental behavior (PEB), the greater the possibility of respondents buying green products.

4. Conclusion

The research found that Indonesian consumers' willingness to pay for green products is positively influenced by pro-environmental behavior and environmental locus of control. The managerial implications for companies manufacturing green products are that firstly, they have to target the customers with a high environmental locus of control (ELOC) and high pro-environmental behavior (PEB). Secondly, they have to increase consumers' environmental locus of control and pro-environmental behavior by devising a major, active campaign about the importance of the environment. Some examples of the messages that the companies can convey are (1) to turn the vehicle off during the red light to increase awareness that even a small act from consumers can save the environment by minimizing the emission; and (2) to promote the benefits of recycled empty product packaging.

4.1. Managerial Implications

The result of this study may give some critical consideration for green products' managers. The critical considerations are first based on the research testing result, Environmental Locus of Control (ELOC) is positively affect Willingness to Pay (WTP). Therefore in order to increase willingness to pay, managers can increase consumers' environmental locus of control first. For example, managers should make campaign to turn vehicle off during red light at traffic point to increase awareness that small act of consumers can save the environment by minimizing the emissions produced.

Second, based on the research testing result, Pro-Environmental Behavior (PEB) is positively affect Willingness to Pay (WTP). Therefore managers can increase pro-environmental behavior first in order to increase consumers' willingness to pay. For example, managers should show consumers that recycling is easy. Directly by showing them how to get it done, indirectly put it in the package, or by promotion where they will get benefits if consumers recycled empty product packaging.

Third, based on the research testing result, consumers' willingness to pay is high therefore marketers should promote the benefits to spend an extra Rp. 20.000,- per week to buy less environmentally harmful products such as, cleaner environment will make a healthier surroundings for citizens. Fourth, based on the research testing result, marketers should let the consumers know that the wrappers are made from recycled materials and therefore it shows that this product is better than the others by trying to conserve nature.

References

[1] Jowit J, 2008, October 29th, Climate Change. Retrieved from The Guardian: https://www.theguardian.com/environment/2008/oct/29/climatechange-endangeredhabitats.

[2] Rinkesh, 2016, Causes and Effects of Environmental Degradation. Retrieved from Conserve Energy Future: http://www.conserve-energy-future.com/causes-and-effects-of-environmental-degradation.php.

[3] Brown J, 2014, August 23rd. Press Release. Retrieved from Asia Today: http://asiatoday.com/pressrelease/consumers-encourage-companies-go-green-companies-profits-increase.

[4] Goldenberg S, 2014, March 31st, Climate Change. Retrieved from The Guardian: https://www.theguardian.com/environment/2014/mar/31/climate-change-food-supply-un.

[5] The Telegraph, 2010, May 30th, News. Retrieved from The Telegraph: http://www.telegraph.co.uk/news/earth/earthnews/7785705/Green-goods-cost-nearly-50-more.html.

[6] BRIEF, 2014, August 1st, Countries, Retrieved from World Bank: http://www.worldbank.org/en/country/indonesia/brief/world-bank-and-environment-in-indonesia.

[7] Isroi, 2014, September 13th, Berbagi Tak Pernah Rugi. Retrieved from Wordpress: https://isroi.com/2014/09/13/menanam-blueberry-blackberry-ghooseberry-strawberry-di-indonesia/.

[8] Le Gall-Elly M. 2009. Definition, Measurement, and Determinants of the Consumer's Willingness to Pay: a Critical Synthesis and Avenues for Further Research. *Recherche et Applications en Marketing*. 24(2):91–112.

[9] Trivedi R H, Patel J D, and Savalia J R. 2015. Pro-environmental Behaviour, Locus of Control and Willingness to Pay for Environmentally Friendly Products. *Marketing Intelligence & Planning*. 33(1):67–89.

[10] Kollmus A, and Agyeman J. 2002. Mind the Gap: Why do People Act Environmentally and What are the Barriers to Pro-environmental Behavior? *Environmental Education Research.* 8(3), (4).

[11] Park J, and Ha S. 2012. Understanding Pro-environmental Behavior. *International Journal of Retail & Distribution Management.* 40(5):388–403.

[12] McCarty J A, and Shrum L J. 2001. The Influence of Individualism, Collectivism, and Locus of Control on Environmental Beliefs and Behavior. *Journal of Public Policy & Marketing: Spring 2001*. 20(1):93–104.

[13] Cleveland M, Kalamas M, and Laroche M. 2005. Shades of Green: Linking Environmental Locus of Control and Pro-environmental Behaviors. *Journal of Consumer Marketing*. 22(4): 198–212.

[14] Albaryak T, Caber M, and Aksoy S. 2010. Clustering Consumers According to Their Environmental Concerns and Scepticisms. *Journal of Trade, Economics and Finance*. 1(1), (5).

[15] Sekaran U, and Bougie R. Research Methods for Business. West Sussex: John Wiley & Sons Ltd. 2013.

[16] Hair J F, Black W C, Babin B J, and Anderson R E. Multivariate Data Analysis (7th ed.). USA: Pearson. 2010.

[17] Allen J B, and Ferrand J L. 1999. Environmental Locus of Control, Sympathy, and Pro-environmental Behavior: A Test of Geller's Actively Caring Hypothesis. *Environment and Behavior*. 31(3):16.

The effect of biomass concentration on polymer alginate in the immobilized biosorbent formation during the sorption processof heavy metal Cu^{2+}

A Rinanti*, D Jonathan, M D S Silalahi, M F Fachrul, R Hadisoebroto

Environmental Engineering Department, Faculty of Architecture Landscape and Environmental Technology, Universitas Trisakti, Jakarta, Indonesia

*Corresponding Author : astririnanti@trisakti.ac.id

Abstract. A research in environmental biotechnology has been done to analysis adsorption of ion Cu^{2+} by biomass of microalgae (*Chlorella sp, Ankistrodesmus braunii, Scenedesmus quadricauda*) and Saccharomyces *cerevisiae* onto alginate polymeras immobilized biosorbent on laboratory scale. The purpose of this study is to achieve the optimum biomass concentration which gives the best biosorption performance. Biosorption of Cu^{2+} was carried out in continuous fixed-bed column reactor system, volume of 1.5 L, equipped with peristaltic pump with a flow rate of 13 mL/min. Biosorption of Cu^{2+} was investigated using immobilized biosorbent with concentration of (g biomass/g polymer) 0.25; 0.5; 1, at pH4,initial concentration Cu^{2+}15 mg/L and 26°C±1. The results of this study showed that the increasing of biomass concentration (0 to 0.5 g/g) would result in better biosorption performance but soon decreased its performance at biomass concentration of 1 g/g. Biosorption capacity and highest removal efficiency of 0.1025 mg Cu^{2+}/g biosorbent and 66.36% occurred by immobilized biosorbent with 0.5 g/g concentration. The connection between the variation of biomass concentration in alginate to the biosorption performance by immobilized biosorbent shown by breakthrough curve, total adsorbed metal mass(q_{total}), efficiency of removal (%R) and biosorption capacity at breakthrough(q_e). Excessive biomass concentrations lead to reduced porosity of the beads thus slowing down the adsorption process.
Keywords: alginate polymer, biosorbent, biosorption, breakthrough curve, green microalgae

1. Introduction

The increasing concern for heavy metals danger in the neighbourhood encourages an alternative technology search to remove dangerous substances of heavy metals which is environmentally friendly. Biosorption is process of binding as well as selected unwanted ions or molecules from heavy metals utilizes biological materials like yeast and microalgae acting as biosorbent. The appliance of biomass microalgae as biosorbent appears as an interesting choice, economical and effective because some of its advantages than others [1]. Those benefits include low needs nutrients, high biomass production because of its autotroph, and unlike others as bacteria and fungi, microalgae generally not producing dangerous compound.

The biomass ability of binding metal ions influenced by several factors, namely temperature, acidity degrees (pH), biosorbent concentration, early sorbate concentration, contact time and the existence of other metal ions [2]. Nevertheless, free or mobile biosorbent when used in a reactor column on a continuous way by low density and size potentially causing blockage (clogging) on reactor so that causes pressure to be drastically decreasing. Biosorbent Immobilization can resolve these problems so that biosorbent can be used in the continuous reactor system to absorb heavy metals. Immobilization can be defined as confinement physical and localization cells microbes at a certain room in such a way as to limit migration microbes freely. Immobilised Cells show different hydrodynamics characteristic from environment but still able to maintain catalyst activity to be reused [3], [4].

One biomass immobilization technique often used to biosorption research is entrapment or coating in polymer matrix. Entrapment is the irreversible method; namely immobilized biomass which is entrapped in a supportive matrix or in fiber. This technique creates protective around immobilized microalgae to ensure the prolonged survival, not just for the process but also the stowage in polymers [5]. Alginate is one of its polymers compounds often used as supportive matrix to biomass immobilization using entrapment technique. Based on this description, this research is aimed at determining the influence of varied biomass concentration addition to polymer alginate in the formation of immobilized biosorbent in efforts to improve efficiency and the absorption capacity of heavy metal Cu2+.

2. Research Method

The heavy metal Cu2+ used as an adsorbate derives from the liquid wastewater of the metal coating industry.

2.1. Biomass cultivation and its production

Biosorbent raw materials are Chlorella sp. microalgae biomass, Ankistrodesmus braunii, Scenedesmus quadricauda and Saccharomyces cerevisiae, which is isolated from the waters of Setiabudi-Jakarta Reservoir. Biomass cultivation is done in vertical column photobioreactor with 10 L capacity, filled with artificial growth medium of PHM (Provasolli Haematococcus Media) to obtain optimum condition for microalgae growth. The operational conditions are controlled at 25^0C with 2500 lux lighting intensity, airflow rate of 800 mL/min, bright periods: dark (24h: 0), to obtain optimum biomass growth [1].

2.2. Mobile Biosorbent Production

Microalgae biomass cultures and Saccharomyces cerevisiae which are in the exponential phase, are harvested by centrifugation method with a rotational speed of 4000 rpm for 15 minutes. Centrifugation is done to separate biomass from its growth medium. The insoluble precipitated biomass is poured into a petri dish then dried in an oven. Dry biomass is crushed with mortar to become a biomass powder. Microalgae powder (MA) and Saccharomyces cerevisiae (SC) were mixed with the MA: SC ratio of 3: 1, forming immobilized biosorbent ready for research.

2.3. Operational Research Condition

The mobile biosorbent was weighed to obtain a load of 2 g, 4 g, and 8 g. The weighed biomass was immobilized with 400 ml of 2% Na-alginate solution to form beads. These beads are called immobilized biosorbents. Immobilized biosorbent was arranged into a continuous fixed-bed column reactor system with a capacity of 1.5 L, to reach a bed height of 5 cm. The feeding solution with an initial concentration of Cu2+ 15 mg/L, pH 4, was flowed into a fixed-bed column reactor vertically up-flow to prevent the channel effect [5], using a peristaltic pump with discharge capacity of 13 ml/min, controlled at room temperature 25°C. Samples were taken from the point of effluent periodically with the capture time range 0, 10, 20, 30, 45, 60, 80, 100, 125 and 150 minutes. Periodic effluent sampling was done to determine the change of concentration of Cu2+ gradually. The percentage of Cu2+ removal in the sample solution was measured using Atomic Absorption Spectrophotometry (AAS).

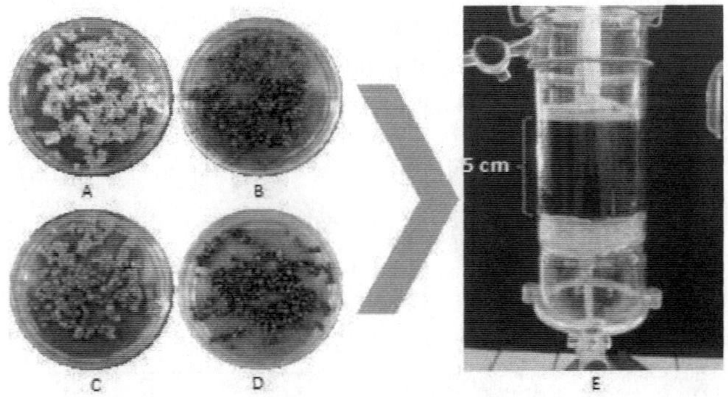

Figure 1. A) Sorbent Ca-alginate without biomass (control); B) immobilized biosorbent with (g biomass/g polymer) 0.25 g/g concentration; C) immobilized biosorbent0.5 g/g; D) immobilized biosorbent 1 g/g; E) immobilized biosorbent is arranged into continuous fixed-bed column reactor system with capacity of 1.5 L, until reach the *bed* height of 5 cm.

2.4. Breakthrough Curve

The breakthrough curve is a curve that describes the relation between changes in the concentration of adsorbate (Cu^{2+}) in effluent with time (t). The change in the concentration of adsorbate begins to be recorded when the solution reaches the effluent point where t = 0. Although the analysis results note that the concentration of adsorbate in the effluent is not equal to zero, but to illustrate the principle of biosorption in the column of adsorbate concentration when ¬t = 0 is considered zero. This concentration change describes the biosorption events occurring in the fixed-bed column reactor, i.e. the absorption of adsorbate at the beginning of the column when the feed solution passes through a column already filled by immobilized biosorbent. The adsorption reaction zone is increasingly moving up along with the saturated immobilized biosorbent at the bottom of the column. When the adsorption zone reaches the effluent point, the breakthrough concentration is reached (10% C_0). The time required to reach the breakthrough point (t_b) becomes an important parameter in this study.

3. Results and Discussion

Cu^{2+} metal sorption results in a continuous system of fixed-bed column reactors are presented in Table 1. The plot of time-effected Cu^{2+} effluent concentration by immobilized biosorbent with different variations of biomass concentration on the polymer will form the breakthrough curve as in Figure 2.

The biomass concentration of polymer alginate in immobilized biosorbent formation influences the adsorption capacity of Cu^{2+} heavy metals occurring in fixed-bed column reactors that operate with continuous system (Figure 1). Based on the graph of the breakthrough curve in Figure 2, the concentration of Cu^{2+} in the effluent to the initial concentration of adsorbate (C_e/C_0) increases with time (t). This means the ability of adsorption by immobilized biosorbent increases with increasing of biomass concentration to polymer alginate. Curves formed by immobilized biosorbent (A) are steeper than those formed by immobilized biosorbent (B). Different performance is shown by immobilized biosorbent (C), where the curve formed is steeper than the curve formed by immobilized biosorbent (B) but not steeper than the curve formed by immobilized biosorbent (A). This suggests that the immobilized biosorbent adsorption capacity decreases with increasing biomass concentration of 0.5 g/g to 1 g/g polymer alginate.

Tabel 1. Parameter of breakthrough curve in biosorption process heavy metal Cu^{2+} by immobilized biosorbent with 5 cm bed height in continuous fixed-bed column reactor system.

immobilized biosorbent (g biomass/g polimer)	M (g)	t_b (menit)	m_{total} (mg)	q_{total} (mg)	q_e (mg/g)	R (%)
0 (control)	171.5	1.79	29.348	9.157	0.0534	31.20
(A) 0.25	183.5	3.38	29.348	16.584	0.0904	56.51
(B) 0.50	190.0	7.34	29.348	19.474	0.1025	66.36
(C) 1.00	203.8	3.75	29.348	17.681	0.0868	60.25

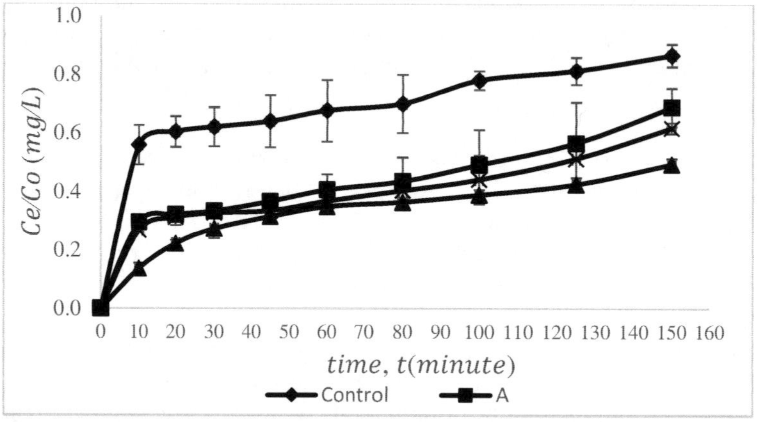

Figure 2. Breakthrough curve in heavy metal biosorption process Cu^{2+} by immobilized biosorbent with bed height of 5 cm in continuous fixed-bed column reactor system.

Based on the breakthrough curve formed, it can be identified that to reach the breakthrough concentration, it takes different (t_b) time. The steeper the curve forms, the faster the (t_b) required time to reach the breakthrough concentration. This is evidenced by the results of the calculation data in Table 1. Based on Table 1, the time required by the immobilized biosorbent column (A) to reach the breakthrough concentration (C_e/C_0 = 0.1C_0) is 3.38 minutes. Columns containing immobilized biosorbent (B) with the gentlest curve form shows the greatest result, i.e. the required time to reach the breakthrough is 7.34 minutes. The breakthrough time (t_b) achieved by the immobilized biosorbent (C) column is 3.75 minutes, greater than the immobilized biosorbent (A) but smaller than the immobilized biosorbent (B). The fastest breakthrough (t_b) time is achieved by control (alginate without biomass) with the steepest curve of 1.79 minutes. Breakthrough time is considered as an important parameter because it shows how fast a fixed-bed column is saturated. The longer breakthrough times indicate that the bed column is better because it takes longer to saturate.

The total metal mass value entered the column (m_{total}) shows a fixed value of 29.348 mg Cu^{2+}. That is, the total mass of Cu^{2+} metal entering the column is the same for each variation of the biosorbent concentration. The result is obtained from the initial concentration of Cu2 + metal multiplied by total reactor operating time of 150 min. The total mass of the adsorbed metal (q_total) also increased by 9.157; 16.584; 19.474 mg Cu2 + for control, immobilized biosorbent (A) and (B.) However, different results are shown by immobilized biosorbent (C) with q_total value decreasing to 17,681 mg Cu^{2+}. In addition, the biosorption capacity (q_e) also shows a change as the total mass of the adsorbed metal (q_{total}) increases.

Biosorption capacity increased respectively 0,0534; 0.0904; 0.1025 mg Cu^{2+}/g biosorbent for control, immobilized biosorbent (A) and (B). However, different results were shown by immobilized biosorbent (C) with q_e value decreasing to 0.0868 mg Cu^{2+}/g biosorbent. The results of this study indicated that the capacity of biosorption (q_e) increased from 0.0534 to 0.1025 mg Cu2 + / g biosorbent as the immobilized biosorbent concentration increased (0 to 0.5 g/g). The decreased biosorption capacity (q_e) of 0.1025 to 0.0868 mg Cu2+/g biosorbent occurred when the immobilized biosorbent concentration was increased from 0.5 g/g to 1 g/g.

The efficiency of heavy metals removal of Cu2+ increased by 31.2%; 56.51%; 66.36% for control, immobilized biosorbent (A) and (B). However, the removal efficiency by immobilized biosorbent (C) appeared to be smaller or decreased to 60.25%. The results of this study indicated that the removal efficiency (%R) increased by 31.2% to 66.36% as the immobilized biosorbent concentration increased from 0 to 0.5 g/g. Decrease in removal efficiency (%R) by 66.36% to 60.25% occured when the immobilized biosorbent concentration was increased from 0.5 g/g to 1 g/g.

This study revealed the existence of interrelation between a variation in the concentration of biomass in the polymer alginate to the performance of biosorption by immobilized biosorbent indicated by the breakthrough curve shape, the total mass of metal adsorbed(q_{total}), the removal efficiency (% R) and a capacity of biosorption at the time of breakthrough (q_e). The greater the concentration of immobilized biosorbent (0 to 0.5 g/g) the steeper the breakthrough curve will be so thatbreakthrough time taken getting bigger, the total mass of the adsorbate which absorbed the greater, the greater removal efficiency and greater biosorption capacity as well. Biosorption capacity and the highest removal efficiency of 0.1025 mg Cu^{2+}/g biosorbent and 66.36% occurred by immobilized biosorbent (B). Lowest biosorption capacity found by immobilized biosorbent (A) with a capacity of biosorption 0.0904 mg Cu^{2+}/g biosorbent and Ca-alginate without biomass as a control with a capacity of 0.0534 mg Cu^{2+}/g biosorbent. The lowest allowance efficiency of 31.2% and 56.51% existed by control and immobilized biosorbent (A)

Ca-alginate beads can remove adsorbate. The ability of Ca-alginate to absorb heavy metals was evidenced by decreasing concentrations of adsorbate passing through fixed-bed column starting initial concentration of Cu2+ 15 mg/L to 9 mg/L at 10 minute. The breakthrough curves formed by Ca-alginate beads looked steeper than those of immobilized biosorbent beads. Based on the previous analysis, the steeper the curve the faster the breakthrough time, the smaller the biosorption capacity. This was evidenced by Ca-alginate biosorption capacity of 0.0534 mg Cu2+/g biosorbent which is smaller than the biosorption capacity of immobilized biosorbents with biosorbent concentration therein. The ability of Ca-alginate beads to absorb heavy metals was obtained from the active side found in alginate polymer compounds Beads.

The concentration of biomass in polymer solution can directly affect the immobilized biosorbent ability to adsorb the adsorbate. Increased biosorption capacity from 0.0534 to 0.1025 mg Cu2+/g biosorbent turned out when the immobilized biosorbent concentration was increased from 0 to 0.5 g/g. This increased capacity was due to the increased surface area of the adsorbent and the availability of more active adsorption side along with the increased concentration of biosorbent[6]. A further increase of immobilized biosorbent concentrations from 0.5 to 1 g/g showed a decrease in biosorption capacity from 0.1025 to 0.0868 mg Cu2+/g biosorbent. This is due to the disruption of the bond between immobilized biomass with Ca-alginate and decreased permeability of beads.

Decreased biosorption capacity at a 0.5 to 1 g / g immobilized biosorbent concentration was also due to the viscosity of the slurry formed by a mixture of biomass with an alginate polymer solution. The quantity of biomass mixed into the alginate solution may affect the viscosity / thickness of the solution. The more the biomass load in the polymer solution the more difficult it is to form beads. Excessive concentrations of biomass can lead to reduced porosity of the beads thus slowing down the adsorption process. The porous structure will provide opportunities for biomass that are within the beads to participate in metal ion absorption. Biosorbent with a concentration of 0.5 g/g of polymer has a lower capacity due to the limited diffusion of metal ions by functional groups on the cell surface. The results of this study are similar to the research of Abdel-Razek [6] which shows that the biosorption capacity increases to the maximum along with increasing the biomass concentration up to 0.75 g/g, then increasing the biomass concentration in the polymer further leads to decreased biosorption capacity. Researchers [6], [7] suggests that this event may be due to an interaction between the active

IOP Conf. Series: Earth and Environmental Science **106** (2017) 012094 doi:10.1088/1755-1315/106/1/012094

side of excessive biomass and the active side of the polymer, so that the free active side of the biomass and the polymer is decreasing.

4. Conclusion

Biomass concentrations trapped by alginate polymers in immobilized biosorbent formation have an effect on the absorption of Cu^{2+} metals. The highest removal efficiency of heavy metals Cu^{2+} of 90.69% occurred by immobilized biosorbent with a concentration of 0.5 g/g. Biomass concentration also influences the biosorption capacity, i.e. the higher the biomass concentration of 0.25 to 0.5 g/g, the greater the biosorption capacity reaches 0.0068 mg/g, but the capacity decreases with increasing biomass concentration of 0, 5-1 g/g of polymer. The result of SEM analysis shows the morphological changes of the immobilized biosorbent surface before and after the process of heavy metal biosorption of Cu^{2+}. Morphological changes occur because of the interaction between heavy metal ions Cu^{2+} and the active side on the immobilized biosorbent surface. These changes cause the pore to surface in the immobilized biosorbent by biomass so that the pore structure appears to be getting smaller.

References

[1] Rinanti A, E Kardena, D I Astuti, K Dewi. 2013. Growth Response and Chlorophyll Content of Scenedesmus Obliquus Cultivated in Different Artificial Media. *Asian Journal of Environment Biology*. 1(1): 1-9.

[2] Abbas H Salman, Ismail M Ibrahim, Mostafa M Tarek, Sulaymon H Abbas. 2014. Biosorption of Heavy Metals : A Review. *Journal of Chemical Science and Technology*. 3(4):74-102.

[3] Amim Jr J, Petri DFS, Maia FCB, Miranda PB. 2010. Ultra thin cellulose ester films: preparation, characterization and protein immobilization. Quím. Nova. 33:2064-2069.

[4] Gorecka E, Jastrzębska M. 2011. Immobilization techniques and biopolymer carriers. *Biotechnol. Food Sci*. 75:65-86.

[5] Godjevargova T, Mihova S, Gabrovska K. 2004. Fixed-bed biosorption of Cu2+ by polyacrylonitrile-immobilized dead cells of Saccharomyces cerevisiae. *World Journal of Microbiology & Biotechnology*. 20:273-279

[6] Abdel-Razek A S. 2011. Removal of chromium ions from liquid waste solutions using immobilized Cunning hamella elegans. *J Nature Science*. 9:211-219.

[7] Covizzi L G, Giese E C, Gomes E, Dekker R F H, Silva R. 2007. Immobilization of microbial cells and their biotechnological applications. *Seminar : Ciências Exatas e Tecnológicas*. 28:143-160.

IOP Conf. Series: Earth and Environmental Science **106** (2017) 012095 doi:10.1088/1755-1315/106/1/012095

Batch leachate treatment using stirred electrocoagulation reactor with variation of residence time and stirring rate

I S Sitorus, W Astono, B Iswanto*

Environmental Engineering Department, Faculty of Landscape Architecture and Environmental Technology, Universitas Trisakti, Jakarta, Indonesia

*Corresponding Author: biswanto@trisakti.ac.id

Abstract. This study aims to reduce pollutant levels of the leachate by electrocoagulation method using a stirred electrocoagulation reactor as the electrochemical water treatment. The release of active coagulants as metallic ions took place in the anode, while in the cathode, the electrolysis reaction in the form of hydrogen gas dischargeoccurred. The source of wastewater is Waste Water Treatment Plant inlet III of Bantar Gebang, Bekasi. Some parameters were analyzed in this research, i.e., Chemical Oxygen Demand (COD), Biochemical Oxygen Demand (BOD), NH_3, NO_3^-, NO_2^-, N-total, and organic substances as well as the microorganism growth before and after electrocoagulation, with variations of detention time (seconds) of 10, 20, 120, 600 and rapid mixing conditions (rpm) of 60, 100 and 200. The results show that the greater the rapid mixing speed and the detention time of electrolysis, the higher the removal of contaminants in liquid waste. The optimum condition of electrocoagulation was encountered at 200 rpm rapid mixing with 600 seconds of processing time. The removal efficiencies of electrocoagulation method for each parameter are TSS of 46.80%, BOD_5 of 71.33%, COD of 73.77%, Pb of 62.5%,and NH_3-N of 57.92%,whereas the pH value has been increased from 8.03 to 8.95. The electrocoagulation method can reduce levels of pollutants, complying with the environmental standards.
Keywords: electrocoagulation, leachate, optimum codition

1. Introduction

Integrated Waste Treatment Site (TPST) is the location of final waste treatment establishedin a region, such as urban areas. One of the main problems in the operation of TPST is the emerging leachate - a liquid waste that arises from the inclusion of water into the pile of waste and dissolves chemical elements including decomposed organic matters [1], [2].

Among various methods of leachate processing, the electrochemical application of electrocoagulation can be applied to reduce organic contents in leachate through the oxidation process [3], [4]. Wholly occurred in water, electrocoagulation requires energy in the form of electricity to change the water content chemically [2], [5]. [6],

An electrocoagulation reactor is an electrochemical cell wherein the anode (usually using aluminum or iron) is used as an agitation agent, according to the following reactions [7], [8].

$$Al(s) \rightarrow Al^{3+} + 3e \qquad (1)$$

$$Cathode: \ 2H_2O \ \rightarrow + 2e - H_2(g) \ + \ 2OH^- \qquad (2)$$

IOP Conf. Series: Earth and Environmental Science **106** (2017) 012095 doi:10.1088/1755-1315/106/1/012095

$$Anode: 2H_2O \rightarrow O_2\,(g) + 4H^+ + 4e^- \quad (3)$$

Total reaction:

$$2Al\,(solid)^* + 6H_2O \rightarrow 3H_2\,(g) + 6OH^- + 2Al^{3+} \quad (4)$$

(*) the formed $O_2$is so small that the decomposition of H_2O in the anode is ignored.

It can be seen in the reactions occurred in the electrocoagulation process that the cathode will produce hydrogen gas and metal ions. Meanwhile, the anode will generate halogen gas and developflocculation deposition.

The electrocoagulation process is performed in an electrolysis vessel in which there is a cathode and an anode as direct electric current conductors called electrodes, immersed in the waste solution as the electrolyte. Sincethe electrocoagulation process produces gas in the form of bubbles, the dirt molded in the water will be lifted up to the water surface.The flocs have a relatively small size, but over the processing time, their size will be increased. After the water undergoes electrocoagulation, the precipitation will take place to precipitate the particles or flocs formed earlier. Subsequently, the effluent produced will be analyzed in the laboratory.

The leachate treatment by batch electrocoagulation method using variations of rotation speed and residence time has been studied deeper in this research.

2. Research Method

2.1. Reactor Description

This laboratory scale research, the electrocoagulation reactor is a container made of PVC pipe with a thickness of 4 mm (Figure 1). Inside the reactor there is a 1mm aluminum that acts as anode and stainless steel stirrer (length = 40 cm; thick = 1 mm). In addition, there is a gear motor (DC motor), connected to the ampere meter, to drive the paddle inside the reactor to process the waste inside. For the outside of the reactor itself consists of inlet and outlet taps as water channels prior to processing and after processing, as well as a gas valve to determine whether any gases are formed from the process proceeds within the reactor. There is also a voltage generator (power supply) that serves as a regulator of the speed of rotation (rpm). Volume of sample that can be processed in reactor as much as 5.5 L.

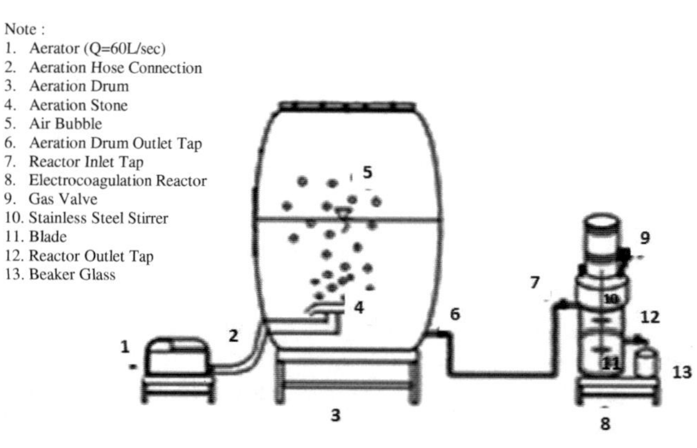

Figure 1. Design of electrocoagulation reactor.

2.2. Characteristics of leachate samples

This research was conducted on a laboratory scale using two variables (residence time and rotation speed) and eight measured parameters of nitrate (NO_3^-) and nitrite (NO_2^-) by colorimetry/spectrophotometry, ammonia (NH_3) by spectrophotometry in phenate, COD by titrimetric

IOP Conf. Series: Earth and Environmental Science **106** (2017) 012095 doi:10.1088/1755-1315/106/1/012095

closed reflux method, BOD by Winkler method, organic matter content ($KMnO_4$) by permanganometric titration, N-total by Kjeldahl method, and the microorganism growth (VSS).

Table 1. Leachate characteristics of TPST Bantar Gebang, Bekasi.

Parameters	Range of concentration (mg/L)
Nitrate	127,92
Nitrite	1.7–2.3
Ammonia	47,83
COD	4.043
BOD	1.462
VSS	738
$KMnO_4$	1.896

2.3. Aeration

The aeration process is the process of adding air oxygen gas into the waste water to remove organic or odorous substances caused by anaerobic organic compounds, eg NH_3, H_2S and CH_4. Before being treated in an electrocoagulation reactor, the first leachate is aerated. Aeration treatment is the traditional way to treat leachate. This method is effective in removing the dissolved organic pollutants present in the leachate. Leachate from IPAS III aerated (60L/min air supply) for 15 hours.

2.4. Electrocoagulation Reactor

Electrocoagulation reactor is a container made of PVC pipe with laboratory scale. The volume of the selected reactor is 5.5 liters. Electrocoagulation is equipped with a stainless steel 40 cm (anode), aluminum (cathode), manometer as a gas pressure gauge and inlet taps and outlets.

2.5. Operation of the Reactor

After the aeration phase is complete, core research is conducted. The purpose of this research is to know the optimum condition of leachate process by using electrocoagulation reactor with variation of rotation speed (60 rpm, 100 rpm and 200 rpm) and residence time (10 seconds, 20 seconds, 120 second and 600 second) (Figure 2).

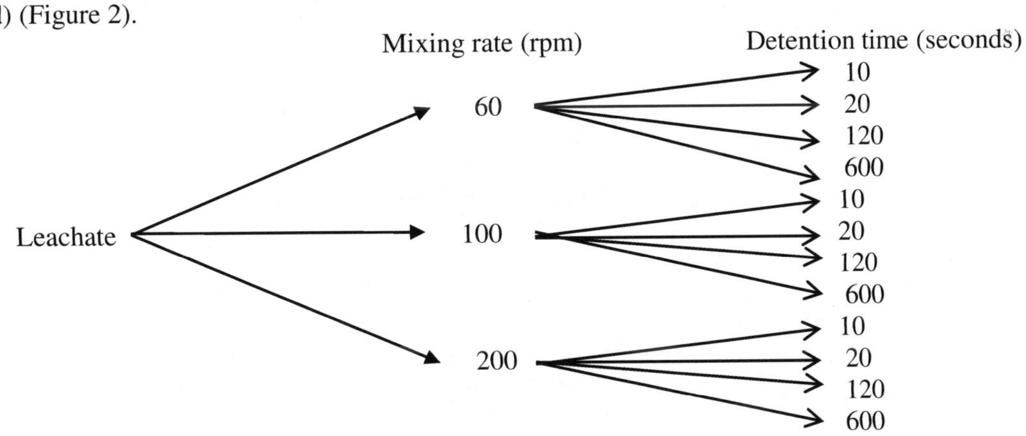

Figure 2. The electrocoagulation plan.

3. Results and Discussion

The wastewater samples were taken from the IPAS III inlet pool of TPST Bantargebang, Bekasi, by reason of location selection and the closer distance than the other IPAS locations. IPAS III inlet pools accommodate leachate produced by waste in Zone 3.

3.1. Effect of Residence Time and Stirring Rate Variation on COD

IOP Conf. Series: Earth and Environmental Science **106** (2017) 012095 doi:10.1088/1755-1315/106/1/012095

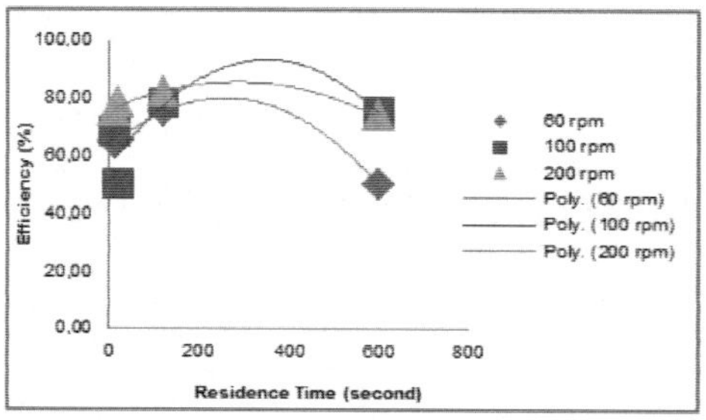

Figure 3. Relation of stirring speed variation and residence time to Efficiency (%) decrease in COD.

Based on the results of Figure 3 it can be concluded that at 200 rpm stirring speed and residence time of 120 seconds, COD concentration decreased to 714 mg/L. When compared to the Decree of Governor No. 69 of 2013 on waste water quality standards for activities and/or businesses, the concentration of COD leachate processed products is still above the standard quality of 100 mg/L. In addition to the most optimum removal efficiency pattern based on the above graph obtained by 93.24% with a stirring speed of 100 rpm and residence time of 354 seconds.

3.2. Effect of Residence Time and Stirring Rate Variation on BOD

Based on Figure 4, it can be concluded that at 200 rpm stirring rate and 60 second stay time decrease BOD concentration by 183 mg/L with removal efficiency equal to 94.67%. In addition to the optimum removal efficiency pattern based on the above graph obtained at 60 rpm stirring rate with residence time of 311 seconds with allowance efficiency of 99.79%.

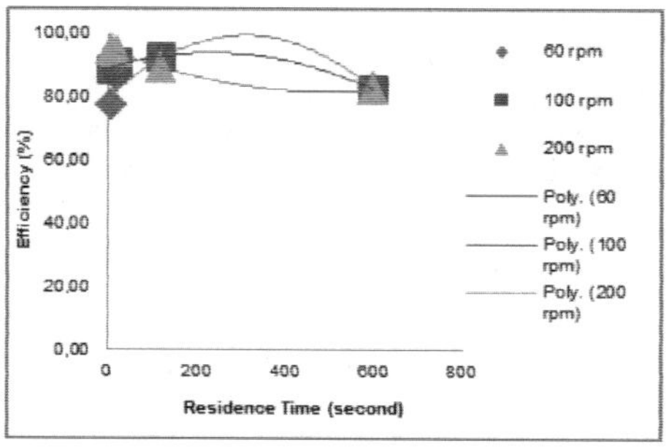

Figure 4. Relation of stirring speed variation and residence time to Efficiency (%) decrease in BOD.

3.3. Effect of Residence Time and Stirring Rate Variation on nitrate (NO_3^-)

Based on Figure 5, it can be concluded that at 60 rpm stirring speed and residence time of 20 seconds can decrease nitrate concentration (NO_3^-) of 92 mg/L with an allowance of 28.41%. For the most optimum removal efficiency pattern on nitrate removal has not occurred with a residence time of 600 seconds. That's because the H_2 gas that is formed is very small or the increase in H_2 gas is very slow. The size of H_2 gas formed is influenced by Al^{3+} formed. So it needs to do the process back with a residence time above 600 seconds to get the optimum condition of the nitrate removal percentage.

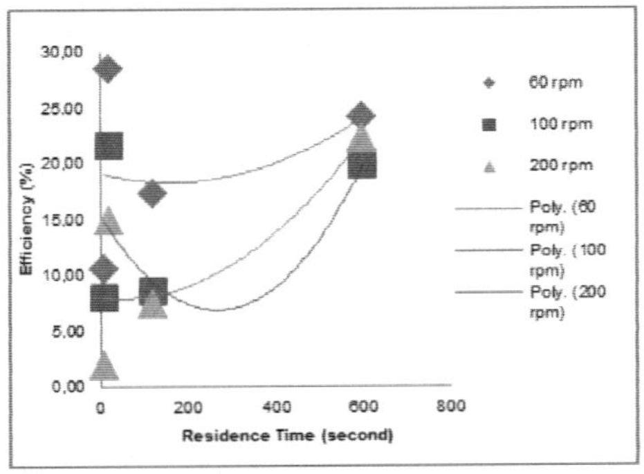

Figure 5. Relation of stirring speed variation and residence time to Efficiency (%) decrease in nitrate.

3.4. Effect of Residence Time and Stirring Rate Variation on ammonia (NH_3)
The aeration process causes ammonia (NH_3) to nitrate. As a product resulting from the aeration process, nitrate is relatively harmless to aquatic life. In detail, the picture of NH_3 change reaction is presented as follows:

$$NH_3 \text{ -------- } NH_4^+ \tag{5}$$

$$NH_4^+ + 2O_2 \text{ ------- } 2H^+ + NO_3^- + H_2O \tag{6}$$

Based on the aeration process performed for 15 hours, the ammonia concentration on effluent produced by air giving at a rate of 60 liters/minute decreased to 36.15 mg/L from an initial concentration of 47.83 mg/L. In the electrocoagulation process, ammonia does not change. That's because O_2 gas in the reactor is very small. In addition, H_2 gas formed in the electrocoagulation process can not react with ammonia, so there is no allowance process.

3.5. Effect of Residence Time and Stirring Rate Variation on organic matter ($KMnO_4$)
In the electrocoagulation process, organic matter ($KMnO_4$) is unchanged or unaffected by stirring rate and residence time. It can be seen from the results of organic matter analysis ($KMnO_4$), the decrease is not significant (Figure 6).

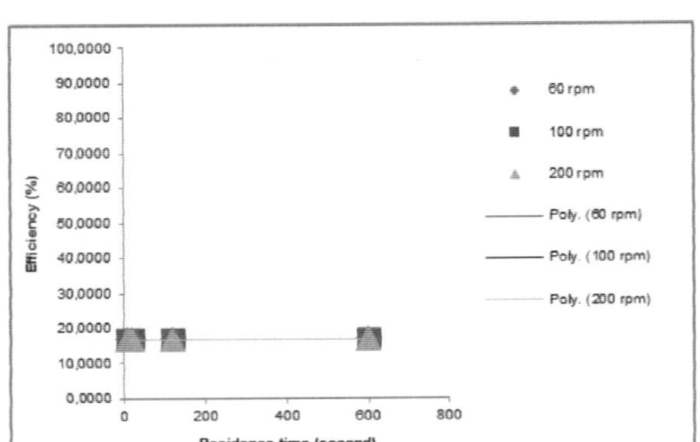

Figure 6. Relation of stirring speed variation and residence time to Efficiency (%) decrease in organic matter ($KMnO_4$).

3.6. Effect of Residence Time and Stirring Rate Variation on organism (VSS)

Based on Figure 7, it can be concluded that at 100 rpm stirring speed and residence time of 120 seconds can decrease VSS concentration of 348 mg/L with allowance removal of 52.85%. While at 200 rpm stirring speed is not likely to occur VSS large allowance because at this speed the microorganisms disrupted and did not have time to develop.

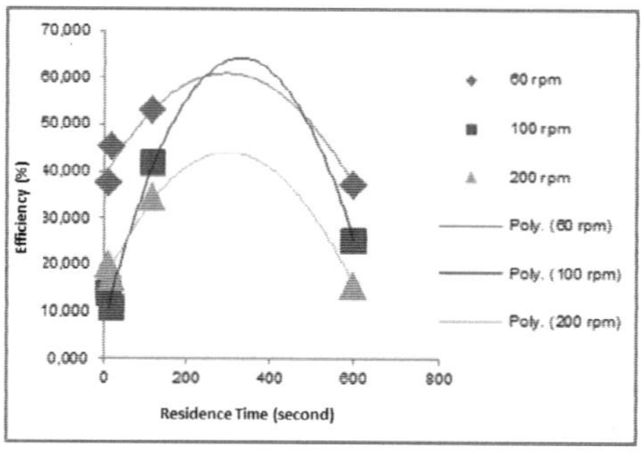

Figure 7. Relation of stirring speed variation and residence time to Efficiency (%) decrease in organism (VSS).

4. Conclusion

The results showed that the characteristic of leachate IPAS 3 TPST Bantar Gebang, Bekasi has concentration of COD, BOD, Nitrate (NO_3^-), Ammonia (NH_3), organic substance ($KMnO_4$) and increase of microorganism (VSS) that is equal to 4,043 mg/L, 1,462 mg/L, 127.92 mg/L, 1,896.17 mg/L, and 738 mg/L. The most optimum removal efficiency pattern with 60 rpm speed can set aside BOD parameter 99.79% for 311 second and nitrate equal to 41.12% for 288 second. For 100 rpm speed COD parameter can be set aside 93.24% for 354 seconds and VSS of 66.77% for 326 seconds.

References

[1] Kementerian Lingkungan Hidup, Peraturan Menteri Lingkungan Hidup Republik Indonesia Nomor 59 Tahun 2016 Tentang Baku Mutu Lindi bagi Usaha dan/atau Kegiatan Tempat Pemroresan Akhir Sampah, Jakarta, 2016.

[2] Un, U T, Koparal, A S, and Ogutveren, U B. January 2009. Electrocoagulation of Vegetable Oil Refinery Wastewater Using Aluminium Electrodes. *Journal of Environmental Management.* 90(1):428-433.

[3] Firmansyah R M, 2015, Pengolahan Air Sampah Tempat Pengolahan Sampah Terpadu (TPST) Bantargebang Bekasi, Laporan Kerja Praktik. Program Diploma Teknik dan Manajemen Lingkungan Institut Pertanian Bogor, 2015.

[4] Riyanto, Elektrokimia dan aplikasinya. Yogyakarta: Graha Ilmu, 1-7, 2013.

[5] Raju, G B, Karuppiah, A T, Latha, S S, Parvathy, S and Prabhakar, S. October 1, 2008.Treatment of Wastewater from Synthetic Textile Industry by Electrocoagulation Electrooxidation.*Chemical Engineering Journal*. 144(1):51 – 58.

[6] Shikoku Industry, and Technology Promotion Center, Januari 26, 2010, Industrial Waste Water Treatment in Japan.

[7] Triwulandari R, Pahlevi MN, Mirwan A. Dec 2012. Pengambilan Logam Cr6+ dan Cr Total dari Limbah Industri Elektroplating secara Elektrokoagulasi. Konversi, 1(1): 46.

[8] Hernandez I L, Diaz C.B, Morales, G.r, Bilyue B, and Nunez, F U. May 1 2009. Influence of the Anodic Material on Electrocoagulation Performance. *Chemical Engineering Journal.* 148(1):97 – 105

Effects of *Clitoria ternatea linn.* flower juice on the elasticity, texture and pores of the denture retainer component

E Erfan[1*], C Cudharma[2]

[1]Oral Biology Department, Faculty of Dental, Universitas Trisakti Jakarta, Indonesia
[2]Clinical Programme, Faculty of Dental, Universitas Trisakti Jakarta, Indonesia

*Corresponding Author : erni.erfan@trisakti.ac.id

Abstract. *Clitoria ternatea* flower has a variety of antioxidant compounds. To analyze the effects of the flower juice on the elasticity, roughness and pore components of partial denture retainer (stainless steel wire). 0.8 cm of retainer wire components were soaked in the juice (1 g/mL) for 4, 6, and 8 hours at room temperature. Polident and mineral water were used as controls. Measurement of the elasticity, texture and pore of each component based on the addition of wire length, number of pixels, and pores average. *C. ternatea* flower juice showed no statistically significant difference effects among all type solutions on the elasticity, texture and pore of the retainer component. *C. ternatea* flower juice did not cause alteration in the retainer component in terms of elasticity, texture and pore.
Keywords: *Clitoria ternatea,* elasticity, pore and wire, roughness

1. Introduction

C. ternatea flower contained phenolic compounds, flavonoid, anthocyanin, saponin, alkaloid, and anthocyanidin.[1, 2] Stainless steel wire is most used in prosthodontic and orthodontic treatment due to its content and excellent corrosion resistance in the oral environment [3]. The wire's components was 17-20%, 8-12%, and 0.15% for was chromium, nickel and carbon respectively [4]. It is imperative to find natural materials which through a simple method can be processed as a cleaning solution for the denture retainer. So that maintenance cost of it can be reduced.

1.1. Conceptual framework

Making artificial teeth has the purpose to improve some functions, the functions are mastication, phonetics, and aesthetic. In addition to maintaining the health of tissues in the oral cavity against further damage. There are a variety of dentures one of which is partially removable partial denture. The denture retainer is a partially removable denture component as part of a denture used for the purpose of obtaining stability or denture fixation [5]. Elasticity, texture, and pores of the denture retainer are the essential factors in determining the denture retainer quality and also affects its corrosion behavior, biocompatibility and supporting in the health oral cavity of a user [3], [6]

2. Research Method

2.1. Used materials
Clitoria ternatea (Ct) flower, Stainless steel wire (AISI 302), mineral water, Polident tablet.

2.2. Preparation of Ct flower juice.
Ct flower, harvested in May where growth in Jakarta. Stainless steel wire with 0.8 mm in length. The flower part used was the crown.

After ten of the crowns are washed cleanly, they are immersed in sterile water for 5 minutes. Then the crowns of the flower are squeezed to get 9 mL its juice. The juice was filtered using filter paper and then filtered through milli pores filter (ϕ = 0.22 μm, Sartorius) to prevent contamination of the juice. The final volume of the juice was 8 mL which was divided into four tubes. The tubes were tube I, II, III as a place of wire which soaked in the water for 4, 6, 8 hours and tube C1 as an internal control. There were two other tubes. They were filled the wire which was soaked in Polident solution and mineral water respectively as the positive control (C2) and negative control (C3).

2.3. Test Performed
The elasticity, texture and pore of the wire were performed by accurately observed using optical an microscope . The microscope oriented at 90 degress to the wire surface.[3] All the images were captured at 50x magnification. The elasticity is measured based on the change in the wire length. The measurements were done before and after (4, 6, and 8 hours) immersion. The imaging data of microstructural features of the wire were analyzed further more by ImageJ plugin version 1.46r to obtaian numerical data in the ratio scale [7]. Furthermore, the data were analyzed statistically using repeated ANOVA ($p>0.05$) to find out the difference between treatment and control group data.

3. Results and Discussion

Tabel 1 shows the length of the tested wire before and after soaked in the juice of *C. ternatea* flower on three repetitions of measurements. The duration of immersion time is used base on the length of time a person eats his food in a day (4-6 hours) (the time when removable orthodontic users do not use their wire) and duration of teens and adult human sleep in a day (± 8 hours) [8]. While sleeping a removable prosthodontic users does not use the dentures. The measurements are intended to know the elasticity of the wires. We did not calculate their modulus of elasticity. Table 1 shows that no tested wire underwent a lengthy alteration. It describes that the three cleaning solutions did not affect the elasticity of the wire for 8 hours immersion period.

Table 1. The length of the wire before and after soaked in the juice of *C. ternatea* flower.

	Length of the wire (cm) after soaked in the juice			
	0 hour	4 hours	6 hours	8 hours
I	0.800	0.800	0.800	0.800
II	0.800	0.800	0.800	0.800
III	0.800	0.800	0.800	0.800
C2	0.800	0.008	0.800	0.800
C3	0.800	0.800	0.800	0.800

The calculated surface roughness values were; arithmetical mean deviation (Ra), root mean square deviation (Rq), kurtosis of the assessed profile (Rku), skewness of the assessed profile (Rsk), lowest valley (given by the min measurements) (Rv), highest peak (given by the max measurements) (Rp), and the total height of the profile (Rt) [9]. The value of the parameters described the surface texture of the tested wire. All values of the seven parameters of each soaked wire are presented in Tables 2.a. and 2.b. The data in both tables are obtained respectively based on value of Gaussian filtering (GF) and Fast Fourier Transform (FTT) bandpass filtering results.

Table 2. Average roughness parameters of the wire at before and after soaked in the juice; (a) Gaussian Filtering (GF); (b) Fast Fourier Transform (FTT) bandpass filtering.

(a) GF for roughness parameters

	hrs	Ra	Rq	Rku	Rsk	Rv	Rp	Rt
I	0	99.426	100.794	1.117	1.042	60	164	224
	4	99.426	100.794	1.117	1.042	60	164	224
	6	141.258	146.076	1.167	1.068	7.333	255	262.333
	8	153.072	157.939	1.133	1.057	0	245.667	245.667
II	0	130.256	153.874	1.198	1.081	1.333	255	256.333
	4	130.256	153.874	1.198	1.081	1.333	255	256.333
	6	148.784	154.017	1.154	1.064	2	255	257
	8	143.733	149.063	1.146	1.063	0	255	255
III	0	132.797	140.521	1.290	1.117	0	255	255
	4	132.797	140.521	1.290	1.117	0	255	255
	6	136.654	143.123	1.233	1.094	0	255	255
	8	134.411	141.399	1.196	1.084	0	254.667	254.667
C2	0	134.113	139.277	1.171	1.071	0	255	255
	4	134.113	139.277	1.171	1.071	0	255	255
	6	136.654	143.123	1.233	1.094	0	255	255
	8	154.256	160.147	1.158	1.068	0.333	255	255.333
C3	0	98.300	111.766	1.818	1.292	2.684	245.129	247.813
	4	98.300	111.766	1.818	1.292	2.684	245.129	247.813
	6	146.419	151.683	1.140	1.062	0.667	253.667	254.333
	8	135.879	145.675	1.283	1.121	0	255	255

(b) FFT bandpass filtering for roughness parameters

	hrs	Ra	Rq	Rku	Rsk	Rv	Rp	Rt
I	0	0.006	0.046	89.700	9.092	0	1	1
	4	0.006	0.046	89.700	9.092	0	1	1
	6	3.929E-4	0.005	698.496	16.191	0	1	1
	8	0.006	0.046	89.600	9.085	0	1	1
II	0	0.006	0.046	89.700	9.092	0	1	1
	4	0.006	0.046	89.700	9.092	0	1	1
	6	3.929e-4	0.005	698.496	16.191	0	1	1
	8	3.929e-4	0.005	698.496	16.191	0	1	1
III	0	0.006	0.046	89.700	9.092	0	1	1
	4	0.006	0.046	89.700	9.092	0	1	1
	6	3.926e-4	0.005	336.370	14.644	0	1	1
	8	0.006	0.046	89.700	9.092	0	1	1
C2	0	3.926E-4	0.005	336.370	14.644	0	1	1
	4	3.926E-4	0.005	336.370	14.644	0	1	1
	6	3.926E-4	0.005	336.370	14.644	0	1	1
	8	0.006	0.046	89.600	9.085	0	1	1
C3	0	3.929E-4	0.005	698.496	16.191	0	1	1
	4	3.929E-4	0.005	698.496	16.191	0	1	1
	6	3.929E-4	0.005	336.370	14.644	0	1	1
	8	0.006	0.046	89.700	9.092	0	1	1

Table 2 showed the surface roughness parameter data based on Gaussian filtering techniques (sigma value = 5) and Fast Fourier Transform (FFT) with bandpass filter size of 10 x 20 pixels with 178.75 pixels/mm. The homogeneity of Table 2 data did not differ significantly.

The pore diameters of the tested wires at before and after immersed in the three cleaning solutions ranged from 30 to 16 μm. The results of statistic test showed that there is no difference of length, surface roughness and pore diameter between wires soaked in the juice of flower of Ct, Polident and mineral water significantly.

The gold standard for morphological measurements for the nano-sized component parts is the image of the electron microscope [10]. However, the use of Scanning Electron Microscope for the purpose of measurement began to be abandoned with increasing the ability of optical microscopes and the increasing functionality of plugins on the software to analyze images such as Image J. Our study proves that the images obtained from optical microscopes can show variations of data although statistically not significantly different. It can be seen from the data contained in Table 2.a., and Table 2.b [11]. Ra is a roughness parameter used to control the quality of roughness in general. Ra is easily defined and measured and provides a general description of high variations [12]. Although statistically insignificant, the data in Table 2.a, shows a change in Ra value on wire immersed in Polident and mineral water after 6 and 8 hours of immersion. While the data in Table 2.b, shows that there is a change of Ra value on the wire soaked in Polident and mineral water, after 8 hours of immersion mass.

Rku parameter illustrates the sharpness of the probability density of the amplitude probability density function profile. For Rku in the GF for roughness parameters, if the value is less than three then the distribution curve is said to be platykurtoic which has relatively few high peaks and low valleys [12]. All of Rku values in Table 2.a, data show that the value of all wires at all immersion periods is less than 1. It can be stated that all the wires tested at all periods of immersion in all types of submersion solutions are not found in parts which have relatively much high peaks and low valleys The largest variation in Rku values occurs in the data group of wires immersed in mineral water.

Table 2.b data for the Rv and Rp parameters, show the same value for all the soaked wire each is zero. However in table 2.a the data for both parameters are more varied although not statistically significant. The sum of Rv and Rp parameters values is Rt [12]. So the same thing happens to the value of the Rt parameter. This shows no 'perforation on all surfaces of the wires after immersion in all periods of immersion.

4. Conclusions

The results of this study proved that Ct flower juice did not cause the alteration in the retainer component in terms of elasticity, texture, and pore. Although the data obtained was the test results performed for 8 hours but enough to be used serve as a scientific base to utilize Ct which is a native Indonesian plant as the main ingredient of a cleaning solution.

Acknowledgment

This study was supported by a Grant-in-Aid from the Ministry of Research, Technology and Higher Education of the Republic of Indonesia.

References

[1] Kaisoon O, Siriamornpun S, Weerapreeyakul N, and Meeso N. 2011. Phenolic compounds and antioxidant activities of edible flowers from Thailand. *Journal of Functional Foods.* 3:88-99

[2] Nair V, Bang W Y, Schreckinger E, Andarwulan N, and Zevallos C. 2015. Protective Role of Ternatin Anthocyanins and Quercetin Glycosides from Butterfly Pea (Clitoria ternatea Leguminosae) Blue Flower Petals against Lipopolysaccharide (LPS)-Induced Inflammation in Macrophage Cells. *J. Agric. Food Chem.* 63:6355-6365

[3] Elayyan F, Silikas N, and Bearn D. 2008. Ex-vivo surface and mechanical properties of coated orthodontic archwires. *European Journal of Orthodontics.* 30:661-667.

[4] Santander SA, and Luna-Ossa C M. 2015. Stainless Steel: Material Facts for the Orthodontic Practitioner. *Revista Nacional de Odontología.* 11:20

[5] Aquino Souza J E, Silva N R F A, Coelho P G, Ferracioli ACZRCSR, and Ricardo Alexandre Zavanelli. 2011. Retention Strength of Cobalt-Chromium vs Nickel-Chromium Titanium vs CP Titanium in a Cast Framework Association of Removable Partial Overdenture. *Journal of Contemporary Dental Practice.* 12:179-186

[6] Vcnab A, Carey J, and Badawic H. 2007. Clinical Variability in Arch Wires: A Preliminary Study Evaluating Mechanical and Surface Characteristics of Two Different Sized Rectangular Stainless Steel Wires. *Open Biomedical Engineering Journal.* 1:13-22

[7] Heima K, Bernierb F, Pelletierb R, and Lefebvreb L P. 2016. High resolution pore size analysis in metallic powders by X-ray tomography Case Studies in Nondestructive Testing and Evaluation. 6:46-52

[8] Watson NF et. al., 2015 Recommended Amount of Sleep for a Healthy Adult: A Joint Consensus Statement of the American Academy of Sleep Medicine and Sleep Research Society. *Journal of Clinical Sleep Medicine*.11:591-592.

[9] Ondimu S, and Murase H. 2008. *Proc. Int. Conf. of the 17th World Congress,* (Seoul : The Int. Federation of Automatic Control) pp 641-646.

[10] Hotaling NA, Bharti K, Kriel H, and Simon Jr CG. 2015. Diameter J: A Validated Open Source Nanofiber Diameter Measurement Tool. *Biomaterials.* 61:327-338.

[11] Rafi K, Development of Porous TiO_2 Layer on Ti6AL4V for Bone Implants. Biotechnology and Medical Engineering National Institute of Technology Rourkela, p 25, 2014.

[12] Gadelmawla E S, Koura M M, Maksoud T M A, Elewa I M, and H H Soliman. 2002. Roughness parameters. *Journal of Materials Processing Technology.* 123:133-145.

Study on financial risk towards individual investor as strategy to improve urban community empowerment

F M Leon*, A Aprilia

Faculty of Economic and Business, Universitas Trisakti, Jakarta

*Corresponding Author: farahmargaretha@trisakti.ac.id

Abstract. Investor will be always influenced by its risk tolerance when investing, each investor has own risk tolerance that differ to another, although this still being questioned until now. This research aimed to know the influence of demography factor in distinguish and classify Financial Risk Tolerance (FRT) and Financial Risk Taking Behavior (FRB) to individual investor. Methodology in this research is data that used as primary data which distributed by offline and online. The sample in this research is 642 respondents in Jakarta. Logistic regression is analyze method that used in this research. The research found that there is influence of gender, marital, status, education and income level to Financial Risk Tolerance (FRT) and Financial Risk Taking Behavior (FRB). For FRT significantly 0.000 for gender and marital status; 0.010 for education and 0.001 for income level. Whereas for FRB significantly 0.000 for gender; 0.003 for marital status and 0.010 for education level. The research contribution is crucial for financial advisor to notice the characteristic investor based on demography factor such as gender, marital, status, education level and income level. Therefore, this research able to give optional decision for appropriate investment to clients as ones of strategy to improve urban community empowerment.
Keywords: demography factor, financial risk taking behavior, financial risk tolerance

1. Introduction

Investments are placements of a certain amount of funds in a given asset in the hope of gaining profit in the future. According to [1], investment is a commitment to a certain amount of money or other resources currently undertaken for the purpose of obtaining benefits in the future. Investments are conducted with the aim of improving the welfare of investment doers. Investors such as employees, entrepreneurs to pensioners invest in different levels and forms of investment. There are so many forms of investment that can be done, from opening savings and/or bank deposits, buying precious metals or gold and jewelry, buying stocks or bonds, buying mutual funds, buying land, homes or many now apartment. But before investing the investor will be faced with various investment options that also have different levels of risk.

In making a decision to invest investors will be faced with uncertainty where the possibility of the outcome of the investment is not in line with expectations or even become a loss. Investors will not only take advantage of investment but also consider the possible risks to be faced from an investment where the level of risk acceptable or tolerated by each investor varies depending on the Financial Risk Tolerance (FRT) level. FRT is the willingness of an investor to accept a negative change or decline in value of investment or loss resulting from investments made that differ from expectations [2]. Every investor has tolerant and attitudes to different risks, where the high risks considered by an investor can be regarded as a small risk to other investors [3]. In making a decision to invest, investors will be influenced by thc FRT level. FRT is a key element and plays an important role to consider in making investment decisions to achieve financial goals for both investors and investment managers [3].

In making a decision to invest an investor can consider various factors of consideration with himself or with the help of others such as financial advisor. Assigning investors into appropriate risk tolerance categories and then showing the portfolio of investments that best suits them is an important task of investment managers and advisers [3]. However, most financial advisors and / or investors often miss-equate Financial Risk Tolerance (FRT) with the Financial Risk Behavior (FRB) of an investor [4]. According to [5] in financial management in general and especially on money management, behavior can be defined as goal-oriented or whereby behavior leads to actions that can lead to positive or negative outcomes. Behavior can be seen from the point of view of the consequences or the outcome of a decision making where it is important to understand the consequences of money management rather than the behavior or actions of money management because if a mismanagement occurs it will increase the likelihood of the occurrence of financial stress [6]. Money management is a daily financial activity (storing and maintaining personal financial records and documents, creating personal financial statements such as balance sheets and cash flows, then creating and executing expenditure and budgeting plans) needed to manage current personal economic resources to get long-term prosperity [2]. The objective of money management is to generate profits and improve welfare [3], but it does not rule out welfare may decrease [3]. A decrease in wealth may cause a person to experience overdue warnings, or even to the point of bankruptcy which is the result or consequence of mismanagement or FRB [6]. It is very important for policy makers and financial advisors to understand how FRTs influence investment decisions in order to improve the quality of their services to clients [10].

Knowing and understanding FRT and financial risk is important in making investment decisions but to know and measure it is not easy to do. According to [3] to measure a person's FRT is not easy because FRT is a complex attitude. Many previous studies have examined demographic factors affecting FRT and financial FRB using variables such as gender, age, marital status, education, occupation and income. Demographics are factors associated with FRT that are often used in research, and there is general agreement among investment managers and researchers that demographic factors can be used to differentiate and classify risk tolerance levels of investors into categories [3]. Yet so far this research aimed to know the influence of demography factor in distinguish and classify Financial Risk Tolerance (FRT) and Financial Risk Taking Behavior (FRB) to individual investor.

2. Research Method

2.1. Financial Risk Taking Tolerance (FRT)

The FRT variable describes the level of ability of an investor to accept risk. This variable is measured by a liker scale of 1 to 4, where 1 = agree, 2 = strongly agree, 3 = disagree, 4 = strongly disagree using 5 question items adopted from [15]. The five items of the question are a) investing is too difficult to understand, b) I feel more comfortable to put my money in the bank than on the stock market, c) when I think about the word risk, then the word loss comes instantly in my mind, d) making money / profits through the purchase of shares and bonds is based on luck factor, and e) in terms of investment, security is more important than the rate of return.

2.2. Financial Risk Taking Behavior (FRB)

FRB variable refers to the result of money management. This variable is measured using 4 items of questions adopted from [5] where the value of 1 if the respondent answered Yes and 0 if the respondent answered No. The items of the question are a) have you ever experienced a wage / salary deduction?, b) have you ever experienced bankruptcy? c) Have you ever received overdue notices from creditors? d) Have you ever experienced vehicle seizure by creditors?

2.3. Demographic Variables, Population and Sample

Demographic variables used include gender, age, status, employment changes affected results. Sampling method in this research is done by purposive sampling that is withdrawal sample based on certain criterion determined by researcher [9]. Here are certain criteria in sampling: a) respondents are domiciled in Jakarta, b) respondents have jobs, c) respondents have at least 2 (two) types of investment. Of the 250 questionnaires distributed only 217 questionnaires were returned but 13 questionnaires were not filled completely so they could not be used for the study. While from the questionnaire online obtained as many as 449 questionnaires so that the total questionnaire received back is as much as 653. Of the total 653 questionnaires returned but only as many as 642 questionnaires that can be used.

2.4. Data Analysis Method

The method of analysis used in this research is to use binary logistic regression analysis using a program of SPSS version 22. The purpose of knowing the influence of independent variable to its dependent variable where the dependent variable is dichotomy variable. In binary logistic regression analysis no normality test is required [11]. The regression equation used in this research is as follows:

$$\text{Ln}\ \frac{p}{1-p} = \beta_0 + \beta_1\ \text{Gender} + \beta_2\ \text{Age} + \beta_3\ \text{Marital Status} + \beta_4\ \text{Education} + \beta_5\ \text{Work} + \beta 6\ \text{Income} \quad (1)$$

or:

$$\text{Ln}\ \frac{p}{1-p} = e\ (-\beta 0 + \beta 1\text{Gender} + \beta 2\text{Age} + \beta 3\text{Marital Status} + \beta 4\text{Education} + \beta 5\ \text{Work} + \beta 6\ \text{Income}) \quad (2)$$

3. Results and Discussion

Regression analysis (Table 1) aims to determine whether there is an influence of independent variables on the dependent variable. The basic decision-making decision are a) if the significance value < 0.05 then there is influence of independent variable to dependent variable, b) if the value of significance > 0.05 then there is no influence of independent variables on the dependent variable.

H_{1a}. There is a gender effect on FRT

From the results of Table 1 obtained value of p-value significance of 0.000 <0.05 indicating that gender has a significant influence on FRT. An odds ratio of 3.884 indicates that men tend to have a higher FRT 3,884 times compared with women. These results are supported by research conducted by [13] who found that men had risk tolerance compared with women. Similar results have also been found by [12] that women have lower levels of risk tolerance than men.

H_{1b}. There is gender influence on FRB

From the results of Table 1 shows a significant value for gender of 0.000 <0.05 which indicates that gender significantly affect FRB. Table results show an odds ratio of 4.395 which means that men tend to have higher FRBs 4.395 times compared with women. This result is reinforced by a study that has been done previously by [6] who found that men had higher FRBs than women.

Table 1. Table of logistic regression analysis result (Source: SPSS data processing).

Dependent Variable	Independent Variable	B	S.E.	Wald	Sig.	Exp (B)
Financial Risk Tolerance (FRT)	Gender	1.357	0.185	53.819	0.000	3.884
	Age	0.016	0.010	2.641	0.104	1.017
	Marital Status	1.066	0.214	24.845	0.000	2.907
	Education	0.840	0.325	6.693	0.010	2.317
	Occupation	0.043	0.233	0.034	0.854	1.044
	Income	0.701	0.219	10.289	0.001	2.017
	Constant	-2.407	0.492	23.967	0.000	0.090
Financial Risk Taking Behavior (FRB)	Gender	1.481	0.177	69.961	0.000	4.395
	Age	0.002	0.010	0.049	0.825	1.002
	Marital Status	0.612	0.204	9.031	0.003	1.844
	Education	0.397	0.321	1.529	0.216	1.487
	Occupation	0.329	0.226	2.122	0.145	1.389
	Income	0.597	0.214	7.812	0.005	1.816
	Constant	-1.768	0.476	13.775	0.000	0.171

H_{2a}. There is an age effect on FRT
From the results of Table 1 it is known that age has a p-value of 0.104> 0.05. This shows that age has no significant effect on FRT. The results of this study differ from the results of research conducted by [6] which in his research found that there is influence of age to FRT. However, the results of this study are supported by previous research results that have been done by [11] who found that there is no significant effect of age on FRT. These results indicate that the elderly or younger a person's age but if not supported by adequate education then it will not have a significant effect on FRT.

H_{2b}. There is an age effect on FRB
From Table 1 obtained significance value of 0.825 > 0.05. This shows that age has no significant effect on FRB. This result is not in accordance with the results of a study that has been done by [6] who found that there is an age effect on FRB. This indicates that a person who is either young or old but if not supported with unmarried status then it will not have a significant effect on FRB. Because a person with an unmarried status has no responsibility to finance household needs, so the more daring to invest in an investment that has a greater risk.

H_{3a}. There is influence of marital status to FRT
From the results of Table 1 obtained the value of significance for marital status of 0.000 < 0.05 indicating that marital status has a significant influence on FRT. The odds ratio of 2.904 indicates that an unmarried person tends to have a higher FRT 2.904 times compared with a married couple. This result is supported by research conducted by study [15] who found that married respondents had lower levels of FRT compared with unmarried respondents. Then the results of this study also supported by the results of research that has been done by study [12] who found that an unmarried person has a higher risk tolerance compared with married couples

H_{3b}. There is influence of marital status to FRB
From the results of Table 1 obtained a significant value for marriage status of 0.003 < 0.05, the marriage status indicates has a significant influence on FRB. The odds ratio of 1.844 indicates that an unmarried person tends to have a higher FRB 1,844 times compared with a married couple. These results are supported by [14] who found that an unmarried person has a high FRB as well.

H_{4a}. There is an influence of education level on FRT
From table 1, there is a significance value for education of 0.010 < 0.05 which indicates that education has an influence on FRT. The odds ratio for education of 2.317 indicates that higher education levels tend to have FRT 2.317 times higher than lower education levels. Similar results have also been found by [5] who found that respondents with higher levels of education had higher risk tolerance than those with lower levels of education. In addition, this result is also reinforced by the results of research conducted by [13] who found that respondents with higher levels of education had higher FRTs than respondents with lower levels of education.

H_{4b}. There is an influence of education level on FRB
From Table 1 obtained significant value for education of 0.216 > 0.05 which means that there is no significant effect of education level on FRB. This result is different from the results of the research that has been done by [16]. However, the results of this study are reinforced by the results of research that has been done by [6] who found that there is no significant effect of education level on FRB. According to [6] this can be attributed to the highly subjective nature of respondents where the likelihood of the respondent being a highly competitive person so as to enhance their competitive advantage tends to avoid FRB.

H_{5a}. There is an influence of the type of work to FRT
From the results of Table 1 for the work obtained a significant value of 0.854 > 0.05. This shows that there is no significant effect of the type of work on FRT. This result is different from the results of previous studies done by [6] who found that there is an influence of the type of work on FRT.

However, the results of this study are supported by the results of a study conducted by [7] who found that there was no significant effect of employment status on FRT. The results of this study indicate that a person who works as a salaried-employee or self-employed but if not supported by income that is considered to have to meet the needs then will not have a significant influence on FRT. Because if the income increases then someone will be more daring to invest in investment products that have a high risk.

H_{5b}. There is an influence of the type of work to FRB

From the results of Table 1 obtained significance value of 0.145> 0.05. This shows that there is no significant effect of the type of work on FRB. This result is different from the results of previous studies that have been done by [6]. The results of this study indicate that a person who works either as a salaried-employee or self-employed but if not supported by marriage status that has not yet married it will not have a significant influence on FRB. Because a person with an unmarried status has no responsibility to finance household needs, so they more daring to invest in an investment that has a greater risk.

H_{6a}. There is an effect of earnings on FRT

From the results of Table 1 obtained significance value for income of 0.001 < 0.05 which means that income has significant effect on FRT. An odds ratio of 2.017 indicates that respondents with higher income levels tend to have FRT 2.017 times greater than respondents with lower income levels. The results of this study are reinforced by the results of studies that have been conducted by [12] who found that risk tolerance increases with increased incomes. According to [6] someone with a high income level is more daring to take a higher risk than someone with lower income.

H_{6b}. There is an effect of earnings on FRB

From the results of Table 1 obtained value significance for income of 0.005 < 0.05 which means that income has a significant influence on FRB. The odds ratio of 1.816 indicates that respondents with higher income tend to have higher FRB at 1.816 times compared with respondents with low income level. These findings are reinforced by the results of research done by [6] who found that there is a significant effect of income on FRB.

4. Conclusion

The research found that there is influence of gender, marital status, education level and income level to FRT and FRB. The managerial implications for investment service providers in assigning investors into the risk tolerance and risk-taking behavior categories should consider the demographic factors of gender, marital status, education and income from investors or clients, as by classifying investors into financial categories risk tolerance and appropriate financial risk-taking can help to produce the most fit and proper investment portfolio for its clients. As for investors to consider the demographic factors, especially income, because the income owned by investors can affect the financial risk tolerance and financial risk-taking behavior. In addition, marital status is also known to have an effect on financial risk tolerance and financial risk-taking behavior. So this study can be implementation as a strategy to improve urban community especially in Indonesia.

References

[1] Bodie K, Kane A, & Marcus A J, Investment, 10th Edition. New York : McGraw-Hill. 2014.

[2] Grable J E, and Lytton R H. FRT 1999. Revisited: The Development of A Risk Assessment Instrument. *Financial Services Review,* 8:163-181.

[3] Anbar A, and Eker M. 2010. An Empirical Investigation for Determining of The Relation Between Personal FRT and Demographic Characteristic. *Ege Academic Review*, 10(2):503-522. Davey G. Measuring Client Risk Tolerance. Reed Business Information Pty Ltd. 16. 2006.

[4] Grable J, and Roszkowski M. 2008. The Influence of Mood on the Willingness to Take Financial Risks. *Journal of Risk Research.* 11(7):905-923.

[5] Kannadhasan M. 2015. Retail Investors FRT and Their Risk Taking Behaviour : The Role of Demographics as Differentiating and Classifying Factors. *IIMB Management Review*, 27:175-184.
[6] Sina P G, and Raturomon LT. 2012. Analisis Stres Finansial Pada Gender. Jurnal Ekonomi & Pendidikan, 9(1):102-112.
[7] Kapoor J R, Dlabay L R, & Hughes RJ. Personal Finance. New York : McGraw-Hill. 2012.
[8] Hermawan Asep. Penelitian Bisnis :Pendekatan Kuantitatif, Kualitatif dan Metode Campuran. Jakarta : Universitas Trisakti. 2013.
[9] Wagner J. An Empirical Analysis Linking A Person's Financial Risk Tolerance and Financial Literacy to Financial Behaviors. Accessed November 02, 2015 from https://www.aeaweb.org/aea/2014conference/ program/retrieve.php?pdfid=764. 2013.
[10] Ghozali I. Aplikasi Analisis Multivariat Dengan Program IBM SPSS 21. Edisi 7. Semarang : BP Universitas Dipenogoro. 2013
[11] Hallahan T A, Faff R W, & McKenzie M D. 2003. An Empirical Investigation of Personal FRT. *Financial Services Review*. 13:57-78.
[12] Yao R. And Hanna, S D. 2005. The Effect of Gender and Marital Status on FRT. *Journal of Personal Finance*. 4(1), 66-85. 2005.
[13] Roussanov, N and Savor, P. Status, Marriage, and Managers Atittudes To Risk. Accessed Januari 03, 2015 dari http://www.usc.edu/schools/business/FBE/seminars/papers/F_4-8-11_ROUSSANOV-Rev.pdf. 2011.
[14] Grable J E. 2000. FRT and Additional Factors That Affect Risk Taking in Everyday Money Matters. *Journal of Business and Psychology*. 14(4):625-630
[15] Cole S, Paulson A. and Shastry G K. Smart Money? The Effect of Education on Financial Outcomes. Accessed Januari 13, 2015 from http://www.hbs.edu/faculty/Publication%20Files /cps-smart%20money%202013%20august_d4fb3555-e078-415a-8b20-aa50dc878205.pdf. 2013.
[16] Chattopadhyay S, and Dasgupta R. 2015. Demographic and Socioeconomic Impact on Risk Attitudes of The Indian Investors – An Empirical Study. Asian Economic and Financial Review, 5(4):601-623.
[17] Yao R, Sharpe D L, & Wang F. 2011. Decomposing the Age Effect on Risk Tolerance. *The Journal of Socio-Economics*. 40:879-887
[18] Sadiq M N, and Ishaq H M. 2014. The Effect of Demographic Factors on the Behaviour of Investors During the Choice of Investment : Evidence from Twin Cities of Pakistan. *Global Journal of Management and Business Research: C Finance*, 14(3):47-56.

Reconception of mandatory-based corporate social and environmental responsibility in Indonesia

S B Yunari[12]*, S Suhariningsih[2], R Syafa'at[2], S Sihabudin[2]

[1]Faculty of Law, Universitas Trisakti, Jakarta, Indonesia
[2]Faculty of Law, Universitas Brawijaya, Malang, Indonesia

*Corresponding Author : sri.by@trisakti.ac.id

Abstract. The Legal Concept of Corporate Social and Environmental Responsibility (CSER) in Law Number 40 Year 2007 (Company Law), as set forth in the general provision Article 1 (3) of Company Law evidently is a definition (*begripsbepalingen*) of voluntary basis, because it comes from the concept of CSR used by western countries based on World Bank's guidelines. Hence, it is certainly contrary to the legal concept of CSER that is perceived to be mandatory in the Company Law. Therefore, the concept of CSER as an implementation of a legal principle in a norm, so as not to cause legal issue, at the normative level as well as at implementation level, must be consistent and need reconception.The purpose of this reconception of CSER is to find a new concept of mandatory-based CSER. The methodology of research used is legal research (doctrinal research), based on secondary legal material acquired analysed presciptively by statute, conseptual and comparative approach.The research outcome is resulting in a discussion of reconception of a legal responsibility-based Corporate Social Liability (CSL) with sustainable local community empowerment oriented, so as to create legal certainty at the normative level and implementation in Indonesia.
Keywords : legal concept, normative level, reconception

1. Introduction

The regulation of Corporate Social Responsibility in Indonesia, as stipulated in Article 15 item b Law No. 25 of 2007 on Investments (hereinafter referred to as Law No 25 of 2007) [1]. The term "Corporate Social Responsibility" (known as CSR), states that : "Every investor shall have obligations: to implement corporate social responsibility". It's mean *a responsibility mounted in every investment company to keep creating relationship which is in harmony, in balance and suitable to the local community's, values, norms, and culture.* If the investor does not perform CSR obligations, the business entity or individual business may be subject to administrative sanctions, in the form of: written warning, restrictions on business activities, freezing of business activities and / or investment facilities; Or revocation of business activities and / or investment facilities.

Similarly, Article 1 item 3, Law No. 40 of 2007 on Limited Liability Company (hereinafter referred to as Law No. 40 of 2007). The term Corporate Social and Environment Responsibility (known as CSER), states that: CSER is the Company's commitment to participate in sustainable economic development in order to improve the quality of life and the environment which is useful to both the Company itself, the local community, and the society in general. Then in Article 74 Law No 40 of 2007, states that : (1) The Company having its business activities in the field of and/or related to natural resources, shall be obliged to perform its Social and Environmental Responsibility. (2) Social and Environmental Responsibility as referred to in paragraph (1) shall constitutes the obligation of the Company which is budgeted and calculated as the cost of the Company, implementation of which shall be performed with due observance to the reasonableness and fairness. (3) The Company which fails to perform its obligation as referred to in paragraph (1) shall be imposed with sanction in accordance with the provision of regulation. (4) Provision regarding Social and Environmental Responsibility shall be further regulated with a Government Regulation. As the implementation regulation of article 74 Law No 40 of 2007, so the government issued the Government Regulation No. 47 of 2012 on Social and Environmental Responsibility for Limited Liability Company (hereinafter referred to as GR No. 47 of 2012) [2].

Based on the three regulations of CSR (Law No. 25 of 2007; Law No. 40 of 2007 juncto the Government Regulation No. 47 of 2012 and Law No. 19 of 2003 juncto the Regulation of the Minister of State-Owned Enterprises No. PER-03/MBU/12/2016), the provisions of CSR in these three regulations in Indonesia have raised some legal issues, namely conflict of norms and vague of norms. Conceptually, CSR in Indonesia has not been interpreted and understood as mandatory, especially CSER concept as stated in Article 1 item 3 Law No 40 Years 2007, states that : "Environmental and Social Responsibility" means a Company's commitment to taking part in sustainable economic development in order to improve the quality of life and environment, which will be beneficial for the Company itself, the local community and society in general.

Philosophically, if the grammatical interpretation is interpreted, it is desired by Article 74 UUPT, that TJSL is a mandatory obligation, which is explicitly stated in Article 74 paragraph (1) containing the word ".... mandatory. ... ", as well as in Article 74 paragraph (2) and paragraph (3) also contains the word" .... obligation .... ". Where the word "must" literally means "to be done; Should not be abandoned ", whereas the word" obligation "is defined as (something) that is required; Something to be done; must". In other words "obligation" is anything that is considered as a necessity that comes from outside to be implemented by an individual as a member of a citizen, to gain recognition of the right in accordance with the implementation of the obligation.

Similarly, the difference between the meaning of the word "obligation" and "commitment" is manifested in the form of legal consequences, where an "obligation" will have legal consequences which may be enforced to the business actor and will be penalized if not exercised by law And the government), whereas "commitment" is an act or a will that has no legal effect, but rather a moral consequence, because the act or intention to execute it comes from within (the business actor) and has no legal effect. Grammatically the word "commitment" has a different meaning from the word "mandatory" or "obligation", in other words the concept and norm of TJSL contained in UUPT, conceptually is an inconsistent, non-synonymous (*contradictio in terminis*), overlapping and has led to a conflict of concepts, which ultimately lead to legal uncertainty within the norm.

Consequently the concepts in the three regulations of CSR contradict each other. Because the understanding of concept of CSR as set forth in the elucidation of Article 15 item b Law No 25 Years 2007 differs from the concept of CSER as regulated in Article 1 item 3 Law No. 40 Years 2007, as well as if compared with the concept of PCPD in Article 1 item 6 and 7 the Regulation of the Minister of State-Owned Enterprises No. PER-03/MBU/12/2016, which according to the authors, tends to lead to conflict of concept. Similarly, the regulations of CSR norms have caused conflict of norm, since they incorporate different norms sectorally. For consistency purposes between CSR Law No. 25 year 2007, Law No. 40 Year 2007 and the Regulation of the Minister of State-Owned Enterprises as mandatory in Indonesia, then a uniform and harmonious concept is needed so that legal norms that regulate CSR for all companies in Indonesia do not cause vague of norm.

Based on the above problems, this paper proposes the need for reconception of CSR as a legal obligation in the form of uniform of concept that will cause harmonious legal norm to govern in the form of CSR norm of mandatory nature in Indonesia. What is actually the idea of reconception of legal obligation-oriented CSR in the sense of *liability* not *responsibility*, will be discussed in this paper as a form of CSR legal obligation with legal consequences because of its mandatory nature.

IOP Conf. Series: Earth and Environmental Science **106** (2017) 012098 doi:10.1088/1755-1315/106/1/012098

2. Research Method

The applied type of research is legal research or doctrinal research secondary legal material obtained from library research of primary, secondary and tertiary legal material, to be prescriptively analysed [2] on any and all laws And regulations governing CSR as well as relevant sectoral laws. The method of approach used in this research is statute, conceptual and comparative approach [3], which is done by analyzing all the concepts related to the legal issues of CSR.

3. Results and Discussion

3.1. The Regulations of CSR in Indonesia.

The phenomenon of mandatory CSR in a global context is not a new thing, especially in European countries, such as France through *Nouvelles Regulations Economique* (NRE) Tahun 2001 [4], then England, through The 2003 Corporate Social Responsibility Bill [5] have applied it. Similarly, in Asian countries, like India, CSR has been mandated in the Indian Companies Act 2013 - section 135[6]. Including the state of Indonesia, has regulated CSR as mandatory since the 1980s specifically for State-Owned Enterprises (SOEs), while CSR for Private-Owned Enterprises has been mandated since 2007, through Law no. 25 of 2007 on Investments (Article 15 item b) and Law No. 40 Year 2007 on Limited Liability Company (Article 1 item 3 and Article 74) juncto Government Regulation No. 47 of 2012 on CSER for Limited Liability Company [7].

The regulations governing CSR in Indonesia are regulated in [7]:

1) Explicitly: Law No. 19 of 2003 on State-Owned Enterprises (Articles 2 and 88) juncto Regulation of The Minister of SOEs No. 03/MBU/12/2016, Law No. 25 of 2007 on Investment (Article 15 and elucidation); Law No. 40 of 2007 on Limited Liability Company (Article 1 item 3 and Article 74) and Government Regulation No. 47 of 2012 on Social and Environmental Responsibility for Limited Liability Company [7].
2) Implicitly: Law No. 4 of 2009 on Mining; Law No. 11 of 2009 on Social Welfare; Law No. 32 of 2009 on Environmental Management; Law No. 41 of 1999 on Forestry; Law No. 22 of 2001 on Oil and Natural Gas; Law No. 27 of 2003 on Geothermal; Law No. 7 of 2004 on Water Resources [8].

3.2. Weakness of CSER Concept in Law No. 40 of 2007.

As mentioned in the background, although the terms and concepts used for CSR in Indonesia vary, but one thing that needs to be underlined is that the terms of CSR, CSER and PCDP are not the same as the term and concept CSR in the western countries. Especially with the CSER in Law No. 40 of 2007, as affirmed in the Decision of the Constitutional Court Number 53/PUU-VI / 2008, stating that: "... The regulation of social and environmental responsibility as referred to Article 74 Law No. 40 of 2007 is different from the concept of CSR, As known and understood and implemented by the business community so far.... ". According to the government statement, there is a very basic difference between the concept of CSER in Law No. 40 of 2007 and CSR which has been applied to the Western world, namely as follows:

Firstly, CSER is legally obliged only to the Limited Liability Company conducting business activities in the field of natural resources and or related to natural resources. However, CSR as a whole is mandatory to all forms of the company in general. *Secondly*, the cost of CSER is charged to the company's operational expenses, the amount of which is determined based on reasonableness and fairness. While the cost of CSR implementation is taken from the company's net profit. *Thirdly*, the violation of the CSER shall be liable to sanctions in accordance with the sanctions as provided in the sectoral laws governing them. While sanctions violation of CSR is only a moral sanction.

In order for consistency to the term, the uniformity of concepts and legal norms that will regulate the CSR mandatory, it is necessary to create a reconception of sustainable CSR that is oriented towards legal responsibility. The idea is called by the author of the term Corporate Social Liability, which is then abbreviated as "CSL", as a renewal of the concept of CSR is known for this. So the context of the word Responsibility in the CSR writer replace with the word Liability to be CSL, as a form of CSR that is mandatory in Indonesia.

IOP Conf. Series: Earth and Environmental Science **106** (2017) 012098 doi:10.1088/1755-1315/106/1/012098

Literally, CSL can be interpreted as a social obligation that connects between companies and all stakeholders, including customers, employees, communities, owners or investors, government, providers, and even rivals. On the basis of this understanding, then logically the appropriate term for mandatory CSR is CSL. So, when it comes to CSR mandatory reconception to CSL, this idea is deemed appropriate because liability is more likely to lead to social responsibility for corporations legally, which the word liability can also be interpreted as legally responsible.

Normatively, the arrangement of CSER in Law No. 40 of 2007 juncto Government Regulation No. 47 of 2012 on Social and Environmental Responsibility for Limited Liability Company, still has some weaknesses, especially related to activity arrangement and budget of CSER and unfussy sanction implementation. However, the formal juridical existence of the rule of law on the CSER has had binding power as a law (genus), although in terms of procedures for the establishment of legislation, the regulation of CSER has not fulfilled the principles of the establishment of good legislation. As well as impact on the implementation of CSER in practice that tends not to be sustainable, so that created a concept that leads to the concept of sustainable [11].

Therefore, The Legislator are willing to make CSR mandatory, then the appropriate paradigm in the mandatory context of the concept of CSR is initiated more precisely when using the word or liability term not responsibility. Therefore, the idea to then make CSR concept into CSL is considered as a reference in fixing the prevailing regulations in the field of CSR in Indonesia, which is more in line with the philosophy of the Indonesian nation based on the principle of kinship economy, this is very different from the concept of CSR Prevalent and adopted by Western countries which tend to be based on the principle of capitalist and liberal economy.

3.3. The CSR Concepts according Laws, doctrine and International institutions.

To support the idea of uniformity of the legal concept of CSR, by applying the statute and conceptual approach, obtained some concepts of CSR as follows:

1) In Laws and Regulations in Indonesia, as follows:
   a) Elucidation of Article 15 item b Law No. 25 of 2007;
   b) Article 1 item 3 and Article 74 Law No. 40 of 2007;
   c) Article 1 item 6 and 7 of the Regulation of the Minister of State-Owned Enterprises Number: PER-09 / MBU / 07/2015 on PCDP.
2) Other concepts of Doctrine and international institutions, such as:
   a) Corporate social responsibility is the commitment of businesses to contribute to sustainable economic development by working with employees, their families, the local community and society at large to improve their lives in ways that are good for business and for development [12], [13].
   b) Corporate Social Responsibility is a business acts in socially responsible manner when its decision and account for and balance diverse stake holder interest [15].
   c) A concept where by companies integrate social and environmental concerns in their business operations and in their interaction with their stakeholders on a voluntary basis [9].
   d) Aligning a company's activities with the social, economic and environmental expectations of its "stakeholders" [16].

Furthermore, by comparative approach, some concepts of corporate social responsibility are used as reference by countries, based on comparative study with Hongkong, India, Netherlands and Taiwan countries, conducted in 2010-2014 as follows:

a) CSR is An organic link between enterprising, profit-making and social development, that is conducted as a win-win solution in areas where a company requires something from society that cannot be bought and requires investment through a business plan [17].
b) CSR maybe defined as the obligation of the companies to protect and enchance the norms and values held by the society within which it operates. CSR is concept whereby corporate organizations consider the interests of society by taking responsibility for the impact of their activities on customers, suppliers, employees, shareholders, community and others direct and indirect stakeholder as well as the environment [18], [19].

c) A firm takes on a visible role in the society which goes beyond the core business and beyond what the law requires and which leads to added value for the company and the society [20].
d) CSR is the way in which business consistently creates shared value in society through economic development, good governance, stakeholder responsive ness and environmental improvement. Put another way, CSR is an integrated, systemic approach by business that builds, rather than erodes or destroys, economic, social, human and natural capital" [21], [22].

3.4. Formulation Reconception of CSR based on Mandatory.

Based on the above concepts, the formulation or definition used as a mandatory CSR concept will be, at least 3 (three) main points that form an understanding or concept of CSL [19] as follows:

a) As an artificial person, the corporation or corporation is not independent and isolated, the company or corporation cannot claim that they have no responsibility for the economic, environmental or social circumstances.
b) The existence and sustainability of the company or corporation is determined by all stakeholders and not just shareholders. The stakeholders consist of shareholders, customers, suppliers, clients, customers, employees and their families, local communities and those involved either directly or indirectly with the company (the local community and general society).
c) Implementing CSL means also carrying out the duties and daily activities of a company or corporation, as a container to gain profit through the business carried on and or managed by it. So this means CSL is an integrated part of business activity, so CSL means also run the company or corporation to gain profit.
d) Thus, the concept of CSR as a legal responsibility must meet the following elements:
e) social responsibility is an obligation to all companies without exception;
f) social responsibility applies to any company established under Indonesian law;
g) social responsibility is aimed at creating a harmonious, balanced relationship between the company and the environment;
h) CSR is implemented by taking into account the values, norms and culture of local communities;
i) CSR should be oriented towards sustainable economic development to improve the quality of life of local communities.

Based on of the idea of reconception, it is expected to create a legal norm of CSR which is sustained as mandated by Articles 33 and 34 of the 1945 Constitution and the purpose of holding CSER in Article 74 Law No. 40 of 2007. Therefore, as a follow-up, it is necessary to make a legal model of CSL that is expected as a follow-up study.

4. Conclusion

The need for uniformity of terminology and reconception of the concept of CSR comprehensively in order to achieve the purpose of the implementation of CSR in Indonesia, as mandated in article 33 and 34 the 1945 Constitution and Law No 40 of 2007. Hence it is expected that a sustainable model of CSR law for company can be established at the level of norms and implementation. The application of the concept of CSR in the regulation of Law No. 25 of 2007, Law No. 40 of 2007 and its implementing regulations, as well as The Regulation of the Minister of State-Owned Enterprises No. PER-03/MBU/12/2016 concerning PCDP mandatorily in Indonesia causes inconsistency, therefore the idea of changing corporate social responsibility (CSR) concept into corporate social liability (CSL) as a form of mandatory corporate social responsibility is deemed important to create a sustainable model of CSR law in accordance with the principle of sustainable economy as the philosophy of the Indonesian nation.

References

[1] A Marina Welker. January 2009. Corporate Security Begins in The Community: Mining, the Corporate Social Responsibility Industry, and Environmental Advocacy in Indonesia, *Cultural Antropology*. 24(1):142-179.
[2] Peter Mahmud Marzuki, Penelitian Hukum. Jakarta: Publisher Kencana Prenada Media Group, p 133, 2011.

[3] Terry Hutchinson, Researching and Writing in Law, New South Wales: Published Lawbook Co, p 35, 2002.

[4] Bantekas Ilias. 2004. Corporate Social Responsibility In International Law, Boston University *International Law Journal*, 22:309

[5] Busyra Azheri, Corporate Social Responsibility: Dari Voluntary Menjadi. Jakarta: Rajawali Pers, 2011.

[6] Monika Kansal, and Shekar Babu Nava Subramaniam. 2017. Governance of Mandated Corporate Social Responsibility: Evidence from Indian Government-owned Firms. *Journal Business Ethic*. 143: 543–563, no. DOI 10.1007/s10551-015-2804-0.

[7] Sri Bakti Yunari, and Siti Nurbaiti. 2012. *Jurnal Akreditasi Nasional,* Regulasi CSR Pasca Berlakunya PP No. 47/2012 tentang Tanggung Jawab Sosial dan Lingkungan Perseroan Terbatas, *Jurnal Hukum Bisnis.* 31(5):570-578.

[8] Lambooy Tineke et.al, CSR in Indonesia-Legislative Development and Case Studies. Jakarta : Konstitusi Press, 2014, p xxvii.

[9] Gunawan Widjaja and Jeremia Ardi Pratama, Risiko Hukum dan Bisnis Perusahaan Tanpa CSR. Jakarta: Publisher Forum Sahabat, 2008.

[10] Maria Farida Indrati S, Ilmu Perundang-undangan – Proses dan Tehnik Pembentukannya, Yogyakarta: Kanisius Publisher, 2008, p 98.

[11] Sonny Keraf, Etika Lingkungan. Jakarta: Penerbit Kompas Gramedia, 2012, p 173.

[12] Piotr Mazurkiewicz, Corporate Environmental Responsibility CSR Framework Possible? Washington: World Bank Group, 2013, p 2.

[13] Djordjija Petkoski and Nigel Twose, Public Policy for Corporate Social Responsibility, in the Private Sector Development Vice Presidency of the World Bank, and the International Finance Corporation. Washington: World Bank Institute, 2012, p 1.

[14] Phil Watts and Rio Tinto, CSR: Meeting changing expectations. Netherlands: WBCSD Publication, 2008, p 2.

[15] Isabelle Maignan, and O C Ferrell. Desember 2004. Corporate social responsibility and marketing : An integrative framework. *Journal of the Academy of Marketing Science*, (32)1:3–19.

[16] Source Watch Website. (2016, 17 July) www.sourcewatch.org. [Online]. http://www.sourcewatch.org/index.php/SourceWatch.

[17] Rio D. Praaning Prawira Adiningrat, Corporate Social Responsibility Basics, Definitions, Practices, in Euro-Asia Economic Forum, Xi'an, People's Republic of China, 2009, p 21

[18] Partha Sarathi Adhya, Company Law. New Delhi: SBS Publisher & Distributors Pvt. Ltd., 2011, p 38.

[19] Ayesha Malhotra, Loren Falkenberg jaydeep Balakrishnan. August 2017. Multi-Level Corporate Responsibility: A Comparison of Gandhi's Trusteeship with Stakeholder and Stewardship Frameworks. *Journal of business Ethics*. 141:133-136.

[20] S.C.W.Eijffinger, N.C.G.M. Stoffele, H.Smid and A.M. Coldeweijer J. J. Graafland, "Benchmarking of CSR: Methodological Problems and Robustness," Journal of Business Ethics, vol. Volume 53, no. Issue 1–2, August 2004, p 137–152

[21] Wen Horng Kao. (2015, 20 Agustus) [Online]. Corporate%20Social%20Responsibility%20in %20Taiwan.html

[22] Maria Lai-Ling Lam. April 2009. Beyond Credibility of Doing Business in China: Strategies for Improving Corporate Citizenship of Foreign Multinational Enterprises in China. *Journal of Business Ethics*. 87:133–146.

Refuse derived fuel potential in DKI Jakarta

H Widyatmoko

Environmental Engineering Department, Faculty of Landscape Architecture and Environmental Technology, Universitas Trisakti, Jakarta, Indonesia

Corresponding Author: widyatmoko@trisakti.ac.id

Abstract. Combustible waste fractions of municipal solid waste (MSW) which can not be easily separated or sorted, reused or recycled, may have a high calorifiv value (CV) that can be used in a fuel for energy recovery. The objective of this study was to explore the Refuse Derived Fuel (RDF) potential of municipal solid waste from DKI Jakarta to produce electricity and to promote it to be socially and politically acceptable. For this purpose, 24 sampels of RDF were taken from Bantargebang, cabonized, molded and pressed to be briquette. All samples were analized for moisture, ash, and calorific value in the physical and chemistry Laboratory of ITB Bandung. The analysis of calorific value (CV) shows the CV difference of 1815.8 cal/g between the briquettes (8051.25 cal/g) and the RDF (9867.12 cal/g. The total waste DKI which can be used as briquettes 5253 ton / day or equivalent with 49154115 kWh / day. If the efficiency of electricity production from RDF was 25%, then Jakarta is able to generate electricity from RDF of 12288529 kWh / day or as much as energy needed by 573,480 middle-class households with energy needs of 642.84 kWh/month.
Keywords: briquette, colorific value, energy, RDF

1. Introduction

Combustible waste fractions of municipal solid waste (MSW) which can not be easily separated or sorted, reused or recycled, may have a high calorifiv value (CV) that can be useful in a fuel for energy recovery (Table 1).

Table 1. Quality standard of RDF.

States	CV	Moisture	Ash	Sulfur	Chlorine
Finland	13-16 kj/kg	25-35 %	5 – 10 %	0.1-0.2 %	0.3-1.0 %
Italy	15 kj/kg	Max. 25 %	20 %	0.6 %	0.6 %
UK	18,7 kj/kg	7-28 %	12 %	0.1-0.5 %	0.3-1.2 %

The MSW can be shredded, dried, baled to produce refuse-derived fuel (RDF) and then burned to generate electricity, thereby making good use of waste that otherwise might have ended up in landfill.

Since the early nineties, producing RDF for energy recovery has been a popular waste management option. However, there is a problem found in RDF contents which sometimes contains hazardous component, its quality varies. Therefore, it should be thoroughly analysed and tested to ensure the RDF is a fuel gained from non-hazardous waste.

The progress of tecnology leads RDF as an important energy alternative in the future. The developments in sensor-based sorting technology enabling the acquisition of RDF material volume to be much higher than ever before. Plastics, fibre, films, paper, cardboard, textiles and wood can all be recovered to an exceptionally high standard material, achieving purity level reaches of about 95% and yields up to 70%. In comparison, manual or semi-automated techniques typically recover just as much as 10-20% [1], [2].

Labour costs are significantly reduced as there is no manual picking anymore and from a quality perspective, it is possible to process and maintain higher throughputs, with consistent quantity and quality achievable within 24 hours a day [1].

DKI Jakarta with population of 9,607,787 [4], has the capability of processing 8291.650 tonnes per day of MSW. The processing of this waste will produce around 8.291650 tonnes of RDF per day [3], [4], [5].

This reseach is aimed at exploring the Refuse Derived Fuel (RDF) potential of municipal solid waste from DKI Jakarta to produce electricity as well as promote it to be socially and politically acceptable and technically feasible.

2. Research Method

Grab sampling was done at Bantargebang, Bekasi, West Java. The total waste volume per day is calculated based on total valume of the transport vahicles [6], [7].

The 24 waste samples were taken randomly and sequentially put into a box and 3 times dropped from a high of 20 cm to determine its density. The samples were than 7 days air dried, sorted into fractions of peper, plastic, glass, tetrapack, Styrofoam, pads/pampers, organic fraction, food scraps, textile and rubbers. After homogenization process, 200 g of all fractions, except metal and glass, were put into perforated cans and carbonized. The carbonized waste was than sieved 50 mesh, mixed with tapioca glue, molded, pressed to be briquette and laboratory analysed for CV, Moistur, ash content.

Water content was determined according to SNI 03-1971-1990 [8]. For this purpose, ± 10 grams of sample in porcelain cup was kept for 3 hours into oven at 105°C and for 30 minutes into desiccator, then finally it was weighed. Residu moisture: $M = (w\text{-}d) \times 100\%$; ($M$ = moisture content (%); w = initial weight (g); d = weight after drying in grams in 105 °C oven)

Ash content was determined according to ASTM E 830-87 [9]. The remaining sample was heated at temperature of 550 ° C and reheated in a furnace at 950 ° C for 7 minutes. Then it was laid into desiccator and weighed. Ash content = $(f / w) \times 100\%$; (E = weight after heating in furnace 550 ° C; f = weight after heating in furnance 950 ° C; w = initial weight in g). Colorific value was analized using calorimeter.

The information on recycling potential of waste components was obtained through interviews with waste picker and waste collector. According to recyclers and waste collector, the damaged paper can not be sold and only 50% of the organic waste and the food waste can be recycled.

3. Results and Discusion

The highest CV is obtained from the residual component of garbage and rubber waste. Components of garbage/pampers waste residues have high CV, but the moisture content in these components is also high. The CV of a material is strongly influenced by the content of the substance in it (Tabel 2).

IOP Conf. Series: Earth and Environmental Science **106** (2017) 012099 doi:10.1088/1755-1315/106/1/012099

Table 2. Recycling Capasity (RC).

Componets	% -RC
Paper	30-40
Plastic Bottles	100
Glass Bottles	100
Tetrapack	-
Styrofoam	-
Pampers	-
Organic waste	50
Food waste	50
Platic bags	-
Textile	-

Table 3. Average of moisture, ash content and CV of DKI Jakarta waste componets.

Component	Moisture (%)	Ash content (%)	CV (cal/g)
Plastic bag	12.43	23.27	6216.6
Tetrapack	15.79	23.88	5951.21
Styrofoam	13.86	25.11	5686.20
Textile	22,72	22.33	7036,90
Organic& Food rest	19.94	24.30	5926.83
Pampers	27.13	23.31	6171.90
Rubber	23.05	26.17	6808.57
Total	134.92	168.37	43798.21
Average	19.27	24.05	6256.89

3.1. Calorific Value of RDF Bricket

Result of laboratory analysis of calorific value, moisture content, and ash content from briquette sample are as follows (Table 3, 4, 5)

Table 4. 3 briquette of waste residue.

Moisture	33.86 %
Ash Content	44.72 %
CV	9867.10 cal/g

Table 5. 4 Briquette CV.

Per Component	6235.39 cal/g
Briquette CV	9867.12 cal/g
Average of CV	8051.25 cal/g
CV-Diference	1815.86 cal/g
Deviation	22.55%

The following data are collected from invoices that describe electricity usage and charge of low income cluster houses in Ciracas (Table 6).

IOP Conf. Series: Earth and Environmental Science **106** (2017) 012099 doi:10.1088/1755-1315/106/1/012099

Table 6. Metering charges of low income cluster houses of R1 450 VA class with tariff Rp. 415,-/kWh in Ciracas, East Jakarta.

Number	PLN(Rp)/month	Customer Number	Name
1	111,227	5471.0239.700	Chaniago
2	236,933	5471.0238.7063	Jalil
3	210,068	5471.0238.7048	Erdi
4	61,363	5471.0238.7089	Erni
5	3,.95	5471.0234.8292	Jalil
6	81,132	5471.0238.7071	Robby
7	88,735	5471.0238.9830	Zubaidi
8	44,319	5471.0237.3790	Tan
9	162,929	5471.0234.0476	Aang
10	53,760	54.71.0238.9814	Hadi
11	385,697	5471.0236.1308	Amri
12	45,143	5471.0234.8306	Suryo
13	8,357	5471.045.3860	Suryo
14	68,46	5471.0245.3878	Ari
15	8,703	5471.0280.5455	Roni
16	16,896	5471.0280.5463	Opik
17	355,369	5471.0245.3886	Petrus
18	54,267	5471.0247.3693	Puspo
19	32,978	5471.0247.3723	Sumari
20	17,633	5471.0247.3731	Sumari
21	58,829	5471.0247.3749	Ratna
22	64,404	5471.0247.3756	Ratna
23	46,664	5471.0247.3685	Laksmono
24	65,925	5471.0247.3707	Wiarso
25	11,.12	5471.0247.3715	Wiarso
26	5,069	5472.0001.7792	Agus
27	18,432	5471.0234.8314	Emilia
28	102,927	5471.0233.6455	Laksmono
29	213,733	5471.0239.0718	Wiarso
30	507,003	5471.0239.0734	Puspo
31	118,641	5471.0234.8322	Ratna
32	5,242	5471.0236.1294	Novi
33	53,253	5471.0238.9806	Hendro
Avgerage	**104,611**		

The analysis of calorific value (CV) shows the CV difference of 1,815.8 cal/g between the briquettes (8,051.25 cal/g) and the RDF (9,867.12 cal/g). This 22.5 % of deviation is related to the complexity of the material homogenization process. The briquette CV average of 8,051,25 cal/g is equivalent to 8,051,250 kCal/ton or 9357,3416 kWh/ton.

DKI Jakarta generates MWS 8,291,650 kg / day. Of that number only 66% ie 5,472,490 kg/day can be transported to Bantargebang Landfill [10]. Therefore the total waste that can be prodused to RDF = 5,472,490 kg/day - (metal + glass component) kg/day = 5,472,490 kg/day - (4% of 5,472,490) kg/day

= 5,253,590.44 kg/day = 5,253,590 kg/day = 5,253 tons/day or equivalent to 5,253 ton/day x 9357.3416 kWh/ton = 49,154,115 kWh/day.

If the efficiency level to produce electricity from RDF is 25%, then Jakarta can produce 12,288,529 kWh/day or as much as energy needed by 573,480 middle-class households with energy consumption of 642.84 kWh/month or 21.428 kWh/day (12288529 kWh/day: 21,428 kWh/day/households = 573479,98 households).

The basic electricity tariff of low income household or the 450 VA of R1 household class is Rp. 450,- per kWh. Based on the data obtained the average electricity needed by the R1 class of 450 VA is 244.512 kWh/month (8.2 kWh/day). Therefore the energy of 12,288,529 is enough to supply 1,498,601 low income households.

4. Conclusion

Jakarta generates MWS 8,291,650 kg/day, but only 66% of that, ie 5,472,490 kg/day can be transported to Bantargebang Landfill [10] and only 5,253 tons/day can be used as RDF to produce electricity of 12,288,529 kWh / day or as much as energy needed by 573,480 middle-class households.

RDF offers a very strong commercial opportunity. It seems to be optimistic, even it's just the tip of the iceberg. DKI Jakarta Government needs to be encouraged to adopt commercially viable alternative fuel such as RDF. The developments in sensor-based sorting technology enabling to achieve optimal purity rates and much higher volumes of high value material than manual picking. Plastics, fibre, films, paper, cardboard, textiles and wood can all be recovered to an exceptionally high standard material. It is possible to process and maintain higher throughputs, with consistent quantity and quality achievable within 24 hours a day.

Creating the RDF processing machinery with a small footprint to make it suitable for small-scale operations is also achievable. Local companies can cooperate with one of the coutries whose having experience with RDF to create sensor-based sorting technology so that after all valuable fractions had been recovered, only the smallest amounts of residue would be destined for the precarious RDF route.

RDF will be an important fuel for the future. RDF can be an alternative energy source as well as a solution to solve recycled waste problem in Jakarta, helping the citizens to reduce their reliance on fossil fuels and avoid landfilling this residual waste. With a high CV, RDF can be used in facilities such as cement kilns.

There should be continous campaign about RDF to make it socially and politically acceptable, to create confidence in the market. It must be tested and evaluated in an appropriate and standardized way and declared that RDF is a fuel made from non- hazardous waste componets and in compliance with environmental regulation. It must be produced to a range of specifications to meet customer requirements. Deeper research of RDF is indispensable and technically feasible.

Acknowledgment

The authors gratefully acknowledge the research grant and financial support provided by Universitas Trisakti.

References

[1] http://www.recyclingwasteworld.co.uk/site/contact-form.aspx? accessed on May 2, 2017 at 7 pm)
[2] Taparugssanagorn K, Yamamoto K, Nakajima F, Fukushi K. Evaluation of waste-to-energy technology: economic feasibility in incorporating into the integrated solid waste management system in Thailand. The IE Network Conference, 2007, 91-96.
[3] Widyatmoko dan Sintorini, 2002. Menghindari, Mengolah dan Menyingkirkan Sampah. Abdi Tandur, Jakarta.
[4] Badan Pusat Statistik [BPS]. (2016). Jakarta dalam angka 2016. Jakarta: BPS
[5] BPS, Bappenas, & UNFPA. (2013). Proyeksi penduduk Indonesia 2010-2035. Jakarta: BPS
[6] Badan Standardisasi Nasional. 1994. SK SNI 19-3694-1994 Tentang Metode Pengambilan Dan Pengukuran Contoh Timbulan Dan Komposisi Sampah Perkotaan. Jakarta : Balitbang DPU

[7] Badan Standardisasi Nasional. 2002. SK SNI 19-2454-2002 Tentang Tata Cara Teknik Operasional Pengolahan Sampah Perkotaan. Jakarta : Balitbang DPU.
[8] Badan Stadardisasi Nasional 2011. SK SNI 185/KEP/BSN/11/2011 SNI 1971:2011 - Cara uji kadar air total agregat dengan pengeringan. BSN: Jakarta.
[9] ASTM International 2004. Standard Test Method for Ash in the Analysis Sample of Refuse-Derived Fuel 1. United State.
[10] Widyatmoko 2011. Pengelolaan Sampah. Universitas Trisakti. Jakarta.

Petroleum residues degradation in laboratory-scale by *rhizosphere* bacteria isolated from the mangrove ecosystem

A Rinanti*, **I J Nainggolan**

Environmental Engineering Department, Faculty of Landscape Architecture and Environmental Engineering, Universitas Trisakti, Jakarta, Indonesia

*Corresponding Author : astririnanti@trisakti.ac.id

Abstract. This research is about petroleum bioremediation experiment to obtain bacterial isolate from mangrove ecosystem which potentially degrade petroleum. It was conducted in an Erlenmeyer batch system filled with growth medium of Stone Mineral Salt Solution (SMSS) plus petroleum residue, placed in an incubator shaker with a rotation speed of 120 rpm, temperature 30^0C, for 14 research days. Indigenous bacteria that have been isolated and identified from the roots of mangrove plants are *Ochrobactrum anthropi* and *Bacillus* sp., *Ralstonia pickettii* and *Bacillus circulans*. Those bacteriain both monoculture and consortium form (mixed culture) are incorporated into erlenmeyer as remediator agents. All bacteria can utilize hydrocarbon compounds, but *Ralstonia pickettii* and *Bacillus circulans* reached exponential phase faster with more cell count than other bacteria. Compared to single cultures, petroleum degradation by a bacterial consortium provides a higher TPH reduction efficiency, i.e. at 5%, 10%, and 15% of initial TPH of 94.4%, 72%, and 80.3%, respectively. This study proved that all bacteria could optimize hydrocarbon compounds up to 15% TPH load.
Keywords: biodegradation, hydrocarbon, petroleum, mangrove, total petroleum hydrocarbon

1. Introduction

Petroleum contamination can occur in mangrove ecosystems due to oil spills occurring in the sea or coast which is swiped by tide to reach the mangrove ecosystem. Oil residues that pollute mangrove ecosystems can influence mangrove plants, especially on their root systems and on microorganisms that live in *rhizosphere*. If mangrove plants cannot adapt to residues, then the sustainability of mangrove forests can be threatened. The composition of residue is a hydrocarbon compound which can still be used by bacteria as a carbon source. Bacteria are groups of microorganisms that readily adapt to their environment, allowing them to use residues as a source of carbon and energy [1].

An effective petroleum removal strategy is required to prevent the destruction of mangrove ecosystems. Bioremediation is an effective and promising way to make quick, cheap remediation with minor health risks [2], [3], [4]. The breakdown of the hydrocarbon compounds by bacteria to become simpler compounds, causing changes in the physical and chemical properties of the compounds. Some of those properties that can be an indicator of the degradation process of hydrocarbon compounds include viscosity, specific gravity, density, water solubility, petroleum residual weight and surface tension [3]. The hydrocarbon splitting process and the breakdown of the residual constituent components can occur both non-biologically and microbiologically. Non-microbiologically for instance is due to evaporation, and dissolution, while microbiologically is due to the activity of microorganisms such as bacteria, yeasts, and fungi.

Rhizosphere of mangrove ecosystems that are covered by petroleum residues allegedly inhabited many bacteria that have been adapted to utilize the carbon source in the residue for its growth [5]. Based on these circumstances, this study was conducted to isolate bacteria from plant *rhizosphere* in the mangrove ecosystem and exploit its ability in the process of petroleum bioremediation.

2. Research Method

2.1. Isolation and purification of indigenous bacteria

Petroleum comes from Kepulauan Seribu (Island's name in Jakarta region that means Thousand Island). The roots of mangrove plant originated from the conservation of the mangrove ecosystem of Pantai Indah Kapuk, Jakarta, Indonesia. Isolate bacteria grown in Erlenmeyer with batch culture containing SMSS (Stone Mineral Salt Solution) liquid medium. Degradation test was done by adding petroleum with initial concentration of Total Petroleum Hydrocarbon (TPH) that is 5%, 15%, 25%, and 50%. Samples in the form of *rhizosphere* were weighed 4.5 grams and inoculated into 45 mL of liquid SMSS medium, then incubated in an incubator shaker for five days at 30^0C with 120 rpm shaking rate. Further diluted to 10^6, and each dilution was taken as 1 mL mixed with SMSS medium until evenly distributed, then poured into a sterile petri dish. After the medium of SMSS in the sterilized petri dish solidified, the culture was incubated at a temperature of 30^0C until bacterial colonies were found to grow.

2.2. Bacteria identification

Identification stages include: morphological testing of colonies, cell morphology, Gram properties, presence or absence of endospores, biochemical testing. Furthermore, to determine the species of bacterial isolate obtained, the test results at the identification stage were matched with **Bergey's Manual of Determinative Bacteriology** [6]. The ability of isolates to degrade petroleum residue is studied by measuring viscosity, specific gravity, residual weight, observing a decrease in pH of the medium and calculating the total number of bacterial cells.

2.3. Residual viscosity measurement

The measurement of the viscosity is carried out at a temperature of 30^0C, which is started by separating the residual phase from the media (water) phase by passing the media phase to the container via a pipe connected to the suction pump. Once separated, the residue is then inserted into a cylindrical glass. A ball of metal material is inserted into a cylindrical glass containing residuals and the length of time the ball travels through the length of the residue in the glass [7]. Afterward, the determination of viscosity used the following equation:

$$\eta = \frac{2r^2 g(\rho - \rho o)\, t}{9d} \quad (1)$$

rho = dynamic viscosity coefficient (Pa.s), r = metal ball radius (m), g = earth gravity force (m/s^2), rho = metal ball density (kg/m^3) po, d = residual density (kg/m^3): the length of the residue through which the metal sphere (m), t = the travel time of the metal ball passes through the residue (s). The value of the viscosity obtained is converted:

$$v = \frac{\eta}{\rho o} \quad (2)$$

In units of m^2/s, then the kinematic viscosity unit used is mm/s equivalent to centistoke (cSt).

2.4. Specific gravitation measurement

Specific gravity [7] was measured by comparing residual density (kg/m^3) with aquadest density (kg/m^3) at 30^0C. Aquadest density at 30^0C temperature measured 1012.2216 kg/m^3.

2.5. Measurement of final residual weight

The final residual weight is measured by weighing the total residual at the end of the observation that has been separated from the media (water) phase in grams.

2.6. Calculation of the number of bacterial cells

Calculation of bacterial cell count is done using total plate count method [2]. Before the number of bacterial cells is calculated, each culture in the Erlenmeyer is shaken first so that the residual phase with the media phase is listed perfectly (homogeneous). Sampling was performed daily for 14 days and the number of bacteria in the sample was calculated in units of cells/mL. Data on the number of bacterial cells obtained from each inoculum made the curve so that it could show the relationship between the number of cells with time.

$$r = \frac{3{,}32[\log x_t - \log x_o]}{t} \quad (3)$$

Xt = the number of bacterial cells at the end time (t_x), *X0* = the number of bacterial cells at the beginning (t_x), *T* = incubation time of t_0–t_x so that it can be determined the increase in the number of cells per unit of time.

3. Results and Discussion

3.1. Selection and identification of bacterial isolates

Bacterial isolates obtained from *rhizosphere* of mangrove plants after purification were as many as 4 species. The results of the first stage selection showed that all the isolated *rhizosphere* bacteria isolates could grow on SMSS medium containing petroleum residue. The first stage selection was performed to determine the isolates of *rhizosphere* bacteria that survive and grow in environmental conditions containing petroleum residues, although the bacterial isolates may not be able to utilize petroleum residue as a source of carbon and energy.

The SMSS growth medium used in the first stage selection was a nutrient-rich medium, because it contained peptone (5 g/L) and yeast extract (1 g/L), in which the *rhizosphere* bacteria were resistant to residue but could not utilize the residue as a carbon sourcewhich allegedly still could grow and survive by utilizing the nutrients of the medium. In the first stage selection, the isolates were not resistant to the presence of petroleum residue and could not utilize the petroleum residue as a source of carbon and energy, the isolates would die and not grow. Selection phase II aimed at determining the isolates of bacteria that are resistant to the presence of petroleum residue and can use it as a source of carbon and energy. Therefore, the medium used is a poor medium of nutrition. The identification results showed that all the bacterial isolates obtained were Gram-negative, non-spore and one isolate was Gram positive, with spore. The complete identification results for each isolate are shown in Figure 1, Table 1 and Table 2.

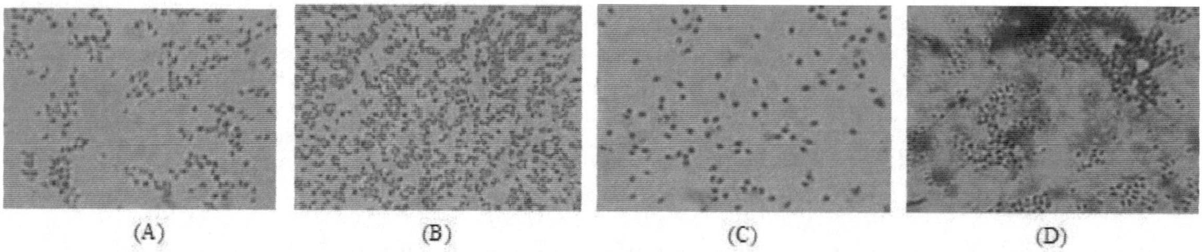

Figure 1. Identified bacterial isolates A) *Ralstonia picketii,* B) *Bacillus circulans* C) *Ochrobactrum anthropi,* D) *Bacillus sp.*

Table 1. Visual observation results colonies of microorganisms.

Bacteria	Reaction to Blood agar	Size	Shape	Elevation	Surface Condition	Colour
Ralstonia pickettii	Himolise	2 mm	Irregular	Convex papillate	Rough	Green
Bacillus circulans	Non-himolise	0.5 mm	Circullar	Low convex	Smooth	Grey
Ochrobactrum anthropi	Himolise	1 mm	Circular	Low Convex	Smooth	Grey
Bacillus sp.	Himolise	2 mm	Irregular	Convex papillate	Rough	Grey

Table 2. biochemical test results.

Bacteria	KIA	A	B	C	D	E	F	G
Ralstonia pickettii	K/A	+	+	-	+	-	-	+
Bacillus circulans	K/K	+	-	+	+	+	-	-
Ochrobactrum anthropi	K/K	-	-	-	+	-	-	+
Bacillus sp.	K/K	+	+	+	+	+	-	+

A = Glucose; B = Maltose; C = Sucrose; D = Arabinose; E = Mannitol; F = Voges-Proskauer; G = Lactose

3.2. Residue viscosity

The residual viscosity of medium that was not given inoculum (control) reached 54.64 cSt. This value is significantly different from all other inoculum factors, either in the form of single culture or its mixtures. The lowest viscosity was shown in the mixed culture of 22.88 cSt (Figure 2).

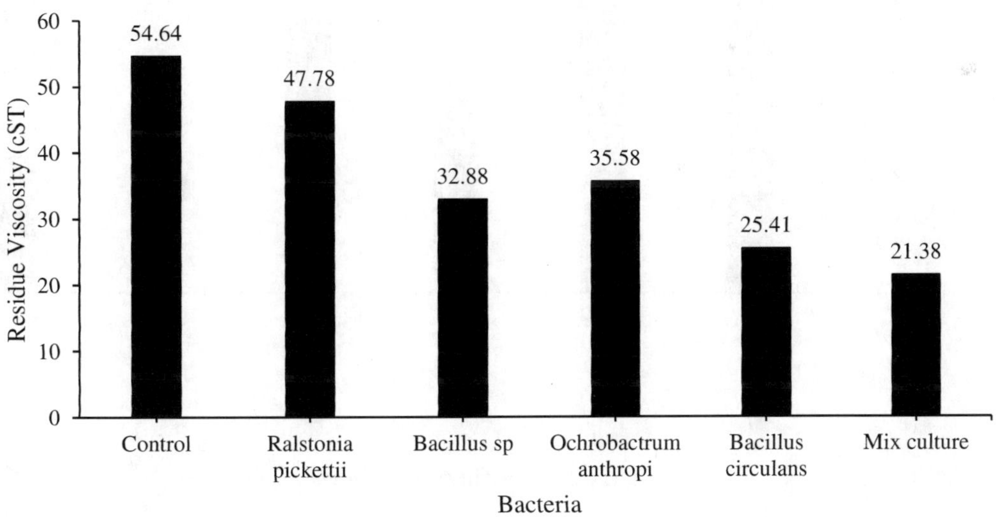

Figure 2. Residue Viscosity (cSt).

Each residual viscosity value due to the bioremediation agent in the form of single culture of *Ralstonia pickettii, Bacillus sp., Ochrobactrum anthropi* ranged between residual viscosity values due to its mixed culture and control. Bioremediation agent in the form of single culture *Bacillus circulans,* showed the lowest viscosity value (28.97 cSt), when compared with other single cultures. Provision of bioremediation agent showed the smallest viscosity of 22.88 cSt. It was suspected that all isolates in the mixed culture worked to degrade residues by utilizing the corresponding residual components for each isolate, resulting in a much simpler fraction in large quantities. The number of these simple fractions causes a decrease in the value of viscosity.

3.3. Specific residue gravitation

The unallocated medium (control) of the residual-specific gravity value reaches 0.6911, the value is significantly different from all the specific gravity values of the residue on the medium given the level of other inoculum factors. The mixed inoculum shows the least specific residual gravity value of 0.6108 which means the highest degradability (Figure 3).

The specific gravity value is closely related to the viscosity value, the hydrocarbon compound having a low specific gravity value, the viscosity is also low. Hence the relationship between specific gravity and viscosity is directly proportional, which means the increase in specific gravity will be followed by increased viscosity and vice versa [3], [7].

The final residual weight at the inoculum level of single culture *Bacillus sp* reached 22.65 g and was not significantly different from the residual weight of the medium without inoculum (control) reaching 22.83 g, but the other single culture inoculum levels and mixtures were all significantly different with control. The final residual weight of the medium inoculated with the smallest mixed culture was 16.88 g compared with the weight of the final residue on the inoculated medium with other control factors. Mixed cultures were the best inoculum factor in reducing the final residual weight compared with all the inoculum factors tested, since the medium inoculated with mixed culture showed the least final residual weight of 16.88 g.

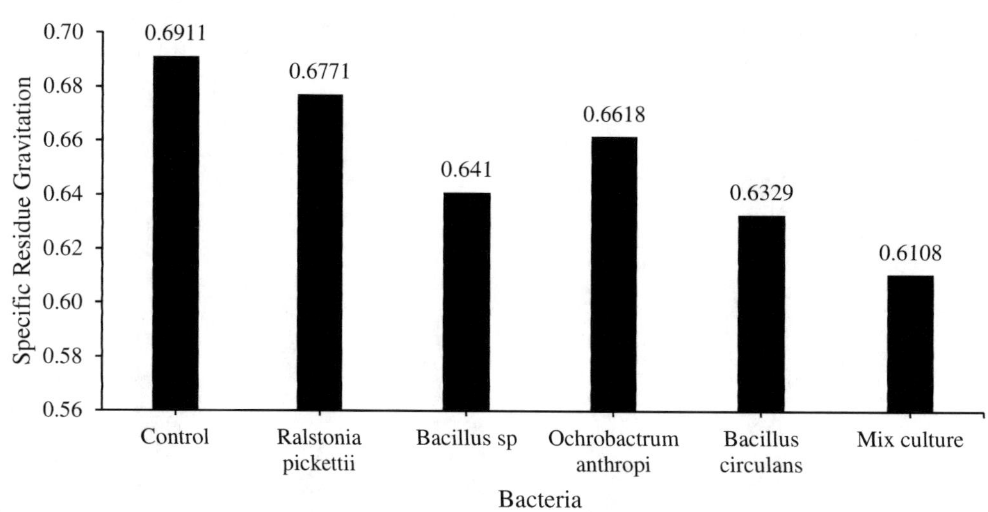

Figure 3. Spesific residue gravitation.

The use of nutrients by bacteria will be directly proportional to the population of growing bacteria, if the number of bacterial populations is high then the nutrients used are also numerous. Residue degradation by bacteria is the process of using residues as nutrients by utilizing the hydrocarbon compounds which contained therein, so that the reduction of residuals used as nutrients will be proportional to the addition of a growing population of bacteria. Thus, there is a relationship between degradation rate of petroleum residue with growth rate of degrading bacteria.

The degradation process was discontinued on day 14 although the degradation process was still going on, but the degradation rate was very slow. Based on the k values of each level of inoculum factor, it

IOP Conf. Series: Earth and Environmental Science **106** (2017) 012100 doi:10.1088/1755-1315/106/1/012100

can also be explained that single culture of *Bacillus circulans* has the highest degradation rate, followed by mixed cultures, single cultures of *Ralstonia pickettii, Bacillus sp., Ochrobactrum anthropic*. The results also showed that the residual degradation process could proceed well, in the form of a single culture and mixed culture. Although mixed cultures consisting of four isolates, residual degradation processes indicated by viscosity, specific gravity, and the weight of the final residue and decreased pH of the medium, all indicated a strong degradation ability.

4. Conclusion

Isolates of *rhizosphere* bacteria in the form of mixed cultures have a higher ability than isolates in the form of a single culture in degrading petroleum residue as indicated by changes in viscosity, specific gravity and residual weight. While in the form of a single culture, each isolate has a specific ability to degrade petroleum residues such as *Bacillus circulans* has the highest ability to decrease viscosity and decreases specific gravity, *Ralstonia pickettii* vesicular has the highest ability to decrease the pH of the medium, and *Bacillus* sp. has the highest ability to reduce the weight of the final residue.

References

[1] Kirchman David L, Processes in Microbial Ecology. Oxford University Press Inc. New York, 2012.

[2] Das N, and Chandran P. 2011. Microbial degradation of petroleum hydrocarbon contaminants: an overview, *SAGE Hindawi access to research Biotechnology Research International*, pp 1-13.

[3] Pao-Wen Grace Liu, Tsung Chain Chang, Liang-Ming Whang, Chun-Hsuan Kao, Po-Tseng Pan. Sheng-Shung Cheng. 2011. Bioremediation of petroleum hydrocarbon contaminated soil: Effects of strategies and microbial community shift. *International Biodeterioration & Biodegradation*. 65(8):1119-1127.

[4] Anwesha Banerjee, Atanu Roy, Suvakshan Dutta, Sandhimita Mondal. 2016. Bioremediation of Hydricarbon – A Review. *International Journal of Advanced Research.* 4(6):1303-1313. DOI : 10.21474/IJAR01

[5] Raj K Gupta, I P Abrol, Charles W Finkl, M B Kirkham, Marta Camps Arbestain, Felipe Macía,Ward Chesworth, James J Germida, Richard H. Loeppert, Maurice G Cook, G O Schwab, Jarosław Pytka, J J Oertli, Arieh Singer, William Joseph Edmonds, Yucheg Feng. 2016. *Soil Microbiology*. Springer Netherlands. DOI 10.1007/978-1-4020-3995-9-544

[6] Buchanan R E, & N E Gibbons (CoE). 1974. Bergey's Manual of Determinative Bacteriology. 8th Ed. S T Cowan J G, Holt J Liston, R G E Murray, C F Niven, A W Ravin, & R Y Stanier (Eds.). Baltimore.

[7] Covino S, Fabianová T, Křesinová Z, Čvančarová M, Burianová E, Filipová A, & Cajthaml T. 2016. Polycyclic aromatic hydrocarbons degradation and microbial community shifts during co-composting of Creosote treated wood. *J. Hazar mater.* 301:17-26

IOP Conf. Series: Earth and Environmental Science **106** (2017) 012101 doi:10.1088/1755-1315/106/1/012101

Characterization of biosurfactant produced by petrofilic bacteria isolated from hydrocarbon impacted soil and its potential application in bioremediation

D M Arsyah[1], E Kardena[2], Q Helmy[2,3]*

[1]Environmental Engineering Department, Faculty of Civil and Enviromental Engineering, Institut Teknologi Bandung, Bandung, Indonesia
[2]Environmental Engineering Department, Water and Wastewater Engineering Research Group, Institut Teknologi Bandung, Bandung, Indonesia
[3]Bioscience and Biotechnology Research Center, Institut Teknologi Bandung, Bandung, Indonesia

*Corresponding Author: helmy@tl.itb.ac.id

Abstract. The study adopts a multi-level perspective in technology transition to analyse how the transition process in the development of geothermal energy in Indonesia is able to compete against the incumbent fossil-fuelled energy sources. Three levels of multi-level perspective are socio-technical landscape (ST-landscape), socio-technical regime (ST-regime) and niche innovations in Indonesia geothermal development. The identification, mapping and analysis of the dynamic relationship between each level are the important pillars of the multi-level perspective framework. The analysis considers the set of rules, actors and controversies that may arise in the technological transition process. The identified geothermal resource risks are the basis of the emerging geothermal technological innovations in Indonesian geothermal. The analysis of this study reveals the transition pathway, which yields a forecast for the Indonesian geothermal technology transition in the form of scenarios and probable impacts.
Keywords: biosurfactants, emulsification index, fatty acids, surface tension

1. Introduction

Bioremediation is a process by which biological agents such as bacteria, fungi, or green plants are used to remove or neutralize contaminants in soil or water. Bioremediation typically involves augmentation of soil or other media, contaminated with pollutants act as nutrient for microorganisms, to improve process for biodegradation of the contaminants. Biodegradation rate of contaminant in soil depends on its bioavailability to the metabolizing organisms, which is influenced by factors such as desorption, diffusion, and dissolution [1]. Addition of biosurfactants as surface tension reduction agents became a promising method to improve bioremediation effectiveness of hydrocarbon-contaminated environments.

Most of persistent contaminants exhibit low water solubility, and hence the addition of emulsifier can increase bioavailability of the contaminants for the metabolizing microorganisms. By reducing surface and interfacial tension among liquids, solids, and gases, biosurfactans are allowed to disperse readily as emulsions so that the intake process of contaminants as a nutrient for microorganism can be improved [2]. Researcher [3] classified biosurfactants into five categories: (i) glycolipids, carbohydrate with the combination of aliphatic acids and long chained hydroxyl aliphatic acids, group of low molecular weight biosurfactants, (ii) phospholipids and fatty acids, containing one or more phosphate groups with phospholipid and glycolipid membranes, (iii) lipopeptide, classified as biosurfactants with high molecular weight, (iv) polymeric surfactants, biopolymer that can be used as thickener or emulsion stabilizer, and (v) particulate biosurfactants, as membrane or extracellular products act in uptake process of hydrocarbon as the substrate. This studies focusing on isolating bacteria that capable to produce biosurfactants, its properties and characteristics of biosurfactants, also its application for biodegradation of petroleum hydrocarbon.

2. Research Method

2.1. *Preparation of Bacteria Isolates*

In the beginning, screening process is carried out to obtain bacteria that potentially capable of producing biosurfactants. Mixed culture from Laboratory of Environmental Biotechnology, Institut Teknologi Bandung was incubated at room temperature 25°C with agitation speed of 110 rpm. In the mixed culture, 0.1% (v/v) of crude oil was added to create a suitable condition and environment for biosurfactants producing bacteria to grow. Furthermore, the culture is inoculated in nutrient agar on petri dish to acquire bacteria colonies that pass the screening process.

2.2. *Emulsification Index*

In emulsification test, 5 ml of supernatant is added with crude oil (1:1 v/v) and mixed in vortex with maximum speed for two minutes, and left for 24 hours. E24 is height measurement of the crude oil emulsion formed, divided with the total of column height after 24 hours, in percentage.

2.3. *Surface Tension Measurement*

In surface tension measurement, supernatants are tested for its surface tension at room temperature with Fisher® Tensiometer model 21, with Du-Nuoy platinum-iridium ring method. 30 ml of sample was put into a clean glass vessel and placed on tensiometer platform. The ring is pulled slowly until it breaks the surface of sample through the liquid-air interface. Between each measurement, the platinum wire ring was rinsed with water and allowed to dry.

2.4. *Measurement of Microorganism Growth*

The growth of biomass is measured through spectrophotometry analysis, with the measurement of optical density (O.D.) and Mixed liquor suspended solids (MLSS). O.D. are measured by Jenway® spectrophotometer model 6305 with 420nm wavelength. MLSS measurement is carried out by measuring its total suspended solid (TSS).

2.5. *16S rRNA Gene Sequencing Analysis*

Polymerase Chain Reaction (PCR) was carried out by Macrogen Inc., South Korea. Taxonomic assignments of sequences were performed using Ribosomal Database Project. The sequences in GenBank were retrieved using Basic Local Alignment Search Tool (BLAST) from NCBI.

2.6. *Isolation of Biosurfactants*

In this research, the measurement of biosurfactants quantity is including fatty acids fraction and exopolysaccharide fraction (EPS). The cell free broth, previously incubated with the same condition mentioned above is precipitated by two different methods. To obtain exopolysaccharide fraction, cold acetone is added to precipitate emulsifiers. Then mixture is allowed to stand overnight at 4°C to get the precipitate exopolysaccharide. Furthermore, the mixture sample is filtered through number 42 Whatman paper, and heated in oven at temperature 105°C for one hour, the measurement is performed by gravimetric method. Cold acetone method also chosen as biosurfactants extraction with organic solvents by [4]. To measure fatty acids fraction, culture supernatant is acid hydrolysed with HCl 2N until it reaches at pH 2, because in this condition, biosurfactants becomes insoluble in aqueous [5]. It then allows to stand for overnight at 4°C to precipitate lipid and protein fraction from biosurfactants. After that, extraction is performed by adding chloroform to the sample to extract fatty acids, because it will be soluble in chloroform. The organic later was transferred to a round-bottom flask and heated in water

IOP Conf. Series: Earth and Environmental Science **106** (2017) 012101 doi:10.1088/1755-1315/106/1/012101

bath at temperature 70°C, allowing the chloroform to evaporate, so then fatty acids will be left for further gravimetric method of measurement [6].

2.7. CMC Values
Determination of the critical micelle concentration (CMC) of rhamnolipids was performed by measuring the surface tension of aqueous solutions. The measurements were carried out with a tensiometer.

2.8. Biodegradation Assay
To determine the performance of biosurfactant on enhancing petrophilic bacteria in degrading petroleum hydrocarbon, a biodegradation assay developed and set up as follows:

- Blank reactor; no petrophilic bacteria were added.
- Control reactor; with addition of petrophilic bacteria but without biosurfactant.
- Reactor with the addition of petrophilic bacteria and biosurfactant.
- Reactor with the addition of petrophilic bacteria and chemical surfactant as comparison

5% (v/v) of biosurfactant was added to 500 ml of Basal Salts Medium (BSM) containing 5% crude oil. The flasks were incubated at room temperature on a rotary shaker (150 rpm). Total petroleum hydrocarbon (TPH) concentration are observed each day for 10-days period.

2.9. Total Petroleum Hydrocarbon
Measurement of TPH was conducted by gravimetric method. Sample was extracted with n-hexane, the organic layer was pooled and dried at 70°C temperature by evaporation of solvents. After evaporation, the amount of residual TPH recovered was weighted [7].

3. Results and Discussion

3.1. Biosurfactant production
From screening process, there are four samples of isolates that undergo selection process while its biosurfactants are tested for its E24 and surface tension value. Table 1 is showing the results of biosurfactants quality from each isolate. The chosen isolates are the one with highest E24 and the lowest surface tension, due to consideration that it will describes the best quality of biosurfactants. From the measurement, isolate A is chosen as isolate that will be inoculum for the next biosurfactants production process.

Table 1. The results of biosurfactants quality parameters of each isolate.

No.	Isolate	Value	
		E24 (%)	Surface Tension (dyne/cm)
1	A	84	43.5
2	B	52	52.5
3	C	61	49.5

It was then revealed by gene sequencing analysis of 16S rRNA that isolate A is identified as *Burkholderia* sp. For further reference, the isolates will be referred as as *Burkholderia* sp. PAU02. Growth and biosurfactant production from *Burkholderia* sp. PAU02 with 2% glucose as sole carbon source is described in Figure 1. It is indicating that biosurfactant production kinetics is categorized as "Growth Associated" [8]. The maximum biosurfactant produced by *Burkholderia* sp. PAU02 is obtained after 42-hour incubation amount to 5.18 g/l.

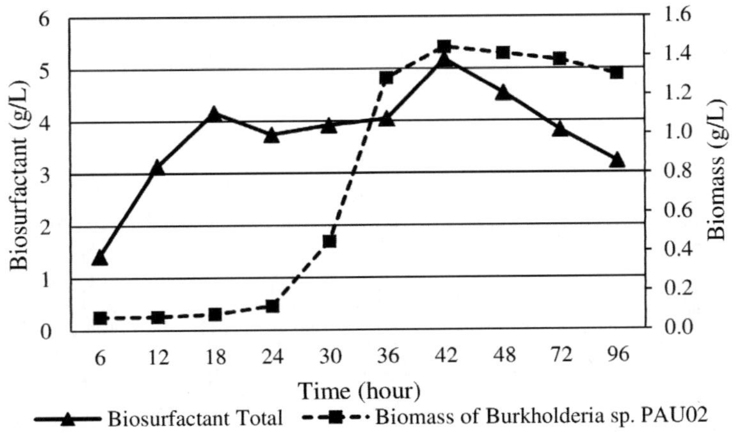

Figure 1. Time course study of biosurfactant production with glucose as a carbon source.

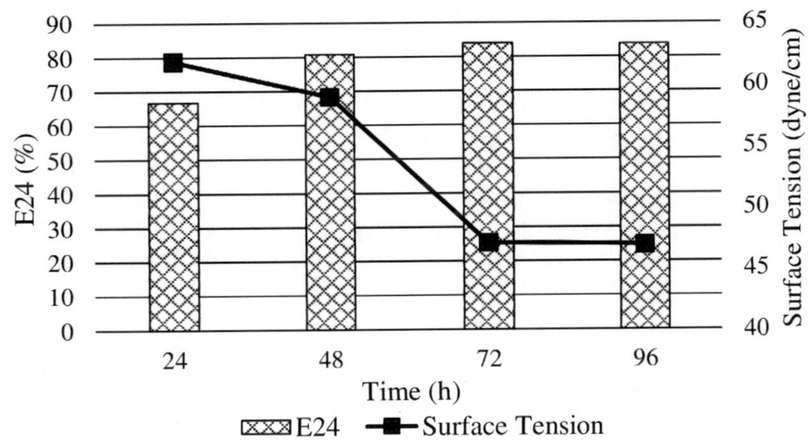

Figure 2. The quality of biosurfactant produced by *Burkholderia* sp. PAU02.

The quality of biosurfactants is measured each day for consecutive four days. Figure 2 shows the data related to quality of the biosurfactant of the respective isolate. *Burkholderia* sp. PAU02 produced optimum biosurfactant at 72-hour incubation period, reducing surface tension to the point of 47.0 dyne/cm and emulsification index of 84%. Based on comparison study of biosurfactant produced by *B. thailandensis* by [9], different method of extraction and purification of biosurfactant can alter the property of biosurfactant itself, including its surface tension. For this research, crude biosurfactant was measured without further purification process. The quantity of biosurfactants produced by each isolate was measured and can be seen in Figure 3. For *Burkholderia* sp. PAU02, the maximum biosurfactant production is achieved at 42-hour. Majority of biosurfactants is a secondary metabolite and some have an essential role in the defence mechanism, either through nutrient transport mechanisms, interactions between the microbial hosts or as biocides. By releasing biosurfactants into the growth medium, microorganisms can survive by using non-polar substrates as a carbon source, to sustain its growth and survival.

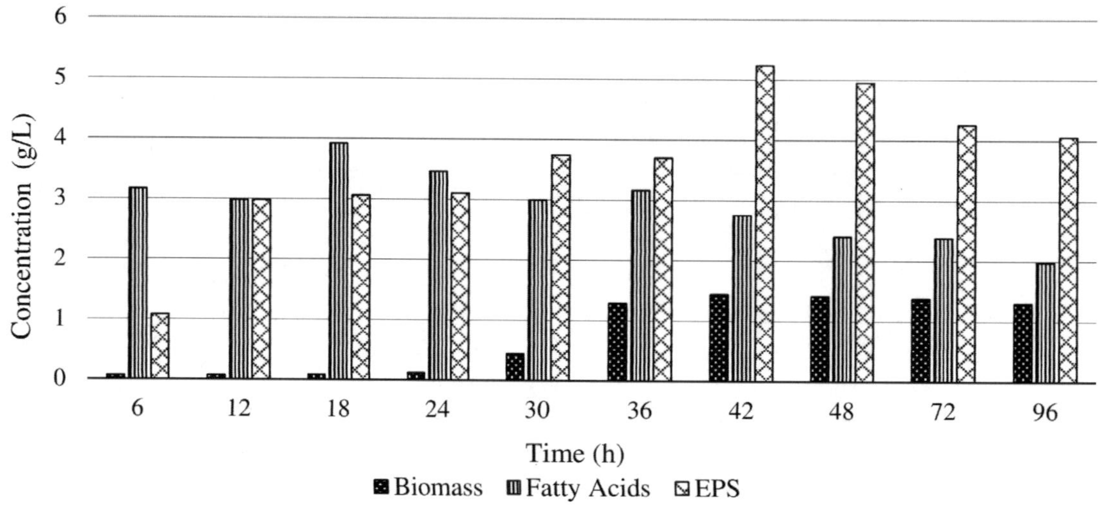

Figure 3. Production of fraction biosurfactant produced and biomass measured.

In contrast to primary metabolites, secondary metabolite is a product of metabolism produced due to change of the environment and inducible. Definition of secondary metabolites is not because the product is produced after growth, but because those products are not involved in the growth of the culture itself [10]. The biosurfactant kinetic production patterns is categorized as type of kinetics with a pattern of "simultaneous formation of an external product with growth" indicating that biosurfactant production rate proportional to the rate of growth. Measurement of CMC value of biosurfactant yielding 279 mg/L. CMC is important for biodegradation mechanism. After reaching CMC, solubilisation process is occured thus forming a micelle and increasing oil solubility.

3.2. Biodegradation Assay
Laboratory scale of biosurfactant enhanced biodegradation of crude oil was conducted. Effect of addition of biosurfactant and surfactant in the biodegradation process is shown in Table 2.

Table 2. TPH removal efficiency of petroleum hydrocarbon after 10-days by petrophilic bacteria in batch reactor with addition of various surfactants.

Sample	Initial TPH (%)	Final TPH (%)	Efficiency (%)
Blank*	5	4.479	10.43
Control** (C)	5	2.626	47.48
C + Biosurfactants *Burkholderia* sp. PAU02 0.34 ×CMC	5	1.728	65.44
C + Tween 80 (3.53 ×CMC)	5	1.687	66.26
C + Tergitol NP-10 (1.18 ×CMC)	5	2.836	43.29

*Blank: Medium + crude oil 5%; **Control: Medium + crude oil 5% + petrophilic bacteria

The efficiency of chemical surfactant Tween 80 is as effective as biosurfactant, due to its low toxicity and biodegradability. However, Tergitol NP-10 is showing even lower efficiency than control parameter. Compared to other non-ionic surfactants, Tergitol NP-10 have non-biodegradable characteristic with higher toxicity. Due to this reason, addition of Tergitol NP-10 is not as efficient compared to other kinds of surfactant. Based on GC/MS results (data not shown), it is known that biosurfactant addition to the system can enhanced the hydrocarbon removal process.

4. Conclusion

Biosurfactants produced by *Burkholderia* sp. PAU02 have a great potential for bioremediation application, due to its emulsifying effects and ability in reducing surface tension. From biodegradation assay studies, it is known that with addition of biosurfactant can enhance the rate of biodegradation efficiency on removing petroleum hydrocarbon up to 65%.

Acknowledgment

The study was performed within the ITB-P3MI research scheme; a research programs on community service, and innovation of ITB research group which is co-funded by the Indonesian Ministry of Research, Technology and Higher Education (Ristek-Dikti).

References

[1] Kardena E, Helmy Q, and Funamizu N. Biosurfactants and Soil Bioremediation. In: N. Kosaric, F. Vardar-Sukan & Sukan, eds. Biosurfactants Production and Utilization-Processes, Technologies, and Economics CRC Press New York pp. 327-360, 2014.
[2] Banat I, Makkar R, and Cameotra S. 2000. Potential commercial applications of microbial surfactants. *Applied Microbiol. Technol.* 53:495-508.
[3] Vater J, Kablitz B, and Wildes C. 2002. Matrix assisted laser desorption ionozation-time of flight mass spectrometry of Lipopetide biosurfactant in whole cell and culture filtrates of Bacillus subtilis C-1 isolated from petroleum sludge. *J.Appl. Environ.Microbiol.* 68:6210-9.
[4] Amrane A. 2008. Evaluation of different carbon and nitrogen sources in production of biosurfactant by Pseudomonas fluorescens. Desalination. 223:143-51.
[5] Mukherjee S, Das P, and Sen R. 2006. Towards commercial production of microbial surfactants. *Trends in Biotechnology*. 24:509-15.
[6] Vermani M, Kelkar S, and Kramat M. 1997. Studies in polysaccharide production and growth of Azotobacter vinelandii MTCC 2459, a plant rhizosphere isolate. *Lett. Appl.Microbiol.* 24: 379-83.
[7] Ellaiah P, Prabhakar T and Sreekanth M 2002 Production of glicolypids containing biosurfactant by Pseudomonas species. *Indian Journal of Experimental Biology*. 40:1083-6.
[8] Mukherjee S, Das P, Sivapathasekaran C, and Sen R. 2009. Antimicrobial biosurfactants from marine Bacillus circulans: extracellular synthesis and purification. *Letters in Applied Microbiology*. 48:281-8.
[9] De Rienzo D M, Kamalanathan I, and Martin P. 2016. Comparative study of the production of rhamnolipid biosurfactants by B. thailandensis E264 and P. aeuginosa ATCC9027 using foam fractionation. *Process Biochemistry*. 51:820-7.
[10] Demain A. Microbial secondary metabolism: a new theoretical frontier for academia, a new opportunity for industry. In: D. J. Chadwik & J. Whelan, eds. Secondary Metabolites: Their Function and Evolution John Wiley and Sons Ltd West Sussex. 1992.

Textile wastewater treatment: colour and COD removal of reactive black-5 by ozonation

I W K Suryawan[1], Q Helmy[2,3*], S Notodarmojo[1,2]

[1]Environmental Engineering Department, Faculty of Civil and Enviromental Engineering, Institut Teknologi Bandung, Bandung, Indonesia
[2]Water and Wastewater Engineering Research Group, Environmental Engineering Department, Institut Teknologi Bandung, Bandung, Indonesia
[3]Bioscience and Biotechnology Research Center, Institut Teknologi Bandung, Bandung, Indonesia

*Corresponding Author : helmy@tl.itb.ac.id

Abstract. Textile industries produced a large amount of highly coloured wastewater containing variety of dyes in different concentrations. Due to the high concentration of organics in the effluents and the higher stability of modern synthetic dyes, the conventional biological treatment methods are ineffective for the complete colour removal and degradation of organics and dyes. On the other hand, physical-chemical treatment are not destructive, mainly just concentrate and separate the pollutants phases. This research paper investigates the removal of colour and chemical oxygen demand/COD from textile wastewater using ozone treatment. Varied ozone dosages of 1.16; 3.81; 18.79; and 40.88 mg/minute were used in the experiment. Varied wastewater containing Reactive Black 5 (RB-5) concentrations of 40 mg/L, 100 mg/L were also applied. Research result showed the highest colour removal efficiency of 96.9 % was achieved after 5 hours incubation time, while the highest COD removal efficiency of 77.5% was achieved after 2 hours incubation time.
Keywords : COD, dye, ozonation, RB-5

1. Introduction

Indonesia is ranked among the top ten largest textile producing countries in the world and second largest in South East Asia with more than 4,500 companies [1]. The textile and garment industry is one of Indonesia's oldest industries and being labor intensive as a large source for jobs. The Indonesian government targets to increase the nation's value of exported textiles and garments to USD $75 billion by the year 2030, implying that this industry would contribute around 5 percent to global exports. Although, the textile sector is the flourishing sector in economics but the most important problem is the side impact of these industries. The textile manufacturing process is characterized by the high consumption of resources like water, fuel and a variety of chemicals in a long process sequence that generates a significant amount of waste. The common practices of low process efficiency result in substantial wastage of resources and a severe damage to the environment. The main environmental problems associated with textile industry are typically those associated with water body pollution caused by the discharge of untreated effluents.

During production process, different type of chemicals are used such as strong acids, strong alkalis, inorganic chlorinated compounds, hypochlorite of sodium, organic compound such as dye stuff, bleaching agent, finishing chemicals, starch, thickening agent, surface active chemicals, wetting and dispensing agents and salts of metals. Various dyes are used during dyeing stage for colouring purposes; multi colours are used to improve best of products [2]. The textile industry consumes large quantities of water and produces large volumes of wastewater from different steps in the dyeing and finishing processes. Wastewater from printing and dyeing units is often rich in colour, containing residues of reactive dyes and chemicals, such as complex components, many aerosols, high COD and BOD concentration as well as much more hard-to-degrade materials [3]. Dye wastewater effluent may lead to environmental problem such as eutrophication in receiving water bodies, and environmental concerns about the possible toxicity and carcinogenicity of some organic dyes. Dyes also decrease light penetration and photosynthetic activity, causing low dissolved oxygen in water bodies [4, 5, 6].

Many methods have been used such as biological, physical and chemical treatment of dye containing wastewater. Among these treatment technique, activated sludge, flocculation, and adsorption are the most commonly applied methods in Indonesia. Biological treatment processes are frequently used to treat textile effluents. However, colour is not readily removed from microbial-based treatment processes. Therefore, due to environmental implications and limitations of each of the above treatment methods, research on combining one of the biological treatment methods with other techniques such as chemical oxidation is needed to ensure a cost-effective and best approach technology for hard-to-degrade dyes/chemicals pollution treatment and control. Biologically treatment organics in textile wastewater must be accomplished via certain methods that enhance the biological degradability of organics, such as ozonation [7], [8].

2. Research Method

2.1. Materials

Azo-reactive black-5 (RB-5) dye was obtained from Bratachem. Dye solutions were prepared by dissolving the dye in distilled water to certain concentration to simulate loaded textile wastewater. Ozone was produced by a corona discharge type ozone generator (Resun RSO25) with dry air used as the feed gas. The ozone was fed into the reactor through a porous glass diffuser located at the bottom of the reactor to produce fine bubbles. The excess ozone leaving the reactor was destroyed by sequential 20 % KI traps incorporated to the reactor set-up as shown in Figure 1. In this study, experiments were conducted at room temperature and no reactor cooling was provided.

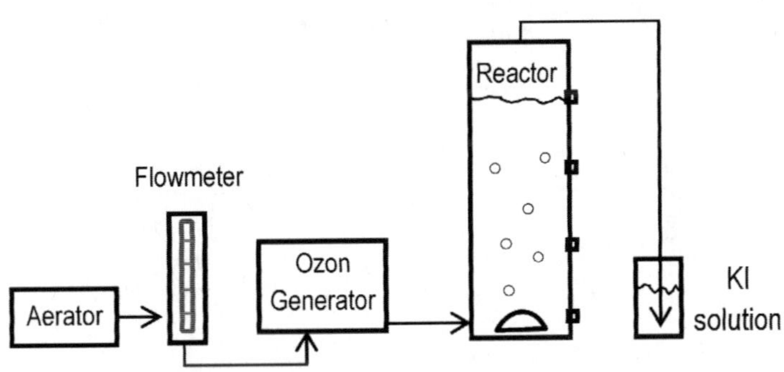

Figure 1. Illustration of the experimental apparatus used for the ozonation.

2.2. Experimental procedure and analysis

The volume of bench scale reactor is 2 L. O_3 is produced from dried air with O_3 output of 1.16; 3.81; 18.79; and 40.88 mg/minute and flow rate of the gas stream of 0.6; 1.4; 2; 4 LPM. Determination of ozone dosage is measured by iodometry method [9]. RB-5 initial concentration used in this study was 40 mg/L, and 100 mg/L.Starch was added to simulate the organic content of textile wastewater. During the experiments, a sample was taken from the sampling port in the batch system at regular time intervals for COD and colour measurements according to the standard methods for the examination of water and wastewater.

IOP Conf. Series: Earth and Environmental Science **106** (2017) 012102 doi:10.1088/1755-1315/106/1/012102

3. Results and Discussion

3.1. Colour removal

The changes in dye concentration and colour of RB-5 initial concentration of 40 mg/L after ozonation, are shown in Figure 2 (upper). Colour removal was quite rapid during the first 60 min except for doses 1.16 mg/min and the initial decolourization rate was higher for higher ozone doses. The treatment time of 15 min; colour removal was about 40% for the ozone doses of 1.16 and 3.81 mg/min, and over 60% for the ozone doses of 18.79 and 40.88 mg/min. Similar result was obtained by [10] with 50 mg/L RB-5 initial concentration and 10 g/h dosage of ozone resulting percentage of decolourization was approximately 31.8% after 60 min of reaction time.

The changes in dye concentration and colour removal of RB-5 initial concentration of 100 mg/L after ozonation, are shown in Figure 2 (lower). With ozone doses of 40.88 mg/min, the best efficiency for the removal of colour is 96.9% at 300 min incubation time. Whilst the treatment with the lowest colour removal efficiency corresponded to the ozone doses of 1.16 mg/min with 87.5%. The decolourization efficiency increased with increasing in ozone dosage.

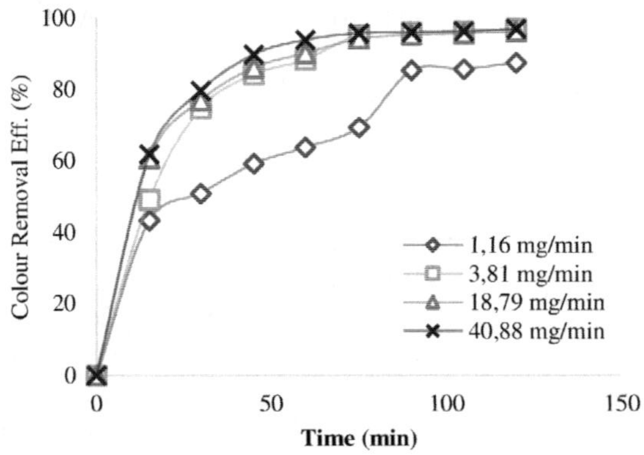

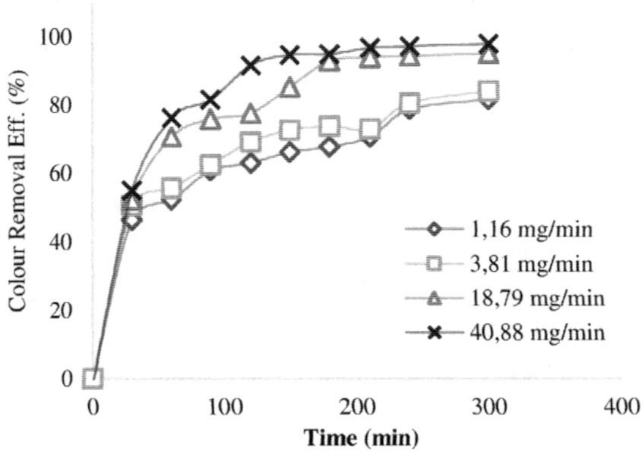

Figure 2. Colour removal efficiency of 40 mg/L (upper) and 100 mg/L (lower) initial concentration of RB-5 as the function of ozonation duration.

Results obtained in this study showed that colour disappears almost completely for 60 min (40 mg/L RB-5 initial conc.) and 150 min (100 mg/L RB-5 initial conc.) of ozonation, making it promising used as raw water for cleaner production application. Analysis from UV–Vis absorption spectrum (data not shown) of initial RB-5 and treated RB-5 showed that chemical structure breakdown was occured during the process. However, generated by-products require further treatment for complete degradation [10,11]. The absorbance decay in the UV region is considered evidence of degradation for the aromatic fragments of the dye molecules and oxidation of their intermediates. Similar result was reported by [12] using photocatalytic reaction in treating Acid Red-4.

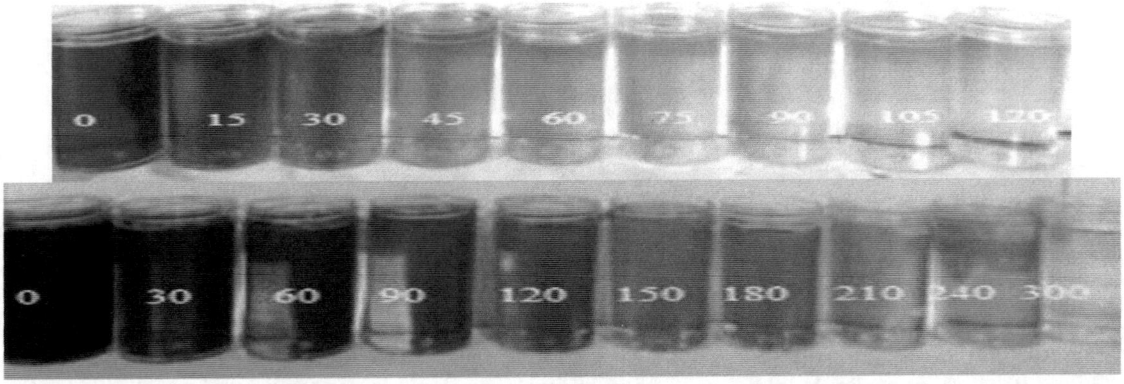

Figure 3. Visual image of colour gradation of RB-5 (upper: 40 mg/L and lower: 100 mg/L initial concentration) as the function of ozonation treatment with dosage of 40.88 mg/min.

The treatment of simulated textile wastewater was conducted by adding 100 mg/L RB-5 with 1 g/L starch into the reactor and ozonated for 120 minute incubation time. After the incubation period, the rate of colour degradation is slower compared to RB-5 without addition of starch. Organic presence in the wastewater makes the rate of colour removal efficiency decrease from 90% to 75% (Figure 4).

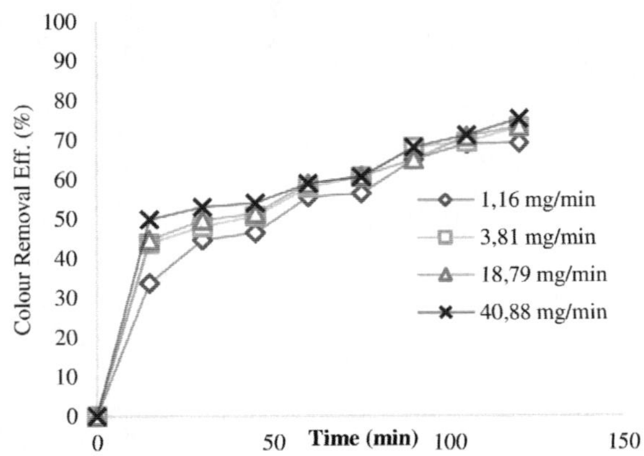

Figure 4. Colour removal efficiency of simulated textile wastewater by adding 100 mg/L RB5 with 1 g/L starch.

3.2. COD removal

Chemical oxygen demand is a common used parameter for the characterization of organic matter present in textile wastewaters; it depends on the dyes used in the production process. COD removal was observed in all ozonation treatment cases. COD reduction observed from 73.5% to 77.5% (40 mg/L RB-5 initial concentrations) was achieved after 120 minutes of treatment process. Similar COD reduction was found from 70.1% to 77.1% (100 mg/L RB-5 initial concentrations) was achieved after 300 minutes of treatment process. Different result occured in the simulated textile wastewater experiments, where COD reduction efficiency decrease to about 43 – 50% after 120 min incubation time indicating that organics breakdown still in the form of intermediate products (see Table 1).

Table 1. COD removal efficiency.

Ozone dose mg/min	Removal Efficiency, %		
	40 mg/L RB-5	100 mg/L RB-5	100 mg/L RB-5 + 1 g/L starch
1.16	77.5	77.1	43.8
3.81	65.0	73.6	50.7
18.79	70.9	64.5	44.9
40.88	73.5	70.1	46.4

3.3. Biodegradability

The biodegradability of the dye was performed for the simulated wastewater containing RB-5 dye and organics. Although the transformation products were not identified, the degradation of RB-5 into other by products was monitored by the change in the degree of oxidation. Research results showed that BOD_5/COD values of RB-5 mixture are increased after ozonation treatments as can be seen in Figure 5. This indicated that partial breakdown of complex into simpler compound is occured after ozonation process, making it more readily or suitable for biological process treatment.

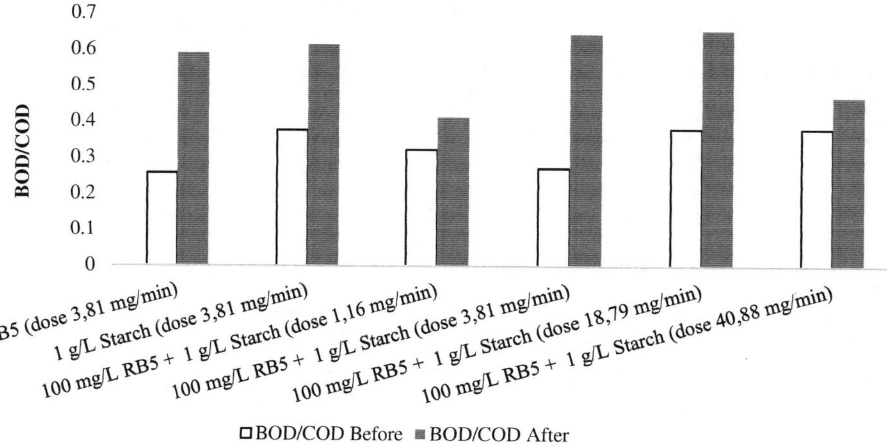

Figure 5. Change of biodegradability index (BOD_5/COD) during the ozonation of RB5.

4. Conclusion

The presented lab-scale ozonation treatment of RB-5 dye was designed to evaluate treatment efficiencies for colour and COD removal system. The results indicated that ozonation shows a promising efficiency in removing both colour and COD. Since ozone system represents a significant cost, instead of ultimate degradation to complete mineralisation of organic compounds, a partial breakdown to more readily biodegradable compounds is more preferable. After partial oxidation by ozone, biological process that utilize microorganisms can be used in a post-treatment process for complete mineralization of dye and organics in the wastewater. Therefore, the use of ozonation combined with biological degradation process may provide more economical and effective process in the treatment of highly coloured wastewater.

Aknowledgment

The study was performed within the ITB-P3MI research scheme; a research programs on community service, and innovation of ITB research group which is co-funded by the Indonesian Ministry of Research, Technology and Higher Education (Ristek-Dikti).

References

[1] CBS, Census report year 2014 Indonesian Central Bureau of Statistics, 2014, p.230.

[2] Dey S, and Islam A. 2015. A Review on Textile Wastewater Characterization in Bangladesh. *Resources and Environment*. 5: 15-44.

[3] Wang Z, Xue M, Huang K, and Liu Z. 2011. Textile Dyeing Wastewater Treatment, Advances in Treating Textile Effluent, Prof. Peter Hauser (Ed.), ISBN: 978-953-307-704-8, In Tech

[4] Fahmi C, and Rahmat N. 2010. Multi-stage Ozonation and Biologial Treatment for Removal of azo dye industrial effluent. *International Journal of Enviromental Science and Development*. 1: 193-8.

[5] Avsar Y, Kabuk H A, Kurt U, Cakmaki M, and Ozkaya B. 2012. Biological treatability processes of textile wastewater using electrocoagulation and ozonation. *J. of Scientific and Industrial Research*. 71: 496-500.

[6] Bilinska L, Gmurek M, and Ledakowicz S. 2016. Comparison between Industrial and Simulated Textile Wastewater Treatment by Aops – Biodegradability, Toxicity and Cost Assessment. *Chemical Engineering J*. 306: 550-9.

[7] Cuiping B, Xianfeng X, Wenqi G, Dexin F, Mo X, Zhongxue G, and Nian X. 2011. Removal of rhodamine B by ozone-based advanced oxidation process. *Desalination*. 274: 84-90.

[8] Khamparia S, and Jaspal D. 2017. Adsorption in combination with ozonation for the treatment of textile waste water: a critical review Front. *Environ. Sci. Eng*. 11: 1-18.

[9] APHA, Standard methods for the examination of water and wastewater, 17th eds. APHA, AWWA and WPCF Washington DC, 1989, 3464 p

[10] Chen H W, Kuo Y L, Chiou C S, You S W, Ma C M, and Chang C T. 2010. Mineralization of Reactive Black 5 in aqueous solution by ozone/H_2O_2 in the presence of a magnetic catalyst. *J. Hazard Mater*. 174: 795–800.

[11] Hoigne J, and Bader, H. 1976. The role of hydroxyl radical reactions in ozonation processes in aqueous solutions. *Water Res*. 10: 377–386

[12] Sugiana D, and Notodarmojo S. 2015. A study of photocatalytic degradation mechanism of acid red 4 azo dye using TiO_2 microparticle catalyst. *Arena Tekstil*. 30: 83-94

Potential of solid waste utilization as source of refuse derived fuel (RDF) energy (case study at temporary solid waste disposal site in West Jakarta)

D Indrawati*, M Lindu, P Denita

Environmental Engineering Department, Faculty of Landscape Architecture and Environmental Technology, Universitas Trisakti, Jakarta, Indonesia

*Corresponding Author : dindrawati@trisakti.ac.id

Abstract. This study aims to measure the volume of solid waste generated as well asits density, composition, and characteristics, to analyze the potential of waste in TPS to become RDF materials and to analyze the best composition mixture of RDF materials. The results show that the average of solid waste generation in TPS reaches 40.80 m^3/day, with the largest percentage of its share is the organic waste component of 77.9%, while the smallest amount of its share is metal and rubber of 0.1%. The average water content and ash content of solid waste at the TPS is 27.7% and 6.4% respectively, while the average calorific potential value is 728.71 kcal/kg. The results of solid waste characteristics comparison at three TPS indicate thatTPS Tanjung Duren has the greatest waste potential to be processed into RDF materials with a calorific value of 893.73 kcal/kg, water content level of 24.6%, andlow ash content of 6.11%. This research has also shown that the best composition for RDF composite materials is rubber, wood, and textile mixtureexposed to outdoor drying conditions because it produced low water content and low ash content of 10.8% and 9.6%, thus optimizedthe calorific value of 4,372.896 kcal/kg.
Keywords: briquette, refuse derived fuel, solid waste

1. Introduction

Related to the increasing energy demand in Indonesia, the current supply of electric power cannot compensate for the high rate of energy demand of 7–8% per year [5] so that an alternative source is needed. Solid waste has a high potential to generate energy and has economic value for society. One alternative to solid waste processing is by applying the concept of Waste-to-Energy, which is a process to produce energy in the form of electricity or heat coming from burning solid waste.

The variety of communal activities in urban areas increases along with the development of technology that further promotes the generation rate of solid waste. The amount of solid waste generated in Indonesia has reached 175,000 tons/day [7]. DKI Jakarta, especially West Jakarta, is one of the areas with the biggest solid waste generation of about 1,574. 92 tons/day [1].

Refuse Derived Fuel (RDF) is a material that has a high calorific value, formed from municipal solid waste that has been separated from its noncombustible materials and metals. Refuse Derived Fuel (RDF) is used as a fuel that is shaped like a pellet/briquette, will not rot even if it is stored for a long time which then makes it quite practical to be transported, and its heating value is greater than the unprocessed waste. Also, the burning results of RDF materials are more environmentally friendly compared to coal use.

IOP Conf. Series: Earth and Environmental Science **106** (2017) 012103 doi:10.1088/1755-1315/106/1/012103

Based on research conducted by [2], there are several types of waste that can be used as raw materials RDF seen from the calorific value. Kara's research [4], shows the specification of waste parameters used in the manufacture of RDF, which has a minimum heat value of 4000 Kcal/kg. Research on the utilization of solid waste from the landfill as a source of energy has also been done by [6] which shows that the potential solid waste reduction of 256 tons/day if processed as RDF. In this study, the potential of municipal waste that is used as a source of energy comes from municipal solid waste in the Temporary Solid Waste Disposal Site (TPS). Related to that, the research aims to analyzing solid waste composition in TPS and examine the potential utilization of municipal solid waste, especially from TPS as a new energy source (RDF).

2. Research Method

2.1. Sampling location

The samplings of solid waste were conducted at three Temporary Solid Waste Disposal Sites (TPS) located in West Jakarta area. They are TPS Arjuna in South Tanjung Duren Village, TPS Tanjung Duren in Tanjung Duren Village, and TPS Makaliwe in Grogol Village, West Jakarta. The samples were then analyzed in Environmental Laboratory, Faculty of Landscape Architecture and Environmental Technology, Trisakti University, and in Physical Chemistry Laboratory, Faculty of Mathematics and Natural Sciences, Bandung Institute of Technology.

2.2. Types of data collected

Types of the primary data collected in this study include data ofsolid waste generation, solid waste composition, water content, ash content, potential calorific value, and data on the quality of the generated RDF briquettes. The secondary data used are data from RDF-related literature studies, as well as the calorific value data of each component derived from the previous research.

2.3. Making of RDF briquettes

Making of RDF briquettes is done by the following steps:

1) Preparation of materials,
2) Shredding,
3) Drying 1,
4) Carbonization process,
5) Sieving process,
6) Mixing raw materials,
7) Molding and compaction,
8) Drying 2,
9) Testing of RDF briquette quality.

Table1. Mixed RDF material treatment.

Treatment	Mixed Composition of Solid Waste		Adhesive
$A1_1$, $A1_2$	Plastic	: 6,5 gr	20 gr
$A2_1$, $A2_2$	Paper	: 4,6 gr	
	Rubber	: 0,2 gr	
	Wood	: 0,5 gr	
	Textile	: 0,3 gr	
	Organic	: 87,9 gr	
$B1_1$, $B1_2$	Rubber	: 20 gr	20 gr
$B2_1$, $B2_2$	Wood	: 50 gr	
	Textile	: 30 gr	
$C1_1$, $C1_2$	Organic	: 100 gr	20 gr

1= Drying in Outdoor Condition; 2= Drying in Indoor Condition

The drying process in making RDF briquettes has been done in two conditions, namely outdoor conditions and indoor conditions. Furthermore, mixtures were made with a total weight of each 120 grams treatment, consisting of 100 grams of solid waste composition treatment and 20 grams of adhesive. The adhesive was prepared by blending 20 grams of starch and 200 ml of water which then boiled. Treatment variation and mixture composition were determined based on the condition of solid waste composition at the field, as well as the heating value of each component. Three experimental treatments were conducted to obtain the best composition variation of RDF briquette material, each of which consists of two replications that can be seen in Table 1.

3. Results and Discussion

3.1. Waste generation and composition

The solid waste generation data were collected based on the number and the capacity of the carrier vehicles entering the TPS so that the mean daily volume at each TPS could be obtained. Table 2 shows the average number of garbage carts, the amount of solid waste, and the density of solid waste in each TPS.

Table 2. Solid waste generation and density at the three TPS.

TPS	Number of Carts (unit)	Carts capacity (m^3)	Solid Waste Volume (m^3/day)	Density (kg/m^3)
Arjuna	19	1.5	28.50	147.91
Makaliwe	33	1.8	59.40	183.17
Tanjung Duren	23	1.5	34.50	189.29
Mean			40.80	173.46

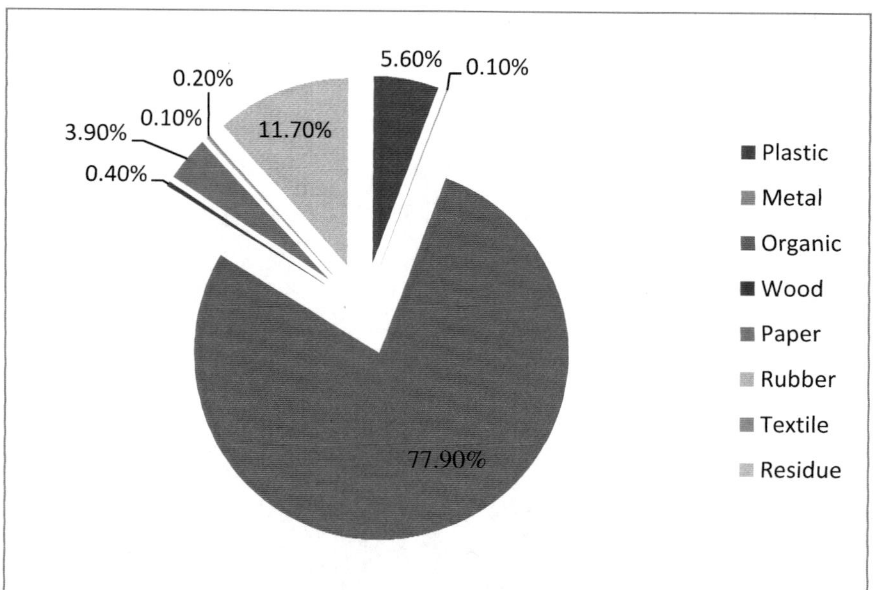

Figure 1. Solid waste composition at TPS.

The solid waste composition in this study was measured using SNI 19-3964-1994 regarding the method of sampling and measuring the generation and composition of municipal solid waste samples. Figure 1 shows the average composition of solid waste on all three TPS. The organic waste component has the highest percentage of 77.9%, followed by the residual waste of 11.7%, plastic waste by 5.6%, paper waste by 3.9%, 0.4% wood waste, 0.2% textile, and the smallest component is metal and rubber waste of 0.1%.

3.2. Water and ash content of solid waste

The average value of ash content for the rubber component in the first order was 1.45%, followed by the plastic component of 1.27%, the paper component of 0.88%, the organic component of 0.84%, wood component by 0.53%, and textile component by 0.43%. The difference of ash measured is due to the content characteristics in the solid waste components, including the high value of ash content since there are inorganic materials that cannot burn at a temperature of 550°C, leaving some remains formed to ash.

Table 3. Average of water content and ash content on TPS.

TPS	Water content (%)	Ash content (%)
Arjuna	29.22%	0.96%
Makaliwe	28.73%	0.90%
Tanjung Duren	23.67%	0.83%
Rata-rata	27.21%	0.90%

Table 3 shows the average moisture content and ash content in each TPS. TPS Makaliwe has the highest mean of the water content of 29.22%, followed by TPS Arjuna of 28.73%, and Tanjung Duren at 23.67%. Meanwhile, TPS Arjuna owns the highest rate of the ash content of 0.96%, followed by TPS Makaliwe of 0.90% and TPS Tanjung Duren with the lowest rate of ash content at 0.83%. The average water content of the TPS is 27.21%, while the ash content is 0.90%.

3.3. Result of solid waste potential analysis based on heat value

Figure 2 shows the ratio of the heating value of each solid waste's component at all three TPS. It can be seen that the calculated RDF briquette's calorific value could not comply with the various RDF standards sinceeach component for each TPS still has high water content and ash content. Thus, a scheme to optimize the yielded calorific value needs to be arranged. Besides, components with high calorific values such as plastic and rubber have relatively fewer quantities than the organic waste component in the TPS.

3.4. Quality of RDF

Solid waste samples used for RDF materials are solid waste originating from TPS Tanjung Duren because it has the most suitable potential in terms of calorific value, moisture content, and ash content. Compared to the other TPS, solid waste at TPS Tanjung Duren has the highest calorific value of 893.73 kcal/kg with water content and low ash content of 24.6% and 6.11% respectively. The produced RDF materials are included in the type of RDF-5, which is combustible. They were then compacted and printed in the form of briquettes.

3.4.1. Water content, ash content, and heat value of briquettes

Table 4 shows the means of moisture content, ash content, and heating value of RDF briquette test results in the laboratory. It can be seen that the average water content of the treated briquettes is ranging from 7.36–7.80%, indicating that the moisture content of RDF briquettes has conformed to the maximum water content standard of 15% based on [6].

IOP Conf. Series: Earth and Environmental Science **106** (2017) 012103 doi:10.1088/1755-1315/106/1/012103

Table 4. Water content, ash content, and heating value of briquette testing results.

Treatment	Moisture Content (%)	Ash Content (%)	Heat Value (Kcal/kg)
A	7.80%	6.10%	4,168.75
B	7.36%	4.9%	4,366.66
C	7.36%	5.9%	4,201.64

For ash content of briquettes, the results show that the ash content of each treatment has also fulfilled the maximum standard of 10% as stated in [4] as well. Contrarily, based on Table 4, the calorific value contained in the briquettes of each treatment could not meet the standard of 4,400 kcal/kg based on the regulation mentioned above. However, when compared to the Finnish, English, and Italian Standards of 3,000–4,000 kcal/kg according to researchers' [2], [6] standard, the briquette calorific value of each treatment can comply well with the criteria [5].

Following the results of moisture content, ash content, and calorific value of each treatment, the best mixture composition can be determined. The treatment that yielded the lowest moisture and ash content while keep maintaining high calorific value is treatment B that is a combination of rubber, textile, and wood with outdoor drying conditions. In 2016, the population in Grogol Petamburan sub-district is 225,038 people. Assuming that the solid waste generation rate is 0.5 kg/person/day, the total solid waste generation in Grogol Petamburan sub-district is 112,519 tons/day. Taking those data into account, if the waste is utilized as briquettes with mixture composition of B1 i.e. rubber, wood, and textile, the potential of the produced energy will be equal to 204.73 kW/hour or 4,913.62 kW/day.

4. Conclusion

The most potential solid waste to be processed into RDF materials is located in TPS Tanjung Duren because compared to the other TPS, it has the highest calorific value of 893.73 kcal/kg and the lowest water and ash content of 24.6% and 6.11% respectively. The best composition mixture for RDF materials concerning the levels of moisture content, ash content, and calorific value isB1 briquette treatment of rubber, wood, and textile with outdoor drying conditions. It produced a moisture content of 6.89%, an ash content of 4.9%, and a heating value of 4,372.896 kcal/kg.

References

[1] Cleanliness Services of DKI Jakarta Provinces. Data on Total Volume and Transported Solid Waste by Cleanliness Services of DKI Jakarta Provinces, Jakarta. 2014.

[2] Dong, T.T.T., & B. K., Lee. 2009, "Analysis of Potential RDF Resources from Solid Waste and Their Energy Values in the Largest Industrial City of Korea", Waste Management, 1725-1731.

[3] Gedebin, A., Leavens, A., Blackmore, K., & al, Refuse Derived Fuel, Current Practice and Prespective Final Report, Sweden: European Commission. e. 2003.

[4] Kara, M., Gunay, E., Tabak, Y., Yildiz, S., 2009, "Prespective for Pilot Scale Studi of RDF in Instanbul, Turkey", Waste Management, 29, 2976-2982

[5] Kuswardono, T., 2007. Alternative Energy Campaign, www.Walhi.or.id/campaign/energy/energy_ info_ 310504

[6] Maschita, D., C. Meidiana., I. R. D. Ari. 2014. Potential processing of solid waste from landfill as refused derived fuel (RDF) with local government-private partnership. Pure Jurnal Vo.3 No.4. E-ISSN: 2338-1760.

[7] Minister of Energy and Mineral Resources. 2006. Regulation of the Minister of Energy and Mineral Resources No. 047 of 2006 on Guidelines on the Making and Utilizing Coal Briquette and Coal-Based Solid Fuel. Ministry of ESDM, Jakarta.

[8] Ministry of the Environment. 2015. HLH 2015 Series - Plastic Waste Handling Dialogue, www.menlh.go.id.

[9] PT. Indocement Tunggal Prakarsa, Tbk. RDF Utilization in Indocement. Indocement, Citeureup. West Java. 2015.

IOP Conf. Series: Earth and Environmental Science **106** (2017) 012104 doi:10.1088/1755-1315/106/1/012104

Increasing urban community empowerment through changing of poverty rate index on the productive zakat impact

M H Zaenal[*], **A D Astuti, A S Sadariyah**

Center of Strategic Studies, The National Board of Zakat (BAZNAS), Jakarta, Indonesia

*Corresponding Author : mohd.hasbi@puskasbaznas.com

Abstract. We show how changes in poverty measures can be applied into growth of islamic philanthropy distribution via zakat, and we use the methodology to zakat community development (ZCD) program in Bantul during the 2016. The purpose of the present paper is to prove zakat is able to be a solution part for the community empowerment. The result is the number of productive zakat program beneficiaries whose income is below the poverty line (poor category) before the program are 244 people (H = 0.171) and after the program change to 168 (H = 0.118), which means the program has succeeded in reducing the number of poor people by 76 people (5.34 percent). The poverty gap (P1) of beneficiaries of productive zakat program in Bantul also decrease. The gap between poverty line and average income of beneficiaries is Rp 63,763 before the program, while the gap after the program is Rp 56,992. The income gap (I) is also decline from 0.197 to 0.169. Poverty severity of beneficiaries of productive zakat program in Bantul seen by Sen Index (P2) decrease from 0.093 to 0.062, while using Foster-Greer-Thorbecke Index (P3), the poverty severity decrease from 0.010 to 0.004. The analysis revealed the zakat community empowerment was significant economically in suppressing the poverty rate, and possible for reducing inequality and ending poverty in Indonesia.
Keywords: community empowerment, income of beneficiarie, poverty line, zakat

1. Introduction

Zakat is one of the primary sector of economic in the Muslim countries. As the third pillar of Islam, payment of zakat is an obligation for an eligible Muslim to purify his wealth by distributing it to the mustahiq based on the specific criteria. Zakat has a huge potential to be economically developed. In the last decade, zakat experienced a rapid growth marked by the increase of total zakat collection fund. However, this zakat growth still has a significant gap compared to its potential. According to Kahf, total potential of zakat in OIC member countries ranged from 1.8 to 4.34 percent of total GDP. If zakat is multiplied by the GDP at current prices in 2010 from the OIC member countries, the global potential zakat reached USD 600 billion [1].

In the context of poverty alleviation in Indonesia, zakat has the huge potential. Since Indonesia has the biggest Muslim population in the world which is 85 percent of total population in Indonesia or 216.66 million population. It can also be portrayed from the increasing of zakat, alms, and sadaqah (zakat, infaq, sadaqah or ZIS) collection fund since 2002 until 2015 (Table 1).

IOP Conf. Series: Earth and Environmental Science **106** (2017) 012104 doi:10.1088/1755-1315/106/1/012104

Tabel 1. Time series of ZIS collected in Indonesia [6].

Year	Billion IDR	Million USD*)	Annual Growth (%)	GDP Growth (%)
2002	68.39	4.98	-	3.70
2003	85.28	6.21	24.70	4.10
2004	150.09	10.92	76.00	5.10
2005	295.52	21.51	96.90	5.70
2006	373.17	27.16	26.28	5.50
2007	740.00	53.86	98.30	6.30
2008	920.00	66.96	24.32	6.20
2009	1200.00	87.34	30.43	4.90
2010	1500.00	109.17	25.00	6.10
2011	1729.00	125.84	15.30	6.50
2012	2200.00	160.12	27.24	6.23
2013	2700.00	196.51	22.73	5.78
2014	3300.00	240.17	22.22	5.02
2015	3700.00	269.29	21.21	4.79

*) 1 USD = IDR 13,740

Table 1 shows that the ZIS collection fund had increased 5310.15 percent since year 2002 until 2015. In year 2005 and 2007, it also increased significantly (almost 100 percent) which was predicted as the implication of tsunami in Aceh and earthquake in Yogyakarta. Both of these tragedies were stated as the national disaster in Indonesia. Table 1. also shows the increasing of ZIS collection fund since year 2002 until 2015 as 39.28 percent in average. This data indicated that the public awareness to pay zakat through the certified zakat institutions (Lembaga Amil Zakat or LAZ) has been increased. The positive trend of this zakat collection also implied the increase of public trust to the zakat institution's performance in managing zakat fund. The annual growth of ZIS collection fund was also higher than the GDP growth year to year. In year 2009, the GDP growth decreased 1.3 percent as a consequence of global financial crisis. On the contrary, zakat growth increased 6.11 percent. The average of zakat growth in 2002 to 2015 (39.28 percent) also showed a higher number than the average of GDP growth which only 5.42 percent. Generally speaking, the zakat growth is not much affected by the global crisis. Therefore, zakat has a huge potential to contribute to the national development.

The zakat development in Indonesia increased significantly when the Zakat Act No. 38/1999 was launched. Based on this act, zakat can be managed by the zakat institutions created by the government (*Badan Amil Zakat*) and also privately created by the public (LAZ). However, a major change in the regulatory framework occurred on the replacement of Zakat Act No. 38/1999 with the Zakat Act No. 23/2011 which brought all major private collectors under the supervision of National Zakat Board (BAZNAS). The Act No. 23/2011 aims to "improve the effectiveness and efficiency of the management of zakat services, and optimise the benefits of zakat for public welfare and poverty alleviation".

Based on this Act, the National Board of Zakat (BAZNAS) as an independent government agency responsible to the President of the republic of Indonesia. It had been given two main obligations: (1) to regulate entire zakat system including planning, implementation, controlling the process, audit, transparencies, collections, and distributions and (2) to coordinate all of the zakat institutions in the country cross-bodies, cross-provinces and cross-regencies.

2. Research Method

To assess and measure changes in material well-being of productive zakat program beneficiaries in Bantul, we have used basic need approach that is ability to fulfill human basic needs, based on the Indonesian Central Bureau of Statistics (BPS) standard. In 2016, BPS has released the poverty line in Bantul in the amount of Rp 360,169 per capita per month. Where, a person is called poor if he/she has average monthly expenditure below the poverty line. When to assess the change in poverty, we use of

IOP Conf. Series: Earth and Environmental Science **106** (2017) 012104 doi:10.1088/1755-1315/106/1/012104

some general poverty index such as Headcount Index, Poverty Gap, Poverty Gap Index (P1), Income Gap Ratio (I), Sen Index (P2), Poverty Severity and FGT Index (P3). In terms of data availability, we conduct the survey to the productive zakat program by BAZNAS in Bantul during 2016.

2.1. General poverty index

The word poverty comes from old French poverté (Modern French: pauvreté), from Latin paupertās from pauper (poor). Poverty is general scarcity or the state of one who lacks a certain amount of material possessions or money (people with $1.25 a day). Absolute poverty refers to the lack of means necessary to meet basic needs such as food, clothing and shelter. Absolute poverty is meant to be about the same independent of location. Relative poverty occurs when people in a country do not enjoy a certain minimum level of living standards as compared to the rest of the population and so would vary from country to country, sometimes within the same country.

The main poverty line used in the OECD and the European Union. When the United States, uses an absolute poverty measure created in 1963–64 and was based on the dollar costs of the U.S. Department of Agriculture's "economy food plan" multiplied. Both poverty measures are usually based on a person's yearly income and frequently take no account of total wealth. Major developments and research in this area suggest that standard one dimensional measures of poverty, based mainly on wealth or calorie consumption, are seriously deficient. This is because poverty often involves being deprived on several fronts, which do not necessarily correlate well with wealth.

The World Bank defines extreme poverty as living on less than US$1.90 per day (PPP), and moderate poverty as less than $3.10 a day. It has been estimated that in 2008, 1.4 billion people had consumption levels below US$1.25 a day and 2.7 billion lived on less than $2 a day.

2.2. Headcount index

The most common method to measure and report the poverty is the headcount ratio, which is given as the percentage of population that is below the poverty line based on the regional standards. For example, The New York Times in July 2012 reported the poverty headcount ratio as 11.1% of American population in 1973, 15.2 percent in 1983 and 11.3 percent in year 2000. The headcount index is one of the most widely-used measurements since it simply measures the proportion of the population that is counted as poor. In this case, we put the headcount index is used to find out the proportion of productive zakat beneficiaries, means the total beneficiaries who are poor categorized (in the sense of the local poverty line) of the total beneficiaries population.

The score describe how many poor based on BPS standard, among the productive zakat beneficiaries in Bantul. The headcount index often denoted by H, the formula as follows:

$$H = \frac{q}{n} \tag{1}$$

H = head count index; q = the number of poor people; n = the number of people in population

2.3. Poverty gap

The poverty gap indicator is produced by the World Bank Development Research Group to measure of the intensity of poverty. It is also defined as the average poverty gap in the population as a proportion of the poverty line. The poverty gap index is an improvement over the poverty measure headcount ratio which simply counts all the people below a poverty line, in a given population, and considers them equally poor. The poverty gap also described the average shortfall of the total population from the poverty line.

In the most cases, poverty line is indicated by the widely accepted international standard for extreme poverty. However, it's been difficult to set a common international poverty threshold since different countries have different thresholds for poverty. Thus, while the headcount index is just to show the proportion of poor people in population, the poverty gap measure how far the average individual income fall below the poverty line.

2.3.1. Poverty gap index
Poverty gap index estimates the depth of poverty by considering how far, on the average, the poor are from that poverty line [5]. Therefore, the sum of the income shortfall of poor people divided by the total number of poor people will show how far the gap is. The formula of P1 is as follows:

$$P = \sum_{t=1}^{q} gi_i \ v_i(z, y) \tag{2}$$

P_1 = poverty gap; g_i = z-y_i (the difference between the i-poor people and the poverty line/income shortfall), z = official poverty line, y_i = income of i-population, where I = 1, 2,..., q

This measure is the mean proportionate poverty gap in the population (where the non-poor have zero poverty gap). Some people find it helpful to think of this measure as the cost of eliminating poverty (relative to the poverty line), because it shows how much would have to be transferred to the poor to bring their incomes or expenditures up to the poverty line (as a proportion of the poverty line). The minimum cost of eliminating poverty using targeted transfers is simply the sum of all the poverty gaps in a population; every gap is filled up to the poverty line. However this interpretation is only reasonable if the transfers could be made perfectly efficiently, for instance with lump sum transfers, which is implausible [2]. It can be seen that the ratio of the minimum cost of eliminating poverty with perfect targeting (i.e.Gi) to the maximum cost with no targeting. Thus this measure is an indicator of the potential saving to the poverty alleviation budget from targeting: the smaller is the poverty gap index, the greater the potential economies for a poverty alleviation budget from identifying the characteristics of the poor – using survey or other information – so as to target benefits and programs.

2.3.2. Income Gap Ratio (I)
In addition to the poverty gap index, there is also a method to measure the gap between the poor people and the poverty line which is Income Gap Ratio (I). The income gap ratio is a relative gap between the poverty line and the average income of the poor [3].

$$I = \sum_{i=\in S(z)} \frac{gi}{qz} \tag{3}$$

I= income gap ratio; g_i = z - y_i (income short-fall of i-poor people, z = poverty line, y_i = i-individual income; q = total people whose is below the poverty line.

2.3.3. Poverty Severity
Despite the ability to see the depth of poverty, poverty gap and income gap cannot capture the inequality between the poor. Therefore, other indexes such as Sen Index as well as Foster, Greer, and Thorbecke Index, are used to see the poverty severity.

2.3.4. Sen Index (P2)
The Sen index sought to show how the headcount and income gap ratio, along with the Gini index of the income distribution of the poor, can give an adequate picture of poverty. Because of this poverty, the Sen index is said to include the three I's of poverty: Incidence, Intensity and Inequality [4].

$$P_2 = \left[I + (1 - I)G_p\right] \tag{4}$$

2.3.5. FGT Index (P_3)
Another method to see the poverty severity by thought of as one of a family of measures proposed by Foster, Greer and Thorbecke (FGT) Index (1984). The general formula for this index depends on a parameter α which takes a value of zero for the headcount, one for the poverty gap, and two for the squared poverty gap. Quite generally, as:

IOP Conf. Series: Earth and Environmental Science **106** (2017) 012104 doi:10.1088/1755-1315/106/1/012104

$$P_\alpha(y,z) = \frac{1}{n}\sum_{i=1}^{q}\left(\frac{g_i}{z}\right) \quad (5)$$

G_i = z-y_i (income short-fall of i-poor people, z = poverty line, q = total people whose is below the poverty line, α = parameter of sensitivity with value >= 0.

2.3.6. Poverty Rate Index Changes Before and After
The next steps after conducting the assessment using these methods and formulas, we noticed changes in poverty level and material well-being of beneficiaries. Table 2 show the result for a year before and after receiving the program.

Table 2. Poverty rate index changes before and after.

Index Score	Before Program	After Program
H	0.171	0.118
P_1	IDR 63,763	IDR 56,992
I	0.197	0.169
P_2	0.093	0.062
P_3	0.010	0.004

3. Results and Discussion

3.1. Headcont index
From the table, show the number of total beneficiaries (356 people) whose income is below the poverty line (poor category) before the program are 244 people (H = 0.171) and after the program change to168 (H = 0.118), which means the program has succeeded in reducing the number of poor people by 76 people (5.34 percent).

This assuming, the indexwhen getting close to 1 means more number of the poors, while when getting closer to 0 means the poors number are decreased. Thus, a good empowerment program is when the H index values after the program are less than the H index before the program.

3.2. Poverty Gap
From the table, show the poverty gap (P_1) among the beneficiaries also decrease. Before the program, the gap between poverty line and the average income of beneficiaries is Rp 63,763, while after the program the gap is IDR 56.992. Meaning that the average distance of the poors income to the poverty line (IDR 360,169), before the program IDR 63,763 and after the program IDR 56,992. Assumed, that the smaller and closer to 0 in *rupiah*, the smaller poverty gap has been generated. Thus, a good empowerment program is when the P_1 indexafter the program are less than the P_1 index before the program.

3.3. Income gap
From the table, show the income gap (I) is also decline from 0.197 to 0.169. Assuming, when the indexgetting close to 0 means the value of income gap is smaller and the poorer the better. Thus, a good empowerment program is when the I indexafter the program are less than the I index before the program.

3.4. Poverty Severity
From the table, show the poverty severity index among the beneficiaries seen by Sen Index (P_2) decrease from 0.093 to 0.062, while using Foster-Greer-Thorbecke Index (P_3) formula, the poverty severity decrease from 0.010 to 0.004. In the Sen Index (P_2) assuming, when the indexgetting close to 0 means the poverty severity was decreased. Thus, a good empowerment program is when the Sen

IOP Conf. Series: Earth and Environmental Science **106** (2017) 012104 doi:10.1088/1755-1315/106/1/012104

indexafter the program are less than the Sen index before the program. As well as in the FGT Index (P3) assuming, when the indexgetting close to 0 means the poverty severity was decreased. Thus, a good empowerment program is when the FGT indexafter the program are less than the FGT index before the program.

4. Conclusion

Based on the above empirical assessment, can be concluded that zakat distribution for the community empowerment in Bantul is possible for reducing inequality and ending poverty. This can be an example program for other regions in Indonesia, including in urban communities. From the calculating of five indexes, we can conclude that the program is effectivefor reducing the poverty, as well as can increase of the welfare level among the beneficiaries.

References

[1] The World Bank, 1.4 Billion Live On Less Than US$1.25 A Day, But Progress Against Poverty Remains Strong, 2015.
[2] The World Bank, Poverty Manual, J H Revision, 2005.
[3] Chakravarty S R, Inequality, Polarization and Poverty, Springer, New York, 2009.
[4] Bellù Lorenzo Giovanni, and Liberati Paolo, Social Welfare Analysis of Income Distributions: Ranking Income Distributions with Crossing Generalised Lorenz Curves. University Library of Munich, Germany in its series MPRA Paper with number 30115, 2005.
[5] David B Grusky, S M Ravi Kanbur, Amartya Kumar Sen, Poverty and Inequality, Stanford University Press, 2006.
[6] National Board of Zakat (BAZNAZ), 2016, Indonesia Zakat Outlook 2017.

Improvement of bagasse become lignosulfonate surfactant for oil industry

R Setiati[1]*, S Prakoso[1], S Siregar[2], T Marhaendrajana[3], D Wahyuningrum[4], S Fajriah[5]

[1]Petroleum Engineering, Faculty of Technology Earth and Energy, Universitas Trisakti, Indonesia
[2,3]Petroleum Engineering, Faculty of Technology Earth and Energy, Institut Teknologi Bandung, Indonesia
[4]Faculty of Mathematic and Natural Science, Institut Teknologi Bandung, Indonesia
[5]Chemical Research Centre, Lembaga Ilmu Pengetahuan Indonesia, Indonesia

*Corresponding Author : rinisetiati@trisakti.ac.id

Abstract. The aim of this research is to use bagasse as a raw material as an effort to enhance the petroleum acquisition. There are two steps of processing bagasse into surfactant lignosulfonate which are the separation process of lignin from bagasse and the sulfonation process of lignin into lignosulfonate. The formation of surfactant lignosulfonate is a result of reaction between ion lignin with bisulfit. The sulfonate group/cluster at lignosulfonate are hydrophilic group which causes lignosulfonate to have amphipathic structure (surfactant). The comparison uses infrared test result against the formed component in surfactant sodium lignosulfonate from bagasse. With the two crude oil samples used in phase test, it turned out that at some light petroleum sample which were mixed with surfactant lignosulfonate formed a middle phase emulsion, with middle phase emulsion stability happening after the second day with a comparison of 10 – 50%. Meanwhile the heavy crude oil did not form a middle phase emulsion at all. Therefore, it can be concluded that bagasse has enough potential to be processed into surfactant lignosulfonate and to be used as injection fluid in the EOR process in oil industries.
Keywords:bagasse, enhanced oil recovery, lignosulfonate, surfactant injection

1. Introduction

Currently in Indonesia there are 63 sugar mills which, 53 factories are state-owned companies while the other ten private-owned sugar factories. So it is necessary for the development of sugarcane plantations and sugar factories in order to meet the needs of the sugar. Actually, based on statistical data, extensive sugar cane crop in Indonesia 395,399.44 hectares, which is spread over an area of 99,383.8 hectares island of Sumatra, Java, covering an area of 265,671.82 hectares, an area of 13,970.42 hectares of Kalimantan, and Sulawesi area of 16,373.4 hectares [12]. It is estimated that each hectare of sugar cane plant capable of producing 100 tons of bagasse. Then the potential of bagasse available from total area of sugarcane plants reached 39,539,944 tons per year. This is taken into consideration for the selection of bagasse processed since sugar cane land in Indonesia is quite large, spread from western Indonesia to East Indonesia, ranging from North Sumatra, Palembang, Lampung, Java, and Sulawesi, so it can be said that natural resources has provided. The development of sugar plantations will support the needs of the sugar industry, which will produce enough sugarcane waste too much. Total production of bagasse annually is relatively abundant, easily available and cheap. Based on data from the Indonesian Sugar

Plantation Research Center (P3GI), bagasse obtained as much as 32% of the weight of cane milled or about 10.2 million tonnes/year or milled per season in Indonesia [13].

Bagasse is the waste from the extraction process cane into sugar products after experiencing five times the milling process. Known as waste because after the mill, there is no more liquid sugar that can be utilized. The rest is sugar cane fibers that are usually considered as goods that are no longer useful are only used as a furnace fuel in the sugar mills or as a mixed fodder. Currently bagasse mostly just as a fuel in boilers, paper-making raw material, or as a source of animal feed. In general, the sugar factory in Indonesia utilizes bagasse as fuel for the plant itself, after the experience bagasse drying process.

Based on the research results done, in the bagasse consist of lignin, cellulose and hemicellulose. In particular, lignin is the basic material of lignosulfonate. Sulfonation processed using bisulfite sodium, so that lignin ion be binded with bisulfite ion to form lignosulfonate. Lignosulfonates containing hydrophilic group (group sulfonate, hydroxyl phenyl and hydroxyl alcohol) and hydrophobic groups (carbon chain) that belong to the group of anionic surfactants [8].

Table 1. The lignin composition at plant waste.

No	Nabati Waste	Cellulose	Hemicellulose	Lignin
1.	The wood of width[1]	40 - 55	24 - 40	18 - 25
2.	The wood of needle[1]	45 - 50	25 - 35	25 - 35
3.	Leave[2]	15- 20	80 - 85	0
4.	Lump Corn[2]	45	35	15
5.	Nut shell[2]	23 - 30	25 - 30	30 - 40
6.	Grass of wheat[3]	30	50	15
7.	Bagasse[3]	50	25	25
8.	Palm empty bunch[3]	41.30 – 46.50	25.30 – 33.80	27.5060 - 32

Source : [5], [3], [14], [9], [15]

Table 2. Reference of bagasse composition.

Reference	Lignin	Cellulose	Hemicellulose
[1]	13	40	29
[2]	18	45	-
[4]	24.2	52.7	20
[11]	21	37	28
[12]	29.46	58.39	17.99

The anionic surfactant formed because of sulfonate group in the molecule lignosulfonates. Lignosulfonate is one type of anionic surfactant which is widely used in the oil industry as an injection fluid to the reservoir in an effort to enhanced oil recovery. This lignosulfonate surfactants have the performance to decrease interfacial tension on the water - oil so that oil can move to be produced. From table 1 shows several other waste which also has a content of lignin, cellulose and hemicellulose. Of these, it appears that one of material, bagasse is containing enough lignin. Table 2 show the result of previous studies of lignin, cellulose and hemicellulose contained in bagasse. Data of components of lignin, cellulose and hemicellulose enough to be reprocessed and these things into consideration that the bagasse can be used as raw material for the manufacture of surfactants with the concept of biomass because it uses vegetable raw materials derived from plants.

Surfactant Lignosulfonate is one of lignosulfonate components that made based raw material lignin with reagent sodium hydroxide and sodium bisulfite, so its called Surfactat Sodium Lignosulfonate. Sulfonate is known with the formula R-SO_3Na as the simplification of sulfat R-O-SO_3Na [6]. The standard formula is shown below.

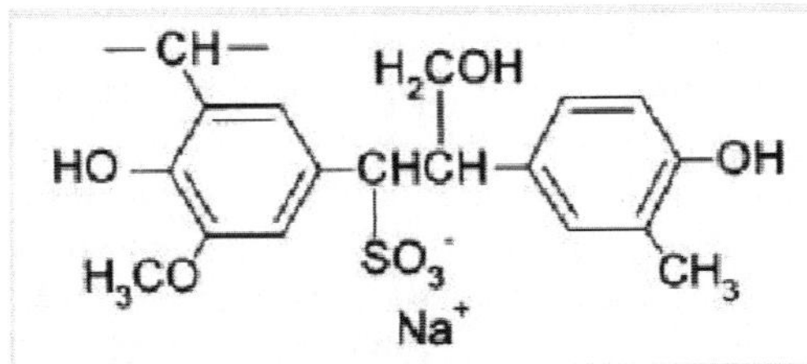

Figure 1. Lignosulfonatestructure [10].

R is the group of carbon atom aromatic $C_8 - C_{22}$ which constitutes the group of hydrophilic whilst group of hydrophobic consists of carbosilat, sulfonate, fosfat or other acid. The Lignosulfonate Surfactant is categorized as surfactant anionic because it has a group of sulfonate and salt ($-NaSO_3-$) which constitutes anion (head) and group of hidrocarbon as the tail. For lignosulfonate, there is propanoid phenyl compounds as the basis of lignosulfonate compound [7]. The Lignosulfonate Surfactants which being used as an injection fluid in EOR process are still commercial petroleum based surfactants (petroleum lignosulfonate). Surfactants are used as injection fluids since surfactants could lower the oil-water interface tension so that the oil could easily move for production. Surfactant is a surface active substance that could lower the oil-water interface tension to a much lower level.

Surfactant–water–oil mixture could form a bottom phase emulsion (dissolve in water), middle phase emulsion/micro emulsion (dissolve in oil and water phase) and top phase emulsion (dissolve in oil). Meanwhile in the EOR process, the expected emulsion is the middle phase emulsion (micro emulsion) because one of the phases in this condition became a continuous phase (external phase) and the others forms granules (discontinuous phase). Due to this condition, the surfactant needs to have a suitable character with petroleum, among others is the ability to form a stable middle phase emulsion. With the ability of forming a stable middle phase emulsion, some of the surfactant could integrate with water and some with oil. This thought is what became a consideration to use bagasse as a raw material in this research as an effort to enhance the petroleum acquisition.

2. Research Method

Materials used are derived bagasse from sugar mills that were previously sifted and dried.Then bagasse already sifted back with a sieve shaker to obtain powder size bagasse with certain mesh. The equipment used is a sieve shaker, reflux (flask and condenser), magnetic stirrer, hot plate, filter paper and oven.

Methods of processing bagasse into lignosulfonate done through two processes, namely the process of isolation lignin from bagasse and processof lignin sulfonation becomes sulfonate. Insulating material for isolation lignin are water, sulfuric acid (H_2SO_4) and sodium hydroxide (NaOH), and for the lignin sulfonation process using sodium bisulfite ($NaHSO_3$).

At this research, sugarcane bagasse has processed into lignosulfonate surfactant. The products have component tested using the infra-red with FTIR instruments and lignosulfonate structure tested using NMR test, then the results were compared with standard commercial lignosulfonate.

3. Results and Discussion

This processing begins by entering the bagasse has been sieved with sieve shaker into the flask and refluxed directly in sodium hydroxide solution at a concentration of 2% for five hours. Then results of reflux NaOH is filtered, diluted and neutralized by titration of concentrated sulfuric acid (H_2SO_4 98%) to pH = 2 and allowed to stand at least eight hours until the precipitate, filtered and dried in an oven at a temperature of 70 °C. The precipitate obtained is lignin isolated from the bagasse.The lignin was further processed by using sodium bisulfit through sulfonation process for five hours at 100°C. The result should be process through oven drying and heating until it got the brown powder shaped. This powder is surfactant sodium lignosulfonate. To validate sulfonation results will be tested by FTIR, LCMS and NMR. Test with FTIR to obtain the existence of the components contained in the sample. With LCMS test to get the components in detail the visible presence by infra red test. And lastly, using NMR test to obtain the sample's molecular structure.

Figure 2 below, there are several differences among the FTIR result of lignin and surfactant, in which the blue line curve as FTIR Surfactant curve is a peak picking primer on the 1,635.34 cm^{-1} wave number as stretching vibration Alkena, at 1,384.64 cm^{-1} wave number as stretching vibration Sulfat, with 1,114.65 cm^{-1} wave number as stretching vibration Carbocylic Acids, at 462.832 cm^{-1} wave number as stretching vibration Ester. From the result of the infra-red test, it appears that there has been peak picking of surfactant curve (blue curve) compared to lignin curve (black curve). There is a curve difference through 3-peak picking curves which proved that the lignin surfactant components consists of alkena, sulfat and carboxylic acid.

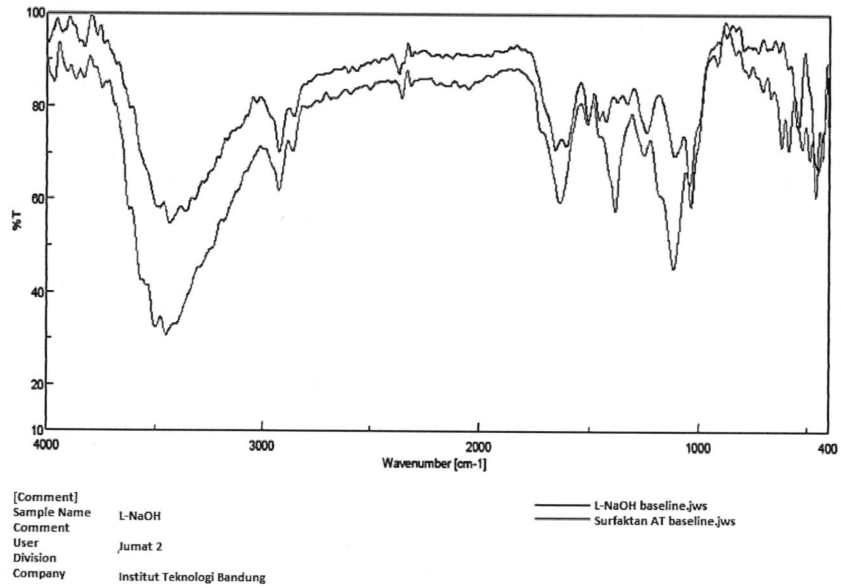

Figure 2. The result FTIR of Lignin and Lignosulfonat (from Bagasse) [13].

The FTIR Lignosulfonate Surfactant curve is a peak picking primer on the 1,635.34 cm^{-1} wave number as stretching vibration Alkena, at 1384.64 cm^{-1} wave number as stretching vibration Sulfat, with 1,114.65 cm^{-1} wave number as stretching vibration Carbocylic Acids, at 462.832 cm^{-1} wave number as stretching vibration Ester. From the result of infra-red test, wave-length spectrum which read as indicates for peak picking especially sulfat at 1384.64 cm^{-1} wave number in which the sulfat has peak picking range within 1,350 – 1,450 cm^{-1} wave number. The appearance of the peak picking indicates that Surfactant LignoSulfonate has completely processed which can be seen on Table 3. The result of infrared spectrum that Lignosulfonate Surfactant has been compared with standart the Lignosulfonate Surfactant of Aldrich products as shown in the Table 3 below.

Tabel 3. Comparison spectrum lignosulfonate bagasse and lignosulfonate standart (Aldrich).

No.	Component of Lignosulfonate	Wave number (cm^{-1})	
		Standart Lignosulfonate (Aldrich)	Bagasse's Lignosulfonate
1.	Alkena C=C	1608.34	1635.34
2.	Sulfate S=O	1365	1384.64
3.	CarboxylicAcids C=O	1187.94	1114.65
4.	Ester S-OR	499.831	462.832

From the NMR test results at Figure 3, can be seen the correlation of the H-3 and H-5 to C-5 / C-3, C-1 and C-7; H-7 correlated with C-8 and H-9 correlated with C-8 and C-7. Figure 4 show propanoid phenyl compounds as the basis of lignosulfonate compound. It could be concluded that the process of sulfonation lignin into lignosulfonates has succed.

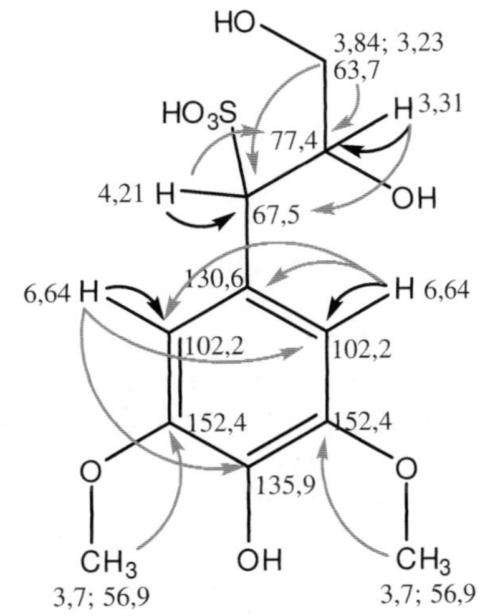

Figure 3. Result of NMR test – Lignosulfonat Surfactant from bagasse.

4. Conclusion

Based on the result, it could be concluded from surfactant sulfonation process, that the product is suitable with surfactant standard components that contain four main components such as Alkena, Sulfate, Carboxylic Acids, and Ester. From NMR test, it is found that propanoid phenyl compounds as the basis of lignosulfonate compound at the lignosulfonate surfactant from bagasse. Based on above findings, bagasse has a good potential to be processed into lignosulfonate surfactant.

Acknowledgment

This research was facilitated by the Research Consortium OGRINDO at ITB, Universitas Trisakti, and the Indonesian Directorate General of Higher Education Decentralization Fund for PUPT Fund 2014 Usakti – Dikti, Decentralization Fund 2014 ITB, PUPT Fund 2015 ITB – Dikti and PUPT Fund 2016 ITB – Dikti. We also thank to ISoSUD 2017 who publishes this article, which in turn will benefit the society.

References

[1] Arora DS, Chander M, Gill PK.2002. Involvement of lignin peroxidase, manganese peroxidase and laccasein the degradation and selective ligninolysis of wheat straw, *Int. Bioterior Biodegrad.* 50:115 – 120

[2] Bon EPS, Ferara MA. 2007. Bioetanol Production via Enzymatic Hydrolysis of Cellulosic Biomass, *FAO. Seminar on the Role of Agricultural Biotechnoligies for Production of Bioenergy in Developing Countries, Rome.*

[3] Boopathy R. 1998. Biological Treatment of Swine Waste Using An Aerobic Baffela Reactors, *Bioresource Technol.* 64:1 – 6.

[4] Brady J W, Himmel ME, Ding S H, Johnson D K, Adney W S, Nimlos M R, and Foust, TD. 2007. Biomass Recalcitrance Engineering Plants and Enzymes for Biofuels Production, *Science* 315:804 – 807.

[5] Cheung SW, Anderson BC. 1997. Laboratory Investigation of Ethanol production from municipal primary wastewater.*Bioresource Technol.* 59:81 – 96.

[6] Fujimoto T. 1985. *New Introducing to Surface Active Agents.* Kyoto, Japan: Sanyo Chemical Industries, Ltd.

[7] Lutnaes B F, Myrvold B O, Lauten R A, Endeshaw MM. ^{1}Hand ^{13}C NMR data of benzylsulfonic acids – model compounds for lignosulfonate, Magnetic Resonance in Chemistry. Willey Interscience. 2007.

[8] Ouyang JM. 2006. Effect of Temperatue on growth aggregation of calcium oxalate in presence of varous carboxylic acids in silica gel system, *Mat. Sci Eng C.* 26:679-682.

[9] Pandey A, Soccol CR, Nigam P, Soccol VT, Vandenberghe LPS, Mohan R. 2000. Biotechnological potential of agro-industrial residues, *Biorsource Technol.* 74(1):81 – 87

[10] Rivai M. 2008. Analisa Kinerja Surfaktan Metil Ester Sulfonat (MES) Ramah Lingkungan Dari CPO, CJO dan CNO, Pusat Penelitian Surfaktan dan Bioenergi, IPB, Bogor.

[11] Samsuri. 2007. Pembuatan Selulosa Bagas untuk Produksi Etanol Melalui Sakarifikasi dan Fermetasi Serentak dengan EnzymXylanase, *Jurnal Makarta Teknologi.* 11.

[12] Setiati, R, Wahyuningrum D, Siregar S, Marhaendrajana T. 2015. Studi Laboratorium Pengolahan Ampas Tebu menjadi Lignin sebagai Bahan Baku Surfaktan, *Prosiding SNITI (Seminar Nasional Inovasi dan Teknologi Informasi)*, Samosir, Sumatera Utara, Indonesia.

[13] SetiatiR, WahyuningrumD, Siregar S, Marhaendrajana T. 2016. Optimasi Pemisahan Lignin Ampas Tebu Dengan Menggunakan Natrium Hidroksida, *Ethos, Jurnal Penelitian dan Pengabdian (Sains dan Teknologi).* 4(2):257 – 264, UNISBA.

[14] Sun Ye, Cheng Jiayang. 2002. Hydrolysis of lignocellulosic materials for ethanol production: a review, *Bioresource Technology.* 83:1 – 11

[15] Syafwina HY, Watanabe T, Kuwahara M. 2002. Pre-treatment of oil palm empty fruit bunch by white-rot fungi for enzymatic saccharification, *Wood Res.* 89:19-20

Petroleum system of Northwest Java basin based on gravity data analysis

E Widianto

Geology Engineering Department, Faculty of Technology Earth and Energy, Universitas Trisakti, Jakarta, Indonesia

*Corresponding Author : ewidianto@trisakti.ac.id

Abstract. Energy management in the upstream oil and gas sector becomes very important for the country's energy security. The renewal of energy resources and reserves becomes necessary and is a must. In the oil and gas industry, gravity data is usually used only for regional surveys, but with the development of instrumentation technology and gravity software development, this method can be used for assessing oil and gas survey stages from exploration to production. This study was conducted to evaluate aspects of petroleum system and exploration play concept in the part of Northwest Java Basin, covering source rock deposition regions (source kitchen area, migration direction), development of reservoirs, structural and stratigraphic trap, based on gravity data. This study uses data from Bouguer gravity anomaly map by filtering process to produce a residual map depicting sedimentation basin configuration. The mapping generated 20 sedimentary basins in Java Island with the total hydrocarbon resources of 113 BBOE (Billion Barrel of Oil Equivalent). The petroleum system analysis was conducted in the Northwest Basin section. The final map produced illustrates the condition of petroleum system and play concept that can be used as exploration direction, expectedly reducing the risk of drilling failure.
Keyword : energy management, gravity data, hydrocarbon resources, petroleum system

1. Introduction

Energy management in the upstream sector of oil and gas becomes very important for the energy security of a country. Indonesia is known today as an oil and gas producer with 60 sedimentary oil basins. The history of the number of oil basins in Indonesia shows the dynamics concerning the technological and geological concepts used [1].

In the petroleum industry from the exploration to the production phase, there are challenges and opportunities that each has specific methods and technologies for its solution (see Figure 1). Each technology has advantages and disadvantages, but in principle must meet the requirements of the economy, can be applied in every field and geological conditions, as well as is environmentally friendly. The gravity method is considered excellent for use in the exploration phase until production from oil and gas fields. The gravity method can be used for surveys in every field and any geological conditions while being environmentally friendly and very economical at the same time [2].

So far, gravity data have been used only for regional surveys in oil and gas industries, but with the development of instrumentation and software technology and interpretation methods, they can be used for evaluation from the exploration phase to reservoir monitoring until the end of life of oil and gas field.

This paper attempts to study the petroleum system and play concepts in part of Northwest Java basin based on sedimentary basin map generated by gravity data. This study is expected to be a reference for the exploration in the area, as well as data base management of oil and gas energy.

IOP Conf. Series: Earth and Environmental Science **106** (2017) 012106 doi:10.1088/1755-1315/106/1/012106

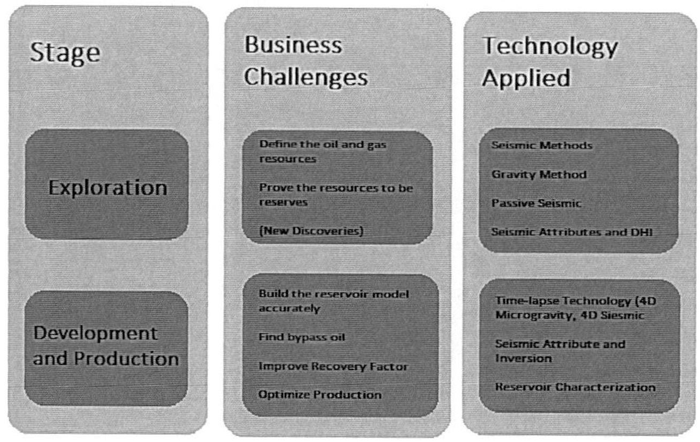

Figure 1. Oil and gas business challenges and technology application.

2. Research Method

The primary data used in this research are Bouguer gravity data. The frequency spectrum contained in the Bouguer gravity data has a quite wide range, covering density contrast layer response from shallow depth to the deepest [2]. Considering the objective of the research is to determine the basement of sedimentary basin, a further process is required. The methodology used in this research is filtering Bouguer gravity anomaly to produce regional and residual maps [3]. The resulting residual map will be an overview of the sedimentary basin configuration of the mapping area. In this research, sedimentation basin mapping is conducted throughout Java Island, but the focus of petroleum system analysis is carried out merely in Northwest Java basin. By combining tectonic elements and stratigraphy, the petroleum system and exploration play concept of Northwest Java basin can be defined. Figure 2 depicts the workflow of this study.

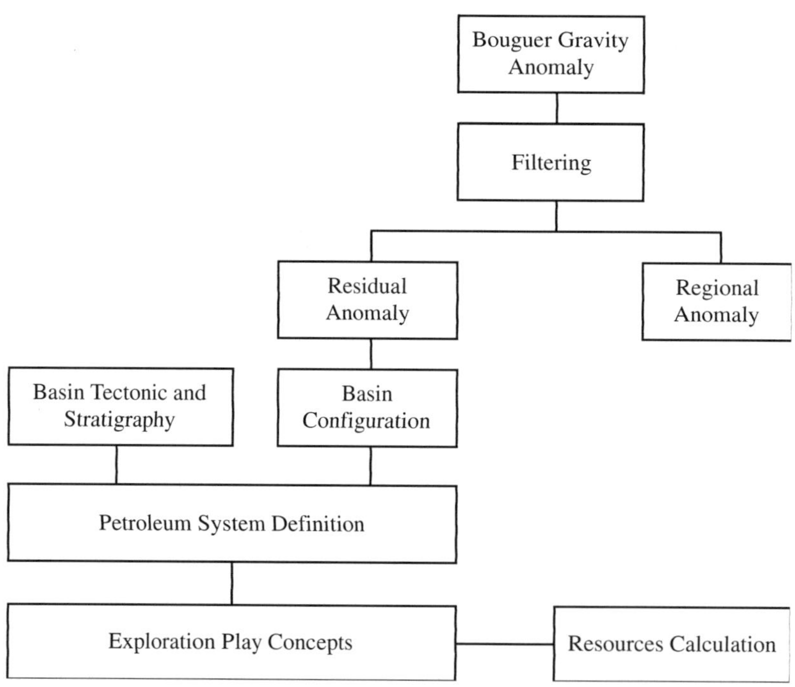

Figure 2. Workflow of study.

3. Results and Discussion

3.1. Gravity maps

Mapping of sedimentary basins in Java Island using gravity data has been done [4]. Figure 3 shows the Java Island's residual gravity map of the Second Vertical Derivative (SVD) process result that describes the configuration of sedimentary basins in Java. On the other hand, in Figure 4 as the depiction and naming of sedimentary basins in Java based on the residual map (Figure 3), it can be seen that there are 20 sedimentary basins in Java Island. This outcome shows that there is always a need for a renewal based on the technological developments and increasing data.

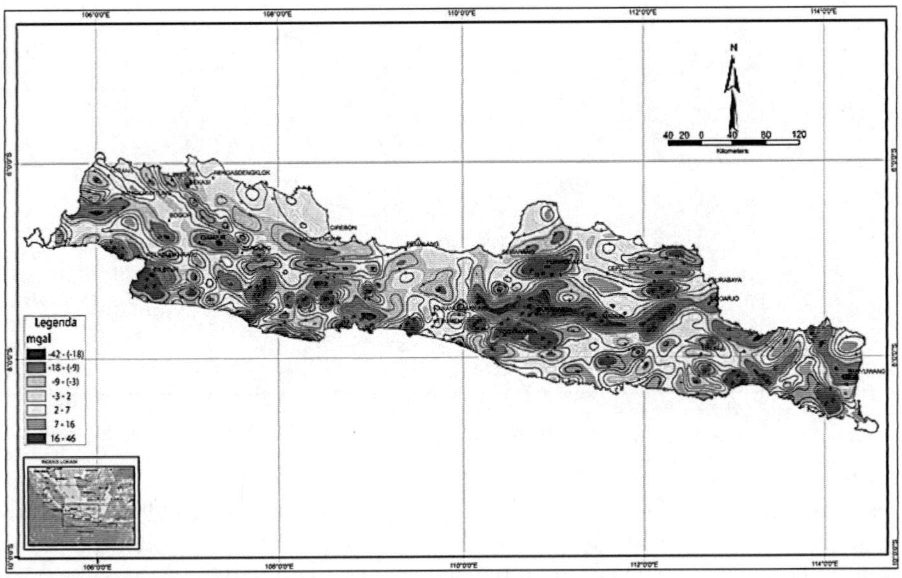

Figure 3. Residual gravity map of Java Island based on SVD process [1].

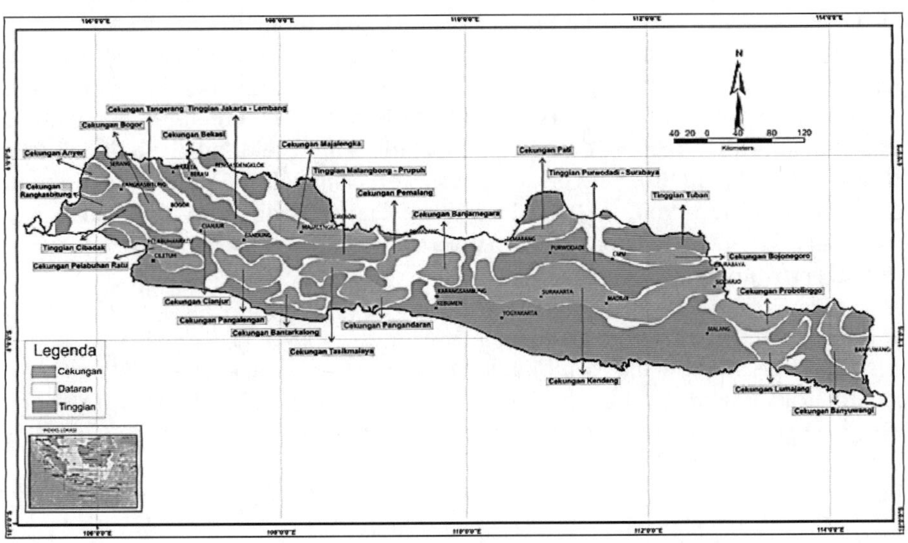

Figure 4. Sedimentary basin map of Java Island [1].

The total speculative hydrocarbon generated in the basin can be calculated using [2]. The average case of total hydrocarbons generated and trapped is 113 BBOE (Table 1).

IOP Conf. Series: Earth and Environmental Science **106** (2017) 012106 doi:10.1088/1755-1315/106/1/012106

Table 1. Hydrocarbon resources calculation of Java Island [1].

AREA	BASIN		SPECULATIVE HYDROCARBON TRAPPED (BBOE)		
			Minimum	Maximum	Average
West Java	ANYER		0.345	1.565	0.936
	BANTARKALONG		0.350	1.755	1.064
	BEKASI		1.264	5.881	3.316
	BOGOR		0.441	1.940	1.243
	CIANJUR		0.657	3.565	2.085
	MAJALENGKA		2.684	14.384	8.935
	PANGALENGAN		1.125	6.991	4.072
	PELABUHAN RATU		1.382	8.037	4.910
	RANGKASBITUNG		1.992	7.034	4.732
	TANGGERANG		1.252	8.086	4.915
	TASIKMALAYA		0.436	2.495	1.672
		Sub Total	***12***	***62***	***38***
Central Java	BANJARNEGARA		0.321	1.187	0.753
	PANGANDARAN		0.712	3.600	2.414
	PEMALANG		1.543	8.131	4.730
		Sub Total	***3***	***13***	***8***
East Java	BANYUWANGI		2.018	7.616	4.449
	BOJONEGORO		0.830	5.528	3.347
	KENDENG		18.463	76.080	45.876
	LUMAJANG		0.274	1.471	0.918
	PATI		0.865	5.001	2.749
	PROBOLINGGO		4.405	16.600	9.695
		Sub Total	***27***	***112***	***67***
		TOTAL	***41***	***187***	***113***

3.2. Petroleum System of Northwest Java Basin
Northwest Java basin is one of the largest oil and gas producing basins in Indonesia. Hydrocarbon resources are present in several sub-basins located both in onshore and offshore of Java Island [1].

A key component of a petroleum system is the presence of an effective source rock. In Northwest Java basin, there are three main types of source rocks: lacustrine (oil prone), coal and fluvial deltaic (oil and gas prone), as well as marine shale (bacterial gas). A brown to gray lacustrine shale, deposited during the syn-rift period, is an excellent source rock as an oil producer.

In the Northwest Java Basin, almost all formations, ranging from the basement to limestone of Parigi Formation, are very potential reservoir rocks. Talangakar and Cibulakan Formations of the upper part are the reservoirs with the largest hydrocarbon production. In some fields, Baturaja Formation also contains significant hydrocarbons. In Jatibarang Field, oil is produced from layers of volcanoclastic rocks of the Jatibarang Formation as well as the clastic sediments of the Banuwati Formation that are good reservoirs [3]. Hydrocarbons can migrate from the source rock either laterally or vertically. Lateral migration occurs within the bedding of sedimentary rocks, which have decent horizontal permeability.

3.3. Exploration play concept of Northwest Java Basin
Exploration play concept is a particular combination of a reservoir, seal, source, and trap associated with proven hydrocarbon accumulations. In the Northwest Java basin, all existing formations including basement have been proven as hydrocarbon producers. From the type of rock view, starting from metamorphic rocks, igneous rocks, volcanic rocks, sandstones, and limestone, all can be a good reservoir. Thus, exploration play concepts in West Java can be categorized as basement play type, volcanic play type, sandstone play, and reef play type (Figure 5).

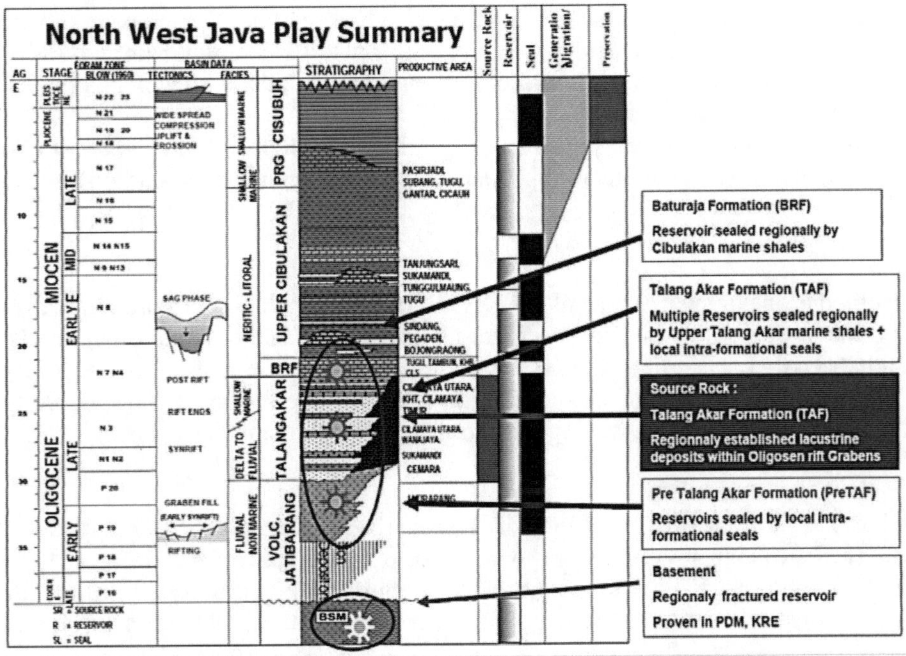

Figure 5. Exploration Play summary of Northwest Java Basin [4].

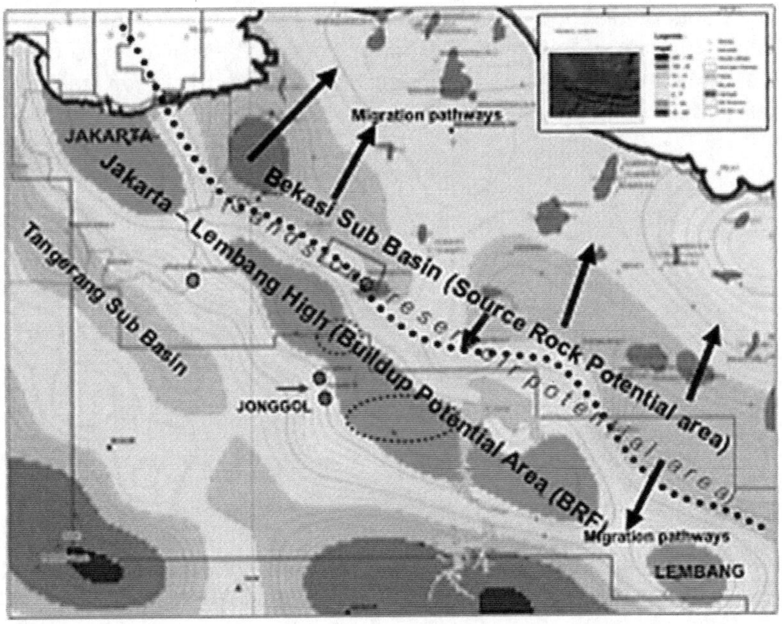

Figure 6. Exploration Play Concepts of Jakarta – Lembang area based on residual gravity map [1].

The case study of gravity map usage for evaluation of exploration play concept in Jakarta – Lembang area, part of Northwest Java basin (Figure 6).

Based on the play type model in Northwest Java Basin, the research area is expected to have sandstone play type and Baturaja reef play type. The development of limestone reefs is strongly affected by environmental factors of sedimentation, one of which is the depth factor that is relatively shallow compared to the surrounding area. Since the sedimentation basin map in Figure 6 is a representation of the depth of basement configuration, the development of the Baturaja Formation reef is still influenced

IOP Conf. Series: Earth and Environmental Science **106** (2017) 012106 doi:10.1088/1755-1315/106/1/012106

by the map. Thus, it can be estimated that the basement high of the Jakarta–Lembang directed NW–SE is a potential area as the development of Baturaja reef. This appraisal has been proven by several drilling wells that were discovered as limestone reef reservoirs such as the Jatinegara well and the excellent reef indication of seismic data in Jonggol area.

The development of sandstone play type can be found on the slopes of the basement height located in the west and the east. This evaluation is expected to be a reference for hydrocarbon exploration in the area to reduce the risk of failure.

4. Conclusion

The gravity method is one of the methods used in the oil industry, from the exploration to the production phase. Other prime properties of this approach are its lower price than seismic methods, as well as its environmentally friendly being. All aspects of petroleum system can be analyzed with gravity map to be used as the reference in exploration activities in the area. For the future, it is advisable to use more gravity methods for oil as well as gas field exploration and development activities.

References

[1] Schmoker, J.W. Volumetric Calculation of Hydrocarbons Generated in Magoon, L.B, W.G. Dow, eds., The Petroleum system – from source to trap AAPG Memoir 60. 1994.

[2] Noble, R.A., Pratomo, K.H., Nugrahanto, K., Ibrahim, A.M.T. Prasetya, I., Mujahidin, N., Wu, C.H., Howes, J.V.C. Petroleum Systems of Northwest Java, Indonesia in Howes, J. V. C., Noble, R. A., Eds., *Proceeding of the Petroleum Systems of SE Asia and Australasia Conference IPA*, Jakarta. 1997.

[3] Pertamina. Pengembangan Potensi Migas Cekungan Sedimentasi Jawa dan Sumatera Berdasarkan Data Gayaberat.Workshop Pertamina EP Technology Centre (Internal Report). 2010.

[4] Eko Widianto. Penentuan konfigurasi struktur batuan dasar dan jenis cekungan dengan data gayaberat serta implikasinya pada target eksplorasi minyak dan gas bumi di Pulau Jawa. (Disertasi S3) Institut Teknologi Bandung. 2008.

The analysis of quantitative methods for renewable fuel processes and lubricant of materials derived from plastic waste

J R Rajagukguk

Faculty of Mechanical Engineering, Universitas Krisnadwipayana, Jakarta, Indonesia

Corresponding Author : jenniria_rajaguguk@yahoo.com

Abstract. Plastic has become an important component in modern life today. Its role has replaced wood and metal, given its advantages such as light and strong, corrosion resistant, transparent and easy to color and good insulation properties. The research method is used with quantitative and engineering research methods. Research objective is to convert plastic waste into something more economical and to preserve the environment surrounding. Renewable fuel and lubricant variables are simultaneously influenced significantly to the sustainable environment. This is based on Fh> Ft of 62.101> 4.737) and its significance is 0.000 < 0.05. Then Ho concluded rejected Ha accepted which means that the variable of renewable fuels and lubricants or very large effect on the environment sustainable, the value of correlation coefficient 0.941 or 94.1% which means there is a very strong relationship between renewable fuel variables and lubricants to the sustainable environment. And utilizing plastic waste after being processed by pyrolysis method produces liquid hydrocarbons having elements of compounds such as crude oil and renewable fuels obtained from calculations are CO_2 + H_2O + C_1-C_4 + Residual substances. Then the plastic waste can be processed by isomerization process + catalyst to lubricating oil and the result of chemical calculation obtained is CO_2, H_2O, $C_{18}H_{21}$ and the rest.

Keywords: fuel processes and lubricant, plastic waste, quantitative methods

1. Introduction

Nowadays, plastic as an adequate new material is increasingly needed. The common black colouring materials for plastic carry some dangerous substances which can cause cancer in a particular hot temperature. In spite of those facts, plastic waste forms will make renewable fuel and lubricants. The plastic cannot be decomposed naturally and have no economic value [1]. It even has negative economic value when we consider the costs of waste management. The increasing number of plastic use will also increase that will give negative impacts to the environment. But the use of plastic is increasing espescially in industry and household [2].

Plastic waste has accumulated up to 6 million tons [3] so that plastic waste management has become an increasingly urgent issue. Technological advances will increase the quantity and quality of waste, as the increasingly diverse use of feedstocks, packing methods and manufacturing products can influence the amount and type of waste. Therefore, if it is not handled properly, it will cause some unexpected changes of environmental balance. It can pollute the soil, water and air.

Inorganic waste is divided into metal waste and its processed products, plastic waste, paper waste, glass and ceramic waste, detergent waste. Most inorganic cannot be parsed by the natural/whole microorganisms (non-biodegradable). Meanwhile, others can only be degrade in a long time. This type of waste at the household level for example plastic bottles, glass bottles, plastic bags, and cans [4].

Table 1. Type of polymer, code and its usage [5].

Code No	Type of Polymer	Usage
1	PET (*polyethyleneterepthalate*)	Bottled mineral water, bottle of cooking oil, juice, chili bottle, medicine bottle and cosmetic bottle; drug,
2	HDPE (*High-density polyethylene*)	Liquid milk, lubricant jerry canandcosmetic bottles
3	PVC (*Polyvinyl Chloride*)	Liquid milk, juice jelly and cosmetic bottles, table cover, shower bottles
4	LDPE (Low-density polyethylene)	plastic bag, plastic cover wrap frozen meat wrap & various other tips
5	PP (*Polypropyleneor Polypropene*)	plastic cup, plastic bottle caps, toys and margarine CD boxes, spoons, plastic and food, Styrofoam, transparent food
	PS (*Polystyrene*)	plastic cup, plastic bottle caps, toys and margarine CD boxes, spoons, plastic and food, Styrofoam, transparent food place
6	Others	Baby bottle, Plastic packaging, and gallons of drinking water, auto parts, household appliances, computers, electronic appliances, toothbrushes, and lego toys.

Qualities Polymers is a full strength, compressive strength, flexural strength, Impact strength. Polymers stairs has a higher thermal stability than open-chain of the polymers, aromatic polymers are characteristically exhibit glass transition temperatures were very high, very high melt viscosity and low solubility. So it is difficult than most other polymer types.

Classification of polymers consisting of natural polymers are polycondensation and polymerization reaction itself and synthetic polymers are polymers that deliberately at the plant as needed and do not exist in nature and synthetic polymer was molded as desired and resistant to microorganisms (Table 1). Polymer based composition microstructure are (1) homopolymers, the polymer formed from the merger of similar monomers with repeating units are the same, (2) copolymers, i.e., polymers that are formed from several different kinds of monomers. These copolymers are subdivided into four groups: random copolymers, regular cross copolymers, block copolymers and copolymer branch/Graft copolymers.

2. Research Method

The tools and materials are plastic waste consist of drink bottles, plastic cups, plastic plate rack, etc. The method used is engineering and quantitative methods while the methodological steps of research preparation of final task proposal described in Figure 1. Alkalinity, detergency properties, dispersancy and oxidation resistance [6].

Research steps are as follows: (1) All types of waste plastic (making mixed oil for diesel fuel apparently does not need to choose what type of plastic to get the best diesel mix, but in the type of waste research is PET (polyethyleneterephthalate), (2) pyrolysis process, as shown in Figure 2. Then the plastic waste put into the pyrolysis machine. Pyrolysis itself is a chemical decomposition process of a material to produce hydrocarbons through high temperature heating with little oxygen, (3) heated waste (once the waste put into the machine, it will be heated until melted) (4) The first gas (when the waste is heated to melt, this plastic waste will produce gas) and (5) condensed into oil (after plastic waste into gas, the gas is condensed through the condensation process so it can produce oil that can be a mixture of diesel fuel) [7].

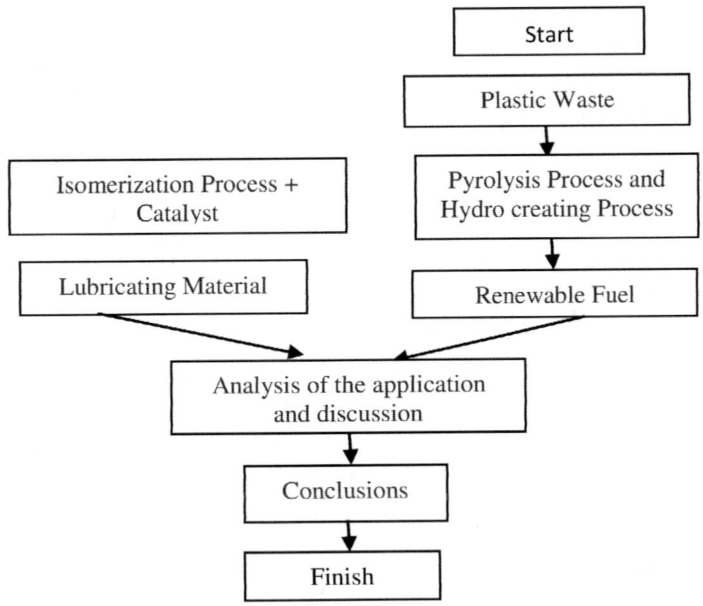

Figure 1. Flow chart research.

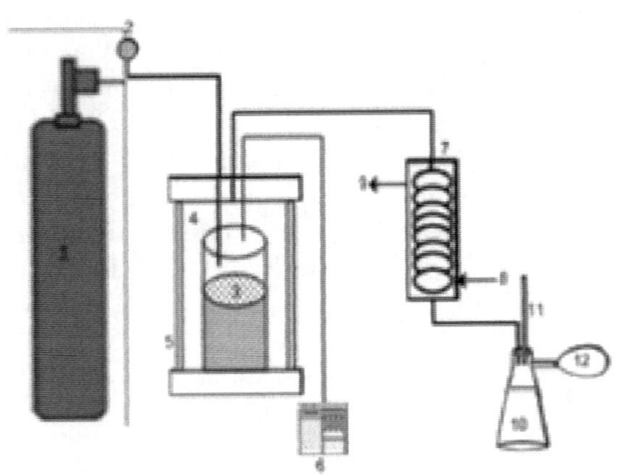

Caption:

1. Nitrogen tube
2. Rotameter
3. Place of catalyst
4. Pyrolysis reactor
5. Electric furnace
6. Thermocouple temperature reading device
7. Reflux condenser
8. Inlet cooling water
9. Water outlet
10. Liquid-gas separator
11. Thermometer
12. Gas container

Figure 2. Schematic of a series of pyrolysis tools [8].

To distinguish the three types of oil are the temperature and distance on the oil flow pipe produced on the machine pyrolysis or cracking machine such as the following:

1) A kind of solar is the result of warming between 250-340°C with the pipe closest to the pyrolysis furnace.
2) Kerosene is the fuel of the oil stove which is generated from the heating process of the pyrolysis furnace with a temperature of 170-250°C up to 2 hours with the oil flow pipe being in the middle or between the diesel and gas pipes.
3) Gasoline is generated from the heating process at temperatures between 350-750°C. It takes 4 hours where the gasoline oil pipeline is farthest from the pyrolysis furnace or at the front.

IOP Conf. Series: Earth and Environmental Science **106** (2017) 012107 doi:10.1088/1755-1315/106/1/012107

Then the process of plastic material into lubricating oil is the result of the pyrolysis process into the condenser to be cooled through the connecting pipes. In this process the plastic material heated using pyrolysis process by burning the garbage at the same time with the distillation at high temperatures. The catalyst (Al_2O_3) is heated with a temperature of 200° - 1000°C CO_2, H_2O, $C_{18}H_{21}$, and Residual. This method is environmentally friendly because it produces CO_2 and H_2O gases. In addition when it heated also produces a liquid hydrocarbon compound ranging from C1-C4 and long chain compounds such as paraffin and olefins which have wax or wax like forms [3].

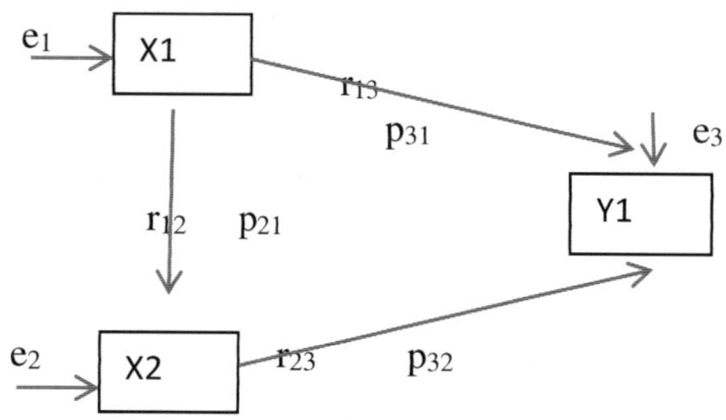

Figure 3. Research model.

The Figure 3 explains that: X1 is an exogenous variable for X2 and Y1; X2 is an exogenous variable for Y1; X1 and Y1 are endogenous variables X1; Y1 is endogenous variable X2; E1 other variables outside the system (residual variables); R12 = the correlation coefficient between X and Y and pij = path coefficient (influence). The data obtained by the test result with the temperature parameter X1 (SP), in the plastic waste forming process resulting in the renewable material X2 (B. BT), the lubricant Y1 ((BP) to the Y2 (LB) environment as Table 2. The picture also explains that X1 is an exogenous variable for X2 and Y1; X2 is an exogenous variable for Y1; X1 and Y1 are endogenous variables X1; Y1 is endogenous variable X2; E1 other variables outside the system (residual variables); R12 = the correlation coefficient between X and Y and pij = path coefficient (influence). The data obtained by the test result is with the temperature parameter X1 (SP), in the plastic waste forming process resulting in the renewable material X2 (B. BT), the lubricant Y1 ((BP) to the Y2 (LB) environment as shown in Table 2.

Table 2. Temperature parameters of each variable.

No	X1 (S.P.)	X2 (B.BT)	Y1((BP)
1	170	200	200
2	175	220	500
3	178	235	600
4	180	240	500
5	190	243	550
6	200	245	600
7	210	250	650
8	220	300	700
9	230	325	800
10	250	350	1000

3. Results and Discussion

The calculation procedure in this hypothesis test using the SPSS Statistics 21.0 Program. Autocorrelation is used to test for the presence of internal correlations among the variables arranged in the time series. To detect the presence or absence of autocorrelation in a regression model using Durbin-Watson method, the autocorrelation test of the linear regression model can be stated no symptoms, if probability Durbin-Watson> 0,05 and from the above table Durbin-Watson value is

1.324. So that shows 1.324> 0.05 concluded regression model did not experience autocorrelation symptoms.

The results of regression coefficient test showed that the nature of probability (significance), when 0.401> 0.05 means experiencing symptoms of heteroscedasticity. It turns out that Plastic Waste variable is 0,000 whereas plastic waste is 0.000. Thus the results above the regression model experienced symptoms of Heteroscedasticity. The linearity will be fulfilled with the assumption that the plot between standardized residual values and standardized prediction values does not form a specific pattern (random).

Hypothesis testing in this research was done by using statistical analysis tools in the form of correlation coefficient, determinant coefficient, F test and t test. In the level of significance or probability (α) is 0.05 or 5%. The value shows that the value of Adjusted R Square r^2 is 0.941, this means simultaneously influenced by renewable fuel and lubricant material 94.1 while the rest of 5.9% influenced by other factor not examined in this research.

The results of regression coefficient test showed that the nature of probability (significance), when 0.401> 0.05 means experiencing symptoms of heteroscedasticity. It turns out that Plastic Waste variable is 0,000 whereas plastic waste is 0.000. Testing equations simultaneously, using between count with the ratio of F arithmetic> F table or by using the value of significance. Basic decision-making is Ho will be rejected or Ha accepted. If the significance value is less than 0.05; below shows the value of F arithmetic is> F table (62.101> 4.737) the value of significance is 0.000 <0.05. Then concluded Ho rejected Ha accepted which means plastic waste materials, renewable fuels and lubricants simultaneously affect the environment sustainable. And partial testing or t test is performed to determine whether partially or individually significantly affects the dependent variable. T-test can be done by comparing t count> with t table (23.738> 2.262), or by using the value of significance. The basis for making a decision is Ho is rejected or Ha is accepted if the significance value t 0,000 <0.05, (1) Variable X1 to Y: The variable of TB on the effectiveness of CO_2 emissions of its significant value is 0.000 <0.05 significant. Then Ho concluded rejected Ha accepted which means that the variable of plastic waste, fuel lubricants to the environment that is sustainable (10.725> 2.262), (2) Variable X2 to Y is 0.784), the variable of lubricant to the environment is sustainable, its significant value is 0.000 < 0.05 significant. Then concluded Ha accepted Ho rejected which means that the variable of renewable materials influential or on the environment sustainable. In multiple linear regression analysis, we can see the constants a = - 461.186; B1 = .784 and the value b2 = 3.580, the linear regression equation is Y = -461.186 + 0.784 X1 + 3.580 X2. For the above multiple regression formula can be concluded through the following statement: If the variable plastic waste increase, an increase of one unit ratio scale, then the renewable fuels and lubricants variables value decreases less with pollution so that a sustainable environment was reduced to contamination.

4. Conclusion

Renewable fuel variables significantly affect the environment sustainably, it is based on the significance value 0.000 <0.05 and the partial variable value; th>tt (23.738> 2.262). Then Ho concluded the rejected Ha accepted which means that the variable renewable fuels have an effect or have big influence on the sustainable environment. Lubricant variable has significant effect on the sustainable environment, it is based on its significance value 0.000 <0,05, but the value of the lubricant variable compared to the continuous environment (10.725> 2.262). Then concluded Ha accepted Ho is rejected which means that the variable of lubricant ingredients have small effect on the environment. Renewable fuel and lubricant variables are simultaneously influenced significantly to the sustainable environment, this is based on Fh> Ft of 62.101> 4.737) and its significance is 0.000 <0.05. Then Ho concluded rejected Ha accepted which means that the variable of renewable fuels and lubricants or very large effect on the environment sustainable, the value of correlation coefficient 0.941 or 94.1% which means there is a very strong relationship between renewable fuel variables and lubricants to the environment Sustainable. With determinant coefficient R = 0,941 influenced by factor of independent variable. Hence Ho concluded Ha refused acceptable which means that the variable of renewable fuels and lubricant ingredients influence or very big influence on environment sustainable. Utilizing plastic waste after being processed by pyrolysis method produces liquid hydrocarbons having elements of

IOP Conf. Series: Earth and Environmental Science **106** (2017) 012107 doi:10.1088/1755-1315/106/1/012107

compounds such as crude oil and renewable fuels obtained from calculations are CO_2 + H_2O + C1-C4 + Residual substances. Then the plastic waste can be processed by isomerization process + catalyst to lubricating oil and the result of chemical calculation obtained is CO_2, H_2O, $C_{18}H_{21}$ and the rest.

References

[1] Harwin, 2011, Environment Product Change Notification Quality About Events, Latest News Brunel University.
[2] Candra Wiguna, 2015, History of Plastic Waste.
[3] Kompas. 2008. 80 percent of plastic waste becomes waste. Retrieved on November 20, 2008 fromhttp://cetak.kompas.com/read/xml/2008/08/06/00444748/80persen.sampah.plastik.jadi.limbah.
[4] Gelbert M et al, 1996, Garbage and Management. Malang: PPPGT/VEDC. Gerungan, W. A. 2009.
[5] Surono, Budi U. 2013. Various Methods of Converting Plastic Samples into Fuel Oil. *Journal of Engineering*. 3(1). Yogyakarta. University of Janabadra.
[6] I Made Arcana. 2010. Structure and morphology of poly ((RS) -b-butyrolactone). Synthesized from aluminoxane catalyst. C Jaimes, M Arcana, A Brethon, A, Mathieu, F Schue.
[7] Social Psychology. Bandung: Refika Aditama. Goleman.
[8] Ramadan, Aprian. 2011. Plastic waste processing into oil using pyrolysis process, *Journal of Environmental Scientific Engineering*. 4(1). University of National Development "Veteran" East Java.

The new management policy: Indonesian PSC-Gross split applied on CO_2 flooding project

S Irham*, S N Sibuea, A Danu

[1]Petroleum Engineering Department, Faculty of Earth Technology and Energy, Universitas Trisakti, Jakarta, Indonesia

*Corresponding Author : syamsul.irham@trisakti.ac.id

Abstract. “SIAD” oil field will be developed by CO_2 flooding. CO_2, a famous pollutant gas, is injected into the oil reservoir to optimize the oil recovery. This technique should be conducted economically according to the energy management policy in Indonesia. In general, Indonesia has two policy contracts on oil and gas: the old one is PSC-Cost-Recovery, and the new one is PSC-Gross-Split (introduced in 2017 as the new energy management plan). The contractor must choose between PSC-Cost-Recovery and PSC-Gross-Split which makes more profit. The aim of this paper is to show the best oil and gas contract policy for the contractor. The methods are calculating and comparing the economic indicators. The result of this study are (1) NPV for the PSC-Cost-Recovery is -46 MUS$, while for the PSC-Gross-Split is 73 MUS$, and (2) IRR for the PSC-Cost-Recovery is 9%, whereas for the PSC-Gross-Split is 11%. The conclusion is that the NPV and IRR for PSC-Gross-Split are greater than the NPV and IRR of PSC-Cost-Recovery, but POT in PSC-Gross-split is longer than POT in PSC-Cost-Recovery. Thus, in this case, the new energy policy contract can be applied for CO_2 flooding technology since it yields higher economic indicators than its antecendent.
Keywords: CO_2 flooding, new energy management policy, PSC-Gross-Split

1. Introduction

"SIAD" oil field is located in the southwest of Cirebon City, West Java, particularly in the basin of the northern West Java. There are two main layers in the "SIAD" oil field, consisting of D1 and F layers. The discussion on this case is limited to the latter only. All reservoir parameters, production, and characteristics refer to solely Layer F [1].

Layer F on the “SIAD” oil field has been produced since 1973. The field “SIAD” has an Original Oil In Place of 55 MMBBL. The latest production status in December 2015 has a recovery factor of 17.8% with a cumulative production of oil (Np) of 9.8 MMBBL and the maximum allowable or so-called estimated ultimate recovery (EUR) of 37 MMBBL, while the remaining reserve that can still be taken is 27 MMBBL [1].

The remaining reserve of Layer F will be developed by using CO2 flooding technology. CO2 flooding is enhanced oil recovery method. The pollutant CO2 gas is injected down into the reservoir to get more oil producing from the well [2], [7]. This technique should be conducted economically according to Indonesia’s oil and gas management policy. Referring to the regulation of the Minister of Energy No. 8/2017 on PSC-Gross-Split [5], the contractor must choose between PSC-Cost-Recovery or PSC-Gross-Split which makes more profit.

2. Research Method

This study needs some data to compute the economic indicators. "SIAD" field has a cumulative production of 175 MBBL for 16 years to be calculated in this research, an investment of 1557 MUS$, an operating cost of 1864 MUS$, and an average oil price of $50/BBL[1], [7].

This project will be evaluated for 20 years. For four years from starting of the project, the oil production is still zero. Yearly incremental oil production of MBBL/year consecutively are : (0,0); (0,0); (0,0); (0,0); 26.8; 23.1; 19.9; 17.1; 14.7; 12,7; 10.9; 9.4; 8.1; 7.0; 6.0; 5.2; 4.4; 3.8; 3.3; and 2.8.

Early in project life, some amount of investment must be available for capital and noncapital expenditure. Investment expenditure in M$/year are : 518; 426; 594; and 19. The total amount of investment is 1557 MUS$ for the four years of the early project [1].

To produce oil from the reservoir into the surface will take production cost which estimated 10 US$/BBL of oil. And to protect the environment from all damage and to do the land restoration at the end of project, some of cost is also needed, which is called Abandon Site Restoration cost (ASR cost). The sum of production cost and Abandon Site Restoration cost is called Operating Cost. Yearly operating cost of M$/year for entirely the CO2 flooding project, consecutively are : (0,0); (0,0); (0,0); (0,0); 275; 238; 206, 178; 154; 134; 116; 101; 88; 77; 67; 59; 51; 45; 40; and 35 [1]. The total amount of operating cost is 1864 MUS$.

Those data will be the same input for both energy management policy : PSC-Cost-Recovery and PSC-Gross-Split.

In PSC-Cost-Recovery, the contractor can calculate the cash flow for 16 years on oil production. The revenue distribution splits after tax are 85% for the government and 15% for the contractor. The outcomes of this analysis are NPV, IRR, and POT. There are some differences between PSC-Cost-Recovery and PSC-Gross-Split. PSC-Gross-Split has a discrete revenue distribution split and tax. There are neither cost recovery nor FTP on PSC-Gross-Split, so the contractor must pay the cost recovery for all their operational costs[3], [4], [5], [6]. The revenue distribution splits for PSC-Gross-Split are 64% for the contractor and 36% for the government since there are additional progressive split and variable split in PSC-Gross-Split . Summary of data input and petroleum fiscal term for both petroleum management policy (PSC-Cost-Recovery and PSC-Gross-Split) is put in Table 1.

Following the results regarding the economic indicators, a sensitivity analysis was conducted for both model managements by changing input data of cumulative production, oil price, investment, and operating cost. Comparison of the output and sensitivity for both models was then carried out to determine the conclusions from the profitability index and sensitivity analysis.

Table 1. Summary of data input, and comparison of fiscal term.

DATA INPUT		OIL AND GAS FISCAL		
NP	9.8 MMBBL	PSC-Cost-Recovery		
Tax	40.5%		Government	Contractor
oil price	50 $/BBL	Base Split	85%	15%
Investment	1157 MUS$			
Opex	1864 MUS$			
		PSC-Gross-Split		
NP	9.8 MMBBL		Government	Contractor
Tax	40.5%	Base Split	57%	43%
oil price	50 $/BBL	Variabel Split	0	8.5%
Investment	1157 MUS$	Progresive Split	0	10%
Opex	1864 MUS$	Final Split	34.5%	65.5%

3. Results and Discussion

3.1. PSC-cost-recovery revenue distribution

PSC-Cost-Recovery is an original Indonesia oil and gas contract [3], [4], [6]. According to the contract terms and conditions, 20% of the gross revenue will take as cost recovery ceiling. The government and contractors shall be entitled to take and receive each year 10% of all Petroleum produced and saved before any deduction for the recovery of Investment Credit and Operating Cost. Such First Tranche Petroleum shall be shared between government and contractor in accordance with sharing splits. The revenue after FTP can be used for cost recovery. Remaining revenue after FTP and cost recovery, which is called Equity To be Split, or ETS, will be shared between government and contractor in accordance with sharing splits. Contractor share is subject of government tax [4], [6].

For the entire 20 years of the assuming CO_2 flooding project, it will give the revenue in MUS$, consecutively from the start of project : (0,0), (0.0); (0.0); (0,0); 1,342; 1,155; 994; 856; 736; 634; 545; 470; 404; 348; 299; 258; 222; 191; 164; and 141. The cumulative gross revenue is 8,759 MUS$.

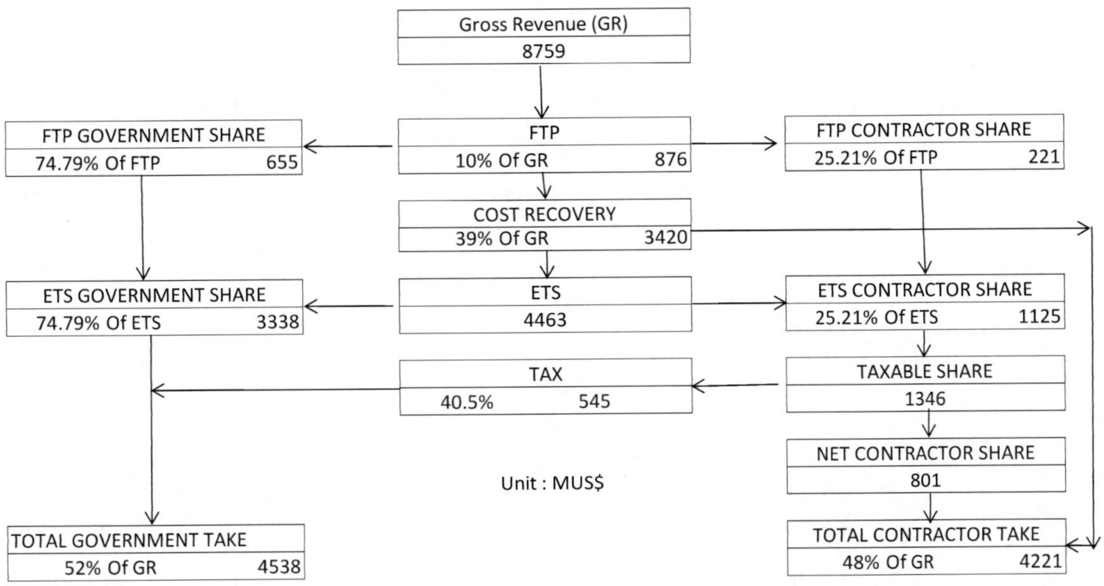

Figure 1. Revenue distribution results of PSC-Cost-Recovery.

The final results for PSC-Cost-Recovery revenue distribution are Gross Revenue of 8,759 MUS$, Contractor Tax of 40.5%, FTP of 876 MUS$ from 10% of Gross Revenue, Cost Recovery of 3,420 MUS$, Total Government Take of 4,538 MUS$ from 52% of Gross Revenue, and Total Contractor of 4221 MUS$ from 48% of Gross Revenue (Figure 1).

3.2. PSC-gross-split revenue distribution

PSC-Gross-Split scheme is simple in calculation of revenue distribution. According to PSC-Gross-Split terms and conditions, 51.5% of the gross revenue will goes to government, and the remaining 38.5% of gross revenue is equity of contractor. Contractor have to pay 40.5% profit tax to the government. There are no First Tranche Petroleum (FTP), nor Cost Recovery. All expenses is under contractor self control [5].

PSC-Gross-Split revenue distribution for Gross Revenue of 8,759 MUS will give Total Government Take is 4,799 MUS$ from 90% of Gross Revenue, Total Contractor Take is 540 MUS$ from 10% of Gross Revenue (Figure 2).

It is clear and simple to see that PSC-Gross-Split cannot be the choice of the contractor of the project. The contractor is receive more money with PSC-Cost-Recovery contract scheme.

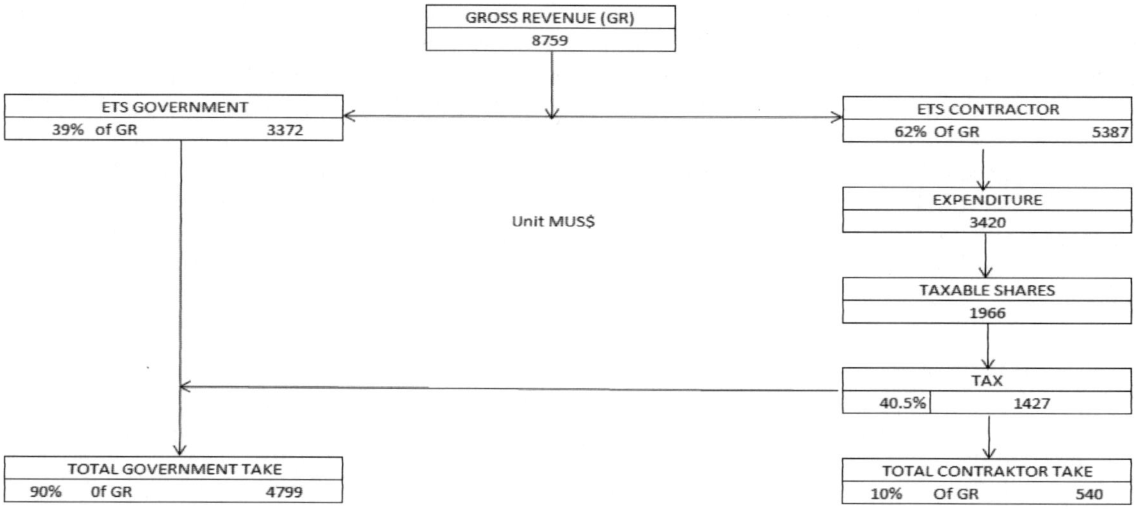

Figure 2. Revenue distribution results of PSC-Gross-Split.

3.3. Economic indicator results

For the entire 20 years of the assuming CO_2 flooding project, although the input data are the same for predicted yearly oil production, assuming oil price, estimated investment , tax rate , and estimated operating cost, the contractor cash flow will be definitely difference for both PSC-Cost-Recovery and PSC-Gross-Split.

Consider of PSC-Cost-Recovery scheme, the contractor cash flow in MUS$, consecutively are : (-518); (-426); (-594); (-19); 952; 650; 136; 102; 87; 75; 64; 55; 47; 41; 35; 30; 26; 22; 19; and 16. The cumulative cash flow is 801 MUS$.

According to PSC-Gross-Split scheme, the contractor cash flow in MUS$, consecutively are : (-518); (-426); (-594); (-19); 327; 281; 241; 207; 178; 152; 131; 112; 96; 82; 70; 60; 51; 43; 36; and 31. The cumulative cash flow is 540 MUS$.

The economic indicator was then calculated from that cash flow. Table 2 shows the economic output comparison between PSC-Cost-Recovery and PSC-Gross-Split.

The results of this comparison are NPV@10% for the PSC-Cost-Recovery of -46 MUS$ and the PSC-Gross-Split of -377 MUS$, while IRR for the PSC-Cost-Recovery of 9% and for the PSC-Gross-Split of 4%. The conclusion is NPV and IRR values of PSC-Gross-Split are smaller than thoseof PSC-Cost-Recovery, and PSC-Gross-Split POT is longer than POT in PSC-Cost-Recovery.

Because of negative value on NPV, and small value of IRR, this CO_2 flooding project project cannot be implemented economically for both energy management policy.

Let us see on the economic indicator of NPV, IRR, and POT. On the government side, the new management policy (PSC-Gross-Split) for the petroleum (oil and gas) exploration and exploitation is more attractive than the old one (PSC-Cost-Recovery). On the contractor side, the new management policy (PSC-Gross-Split) for the petroleum (oil and gas) exploration and exploitation is less attractive than the old one (PSC-Cost-Recovery). For this case of CO_2 flooding project, if there is opportunity to choose the type of government management policy contractor will tend to prefer the PSC-Cost-Recovery rather than PSC-Gross-Split.

From the economic calculation, PSC-Cost-Recovery contract model provides better results compared to the PSC-Gross-Split or the new petroleum management policy.

Table 2. Economic output comparison between PSC-Cost-Recovery vs PSC-Gross-Split.

NO	Indicator	Unit	PSC-CR	PSC-GS
1	Oil Production	MBBL	175	175
2	Oil Price	US$/BBL	50	50
3	Total Gross Revenue	MUS$	8759	8759
4	FTP	MUS$	876	
	Total Cont. FTP	MUS$	221	
	Total Gov. FTP	MUS$	655	
5	Investment	MUS$	1557	1557
6	Invesment Credit (IC)	MUS$		
7	Total Operating Cost	MUS$	1864	1864
	Operating Cost	MUS$	1752	1752
	Abandonment & Site Restoration Cost	MUS$	112	112
8	Cost Recoverable	MUS$	3420	3420
	(% Of Gross Revenue)	%	39%	39%
	Unrecovered Cost	MUS$	0	0
	(% Of Gross Revenue)	%	0%	0%
9	Equity to be Split	MUS$	4463	8759
	Cont. Equity	MUS$	1125	5387
	Gov. Equity	MUS$	3338	3372
10	Contractor			
	Net Cash Flow	MUS$	801	540
	(% thd. Gross Rev)	%	9%	6%
	IRR	%	9%	4.3%
	NPV@10%	MUS$	-46	-377
	POT	years	5.0	11.6
11	Government			
	FTP	MUS$	655	
	Equity	MUS$	3338	3372
	Tax	MUS$	545	1427
	Net Cash Flow	MUS$	4538	4799
	(% Of Gross Revenue)	%	52%	55%
	Gov. PV @ 10%	MUS$	1731	2061

3.4. Sensitivity analysis of PSC-cost-recovery and PSC-gross-split

Sensitivity analysis was performed to figure out the change of project economic indicators as well as changes of the input data, such as production, investment, and operating cost. The result of sensitivity analysis are presented in Figure 3A, and Figure 3B.

There is no discrepant sensitivity in price, investment, and operating cost between PSC-cost-recovery and PSC-gross-split (Figure 3).

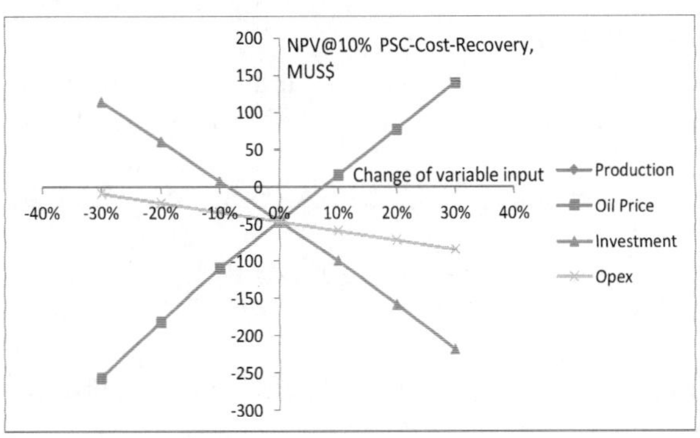

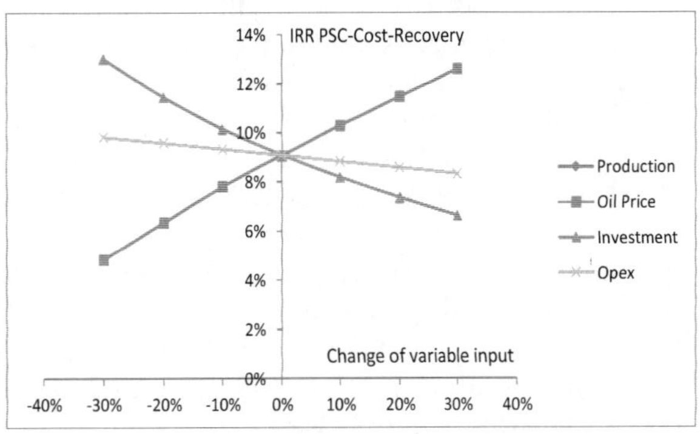

Figure 3A. PSC-Cost-Recovery : Sensitivity Analysis of NPV and IRR.

The line curve of production is the same as the line curve of oil price. The NPV and IRR of PSC-Cost-Recovery scheme of contract are very sensitive of investment and of oil price (Figure 3A).

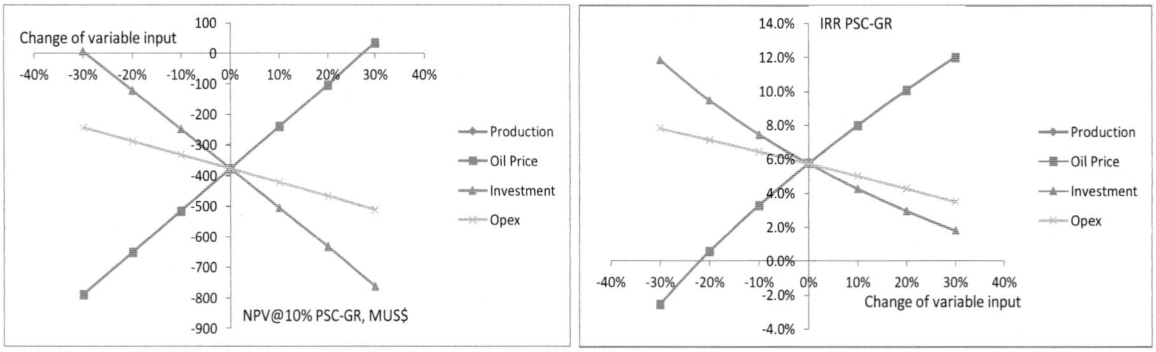

Figure 3B. PSC-Gross-Split : Sensitivity Analysis of NPV and IRR

The NPV and IRR of PSC-Cost-Recovery scheme of contract are very sensitive of investment and of oil price. It is must be noted that the line curve of production is the same as the line curve of oil price. (Figure 3B).

4. Conclusion

NPV for the PSC-Cost-Recovery is negative (-46) MUS$ and for the PSC-Gross-Split is also negative (-377) MUS$, while IRR for the PSC-Cost-Recovery is 9% and for the PSC-Gross-Split is 4.3%. NPV and IRR of PSC-Gross-Split are greater than of PSC-Cost-Recovery, and POT in PSC-Gross-split is longer than POT in PSC-Cost-Recovery. The new energy policy contract cannot be applied for this CO_2 flooding project technology because it yields lower economic indicators than the old one (PSC-Cost-Recovery).

Acknowledgment

The authors gratefully acknowledge to Mr. Danang Sismartono PPPTMGB LEMIGAS in obtaining the data for this paper.

References

[1] Field Report Data CO2 Flood. PPPTMGB LEMIGAS, Jakarta.
[2] H.K Van Pollen and Associates, Inc., Fundamental of Enhanced Oil Recovery, Tulsa, Oklahoma, 1980.
[3] Harimurti D, Indonesian Milestone in Production-Sharing Contract in Perspective of Government Take, Contractor Take, Cost Recovery and Production Target, SPE-187008-MS, Society Of Petroleum Engineer, 2017
[4] Irham S, Pengelolaan Lapangan Migas, Universitas Trisakti, Jakarta, 2017.
[5] Kurniawan T and Jaenudin J, Proposed Modification of Abandonment and Site Restoration Mechanism in Gross Split PSC for Marginal Field in Indonesia, SPE-186378-MS, Society of Petroleum Engineer, 2017
[6] Lubiantara B, Ekonomi Migas: Tinjauan Aspek Komersil Kontrak Migas, Gramedia, Jakarta, 2012.
[7] http://www.migas.esdm.go.id/post/read/permen-esdm-nomor-08-tahun-2017-tentang-kontrak-bagi-hasil-gross-split
[8] Nasir A, Sejarah Sistem Fiskal Migas Indonesia, Grasindo, Jakarta, 2014.
[9] Patmosukismo S, Oil and gas : Politic, law and industry (MIGAS Politik, Hukum & Industry). ISBN 978-979-823189-6, 2011.
[10] Sismartono D, Studi Simulasi Injeksi CO_2 Dari Hasil Sampling Gasifikasi Low Rank Coal Pada Lapangan X dan Keekonomiannya, Tesis Program Studi Teknik Perminyakan Institut Teknologi Bandung, Bandung, 2013.

The new energy management policy: Indonesian PSC-gross-split applied on steam flooding project

S Irham*, P Julyus

Petroleum Engineering Department, Faculty of Earth Technology and Energy, Universitas Trisakti, Jakarta, Indonesia

*Corresponding Author : syamsul.irham@trisakti.ac.id

Abstract. "SIPY" oil field has been producing oil using steam flooding technology since 1992 under the PSC-Cost-Recovery policy. In 2021, the contract will be finished, and a new agreement must be submitted to the Indonesian government. There are two applied fiscal policies on oil and gas management: PSC-Cost-Recovery and PSC-Gross-Split (introduced in 2017 as the new energy management plan). The contractor must choose between PSC-Cost-Recovery and PSC-Gross-split which makes more profit. The aim of this research is to determine the best oil and gas contract policy for the contractor. The methods are calculating contractor cash flow and comparing the Profitability Indexes. The results of this study are (1) Net Present Values for the PSC-Cost-Recovery and the PSC-Gross-Split are 15 MMUS$ and 61 MMUS$, respectively; and (2) Internal Rate of Return values for the PSC-Cost-Recovery and PSC-Gross-Split are 10% and 11%, respectively. The conclusion is that the Net Present Value and Internal Rate of Return of PSC-Gross-Split are greater than those of PSC-Cost-Recovery, but in Pay Out Time of PSC-Gross-split is longer than Pay Out Time in PSC-Cost-Recovery. Thus, the new energy management policy will be more attractive than PSC-Cost-Recovery.

Keywords: new energy management policy, oil steam flooding, PSC-Gross-Split

1. Introduction

Oil field "SIPY" is a reservoir structure located in the middle of Sumatra, approximately 120 kilo meters to the north of Pekanbaru and about 75 kilo meters south of Dumai Administrative City in Riau Province. This oil field stretches from the north to the south, having a symmetrical shape with an estimated width of 10 kilo meters from the west to east as well as 18 kilo meters length from the north to south [1]. This "SIPY" field was discovered in 1958, and its production began in 1958 along with the completion of the drainage channel construction from "SIPY" Field to Dumai harbour. The field has an Original Oil In Place of 6,827 MMBBL and an estimated cumulative production in 2021 of 2,116 MMBBL or about 31% of the total Original Oil in Place. The estimated ultimate recovery (EUR) is 60% for Steam Flooding. The maximum oil production that can be taken is 4,096 MMBBL, while the estimated remaining reserves that can be acquired until the year of 2021 is 1,980 MMBL [1]. The PSC Recovery agreement of this field will be expired in 2021, but the contractor still have an opportunities to propose a new period of contract extension. Referring to researchers [5] the contractor may choose a contract management policy on the field to extend the old agreement. The contractor may apply for a contract extension before two years or at the earliest ten years prior to the expiry date according to the ESDM Minister Regulation No. 15 of 2015.

Recently in Indonesia are experiencing various of revision and improvement in the regulations, related with the oil and gas industry activity, this kind of situation certainly will impact to the investment atmosphere in the Indonesia oil and gas upstream.

The new regulation is PSC-Gross Split [7]. The Contractor should use this new regulation, in order to propose the contract extension.

This study purpose is to analyze if the Gross Split mechanism is more attractive or equal the old PSC Cost Recovery.

2. Research Method

The calculation of economic aspects related to the new energy management policy of the oil field includes the computations of gross revenue, First Tranche Petroleum, the amount of capital investment that can be returned in the cost recovery scheme, profit sharing in the form of equity to be split, taxes to be paid to the government, Profitability Index (for PSC-Gross-Split there is no First Tranche Petroleum), cost recovery, and different tax to the contractor. Prior to the economic calculations, some supporting data were needed, such as oil price, capital expenditure (CapEx), Operating Expenses (OpEx), cumulative production, as well as tax and revenue split distribution to yield the Profitability Index. In the economic calculation of an oil field, it takes at least three economic indicators to determine whether a field of oil and gas is to be developed, including the NPV, IRR, and POT.

This study needs some data to compute the profitability index. "SIPY" field has a cumulative production of 132 MMBBL until 20 years for the calculation in this research, as well as an average oil price of $50/BBL, an investment of 452 MMUS$, and an operating cost of 2,574 MMUS$. For three years from starting of the project, the oil production is still zero. Yearly oil production of MBBL/year consecutively are : (0,0); (0,0); (0,0); 1,500; 4,600; 5,800; 5,100; 5,400; 7,100; 10,500; 12,100; 12,500; 12,300; 12,000; 12,700; 14,300; 16,100; 19,100; 20,800; and 24,900.

Steam flooding project is one of enhanced oil recovery method [1], [2], [3]. The implementation of this technique in SPY field will need three years capital and noncapital investment from the beginning of the project. Yearly investment in M$/year are : 319,330; 66,260; and 66,660. The total amount of investment is 452,250 MUS$ for the three years of the early project. Producing oil from the reservoir into the surface will take production cost which estimated 15 US$/BBL of oil. And to protect the environment from all damage and to do the land restoration, some of cost is also needed, which is called Abandon Site Restoration cost (ASR cost). The sum of production cost and Abandon Site Restoration cost is called Operating Cost. Yearly operating cost of M$/year for entirely the steam flooding project, consecutively are : (0,0); (0,0); (0,0); 25,327; 97,153; 186,980; 266,306; 350,133; 459,459; 619,786; 804,113; 994,439; 1181,766; 1364,592; 1557,919; 217,327; 244,327; 533,633; 848,480; and 1224,806. Those data will be the same input for both energy management policy: PSC-Cost-Recovery and PSC-Gross-Split.

Under PSC-Cost-Recovery, the contractor can calculate the cash flow for 17 years on oil production. The revenue distribution splits after tax are 85% for the government and 15% for the contractor. The outcomes of this analysis are Net Present Value, Internal Rate of Return, and Pay Out Time. There are some differences between PSC-Cost-Recovery and PSC-Gross-Split. PSC-Gross-Split has a discrete revenue distribution split and tax. There are no cost recovery and First Tranche Petroleum on PSC-Gross-Split, so the contractor must pay the cost recovery for all their operational costs. The revenue distribution splits for PSC-Gross-Split are 53% for the contractor and 47% for the government since there are additional progressive split and variable split in PSC-Gross-Split.

Following the profitability index result, a sensitivity analysis was conducted for both model managements by changing input data of cumulative production, oil price, investment, and operating cost. Comparison of the output and sensitivity for both models was then carried out to determine the conclusions from the profitability index and sensitivity analysis.

3. Results and Discussion

3.1. PSC-Cost-Recovery Revenue Distribution

The oil production which predicted from the past historical production, the estimated investment, and the estimated production cost, are presented as seen in table below. The revenue come from oil production multiply by oil price. According to the contract term arrangement, the revenue distributed to government and contractor as simple as show in figure 1. Contractor cash flow was calculated from revenue and expenses (all investment and operating cost). Finally, the profit indicator was calculated from the cash flow.

PSC-Cost-Recovery scheme is a little bit complicated in calculation of revenue distribution. It is also need time consumption for Plan of Development approval, Work Program and Budgeting control, Authorization of Expenditure control, which are the implication of cost recovery approval by government.

According to PSC-Cost-Recovery terms and conditions, 20% of the gross revenue will take as cost recovery ceiling. The Parties shall be entitled to take and receive each year 20% of all Petroleum produced and saved ...before any deduction for the recovery of Investment Credit and Operating Cost. Such First Tranche Petroleum shall be shared between government and contractor in accordance with sharing splits. The revenue after FTP can be used for cost recovery. Remaining revenue after FTP and cost recovery, which is called Equity To be Split, or ETS, will be shared between government and contractor in accordance with sharing splits. Contractor share is subject of government tax [4], [6].

For the entire 20 years of the assuming steam flooding project, it will give the revenue in MUS$, consecutively from the start of project : (0,0), (0.0); (0.0); 75,000; 230,000; 290,000; 255,000; 270,000; 355,000; 525,000; 605,000; 625,000; 615,000; 600,000; 635,000; 715,000; 805,000; 955,000; 1,040,000; 1,245,000. The cumulative gross revenue is 9,840,000 MUS$.

The final results for PSC-Cost-Recovery revenue distribution are Gross Revenue of 9,840 MMUS$, Contractor Tax of 44%, First Tranche Petroleum of 1,968 MMUS$ from 20% of Gross Revenue, Cost Recovery of 3,452 MMUS$ from 35% of Gross Revenue, Net Government Take of 5,429 MMUS$ from 55% of Gross Revenue, and Net Contractor Take of 958 MMUS$ from 10% of Gross Revenue (Figure 1).

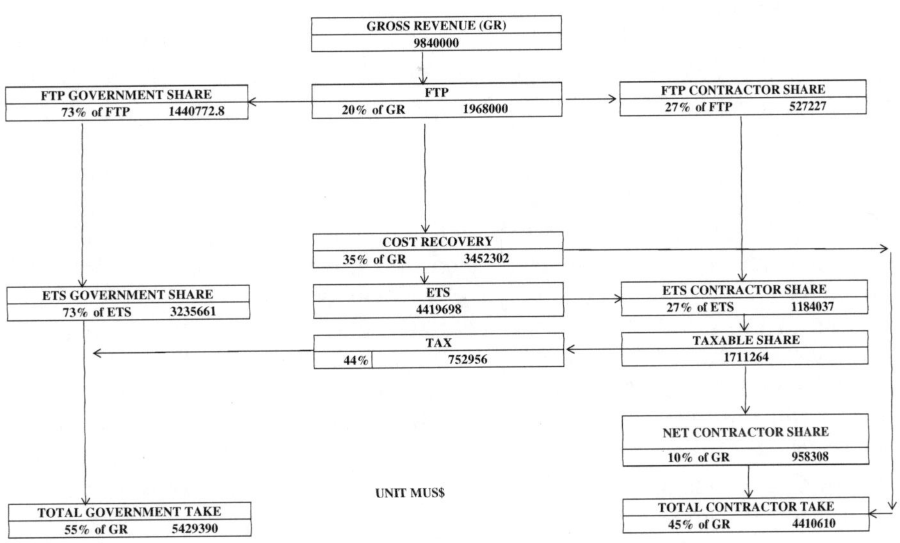

Figure 1. Revenue distribution results of PSC-Cost-Recovery.

3.2. PSC-Gross-Split Revenue Distribution

PSC-Gross-Split scheme is more simple in calculation of revenue distribution rather than PSC-Cost-Recovery. And there are no time consumption for control of expenses nor cost recovery approval by government.

According to PSC-Gross-Split terms and conditions, 47% of the gross revenue will goes to government, and the remaining 53% of gross revenue is equity of contractor. Contractor have to pay his profit tax to the government. There are no First Tranche Petroleum (FTP), nor Cost Recovery. All expenses is under contractor self control [5].

The final results for PSC-Gross-Split revenue distribution are: FOR Gross Revenue of 9,840 MMUS$, total Government Take is 5,593 MMUS$ (57% of Gross Revenue), and total Contractor Take is 4247 MMUS$ (43% of Gross Revenue) (see Figure 2).

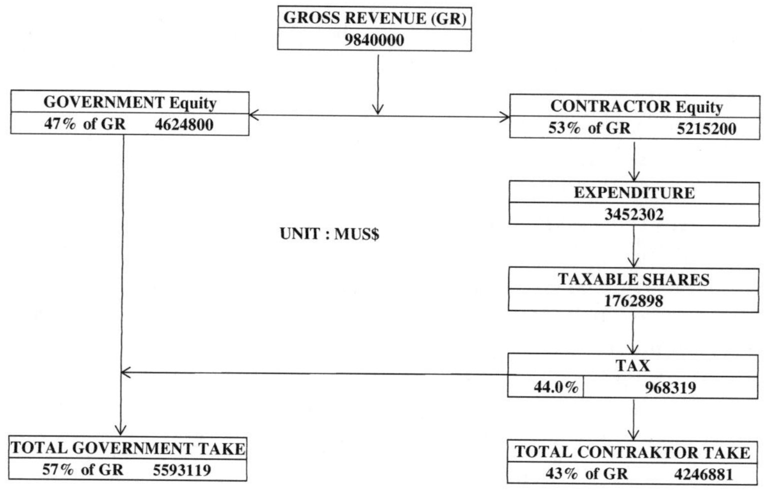

Figure 2. Revenue distribution results of PSC-Gross-Split.

3.3. Profitability Indicators Result

For the entire 20 years of the assuming steam flooding project, although the input data are the same for revenue, investment , tax rate , and operating coast, the contractor will have difference cashflow for both PSC-Cost-Recovery and PSC-Gross-Split.

On the side of PSC-Cost-Recovery scheme, the contractor cash flow in MMUS$, consecutively are : (-319.33); (-66.26); (-66.66); 36.92; 119.07; 150.87; 132.32; 60,70; 36.86; 54.71; 63.11; 65.21; 64.16; 62.59; 66.26; 74.66; 84,11; 99.87; 108.79; and 130.32. The cumulative cash flow is 103.96 MMUS$.

On PSC-Gross-Split scheme, the contractor cash flow in MMUS$, consecutively are : (-319.33); (-66.26); (-66.66); 14.42; 28.04; 35.77; 31.26; 33.19; 14.14; 66.04; 76.34; 78.92; 77.63; 75.70; 80.21; 90.51; 102.10; 121.42; 132.37; and 158.77. The cumulative cash flow is (-51.48) MMUS$.

The economic indicator was then calculated from that cash flow. Figure 3 shows the economic output comparison between PSC-Cost-Recovery and PSC-Gross-Split.

Look at the profitability index of NPV, IRR, and POT. On the government side, the new management policy (PSC-Gross-Split) for the petroleum (oil and gas) exploration and exploitation is more attractive than the old one (PSC-Cost-Recovery). On the contractor side, the new management policy (PSC-Gross-Split) for the petroleum (oil and gas) exploration and exploitation is less attractive than the old one (PSC-Cost-Recovery). For this case of steam flooding project, if there is opportunity to choose the type of government management policy, contractor will tend to prefer the PSC-Cost-Recovery , rather than PSC-Gross-Split.

3.4. Sensitivity Analysis PSC-Cost-Recovery and PSC-Gross-Split

All data value that used in this project are estimated or predicted data. Of course, those data could be difference in real time the project implementation. It may be under or over estimated value. The impact of all change data input will change the profitability of the project. Sensitivity analysis was performed to figure out the change of project profitability inline with changes of the input data, such as production, investment, and operating cost.

No	Indicator	Unit	PSC-Cost-Recovery	PSC-Gross-Split
1	Oil Production	MBBL	196,800	196,800
2	Oil Price	US$/BBL	50	50
3	Years Oil Production	years	17	17
4	Total Gross Revenue	MUS$	9,840,000	9,840,000
5	FTP	MUS$	1,968,000	0
	Total Cont. FTP	MUS$	527,227	0
	Total Gov. FTP	MUS$	1,440,773	0
6	Investment	MUS$	452,250	452,250
	Intangible	MUS$	52,000	52,000
	Tangible	MUS$	400,250	400,250
7	Invesment Credit (IC)	MUS$	0	0
8	Total Operating Cost	MUS$	3,000,052	3,000,052
	Operating Cost	MUS$	2,952,000	2,952,000
	Abandonment & Site Re	MUS$	48,052	48,052
9	Cost Recoverable	MUS$	3,452,302	3,452,302
	(% of. Gross Revenue)	%	35%	35%
	Unrecovered Cost	MUS$	0	0
	(% of. Gross Revenue)	%	0	0
10	Equity to be Split	MUS$	4,419,698	9,840,000
	Cont. Equity	MUS$	1,184,037	5,215,200
	Gov. Equity	MUS$	3,235,661	4,624,800
11	Contractor			
	Net Cash Flow	MUS$	958,308	794,579
	(% of. Gross Rev)	%	10%	8%
	IRR	%	13.3%	8.6%
	NPV@10%	MUS$	103,962	(51,479)
	POT	years	7.29	12.44
12	Government			
	FTP	MUS$	1,440,773	0
	Equity	MUS$	3,235,661	4,624,800
	Tax	MUS$	752,956	968,319
	Net Cash Flow	MUS$	5,429,390	5,593,119
	(% of. Gross Revenue)	%	55.2%	56.8%
	Gov. PV @ 10%	MUS$	1,379,884	1,535,326

Figure 3. Economic output comparison of PSC-Cost-Recovery vs PSC-Gross-Split.

The result of sensitivity analysis are presented in Figure 4A, Figure 4B, and Figure 4C. Figure 4A is sensitivity NPV and IRR of PSC-Cost-Recovery scheme as function of change in data input of investment, production, oil price, and operating cost. The line of production is the same with line of oil price. Profitability is more sensitive to investment rather than three others variable data input.

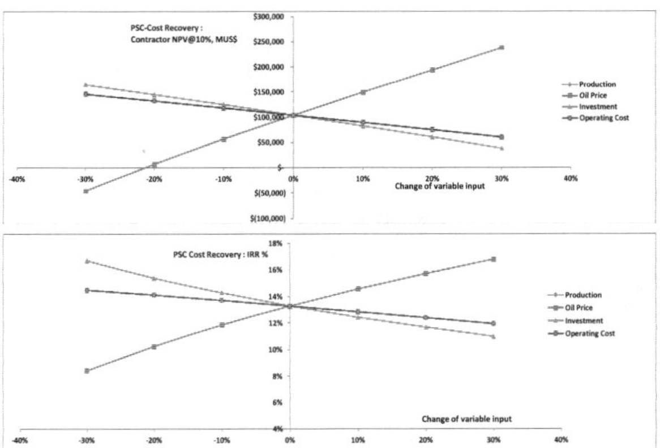

Figure 4A. PSC-Cost-Recovery : sensitivity analysis of NPV and IRR

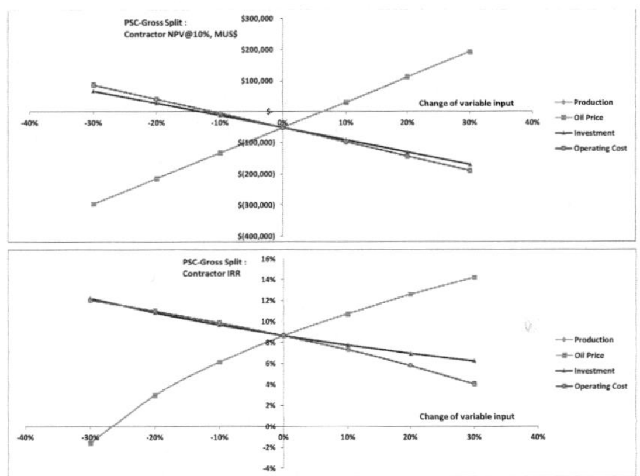

Figure 4B. PSC-Gross-Split : sensitivity analysis of NPV and IRR.

Figure 4B is sensitivity NPV and IRR of PSC-Gross-Split scheme as function of change in data input of investment, production, oil price, and operating cost. The line of production is the same with line of oil price. Profitability is also more sensitive to investment rather than three others variable data input. So that, it is clearly figure out by figure 4, both of the petroleum management policy are the same sensitive to changes of investment.

The new petroleum management policy which represent by "Peraturan Menteri ESDM Republik of Indonesia No. 8 Tahun, 2017" is look like tend to not attract the investor or contractor on this steam flooding project. One of the tool to solve this condition is by tax incentive. If the government give tax incentive to the contractor, the project profitability of PSC-Gross-Split scheme management policy would become the same as profitability of PSC-Cost-Recovery scheme management policy (See Figure 4C).

If NPV is important point of profitability, PSC-Gross-Split tax rate should be decrease from 44% to 17%, so the contractor NPV@10% will become 103,962 MMUS$, that is the same with the result of PSC-Cost-Recovery.

If IRR is important point of profitability, PSC-Gross-Split tax rate should be decrease from 44% to 10%, so the contractor IRR will become 13.3%, that is the same with the result of PSC-Cost-Recovery.

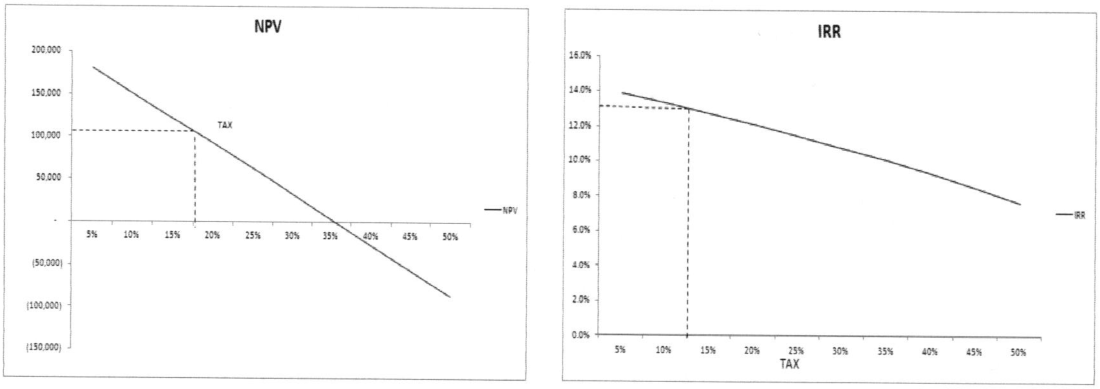

Figure 4C. PSC-Gross-Split : Sensitivity Analysis of NPV and IRR as changes of Government Tax.

4. Conclusion

Contractor Net Present Value for the PSC-Cost-Recovery is 103,962 MUS$ and for the PSC-Gross-Split is minus 51,479MMUS$. Internal Rate of Return for the PSC-Cost-Recovery is 13.3% and for the PSC-Gross-Split is 8.6%. Contractor Net Present Value and Internal Rate of Return of PSC-Gross-Split are greater than of PSC-Cost-Recovery, but Pay Out Time in PSC-Gross-split is longer than Pay Out Time in PSC-Cost-Recovery. The new energy management policy is considered as less attractive scheme than PSC-Cost-Recovery, for the contractor side.

Acknowledgment

The authors gratefully acknowledge the permission given in publishing this article from the Faculty of Mineral Technology, Universitas Trisakti, as well as the facilities provided by Danang Sismartono PPPTMGB LEMIGAS, Jakarta, in obtaining the data for this paper.

References

[1] Field Report Data Steam Flood Duri. PPPTMGB LEMIGAS, Jakarta, 2017.
[2] H.K Van Pollen and Associates, Inc., Fundamental of Enhanced Oil Recovery, Tulsa, Oklahoma, 1980.
[3] Harimurti D, Indonesian Milestone in Production-Sharing Contract in Perspective of Government Take, Contractor Take, Cost Recovery and Production Target, SPE-187008-MS, Society Of Petroleum Engineer, 2017
[4] Irham S, Pengelolaan Lapangan Migas, Trisakti University, Jakarta, 2017.
[5] Kurniawan T and Jaenudin J, Proposed Modification of Abandonment and Site Restoration Mechanism in Gross Split PSC for Marginal Field in Indonesia, SPE-186378-MS, Society Of Petroleum Engineer, 2017
[6] Lubiantara B, Ekonomi Migas: Tinjauan Aspek Komersil Kontrak Migas, Gramedia, Jakarta, 2012.
[7] http://www.migas.esdm.go.id/post/read/permen-esdm-nomor-08-tahun-2017-tentang-kontrak-bagi-hasil-gross-split
[8] Nasir A, Sejarah Sistem Fiskal Migas Indonesia, Grasindo, Jakarta, 2014.
[9] Patmosukismo S, Oil and gas : Politic, law and industry (MIGAS Politik, Hukum & Industry). ISBN 978-979-823189-6, 2011.

IOP Conf. Series: Earth and Environmental Science **106** (2017) 012110 doi:10.1088/1755-1315/106/1/012110

Assessment of existing local houses condition as analysis tools for shore housing improvement program in Weriagar district, Bintuni Bay

F Firmansyah*, A Fernando, I P R Allo

School of Architecture, Planning and Policy Development (SAPPK) Bandung, Institut Teknologi Bandung, Bandung, Indonesia

*Corresponding Author: firmansyah@ar.itb.ac.id

Abstract. The housing assessment is a part of the pre-feasibility study inThe Shore Housing Improvement Program in Weriagar District, West Papua. The housing assessment was conducted to identify the physical condition of existing houses. The parameters of assessment formulated from local references, practices and also national building regulation that covers each building system components, such as building structure/frame, building floor, building cover, and building roof. This study aims to explains lessons from local practices and references, used as the formula to generate assessment parameter, elaborate with Indonesia building regulation. The result of housing assessment were used as a basis to develop the house improvement strategy, the design alternative for housing improvement and further planning recommendations. The local knowledges involved in housing improvement program expected that the local-based approach could respect to the local build culture, respect the local environment, and the most important can offer best suitable solutions for functional utility and livability.
Keywords: housing assessment, housing improvement, local knowledges, vernacular, weriagar district

1. Introduction

Weriagar district is a remote area away from Bintuni bay regency, only accessible by boat for 3-4 hours. It has a distinct character of change-sensitive estuary area and local culture owned by its Indigenous People. The living community has lived for centuries and inherited custom land law as well as livelihood in fisheries. Due to its location, it had been revealed that coastal and river erosion threats Weriagar District since fisheries and settlement activities interfere its natural function.There is also a problem in settlement area, that always submerged by sea tides, it affects the houses feasibility, as well as the health of the living community. Based on the issue, The Shore Housing Improvement Program was developed to improve settlement and environment condition in Weriagar, in order to increase life quality of the living community. The program was first initiated with pre-feasibility study, to identify the main alternative solution and recommends the most prospective solution.

Generally, the housing assessment was conducted to a building that used familiar structure system and assessed with common building regulation. The Weriagar district community still hold strong traditional values and very dependent on nature, so the parameters used to housing assessment should be adjust based on it. So that, the assessment results can be balance and fit with local wisdom.

In addition to the assessment parameter derived from local knowledge, it is needed to elaborate with other parameter that have detailed and specific basis of building. In this case is used Indonesian national regulation on public housing, it contains the basis of feasibility standard for a health and safety in every building structure elements, covering the building foundation, building frame, roof structure, lighting, air circulation, sanitation, the space ratio per person, and the usability of materials and the maintenance which are based on local knowledge [1].

In order to strategy development of housing improvement, It should be done with a local-based approach, it has respect to the community and respect to the process and building culture [2] and also leave minimum harm impact to environment. The local knowledge of houses were developed by living community to adapt with their environment characteristic and culture, with basic construction techniques which people have learned from over centuries, local labour availability, and also environmentally friendly materials [3].

The objective of this paper, specifically explains the housing assessments analysis tools to identify existing local houses condition for development strategy, and basis of design alternatives for housing improvement program, also to explain lessons from local practices and references could be used to formulating assessment parameter.

2. Research and Method

2.1. Scope

Weriagar District settlement located in Bintuni Bay Regency, West Papua.2°16'22.3"S 132°53'38.3"E. The housing assessment scope are 33 pre-identified random samples from total 131 houses, based on The Inhabitant Survey Report of Centre of Population and Policy Study in 2013.

2.2. Method

There were 3 steps in physical house assessment, (1) data collecting; (2) data analyzing; and (3) providing recommendations.

2.3. Assessment Parameter

The following housing assessment parameters adapted from Indonesia national regulation about house feasibility criteria which has adjusted with local wisdom. Buildings are divided into observation 3 elements. They are building structure/frame, building floor, and building cover.

Building structure/frame consists of foundation, bottom column, upper column, floor, roof and wall. In the assessment of foundation, a maximum score is given if there is no differential settlement or settlement of the building and if the building is not tilting. In the assessment of bottom column (from foundation to floor), a maximum score is given if columns are made of processed wood (instead of logs) with sufficiently close spacing among each other and bound with bracings between the bottom poles to form a rigid system. In the assessment of upper column (from floor to roof structure), a maximum score is given if columns are made of processes wood (instead of logs) and have connections with roof frame and house's sheathing frame. In the assessment of Floor a maximum score is given if floors are made of processed wood (instead of logs) with tight spacing so that floorboards do not sag, shift, nor tilt, and have different elevations between room. In the assessment of roof, a maximum score is given if the roofs are made of processed wood (instead of logs), wood dimensions are sufficient to bear loads, and have correct wooden roof construction system (roof stands). In the assessment of wall, the maximum score is given if walls are made of processed wood (instead of logs), have correct wall support system and connected to the main frame/roof system.

Building floor consists of floor frame and floorboard. In the assessment of floor frame, a maximum score is given if floor frameworks are made of processed wood (instead of logs), have tight spacing so the floorboards do not sag, shift, and tilt, and have different elevations between rooms. In the assessment of floorboard, a maximum score is given if floors are made of processed wood (instead of logs), have tight floorboard spacing and not undulated.

Building cover consist of wall, door, and window. In the assessment for wall, a maximum score is given if wall covers are made of wood plank, processed neatly and uniform in size. If *gaba-gaba* is used, it should have a uniform shape and size, tightly closed and fixed on the wall framework solidly. In the assessment of door, a maximum score is given if doors are made of processed wood (instead of logs) with finishing, and have regular shape and size and a minimum score is given ifdoors consist merely of frames/sillsclosed only by curtains and door leaves are absent. In the assessment of window, the maximum assessed if using wood materials that already processed (not from logs) with finishing, has a definite shape and size. Average assessed if window not only have frame/sills but also have shutters made from unprocessed wood and not finishes. Minimum assessed if window only have frame/sills, closed only by a curtain and don't have shutters.

Building Roof consist of roof frame and roof cover. In the assessment for roof frame a maximum score is given if roofs are made of processed wood (instead of logs), with sufficient wood dimensions to bear loads, and have correct wooden roof construction system (roof stands). In the assessment of foof cover, a maximum score is given if roof coverings are made of zinc, which is used to cover all parts of the building, and already has ceilings made of plywood/gypsum and a minimum score is given if roof coverings are only made of unsealed *rumbia*(still leaking), if there are still parts of the roof that arenot covered by roofing materials, and if there is no ceiling.

3. Results and Discussion

3.1. Data Collecting

The results of data collecting from33 pre-defined house sample consist of size, room type, and interior facilities; Existing interior visual data and exterior condition of houses; Number of householders and inhabitants; Daily and periodically of routine activity pattern in each room; Space requirements references and activity type. The data collected through interview and observation.

The following sample of data result are collecting to explain the condition of sample house (see Figure 1).

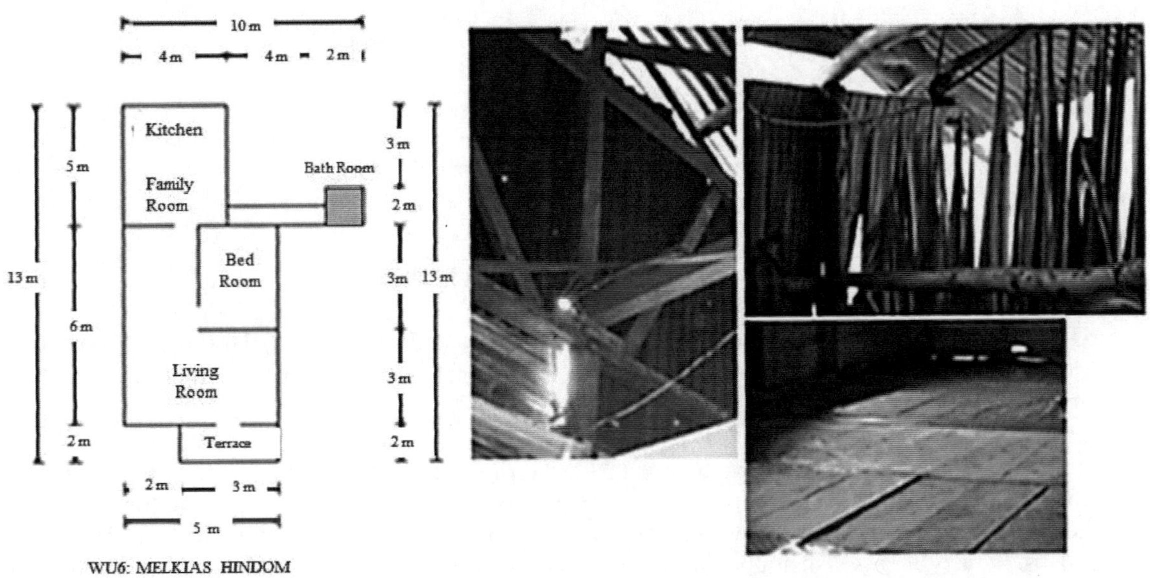

Villages	Householder	Householder number	Inhabitants	Status	Area
North Weriagar	Melkias Hindom	1	4	Indigeneous	72

Figure 1. Sample of result of data collecting.

IOP Conf. Series: Earth and Environmental Science **106** (2017) 012110 doi:10.1088/1755-1315/106/1/012110

3.2. Data Analyzing

Based on collected data of houses in every village in Weriagar District, the analysis process is conducted which consist assessment of physical condition percentage using predefined parameter. The following sample of the housing assessment of physical condition percentage.

Table 1. Sample of House Assessment form.

Building Component System	Building Component	Building Condition Percentage			
		0-25 %	26-50%	51-75 %	76-100 %
Frame/Structure	Foundation		v		
	Bottom Column	v			
	Upper Column		v		
	Floor		v		
	Ceiling		v		
	Wall		v		
Floor	Floor Frame		v		
	Floor Board		v		
Building Cover	Wall		v		
	Door	v			
	Window	v			
Roof	Roof Frame			v	
	Roof Tile			v	

3.3. Recommendation

Based on observation and assessment results, the following recommendations could be made:

- Elevated Building
  All of the buildings in Weriagar District need to be elevated in order to adapt to local conditions (tidal area). The minimum building floor height is higher than the maximum height of the tide that occurs periodically every year.
- Frame Construction System
  Frame construction system needs to use in the elevated building. In this system, the foundation structure does not rely on soil type or bearing capacity but on the bond between the foundation and the upper framework which ensures the overall stability of the structural system. This frame construction system is also applied in the overall structure of the system including structural floors, pillars, and roof of the building.
- Local Materials
  The usability of local materials becomes dominant because of limited availability of materials from the outside and lack of supporting system to transport goods. The local materials have been processed beforehand to ensure their strength, durability and to facilitate their maintenance. Local materials are used both for building structure and building envelope.
- New Materials
  The use of new materials needs to be introduced mainly for roofing to increase durability against rain and to facilitate maintenance. Moreover, additional materials such as glass for windows and openings of the buildings are also needed.

Based on the analysis result of space usage preferences, inhabitant activities, and existing house layout, the following recommendations on house layout are made:

- House layout should follow daily activity pattern.
- House layout should follow room usage and function pattern which may be different to common building standards.
- Room and house layout should consider the growing building type to anticipate additional room needs which have not been fulfilled in the initial house construction stage, so that the house shape and character can still be maintained. Furthermore, the original and additional parts would be integrated.

4. Conclusion

Each location has different natural challenges, the manifestation of human resistance to living in natural landscape reflected into the type of the settlement and the manner the houses are built. Local knowledge can be used as a strategy in the strategy development of housing improvement in Weriagar District due its offer best suitable solutions for functional utility and liveability. [3] Lesson from local knowledge elaborated with building regulation, generating the balanced assessment parameter for achieving contextually appropriate and sustainable development [4].

Visual physical assessment is effective and easy to conduct, due the houses has a very simple structure, but visual assessment can only detect major damage of structural or building physic. Assessment should be done by people who are competent in their fields, since visual assessment has no clear basis, and rely on the experience and knowledge of observer.

References

[1] Peraturan Menteri PU. Standar Pelayanan Minimal Bidang Perumahan Rakyat Daerah Provinsi dan Daerah Kabupaten/Kota, 2008.

[2] Asquith, Lindsay, and Marcel Vellinga, Eds. Vernacular architecture in the 21st century: Theory, education and practice. Taylor & Francis, 2006.

[3] Kumar, Ashwani. 2013. Vernacular practices: as a basis for formulating building regulations for hilly area. *International Journal of Sustainable Built Environment.* 2(2):183-192.

[4] Kumar A, and N Munoth. 2011. Vernacular architecture–a prerequisite for sustainable development. Arch. *Time Space People*. 11(7):16-22.

Effect of iron and magnesium addition for ethanol production from the conversion of palm oil mill effluent by anaerobic processes

M Handajani*, A Gumilar, M Syafila

Environmental Engineering Department, Faculty of Civil and Environmental Engineering, Institut Teknologi Bandung, Bandung, Indonesia

*Corresponding Author: marisa.handajani@ftsl.itb.ac.id

Abstract. Nowadays, crisis of the energy is the main problem in the world. Currently, most the energy resource derived from the fossil material that cannot be refurbished. Ethanol is an alternative fuel that content as a fossil fuels. Wastewater with the high concentration of the organic can be used for the ethanol production to replace foodstuff as a raw material. In this study, palm oil mill effluent (POME) with the concentration of COD is 24,500 mg/L has been used as a substrate. The purpose of this study was to determine the effect of the metal addition in the substrate metabolic pathways. Circulating batch reactor (CBR) is used with the flushing N_2 1L/min for 24 hours and continued operates for 72 hours by internal biogas. The additional variation concentration of Fe(II) ion are 0.5; 1.0 and 2.5 mg/L, and Mg(II) are 0.5 and 1.5 mg/L were added by combination. The results showed that the combination of Fe (II) 2.5 mg/L and Mg(II) 1.5 mg/L produced the highest ethanol concentration is 715.8 mg/L and degree of acidification (DA) 0.284-0.357. Another combination of Fe(II) and Mg(II) provide results for the ethanol production 463.7-689.7 mg/L with the rate of ethanol production is 1.09–26.5 mg/L/hour.
Keywords: anaerobic process, circulating batch reactor, palm oil mill effluent

1. Introduction

The limitations of fossil fuel being a serious problem while the world energy demand were increase every year. Biofuels such as biogas, bioethanol, biodiesel, bio-methanol, bio-oil, bio-hydrogen and bioelectricity are renewable energies with high quality fuel properties and can be utilized for various energy services [1]. Bioconversion of the wastewater with the high concentration of organic by anaerobic process can be produced biofuels which is can instead of fossil fuels [2]. Renewable resources including agricultural and industrial residues have been studied through microbial bioconversion processes.

Indonesia is the world's largest producer and exporter of crude palm oil (CPO). CPO's are one of the main commodities on agro industrial sector in Indonesia. The land area of palm trees increased every year to meet the Government target for crude palm oil production to 40 million metric tons per year in 2020 [3]. In 2014, the production of palm oil in the world reached 62.34 million tons, mostly from Indonesia (30.5 million tons) and Malaysia (19.9 million tons) [4]. The process of palm oil can produce the wastewater with the high concentration of organic. Nowadays, the potential of the wastewater are not utilized optimally, for example the treating of the wastewater released the methane into the environment about 32.9 kg/ton palm oil produced [5].

Wastewater from the palm oil mill effluent has a very high concentration of organic. BOD and COD concentration of the wastewater are 25,000-65,714 and 44,300-102,696 mg/L respectively [6]. Characteristic of the palm oil mill effluent shown in Table 1.

Table 1. Characteristic of raw palm oil mill effluent (POME) [6].

Parameter	Unit	Range
pH	-	4-5
Biological oxygen demand (BOD)	mg/L	25,000-65,714
Chemical oxygen demand (COD)	mg/L	44,300-102,696
Total solids (TS)	mg/L	40,500-72,058
Suspended solid (SS)	mg/L	18,400-46,011
Volatile solids (VS)	mg/L	34,000-49,300
Oil and grease (O and G)	mg/L	4,000-9,341
Ammonia nitrogen (NH_3-N)	mg/L	35-103
Total nitrogen (TN)	mg/L	750-770

Anaerobic treatment using microbe can produce the biofuels. In general, this processes divided into four main stages, namely hydrolysis, acidogenesis, acetogenesis and methanogenesis [7]. In every stages depend on the condition such as pH, microbe, temperature and so on. The presence of micro nutrients (metal addition) can increase the activity of alcohol dehydrogenase enzyme [8]. During the first stage, organic compounds (carbohydrates, proteins and lipids) are hydrolyzed and utilized by hydrogen producing bacteria to produce ethanol, H_2 and volatile fatty acids (VFAs) [9].

The aim of this study was to determine the effect of metal addition for the formation of ethanol and other product in acidogenic step. Ethanol and other intermediates products are formed during acidogenic fermentation by anaerobic mixed culture bacteria. The alcohol dehydrogenase (ADH) is the main enzyme for the ethanol production [10].

2. Research Method

2.1. Organic wastewater and biomass

The fresh raw palm oil mill effluent (POME) was collected from the influent basin of PT. Condong, Garut, Indonesia. In this study, biomass was taken from the sludge of palm oil mill effluent mixed by cow rumen with the ratio of 50:50 and has been acclimated to the POME. The acclimated biomass concentration used for the anaerobic batch reactor in terms of mixed liquor suspended solids (MLSS) and mixed liquor volatile suspended solids (MLVSS), were about 5.2 g SS/L and 3.98 g VSS/L respectively.

2.2. Analytical methods

Biological oxygen demand (BOD), chemical oxygen demand (COD), TSS, oil and grease, total nitrogen and pH were analyzed according to standard methods [11]. Analysis of volatile fatty acids by using HPL Ctype Hitachi inter faceD-7000 HSM with analytical column Aminex HPX-87H, ethanol by using GC Shimadzu17-A with column Analytic DB-Wax.

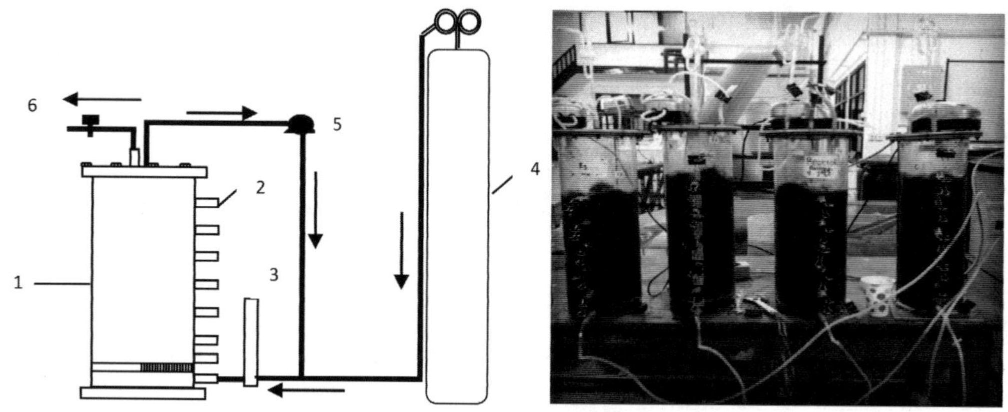

Note: 1. Anaerobic reactor; 2. Sampling port; 3. Flowmeter; 4. Gas N_2; 5. circulation pump; 6. Gas ventilation

Figure 1. Anaerobic batch reactor.

2.3. Bioreactor and experimental design

Anaerobic batch reactor with the working volume of 5L were operated with the flushing N_2 1L/min for 24 h and continued operates for 72 h by internal biogas (Figure 1). The reactor filled with a mixture of consisted of 20% (v/v) mixed culture bacteria and 80% (v/v) of POME as a substrate. The additional variation concentration of Fe(II) ion are 0.5; 1.0 and 2.0 mg/L, and Mg(II) are 0.5 and 1.5 mg/L were added by combination as shown in Table 2. Sample was collected every 6 h and analyze for ethanol and volatile fatty acid.

Table 2. Combination of magnesium (Mg) and iron (Fe) addition.

Metal addition (mg/L)		Reactor ID
Mg	Fe	
0	0	Control
	0.5	A
	1	B
	2.5	C
0.5	0	D
	0.5	E
	1	F
	2.5	G
1.5	0	H
	0.5	I
	1	J
	2.5	K

2.4. Calculation

The determination of the optimum conditionsbased onthe efficiency and the formation of ethanol and a few other variables, namely :

$$\text{Degree of acidification (DA): DA} = \frac{(\text{TAV}) \quad [\text{mgCOD/L}]}{\textit{influent of soluble COD}\ [\text{mgCOD/L}]} \tag{1}$$

$$\text{Degree of ethanofication (DE) : DE} = \frac{(\text{Ethanol}) \quad [\text{mgCOD/L}]}{\textit{influent of soluble COD}\ [\text{mgCOD/L}]} \tag{2}$$

Acidogenic product equation Van Haandel and Van Der Lubbe (2012) :

$$COD_{theoritical} = 8\cdot(4x+y-2z)/(12x+y+16z)\ gCOD.g^{-1}CxHyOz \quad (3)$$

where : x = C; y =H and z = O

Formation rate product (mg COD/L/hour) :

$$= \frac{[Product\ concentration_{end} - Initial\ concentration_{in}]}{time}\frac{[mgCOD/L]}{[hour]} \quad (4)$$

3. Results and Discussion

Palm oil mill effluent (POME) was characterized as shown in POME also contains considerable amounts of nutrients such as nitrogen, potassium Table 3., magnesium, calcium, cadmium, copper, chromium and iron. The high concentration of carbohydrate, nutrients, lipids and minerals also found in POME [12].

Table 3. Characteristic of palm oil mill effluent (POME).

Parameter	Unit	Results
BOD	mg/L	12,450
Soluble COD	mg/L	17,300
Total COD	mg/L	24,500
TSS	mg/L	15,000
Oil and Grease	mg/L	185
Nitrogen Total	mg/L-N	250
Ethanol	mg/L	39.05
Volatile fatty acids	mg/L	361.07
pH	-	5.85

Ethanol production during the process shown in Figure 2. Ethanol was start to produce after 12 hr for each reactor with the combination of metal addition. Highest ethanol production occurs at the reactor with combination of 2.5 mg/L-Fe + 1.5 mg/L-Mg is 715.8 mg/L and the maximum rate of ethanol production is 33.33 mg/L/hr. The concentration of ethanol increased until 42 hr, then after 42 to 72 hr ethanol concentration are stable at 628.8 to 678.1 mg/L by the rate of ethanol production is 5.42 to 25.57 mg/L/hr. In other variations, typical of ethanol production is the same, ethanol concentration was increased until 42 to 48hr and then stable until the end of experiments. The highest ethanol concentration for each combination 0.5 mg/L-Mg + 0.5 mg/L-Fe; 0.5 mg/L-Mg + 1.5 mg/L-Fe; 0.5 mg/L-Mg + 2.5 mg/L-Fe; 1.5 mg/L-Mg + 0.5 mg/L-Fe and 1.5 mg/L-Mg + 1.5 mg/L-Fe are 463.7; 572.3; 689.7; 480.6 and 555.8 mg/L respectively. Ethanol production for the reactor without metal addition (control) has lower concentration than the reactor with metal addition, either individual addition of metal and combination. The highest concentration of ethanol in control is 312.7 mg/L. Highest concentration of ethanol production at the reactor with the individually metal addition are 292.3; 310.6; 422.4; 532.2 and 674.5 mg/L for the metal addition of 0.5 mg/L-Mg; 1.5 mg/L-Mg; 0.5 mg/L-Fe; 1.5 mg/L-Fe and 2.5 mg/L-Fe respectively. Iron (II) addition can enhance of ethanol production. The addition of Fe (II) 200 mg/L will produce ethanol with the same amount without addition of Fe (II), addition 400 mg/L will produce ethanol 122 times lower than without addition of Fe (II) and addition of 600 mg/L will not produce the ethanol [13]. Nitrogen flushing can shift metabolism pathway to be acetyl CoA production then it will be oxidized to be ethanol by alcohol dehydrogenase enzyme. Addition of Fe (II) 10 times that of the control can increase the activity of alcohol dehydrogenase enzyme and can enhance ethanol production [8].

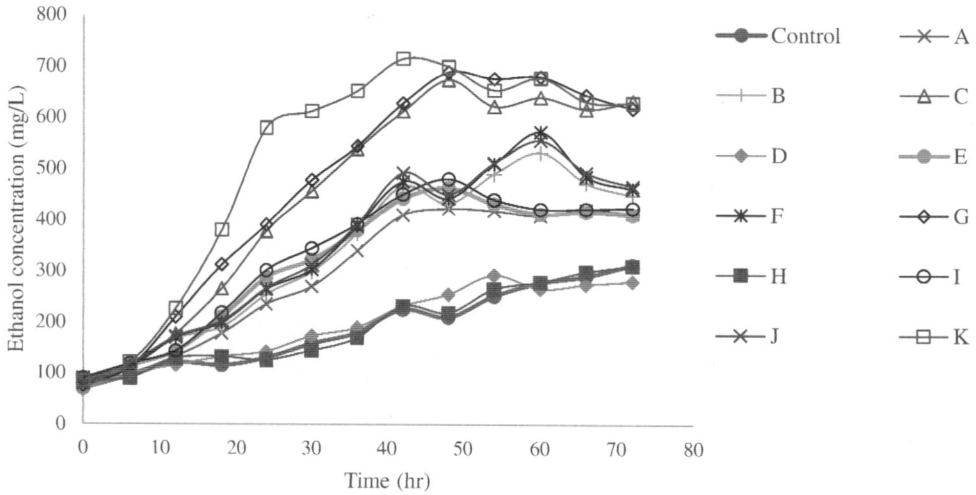

Figure 2. Ethanol formation from the variation of metal addition.

Volatile fatty acids (VFA's) such as acetate, propionate, butyrate, valerate and biogas (H2, CO2 and CH4) was produced during the experiments. Acetate and propionate is acidogenic product most widely produced in the anaerobic processes.Degree of acidification (DA) was calculated by the comparison of VFA's production and soluble COD in the influent. In appropriate concentration of metal addition, micronutrient such as Fe, Mn, Zn, Co, Cu, Mo, Se and Ni are required by enzyme as co-factor, but at the concentration does not exact this is can be toxic for microorganism to convert the substrate [14]. In Table 4 shows the acidogenic product in degradation of wastewater by anaerobic processes.

Table 4. Acidogenic product in degradation of wastewater by anaerobic processes.

			Metal addition (mg/L)					
Parameter	**Unit**	**Mg (II)**	**0.5**	**0.5**	**0.5**	**1.5**	**1.5**	**1.5**
		Fe (II)	**0.5**	**1.5**	**2.5**	**0.5**	**1.5**	**2.5**
Ethanol *	mg/L		463.7	572.3	689.7	480.6	555.8	715.8
Acetate *	mg/L		1885.7	1765.4	1729.0	1766.2	1810.1	1730.2
Propionate *	mg/L		1007.8	1229.5	1182.0	1753.7	1410.0	1170.2
Butirate *	mg/L		533.5	452.4	801.8	653.2	662.7	702.6
Valeric acid *	mg/L		201.8	236.0	242.8	219.7	235.8	188.0
Ethanol formation rate *	mg/L/ hr		14.80	17.62	27.08	21.60	28.81	33.33
Degree of Acidification (DA)*	-		0.284	0.292	0.323	0.357	0.332	0.305
Degree of Ethanofication (DE)*	-		0.056	0.069	0.083	0.058	0.067	0.086

* maximum results

4. Conclusion

The addition of Mg (II) and Fe (II) are influenced to the ethanol production. The addition of 1.5 mg/L-Mg + 2.5 mg/L-Fe has produced ethanol with the highest concentration is 715.8 mg/L. Volatile fatty acid such as acetate and propionate are the highest product formation from the anaerobic process.

Acknowledgment
This research was funded by Program Penelitian, Pengabdian kepada Masyarakat dan Inovasi (P3MI) Tahun 2017.

References

[1] Mshandete A M, & Parawira W. 2009. Biogas technology research in selected sub-Saharan African countries - A review. *African Journal of Biotechnology*. 8(2):116-25.

[2] Salihu A, & Alam M Z. 2012. Palm oil mill effluent: a waste or a raw material? *Journal of Applied Sciences Research.* 8(1):466-73.

[3] Paoli G D, Gillespie P, Wells P L, Hovani L, Sileuw A, Franklin N., et al. 2013. Oil Palm in Indonesia: Governance, Decision Making and Implications for Sustainable Development. Summary for Policy Makers & Practicioners. Jakarta, Indonesia: The Nature Conservancy Indonesia Program.

[4] Ahmed Y, Yaakob Z, Akhtar P, & Sopian K. 2015. Production of biogas and performance evaluation of existing treatment processes in palm oil mill effluent (POME). *Renewable and Sustainable Energy Reviews*. 42:1260-78.

[5] Schuchardt F, Wulfert K, Darnoko D, & Herawan T. 2007. Effect of new palm oil mill processes on the EFB and POME utilisation. *Proceedings of Chemistry and TechnologyConference.* PIPOC 2007: 44-57. Kuala Lumpur.

[6] Ji C M, Eong P P, Ti T B, Seng C E, & Ling C K. 2013. Biogas from palm oil mill effluent (POME): Opportunities and challenges from Malaysia's perspective. *Renewable and Sustainable Energy Reviews*. 26:717-26.

[7] Demirel B, & Scherer P. 2008. The roles of acetotrophic and hydrogenotrophic methanogens during anaerobic conversion of biomass to methane: a review. *Rev Environ Sci Biotechnol*. 7:173-90.

[8] Saxena J, & Tanner R. 2010. Effect of trace metals on ethanol production from synthesis gas by the ethanologenic acetogen, Clostridium ragsdalei. *Journal of Industrial Microbiology and Biotechnology*. 38:513-21.

[9] Rafieenia R, Girotto F, Peng W, Cossu R, Pivato A, Raga R, et al. 2017. Effect of aerobic pre-treatment on hydrogen and methane production in a two-stage anaerobic digestion process using food waste with different compositions. *Waste Management.* 59:194-9.

[10] Li Y F, Ren N Q, Chen Y, & Zheng G X. 2007. Ecological mechanism of fermentative hydrogen production by bacteria. *International Journal of Hydrogen Energy*. 32:755-60.

[11] APHA, AWWA, WEF. 2012. Standard Method for the Examination Water and Wastewater (22nd Edition). Washington DC, United States of America: American Public Health Association.

[12] Ohimain E I, & Izah S C. 2017. A review of biogas production from palm oil mill effluents using different configurations of bioreactors. *Renewable and Sustainable Energy Reviews*. 70:242-253.

[13] Yamashita Y, Kurosumi A, Sasaki C, & Nakamura Y. 2008. Ethanol production from paper sludge by immobilized Zymomonas mobilis. *Biochemical Engineering Journal*. 42:314-319.

[14] Barton L L. 2004. Structural and Functional Relationship in Prokaryotes. Albuquerque, USA: Springer.

Design of combination biofilter and subsurface constructed wetland-multilayer filtration with vertical flow type using *Vetiveria zizanioides* (akar wangi)

A D Astuti*, M Lindu, R Yanidar, M Faruq

Environmental Engineering Department, Faculty of Architecture Landscape and Environmental Technology, Universitas Trisakti, Jakarta, Indonesia

*Corresponding Author : ariani_da@trisakti.ac.id

Abstract. As environmental regulation has become stricter in recent years, there is an increasing concern about the issue of wastewater treatment in urban areas. Senior High School as center of student activity has a potential source to generated domestic wastewater from toilet, bathroom and canteen. Canteen wastewater contains high-organic content that to be treated before discharged. Based on previous research the subsurface constructed wetland-multilayer filtration with vertical flow is an attractive alternative to provide efficient treatment of canteen wastewater. The effluent concentration complied with regulation according to [9]. Due to limited land, addition of preliminary treatment such as the presence of biofilter was found to improve the performance. The aim of this study was to design combination biofilter and subsurface constructed wetland-multilayer filtration with vertical flow type using *vetiveria zizanioides* (akar wangi) treating canteen wastewater. *Vetiveria zizanioides* (akar wangi) is used because from previous research, subsurface constructed wetland-multilayer filtration (SCW-MLF) with vertical flow type using *vetiveria zizanioides* (akar wangi) can be an alternative canteen wastewater treatment that is uncomplicated in technology, low cost in operational and have a beautiful landscape view, besides no odors or insects were presented during the operation.
Keywords: biofilter, constructed wetland, domestic wastewater

1. Introduction

In Jakarta, there are more than 176 Senior High School and Vocational School, but most of them do not have a wastewater treatment plant (WWTP), so wastewater generated directly discharged into receiving water bodies. Wastewater generated by the schools not only wastewater from toilets or bathrooms but also from the canteen/cafeteria. Canteen wastewater contains high-organic content that have to be treated before being discharged. Constructed wetland is planned or controlled system which has been designed and constructed using natural processes that involve vegetation, media, and microorganisms to treat wastewater [1], [2], [3]. Constructed Wetlands are widely used as advanced treatment of wastewater, in various countries to treat domestic and non-domestic wastewater, but not for canteen wastewater treatment that containing high concentrations of organic compound. Some researches in constructed wetlands conducted to process wastewater containing low concentrations of heavy metals and do not use multilayer filtration. Vertical wetlands treated domestic wastewater achieved 99% COD removal [4], [5], [6]. Therefore, this study was carried out to design combination of biofilter and subsurface constructed wetland multilayer filtration (SCW-MLF) vertical flow treated

IOP Conf. Series: Earth and Environmental Science **106** (2017) 012112 doi:10.1088/1755-1315/106/1/012112

high-organic wastewater. This treatment unit expected to be a canteen wastewater treatment alternative which is simple since it does not require complicated technology. It is cheaper in terms of cost of manufacture and operation as well as maintenance, and more natural [7], [8]. *Vetiveria zizanioides* can reduce dissolved P by 99% after three weeks and 74% N dissolved after five weeks. Due to limited land, addition of preliminary treatment such as the presence of biofilter was found to improve the performance.

2. Research Method

This study was carried out by implementing step by step data analysis of the quantity and quality of wastewater from the school canteen of Senior High School, West Jakarta, to design and determine the unit treatment. The data related to the quantity and quality of wastewater is obtained from domestic processing activities of the canteen, and are used as the basic design the unit of biofilter and vertical-flow subsurface constructed wetland multilayers filtration with Akar Wangi (*Vetiveria zizanoides*)

2.1. Data of wastewater quantity

Data related to the quantity of domestic wastewater from canteen of Senior High School, West Jakarta, are obtained from the direct measurement of the flow rates of wastewater collected in the collecting tank. This measurement is conducted during the school periods started from 7 a.m. to 4 p.m. (with 9 hours duration) for 5 days (weekdays) in order to calculate the daily waste flow rates started from Monday to Friday. The results of the measurement indicate that the wastewater flow rates are typically varied, as observed from both daily and hourly measurement of wastewater flow rates within the same day. The highest rate of daily measurement was 3,459 m^3/day.

2.2. Data of wastewater quality

The data related to the wastewater quality from the canteen of Senior High School, West Jakarta, is conducted by laboratory analysis using the sample of the wastewater taken from school canteen's wastewater outlet in order to investigate the characteristics of wastewater. Table 1 below presents information related to the results of laboratory analysis of the wastewater.

Table 1. Canteen wastewater quality analysis.

No	Parameter	Unit	Range of Concentration	Dom WW Standard
1	pH	-	6.8 – 7.4	6-9
2	COD	mg/L	665.6 – 3200	100
3	BOD	mg/L	208.1 – 1,208.1	30
4	Total Nitrogen	mg/L	3.1 – 26.6	-
5	Total Fosfat	mg/L	1.8 – 51.8	-
6	Oil and Fat	mg/L	8 – 146.8	5

2.3. Effluent standard

Quality effluent standard used is the effluent standard based on regulation of Ministry of Environment and Forestry's Standard for Domestic Wastewater Nr. P.68/2016 [9].

3. Results and Discussion

The units of wastewater treatment of school canteen, consist of wastewater reservoir, an aeration tank, and a vertical flow type constructed wetland multilayers filtration unit with vetiver grass, precisely located on the west side of school. This location was selected as the area is next to the school canteen, so the researcher to place and conduct treatment phase easily.

3.1. The basic design

The design of the building dimension of the whole units is based on the selected design criteria, with the help of key and empirical formulas in the planning of a vertical flow type of subsurface constructed wetland multilayers filtration unit. The dimensions are explained as follows:

IOP Conf. Series: Earth and Environmental Science **106** (2017) 012112 doi:10.1088/1755-1315/106/1/012112

1. The construction of aeration tanks and influent pipes are calculated using design criteria.
2. The construction of the vertical flow type of subsurface constructed wetland multilayers filtration unit with vetiver grass is calculated based on design criteria with the organic loading rate Kg BOD/Ha/Day.
3. The height of media and the diameter of gravel, the mixture of soil and sand as multilayer filtration are based on design criteria.
4. The pipes of wastewater distribution are calculated based on design criteria (volume) in accordance with closed/pressurized hydraulic formulas.
5. The pipes of wastewater are calculated based on the design criteria and pump formula.

3.2. The calculation of wastewater treatment plants

The following presents the results of treatment unit calculation for the treatment of wastewater of the school which is conducted in sequences started from preliminary treatment that consist of the collecting tank (aeration tank), and biofilter. And wastewater flow in vertical flow type subsurface constructed wetland multilayers filtration unit.

3.2.1. Dimensions of preliminary treatment unit

The collecting tank is not only useful for collecting wastewater removed from the canteen and then flowed through biofilter, but also for sedimentation of the suspended solids, garbage, oil and fat. The collecting tank equipped with an aerator as oxygen supply as part of biological treatment, but the supply from the aerator is also expected as water flotation which is able to remove the oil and fat in the wastewater so they will not be emulsified.

3.2.2. The Dimensions of vertical-flow subsurface constructed wetland multilayers filtration unit

This unit is designed to optimally remove the parameters of pH, COD, BOD, total nitrogen, total phosphate and oil and fat contained in domestic wastewater. The following presents dimensions of the unit and its equipment.

The Calculation is based on the results of calculation of BOD surface loading 309,78 - 850,73 kgBOD / Ha / Day [10]

Given that:

BOD *surface loading* = 600 kg/Ha/Day

BOD loading (kg/day) = 0.644 kg/day
Efficiency Removal at Biofilter 30%, thus BOD loading = 0.45 kg/day
Dimension of *Constructed Wetland*
1 Ha = 600 kg/day
With BOD concentration at 0.45 kg/day, so that the total wetland area (X) is 600 kg/Ha/day = 0.45 kg/day/X (Ha)

$$X\ (\text{Ha}) = \frac{0.45\ \text{kg/day}}{600\ \text{kg/Ha/day}} = 0.000751\ \text{Ha} = 7{,}51\ \text{m}^2 \cong 8{,}1\ \text{m}^2\ (10\%)$$

The total area of IPAL should be equal to or bigger than the result of calculation to meet the safety standard.

IOP Conf. Series: Earth and Environmental Science **106** (2017) 012112 doi:10.1088/1755-1315/106/1/012112

Table 2. Multifiltration layer thickness and the total depth of the reactor.

No	Layers	Filter Types	Thickness design criteria	Thickness Selected	Diameter
1.	Space of distribution pipelines and plants	-	-	35 cm	-
2.	Surface	90% soil 10% quartz sand	0 – 30 cm	20 cm	-
3.	Second layer	Compost	0 – 40 cm	30 cm	0.03 cm
4.	Third layer	Silica sand	0 – 70 cm	20 cm	0.03 cm
5.	Forth layer	Gravel	0 – 40 cm	30 cm	1 – 2 cm
6	Collecting systems	-	-	15 cm	3 – 5 cm
			Total depth of reactor	150 cm	-

Table 3. Design of unit system of domestic wastewater treatment.

No	**Unit**	**Unit**	**Shape**	**Capacity**	**Dimension**	**Remarks**
1.	Collection Piping	1	Cilinder	-	-	-
2.	Collection Tank	1	rectangular	0,6 m^3	P = 1 m L = 1 m T = 0.6 m	-
3.	Aerator	1	Oval	-		-
4.	Sewage pumps	1	Oval	-		Submersible
5.	Biofilter	1	rectangular	1.4 m^3	P = 1 m L = 1 m T = 1.4 m	-
6.	Equalization Tank	1	rectangular	2.1 m^3	P = 1.5 m L = 1 m T = 1.4 m	-
7.	Constructed Wetland	1	rectangular	3.459 m^3/day	P = 8.1 m L = 1 m T = 1.5 m	-
8.	Distribution Piping	3	PVC	3.459 m^3/day	½ inchi	-

The available are to build the unit in this study is very limited compared to the volume of the wastewater resulted from the canteen activities and high influent concentrations, with a total area at 11.5 m^2 for the treatment unit and 8.1 m^2 for the vertical flow type of the subsurface constructed wetland multilayers filtration unit with *vertiveria zizanioides*. Therefore, the vertical flow type is considered as an effective alternative to be implemented in the limited area since it requires shorter retention time compared to the horizontal flow type. Meanwhile, the use of filtration media consisting of 3 types with 3 layers in certain thickness standard is expected to maximize the treatment as it provides optimum retention time. Moreover, using the vertical flow type of constructed wetland multilayers filtration with vetiver grass and the subsurface flow is considered safer especially related to environmental health condition, because it will prevent the vector breeding in wastewater, considering the characteristics of wastewater in the canteen and its location which is close to the people's activities. The retention time (td) of wastewater in this unit is calculated based on the porosity of each filter obtained from the experimental results in the field, with the assumption that the wastewater is 20 cm below the surface of compost layer. The calculation of wastewater retention time in the vertical flow type constructed wetland multilayers filtration unit.

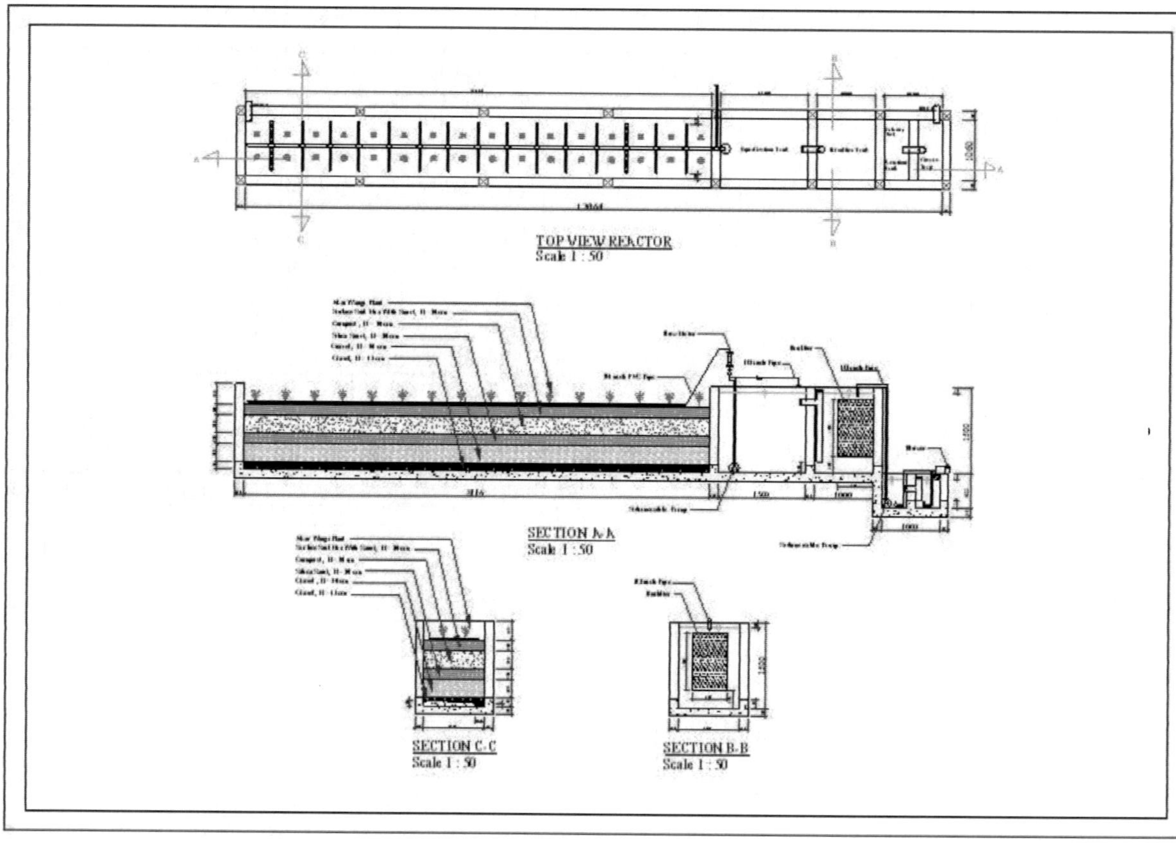

Figure 1. The design of domestic wastewater treatment from canteen of senior high school West Jakarta.

3.3. The distribution piping

The distribution pipes supply water from the pump in the aerator tank to the vertical flow type constructed wetland multilayers filtration with vetiver grass and multilayers filtration unit. The criteria of the pipe are presented as follows.

a. Pipes with ¾ inch diameter is used as the connector of the pump to the distribution pipes with the length of the pipes is 1.5 meters.
b. Tee with ¾ x ½ inch

Pipe with ½ inch diameter is used as a distribution pipe to the vertical flow type constructed wetland multilayers filtration unit with vetiver grass and multilayers filtration unit with size about 8 meter.

4. Conclusion

The design of domestic wastewater treatment from canteen of Senior High School West Jakarta using a combination biofilter and vertical-flow subsurface constructed wetland multilayers filtration with vetiver grass, with the collection tank and aeration tank in the prior stage of the preliminary treatment, has domestic wastewater removal rates at 3,459 L/day in area of 11.5 m^2.

Acknowledgment

This work was supported by grants from Directorate for Research and Community Services Directorate General of Strengthening Research and Development Ministry for Research, Technology and Higher Education In accordance with the Agreement Assignment Implementation Research Grant.

References

[1] Mengzhi Chen, Yingying Tang, Xianpo Li, Zhaoxiang Yu. 2009. Study on the Heavy Metals Removal Efficiencies of Constructed Wetlands with Different Substrates. *J. Water Resources and Protection*. 1:1-57

[2] Vymazal J, Constructed Wetlands for Wastewater Treatment. Water 2010, 2:530-549, www.mdpi.com/journal/water, 2008.

[3] Konnerup D, Koottatep T, & Brix H, Treatment of domestic wastewater in tropical, subsurface flow constructed wetlands planted with Canna and Heliconia, 5:248–257, 2008.

[4] Tanner C C, et al. 2012. Constructed wetlands and denitrifying bioreactors for on-site and decentralised wastewater treatment : Comparison of five alternative configurations. *Ecological Engineering*. 42:112–123. Available at: http://dx.doi.org/10.1016/j.ecoleng.2012.01.022.

[5] Abou-elela S I. 2013. Municipal wastewater treatment in horizontal and vertical flows constructed wetlands. *Ecological Engineering*, 61:460–468. Available at: http://dx.doi.org/10.1016/j.ecoleng.2013.10.010

[6] Chang J et al. 2012. Treatment performance of integrated vertical-flow constructed wetland plots for domestic wastewater. *Ecological Engineering*. 44:152–159. Available at: http://dx.doi.org/10.1016/j.ecoleng.2012.03.019

[7] US-EPA. Constructed Wetland Treatment for Municipal Wastewater. United States Environmental Protection Agency, EPA/625/R-99/010. Office of Research and Development. Cincinnati. 2000.

[8] Moshiri G A, 1993, Constructed Wetland for Water Quality Improvement, Lewis Publishers, London.

[9] Ministry of Environment and Forestry's Regulation for domestic wastewater standard Nr. P.68/2016.

[10] Astuti D A, et. Al, 2017.. Effect of Plants *Vetiveria Zizanioides* (Akar Wangi) on Performance of Field Plant Subsurface Constructed Wetland-Multilayer Filtration With Vertical Flow. *Proceeding QIR 2017*. Universitas Indonesia, Jakarta, Indonesia.

Effect of immobilized biosorbents on the heavy metals (Cu^{2+}) biosorption with variations of temperature and initial concentration of waste

W P Siwi, A Rinanti, M D S Silalahi, R Hadisoebroto, M F Fachrul*

Environmental Engineering Department, Faculty of Landscape Architecture and Environmental Technology, Universitas Trisakti, Jakarta, Indonesia

*Corresponding Author: melati@trisakti.ac.id

Abstract. The aims of research is to studying the efficiency of copper removal by combining immobilized microalgae with optimizations of temperature and initial Copper concentration. The research was conducted in batch culture with temperature variations of 25°C, 30°C, and 35°C, as well as initial Cu^{2+} concentrations (mg/l) of 3, 5, 10, 15 and 20 using monoculture of *S. cerevisiae*, *Chlorella* sp., and mixed culture of them both as immobilized biosorbents. The optimum adsorption of 83.4% obtained in temperature of 30°C with an initial waste concentration of 17.62 mg/l, initial biomass concentration of 200 mg, pH of 4, and 120 minutes detention time by the immobilized mixed culture biosorbent. The cell morphology examined using Scanning Electron Microscope (SEM) has proved that the biosorbent surface was damaged after being in contact with copper (waste), implying that heavy metals (molecules) attach to different functional cell surfaces and change the biosorbent surface. The adsorption process of this research follows Langmuir Isotherm with the R^2 value close to 1. The immobilized mixed culture biosorbent is capable of optimally removing copper at temperature of 30°C and initial Cu^{2+} concentration of 17.62 mg/l.

Keywords: biosorbent, biosorption, heavy metals, immobilized microalgae, Langmuir Isotherm

1. Introduction

The development of industrial sectors such as mining, electroplating, agriculture, fisheries, and others increases heavy metal pollution in the environment. One of the heavy metal waste that pollutes the environment is copper (Cu). Cu is a heavy metal widely utilized in the power industry, also as fungicides and anti-pollutant paint. One of the alternative methods that can be applied to overcome the pollution problem is biosorption, a term to describe the removal of heavy metals by passive binding in biomass plants or microorganisms in a solution through the metabolic or chemical-physical steps including the removal of toxins from hazardous materials. The advantages of using biosorption method are economical, easily available, as well as environmentally friendly because it utilizes biological materials. The biosorption process is influenced by the initial concentration of metal-containing waste, temperature, pH, and biomass concentration in the solution [1]. This research was carried out by optimization of variations of temperature and initial concentration of the waste.

In general, biomass can be used as biosorbent in the form of microorganisms such as bacteria, microalgae, and fungi, whether living or dead. However, microalgae's weak structure and tiny size are not suitable for continuous operation of metal elimination. Immobilization of biomass is then implemented to overcome this problem. Such immobilization is beneficial in improving the mechanical strength and resistance to chemicals, as well as facilitating the separation of biomass and generated effluent [2]. This Cu^{2+} metal biosorption study was conducted in a batch reactor using the immobilized consortium of *Chlorella* sp. and *S. Cerevisiae* with an optimization of temperature and initial concentration of waste to treat wastewater containing Cu^{2+}.

2. Research Method

The optimal biosorption was achieved through three stages of testing, i.e., the determination of mobile and immobilized biosorbents with a concentration of 200 mg and 500 m; followed by the evaluation of the effects of temperature using variations of 25°C, 30°C, and 35°C; and continued with the optimization of initial waste concentration using variations of 3 mg/l, 5 mg/l, 10 mg/l, 15 mg/l, and 20 mg/l.

2.1. Optimization Testof Temperature

The immobilized biosorbent was contacted with 100 ml of waste containing Cu^{2+} ions with a concentration of 10 mg/l at pH 4 and in 120 minutes detention time, by aeration and regulated with temperature variations 25^{0}C, 30^{0}C and 35^{0}C. Afterwards, the solution was measured using AAS, and the final concentration was calculated to obtain the removal efficiency.

2.2. Optimization Test of Initial Waste Concentration

The immobilized biosorbent was contacted with 100 ml of waste containing Cu^{2+} ions with various concentrations of 3 mg/l, 5 mg/l, 10 mg/l, 15 mg/l, and 20 mg/l at the optimum temperature and pH 4, with a contact time of 120 minutes and aeration treatment. Afterwards, the solution was measured using AAS, and the final concentration was calculated to attain the removal efficiency.

2.3. Analysis of Removal Efficiency and Adsorption Capacity

The removal efficiency and metal adsorption capacity of Cu^{2+}can be calculated by the following formula [3]:

$$\text{Removal Efficiency (\%)} = \frac{Co-Ce}{Co} \times 100\% \quad (1)$$

$$q\ (mg/g) = \frac{(Co-Ce)\times V}{W} \quad (2)$$

3. Results and Discussion

The optimization of Cu^{2+} metal biosorption was carried out using a single culture of *S. cerevisiae*, monoculture of *Chlorella* sp., and a mixed culture consisting of *S. cerevisiae* and *Chlorella* sp. (1:1). The characterization results of electroplating industrial wastewater samples were used to acquire the initial concentration of waste by 53.674 mg/l. It was then diluted to get the intended variations of initial waste concentration for this research.

3.1. Determination of Mobile and Immobilized Biosorbents

The optimization test of Cu^{2+} metal biosorption was carried out using a single culture of *S. cerevisiae*, a monoculture of *Chlorella* sp.,a and mixed culture of *S. cerevisiae* and *Chlorella* sp. (1: 1), treated with mobile and immobilizedbiosorbents. The experiment was conducted at pH 4, a contact time of 120 minutes, an initial concentration of Cu^{2+}wastewater of 2.99 mg/l,a temperature of 30^{0}C, as well as initial biosorbent concentrations of 200 and 500 mg.

Based on the removal efficiency of Cu^{2+} as shown in Figure 1A, the biosorption using abiosorbent concentration of 200 mg indicated that mobile and immobilized biosorbents could adsorb Cu^{2+}ions, but the higher adsorption was delivered by immobilized biosorbent. It implies that both single culture and mixed culture are capable of performing adsorption, but higher adsorption is achieved by the mixed culture. The maximum adsorption of 71.91% was attained by the immobilized mixed culture biosorbent. Compared with the removal efficiency value shown in Figure 1B, the biosorption using a biosorbent concentration of 500 mg showed its capability to absorb even higher, reaching 81.05% using the immobilized mixed culture. It reveals that the higher the initial biosorbent concentration, the greater the yielded removal efficiency. It is in line with [3] who reported that the biosorption of Uranium (VI) by immobilized biosorbent of *A. niger* with initial concentrations ranged from 0.1 to 1.2 g/l showed that the removal efficiency of Uranium (VI) increased from 55.2% to 93.4% for an increment of initial biosorbent concentration of 0.1–0.5 g/l.

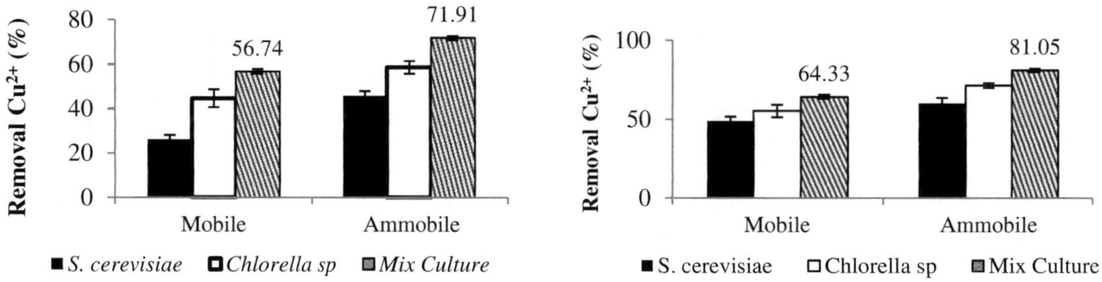

A) B)

Figure 1. The relationship of Cu^{2+}removal efficiency with mobile and immobilized biosorbents at pH 4, 120 minutes contact time, initial Cu^{2+}concentration of 2.99 mg/l, as well as a biosorbent concentration of A) 200 mg and B) 500 mg

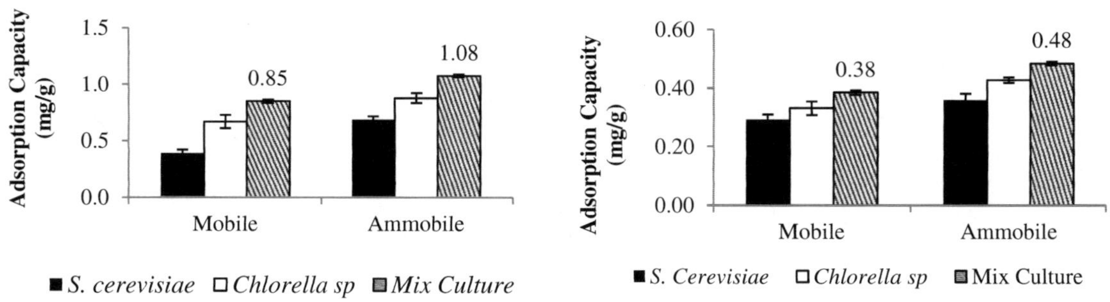

A) B)

Figure 2. The relationship of Cu^{2+}adsorption capacity with mobile and immobilized biosorbents at pH 4, 120 minutes contact time, initial Cu^{2+}concentration of 2.99 mg/l, as well as initial biosorbent concentration of A) 200 mg and B) 500 mg

Adsorption capacity values in Figure 2A and Figure 2B are opposed to removal efficiencies in Figure 1 and Figure 2; the greater the initial concentration of biosorbent used, the smaller the adsorption capacities. The initial biosorbent concentration of 200 mg with mixed culturehad a higher adsorption capacity compared with the initial concentration of 500 mg biosorbent. The maximum adsorption capacity was demonstrated by the immobilized mixed culture biosorbent with a concentration of 200 mg or equal to 1.08 mg/g.

The results obtained in this study are consistent with [4] which observed that the adsorption capacity of immobilized *Chlorella* sp. was greater than the mobile biomass of *Chlorella* sp., reaching a removal capacity of Cu^{2+}ions of up to 24.91 mg/g. The study of [5] also reported using immobilized biosorbent of *Chlorella* sp. that had an adsorption capacity of Cu^{2+}ions of 28.5 mg/g.

Based on these matters, it can be concluded that immobilized biosorbents havea better ability to adsorb Cu^{2+} than the mobile ones. The initial concentration of 200 mg biosorbent has a higher adsorption capacity than the initial concentration of 500 mg biosorbent.

3.2. Temperature Optimization Test

At the temperature of 25°C, immobilized biosorbent could present a removal efficiency of 60.22–73.58%, while at the temperature of 30°C, it was capable of increasing the Cu^{2+}removal of up to 79.4%. At the temperature of 35°C, the adsorption efficiency gradually decreased of only up to 66.97%. It shows that the higher the temperature, the biosorbent becomes less capable of performing adsorption.

The results of this study correspond to [6] which researched the optimum adsorption temperature of heavy metals-Cr (VI), Cd, Cu^{2+}, and Ni-using immobilized mixed culture biosorbent and temperature variations ranging from 10–50°C. It observed that the efficiency of metal uptake continued to increase at a temperature of 10–30°C, then the reduction of adsorption efficiency occurred along with the temperature increase. The optimum adsorption efficiency of the contained metal of 90% was achieved at the temperature of 30°C. According to [6], a decrease in the adsorption efficiency at 40–50°C was led by the increased absorption of heavy metals from the solid to liquid phase, the surface deactivation of biosorbent, as well as the damage to the active site on the surface of biosorbent due to the disruption or the weakening bonds of active site in binding heavy metals.

3.3. Optimization of Initial Concentration of Waste Containing Cu^{2+} Ions

The optimization test of initial concentration of waste containing Cu^{2+} by immobilized biosorbent was performed at pH 4 with concentration variations of 2 mg/l, 5 mg/l, 10 mg/l, 15 mg/l and 20 mg/l; using the optimum temperature obtained at the previous test of 30°C, 120 minutes contact time, and an initial biosorbent concentration of 200 mg.

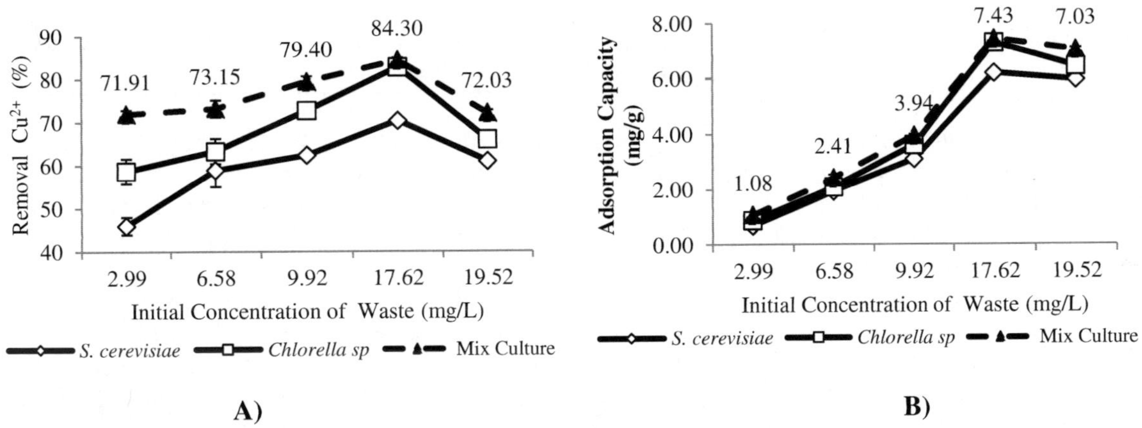

Figure 3. Effect of initial waste concentration optimization on A) the removal efficiency (%) and B) adsorption capacity (mg/g) at pH 4,120 minutes contact time, 30°C temperature, and initial biosorbent concentration of 200 mg.

Figure 3A shows that the removal efficiency of Cu^{2+} by immobilized biosorbent was increasing and the highest removal efficiency of 84.30% was obtained at a concentration of 17.62 mg/l. Figure 3A and Figure 3B show that the immobilized mixed culture biosorbent could adsorb better than the single culture due to the active site discrepancy of the biosorbents. The mixed culture has more active sites used to carry Cu^{2+} adsorption than that owned by the monoculture.

The adsorption capacity of immobilized biosorbent had the range of 0.69–7.43 mg/g on the optimization of the initial concentration of Cu^{2+} waste with variations of 2.99 mg/l, 6,58 mg/l, 9.92 mg/l, 17.62 mg/l, and 19.52 mg/l. The highest adsorption capacity of 7.43 mg/g was generated by the immobilized mixed culture biosorbent with the initial Cu^{2+} waste concentration of 17.62 mg/l.

Researchers [7] stated that the higher the concentration of metal ions, the more the amount of the substance adsorbed to the achievement of the optimum concentration. Once the optimum concentration

IOP Conf. Series: Earth and Environmental Science **106** (2017) 012113 doi:10.1088/1755-1315/106/1/012113

is obtained, the biosorbent has surely been saturated. According to [8], the increase of the biosorption ability is directly proportional to the increase in concentration. It relates to the effect of stresses that occur, ergo increasing all ionic transfers and leading to a higher adsorption of metal ions.

3.4. Analysis of Fourier Transform Infra-Red (FTIR)
The analysis of Fourier Transform Infra-Red (FTIR) was conducted to determine changes in functional groups of biosorbent cells before and after being contacted with Cu^{2+} as they were estimated binding the Cu^{2+} ions. FTIR results of different wave lengths indicate a change of functional groups as listed in Table 1 which describes the wave lengths used, the functional groups, and the content changes of the functional groups of immobilized biosorbent cells.

Table 1. FTIR analysis of biosorbent before and after being in contact with Cu^{2+}.

Functional Groups	***S. cerevisiae* Before (%)**	***S. cerevisiae* After (%)**	***Chlorella* sp. Before(%)**	***Chlorella* sp. After (%)**	**Mixed Culture Before(%)**	**Mixed Culture After (%)**
C-H (alkene)	97.71	95.86	97.60	94.45	96.03	94.45
C-O (carboxylix acid)	98.19	-	98.44	-	96.89	-
C-N (amine-amide)	98.19	96.73	98.44	95.84	96.63	95.84
C-H (alkene)	97.84	95.07	98.25	95.77	98.94	98.20
C=C (alkene, aromatic ring)	97.67	94.28	98.06	95.76	96.88	93.12
C=O (aldehyde, carboxylic acid)	97.62	-	98.48	98.43	98.45	95.83
O-H (phenol)	-	98.26	-	98.91	-	98.17
C≡C (alkalo)	-	98.68	-	98.79	-	98.79

Changes in functional groups that occurred to the immobilized biosorbents of *S. cerevisiae*, *Chlorella* sp., and mixed culture showed that the functional groups were involved in the adsorption process of Cu^{2+}ions. Changes of functional groups of biosorbents were present either in the form of losses of both formerly existed and non-existed functional groups or changes in the levels of biosorbent functional groups.

Previously, alterations of the biosorbent surface observed through SEM has explained that the occurred changes were due to the involvement of functional groups attached to the cell wall of the biosorbent during the biosorption process. It was reinforced by the FTIR results afterward that indicated a change of functional groups before and after the biosorbent being in contact with Cu^{2+}. The aspect affecting the loss or reduction of functional groups contained in the biosorbent as described by [9] is the chemical adsorption by bonding with a number of functional groups (hydroxyl, carboxyl, carbonyl, and amino groups) contained on the surface of the microalgal cell, ensuing exchanges of monovalent and divalent ions such as Na, Mg, and Ca in the cell walls to be replaced by heavy metal ions.

4. Conclusion
Both of mobile and immobilized biosorbents were capable of absorbing Cu^{2+}, but the higher adsorption of 71.91% was delivered by the immobilized biosorbent. Both single culture and mixed culture biosorbents were capable of performing adsorption as well, but the higher adsorption of 71.91% was achieved by the mixed culture. Temperature and initial waste concentration affect the adsorption of Cu^{2+} ions by single culture biosorbents of *S. cerevisiae* and *Chlorella* sp., as well as the immobilized mixed culture biosorbent. The optimum Cu^{2+} adsorption efficiency of 83.4% occurred at a temperature of 30°C, an initial waste concentration of 17.62 mg/l, an initial biosorbent concentration of 200 mg, pH 4, and 120 minutes contact time by the immobilized mixed culture biosorbent. FTIR test results indicated changes of functional groups of both formerly existed chains such as carboxylic

acids and formerly non-existed chains as phenols and alkynes, as well as reduced levels of functional groups after the immobilized biosorbent had been being in contact with Cu^{2+} ions.

Acknowledgment

The authors would like to thank Directorate for Research and Community Services as well as Directorate General for Strengthening Research and Development at the Ministry of Research, Technology, and Higher Education Indonesia for funding this study through Universities Prime Research Grant Program (Penelitian Unggulan Perguruan Tinggi-PUPT) 2016.

References

[1] Nilanjana D, Vimala R, and Kartika P. 2007. Biosorption of heavy metals—An overview. School of Biotechnology, India: Chemical and Biomedical Engineering.
[2] Kurniasih, Ariesyady H D, Sulaeman A, and Kardena E. 2013. Biosorption of chromium (VI) using immobilized algal-bloom biomass: Kinetics and equilibrium studies. Faculty Civil and Environmental Engineering. Institut Teknologi Bandung. *International Journal of Environmental and Resource* (IJER). 2(1).
[3] ZhangY, Dai Y, Liu B, Zheng F M R, and Dong L. 2011. Biosorption of lead from aqueous solution by using chemically treated Saccharomyces cerevisiae. *Environmental Engineering & Management Journal (EEMJ).* 10(6):729–732.
[4] Hastuti Rum, and Gunawan. 2006. Amobilisasi biomassa *Chlorella* sp. pada silika gel sebagai adsorben tembaga. *JSKA*. 9(2).
[5] Wan Maznah W O, Al-Fawwaz A T, and Surif, M. 2011. Biosorption of copper and zinc by immobilized and free algal biomass, and the effects of metal biosorption on the growth and cellular structure of *Chlorella* sp and *Chlamydomonas* sp. isolated from rivers in Penang, Malaysia. *Journal of Environmental Sciences*. 2012. 24(8):1–8.
[6] Ilamathi R, Nirmala G S, and Muruganandam L. 2014. Heavy metals biosorption in liquid-solid fluidized bed by immobilized consortia in alginate beads. *J. Bioprocess Biotech*. 4:145.
[7] Sembiring Z, Suharso B, and Sumadi. 2009. Isoterm adsorpsi ion Pb (II), Cu (II) dan Cd (II) pada biomassa Nannochloropsis sp.dienkapsulasi akuagel silika. Indo. *J. Chem.* 9(1):1–5.
[8] Davis J A, Volesky B, and Vierra R H S F. 2003. Sargassum seaweed as biosorbent for heavy metals. *Water Res,* 34(17):4270–4278.
[9] Chojnacka K, Chojnacki A, and Gorecka H. 2005. Biosorption of Cr^{3+}, Cd^{2+}, and Cu^{2+}ions by blue-green algae *Spirulina* sp.: Kinetics, equilibrium, and the mechanism of the process. *Chemosphere*.59(1):75–84.

The multi-level perspective analysis: Indonesia geothermal energy transition study

A Wisaksono[1*], J Murphy[2], J H Sharp[1], P L Younger[1]

[1]School of Engineering, University of Glasgow, Glasgow, UK
[2]School of Interdisciplinary Studies, University of Glasgow, Dumfries, UK

*Corresponding author: a.wisaksono.1@research.gla.ac.uk

Abstract. The study adopts a multi-level perspective in technology transition to analyse how the transition process in the development of geothermal energy in Indonesia is able to compete against the incumbent fossil-fuelled energy sources. Three levels of multi-level perspective are socio-technical landscape (ST-landscape), socio-technical regime (ST-regime) and niche innovations in Indonesia geothermal development. The identification, mapping and analysis of the dynamic relationship between each level are the important pillars of the multi-level perspective framework. The analysis considers the set of rules, actors and controversies that may arise in the technological transition process. The identified geothermal resource risks are the basis of the emerging geothermal technological innovations in Indonesian geothermal. The analysis of this study reveals the transition pathway, which yields a forecast for the Indonesian geothermal technology transition in the form of scenarios and probable impacts.
Keywords: geothermal, multi-level perspective, renewable energy, transitions

1. Introduction

Energy sustainability requires innovation, planning and development to be conducted carefully in order to ensure energy sustainability. Hence, this paper suggests using the socio-technical multi-level perspective approach, which is the main theory of the socio-technical transition [1], to study the ways in which geothermal energy might replace mainstream energy in electricity generation. The levels are socio-technical landscape (ST-landscape), socio-technical regime (ST-regimes) and niche innovations [2], [3], [1]. ST-landscape includes prices, economic growth, conflicts, political status, environmental issues, etc. There are exogenous factors that exhibit high degrees of proximity and power across incumbent socio-technical systems [2]. The ST-regime is the set of rules or regulation that govern the activities within the broad communities [3]. Niche innovations represent radical innovations, which represents fundamental change that often depends on the integration of many interdependent systems to succeed [3], [1].

2. Research Method

As developed by [1], the multi-level perspective is a non-linear process and evolutionary framework in the process of technological transition [3]. This involves interactions between three levels across which society and technology co-evolve. The technique is also being used in other case studies, for example in car technology transitions, climate change responses, energy transitions, etc. [4], [5], [6] [7], [8]. There are many external factors included in a multi-level perspective framework, which are heterogeneous and related to the transition technology process. These factors fall within the ST-landscape at the highest level within a multi-level perspective framework. These are then followed by ST-regime while niches are at the lowest level of the perspective. In multi-perspective analysis, the success of a new technology is not only influenced by processes in niche innovation, but also by movements of, or changes in, the level of ST-regime and ST-landscape. However, in theory, resistance to change will be primarily at the ST-landscape level, since ST-landscape factors are less likely to change [3]. Any change within the ST-landscape may put pressure on the ST-regime and to open a new path for new innovations to enter and influence both ST-regime and ST-landscape. The basis of the niche innovation in geothermal exploration is to deal with resource risk that may arise during the process. Innovations in geothermal drilling and exploration attempt to mitigate this [9]. Some self-reinforcing innovations may initially require some degree of protection from external pressures (a protected environment: e.g. research universities, government research institutions, etc.). From here, niche actors can then nurture the path-breaking innovation so it becomes more robust through performance improvements and expansions in supportive socio-technical networks [10]. Examples of protected space are research institutions and research universities. In the past, society created technology to help itself, but some argue that technology has become so powerful that it has been able to "lock-in" societal activities [11], [12].

Path dependency is the way in which critical junctures at an early stage in the development of a socio-technical system persist through self-reinforcing means [13]. In multi-level perspectives, path dependency and technological lock-in are important in ST-regime level [15], [16], [17].

Controversy may be regarded as one of the side effects of technology transitions, especially if the transition implies societal change. Technological systems are dynamic and unstable, and any change in the components in the system (e.g. new entrant or a change in the institutional setup) may trigger actions and reactions within the system [7] at the ST-landscape development level.

3. Results and Discussion

3.1. ST-regime analysis

Published laws from 2007 and 2009 make it clear that renewable sources expected to be the main sources of energy in the future. However, the law of 2009 can be regarded as self-contradictory since it also requires optimisation of primary energy both local (preferred) and imported. The wording of the law indicates priority be given to primary resources such as coal, oil and gas.

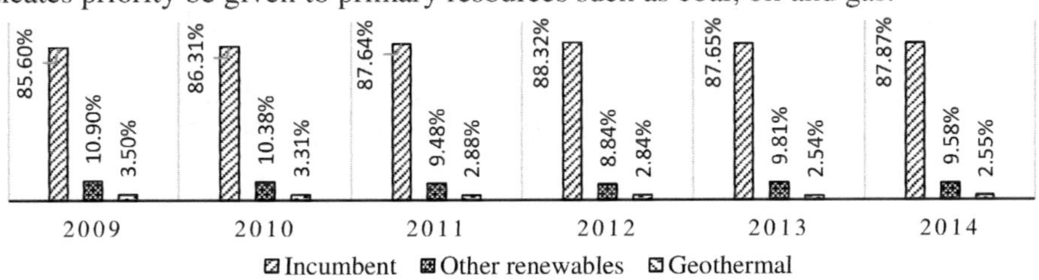

Figure 1. Indonesia geothermal installed capacity power plant distribution versus other resources.

Meanwhile, Figure 1 demonstrates this, where the planned use of fossil fuel as incumbent fossil energy still increases in comparison to renewables. Table 1 explains the currently applicable regulation and policies. The regulation basis pertaining to energy and electrification in Indonesia are [14] and [15] for electrification, followed by [16] and [17]. Indonesian Laws are the highest in the hierarchy and verified by the house of representative. Further examination of government regulations explains this anomaly. In table 1, regulations from the year 2012 show that economic values is the factor that the most difficult to change within ST-regime. Article [14] also underlined that renewable energy development depends on the economic value.

Table 1. Meanings of current applied national energy and electrification regulation and policy pertaining to source of energy.

Law no. 30 year 2007 *pertaining to energy*	**Law no. 30 year 2009** *pertaining to electrification*
Chapter 5, energy management. In article 20 (Energy supply), sentence no: (2) Renewable energy are preferred for energy development (4) Government must increase the use of renewables (5) Renewable energy suppliers will get convenience and incentives by the government, until it meets the economic valu In article 21 Energy utilization, sentence no: (2) Re-phrase article 20 sentence 4 (3) Re-phrase article 20 sentence 5, for energy utilization	Chapter 5, the primary energy utilisation. In article 6, sentences no: (1) The utilization of primary energy, which are taken from local resources and imported, must be optimised. (2) Renewable energy are preferred for energy utilisation. (3) Refer to sentence no. 1, local resources are preferred.
Government regulation No. 14 year 2012 *pertaining to electricity*	**Government regulation No. 79 year 2014** *pertaining to energy mix*
Chapter 5, electricity pricing-network-tariff. The articles in the chapter, explains the mechanism of pricing, network and tariff (economic value), which are determined by the decision and mutual agreement between the related governmental parties. Includes the minister, governor, mayor/residents, and house of representative.	Chapter 2, goals and target. On article 9, sentence (*f*) for energy mix: 1. Year 2030 Renewables increased 8% from year 2025. 2. Year 2030 oil decreased 5% from year 2025. 3. Year 2030 coal decreased 5% from year 2025. 4. Year 2030 natural gas increased 2% from year 2025. Chapter 3, national energy policy. On article 10, are mention the increasing exploration of the potential or proven reserves of fossil energ and renewables, decrease gas and coal export, to be used domestically, and make sure the environmental capacity support for renewables. Chapter 3, national energy policy. On article 11, are mention the maximisation of renewables, in regard to economic value; minimisation of oil, optimisation of renewables and natural gas; determination of coal as energy national supply.

That is why in [17], the government keen on energy mix policy, in which fossil fuels are still the main energy, and even coal being determined as the national energy supply. It shows the argument by Arthur, when usual policy of letting the superior technology reveal itself in the outcome that dominates, is proven [11]. The actors involved in this case study are shown in Figure 2. Central to the system are energy providers, the Indonesian national electricity company and consumers. Energy providers include multi-national companies, private local companies and the government-owned geothermal company. The other groups, which are government organizations, non-governmental organizations, environmental activist groups, and the media, also attempt to influence the system. The co-evolvement of society and technology demonstrated here at the ST-regime level.

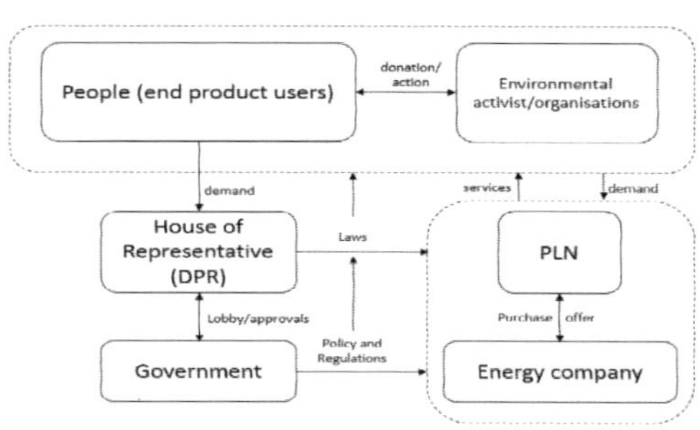

Figure 2. Identified actors in the socio technical regime of geothermal energy in Indonesia.

3.2. ST-landscape developments

Developments in the ST-landscape could affect and pressurize the socio-technical ST-regime; however, the impact and pressure from the ST-landscape is independent from (i.e. not influenced by) the ST-regime. The Indonesian primary national energy supply remains dominated by fossil fuels. The high levels of fossil energy supply reflects high demand for oil and gas products such as diesel for electricity and industry. In these cases, the fuel is a form of final energy that is relatively easy to use and reaches broad categories of consumer. Indonesia has one of the lowest electrification rates in Asia, where frequent rolling power blackouts are experienced in Sumatra, Kalimantan, Java (excepting the Jakarta rural area) and eastern part of Indonesia. Optimised use of geothermal energy could readily increase the electrification ratio (see Figure 3), especially in the eastern part of Indonesia. This is the main objective of government plans to increase the installed capacity of the energy mix. The problem of the demand for electrification directly pressurises the ST-regime.

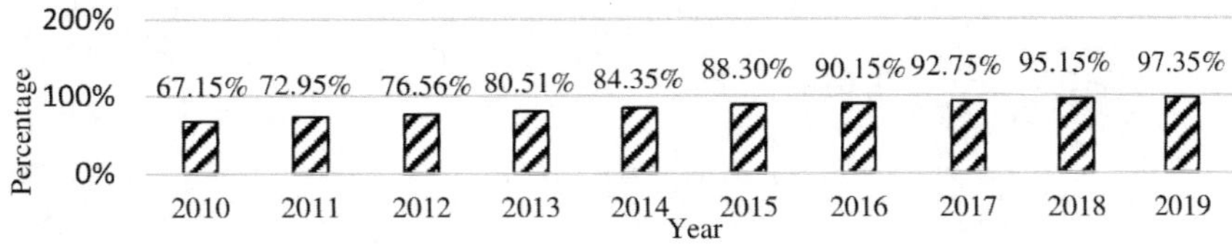

Figure 3. Electrification ratio (2010-2015 = actual and 2016-2019 = goals).

Regulatin [25] permits geothermal exploration inside the protected forest and conservation area (mostly located near volcanic areas) [18]**,** which allows geothermal exploration in conservation areas. However, controversies relating to geothermal development have arisen from cultural reasons and constraints, even though the law protects these projects. Despite the problem of increasing electricity demand, government policy to overcome energy deprivation is not always supported by a consensus from the broader cultural society. As evidenced by the examples above and argued by [8], public acceptance is needed if the aim to socially embed new technology is to be achieved. The existing policies and regulations were not publicly accepted within this Indonesian cultural context.

3.3. Technological niches

The identified geothermal resource risks are the basis of the emerging geothermal technological innovations in Indonesian geothermal development such as Enhanced Geothermal Systems (EGS), subsurface analysis and cross-field innovations. EGS is a technology that is able to perform geothermal exploration in areas that have technically and economically less potential, according to conventional geothermal systems by enhanced drilling techniques and reconditioning of the reservoir. There are very few activities evident in terms of EGS in Indonesia now, principally due to fewer hot dry rock resources and too many unexplored hydrothermal systems. EGS yet to be implemented in Indonesia since geothermal systems in Indonesia are generally convective hydrothermal [19] and the majority of the resources are not hot dry rock with low levels of water or vapour. Subsurface data modelling would be extremely useful to model the reservoir conditions during exploration. It would enable more efficient reservoir maintenance and management and serve as important data with regard to planning further developments, e.g. drilling new boreholes. According to Younger, reservoir modelling and management approaches need to be widely developed as common innovations in the geothermal industry [20]. Indonesian researchers have developed their own research studies in geothermal modelling fields as presented at a recent geothermal conference at Stanford University in early 2017 [21] [22]. This shows strong evidence that niche innovation in subsurface modelling is growing in Indonesia. Meanwhile, the most relevant potential cross-field technology is at the electricity production level, specifically hybrid power plants. There is only one new hybrid power plant in Indonesia (Yogyakarta area), which does not involved geothermal, and instead uses a combination of solar PV and wind turbines.

3.4. Discussion: Transition forecast

Based on the multi-perspective theory and analysis from the previous sections, the socio-technical pathway would be the transformation pathway, with incremental change [1]. In Figure 4, electricity demands and the environmental issue both create resistance in the ST-regime's path-dependence, which also supports the indication for a longer time for the transition. Based on the empirical data, ST-regime analysis and ST-landscape developments at the transition pathway:

a. ST-landscape development pressure from electrification demands, however environmental issues reduce this pressure. The Indonesian government plans to build a geothermal power plant totalling 7200 MW in order to increase the electrification ratio from 88.30%. Due to high demand, growth of renewables has previously been neglected in order to build and explore quick, cheap and powerful coal powered energy resources [23].

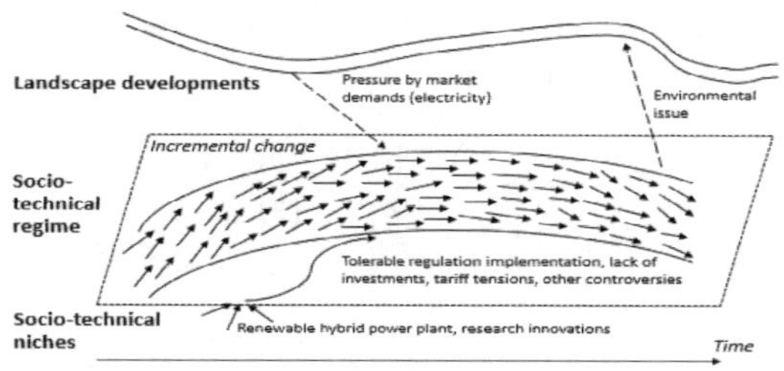

Figure 4. Geothermal energy transformation pathway, with incremental change.

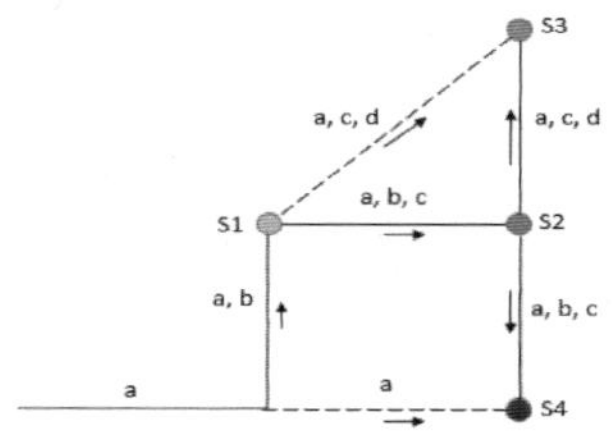

Figure 5. Four possible scenarios, with their transition seeds

b. Financially strong incumbent ST-regime. Funding and investment constraints has allowed the increase in the use of cheap and efficient fossil fuel energy sources, namely natural gas and coal, supported by numerous local reserves.
c. Policy change influence at ST-regime. Energy mix government regulation [17], creates both lock-in to fossil energy, but opens a path to incremental change with a very small number of niches in renewable and geothermal. The government power resources plan consists 23% renewables in 2025 [24].
d. Few niche innovations. Excitement for environmental friendly energy (e.g. solar, wind, hydro, geothermal) aligned with laws and government regulations will open the door for investment in renewables. With support from specific government regulation relating to geothermal sources [25], [26].

Based on the transition seeds, the potential scenarios for the next 20 years are:

S1. Geothermal energy with installed capacity of 5% of the total energy mix policy. Incumbent energy resources still leading, while geothermal energy development is stagnant. The instant solution to increase the electrification ratio is to push the usage of the local natural resources. This scenario is aligned with the current government decision to develop and commence several coal powered steam power plants in Indonesia between 2015 and 2019 [27], [28].

S2. Geothermal energy with installed capacity of 10% of the total energy mix policy. S2 occurs, when S1 fails to meet the target. Geothermal energy and other renewables still take a long time to grow in Indonesia. S2 only starts after S1 ends. The current local reserves of natural gas are extensive. Natural gas now leads the energy source.

S3. Geothermal energy with installed capacity of 20% of the total energy mix policy. Government gives more investment and support in renewables research to comply with the law and the government's recent plan [23]. The latest policy to support this scenario is government support in geothermal project funding by injection of project capital, allowing geothermal developers to seek financial support from state owned financing company [29].

S4. Geothermal energy with installed capacity of 5% of the total energy mix policy. Nuclear power plants alone also could raise the electrification ratio to 100% in combination with existing fossil and renewables. Controversies within the ST-regime will become constraints in this development. However, due to urgency, it becomes reality [30].

4. Conclusion

Based on the theory of the multi-level perspective framework as applied to technology transitions, conflicts were seen to be part of the ST-landscape development. Conflicts can act to trigger change in the ST-landscape, which may then put pressure on the ST-regime. However, conflicts within the ST-regime will only make the incumbent ST-regime path dependency stronger. Further study is needed to analyse this finding. The ST-landscape development on the other hand, helps to identify the potential changes in the level, which, in this case, are the demands for electricity. Based on the data, we conclude that these demands are due to electricity deprivation in large parts of Indonesia. According to the latest popular niche innovation in the global geothermal industry, the study points to multiple evidences of innovation in Indonesia. Unfortunately, at the current time, it still unable to influence the incumbent ST-regime. Overall, the framework smoothly connects the ST-regime with ST-landscape and niche innovations. The pathway analysis was able to help the study examine several scenarios, which forecast the outcome of technological transition in geothermal development in Indonesia.

Acknowledgment
The authors would like to gratefully acknowledge that this work has been made possible by financial support from the Indonesia Endowment Fund for Education (LPDP), Republic of Indonesia.

References
[1] Geels F W. 2005. Tech. Forecast. & Soc. Change. 72: 681.
[2] Raven R, Schot J and Berkhout F. 2012. Env. Innov. and Soc. Transitions. 4: 63.
[3] Geels F W. 2002 Res. Pol. 31: 1257.
[4] Marletto G. 2014. Tech. Forecast. & Soc. Cha. 87: 164.
[5] Van Bree B, Verbog G and Kramer G. 2010. Tech. Forecast. & Soc. Change. 77: 529.
[6] McDowall W. 2014. Futures. 63: 1.
[7] Jacobsson S, and Bergek A. 2004. Indust. and Corp. Change. 13:815.
[8] Geels F W, and Verhees B. 2011. Tech. Forecast. & Soc. Change. 78:910.
[9] Blankenship D, and Finger J. Handbook of Best Practices for Geothermal Drilling (Albuquerque: Sandia National Laboratories). 2010.
[10] Smith A, and Raven R. 2012. Res. Policy. 41: 1025.
[11] Arthur B W. 1989. The Eco. J. 99. 49: 116.
[12] Unruh G C. 2007. Energy Policy. 28: 817.
[13] Upham P, Virkamäki V, Kivimaa P, Hildén M and Wadudd Z. 2015. J. Transp. Geo. 46: 210.
[14] Law of Republic of Indonesia. 2007. Law no. 30. (Jakarta: House of Representative).
[15] Law of Republic of Indonesia. 2009. Law no. 30. (Jakarta: House of Representative).
[16] Government regulation of Republic Indonesia. 2012. Government regulation no. 14. (Jakarta: Presidential office).
[17] Government regulation of Republic Indonesia. 2014. Government regulation no. 79. (Jakarta: Presidential office).
[18] Law of Republic of Indonesia. 2009. Law no. 21. (Jakarta: House of Representative).
[19] Kasbani. 2009. *Buletin Sumber Daya Geologi* 4 (3):64-73 (in Indonesian).
[20] Younger P L. 2015. Energies. 8:11737.
[21] Eljafrie I, and Syihab Z. 2017. *Proc. 42th Stan. Geo. Workshop* (Stanford: Stanford University).
[22] Priyangga H, Soedarmo A, Ogra K, and Zhabagina A. 2017. *Proc. 42th Stan. Geo. Workshop* (Stanford: Stanford University).
[23] Jensen F, and Asmarini W. 2016. Indonesian power plan delays fan industry frustrations. URL: http://www.reuters.com/article/indonesia-coal-idUSL4N18Q07U.
[24] PLN. 2015. 35000 MW for Indonesia. (Jakarta: PLN).
[25] Law of Republic of Indonesia. 2014. Law no. 21. (Jakarta: House of Representative).
[26] Ministry of ESDM Republic Indonesia. 2014. Ministry regulation no. 17. (Jakarta: Ministry of ESDM).
[27] Ministry of Finance Republic Indonesia. 2016. URL: http://www.kemenkeu.go.id/en/Berita/inaugurating-pltmg-arun-president-expects-multiplier-effects-economy-aceh
[28] Amindoni A. 2016. URL: http://www.thejakartapost.com/news/2016/06/09/ri-finally-ready-to-commence-batang-mega-project.html.
[29] Ministry of Finance Republic of Indonesia. 2017. Ministry regulation No. 62/PMK.08/2017 (Jakarta: Ministry of Finance).
[30] Yap S. 2016. Inside Indonesia's nuclear dream. URL: http://www.channelnewsasia.com/news/asiapacific/inside-indonesia-s/2519384.html.

Optimization biogas management as alternative energy from communal scale dairy farm

R Ruhiyat*, L Siami

Environmental Engineering Department, Faculty of Architecture Landscape and Environmental Technology, Universitas Trisakti, Jakarta, Indonesia

*Corresponding author: ratnaningsih@trisakti.ac.id

Abstract. Cow Slurry can be the main pollution source in most villages in Indonesia. In this study, treatment of cow slurry intended to reduce pollution in Citarum river and greenhouse gases effect of CH_4 and CO_2. As a part of renewable energy, biogas can be one of solution to be implemented in small-scale and remote area. In Pejaten, Tarumajaya Village, the cost-effective reached when 7cattleman united to treat cow slurry in one biodigester. The breed varies cow from calf, veal to adult cattle. The installation of anaerobic-bio digester that produce biogas 28 m^3/day equivalent with Rp 168,000 to be consumed for 14 households. In addition, villager also benefitted manure as 42.5 ton monthly. As a whole, the highest profit comes from adult cattle that produce 900 kg/month slurry as Rp 59,919 monthly. Furthermore, this system gives job opportunity for villagers to be biodigester operator is the main beneficial with the higher income compare to mower that only Rp 600.000 monthly as Rp 1.065.000.
Keywords: anaerobic-biodigester, biogas, cow-slurry, cost-effective, job opportunity, manure

1. Introduction

Dairy Commodity is a means of empowering the community because the characteristics of its products can be harvested daily, thus allowing farmers to earn a sustainable income [1]. Although cattle farm provide economic benefits and community empowerment but it have a negative impact such as animal waste in the form of livestock manure, the remaining feed and others.Livestock manure chemically contains complex and high organic materials. Livestock manure discharged into waters can contaminate water quality due to high organic material, which can cause algae blooming, development of pathogenic bacteria and decrease oxygen content in waters due to increased microorganisms, thereby threatening fish life in these waters. Meanwhile, if stacked can pollute the land and seep into the groundwater or carried away "run off" into the surface water. The accumulation of livestock waste over a long period of time, will result in an anaerobic process in the decomposition of organic matter and will lead to the release of methane (CH_4) and CO_2 into the air, which can contribute to global warming. Good livestock waste management is essential to minimize negative environmental impacts, such as pollution to soil, water and air and infectious diseases [2]. The previous study showed that biogas also have possibility to be utilize as power generation source [3],[4] and as daily needs consumption energy [5]. In addition, the sludge output from Biogas reactor can be reused as raw material of organic fertilizer. Therefore, processing of cattle livestock wastes with biogas technology or anaerobic digesters can provide benefits in terms of economic and environmental aspects.This is one of the cause of the slacken gait of biogas development as [6]. Biogas development opportunity as an alternative energy source and energy decentralization is very big considering the decrease of fossil fuel and increasing of pollution caused by the use of petroleum fuels and the increasing of oil price. As alternatif energy, compare to other energy sources Indonesia government have just pay more attention lately. Even biogas technology itself has been introduced since 1962 in Indonesia. Research on the use of livestock waste, especially cow feces to become biogas has been widely done both inside and outside the country, some research suggests that biogas has the potential to be developed to become a renewable energy source both from technical aspect, environmental aspect and economic aspect. But until now the utilization of biogas technology in cow breeding, especially in Indonesia has not been running sustainably [7], [8], [9], [10], [2]. Study on existing condition of biogas technology utilization in study area is a first step to know coefficient of coefficient in arranging optimiation model of biogas technology utilization in dairy farm in sustainable study area. The optimization model approach will be done using Linear programming.

2. Research Method

The wide focus research of biogas usage has been done by in the previous study by qualitative method to get the affected and interrelation component for sustainable biogas production in cow farmer. Thus, this research conduct to examine area of study that related wiith farming activity and biogas production, existing condition of biogas utilization and to formulate optimation model of biogas utilization community based by linear programming in resource allocation and pollution control. .The research object, is Biogas Reactor, farmer and non breeder in the study area. Furthermore, this study will allocate maxium number of farmer and biogas consumption with natural and human resource, revenue of biogas farmer/worker as the constrain. Data were obtained by direct observation in the study area, distributing questionnaires and direct interviews, and laboratory analysis. The number of respondents for breeders is 120 kk. Optimum condition in reaching sustainable development require:

1) In order to avoid pollution in the river then, all cow manure produced in the study area is processed into biogas (not disposed of the river)
2) Biogas price should be $\leq$ LPG (Liquid Petroleum Gas) price
3) Land requirements for houses, cattle pens and biogas reactors $\leq$ 40% of total residential area
4) Total grass fields $\leq$ 60 Ha (available with permission from Perhutani)
5) River water usage for livestock and biogas $\leq$ 10% river water discharge
6) The income of dairy cattle worker, worker of biogas and worker of fodder and workers $\geq$ Rp 1,060,500 / month (Regional minimum wage of Kab. Bandung).

The Scenario to be applied are :

1) The maximum income can be obtained by varying the number of livestock by age category (calf, veal and adult cattle), grass management and biogas business management as an interconnected and bonded sustainable supporting business. To run a dairy farm business, cultivation of grass and biogas effort required limited land resources and water resources.
2) The maximum revenue of biogas exploitation can be achieved by varying the number of livestock manure being treated and the pattern of livestock management

The revenue of dairy farms (Y) can be counted as:

$$MaksY = aX_1 \quad + bX_2 \quad + cX_3 \tag{1}$$

Whereas a is the average profit for 1 calf caring. X_1 is number of calf . b is the average profit for 1 veal caring. X_2 is number of veal. c is the average profit for 1 adult cattle caring. X_3 is number of adult cattle.For obstacle of lawn formulated as:

$$dX_1 + e\, X_2 + f\, X_3 \leq 600{,}000 \text{ m}^2 \tag{2}$$

Whereas $X_1$is the lawn need for X_1, X_2 is the lawn need for X_2 and X_3 the lawn need for X_3.While for land formulated as :

$$(g+h)\, X_1 + (i+j)\, X_2 + (k+l)\, X_3 \leq 40\ \%\text{land-yard} \tag{3}$$

Whereas g, i, k is land need for cow pen of X_1, X_2, andX_3 and h, j, l is land need for biogas reactor of X_1, X_2, andX_3. Water resources formulated as :

$$mX_1 + n\,X_2 + q\,X_3 \leq 10\%\ \text{water flow} \tag{4}$$

Whereas m, n, q is water need for cattle and biogas reactor.Total biogas production formulated as :

$$P_{bg} = (T * M * C_{bg}) \tag{5}$$

Whereas T is number of cattle in study area and P_{bg} is biogas production (m^3/yr). $M$is number of manure every cattle (kg/head/yr).C_{bg} is coefficient of biogas forming from cow manure (m^3/kg) = 0.03 m^3/kg Manure [2].To obtain optimal results, it is necessary to analyze the sensitivity of the model to the variation of the number of cattle based on the age category. Based on the analysis of the long-term opportunity of raising livestock (calf 4 months, veal 8 months, and adult cattle 24 months), then made some alternative combination of livestock breeding between amount of calf, veal and adult to choose the most optimal combination.

This study located in Pejaten Village, Tarumajaya, Kertasari sub-district, Bandung, West Java (7°12'11.1"S 107°39'50.2"E). The domain system is village of cow-farming community based where residential area and cow farming along the river. Besides, it also nearby Cisanti wellspring and conservation area, which cow-farmer activities have direct impact to environment. Even several biogas reactor either individual or communal installed in this area but obstruction occurred like unoptimum function, unmanaged and abandon. Dairy farming in Pejaten comprises of 5 group with different number of cattleman as seen in the Table 1.

Table 1. Dairy Farming in Pejaten

	Unit	Group				
		Pejaten 1	Pejaten 2	Pejaten 3	Pejaten 4	Grand Total
Number of Cattleman	Pop.	35	35	25	44	120
Number of dairy cows	Head	69	73	58	89	477
Average ownership lactacy cattle per Cattleman	Head	2	2	2	2	2
Total Revenue	Rp/month	35,561,200	46,155,300	42,965,200	61,485,000	186,166,700
Average revenue per Cattleman	Rp/month	1,016,034	1,318,723	1,718,608	1,397,386	1,339,329
Total MilkProduction	liter/month	28,633	32,887	28,078	38,062	127,661
AverageMilkProductionper Cattleman	liter/ month	895	940	1,276	928	982

Source: Report of unit economic KPBS, 2009
* The Revenue excluded with cost of fodder

As KPBS (Koperasi Peternakan Sapi Perah Bandung Selatan) become the second biggest unit economic in Indonesia wheras Kertasari Sub-district is one of member. Hence, the number of Dairy cows greater than beef cattle. Every cattleman has 1 to 6 head of cows.

3. Results and Discussion

3.1 Survey Result

Based on the survey results, it can be drawn up several point. Manure livestock aslivestock byproduct need systemic management to reduce its effect to greenhouse effect and other seriousenvironmental pollution [11]. The population of livestock in the study area reached 477 animals, if the average produce 20 kg of dirt per day, it will produce 9.40 kg/day manure.

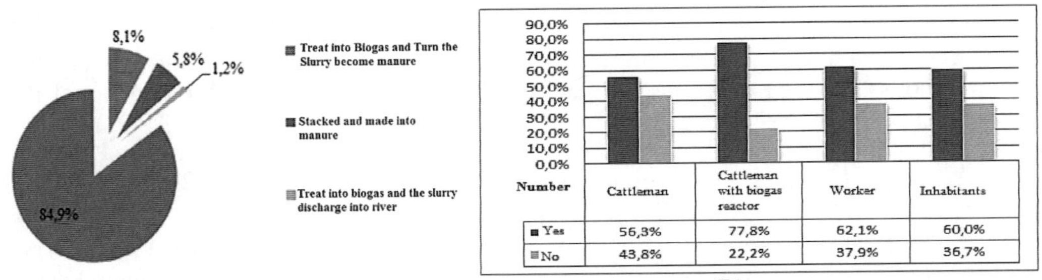

(A) (B)

Figure 1. A) The result of Survey Habitual Cattleman in waste treatment and B) Management of Communal Biogas Reactor

Most of cattlement directly discharged livestock manure into the river (Figure 2). As many as 86.3% dispose directly into the river stream, 5% of farmers accumulate to be used as fertilizer and only 3.8% breeders cultivate into biogas. Several causes of cattlement did not treat the manure is limited land provide for reactor (57.5%) and funding issues (21.3%). Thus, the development of biogas reactor should be done through group system and the existence of stimulus fund for farmers as. In fact, only 22.8% of farmers understood that livestock manure could be processed into biogas and fertilizer and 25% of farmers understood to be fertilizers. As many as 12.7% knew that livestock manure could be made as biogas and slurry discharged into the river and 39.2% of farmers state that livestock manure can be discharged into the river. While the reactor owner's understanding of livestock manure management answered 88.9% biogas owners understand that cattle manure can be made biogas and its effluent is used as fertilizer, and only 11.1% of respondents stated that livestock manure can be discharge into the river. Main reason of undevelop biogas utilization are inhabitants did not know the usefulness of biogas (67%). Unsuccessful biogas reactor development due to the operator absence's, therefore it is necessary to do further study the possibility of communal biogas reactor management and fully managed by operator who benefit financially from management of biogas reactors. Regarding with the readiness of the community in utilized biogas communal showed from 148 respondents, 56.3% of farmers agreed while 43.8% disagreed (Figure 3).

If biogas reactors are communally administered, a management officer must be required to obtain service fees, so that biogas reactor can operate sustainably. In Figure 4 it can be seen the percentage of respondents willingness to pay.

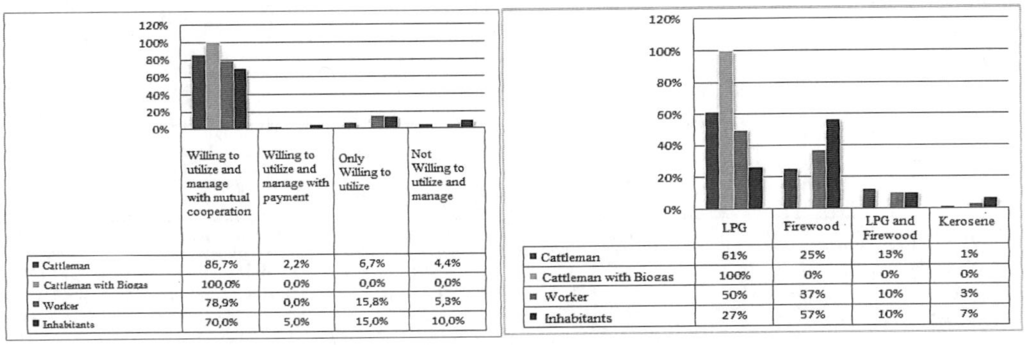

Figure 2. A) Community Availability Partitioned in the Management of Biogas Reactorsand B) Type of Cooking Fuel Consumption

As 82.4% of the inhabitants is willing to use and manage biogas with mutual cooperation even in fact, there are no biogas reactors that function individually or communally, this is probably due to the lack of understanding of biogas reactor management, Organization of biogas managers. In addition, this biogas reactor given to farmers who have busy activity, so it is not possible to do sidejob. Of the 135 respondents, only 57 respondents were willing to pay contributions to use biogas. The ability to pay the biogas usage fee monthly are varied from Rp 100,000 (1.8%), Rp 50,000 (7%), Rp 25,000 (73.3%). Inhabitant are common to use LPG and firewood, as shown in Figure 5 whereas half inhabitant (53%) useLPG, quarter of still using firewood (25%) and 12.5% use combination for firewood and LPG. The monthly expenditure on fuel per head of family was Rp 78,000, equivalent to 6 kg of LPG and equivalent to 12 m^3 biogas, so the biogas price equal to Rp 6,500/m^3.

3.2 Model Result

By optimation model, it found that variation 5 reach monthly optimum profit by Rp 1,432,291,54. The result of constrain showes that lawn is the component which always run out for every variation as seen in the Table 2.

Table 2. Constrain in Optimation Model

Variation	**Slack-/Surplus+**		
	Constraint 1 (lawn)	**Constraint 2 (land)**	**Constraint 3 (water resource)**
1	0	30,801	571,200
2	0	31,287	577,400
3	0	31,598	581,374
4	0	31,815	584,139
5	0	31,898	582,870
6	0	29,192	550,675

Farm size that include in availability land is the constrain of manure processing into biogas [12] and it's stated as constrain 2, land remind between 30.801 – 29.192 m^2 exclude population growth. While constrain 3, Pejaten village is water surplus area. Profitability index (PI) of variation 6 are shown in Table 3.

Table 3. Profitability Index (PI) for Dairy Cow with Variation 1:1:6

Year	**CF**	**IR (12%)**	**Present Value**	**IR (15%)**	**Present Value**
1	10,241,622	0.892857	9,144,304.06	0.869565	8,905,756.20
2	4,055,733	0.797194	3,233,206.24	0.756144	3,066,718.39
3	2,296,093	0.71178	1,634,312.72	0.657516	1,509,717.56
4	723,606	0.6355518	459,864.90	0.571753	413,724.14
5	7,336,117	0.567427	4,162,710.62	0.497177	3,647,348.43
6	14,154,594	0.508831	7,171,156.31	0.432326	6,119,399.17
7	26,470,998	0.452349	11,974,129.53	0.375937	9,951,427.62
Asset in 80th month	113,000,000	0.352347	39,815,211.00	0.27859	31,480,670.00
Sum PV	65,278,763		77,594,895.38		65,094,761.51
OI			54,000,000.00		54,000,000.00
NPV			23,594,895.38		11,094,761.51
PI			1.44		1.21

Table 4. Optimum Condition of Cow Cattle (variation 1 : 1 : 6)

Description	Unit	Calf	Veal	Adult Cattle	Total
Number of Cow	Head	269	269	1,617	**2,155**
Fodder need	kg/month	40,350	121,050	1,697,850	**1,859,250**
Pen cow width	m^2	12,878	38,633	541,873	**593,384**
Pen cow width and Biogas Reactor	m^2	678	1,423	9,427	**11,528**
Water need	liter/day	14,795	14,795	88,935	**118,525**
Profit	Rp/month	43,068,245	66,770,104	1,309,064,988	**1,418.903,337**
Biogas Production	m^3/ month	1,816	3,632	43,659	49,106

And the optimum one is for variation 4 with profit 1,418.903,337 and PI more than 1 for IR 12% and 15%. By this variation, need 270 cattleman with the monthly revenue Rp 4.853.424. It means 2.25 times than those number of cattleman (the average regional wage Rp 1,0605,000) with monthly revenue

IOP Conf. Series: Earth and Environmental Science **106** (2017) 012115 doi:10.1088/1755-1315/106/1/012115

Rp 1,339,329 (excluded cost of lawn worker). Biogas production from 2,155 cow is 49,106 m^3/month equal with 19,553 kg LPG as Rp 254,189,000 /month (see Table 4). At the moment, the number of cow is reach 477 head, with 120 cattleman. Means the utilize of natural resources is sufficient for farming needs. Even cattleman often facing less fodder since catleman do not varry on the fodder intensively.

4. Conclusion

To reach sustainable biogas management, it need to consider the interrelation component compose of farmers,cattle, land farming, food, water resource, biogas, non-biogas fuel in Pejaten Village, Tarumajaya Bandung West Java connected with profit, farmer welfare and pollution control. The farmers get used to throw cow waste into Citarum river and make the decrease of DO from 5.8 into 4.9 mg/L and BOD increase from 15 into 101 mg/L. Biogas potensial of 477 cattle is 285 m^3/day equal with 142 kg LPG for 711 household consumption. By linear programing, it shows the maximum profit able to reach if the combination of calf: veal : adult cattle is 1 : 1: 6 with the total number 2.155 cows. The constrain as maximum revenue is availability of land farming for 60 ha. The other constraint is water resource and homeland.

References

[1] Thawaf Rochadi, MS., Firman, A., & Sugandi D, MS. 2006. Analisis Kinerja Usaha Ternak Sapi Perah Rakyat pada Tiga Kondisi Usaha Koperasi/KUD Susu di Kabupaten Bandung. *Seminar Kinerja Usaha Peternakan Sapi Perah di Jawa Barat*, Bandung, Indonesia

[2] Wikan, W. T., Agung, H. 2005. Development of Biogas Processing for Small Scale Cattle Farm in Indonesia. *International Seminar on Biogas Technology for Poverty Reduction and Sustainable Development (ICAERD),* Beijing, China

[3] Electrigaz Technologies Inc., Feasibility Study – Biogas Upgrading and Grid Injection in the Fraser Valley, British Columbia, prepared for BC Innovation Council, Electrigaz Final Report, 2008

[4] Lietuvos Zemes UKIO Konsultavimo Tarnyba, Biogas Feasibility Study-UAB Dotnuva Experimental Farm, Substantiation of Biogas PowerPlant Construction, Baltic Sea Region Programme, 2013

[5] Teune, B., Orprecio, J., Dalusung, A., Yeneza, G., Feasibility Study of A National Biogas Programme on Domestic Biogas in The Philippines, SNV Netherlands Development Organisation and Winrock International, 2010.

[6] Wresta, A., Andriani, D.,Saepudin, A., Sudibyo, H. 2015. Economic analysis of cow manure biogas as energy source for electricity power generation in small scale ranch. *Energy Procedia*. 68:122-131.

[7] Anspach V., Moller D, 2008. Biogas and Organic farming : Empirical evidence on production structure and economics in Germany. *16th IFOAM Organic World Congress*. Modena, Italy.

[8] APCAEM, Recent Developments In Biogas Technology For Poverty Reduction and Sustainable Development : United Nations Economic and Social Comission for Asia and The Pacific (ESCAP), 2007

[9] Glory Brent, A. Inputs for Biogas Economic Assessment, Cornell University, New York, p.14853-7801. 2008.

[10] Murphy, J.D.G., Belward, A., & Burrage, K. Optimisation of a Dairy Farm Model-Comaprison of simulated annealing, Simulated Quenching and General Algorithms, Elsevier Ltd. 1999.

[11] M.Y, Nurul Aini., C.E. Ibrahim, T. Siti Salmiyah, S.A. Syed Hussein, D. Kamaruddin. 2012. Small-Scale Biogas Plant In A Dairy Farm. *Malaysian Journal of Veterinary Research*. 3: 49-54.

[12] Mattes, S., Daniela, T. 2016. Unlocking the Energy Potential of Manure—An Assessment of the Biogas Production Potential at theFarm Level in Germany. Agriculture. http://www.mdpi.com//[2017 June 25].

Development of weighting value for ecodrainage implementation assessment criteria

S Andajani*, D P A Hidayat, B E Yuwono

Hydraulic Laboratory, Faculty of Civil Engineering and Planing, Universitas Trisakti, Jakarta, Indonesia

*Corresponding author: sandajani23@gmail.com

Abstract. This research aim to generate weighting value for each factor and find out the most influential factor for identify implementation of ecodrain concept using loading factor and Cronbach Alpha. The drainage problem especially in urban areas are getting more complex and need to be handled as soon as possible. Flood and drought problem can't be solved by the conventional paradigm of drainage (to drain runoff flow as faster as possible to the nearest drainage area). The new paradigm of drainage that based on environmental approach called "ecodrain" can solve both of flood and drought problems. For getting the optimal result, ecodrain should be applied in smallest scale (domestic scale), until the biggest scale (city areas). It is necessary to identify drainage condition based on environmental approach. This research implement ecodrain concept by a guidelines that consist of parameters and assessment criteria. It was generating the 2 variables, 7 indicators and 63 key factors from previous research and related regulations. the conclusion of the research is the most influential indicator on technical management variable is storage system, while on non-technical management variable is government role.
Keywords: criteria, cronbach alpha, ecodrain, loading factor, sustainable

1. Introduction

Sustainable development is a concept that emphasizes the conservation and preservation of natural resources. It is comprehensive and is therefore not limited to certain activities, and it is closely related to the continuation of human life on earth. Sustainable development concerns three primary categories: economic, social, and environmental. These three elements are interconnected to create security, harmony, and prosperity. Sustainable development seeks to balance human needs and demands with the environment's capacity to cope with human consumption and industry [1]. For assuring local sustainability in the form of shaping sustainable human settlements an appropriate sustainable management on a local level is necessary to be carried out. In the process of operationalization of sustainable development an important part is represented by developing innovative models for sustainable land management [2].

Water is the source of life, and anyone can't survive without water. Urban drainage presents a classic set of modern environmental challenges: the need for cost-effective and socially acceptable technical improvements in existing systems, the need for assessment of the impact of those systems, and the need to search for sustainable solutions. There are numerous techniques and approaches known around the world how to support sustainable rainwater management, especially in the urban areas, where the storm water can cause significant damages [3].

IOP Conf. Series: Earth and Environmental Science **106** (2017) 012116 doi:10.1088/1755-1315/106/1/012116

In urban areas, the drainage problem consist of flood, drought and water pollution are getting more complex and need to be handled as soon as possible. Flood caused by incapability of river in collecting actual water discharge (river overflow) while inundation occurs when runoff and local inflow cannot be drained into the nearest drainage channel or river. Drought caused by the decreasing of groundwater level and water supply especially in the dry season. There are 2 method of drainage on water resources engineering, the conventional method adopt "hard engineering principle" and the recent one known as "ecodrain". The principle of conventional method is to drain runoff flow as faster as possible to the nearest drainage area while the principle of Ecodrain is increase water infiltration and storage to decrease runoff discharge. The conventional drainage design approach does not address sustainability issues. Moving forward, an ecodrain approach using green infrastructures is recommended. In addition to flow and flood management provided by the conventional methods, green infrastructures can bring multiple benefits such as increased amenity value and groundwater recharge. The main advantage of the sustainable approach is the additional benefits such as environmental improvement,natural groundwater recharge, runoff reduction as well as energy savings [4].

For the optimal result, ecodrain should be applied in smallest scale (domestic scale) to the biggest scale (city areas). Ecodrain can be implemented if the drainage infrastructure meets the criteria and indicators. Therefore, it is necessary to identify drainage condition based on environmental approach. For identify the implementation of ecodrain concept, it use a guidelines that consist of ecodrain criteria and indicators. According to researchers [5], [6], there are 2 variables, 7 indicators and 63 key factors for identify the implementation of ecodrain concept, consist of: technical and nontechnical management as variables, technical management consist of infiltration system, drainage system, complementary building system, storage system, and pump system as indicators, while non-technical management consist of government role and community participation as indicators. From those indicators, 63 key factors was generated (type and channel dimension, water quality, sediment transport, monitoring evaluation existence, etc [5], [6].

Based on researcher [7] and Indonesia's related regulation, there are several factors that influenced implementation of ecodrain concepts, such as:

- Separation between drainage and waste water network for urban areas.
- Masterplan with area more than 10 hectare should be built retention ponds, rainwater subreservoir and absorption wells
- Groundwater recharge, rainwater reuse and recycle and low impact development concept.
- River has function as natural drainage channel.
- Development area plan should have at least 30% of total area which used as retention areas
- etc

For non technical management variables, researcher [8] was mentioned that public perception toward cultural attraction will give the tremendous boost to any development. The groups identified as the most influence and affected individual are the local authorities, local communities, and the tourists [8]. Therefore, government role and community participation is used as non technical management variables. This research aim to develop the weighting value for 2 variables (technical and non technical), 7 indicators and 63 key factors using SPSS. From this research, the most influenced indicator and factor can be determined. Further, the most influenced factor should be prioritized by government and community to implement ecodrain concept.

2. Research Method

Methods used in this research are:

a) Developing factors validity and indicators reliability using SPSS. The validity of all factors has been tested by determining loading factor meanwhile the reliability of all indicators has been tested by using Cronbach alpha. All factors and indicators are valid if the value of loading factor and Cronbach alpha close to 1.
b) Developing weighting value for all factors and indicators from percentage of the loading factors and Cronbach alpha value. From the weighting value, it can be conclude the most influential factor and indicator to identify the implementation of ecodrain concept in an area.

3. Results and Discussion

The result of this research consist of:

a) All indicators (Infiltration system, drainage channel system, complementary system building, storage system, pump system, government role and community participation) from 2 variable (technical and non technical) meets the reliability requirement with Cronbach Alpha value 0.732-0.838

b) All key factors from 7 indicators fulfil the validity test with loading factor value 0.428-0.748.

The recapitulation of evaluation results on validity and reliability evaluation can be seen in Table 1.

Table 1. Validity and reliability test for indicator and factor of evaluation.

Indicator	Code	Validity *Loading Factors*	Reliability *Cronbach Alpha*	Description	Weighting Value
Infiltration System (A1)	***A1***		*0.732*	*Reliable*	*13.19*
Type of infiltration system selection	A11	0.553		Valid	1.55
Determination of infiltration system number and capacity	A12	0.561		Valid	1.57
Determination of infiltration system location	A13	0.534		Valid	1.5
Waste management in infiltration system	A14	0.617		Valid	1.73
Water quality management in infiltration system	A15	0.597		Valid	1.67
Sediment management in infiltration system	A16	0.683		Valid	1.91
Vegetation management in infiltration system	A17	***0.695***		Valid	***1.95***
The existence of monitoring evaluation system	A18	0.468		Valid	1.31
Drainage Channel Systems (A2)	***A2***		*0.747*	*Reliable*	*13.46*
Type of drainage channel selection	A21	0.524		Valid	1.46
Determination of drainage channel dimension and capacity	A22	***0.688***		Valid	***1.92***
Determination of drainage channel location	A23	0.638		Valid	1.78
Waste management in drainage channel system	A24	0.535		Valid	1.49
Water quality management in drainage channel system	A25	0.532		Valid	1.48
Sediment management in drainage channel system	A26	0.665		Valid	1.85
Vegetation management in drainage channel system	A27	0.678		Valid	1.89
The existence of monitoring evaluation system	A28	0.571		Valid	1.59
Complementary System Building (A3)	***A3***		*0.789*	*Reliable*	*14.21*
Type of complementary system building selection	A31	0.517		Valid	1.45
Determination of complementary system building dimension and capacity	A32	***0.711***		Valid	***1.99***
Determination of complementary system building location	A33	0.688		Valid	1.93
Waste management in complementary system building	A34	0.582		Valid	1.63

Indicator	Code	Validity *Loading Factors*	Reliability *Cronbach Alpha*	Description	Weighting Value
Water quality management in complementary system building	A35	0.63		Valid	1.76
Sediment management in complementary system building	A36	0.67		Valid	1.88
Vegetation management in complementary system building	A37	0.669		Valid	1.87
The existence of monitoring evaluation system	A38	0.608		Valid	1.7
Storage Systems (A4)	***A4***		*0.829*	*Reliable*	*14.93*
Type of storage system selection	A41	0.596		Valid	1.64
Determination of storage system number and capacity	A42	0.672		Valid	1.85
Determination of storage system location	A43	0.72		Valid	1.99
Waste management in storage system	A44	0.673		Valid	1.86
Water quality management in storage system	A45	0.624		Valid	1.72
Sediment management in storage system	A46	***0.743***		Valid	***2.05***
Vegetation management in storage system	A47	0.717		Valid	1.98
The existence of monitoring evaluation system	A48	0.667		Valid	1.84
Pump Systems (A5)	***A5***		*0.81*	*Reliable*	*14.59*
Type of pump system selection	A51	0.705		Valid	1.95
Determination of pump system number and capacity	A52	0.725		Valid	2
Determination of pump system location	A53	***0.73***		Valid	***2.02***
Waste management in pump system	A54	0.688		Valid	1.9
Water quality management in pump system	A55	0.435		Valid	1.2
Sediment management in pump system	A56	0.687		Valid	1.9
Vegetation management in pump system	A57	0.626		Valid	1.73
The existence of money system	A58	0.685		Valid	1.89
Government Role (B1)	***B1***		*0.838*	*Reliable*	*15.1*
Determination of O&M organization	B11	0.509		Valid	0.92
Determination of supporting personnel O&M organization	B12	0.61		Valid	1.1
Determination of supporting personnel number	B13	0.653		Valid	1.18
O&M drainage infrastructure on regular schedule	B14	0.537		Valid	0.97
Determination of local government support on O&M budget	B15	0.508		Valid	0.91
Determination of local government support on O&M development and rehabilitation budget	B16	0.597		Valid	1.08
Determination of development and rehabilitation priority	B17	***0.668***		Valid	***1.2***
Formation coordination organization among stakeholder	B18	0.572		Valid	1.03
Giving reward to local government	B19	0.428		Valid	0.77

Indicator	Code	Validity *Loading Factors*	Reliability *Cronbach Alpha*	Description	Weighting Value
The existence of local laws and regulations & policies that support ecodrain concept	B110	0.667		Valid	1.2
Application of punishment/penalty	B111	0.503		Valid	0.91
Socialization of local laws / regulations, policies and penalty to the community	B112	0.657		Valid	1.18
Involvement of customary institutions in socialization	B113	0.513		Valid	0.92
Follow up on community complaints	B114	0.494		Valid	0.89
Determination of ecodrain master plan	B115	0.467		Valid	0.84
Community Participation (B2)	***B2***		*0.806*	*Reliable*	*14.52*
Community Participation on drainage infrastructure regular maintenance	B21	0.6		Valid	1.65
The community's active role on reporting drainage infrastructure breaking/failure	B22	0.68		Valid	1.87
The community's active role on reporting inundation	B23	0.676		Valid	1.86
The community's active role on giving post development impact data	B24	***0.748***		Valid	***2.06***
Community Participation on drainage infrastructure planning and construction process	B25	0.685		Valid	1.89
The community's active role on infiltration wells construction	B26	0.629		Valid	1.73
Community understanding about ecodrain concept	B27	0.65		Valid	1.79
Community's ability to finance infrastructure damage	B28	0.606		Valid	1.67

Based on Table 1 (weighting value result), it can be seen that the most influential indicator on technical management variable is storage system while on non-technical management variable is government role. The most influential factor from each indicator can be seen on Figure 1.

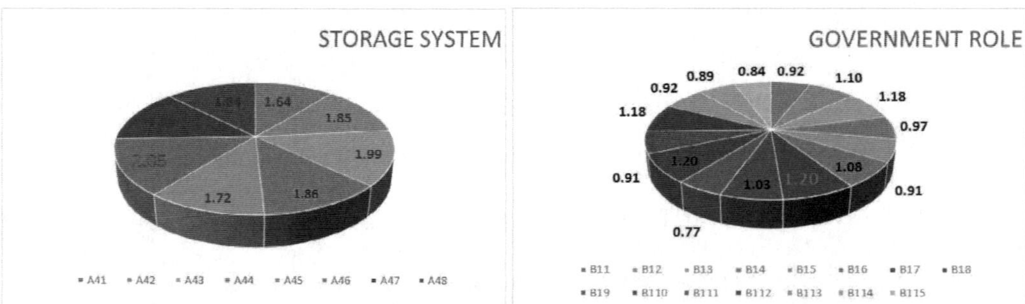

Figure 1. Most influential indicator and factor for ecodrain assessment.

4. Conclusion

It can be conclude that the most influential indicator on technical management variable is storage system while on non-technical management variable is government role. For storage system, the most influential factor for ecodrain assessment is Sediment management in storage system (with weighting value of 2.05) and for government role the most influential factor for ecodrain assessment is Determination of development and rehabilitation priority (with weighting value of 1.20). Therefore, the most influential factors should be prioritized to identify the implementation of ecodrain concept in an area.

References

[1] Shahrom Md Zaina, Wan Hamidon Wan Badaruzzamana, Riza Atiq O K Rahmata, Othman Jaafara b, Noor Ezlin Ahmad Basria, b, Hassan Basria. 2012. Learning Outcome Measurement for Environmental and Sustainable Development Component in the Field of Civil Engineering. *Procedia - Social and Behavioral Sciences.* 60:90 – 97.

[2] Jurijs Holmsa, Irina Arhipovaa, Ildiko Tulbureb, Gatis Vitolsa. 2017. Ecosystem Provisioning Services Automated Valuation Process Model for Sustainable Land Management. *Procedia Computer Science*. 104:65 – 72

[3] M. ZeleĖákováa, G. Markoviþb, D. Kaposztásová, Z. Vranayová. 2014. Rainwater Management in Compliance With Sustainable Design of Buildings. *Procedia Engineering*. 89: 1515 – 1521

[4] J.-f. Chowa,b,c, D. Savić, D. Fortunec, Z. Kapelan, N. Mebratec. 2014. Using a systematic, multi-criteria decision support framework to evaluate sustainable drainage designs. *Procedia Engineering*. 70:343 – 352

[5] Andayani, Yuwono.Key Factor for Regional Urban Eco-Drainage Evaluation. IRES 2015

[6] Andayani, Yuwono. Evaluation of Model Development of Urban Eco Drainage in Region Scale. ICSBE 2016

[7] Sarbidi. 2014. Design Criteria of the Urban Area Sustainable Drainage For Human Settlements. *Jurnal Permukiman* 9(1) 1-16. No. 1 April 2014.

[8] Sabrina Idilfitria, Nur Izzati Mohd Rodzia , Nik Hanita Nik Mohamad, Suria Sulaiman. 2015. Public Perception of the Cultural Perspective towards Sustainable Development. *Procedia - Social and Behavioral Sciences*. 168:191 – 203

IOP Conf. Series: Earth and Environmental Science **106** (2017) 012117 doi:10.1088/1755-1315/106/1/012117

The effect of mixing ratio variation of sludge and organic solid waste on biodrying process

A C Nasution, G A Kristanto*

Civil Engineering Department, Environmental Engineering Study Program, Universitas Indonesia, Depok, Indonesia

*Corresponding Author: andari@eng.ui.ac.id

Abstract. In this study, organic waste was co-biodried with sludge cake to determine which mixing ratio gave the best result. The organic waste was consisted of dried leaves and green leaves, while the sludge cake was obtained from a waste water treatment plant in Bekasi. The experiment was performed on 3 lab-scale reactors with same specifications. After 21 days of experiment, it was found that the reactor with the lowest mixing fraction of sludge (5:1) has the best temperature profile and highest moisture content depletion compared with others. Initial moisture content and initial volatile solid content of this reactor's feedstock was 52.25% and 82.4% respectively. The airflow rate was 10 lpm. After biodrying was done, the final moisture content of the feedstock from Reactor C was 22.0% and the final volatile solid content was 75.9%.The final calorific value after biodrying process was 3179,28kcal/kg.
Keywords: co-biodrying, mixing ratio, organic waste, sludge cake

1. Introduction

Rapid urbanization followed by the industrial development and construction activities is one of the reasons for solid waste (SW) accumulation [1]. SW accumulation will produce odor and leachate during itscollectionandtransportation [2]. Solid waste accumulation in landfill will lead to overcapacity and difficulty to find new site. Moreover, SW management in Indonesia was still dominated with open dumping system which lacked with methane gas control system [3].

Moreover, sludge from wastewater treatment (domestic and industrial) is continuously produced in large number and becoming one of the environmental problems [4]. High moisture content makes it difficult to processed sludge directly [5]. Therefore, waste begins to be used for the recycling of materials and energy production from solid waste. Waste treatment into new energy sources is expected to replace fossil fuels [6].

Biodrying, which is based on a process similar to composting, aims at removing water from biowastes with high water content using the heat generated during the aerobic degradation of organic substances, in addition to forced aeration [4], [7], [8]. The removal of excess moisture content on waste may increase the potential for thermal energy recovery [9]. Moreover, biodrying results can be utilized as RDF (refused derived fuel) which is one of the most excellent sources of renewable energy [10].

When organic solid waste and sludge were mixed, there will be a reduction in the number of microorganisms from the sludge, and enhancement of biodegradable organic matter in the mixture. Therefore co-biodrying of organic solid waste and sludge can increase the profile of temperature and stability of the system by utilizing microorganisms contained in sludge and solid waste that are readily biodegradable and rich in nutrients for microorganisms [2].

2. Research Method

2.1. Preparation of material

Sample used in this study is artificial organic solid waste and sludge cake. The organic solid waste is yard waste from UPS FTUI, while the sludge cake comes from Waste Water Treatment Plant, PT. Jababeka Infrastructure. The sample was put as much as 17 kg, adjusted to the inner volume of the reactor. The mixing ratio is varied to 3: 1, 4: 1, and 5: 1, for solid waste and sludge respectively. The solid waste itself consists of green and dry waste and the fraction used is 2: 1, for each reactor. The selection of fractions refers to the research that has been done [2] as well as based on preliminary experiments conducted before the study began. The feedstock that has been collected is then homogenized and weighed according to a predetermined fraction. Here is the fraction for each composition on each reactor (Table 1).

Table 1. Composition of material used for the study.

Composition	Reactor name		
	Reactor A (kg)	Reactor B (kg)	Reactor C (kg)
Dry yard waste	8.5	9	9.5
Green yard waste	4.25	4.6	4.7
Sludge	4.25	3.4	2.8

2.2. Experimental equipment

The biodrying process was performed on a reactor made of polystyrene foam with volume of 97.5 L (64 cm x 44 cm x 34 cm) with wall thickness of 3 cm. In this study, there were 3 lab-scale reactors, each filled with 17 kg of material mixture. The reactor lid was coated with a 2 cm thick sponge to prevent energy loss and steam condensation.

Table 2. Physicochemical properties of the raw materials.

Parameter	Solid waste : Sludge		
	Reactor A (3:1)	Reactor B (4:1)	Reactor C (5:1)
Moisture content (%)	61.81	56.85	54.25
Volatile solid (%)	88.88	83.18	82.37
C/N	9.7	10.9	11.8
FAS (%)	67.81	68.54	72.91

The reactor was connected to the air pump (Multipro Mini Air Compressor) used for aeration, and the pump was equipped with a flowmeter (Wiebrock capacity of 15 lpm) to adjust the airflow. The airflow used in this experiment is 10 lpm (air flow is supplied to the reactor using a hose with diameter of 0.8 mm. Leachate which formed during the process will be channeled into the leachate collector box through the hole provided in the reactor. Temperature sensors were placed in the center of the reactor. Temperatures were recorded using an acquisition data system consist of Arduino and thermocouples. The following were the reactor design used in this study.

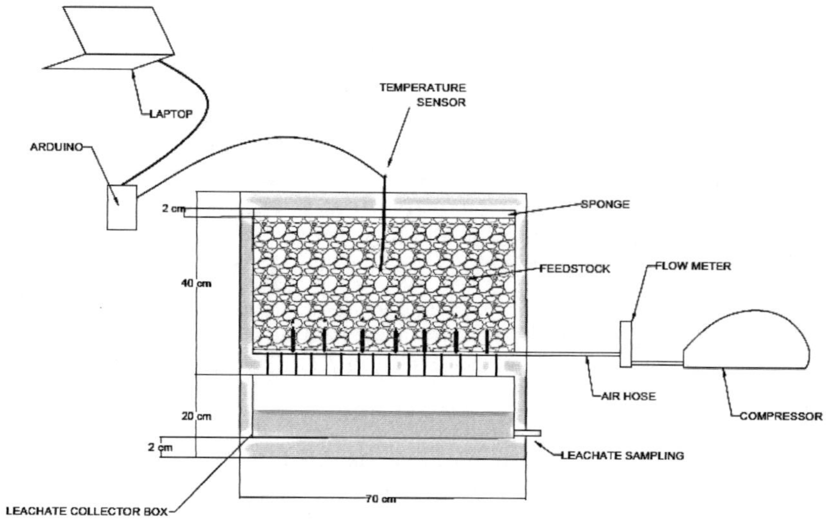

Figure 1. Schematic diagram of biodrying reactor.

3. Result and Discussion

3.1. Temperature profile on biodrying process

The high temperature with a long duration was the key point to achieve the good biodrying performance [2]. Figure 2 shows the daily temperature profile of co-biodrying process. The initial temperature for each reactor ranges in 30°C. In Reactor A (3:1), the highest temperature reached 38.9°C on the third day. The average temperature in reactor A is 34.4°C. While at Reactor B (4:1), the highest temperature reached 44.4°C on the second day, with average temperature of 36.1°C. In reactor C (5:1) the highest temperature reached 52.4°C on the third day and the average temperature was 37.6°C.

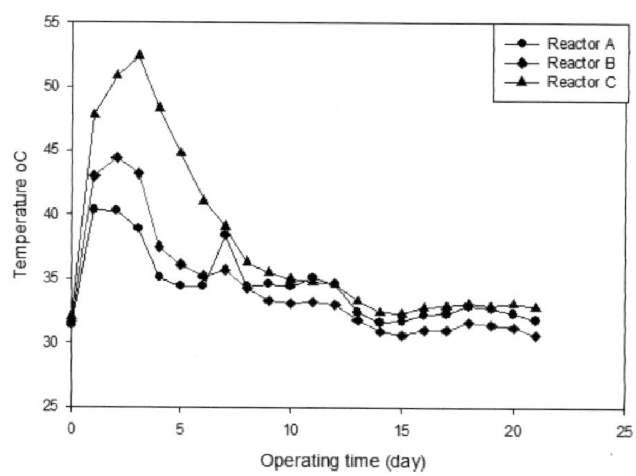

Figure 2. Temperature profile during biodrying process.

The rapid increase in temperature is an indication of microbiological degradation of organic matter and heat release from metabolic activity. This phase is called the warming-up phase, where in this experiment occurs on the 1st day until the 4th day. The depletion of temperature in each reactor might be caused by two factors. The first factor is caused by the high temperature formed causing the death of mesophile (15-35°C) microorganisms initially present in the sludge, thus forming a long lag phase for the microorganism enough to start its activity again. Another factor is the lack of biodegradable materials that lead to the death of microorganisms. Based on the temperature range generated in this experiment,

IOP Conf. Series: Earth and Environmental Science **106** (2017) 012117 doi:10.1088/1755-1315/106/1/012117

it can be seen that microorganisms that work during the biodrying process are mesophiles (active at 8-48°C) and thermophiles (active at 42-68 °C).

3.2. Loss of moisture content

The initial moisture content on reactor A was 61.8%. The moisture content decrease that occurred in reactor A was 32.4% where the final content moisture after process was 29.4%. Whereas in reactor B there was a moisture content decrease of 31.8% where the initial moisture content was 56.9% and the final content moisture was 25.1%. In reactor C, the initial moisture content is 54.3%. After biodrying process was done, the final moisture content was 22.0%, where in this reactor there was a decrease of moisture content of 32.3%.

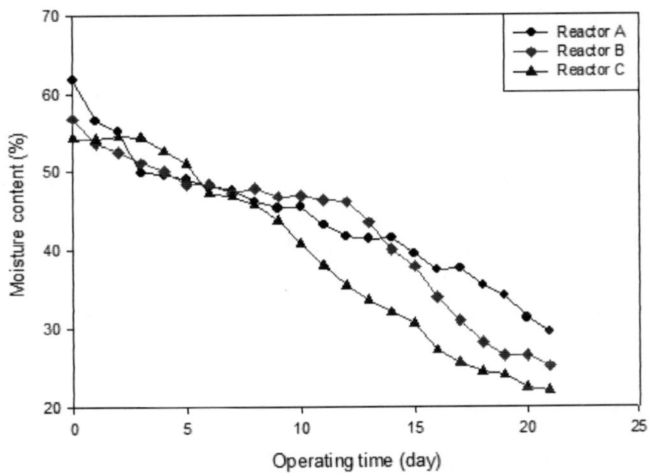

Figure 3. Change of moisture content in each feedstock during co-biodrying process.

Figure 3 shows the relationship between decreasing of moisture content and the operating time of the biodrying reactor. The higher the temperature formed during biodrying process, the greater value of the moisture content loss that occurs. High temperatures convert water to steam, increasing the water holding capacity of the air contained in the feedstock [2]. In biodrying process, the loss of water is greater than the addition of water from metabolic process resulting in a dry material [11], [12].

3.3. Volatile solid depletion

Volatile solid (VS) determines the number of biodegradable organic matter present in the material [15]. In Reactors A, B and C, the initial solid volatile percentage was 88.8%, 83.2%, and 82.4% respectively. After the biodrying process takes place, the percentage of VS in reactor A became 74.3%. So during the biodrying process there was decrease in VS by 14.6%. In reactor B there was a decrease of VS by 10.4%, where the final percentage of VS was 72.8%. While at reactor C, the final percentage of VS was 75.9%, where there was a decrease of VS by 6.4%.

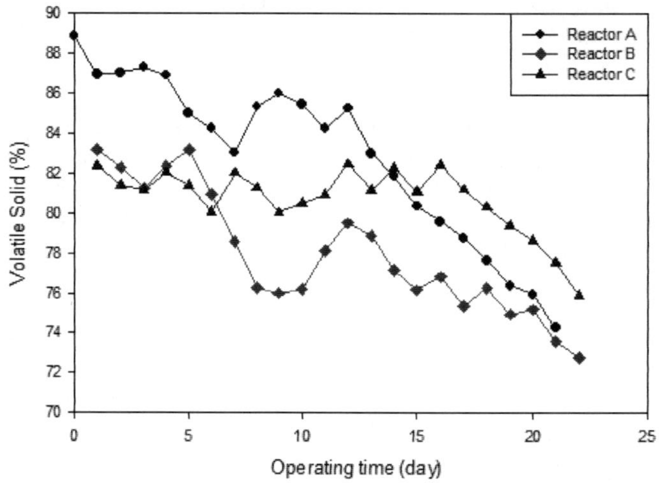

Figure 4. Change of volatile solid content in each feedstock during co-biodrying process.

Based on Figure 4, the larger the sludge fraction added to solid waste, the greater the VS content of the feedstock. In the biodrying process, the less VS is degraded, the better the output is produced. This is because the purpose of the biodrying process is to reduce the biodegradation of VS in order to maintain the heating value of the waste [13], [14]. The decrease of VS in each reactor describes the amount of organic used by microorganisms during the metabolism process.

3.4. Calorific value

In the biodrying process, it is expected that the product has an optimal calorific value with low moisture content. The calorific value calculation is performed based on the following equation [15].

$$LCV = HCV/\,((100\% \,/\, (100\% - moisture\ content\ RDF)) \tag{1}$$

Based on the equation (1), we get the caloric value of feedstock for each reactor as follows.

Table 3. The calorific value of the feedstock in each reactor before and after biodrying process.

Sample	Before *biodrying*			After *biodrying*			ΔLCV (kcal/kg)
	HCV (kcal/kg)	*Moisture (%)*	LCV (kcal/kg)	HCV (kcal/kg)	*Moisture (%)*	LCV (kcal/kg)	
Reactor A	3956	61.81	1510.72	4087	29.40	2885.42	1374.70
Reactor B	4064	56.85	1753.45	4043	25.07	3029.24	1275.79
Reactor C	4164	54.25	1904.89	4076	21.99	3179.28	1274.39

Higher Calorific Value (HCV), is the upper heat value determined when the moisture content value of the material is equal to 0, while LCV (Lower Calorific Value) is the lower heat value determined when H_2O in the material is on gas phase. LCV is the real calorific value of a material. Based on data obtained, there is an increase in LCV value from feedstock that has been bio-dried. The biodrying process increases the energy content of the material by maximizing the removal of moisture content and maintaining the calorific value (HCV) from organic matter by minimizing biodegradation [10].

4. Conclusion

The mixing of organic solid waste and sludge can improve the performance of biodrying process. In this study, Reactor C with mixing ratio of solid waste and sludge of 5:1 gave better results than other reactors.

The final moisture content of the feedstock from Reactor C was 22.0% with final volatile solid content was 75.9%. The final calorific value of the feedstock in Reactor C was 3179.28kcal/kg.

Acknowledgments
This project was financially supported by Universitas Indonesia under the project of Publikasi Internasional Terindeks for Thesis of UI (PITTA).

References

[1] Chattopadhyay S, Dutta A, Ray S. 2007. Existing municipal solid waste management of Kolkata – deficiencies and solutions. *J. IAEM.* 34(3):161–167.
[2] Ma J, Zhang L, and Li A. 2016. Energy-efficient co-biodrying of dewatered sludge and food waste: Synergistic enhancement and variables investigation. *Waste Management*. 56:411-422.
[3] Ciptakarya P, Profil Kabupaten / Kota : Kota Mojokerto, Jawa Timur, Mojokerto: Pemerintah Kota Mojokerto. 2002.
[4] Tambone F, Scaglia B, Scotti S, Adani F. 2011. Effects of biodrying process on municipal solid waste properties. *Bioresour. Technol.* 102(16):7443–7450.
[5] Villegas M, and Huiliñir C. 2014. Biodrying of sewage sludge: Kinetics of volatile solids degradation under different initial moisture contents and air-flow rates. *Bioresource Technology*. 174:33-41.
[6] Tambone F, Scaglia B, Scotti S, Adani F. 2011. Effects of biodrying process on municipal solid waste properties. *Bioresour. Technol.* 102(16):7443–7450.
[7] Frei K M, Cameron D, Stuart P R. 2004. Novel drying process using forced aeration through a porous biomass matrix. *Dry. Technol.* 22(5):1191-1215.
[8] Navaee-Ardeh S, Bertrand F, and Stuart P. 2010. Key variables analysis of a novel continuous biodrying process for drying mixed sludge. *Bioresource Technology*, 101(10): 3379-3387.
[9] Rada E C, Franzinelli A, Taiss M, Ragazzi M, Panaitescu V, Apostol T. 2007. Lower heating value dynamics during municipal solid waste bio-drying.*Environmental Technology*. 28:463–469.
[10] Adani F, Baido D, Calcaterra E, Genevini P. 2002. The influence of biomass temperature on biostabilization-biodrying ofmunicipal solid waste. *Bioresour. Technol.* 83(3):173-179.
[11] Nakasaki K, Nakano Y, Akiyama T. 1987b. Oxygen diffusion and microbial activity in the composting of dehydrated sewage sludge cakes. *J. Ferment. Technol.* 65:43–48.
[12] Richard T L, Veeken, A H M, De Wilde V, Hamelers H V M. 2004. Air-filled porosity and permeability relationships during solid-state fermentation. *Biotechnol. Progr.* 20:1372–1381.
[13] Sadaka S, Vandevender K, Ph D, and Costello T. 2011. Partial Composting for Biodrying Organic Materials. *Agriculture and Natural Resources.*
[14] Huilinir C, Villegas M, 2014. Biodrying of pulp and paper secondary sludge: kinetics of volatile solids biodegradation. *Bioresour. Technol.* 157:206-213.
[15] G. Tchobanoglous, H. T. (1993). Integrated Solid Waste Management. United States: McGraw-Hill, Inc.

Microclimate and architectural tectonic: vernacular floating house resilience in Seberang Ulu 1, Palembang

P Puspitasari*, T Kadri, I Indartoyo, L Kusumawati

Architecture Department, Faculty of Civil Engineering and Planning, Trisakti University, Jakarta, Indonesia

*Corresponding Author : popi@trisakti.ac.id

Abstract. This paper aims to describe the results of preliminary research on floating houses on the Musi River, Seberang Ulu 1, Palembang, focused on studying the influence of microclimates to the tectonics of *Rumah Rakit* (Floating House).The increase of water surface due to global warming will increase the need of using floating house typology in the future. The description of the inhabitants' experiences on applying technics to create vernacular floating houses is considered as significant knowledge to develop advance technology on the basis of local characteristic. Vernacular floating houses resilience consists of natural experiences of inhabitants in adapting their daily activities to the characteristic of local climate. By using qualitative approach, the *Rumah Rakit* inhabitants' verbal information in this article becomes the main aspect in exploring local knowledge. At the end, the conceptual model of vernacular *Rumah Rakit* in Seberang Ulu 1, Palembang is formulated, in terms of building architectural tectonic that is closely related to the local climate characteristic. The knowledge can be utilized in the context of rehabilitation or preservation of such architectural objects that are their existences tend to be extinct at this time.
Keywords: floating house-*Rumah Rakit*, microclimates, building tectonic, vernacular architecture

1. Introduction

Theoretically, the house is a human response to the characteristics of climate, geography, topography and culture by using technology that is enriched by normative ritual and art. Today, humans face global climate change. The change is caused by the fluctuating increase of earth's surface heat that impact on the uncertainty of weather condition, the increasing of sea level and the occurrence of various natural disasters. It is predicted that the expansion of water area will occur in the next 100 years. The inward expansion will be going around 1-1.5 km towards the mainland. Therefore, the demand for floating settlements is possible increased in the future.

Architecturally, the above conditions are pose challenges to prepare various capabilities, especially in the development of structural technology and building construction. What to do right now is to learn from the occupants' experiences of floating houses. The occupants inhabit the vernacular houses and survive for a long time by trial and errors. The occupants build their house based on the experiences of tackling local climate change without the use of high technology. By applying simple techniques, they can survive and solve the problem of the influence of microclimate fluctuations to form a reliable building system.

Some studies on the same location have been conducted by other researchers. They concern on the phenomenological study of floating house transformation in relation to the history of the existence of

Chinese in Palembang [1]. The study of floating house forms on its structure and construction concerned more on the study of the relation of Chinese architectural proportion principles to traditional floating house in Palembang by using deductive analysis [3], [6]. While researcher [2] try to explore the potential of *Rumah Rakit* as tourism object. The research conducted by [7] contributes to enrich the insight of vernacular waterfront house related to microclimate characteristic.

Related to the topic of writing, the important key words that should be comprehended are vernacular architecture and its endurance, local climate, architectural tectonic and floating house of *Rumah Rakit*. Vernacular architecture (Latin: *vernaculus* = native = indigenous) is a pragmatically constructed architecture, based on local or traditional knowledge by using local materials [8], [9]. Tectonics (Greek, tectonics = *tekton* = carpenter/builder) is the art of construction or the art of linking material by applying certain techniques. The subjects of architectural tectonic are focused on: the correlation between interconnected materials, construction technology, technical mechanisms and procedures, climate behaviour, geographical conditions, topographic and environmental settings [5]. The resilience of vernacular architecture relies on its ecological-based technology, its resistance to the climate change fluctuations, the use of local materials and the techniques with simple equipment [9].

The microclimate is a specific weather condition that is different from its zonal climate [4]. The microclimate affects the design of building in terms of: its orientation, form, volume, facade, opening location, building construction and site plan [8], [11]. Floating buildings require a floating structure system that capable to buttress free vertical movement to follow the changes of water level and its movement in all directions. The movements influenced by wind speed, speed of water flow and weight of load [10]. *Rumah Rakit* is a house that is set up on a bamboo or wooden circuit and float on the water [3]. This paper is for enriching the above findings by emphasizing on the influence of local climate to the tectonics of *Rumah Rakit* based on the empirical experiences of the occupants and local expert.

2. Research Method

This paper is descriptive-qualitative, aimed to describe the results of field observations and interviews. The interviewees are conducted among local experts who have the expertise to build a floating house. The technique for determining the interviewees conducted in a purposive manner by following the instruction of local informants who has been living for more than 20 years in the target location. Local experts' verbal information was recorded by tape recorder. Observation units determined purposively. Data and information analysed inductively to produce an integrated conceptual model. This conceptual model is temporary considering that the perfection of the model depiction depends on the comprehensiveness of data and information.

3. Results and Discussion

The location of research is the area of *Rumah Rakit* on Musi River, Seberang Ulu 1, Central Palembang Old City. At that location there are a number of jetties for boats transporting goods and passengers that cross between riverbanks or inter-districts traffics. Based on the interviews of local experts and occupants, the findings are described as follows.

3.1. The determination of Rumah Rakit location in relation to the characteristic of topography, geography and water surface behavior

The location, where *RumahRakit* will be located, depends on the slope level of river base ground. The angle of ground slope is the angle of riverbank surface slope that leads to the river base. The facade of *Rumah Rakit* is facing to river body while the rear side is accessible from the mainland. A sloping ground surface considered unfavourable to the position of building during the low tide season. The rear side of foundation will hit the ground slope as well as cause the front side of building's floor to be submerged.

The local expert said that the position of *Rumah Rakit* to the mainland is determined by the results of observations on the rise and low tides of river water surface. If the river base surface is relatively slanting, the distance of *Rumah Rakit* pile should be approximately erected 2 meters from the base river surface in a maximum low tide. The tight distance will cause the foundation of the *Rumah Rakit* hit the river base surface because of the waves force caused by the movement of passing boats. This force influences to the back-forth movement of buildings. However, this distance will vary by geographic

IOP Conf. Series: Earth and Environmental Science **106** (2017) 012118 doi:10.1088/1755-1315/106/1/012118

location. *Rumah Rakit* on steep topographic locations may be located less than 2 meters from the river base surface in maximum low tide. The piles that control the position of *Rumah Rakit* are possible shifted in case of collision.

The *Rumah Rakit* location is also determined by the length of pile. Commonly, the length of pile is 12 m and it planted into the river base ground at a depth of 3-4m. The piles appear above the water level at an altitude of 3 m or more, corresponding to the height of building. A certain location considered not selected when the pile does not appear above the water surface.

3.2. The movement of Rumah Rakit

3.2.1 The movements due to the changes cycle of day-night and rain-tide seasons

The changing tides of Musi River occur according to the cycle of day and night. The surface of river water recedes in the morning until noon and it risings in afternoon towards the morning. The river water surface tides behaviour is influenced by the downstream sea level. Day-night changes determine the direction of wind flow. The wind flows from the sea to land during the day while it flows from land to the sea at night. Therefore the change of day-night cycle causes the up-down movement of *Rumah Rakit* due to tidal behaviour, and the left-right movement due to the cycle of wind flow from the sea to land and vice versa.

Based on the experience of local expert, the rainy season occurs on the 11th (November) to fourth (April) months and the rising tide season occurs on the fifth (May) to 10th (October) months. The highest rising tide season occurs in May-June. During rainy season, the water flows from the upstream mountain and this causes flooding in the downstream area. While in the highest rising tide season, the water flows from downstream to upstream. This condition affects to the change of water currents direction and the left-right building movement or vice versa.

A countermeasure the movements of building related to flexibility connection between foundation and piles. The foundation tied by a rope and the rope looped around piles. The rope will rise and fall following the ups-downs movements of building. The distance between foundation and pile is about 1 m. In order to make foundation saves from the hard crash, then two or more rubber tires mounted on both left and right sides of foundation.

3.2.2 The movements related to the passing rate and anchoring activity of boats and ships

The movements of traffic (boats or ships) passing around *Rumah Rakit* affect to the occurrence of waves. The waves pushed the *Rumah Rakit* in all directions: back-forth, left right, and up-down. The power of thrust generated by the waves depends on the speed and load of boats/ships. The local expert stated that they tackle the movements by: 1) keeping the position of building at least 2m from the river ground in maximum low tide; 2) looping the rope in triangle position to piles and binding it to foundation of building. A straight bond rope causes the building unstable and rolling down to the front side. It is possible to set the piles in four sides of building; however the occupants argue that it is not profitable in cost and the force of rear reflective waves is relatively insignificant.

3.3. Building components and construction

3.3.1 Foundation

The foundation of *Rumah Rakit* is constructed by a series of bamboos arranged in several layers. The bamboo series called as *Rakit* (rafts). The bamboo commonly has a diameter of 15 cm and a length of 12 m. The width of foundation is the same as the width of building plus a minimum extension of 20 cm on each side. Terraces on all four sides are usually on the size of 10 bamboos. The width of terrace should be 1.5 m when the used bamboo has diameter of 15 cm, while the length of terrace corresponds to the length of *Rumah Rakit* (about 8-10 m). The front terrace (which faces to the river) is used as an access of boat. One of the left and right side terraces is used for bathing and washing. While the rear terrace (which faces to the mainland) is for a wooden stair leading to the mainland.

The foundation is tied to wooden piles planted into the bottom of river ground as deep as 3m on the left and right side foundation. To keep the foundation from drifting, the tied ropes wrapped around the piles forming a triangle. The rope can move in accordance with the ups-downs of water surface and the left-right as well as front-rear movements of building. The number of piles depends on the affordability of the owner to pay for the cost; at least two piles on each left and right side of building.

The pile placement is arranged on the division of axes on the length size of foundation. For example, if using 2 piles, one pile placed at a distance of ¼ x the side length from one corner of the foundation, and the second pile is ¼ x the side length from the other foundation corner. The rope tied at the bottom corner of foundation then it wrapped around the pile and tied again in the middle of the side length of foundation in order to form a triangle. The total length of rope is about 1,5x2 m, so if the length of foundation is 12 m then the length of required space on the river for one *Rumah Rakit* is 15 m.

The length of pile is 12m, and it plugged into the ground as deep as 3 m. If the depth of water is 6 m then the rest 3 m of pile appears on the water surface. The piles joined and clamped by two woods at the top. Before the piles fixed, the location marked first by using woods. When the actual piles fixed then the marker removed. The piles first placed in the direction of the flow of water in order to prevent the foundation from drifting. The winding rope will be tighter to the piles in the direction of the flow of water.

3.3.2 Walls and openings

Columns are set on the transverse woods by using bolts, pegs or clamped by that transverse woods. The next is to install the wooden frame for walls, openings and roof. The wooden frame arranged according to the size of doors and openings. A large door is set at the middle of front door and small door is at the rear, while the wooden or glass windows installed at the left and right side of the doors. According to the local expert, the doors position placed not directly opposite to the windows. It is because the reason to keep a good fortune. Anyhow, it is a right consideration because a cross ventilation is on that position. Another philosophical consideration is to keep the bedroom in the upstream direction in order to avoid the drifting position.

3.3.3 Roof

Zinc is used for roof cover material, in addition to shingles. The shingle roof cover is to be rare today and it is relatively not last longer than zinc. The shingle roof cover needs to change every 6 months, so the actual cost of treatment is higher than zinc. However, the room temperature is relatively cooler by using the shingle roof cover. Both material zinc and shingle considered suitable for *Rumah Rakit* because they are relatively lighter. Zinc is a better conducting material than shingles. Therefore, the room temperature will be relatively higher with zinc roof cover. In order to flow out the heat, the sloping wooden planks ventilations created on the roof wall. In this position, the rainwater does not flow into the room.

The roof slope angle is in the ratio 2 m of skoor length for 4m of beam length. The slope angle could rationalized be 30°. But the owner was often shortening the length of skoor (hanging pole on the roof horse frame) into 1.5 m for saving the cost, so the slope of the roof is relatively more sloping. A sharper roof slope angle will reduce the room temperature. The heat level of room temperatures is determined by the feelings of occupants.

Based on the above research results, it can be formulated a conceptual model as follows: the architectural tectonics of *Rumah Rakit* is related to three important aspects, namely the determination of buildings location; buildings movement; building components and construction system.The location of building determined by water surface behaviour; as well as geography and topography conditions. The movement of buildings occurs due to day-night cycles; the cycles of rainy and tides seasons; the speed movement of passing and anchoring boats/ships; and reaction to the weight of the load. Building component and construction system of *Rumah Rakit* is required to be flexible against movement in all directions; responsive to the nature of local climate; it built on trial and error and empirical experiences (Figure 1).

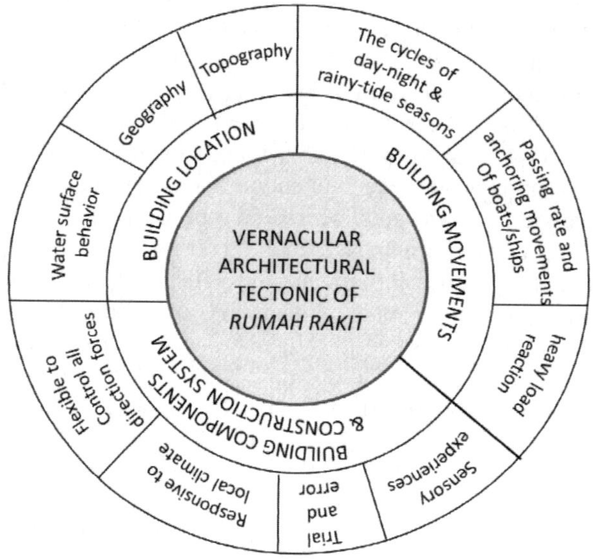

Figure 1. The conceptual model of vernacular architectural tectonic of *Rumah Rakit*, Seberang Ulu-1, Palembang.

4. Conclusion

The result of the research is a conceptual model, which can be used as reference for the architects and observers on the settlement environment so they may be able to develop the design model of floating building. The characteristics of local climate determine the form and system of building tectonics. Different geographical conditions influence to the differences of water tide behaviour and the stability of building position against topographic conditions. The characteristic of local climate causes the various movements of buildings in all directions. The movements caused due to the cycle of day-night and the cycle of rainy-tide seasons. Other factors are the movements originated by the surrounding in the form of the passing rate and anchoring movement of boats/ships; in addition to the heavy load reaction beard by foundation. These types of movements require a flexible construction system that has high capabilities to control the various forces from all directions and responsive to local climatic characteristics. Such tectonic system of *Rumah Rakit* based on the vernacular experimentation and sensory experiences.

Acknowledgment

Appreciation is directed to the Faculty of Civil Engineering and Planning, Universitas Trisakti and Kementerian Riset Teknologi dan Pendidikan Tinggi Republik Indonesia which have provided research funding support through regular research faculty program and Program Penelitian Unggulan Perguruan Tinggi 2017. The great honour delivered for local experts and informants.

References

[1] Adiyanto, Johannes. 2008. Transformation of Architectural Shape of Resident of Descendants China in Palembang (Architectural Reading with Hermeneutics Method Phenomenology, *Proc. on Ke-Bhinekaan Bentuk Arsitektur Nusantara*, 12-13 September 2008. Surabaya: Jurusan Arsitektur FTSP, ITS.

[2] Aniza, Fiatria. 2015. Designing Rumah Rakit as Culinary Tourism Object. Palembang: Politeknik Negeri Sriwijaya

[3] Iskandar, Yulindiani and Lahji, Khotijah. 2010. Local Wisdom in the Settlement of Structure and Construction of Raft House on the Musi River, Palembang. *Lcl.Wsd.* II2. p.37 – 45.

[4] Latifah, Latifah, Nur Laela. Building Physics 1: Natural Penghawaan & Natural Illumination, Thermal Control (Silar Chart & SPSM). Jakarta: Griya Kreasi. 2015.

[5] Mangunwijaya. Wastu Citra. Jakarta: Gramedia Pustaka Utama. 2013.
[6] Perkasa Jaya, Anjuma. 2012. Proportion in The Architecture of Palembang Traditional Rafts House, Thesis.Yogyakarta: Universitas Gadjah Mada.
[7] Priya, R. Shanti, Sundarraja, Radhakrishna, Vijayalakshmi. Solar Passive Techniques in the Vernacular Buildings of Coastal Regions in Nagapattinam, Tamilnadu, India: A Qualitative and Quantitative Analisys. Engy Bld.49 p.50-61. 2012.
[8] Susetyarto, M. Bambang. Vernacular Architecture: Cultural Sustainability at Bena Flore Villages. Sukoharjo: Padepokan Seni Djayabhinangun. 2013.
[9] Tipnis, Aiswarya. Vernacular Traditions: Contemporary Architecture. New Delhi: The Energy and Resources Institut (TERI). 2012.
[10] Wang CM, Wang BT. Large Floating Structure. New York: Springer. 2015.
[11] Wiryomartono, Bagus. Perspectives on Traditional Settlements and Communities: Home, Form and Culture. Singapore: Springer. 2014.

IOP Conf. Series: Earth and Environmental Science **106** (2017) 012119 doi:10.1088/1755-1315/106/1/012119

The effect of leachate recirculation with enzyme cellulase addition on waste stability in landfill bioreactor

N Saffira[*], **G A Kristanto**[**]

Environmental Engineering Study Program, Civil Engineering Department, Universitas Indonesia, Depok, Indonesia

Corresponding Author: [*]andari@eng.ui.ac.id, [**]gakristanto@gmail.com

Abstract. Landfill bioreactor with leachate recirculation is known to enhance waste stabilization. However, the composition of waste in Indonesia is comprised by organic waste which is lignocellulosic materials that considered take a long time to degrade under anaerobic condition. To accelerate the degradation process, enzyme addition is ought to do. Cellulase is an enzyme that can catalyse cellulose and other polysaccharide decomposition processes. Therefore, operation of waste degradation using leachate recirculation with a cellulase addition to enhance waste stabilization was investigated using anaerobic bioreactor landfill. The experiment was performed on 2 conditions; leachate recirculation with cellulase addition and recirculation only as a control. The addition of cellulase is reported to be significant in decreasing organic content, represented by volatile solid parameter. The volatile solid reduction in the cellulase augmented reactor and control reactor was 17.86% and 7.90%, respectively. Cellulase addition also resulted in the highest cellulose reduction. Settlement of the landfill in a bioreactor with enzyme addition (32.67%) was reported to be higher than the control (19.33%). Stabilization of landfill review by the decreasing rate constant of the cellulose and lignin ratio parameter was more rapidly achieved by the enzyme addition (0.014 day^{-1}) compared to control (0.002 day^{-1}).
Keyword: landfill bioreactor, leachate, lignocellulosic

1. Introduction

Limitation of land and the high volumes of waste disposal to landfill are the problems faced in waste management in Indonesia. Coupled with the operation of the landfill in Indonesia almost entirely using open dumping system which gives rise to a variety of environmental problems [1]. Accumulation of waste disposal that degraded for a long time can lead to increased leachate potential percolate into the soil and affect the quality of the ground water around the landfill area [2]. Thus, the alternative solution in an attempt accelerate the degradation process and stabilization of waste is needed.

Composition of waste in Indonesia is comprised by organic waste, paper, and wood which is lignocellulosic materials [2]. Lignocellulosic materials are considered takes a long time to degrade under anaerobic condition [5]. To accelerate the degradation process, enzyme addition is ought to do. The addition of enzyme cellulase will stimulate chemical processes based on biological catalyst reaction [5].

Landfill bioreactor with leachate recirculation is known to enhance waste stabilization [5]. However, the operation of waste degradation using leachate recirculation with a cellulase addition to enhance waste stabilization is a relatively new application technology that still needs to be investigated further. Recent development in the biotechnology industry in expanding the production of cellulase with the reduced cost manufacturing has become one of the drivers to the use of cellulase on landfills. Cellulase enzyme is an enzyme that can catalyze cellulose and other polysaccharide decomposition processes [7]. Based on research conducted by [8], addition of cellulose on leachate recirculation enhance the waste stabilization process and doubled the biogas production. The addition of cellulase on leachate recirculation aims to optimize the degradation process of lignocellulosic material in landfill. Acceleration of waste degradation increases the capacity and expand the operational life of the landfill. However, research using cellulase in leachate recirculation has not been performed especially in Indonesia.

The study was conducted using anaerobic bioreactor. The experiment was performed on 2 conditions; leachate recirculation with cellulase addition and recirculation only as a control. The effect of landfill stabilization by the addition of cellulase on leachate recirculation was investigated by a degradation process of cellulose, lignin and volatile solid (VS) content in waste as well as settlement of waste in landfill bioreactor.

2. Research Method

2.1. Feedstock Preparation

Composition of waste used in this study was 100% organic waste, aims to represent lignocellulosic waste derive from plant biomass. Solid waste samples were collected from Pasar Kemiri Depok, West Java. Both reactors used were filled with 53 kg with the same composition of waste in each. Feedstock was manually shredded into 5 – 10 cm considerate the dimension of the bioreactor used. The waste degradation process can be accelerated by the shred process [9].

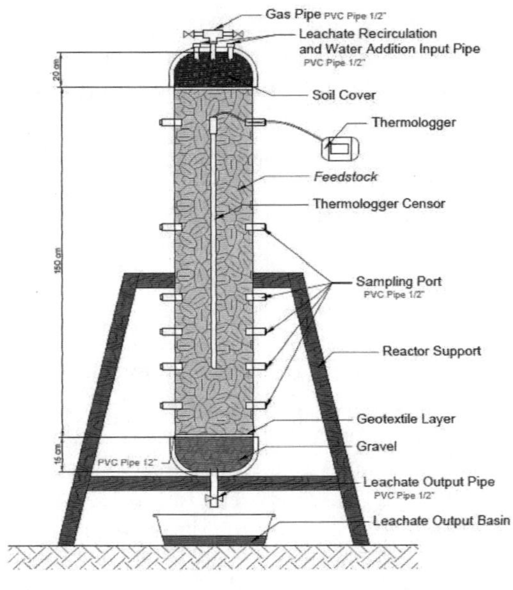

Figure 1. Anaerobic bioreactor landfill simulator.

2.2. Reactor Assembly

The study was conducted using anaerobic bioreactor operated on different conditions; leachate recirculation with cellulase addition and recirculation only as a control. Two bioreactors made from acrylic (PVC) cylinder pipe. The dimensions of the bioreactors were as follows, diameter = 31.5 cm, height = 185 cm, internal diameter = 30 cm. One side of the pipe wall was cut and covered using transparent acrylic for the measurement of reduced waste in the settlement process. Also on both sides of the cylinder pipe connected using ½" diameter pipes as a port for sampling requirement. At the top of the reactor installed pipes served as leachate inlet and gas outlet pipe connected to the 5 L gas bag. One port at the bottom served as a leachate outlet pipe. This reactor was modified from a simulated reactor study conducted by [8]. Approximately 15 cm gravel layered at the bottom of the reactor, acted to hold the sample into the leachate outlet to prevent clogging. Above the gravel layer, geotextile layer was placed. On top of that, 150 cm feedstock was placed on with weight 53 kg and density 500 kg m^{-3}. The reactor was equipped with a temperature sensor to record temperature data connected with the thermologger. The soil was used as a cover above the waste layer.

2.3. Leachate Recirculation and Cellulase Enzyme

Leachate recirculation reduces the organic content by utilizing the activity of microorganisms in leachate. The design volume of leachate recirculation in landfills is determined by geographic location, seasonal rainfall and input volume waste disposal in landfill [10]. Volume leachate recirculation at the anaerobic bioreactor used the low recirculation rate equilibrium flow rate of 5 L.ton^{-1} [11], so leachate recirculation performed on both reactors was 270 ml. Addition of water performed to simulate rainfall in landfill (600 ml day^{-1}). Treatment of lignocellulosic waste was achieved by enzymatic hydrolysis using cellulase enzyme. Cellulase enzyme used in this study was Cellulase CEL 150 produced by Trichoderma reesei, purchased from Sinobios (China) [12]. The CEL 150 enzyme activity consists of several cellulase enzyme including Endo-ß-1.4-glucanase (CMC) /1,500,000 U/g, FPA, Xylanase, ß-glucanase. According to study [11], the amount of enzyme added was equivalent to 15 million u ton^{-1}. As for this study, the 0.53gram cellulase was added on leachate recirculation.

2.4. Analytical Method

The moisture content was measure using the gravitational method of Standard Method 2540-B. Samples are dried to constant mass in an oven at (105 ± 5)°C at an interval of minimum 1 hour, then cooled at desiccator and weighed using an analytical balance. A modified Standard Method APHA Method 2440-E was used to measure volatile solid. The constant weight of samples (105 ± 5) °C drying result was combust in the furnace 550°C for 1 hour. The lost weight of combustion results from weight of total solid is content of volatile solid. Waste temperature was measured using temperature sensor connected to the thermologger unit. Settlement of waste in bioreactor determined using a measuring tape mounted near the transparent acrylic side of the reactor. Measurements are made daily to illustrate the ongoing waste stabilization process. Lignin and cellulose content were measured using Chesson–Datta Method and reported as percentage w/w.

3. Results and Discussion

3.1. Characteristics of Feedstock.

Physical and chemical characteristics of feedstock were required to determine the initial condition of waste and as a reference of degradation process due to treatment in this study. Parameters measured in this study were waste settlement, temperature, C:N ratio, moisture content, volatile solid, and lignin:cellulose ratio. The initial of waste layer was 150 cm on both reactor. The settlement of waste at both reactors at the end of the study were compared. The initial feedstock temperature was 32.6 °C. In general, temperature in landfill equipped with water addition is in the range of 30 – 45°C and will increase due to microorganism activity and the characteristics of the waste itself (Kim, 2004). The C:N feedstock ratio was 17:1, indicated that the feedstock has adequate nutrient availability for the degradation process by microorganisms. Volatile solid can describe the cellulose content of waste and estimate the readily degradable organic content. Volatile solid of the feedstock was 73.73%TS. The C:L ratio in feedstock was 4.16 with 50.30 %w/w cellulose content and 12.09 %w/w lignin content.

3.2. Settlement of Waste

Settlement of waste described the degradation process in the reactor. Based on Figure 2, settlement in the reactor with augmented cellulase was considerably different than settlement in the reactor with leachate recirculation only. Settlement of waste in a reactor with augmented cellulase was 13.3% larger than the control. Settlement that occurred until day 91st in the cellulase augmented reactor and the control reactor was 32.67% and 19.33%, respectively. The first order settlement rate for the reactor with cellulase addition was 0.0028 day^{-1} and reactor with leachate recirculation only was 0.0016 day^{-1}.

The settlement behavior of landfill is due to the consolidation and degradation that occur in landfill [13].

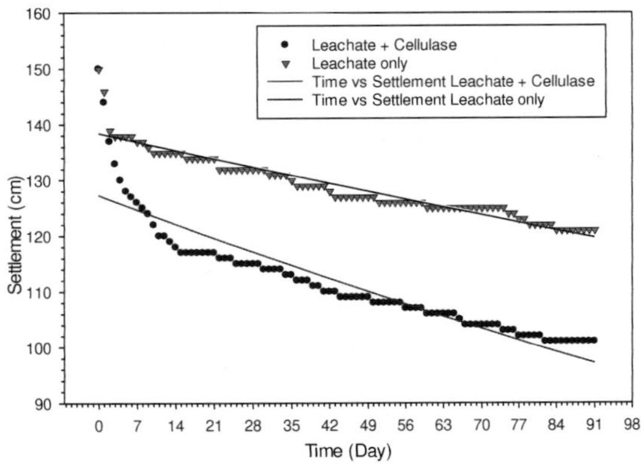

Figure 2. Settlement in bioreactors over 90 days.

The degradation of organic matter in landfill caused a major amount of settlement as the organic material is converted into different phase as decomposition product consist of liquid form and the landfill gas. Settlement in this study has a strong correlation with the reduction in organic content indicated by volatile solid parameter. Statistical analysis with significance level set at 0.05, indicated that there was a significant difference in the cumulative percentage of the settlement between the reactor operated with cellulase addition on the leachate recirculation and the reactor operated with recirculation only ($t(182)=14.812$, $p=0.000$).

3.3. Volatile Solid, Cellulose and Lignin Content

Chemical characteristic parameters of the waste decomposition process used in this study are volatile solid (VS), lignin and cellulose. Volatile solid is commonly used to estimate the organic content of waste [14]. The major organic content of waste is composed of cellulose. Volatile solid should not decrease to zero as a point of stabilization [16], but low cellulose and low volatile solid may be possible parameters of stabilization [17].

Both reactors showed the decreasing of volatile solid content, indicated that decomposition of organic matter occurs. At the end of this study, volatile solid reduction in the cellulase augmented reactor and the control reactor was 17.86% and 7.90%, respectively. Reduction of volatile solid indicated organic content had been degraded during the operation. The addition of cellulase is reported to be significant in decreasing organic content represented by volatile solid parameters ($t(11)=-7.712$, $p=0.000$). The average difference of volatile solid content on both reactors was 7.19%TS.

Based on study [18], when the cellulose is degraded, the lignin content will remain and the percentage of its contents on the waste will increase. Cellulose content on both reactors decreased. Cellulose content reduction in the cellulase augmented reactor and the control reactor was 24.50% w/w and 18.40% w/w, respectively. While lignin until the end of the study was increased. This occurrence caused by lignin considerably difficult to degrade under anaerobic condition [8]. Lignin is not decomposed until cellulose reaches level below 25% [17]. In the study [16], the lignin level at the beginning of the study increased and began to decrease slowly after cellulose has been degraded.

Cellulase enzyme is an enzyme that can catalyze cellulose and other polysaccharide decomposition processes [7]. Statistical analysis determined there was a significant difference in cellulose content in the bioreactor with the cellulose addition, compared to control ($t(5)=-3.8000$, $p=0.013$). Cellulose to lignin ratio can be used as an indicator of waste stability in landfills. The cellulose to lignin ratio of less than 0.2 may indicate that the waste has reached a stabilization point (Kim, 2004). The initial of the ratio C:L in the feedstock was 4.16 and continued to decrease until the end the study in a reactor with cellulase addition and control; 0.96 and 1.41, respectively. Until the end of this study, based on C:L parameters the waste in both reactors has not reached stable condition. The higher constant of the first degradation rate of ratio C:L showed in a reactor with cellulase addition than control. The first order degradation rate of ratio C:L for the reactor with cellulase addition was 0.014 day^{-1} and reactor with leachate recirculation only was 0.012 day^{-1}. Cellulase significantly enhanced the stabilization of waste in this study based on a ratio C:L parameter ($t(5)=-3.280$, $p=0.022$).

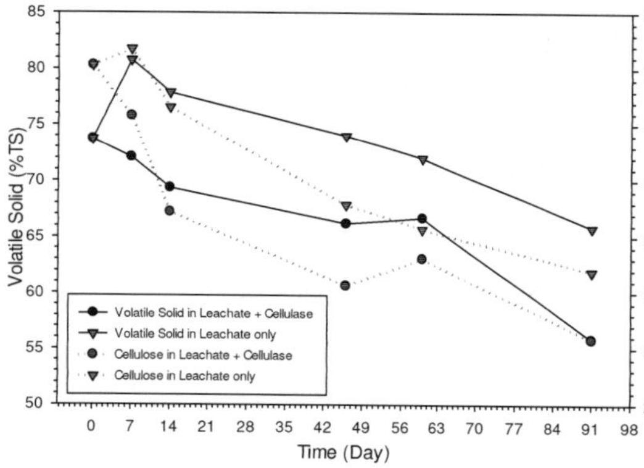

Figure 3. Volatile solid and cellulose in bioreactors.

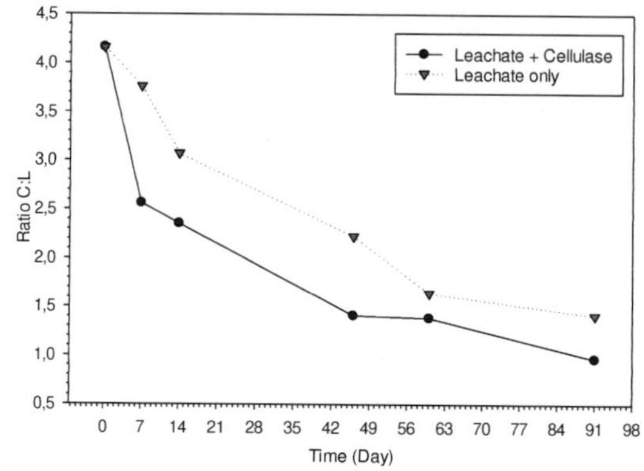

Figure 4. Ratio C:L in bioreactor.

3.4. Output of the treatment with cellulase addition to leachate recirculation

The cellulase enzyme added to leachate recirculation significantly affects organic degradation as indicated by parameters such as settlement, cellulose, ratio cellulose to lignin and volatile solid. The highest reduction of organic matter was showed in a reactor with cellulase addition and this also indicated shorten the waste stabilization process. Application of cellulase enzyme in landfill based on this result can accelerate the stabilization process, thus increase the capacity and age of the landfill. However, further research is needed to improve efficiency, especially in terms of operational costs. It is important to acknowledge in advance the optimum enzyme addition required at the time of application in landfill operation.

Table 1. Output of the treatment results.

Parameters	Initial	Post Treatment		Δ (%)	
		Leachate+ cellulase	Control	Leachate+ cellulase	Control
Settlement (cm)	150	101	121	32,67	19,33
Cellulose (%w/w)	50,30	25,80	31,90	48,71	36,57
Ratio C:L	4,16	0,97	1,41	76,78	66,01
Volatile Solid (%TS)	73,73	55,86	65,82	24,23	10,72

4. Conclusion

Degradation of organic content in waste significantly increased due to the effect of the addition of enzyme cellulase on leachate recirculation investigated by a degradation process of cellulose, lignin and volatile solid content in waste as well as settlement of waste in landfill bioreactor. Stabilization of landfill review by the decreasing rate constant of the cellulose and lignin ratio parameter was more rapidly achieved by enzyme addition compared to control. It is important to identify the optimal quantity of the enzyme cellulase to enhance stabilization and improve the efficiency of the operation in landfill.

Acknowledgment

The authors gratefully thanked Universitas Indonesia as sponsor for this research under the project of Publikasi Internasional Terindeks untuk Tugas Akhir Mahasiswa UI (PITTA).

References

[1] Damanhuri E. 2008. A Future prospect of municipal solid waste management in Indonesia. *In Keynote Lecture in the 5th Asian-Pacific Landfill Symposium,* Sapporo, Japan.

[2] Aboyeji O S. 2016. Evaluations of groundwater contamination by leachates around Olusosun open dumpsite in Lagos metropolis, southwest Nigeria. *Journal of Environmental*

Management. p.1-2.
[3] Godliving Y. 2009. Recent Advances in Pretreatment of Lignocellulosic Wastes and Production of Value Added Products. *African Journal of Biotechnology*. 8(8).
[4] Eriksson K, Blanchette R, & Ander P, Microbial and Enzymatic Degradation of Wood and Wood Components. 1 ed., Springer-verlag Berlin Heidelberg, Berlin, 1990.
[5] Karam J, and Nicell J A. 1997. Potential applications of enzymes in waste treatment. *Journal of Chemical Technology & Biotechnology*. 69(2):141-153.
[6] Qasim S R, and Chiang W, Sanitary landfill leachate: generation, control and treatment, CRC Press, 1994.
[7] Worthington Biochemical Corporation, Cellulase. s.l.:s.n., 2014.
[8] Frank R R, Davies S, Wagland S T, Villa R, Trois C, and Coulon F. 2016. Evaluating leachate recirculation with cellulase addition to enhance waste biostabilisation and landfill gas production, Waste management. 55:61-70.
[9] Valencia Vázquez R, Enhanced stabilisation of municipal solid waste in bioreactor landfills, 2008.
[10] Christensen T H, Sanitary Landfilling: Process, Technology and Environmental Impact. 1st ed., Academic Press Limited, London, 1989.
[11] Jayasinghe P A, Hettiaratchi J P A, Mehrotra A K, and Steele M A. 2012. Enhancing gas production in landfill bioreactors: flow-through column study on leachate augmentation with enzyme. *Journal of Hazardous, Toxic, and Radioactive Waste*. 17(4):253-258.
[12] Sinobios Imp. & Exp. co., Ltd, Cellulase Product Information CEL150, Sinobios, Shanghai, 2016.
[13] Tchobanoglous, G., Integrated Solid Waste Management, Mc. Graw Hill, New York, 1993.
[14] Townsend T G, Miller W L, Lee H J, and Earle J F K. 1996. Acceleration of landfill stabilization using leachate recycle. *Journal of environmental engineering*. 122(4): 263-268.
[15] EPA, *A technical assessment of leachate recirculation*, Environment Protection Agency, Bristol, 2009.
[16] Kim, J. Effect of plastics on the lignin results for MSW and the fate of lignin in laboratory solid waste reactors. Doctoral dissertation. Virginia Polytechnic Institute and State University, 2004.
[17] Kelly R J, Shearer B D, Kim J, Goldsmith C D, Hater G R, and Novak J T. 2006. Relationships between analytical methods utilized as tools in the evaluation of landfill waste stability. *Waste Management*, *26*(12):1349-1356.
[18] Bricker G D, Novak J T, Randall C W, & Goldsmith C D, Analytical Methods of Testing Solid Waste and Leachate to Determine Landfill Stability and Landfill Biodegradation Enhancement, Blacksburg, VA: Virginia Polytechnic Institute and State University, 2009.

Phytoremediation of heavy metal copper (Cu^{2+}) by sunflower (*Helianthus annuus l.*)

G Mahardika, A Rinanti, M F Fachrul*
Environmental Engineering Department, Faculty of Landscape Architecture and Environmental Technology, Universitas Trisakti, Jakarta, Indonesia

*Corresponding Author: melati@trisakti.ac.id

Abstract. A study in microcosmic condition has been carried out to determine the effectiveness of *Helianthus annuus* as a hyperaccumulator plant for heavy metal, Copper (Cu^{2+}), that exposed in the soil. Artificial pollutants containing Copper (Cu^{2+}) 0, 60, 120, 180 ppm are exposed to uncontaminated soil. The 12-weeks old *H. annuus* seedling were grown in Cu^{2+} contaminated soil, with variations of absorption time 3, 6, and 9 weeks. Analysis of Cu^{2+} concentration on soil and *H. annuus* (root, stem, leaf) was analised by Atomic Absorbtion Spectrometry (AAS). *H. annuus* are capable for Cu^{2+} removal, and the highest removal of Cu^{2+} is 85.56%, the highest metal accumulation/bioconcentration factor (BCF) is 0.99 occurred at roots with 9 weeks of exposure time and the highest translocation factor (TF) is 0.71. This highest removal is five times better than absorption by stems and leaves. The results concluded, the use of *H. annuus* for phytoextraction of heavy metals Cu^{2+} in contaminated soil can be an alternative to the absorption of heavy metal Cu^{2+} with low concentration metals which is generally very difficult to do in physical-chemical removal.
Keywords: bioconcentration factor, *helianthus annuus*, hyperaccumulator, phytoremediation, translocation factor

1. Introduction

Phytoremediation is a method to recover contaminated soil by heavy metals. Phytoremediation through phytoextraction process is a biotechnology that utilizing plants to clean the environment [1]. This technology is suitable to clean up the environment contaminated by toxic metals in-situ [2]. Unlike organic compounds, metal compounds could not be degraded. Therefore, this phytoremediation method aims to remove or eliminate pollutant components.

Phytoextraction is one of the phytoremediation's strategies. The terms of phytoremediation and phytoextraction are sometimes incorrectly used as synonyms, but phytoremediation is a concept while phytoextraction is a specific cleanup technology [1]. Phytoextraction involves the absorption of contaminants by roots followed by translocation and accumulation in the aerial parts [3]. This method using hyperacumulation plants that have a natural ability to extract metals from the soil, translocating metals from roots to shoots, and accumulate and tolerate high metal concentrations within the plant tissues [4].

The application of this technology required a type of heavy metals hyperaccumulator plant. Hyperaccumulator plants are hypertolerant, able to absorb heavy metals from high absorption soil solution, and have the ability of translocation from root to stem and to leaf. Hyperacumulator does not refer to plants that have familial relations, but both have the ability to live in metal-contaminated soil and are able to accumulate high amounts of metals in their organs far beyond the accumulation rate at other plants without experiencing phytotoxic effects.

The process of phytoextraction occurs with the mechanism as shown in Figure 1. In process 1 showing the absorption of the metal by the root (root uptake), process 2 showing translocation of heavy metals from the roots to other plant parts, and process 3 showing localization of metals in plant parts. Research on a microcosmic scale aims to exclude Cu^{2+} heavy metals exposed in soil by Sunflower (*Helianthus annuus* L.) as hyperacumulators.

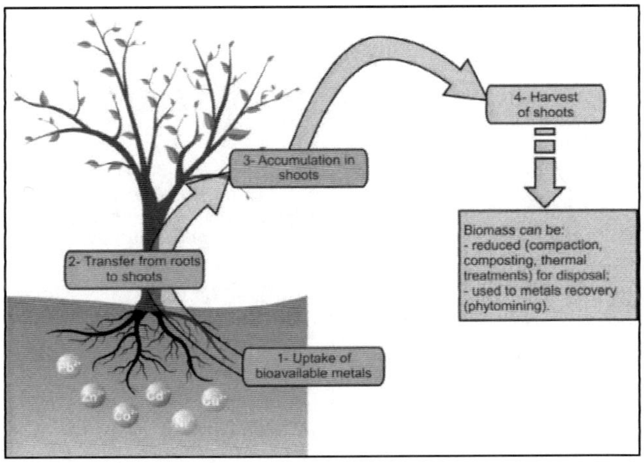

Figure 1. Mechanism of Phytoextraction [3].

2. Research Method

2.1. Planting H. annuus
H. annuus is grown on sterile (heavy metal-free) soil with polybag container. The soil media used for each plant is 5 kg. *H. annuus* was grown to 12-weeks old, to be further used as a hyperaccumulator plant in this study.

2.2. Contamination of Cu^{2+} on Soil
Solution of Cu^{2+} concentration 60, 120, 180 ppm was exposed to sterile soil. The research design can be seen in Table 1.

Table 1. Research design.

Cu^{2+} concentration (ppm)	Exposure Time		
	3-weeks	6-weeks	9-weeks
0	A_0 R_1	A_0 R_2	A_0 R_3
5	A_1 R_1	A_1 R_2	A_1 R_3
10	A_2 R_1	A_2 R_2	A_2 R_3
100	A_3 R_1	A_3 R_2	A_3 R_3

2.3. Atomic Absorption Spechtrophotometer (AAS)
This method is used to analyse Cu^{2+} absorbed by roots, stems, leaves, and soil, at week 3, 6, and 9.

2.4. Removal Percentage
Determination of metal percentage that removed can be calculated by the following equation:

$$\% \, removal = \left[\frac{Co - Ce}{Co}\right] \times 100 \, \% \tag{1}$$

Co = the concentration of the initial metal solution (mg/L),
Ce = final concentration in the soil (mg/L)

IOP Conf. Series: Earth and Environmental Science **106** (2017) 012120 doi:10.1088/1755-1315/106/1/012120

2.5. Bioconcentration Factor (BCF)

The value of BCF is the effectiveness of the plants ability to accumulate metals in their tissues. BCF can be calculated by the following equation [5]:

$$BCF = \frac{Cu^{2+}\ concentration\ on\ plant\ \left(\frac{mg}{L}\right)}{Cu^{2+}\ concentration\ on\ soil\ \left(\frac{mg}{L}\right)} \quad (2)$$

2.6. Translocation Factor (TF)

TF is the value of Cu2+ translocation from soil to a particular part of the plant. TF can be calculated by the following equation [6]:

$$\mathrm{TF} = \frac{\mathrm{Cu^{2+} concentration\ on\ the\ part\ of\ the\ plant}\ \left(\frac{mg}{L}\right)}{\mathrm{Cu^{2+} concentration\ on\ soil/root}\ \left(\frac{mg}{L}\right)} \quad (3)$$

3. Results and Discussion

3.1. Cu^{2+} Removal from Soil

In Table 2 show that Cu^{2+} concentration in soil is decreasing every week. This is happening due to the absorbtion process by *H. annuus,* resulting in the translocation Cu^{2+} into the plant's organ.

Table 2. Cu^{2+} concentration in soil.

Initial Concentration Cu^{2+} in soil (ppm)	Concentration of Cu^{2+} (ppm) 3 weeks	6 weeks	9 weeks
0	0	0	0
60	21.61	21.57	21.46
120	21.28	21.12	21.01
180	27.37	26.62	26.00

Based on Table 2 percentage of Cu^{2+} absorbtion could be measured in soil (Table 3). The highest elimination percentage (reaching 85.56%) happen at the concentration of 180 ppm with the exposure time Cu^{2+} for 9 weeks. It means that the heavy metal availability in the soil affects the capability of *H. annuus* to absorp the copper (Cu^{2+}).

Table 3. % Removal of Cu^{2+} from soil.

Initial Concentration Cu^{2+} in soil (ppm)	Absorption of Cu^{2+} 3-weeks	6-weeks	9-weeks
0	0.00%	0.00%	0.00%
60	63.98%	64.05%	64.24%
120	82.27%	82.40%	82.50%
180	84.79%	85.21%	85.56%

Based on Figure 2 the longer the exposure time, the higher percentage of removal ($R^2 = 0.9$).

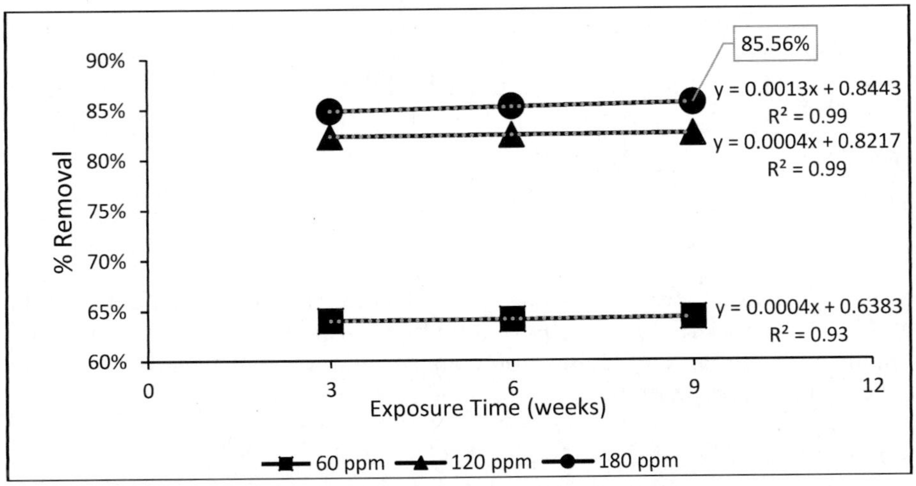

Figure 2. % Removal of Cu^{2+} from soil.

3.2. Cu^{2+} Absorbtion by H. annuus

The measurement of Cu^{2+} absorbtion was done with *H. annuus*, to make sure that the absorbtion is really happening with the plant (Table 4). Copper (Cu^{2+}) that goes in to the plant's tissue, cannot be degraded, however it will be accumulated in the plant tissues [7]. In this research, absorption that happened at concentration of contaminated soil 60 ppm, not yet reached in stem or in leaf..

Table 4. Cu^{2+} Concentration on *H. annuus.*

Initial Concentration of Cu^{2+} in *H. annuus*	**Part of Flower**	**Concentration of Cu^{2+} in *H. annuus***		
		3-weeks	**6-weeks**	**9-weeks**
60 ppm	roots	34.79	35.68	42.36
	shoots	0.00	0.00	0.00
	leaves	0.00	0.00	0.00
120 ppm	roots	57.59	58.40	70.78
	shoots	22.36	22.50	22.85
	leaves	18.51	19.11	19.28
180 ppm	roots	79.65	89.28	128.84
	shoots	24.74	25.17	25.37
	leaves	23.47	23.63	24.09

Referring to Table 5, the highest Cu^{2+} content is at the root followed by shoots and leaves of *H. annuus*. This is in accordance with previous studies where it was found that the metal content was accumulate more in the root than the shoots and leaves because the roots were the first to utilize nutrient uptake and the metal content [8]. Roots of the plants act as a barrier against heavy metal translocation, and it causes the concentration in the roots are higher.

After knowing the concentration of Cu^{2+} absorbtion with *H. annuus* then the calculation could be done with BCF to know the effectiveness of *H. annuus* to accumulate metals in the tissue (root, stem, leaf) (Table 5). The highest BCF is 8.1 obtain in the initial concentration of 50 ppm with absorbtion time of 9-weeks.

Table 5. *H. annuus* Bioconcentration Factor (BCF).

Initial Cu^{2+} Concentration in Soil	Bioconcentration Factor (BCF)		
	3-weeks	6-weeks	9-weeks
60 ppm	0.58	0.59	0.71
120 ppm	0.82	0.83	0.94
180 ppm	0.71	0.77	0.99

Compared to the previous research [9], the largest bioconcentration factor found in heavy metal absorption by *H. annuus* was 1.78 in the absorption of heavy metals Zn. The bioconcentration factor obtained is quite high due to the addition of compost and vermicompost that can increase the absorption of heavy metals in the study.

3.3. Translocation Factor (TF)

The translocation of Cu^{2+} from soil to specific parts of H. annus can be seen in Table 6. From the table it can be seen that the largest translocation factor is 0.71 at 6-weeks absorption. According with previous studies [9], translocation factors obtained greater than or equal to (≥) 1 indicate that the plant includes hyperacumulator plants and is able to perform phytoextraction.

Table 6. H. annuus Translocation Factors (TF).

Initial Cu^{2+} Concentration in Soil	Translocation Factors (Roots → Shoots and Leaves)		
	3 -weeks	6-weeks	9 -weeks
60 ppm	0.00	0.00	0.00
120 ppm	0.71	0.71	0.60
180 ppm	0.61	0.55	0.38

However, that result not happened in this study, due to the heavy metals concentration of Cu2+ tend to be low. The translocation factor is less than 1.0, does not mean that H. annuus is unable to absorb heavy metal Cu2+ but at that concentration, translocation will run more slowly or take more than 9-weeks.

4. Conclusion

Phytoremediation by H. annuus as a hyperaccumulator has been shown to be effective in removing heavy metals from the soil. The removal efficiency of Cu2+ by H. annuus reached 85.56% in the soil with Cu2+ exposure concentrations 180 ppm for 9 weeks. Measurement of TF and BCF can be concluded that the highest accumulation of Cu2+ occurs in the root, which is 5 times better compared to shoots and leaves.

References

[1] Prasad M, & Freitas H O. 2003. Metal hyperaccumulation in plants-Biodiversity prospecting for phytoremediation technology. *Electronic Journal of Biotechnology.* 6(3). http://dx.doi.org/10.2225/vol6-issue3-fulltext-6

[2] Vashegyi A, Mezôsi G, Barta K, Farsang A, Dormány G, Bartha B, & Erdei L. 2005. Phytoremediation of heavy metal pollution: A case study. Acta Biologica Szegediensis, 49(1-2), 77-79. Retrieved from http://www2.sci.u-szeged.hu/ABS/2005/Acta%20HP/4977.pdf

[3] Favas P J, Pratas J, Varun M, D'Souza R, Paul M S. 2014. Phytoremediation of Soils Contaminated with Metals and Metalloids at Mining Areas: Potential of Native Flora, Environmental Risk Assessment of Soil Contamination, Dr. Maria C. Hernandez Soriano (Ed.), InTech, DOI: 10.5772/57469. Available from: https://www.intechopen.com/books/environmental-risk-assessment-of-soil-

contamination/phytoremediation-of-soils-contaminated-with-metals-and-metalloids-at-mining-areas-potential-of-nativ.

[4] Mahmood T. 2010. Phytoextraction of Heavy Metals – The Process and Scope for Remediation of Contaminated Soils. *Soil & Environ*. 29(2): 91-109.

[5] Snyder K V W. 2006. Removal of Arsenic from Drinking Water by Water Hyacinths (*Eichhornia crassipes*). J. U.S. S J W P, Volume 1.

[6] Gupta S, S Nayek, R N Saha, and S Satpati. 2008. Assessment of Heavy Metal Accumulation in Macrophyte, Agricultural Soil and Crop Plants Adjacent to Discharge Zone of Sponge Iron Factory. *Environ. Geol*. 55, 731-739.

[7] Noer L F, 2006, Fitoremediasi Logam Krom dalam Limbah Cair dengan menggunakan Tumbuhan Akuatik *Valisneria spiralis* auct. non L dan *Cabomba caroliniana* Gray. SITH ITB : Bandung.

[8] Fulekar M H. 2016. Phytoremediation of Heavy Metals by *Helianthus annuus* in Aquatic and Soil Environment. *International Journal of Current Microbiology and Applied Sciences*. 5(7): 392-404.

[9] Angelova V R, Nemska M N, Uzunova G P, Ivanov K I, Lee H Q. 2016. Potential of Sunflower (*Helianthus annuus*) for Phytoremediation of soils Contaminated with Heavy Metals. *International Journal of Biological, Biomolecular, Agricultural, Food, and Biotechnological Engineering*. 10(9): 576-583.

Institute of Physics
Dirac House, Temple Back
Bristol BS1 6BE UK

ISSN: 1755-1307
ISBN 978-1-5108-5772-8

9 781510 857728

CURRAN ASSOCIATES INC.
proceedings.com